POSTHARVEST PATHOGENS AND DISEASE MANAGEMENT

POSTHARVEST PATHOGENS AND DISEASE MANAGEMENT

P. NARAYANASAMY
Former Professor and Head
Department of Plant Pathology
Tamil Nadu Agricultural University
Coimbatore, India

WILEY-INTERSCIENCE
A JOHN WILEY & SONS, INC., PUBLICATION

Published by John Wiley & Sons, Inc., Hoboken, New Jersey
Published simultaneously in Canada

Library of Congress Cataloging-in-Publication Data:
Narayanasamy, P., 1937–
Postharvest pathogens and disease management / P. Narayanasamy.
p. ; cm.
Includes bibliographical references and index.
ISBN-13: 978-0-471-74303-3 (cloth : alk. paper)
ISBN-10: 0-471-74303-8 (cloth : alk. paper)
1. Fruit—Postharvest diseases and injuries. 2. Vegetables—Postharvest diseases and injuries. 3. Phytopathogenic microorganisms. I. Title.
[DNLM: 1. Food Microbiology. 2. Plants, Edible—microbiology. 3. Food Contamination—prevention & control. QW 85 N218p 2006]
SB608.F8N35 2006
634′.0468—dc22

2005009866

Printed in the United States of America

10 9 8 7 6 5 4 3 2 1

TO MY PARENTS
FOR THEIR
LOVE AND AFFECTION

CONTENTS

PREFACE

Domestication of various plant species has provided the innumerable plant products required to satisfy the needs of humans and animals. As the plant species were selected for higher yield potential and better quality, the levels of resistance to biotic and abiotic stresses were depleted over long periods. Consequently, losses due to microbial pathogens—fungi, bacteria, mycoplasma, viruses, and viroids—increased, resulting in food famines, when conditions conducive for the development of epidemics prevailed. Losses in quantity and quality of agricultural produce due to postharvest diseases have been proposed to be as high as that from field crop diseases, though precise estimates have not been made. Yet, due attention to the study of various aspects of postharvest pathogens and the diseases caused by them has not been bestowed. The need for precise assessment of losses based on techniques to detect and identify the pathogens rapidly and accurately has been realized. Detailed studies on various aspects of microbial pathogens that will be useful to develop effective management systems for postharvest diseases of plant produce have been taken up in recent years.

This book aims to provide comprehensive information to understand the potential of various kinds of microbial pathogens capable of causing serious losses, on a variety of harvested produce, and the conditions favorable for their development and spread. The physiological functions of the harvested produce are very different from those that are attached to the mother plants. The level of natural resistance of the produce to diseases caused by microbial pathogens is progressively reduced as the process of ripening/senescence sets in. Furthermore, the storage conditions may also predispose the stored products to diseases. The efficacy of different disease management strategies based on cultural, physical, chemical, and biological methods has been shown to depend on the pathogen ecology and population dynamics, host plant species, produce and handling, and storage conditions. Some of the synthetic chemicals used against postharvest pathogens have been recommended for restricted use or replacement because of the development of resistance in pathogens to such chemicals. In addition, the persistence of chemicals in the commodities beyond permissible limits has also contributed to the realization of the necessity for alternative strategies. In this context, the uses of microbial biocontrol agents and/or compounds of plant and animal origin that can act directly on the pathogens and/or enhance the natural disease resistance of produce hold promise as effective alternatives. Moreover, they are natural and ecofriendly, helping the preservation of the environment. The effectiveness of protection to harvested produce may be markedly enhanced by integrating two or more effective strategies to derive maximum benefit.

The information presented in this book, reflecting an extensive literature search, will be useful for teachers, researchers, graduate students in the departments of plant pathology, microbiology, food technology, environmental sciences; personnel of commercial production centers and plant quarantine and certification agencies; and policy planners who are concerned with the production and supply of agricultural produce of high quality and acceptable for human and animal consumption. The protocols presented in different chapters and basic methods can form useful aids for teachers and for development of research investigations. It is hoped that this book will stimulate research and developmental efforts resulting in the supply of safe and healthy foods and feeds to consumers and preservation of the natural environment.

P. NARAYANASAMY

ACKNOWLEDGMENTS

The author is delighted to express his sincere appreciation to all colleagues in the Department of Plant Pathology for the useful suggestions, especially to Dr. T. Ganapathy for his technical help in the preparation of figures and Mrs. K. Mangayarkarasi for the excellent secretarial assistance. Permission given by editors and publishers to reprint the figures has been individually acknowledged in the text.

It is with a sense of pride and pleasure I record my profound appreciation to my family members—Mrs. N. Rajakumari, Mr. N. Kumara Perumal, Mrs. Nirmala Suresh, Mr. T.R. Suresh, and Mr. S. Varun Karthik—for their affectionate encouragement and loving care that allowed me to devote my time to the preparation of this book.

PART I

CAUSES AND EFFECTS

1

INTRODUCTION

Nature has blessed mankind and other living organisms by providing a wide range of plant products for our consumption. However, human beings, in their desire to domesticate plants to satisfy their need for food, feed, fiber, timber, and other plant products, have selected plants with high yield potential. Unfortunately, most of the plant species and crop cultivars with high yield potential became increasingly susceptible to diseases and pests that can cause severe epidemics and food famines. Diseases caused by microbial pathogens – fungi, bacteria, and viruses – account for substantial losses of grains, fruits, and vegetables at both pre- and postharvest stages of crop production. Stakman and Harrar (1957) emphasized that "the responsibilities of the plant pathologists do not end with the harvest of satisfactory yields of plant products and that harvesting marks the termination of one phase of plant protection and the beginning of another". This statement clearly indicates that the second phase of plant protection – of seeds, fruits, vegetables, and other economic plant parts, from the time of harvest until they reach the consumer – is equally important. Postharvest pathology, earlier termed "market pathology", deals with the science of, and practices for, the protection of harvested produce during harvesting, packing, transporting, processing, storing, and distribution.

Several studies have shown that losses occurring between harvesting and ultimate utilization of agricultural/ horticultural products by the consumer are substantial, though there is no precise estimate. The harvested produce may be classified as: (i) durables, consisting of cereal grains, oilseeds, and grain

Postharvest Pathogens and Disease Management, by P. Narayanasamy

legumes; and (ii) perishables, comprising succulent storage organs such as fleshy fruits, vegetables, rhizomes, and tubers. The harvested commodities are essentially dormant plant organs with physiological functions quite different from tissues of the mother plant. As the ripening/senescence process sets in, the susceptibility of the stored products to decay caused by microbial pathogens increases progressively. Furthermore, storage is an abnormal state for living plants or plant organs and it requires crowding of high volumes of plant produce in intimate contact in the limited available space. These unnatural conditions are likely to predispose them to different kinds of diseases which may be due to physiologic or pathogenic causes. The discussion in this book is confined to diseases caused by microbial pathogens.

1.1 IMPORTANCE OF POSTHARVEST DISEASES

The postharvest diseases that cause spoilage of both durable and perishable commodities are widespread. Greater losses occur in developing countries due to nonavailability of proper storage and transportation facilities and improper handling methods, resulting in greater levels of injuries or wounds during harvesting and transit. Durable commodities are generally stored in a dry state with moisture level below 12%, whereas perishable products have higher moisture levels (about 50% or more) at the time of storage. The harvested produce might have been infected by pathogens prior to harvest under field conditions or they may get infected during transit and storage. It is estimated that, in the tropics, about 25% of all perishable food crops harvested are lost between harvest and consumption. Losses in durable commodities, such as cereals, oilseeds, and pulses, may be about 10% on a worldwide basis (Waller, 2002). The losses may be both quantitative and qualitative. Some of the assessments of losses in different commodities are presented in Table 1.1. Under conditions favoring pathogens, loss caused by postharvest diseases may be greater than the economic gains achieved by improvements in primary production. Studies on postharvest diseases are primarily directed at preventing economic loss from spoilage of harvested commodities during transit and storage, and at eliminating the adverse effects of mycotoxins produced by fungal pathogens contaminating both durables and perishables. The mycotoxins are known to be carcinogenic, causing several serious ailments in humans and animals (Narayanasamy, 2005). There is an imperative need to gather information on the microbial pathogens involved in various postharvest diseases, conditions favoring disease development, and methods of developing effective systems of disease management.

1.2 THE STUDY OF POSTHARVEST DISEASES

The rapid detection and identification of pathogens is the basic requirement for the study of various aspects of the diseases and to develop effective systems

TABLE 1.1 Assessment of Losses Caused by Postharvest Diseases

Crop/Country	Pathogens	Loss (%)	Reference
Apple/England	*Gloeosporium* spp. *Monilinia fructigena*	0.3–21.03	Preece, 1967
Apple/France	*Gloesporium* spp. *Botrytis cinerea* *M. fructigena* *Penicillium* spp.	7.58–22.6	Bondoux, 1967
Mango/India	*Colletotrichum gloeosporioides*	29	Sohi et al., 1973
Strawberry/USA	*Botrytis cinerea*	10–15	Legard et al., 2000
Kiwifruit/USA, New Zealand	*Botrytis cinerea*	20–30	Michailides and Elmer, 2000
Avocado/ New Zealand	*Botrydiplodia theobromae*	23	Ledger et al., 1993
	Colletotrichum gloeosporioides	28	

of management. The pathogen population has to be quantified by employing specific and sensitive techniques that will provide results rapidly (Chapter 2). The environmental factors that may influence the development, survival, and overwintering of microbial pathogens causing postharvest diseases have to be studied to explore the possibility of controlling them during storage (Chapter 3). When the conditions are favorable for the development of microbial pathogens, the phenomenon of pathogenesis is initiated and the characteristic symptoms are induced. The names of postharvest diseases are commonly based on the chief symptoms induced in the produce (Chapter 4). The influence of cultivation practices (Chapter 5), handling, and storage conditions have to be assessed carefully in order to avoid or minimize the incidence of postharvest diseases. The storage conditions markedly influence the development of microorganisms causing diseases and producing mycotoxins capable of inducing mycotoxicoses in humans and animals, if the contaminated foods and feeds are consumed (Chapter 6).

1.3 DEVELOPMENT OF DISEASE MANAGEMENT STRATEGIES

Depending on the virulence (pathogenic potential) of the microorganisms, level of susceptibility/ resistance of the crop cultivar and the environment to which both the host and pathogen are exposed, a combination of short-term and long-term strategies to contain the postharvest diseases have to be developed. Some strategies are preventive in nature and they may be effective only if they are adopted prior to infection by microbial pathogens. Exclusion of microbial pathogens at all stages after harvest is necessary to prevent access to the harvested produce. The effects of different physical agents, such as ultraviolet-C (UV-C), different forms of heat, and modification of storage atmosphere, are discussed in Chapter 7. Production of crop cultivars resistant to postharvest pathogens may be the most desirable strategy. However, the non-availability of dependable sources of resistance to diseases and the long period of time required for incorporating disease resistance genes into susceptible cultivars limit the wider application of this approach (Chapter 8).

The increasing concern of the public regarding health hazards and environmental pollution following indiscriminate use of agrochemicals have necessitated the intensive search for alternative strategies for the control of postharvest pathogens. A wide range of fungi and bacteria occurring in the natural environments in which the agricultural and horticultural crops are grown has been screened and selected microorganisms have been employed for the control of the postharvest pathogens. These biocontrol agents are particularly suited for the control of postharvest diseases, since the area of treatment is limited and the conditions in which the commodities have to be treated can be controlled. The effectiveness of the biocontrol agents in providing protection against the pathogens, the specific requirements of the biocontrol agents and the methods of improving their efficiency are highlighted in Chapter 9. Developments in molecular biology and genetic engineering have opened up new vistas for the development of cultivars with resistance to diseases by incorporating genes from wild relatives, microbes, and even animals. In addition, the levels of resistance of susceptible cultivars can be enhanced by employing physical and chemical elicitors of disease resistance (Chapter 10).

Among the disease management strategies, application of various chemicals has been widely practiced with varying degrees of success. However, their use has to be minimized or avoided, because of two important drawbacks – namely development of resistance to chemicals in several microbial pathogens and the levels of residues of chemicals persisting in the harvested produce leading to health hazards in humans and animals. These factors were the primary considerations for the restricted use of chemicals approved specifically for use against certain diseases of fruits and vegetables (Chapter 11). Most of the strategies that have been tested could provide satisfactory control only under particular set of conditions. Hence, it is considered that safe postharvest treatments should be integrated to enhance the level of control of postharvest diseases. The possibility of integrating physical, biological and chemical methods to effectively control postharvest pathogens and to provide pathogen-free and chemical-free products for the consumer is discussed in Chapter 12.

As the management of postharvest diseases has to commence from the field where from the produce is harvested and extend during the storage and transit, the cooperation of the producer and trader is essential. The cultivation practices and plant protection measures should be carefully monitored so that pathogen inoculum reaching the produce is minimized, if not eliminated entirely. The nature, amount, and time of nutrient application should be properly determined to avoid enhancement of host plant and produce susceptibility to microbial pathogens, both in the field and in storage. Furthermore, the use of fungicides in the field should be regulated to prevent the development of resistance to chemicals that are to be applied later, prior to storage. For example no preharvest application of dicarboximide was permitted in New Zealand because of the widespread emergence of dicarboximide resistance in *Botrytis cinerea* that causes gray mold disease of kiwifruit (Michailides and Elmer, 2000). Hence, the realization of the possible occurrence of fungicide-resistant strains of fungal pathogens and the importance of adopting effective corrective measures can be expected to pave the way for effective management of postharvest diseases. Further, another dimension to disease management has arisen, as the preference of consumers for minimally-processed fruits and vegetables is steadily increasing. Outbreaks of human illness seem to have originated from food-borne pathogens such as *Escherichia coli* and *Salmonella* present in fresh-cut fruits and vegetables (Chapter 6). The need to take appropriate measures is emphasized to enhance the level of food safety (Chapter 12).

This book aims to provide comprehensive information to understand the potential of the different kinds of postharvest microbial pathogens that cause serious losses in a wide range of harvested produce, when conditions conducive for their development exist. The options that can be exercised at different stages are discussed in detail, with an emphasis on the coordination of efforts and cooperation of all agencies involved in the production, transport and storage. The teachers, researchers, and graduate students in the Departments of Plant Pathology, Microbiology, Food Technology and Environmental Sciences, Commercial Production Centers, and Plant Quarantine and Certification agencies will find the information presented in this book helpful in their efforts to contain the postharvest diseases. The protocols described in different chapters will assist in the development of research investigations for monitoring disease incidence followed by the application of effective management systems.

SUMMARY

The importance of postharvest microbial pathogens – fungi, bacteria, and viruses – with potential to inflict substantial quantitative and qualitative losses of harvested produce has been recognized. The factors favoring the development of postharvest diseases and various strategies available for the management of these diseases are outlined.

REFERENCES

Bondoux, P. 1967. Les maladies cryptogamiques des poires et des pommes au cours de l'entreposage. *Annales des Épiphyties* **18**: 509–550.

Ledger, S.N., Campbell, T.P., Banks, A.G., Atkinson, I., Kernot, I.I. and Fullelove, G.D. 1993. Internal quality of avocados in retail shops. *Queensland Department of Primary Industries Report*, 27 pp.

Legard, D.E., Xiao, C.L., Mertely, J.C. and Chandler, C.K. 2000. Effects of plant spacing and cultivar on the incidence of *Botrytis* fruit rot in annual strawberry. *Plant Disease*, **84**: 531–538.

Michailides, T.J. and Elmer, P.A.G. 2000. *Botrytis* gray mold of kiwifruit caused by *Botrytis cinerea* in the United States and New Zealand. *Plant Disease*, **84**: 208–223.

Narayanasamy, P. 2005. *Immunology in Plant Health and its Impact on Food Safety.* Haworth Press, New York.

Preece, T.F. 1967. Losses of Cox's Orange Pippin apples during refrigerated storage in England, 1961–1965. *Plant Pathology*, **16**: 176–180.

Sohi, H.S., Sokhi, S.S. and Tewari, R.P. 1973. Studies on the storage rot of mango caused by *Colletotrichum gloeosporioides* and its control. *Phytopathologia Mediterranea*, **1**: 114–116.

Stakman, E.C. and Harrar, J.G. 1957. *Principles of Plant Pathology*. Ronald Press, New York.

Waller, J.M. 2002. Postharvest diseases. In: *Plant Pathologist's Pocketbook* (eds) J.M. Waller, J.M. Lenné and S.J. Waller, pp. 39–54. CAB International, U.K.

2

DETECTION AND IDENTIFICATION OF POSTHARVEST MICROBIAL PATHOGENS

Postharvest diseases have been recognized to be important, not only because of their potential to inflict greater losses than field crop diseases directly, but also due to their ability to cause health hazards in humans and animals when the contaminated commodities are consumed. The imperative need for the rapid detection, accurate identification, and precise quantification of the microbial pathogen(s) causing the disease(s) in question has been underscored by researchers and technologists. Furthermore, development of effective systems of management of postharvest diseases essentially depends on the rapid identification of the pathogen(s) causing such diseases. Assessment of disease incidence and intensity and the extent of losses requires suitable detection techniques. In addition, prevention of disease carry-over through seeds and propagative materials rests primarily on the detection techniques employed to identify and eliminate the plant materials carrying the microbial pathogens, leading to significant reduction of pathogen population capable of infecting fruits and vegetables in the ensuing seasons.

Agricultural produce may be exposed to infection by microbial pathogens both prior to harvest in the field and also after harvest during transit, handling, and storage. Postharvest diseases are primarily caused by fungal and bacterial pathogens, whereas viral pathogens may cause progressive reduction in yield and quality of the produce. The detection techniques – conventional and molecular – that have been widely employed to detect postharvest pathogens are discussed in this chapter, with an indication of their usefulness and

Postharvest Pathogens and Disease Management, by P. Narayanasamy

TABLE 2.1 Priority (Based on Economic Importance and Frequency of Occurrence) of Microbial Pathogens

Crop	Pathogen	Priority
Beans (*Phaseolus* spp.)	*Pseudomonas syringae* pv. *phaseolicola*	1
	Xanthomonas campestris pv. *phaseoli*	1
	P. syringae pv. *syringae*	2
Carrot (*Daucus carota*)	*Alternaria dauci*	1
	A. radicina	1
	X. campestris pv. *carotae*	2
Celery (*Apium graveolens*)	*Septoria apicola*	1
	Phoma apiicola	2
Cucumber (*Cucumis sativus*)	*Cucumber green mottle mosaic virus* (CGMV)	1
	P. syringae pv. *lachrymans*	3
Lettuce (*Lactuca sativa*)	*Lettuce mosaic virus* (LMV)	1
Pea (*Pisum sativum*)	*Pea seed-borne mosaic virus* (PSbMV)	1
	P. syringae pv. *pisi*	2
	Asochyta spp.	3
Pepper (*Capsicum annuum*)	*Tobamo viruses*	1
Tomato (*Lycopersicon esculentum*)	*Clavibacter michiganensis* pv. *michiganensis*	1
	P. syringae pv. *tomato*	2
	Tobamo viruses	2
	P. solanacearum	3
	Potato spindle tuber viroid (PSTV)	3

Source: Meijerink, 1997.

limitations. Information on various other methods of detecting the microbial plant pathogens in plants is available in other publications (Fox, 1993; Narayanasamy, 2001, 2005).

Microbial pathogens may be present in various agricultural commodities such as seeds, fruits, and vegetables and processed foods and feeds (Table 2.1). Different techniques have to be adopted for accurate detection and precise quantification of the pathogens and their toxic metabolites such as mycotoxins.

2.1 MICROBIAL PATHOGENS IN SEEDS

Healthy seed is acknowledged as a prerequisite for success in efficient crop production. Hence assurance regarding the supply of healthy seeds is demanded by growers. The interest in seed health testing has increased substantially with expansion of international trade and tighter restrictions on the movement of seeds. Production and distribution of healthy seeds has become an international issue. International and national plant protection and quarantine agencies, such as the International Seed Testing Association (ISTA), have prescribed seed health management methods and standards based on sound scientific and practical technology. Descriptions of various methods are presented in the working sheets published by ISTA. The working sheets provide one or more tests for each host–pathogen combination and relevant literature in addition to the details regarding the number of seeds, size or weight of working samples, and recommendations for chemical or physical treatment applied prior to inoculation of seeds.

2.1.1 Conventional Methods of Detection

Conventional methods, requiring isolation of microbial pathogens from test seeds, require a long time to yield results and may be influenced by the conditions under which the tests are performed. However, they form the important, basic laboratory methods. The sensitivity and specificity of conventional methods for detection of microbial pathogens are generally low. These methods are described briefly and for additional information Neergaard (1977), Agarwal and Sinclair (1996), Maude (1996), and Narayanasamy (2002) may be consulted. Direct examination of seeds after immersion in water, using a stereomicroscope, may reveal the symptoms on the seeds more clearly or encourage the production and liberation of spores. It may be possible to identify some fungal pathogens by direct examination. But further tests may be required for accurate identification of the fungi.

2.1.1.1 Detection of Fungal Pathogens

A. Blotter Test: Three layers of moistened absorbent (blotting) paper are placed in sterilized Petri plates and test seeds (5–10) are arranged equidistant from one another on the paper layers. The Petri plates with the contents are incubated at $22 \pm 2°C$ under near ultraviolet (NUV) light with alternate cycles of 12 h of light and darkness for 7 days. Using a stereomicroscope, the developing fungal colonies are examined and their characteristics are used for the identification of the fungi. In some cases, deliberate suppression of seed germination using herbicide 2,4-dichlorophenoxyacetic acid (2,4-D) solution can be used to permit the fungal structures to develop. Seed-borne infection of brassica seeds by *Phoma lingam* could be detected with more certainty using this method (Hewett, 1977; Maguire and Gabrielson, 1983). Later, freezing carrot seeds for the detection of *Alternaria radicina* was found to be effective (Pryor et al., 1994).

B. Direct Plating or Agar Test: Plating the required number of seeds, after surface sterilization, on a suitable medium for detection of seed-borne pathogens may provide more reliable information, since the colony characteristics of the pathogen growing on the medium may facilitate identification. Selective media have been used for the detection of slow-growing pathogens to preferentially encourage their growth, while the development of saprophytic fungi is suppressed. A semiselective medium (CWmedium) consisting of galactose (30g), calcium nitrate (3g), dipotassium hydrogen phosphate (1g), magnesium sulfate (1g), benomyl (100ppm), chloramphenicol (100ppm), agar (20g), and water (1L) was developed for the detection of *Alternaria brassicola* in cruciferous seeds. The growth of some common saprophytic fungi (such as *Aspergillus* spp., *Cladosporium* spp., *Fusarium* spp., *Penicillium* spp. and *Ulocladium spp.*) is effectively suppressed, facilitating the development of *A. brassicola* present in the seeds. The morphology and color of the colony are the criteria used for the identification of the pathogen, which forms a dark brown to black colony with black powdery fruiting structures after 7 days of incubation at 24°C with 12h diurnal light. Several international laboratories have tested and confirmed the efficiency and sensitivity of the CW medium for the detection of *A. brassicola* in crucifer seeds (Wu and Chen, 1999). Fungal pathogens infecting vegetatively propagated material may be detected by plating the macerated tissue in a semiselective medium as in the case of *Fusarium oxysporum* in corms of gladiolus (Roebroeck et al., 1990).

C. Growing-on Test/Seedling Symptom Test: A specified number of seeds, as per the prescribed standards for each crop, is planted in soil (preferably sterile) and the percentages of germination and plants infected are determined in the greenhouse. The pathogens, such as *Sclerotium* sp., that could not be detected in soybean seeds were detected by this test (Dhingra and Muchovej, 1980). The need for large greenhouse space and the long time required for obtaining results are the limitations of this test. This method was, however, reported to be very effective in the case of obligate pathogens such as downy mildews carried by seeds.

D. Chemical Methods: The presence of the mycelium of fungal pathogens in the embryos of infected cereal seeds can be observed by staining. The embryos of wheat seeds infected by loose smut pathogen *Ustilago nuda* are extracted in NaOH (5%) and then stained with trypan blue (Khanzada et al., 1980; Khanzada and Mathur, 1988). In a modified method for the detection of *U. tritici* and *U. nuda*, the wheat and barley seeds are incubated in NaOH (10%) containing trypan blue (1g/L) for 12 to 16h. The embryos are separated from endosperm by passing the seeds through different sieves followed by boiling them in alkali solution for 15min. The mycelium stained by trypan blue can be observed after washing the embryos and boiling again in acetic acid or lactic acid (45–50%) for 1min (Feodorova, 1997). Treatment of rice seeds with NaOH (0.2%) was shown to be effective for the detection of *Trichoconiella padwickii* in rice seeds in which the diseased portion of infected seed turned black (Dharam Singh and Maheshwari, 2001). The possibility of using isozyme patterns for the detection of and identification of fungal pathogen present in seeds has been explored. Bonde et al., (1989) showed that the teliospores produced by *Tilletia indica* and *T. barclayana* in stored grains could be differentiated based on the isozyme patterns of these pathogens.

The fungal pathogens cause deterioration of seed quality. Production of characteristic volatiles and odor from infected cereal grains may be considered as a rapid, early indication of deterioration of grain quality. The odor of wheat grains infected by the common bunt pathogen (*Tilletia caries*) was assessed by an electronic nose, which was found to be more efficient than a panel of grain assessors. The electronic nose seems to sense a different characteristic not related to common bunt odor (Börjesson and Johnson, 1998). The use of electronic nose technology for early detection of grain spoilage, based on volatiles as an indicator of fungal activity and differentiation between species, was tested in wheat, maize, and other cereals. The range of volatiles produced by spoilage fungi in vitro and on grain has been determined and the key volatile groups indicative of spoilage have been identified (Magan and Evans, 2000).

E. Physical Methods: Immersion of seeds in sodium chloride solution (20%) is a simple method to remove lighter, infected seeds and fungal sclerotia that will float on the surface of the salt solution. A color image analysis of wheat kernels inoculated with *Fusarium* sp. was performed using the Multi Scan® 4.01 and calculations were made using MS Excel® to determine the extent of colonization by the fungal pathogen. This method may be used for rapid evaluation of kernel colonization by fungal pathogens and it may aid in the programs to control grain quality for food and feed purposes (Wiwart and Korona, 1998). It is possible that visual inspection of large wheat seed lots may result in harvest delays and missed wheat kernels infected by *Tilletia indica*, causing Karnal bunt disease, due to inspector fatigue. A high speed optical sorter developed by Dowell et al., (2002) efficiently removed the infected kernels rapidly and processed up to 8800 kg/h. This technology has the potential for use to remove bunted grains from seed wheat and wheat for food or feed use.

2.1.1.2 Bacterial Pathogens Examination of seeds using a light microscope or stereomicroscope and blotter test applicable for the detection of fungal pathogens are generally not useful in the case of bacterial pathogens. However, by using microscope with an ultraviolet light source, the fluorescing spots produced by *Pseudomonas syringae* pv. *phaseolicola* on the seed coat of beans may be recognized. But the applicability of this test is limited to white-seeded bean cultivars only (Wharton, 1967). The methods are employed for the detection of bacterial pathogens in appropriate media, followed by various biochemical tests to determine the characteristics of the pathogens isolated. Various selective media have been used for the isolation of seed-borne bacterial pathogens. Washed and surface-sterilized seeds, or their extracts at an appropriate dilution, are plated in the specific medium. Washed rice seeds are placed on King's medium B and fluorescent colonies may be observed after 1 or 2 days if the seeds are infected by *Pseudomonas fuscovaginae* (Agarwal et al., 1989).

A rapid diagnosis (RD) medium (Merck) was demonstrated to be effective for the detection of *Erwinia amylovora* causing fire blight disease of pear and apple. The medium consists of peptone (3.0 g), sodium chloride (5.0 g), sodium dihydrogen phosphate (2.2 g), sodium monohydrogen phosphate (2.7 g), tryptophan (1.0 g), sodium pyruvate (1.0 g), tergitol 7 (0.15 g), sorbitol (2.0 g), chromogenic mix (0.4 g), and agar (10.0 g) in one liter of distilled water. The diagnosis of *E. amylovora* could be completed within 36 h. The results were validated by BIOLOG and the fatty acid profile methods. The BIOLOG microplate system provides accurate identification of bacterial genus, species, and pathovars, including *E. amylovora*. The reactions of the isolates are determined by their ability to utilize a series of 95 carbon sources. A color reaction indicates the positive use of the carbon source by the isolate concerned and this pattern of utilization of carbon sources is known as the "metabolic fingerprint" of the isolate. Fatty acid profiles of bacterial species/isolate constitute a stable identification system. The test bacteria grown under standard conditions are extracted and the purified fatty acids are subjected to HPLC analysis. Similarities in the fatty acid profiles of the known and test species are worked out (Kritzman et al., 2003).

The dilution plating (DP) test is conducted using various dilutions of the seed extract and colony forming units (CFU) formed in different dilutions of extracts are determined to give an estimate of living bacterial cells. This method has the advantage of estimating living bacterial cells and hence it may be preferred to other methods that cannot differentiate living and dead bacterial cells.

The growing-on test has been reported to be effective for the detection of some bacterial pathogens. Seed-borne infection of French bean (*Phaseolus vulgaris*) by *Pseudomonas syringae* pv. *phaseolicola* (Grogan and Kimble, 1967), garden pea by *P. syringae* pv. *pisi* (Watson and Dye, 1971), soybeans by *P. syringae* pv. *glycinea* (Parashar and Laben, 1972), *Xanthomonas axonopodis* pv. *malvacearum* (Halfon-Meiri and Volcani, 1977), and rice by *X. oryzae* pv. *oryzicola* (Agarwal et al., 1989) have been detected with varying levels of accuracy. The possibility of gathering epidemiologically important information on the extent of seed-to-plant transfer of bacterial pathogen, and consequent incidence of diseases in the field, is a distinct advantage of this test.

Isozyme profiles of bacterial pathogens have been shown to form a reliable basis for their identification. The usefulness of patterns of enzymes esterase (EST) and superoxide dimutase (SOD) produced by *P. syringae* pv. *pisi* strains for its identification was assessed by Malandrin and Samson (1998). The zymotypes of EST could be used to identify not only this bacterial species, but also to differentiate the races of *P. syringae* pv. *pisi*.

Bacteriophages infect specific bacterial plant pathogens and hence the specificity of the relationship between the bacteria and respective bacteriophages can be a basis for the identification of the bacteria. Pure cultures of plant bacterial pathogens are seeded in appropriate nutrient medium suitable for development. Suspensions of selected, specific phages are spotted onto the petriplates with bacterial pathogen culture. Formation of plaques (clear zones formed due to the lysis of bacterial cells) may be observed after incubation of the petriplates at 25°C for 24h. Detection of *Pseudomonas phaseolicola*, causing haloblight of beans, using specific bacteriophage was reported by Taylor (1970). *Xanthomonas campestris* pv. *phaseoli* was detected using the phage test (Sheppard et al., 1989). The possibility of employing bacteriophages for the detection and identification of seed-borne bacteria was indicated by Schaad (1989). The bacterial pathogens *X. campestris* pv. *oryzae* infecting rice and *X. axonopodis* pv *malvacearum* infecting cotton could be detected by employing specific bacteriophages. Furthermore, *X. campestris* pv. *oryzae* was differentiated into 15 lysotypes based on their sensitivity to the OP_1 phage (Ou, 1985). Two races of *X. axonopodis* pv. *malvacearum* showed distinct differences in their susceptibility to the bacteriophage (Freigoun et al., 1994). The unavailability of truly species-specific bacteriophages, development of resistance to phages, and frequent formation of false negatives appear to be the limitations of this detection technique, restricting its wider applicability.

Human bacterial pathogens, such as *Escherichia coli*, are known to be carried by seeds on their surface. *E. coli* was isolated from alfalfa (lucerne) seeds used to produce sprouts and it is implicated in human illness. Isolation of *E. coli* has been found to be difficult, probably due to inaccessibility of the pathogen entrapped in protected areas of the seed coat. Adoption of a combination of optimal conditions such as soaking seeds for 1h and pummeling seeds for 1min, an enrichment step in modified tryptic soy broth, and use of immunomagnetic beads for separation of *E. coli* cells resulted in effective isolation of this bacterial pathogen (Wu et al., 2001).

2.1.1.3 Viral Pathogens Generally, seeds infected by viruses may not exhibit distinct external symptoms as in the case of fungal or bacterial infections. The size of infected seeds may be reduced and in certain cases filling of seeds may be affected, resulting in a significant reduction in 1000-grain weight. Since the isolation of viral pathogens in pure cultures is not possible, the seeds extracts in appropriate buffers may be used as inoculum and the virus may be transmitted to the local lesion hosts or systemic hosts, as the case may be. The detection techniques based on the biological properties, such as virus dilution end point (DEP), thermal inactivation point (TIP), and longevity in vitro (LIV) are time-consuming and subject to errors due to several factors. Growing-on test for the detection of certain viruses, such *Lettuce mosaic virus*, involves growing large numbers of lettuce seedlings in insect- (aphid-) proof glasshouses and periodical inspection by competent personnel to identify the infected seedlings (Grogan, 1983). The requirement of large space in a glass house, adequate manpower, and lower levels of reliability and sensitivity of the tests based on biological properties of viruses have necessitated the use of more sensitive and rapid techniques based on the intrinsic properties of viruses by using immunological and nucleic acid-based techniques.

2.1.2 Molecular Methods of Detection

Detection of seed-borne plant pathogens by employing molecular techniques are preferred because of their greater sensitivity and reliability. The results of molecular techniques may be obtained within few hours, whereas the conventional methods need several days or weeks even. Molecular detection techniques are based on the serological properties and/or nucleic acid sequences of the pathogen genomes. Molecular techniques have become indispensable for the specific detection and identification of seed-borne viruses, since the characteristics of viruses that can be studied by other techniques are limited and cannot be used for differentiating viruses and their strains. These methods are particularly useful for the detection of pathogens that cannot be isolated in pure cultures.

2.1.2.1 Fungal Pathogens

A. Immunodiagnostic Techniques: Detection and identification of fungal pathogens by conventional methods have been relatively easier compared to bacterial and viral pathogens, for which the usefulness of morphological characteristics was limited or insufficient for establishing their identity. Fast-growing saprophytes contaminating the seeds overgrow the fungal pathogens, posing considerable difficulty in the isolation of the pathogens in pure cultures. Such a situation was observed in the case of rice seed-borne fungal pathogens. More than half of 56 fungal pathogens infecting rice have been reported to be seed-borne (Mew et al., 1988). Hence, the need for alternative, more efficient methods was realized to overcome such problems. The fungi are complex antigens and the lack of characterization of specific antigens did not hamper the use of immunological techniques for the detection and identification of fungal pathogens by the concerted efforts of researchers.

Polyclonal antisera were developed using culture filtrate, cell fractions, whole cells, cell walls, and extracellular components as immunogens. As the specificity and sensitivity of assays using polyclonal antibodies (PABs) were less species–specific, subspecies-specific monoclonal antibodies (MABs) were successfully produced. Methods of preparing polyclonal and monoclonal antisera are available in other publications (van Regenmortel, 1982; Narayanasamy, 2001; Narayanasamy, 2005). *Humicola lanuginosa* and *Penicillium islandicum* associated with discoloration of stored rice could be consistently detected with significant specificity by using monoclonal antibodies developed by Dewey et al. (1989, 1990). *P. islandicum* has to be detected rapidly and accurately, since the mycotoxin produced by this fungus causes liver lesions, cirrhosis, and primary liver cancer if contaminated rice grains are consumed. Enzyme-linked immunosorbent assay (ELISA) was shown to detect *P. islandicum* in a substantially greater number of discolored rice grains compared with direct plating of surface-sterilized grains (Dewey et al., 1990).

Several other seed-borne fungi have also been reported to produce mycotoxins capable of causing illness in human and animals when contaminated foods/feeds are consumed. It is essential, therefore, to detect fungi such as *Aspergillus* spp., *Penicillium* spp., and *Fusarium* spp. to eliminate contaminated foods and feeds. Two MABs that could react with the antigens of 12 field and 27 storage fungi were raised by Banks et al. (1992). By employing indirect ELISA and PABs raised against *P. aurantiogriseum* var. *melanoconidium*, the presence of the fungal pathogen was detected in barley seeds and a positive correlation between absorbance values and the pathogen population (growth) was observable. Chang and Yu (1997) applied DAS-ELISA for the detection of the mold fungi *Aspergillus parasiticus, Penicillim citrinum*, and *Fusarium oxysporum* in rice and corn. The amount of mold growth was strongly related to the absorbance values determined in the assay, the detection limit being 1 μg/ml. The effectiveness of DAS-ELISA for the detection of nine toxigenic *Aspergillus* spp. in rice and corn was further confirmed by Wang and Yu (1998). The usefulness of a direct competitive ELISA as a postcolumn monitoring system after liquid chromatography (LC) for the detection and quantification of fumonisins in maize was shown. This procedure was able to detect 0.1 ng of FmB_1 in maize samples and *Alternaria alternata* (AAL) toxin in culture extracts (Yu and Chiu, 1998). Immunoassays have also been demonstrated to be useful for the detection of mycotoxin produced by seed-borne fungi such as *Aspergillus* spp., *Claviceps* spp., and *Fusarium* spp. (Chapter 6).

The relative effectiveness of ELISA and immunoblotting methods for the detection of *Phomopsis longicolla* in soybean seeds was assessed by Gleason et al. (1987). The nonspecific interference in ELISA formats could be overcome by a new procedure designated as seed immunoblot assay (SIBA). The mycelium of *P. longicolla* growing from infected soybean seeds placed on nitrocellulose paper forms a conspicuous colored blotch after the paper is assayed. The advantage of SIBA is that this test detects viable *P. longicolla* propagules, whereas ELISA does not differentiate live and dead fungus. SIBA test was found to be effective for the detection of *Tilletia indica* causing Karnal bunt disease in wheat grains. Formation of colored imprints on nitrocellulose on which infected seeds were placed indicated the seed infection (Anil Kumar et al., 1998).

B. Nucleic Acid-based Techniques: Nucleic acid-based techniques using specific DNA probes have been employed for the detection of fungal pathogens in seeds, propagative materials, or plants and soil. The conventional methods of detection and identification of seed-borne fungal pathogens have been commonly used, especially when large scale seed testing is needed. However, nucleic acid-based techniques are required to investigate genetic diversity within fungal species and to precisely identify specific groups (races/pathotypes/ strains) within a taxonomic species. The nucleic acid-based techniques are more specific, sensitive, and reliable than immunodiagnostic methods and very small quantities of samples are sufficient.

Dot-blot hybridization was used for the detection of *Peronosclerospora sorghi* causing downy mildew of sorghum. The target DNA from this fungal pathogen was extracted by grinding up sorghum seeds and it was shown that single or bulked samples could be tested for the presence of *P. sorghi*. The DNA of no other fungi associated with sorghum seeds hybridized with probes, indicating the specificity of the assay (Yao et al., 1990). Sensitivity of the assay was further improved by employing probes generated from the mitochondrial DNA rather than from chromosomal DNA. These probes were highly specific and did not hybridize with DNA of other fungi tested (Yao et al., 1991).

Restriction fragment length polymorphism (RFLP) analysis provides information on the natural variations in the genomes of different groups or strains of one organism. Changes in restriction endonuclease recognition sites, due to deletion or insertion in the DNA sequences, may result in variations (polymorphisms) in fragment sizes. Following digestion of the DNA of the test organism with specific restriction enzymes such as *Eco* RI and *Cla* I, the resultant DNA fragments are separated by electrophoresis in agarose or polyacrylamide gel to determine the differences in the size of the fragments. The specific set of fragments formed due to the combination of each restriction enzyme and target DNA is considered as the fingerprint for the test strain or variety of the organism. The DNA fragments are transferred to a nitrocellulose or nylon membrane and, by using an appropriate probe, hybridization with the target DNA is performed.

Polymerase chain reaction (PCR) primarily involves heat denaturation of the target dsDNA and hybridization of a pair of synthetic oligonucleotide primers to both strands of the target DNA by an annealing step. Then, using a thermostable *Taq* DNA polymerase enzyme from *Thermus aquaticus*, new DNA is synthesized on templates to produce twice the number of target DNA.

Substantial loss of export market for wheat was the greatest threat of Karnal bunt to the U.S. wheat industry, because of either prohibition or restriction of wheat imports by other countries. The available PCR assays could not differentiate *Tilletia indica*, causing Karnal bunt disease, from *T. walkeri* infecting ryegrass. The nucleotide sequences of a 2 to 3 kb region of mitochondrial DNA, earlier amplified by PCR only from *T. indica* was determined for three isolates of *T. indica* and three isolates of *T. walkeri*. By using five sets of PCR primers specific to *T. indica*, the Karnal bunt pathogen could be effectively detected in wheat samples. Furthermore, a 212-bp amplicon was developed as a target sequence in fluorigenic 5 (prime) nuclease for PCR assay by employing the TaqMan® system to detect and differentiate *T. indica* and *T. walkeri* (Frederick et al., 2000). Differentiation of *Tilletia* spp. by repetitive-sequence-based polymerase characterization (rep-PCR) assay was demonstrated. Computer-based analysis of the database of combined fingerprints clearly distinguished each taxon and indicated the phylogenetic relationship among the isolates of *Tilletia* spp. Three main clusters (groups) were recognized. Group 1 included *T. indica* and *T. walkeri*. The results showed that rep-PCR has the potential for application as a diagnostic tool (McDonald et al., 2000).

Phomopsis longicolla causing seed decay is invariably associated with *Diaporthe phaseolorum*, both inflicting significant yield losses. The conventional methods have not been found to be useful in differentiating seeds infected by these two pathogens. RFLP analysis of polymerase chain reaction (PCR) amplification products could be used to differentiate these pathogens (Zhang et al., 1997). *Aspergillus flavus* infects peanut plants causing aflaroot disease and peanut kernels. It produces a mycotoxin, known as aflatoxin, capable of inducing serious ailments in humans and animals. A specific probe (pAF 28) could be employed to detect and distinguish *A. flavus* vegetative compatibility groups (VCGs) based on DNA fingerprinting analysis and also to predict the approximate number of VCGs present in a sample population (McAlpin et al., 2002). By using specific primers, *Plasmopara halstedii* causing sunflower downy mildew disease was detected in all infected sunflower seeds by employing PCR assay, especially in shell fraction (Says-Lesage et al., 2001).

Rhynchosporium secalis causing barley scald disease generally overwinters in barley residues; it can infect seeds symptomlessly or inducing the typical scald symptoms. *R. secalis* could be isolated from seeds on lima bean agar supplemented with Bengal rose and streptomycin. To detect symptomless infections, a species-specific primer set based on sequence analysis of the internal transcribed spacer (ITS) region of *R. secalis* was employed. The primer set did not amplify the diagnostic band specific for *R. secalis* from seed DNA extracts of resistant cv. Seebe, but the pathogen could be detected in symptomless seed DNA extracts of susceptible cv. Harrington. (Lee et al., 2001). In a further study, a primer set, RS8 and RS9, capable of amplifying a 264-bp fragment from the DNA of all isolates of *R. secalis*, was employed. This primer set did not amplify the DNA from other species used for validation of the specificity of the primers selected. This PCR assay could detect as little as 1 to 10 pg of *R. secalis* DNA and required only 1, day as against 10 days needed for the culture method (Lee and Tewari, 2001).

Isolates (87) of sorghum ergot pathogen *Claviceps africana* from diverse geographic locations were analyzed by amplified fragment length polymorphism (AFLP) primer combinations to assess the extent of genetic relationships among isolates. Two major groupings of isolates were recognized and the isolates within the two major groups showed between 75 and 100% homology. The results indicated that the international trade and / or movement might have led to these observed relationships between isolates of *C. africana* (Tooley et al., 2002).

Quantification of fungal mass of *Rhynchosporium secalis* (Rs) in barley seeds was achieved by using a competitive PCR assay. A primer set (RS1 and RS3) derived from the internal transcribed spacer (ITS) region of ribosomal RNA genes was tested. The DNA prepared from infected seeds with different intensity of infection was subjected to competitive PCR with a heterologous internal control which could compete for the same primer set in conventional PCR assay, allowing the quantification of pathogen biomass. The resulting PCR product ratio for each PCR (Rs-amplified DNA/internal control template-amplified DNA) registered increases proportionally with increasing levels of infected seeds. Correlation between the results of the competitive

PCR assay and visual disease assessment was observed in tests using artificially mixed seed samples (Lee et al., 2002).

Various fungicides have been used for treatment of infected seeds in order to eliminate seed-borne fungal pathogens. To assess the efficacy of fungicides, a PCR–RFLP technique has been employed. Treatment of rice seeds with thiram was carried out to determine its effect on *Fusarium* spp. responsible for poor seed germination. Differences in the fungal communities present in treated and untreated seeds were significant soon after treatment with the fungicide. However, no significant difference in the fungal communities of treated and untreated seeds was discernible with increasing interval after treatment. The dominance percentage of *Fusarium* spp. increased with duration of the incubation period (Wang et al., 2001).

The effect of seed treatment with fungicides on the incidence of seedling blight caused by *Fusarium* spp. and *Microdochium nivale* was assessed by quantitative PCR assay. Seed treatment with bitertanol + fuberidazole (375 + 23 g/l) or fludioxonil (24.3 g/l) was found to improve seed germination at 4° and 18°C and this might be due to significant reduction in pathogen DNA, as determined by PCR assay (Glynn et al., 2000). In a further study, it was observed that the incidence of the disease (pathogens) on seedlings from fungicide-treated seeds was low as compared with seedlings from untreated seeds (Glynn et al., 2001).

Wheat is infected by several fusaria, such as *Fusarium culmorum, F. graminearum* (*Gibberella zeae*), *F. poae, F. crookwellense, F. sporotrichoides, F. sambucinum* (*G. pulicaris*), *F. avenaceum* (*G. avenacea*), *F. trinctum*, and *Microdochium nivale* (*Monographella nivalis*), causing fusarium head blight, also known as scab disease The pathogens produce mycotoxins, trichothecenes, among which deoxynivalenol (DON) is very important. A competitive PCR-based assay to quantify trichothecene-producing *Fusarium* spp. based on primers derived from the trichodiene synthase gene (*Tri5*) was developed. The primers specifically amplified a 260-bp product from 25 isolates of the trichothecene-producing *Fusarium* spp. Wheat grains harvested from field trials to determine the efficacy of the fungicides metaconazole, azoxystrobin, and tebuconazole, applied for the control of Fusarium head blight disease in winter wheat, were tested. There was no relationship between Fusarium head blight severity and DON concentration in the wheat grains. However, a significant correlation was evident between the amount of trichothecene-producing *Fusarium* spp. and DON concentration in the grains. Metaconazole and tebuconazole significantly reduced the population of trichothecene-producing *Fusarium* spp. in the grains from treated field plots (Edwards et al., 2001).

Different *Fusarium* spp. may tend to predominate in different geographical locations and environmental conditions. By employing species-specific primers in PCR assays, the component *Fusarium* species could be determined (Nicholson et al., 2004). For the detection of *F. proliferatum, F. subglutinans*, and *F. verticilioides* commonly infecting maize and producing mycotoxins, the ribosomal internal transcribed spacer (ITS) and a portion of calmodulin gene were sequenced and analyzed to design species-specific primers. Three pairs of primers (PRU 1/2, SUB 1/2, and VER 1/2) produced PCR products of 585, 631 and 578 bp for *F. proliferatum, F. subglutinans*, and *F. verticillioides* respectively. This technique has the potential for the specific detection of these toxigenic fungi in maize kernels (Mulé et al., 2004). A PCR assay based on DNA markers unrelated to fumonisin production or on the sequences of the genes involved in fumonisin production was used to discriminate between two populations occurring within *F. verticillioides*, fumonisin-producing and fumonisin-nonproducing strains (González-Jaen et al., 2004).

Another specific, sensitive, and robust PCR assay was developed to detect *F. verticillioides* strains, using two pairs of specific primers based on the intergenic spacer region of the rDNA units. The first pair was species-specific to *F. verticillioids*. On the other hand, the second pair of primers could discriminate the major fumonisin-producing *F. verticillioides* strains that are mainly associated with crops and a minor group of strains, nonfumonisin-producing strains that are associated with bananas. The specificity of the primer sequences provided the basis for a simple, rapid, precise, and sensitive detection, identification, and differentiation of *F. verticillioies*, representing a potential risk for animal health (Patiño et al., 2004).

A fluorginic (TaqMan®) real-time PCR assay was used for the group-specific detection of trichothecene- and fumonisin-producing *Fusarium* spp. and for the identification of *F. graminearum* and *F. verticillioides* in field-collected barley and corn samples. Primers and probes were designed from genes involved in mycotoxin biosynthesis (*TR16* and *FUM1*). In addition, primers and probes were prepared based on the rDNA sequences of *Fusarium* to provide a genus-specific internal positive control. Barley samples (20) and corn samples (5) infected by *F. graminearum* and *F. verticilioides* tested positive for the presence of trichothecene and fumonisin (Bluhm et al., 2004). A real-time PCR was developed to monitor and quantify the major *Fusarium* species in the Fusarium head blight complex occurring during different seasons and at various locations. TaqMan® primers and probes showing high specificity for *F. avenaceum, F. culmorum, F. graminearum, F. poae*, and *Michrodochium nivale* var *majus*, were designed. By maintaining an internal PCR control and using proper dilutions of pure genomic DNAs of the pathogens it was possible to determine the concentrations of fungal pathogen DNA for each of the *Fusarium* spp. in leaves, ears, as well as in harvested grains of winter wheat. The results indicated the possibility of applying the *Fusarium* TaqMan® technology to quantify and monitor the dynamics of individual species of the complex causing Fusarium head blight in cereals during the growing season (Waalwijk et al., 2004).

The Light Cycler™ technology involving combination of rapid in vitro amplification of DNA with real-time detection and quantification of the amount of target molecules present in test samples was employed for the detection of *F. graminearum*. Based on the PCR primers specific to the *tri5* gene, a quantitative group-specific assay was developed. This system enables a 35-cycle PCR with 32 samples to be completed in 45 min, including quantification and identification of the product. The assay provided reproducible results (98%) in the range between 0.05 and 6.0 ng of purified *F. graminearum* DNA. This report appears to be the first on the use of the Light Cycler™ system in combination with SYBR Green I, a fluorescent dye enabling real-time detection of PCR products, for the quantification and identification of *F. graminearum* DNA in pure cultures as well as in contaminated wheat samples (Schnerr et al., 2001). Another rapid and sensitive method using DNA Detection Test Strips™ for the detection of *F. graminearum* has been developed. This assay could detect a minimum of 0.26 ng of *F. graminearum* DNA. The PCR products were detected without the need for any special equipment or hazardous fluorescent dyes (Knoll et al., 2002).

Carrot seeds are the primary sources of infection for *Alternaria alternata, A. radicina*, and *A. dauci*, the former two pathogens having a high toxigenic potential. The conventional deep-freezer blotter method and plating on selective medium are both time consuming and laborious. Hence, a PCR assay utilizing specific primers based on sequences of the internal transcribed spacer (ITS) regions of the ribosomal repeat (rDNA) was developed for the identification of the three *Alternaria* species on carrot seeds and roots. The primers were highly specific, sensitive, and capable of differentiating the fungal pathogens. Results of PCR assays applied to naturally infected seed samples and root tissues were similar to those obtained with the conventional methods, indicating that the PCR assay could be preferred as alternative, especially when results are required rapidly (Konstantinova et al., 2002).

Black spot disease of crucifers caused by *Alternaria brassicae* is an important seed-borne disease. Production of disease-free seeds is considered as the effective disease management strategy. Detection and identification of the pathogen using conventional isolation and identification based on morphological characteristics are time-consuming and laborious. Hence a PCR-based diagnostic procedure involving the use of specific primers designed from DNA sequence in the ITS region of nuclear rDNA was employed to detect and differentiate *A. brassicae, A. brassicola, and A. japonica*, causal agents of black spot of crucifers. These pathogens were detected in DNA extracted from seed macerates (Iacomi-Vasilescu et al., 2002). In a later investigation, a PCR assay employing pathogen-specific primers was developed. The primers were designed on the basis of the sequence of two clustered genes potentially involved in pathogenicity. By employing two different sets of primers, the conventional and real-time PCR, *A. brassicae* was specifically detected in the DNA extracted from seed. The real-time PCR provides two advantages, that is automation for large scale application and quantitative estimation of seed infection. Furthermore, the presence of seed-borne pathogens such as *A. brassicola*, and *A. japonica* in radish, *A. alternata* in radish and cabbage, *Stemphylium botryosum, Penicillium* sp., *and Aspergillus* sp. in cabbage, and *Verticillium* sp. in tomato seeds was detected by the quantitative real-time

TABLE 2.2 Level of Seed Infection by Fungal Pathogens in Cabbage and Radish, as Determined by Quantitative Real-Time PCR

Fungal pathogen	Infection level (%)					
	Cabbage	Radish				
	1	1	2	3	4	5
Alternaria brassicae	0	0	2	3	9.5	10
A. brassicola	8	0	0	0	0	0
A. japonica	0	1	0	0	0	0
A. alternata	19	39	11	38.5	27.0	13.5
Fusarium sp.	2	0	0	1	0	0.5
Phoma sp.	1	0	0	1.5	0	0

Source: Guillemette et al., 2004.

PCR. This diagnostic technique was applied to check the natural infection levels in cabbage and radish seeds (Table 2.2) (Guillemette et al., 2004).

2.1.2.2 Bacterial Pathogens

A. Immunodiagnostic Techniques: A variety of specific antigenic determinants capable of reacting with antibody probes is readily accessible. When compared to fungal pathogens, bacterial pathogens are more amenable for detection by immunoassay. Both polyclonal and monoclonal antibodies specific to bacterial species/pathovars have been used for the detection of seed-borne bacterial pathogens. Monoclonal antibodies provide greater sensitivity and specificity for the immunological techniques that are particularly useful in detecting latent infections.

Tube/slide agglutination tests have been adopted for the detection of *Pseudomonas syringae* pv. *phaseolicola* (Taylor, 1970; van Vuurde and van den Bovenkamp, 1981), *Xanthomonas campestris* pv. *phaseoli* (Trujillo and Saettler 1979) in bean seeds and *P. syringae* pv. *phaseoli* in peas (Ball and Reeves, 1992). The virobacterial agglutination (VBA) test was performed by sensitizing *Staphylococcus aureus* cells with specific PABs. *P. syringae* pv. *phaseolicola* and *P. syringae* pv. *pisi* could be detected in bean and peas respectively (Lyons and Taylor, 1990). The possibility of differentiating the pathovars of *P. syringae* by Ouchterlony double-diffusion test was reported by Mazarei and Kerr (1990). By using the antiserum raised against glutaraldehyde-fixed bacterial cells of *P. syringae* pv. *pisi*, this pathogen could be specifically detected in pea seeds.

Among the immunodiagnostic tests, enzyme-linked immunosorbent assay (ELISA) and its variants have been used widely for the detection of seed-borne bacterial species and for their differentiation. By using specific MABs, *P. syringae* pv. *pisi* could be detected in pea seeds by employing both indirect and competitive ELISA formats. Furthermore, the strains of this pathogen were also differentiated by these tests (Candlish et al., 1988). The presence of

Erwinia stewartii in corn (maize) seeds was detected consistently by applying ELISA tests (Lamka et al., 1991). The PABs generated against virulent cells of *Ralstonia solanacearum* encapsulated with mucin were used for the detection of *R. solanacearum* in tomato seeds. The detection limit of the ELISA format employed was as few as 100 cells/ml (Rajeshwari et al., 1998). *X. campestris* pv. *oryzae* was detected in seeds and leaves of 60 rice accessions by Zhu et al. (1988). The MABs generated specifically against *X. campestris* pv. *oryzae* were able to react with all 178 tested strains of the pathogen obtained from diverse geographical locations. These strains were classified into four groups based on their reaction with four MABs. Later Huang et al. (1993) grouped 63 strains of *X. campestris* pv. *oryzae* into nine reaction types consisting of four serovars and seven subserovars. Detection and identification of *X. oryzae* pv. *oryzae* in rice seed contaminated with other microflora could be done reliably by using pathovar-specific MABs (Gnanamanickam *et al.*, 1994). Seed-borne infection of wheat by *X. campestris* pv. *undulosa* could be recognized with high precision by using a semiselective enrichment broth (SSEB) for concentrating the bacterial pathogen followed by an ELISA test. The SSEB-ELISA protocol enhanced the sensitivity of detection and a positive reaction could be observed even when the bacterial population was less than 5×10^2 CFU/ml which is 10 times less than the population required for detection if an ELISA test without enrichment was followed (Frommel and Pazos, 1994).

The suitability of immunofluorescence tests has been examined for the detection of seed-borne bacterial pathogens *X. campestris* pv. *campestris* by using PABs or MABs, either individually or as a mixture of both kinds of antibodies. Cross reactions with other pathovars, that is *vesicatoria* and *amoraciae*, reduced the specificity of detection (Franken, 1992). On the other hand, an indirect immunofluorescence test was reported to be useful for the detection of *X. campestris* pv. *undulosa* in wheat seeds by Bragard and Verhoyen (1993). Specific identification of *X. oryae* pv. *oryzae* in highly contaminated rice seeds was possible by employing specific MABs (Xco-1 and Xco-2) which reacted with the epitopes of bacterial cell surface antigens (Gnanamanickam et al., 1994).

B. Nucleic Acid-based Techniques: The sensitivity and specificity of nucleic acid-based diagnostic techniques have been shown to be relatively greater than immunoassays in general. However, these methods are expensive and involve handling radioactive probes in certain cases. With the development of nonradioactive probes labeled with digoxigenin, the possibility of wider use of these tests has improved to a great extent. By using specific DNA probes it is possible to detect, identity, differentiate, and quantify the seed-borne bacterial pathogens very rapidly and reliably. Nucleic acid hybridization techniques have been adopted for the detection of *Pseudomonas syringae* pv. *phaseolicola* in beans (Schaad et al., 1989; Prossen et al., 1991), *Xanthomonas campestris* pv. *phaseoli* in beans (Gilbertson et al., 1989), *X. oryzae* pv. *oryzae* and *X. oryzicola* (Cottyn et al., 1994), and *P. glumae* in rice seed (Tsushima et al., 1994). In actively growing bacterial cells, a large number of rRNA molecules. (>10,000 copies/cell) is present and these rRNA molecules have diverse sequence regions that may be correlated with phylogenetic relatedness between bacterial species/ genera. A DNA sequence present in the rRNA of *X. oryzae* pv. *oryzae* that hybridized only with this pathovar was identified and employed for its detection (De Parasis and Roth, 1990). Likewise Tsushima et al. (1994) developed a probe specific for *P. glumae* causing bacterial grain rot of rice and this probe hybridized to all strains of *P. glumae*, but not to *X. oryzae* pv. *oryzae* and several other bacterial species, indicating the specificity of this detection technique.

Detection and differentiation of strains/pathovars of bacterial pathogens may be possible by adopting restriction fragments length polymorphism (RFLP) analysis, using specific restriction enzymes. RFLP patterns observed after digestion with *Pst* 1 showed specificity for a group of strains within a pathovar. Thus the strains of *X. oryzae* pv. *oryzae* (Xoo) occurring in the Philippines were distinguished and the strains prevalent in the United States appeared to be not closely related to the Asian strains based on the differences in RFLP patterns (Leach and White 1991). The strains of Xoo found in 16 of 18 locations in India belonged to the pathotype 1b as determined by the RFLP analysis (Yashitola et al., 1997).

Polymerase chain reaction (PCR) alone or in combination with other molecular diagnostic methods has been very useful for the detection of seed-borne bacterial pathogens. By using the primers from DNA sequences of the phaseolotoxin gene of *P. syringae* pv. *phaseolicola* causing haloblight disease, this pathogen could be detected even though the seeds were contaminated with saprophytic bacteria. The detection limit in seed-soak wash was 1 to 5 CFU/ml (Prossen et al., 1991). Later a rapid and sensitive PCR-based protocol was developed for concurrent detection of *Xanthomonas campestris* pv. *phaseoli* (*Xcp*) causing bean common blight disease, in addition to *P. syringae* pv. *phaseolicola* (*Psp*) in bean seed. The primer X4 specific for *Xcp* and HB14 specific for *Psp*, in combination, amplified DNA fragments from the respective bacterial pathogens present alone or in mixed infections in bean seeds. The PCR assay could detect these pathogens in seeds lots as few as one infected seed in 10,000 seeds (Audy et al., 1996). For the detection of *Alternaria alternata (*Aa*)* and *A. radicina (Ar)* with high toxigenic potential, sequence-specific primers (based on ITS of ribosomal gene repeat, rDNA) were designed. The primers were highly sensitive and were able to differentiate between these two species, in addition to *A. dauci.* There was a positive correlation between the percentages of seed infection assessed by blotter method and intensity of the amplified specific product. This PCR assay is a good alternative for the deep freeze blotter method when the results have to be obtained within a short time (Konstantinova et al., 2002).

Xylella fastidiosa, a xylem-limited bacterium causing citrus variegated chlorosis disease, was detected by PCR in all main fruit vascular bundles as well as in seed and dissected seed parts, although no visible abnormalities in the seeds could be observed. The identity of the pathogen in seeds was established by cloning and sequencing the specific amplification product obtained by standard PCR assay using specific primers. *X. fastidiosa* was transmitted from infected seeds to the seedlings and this appears to be the first report of seed infection by and transmission of *X. fastidiosa* to seedlings (Li et al., 2003).

The BIO-PCR technique developed by Schaad et al. (1995) involves a combination of biological and enzymatic amplification of DNA sequences of target bacteria. The seed-borne bacteria are isolated on a general agar medium by plating aqueous extract obtained after soaking bean seeds overnight. The bacterial cells are harvested after incubation for 45 to 48 h and subjected to two consecutive cycles of PCR without prior DNA extraction. Nested pairs of primers required to amplify the phaseolotoxin gene (*tox*) region of *P. syringae* pv. *phaseolicola* DNA were used as probes. The sensitivity of the technique could be enhanced by allowing the bacteria in the seed extract to multiply in a semisolid medium for 18 h prior to PCR assay. It is possible to detect *P. syringae* pv. *phaseolicola*, even if only one seed in a lot of 400 to 600 seeds is infected (Mosqueda-Cano and Herrera-Estrella, 1997). The BIO-PCR technique provides certain advantages over direct PCR assay: (1) absence of false positive results due to the presence of dead bacterial cells; (2) noninterference by the saprophytic bacterial contaminants; (3) absence of false negative results due to the presence inhibitors of PCR in the seeds; and (4) no need for extraction of DNA of target bacteria before amplification.

Incorporation of an immunological step prior to PCR assay significantly improves the sensitivity of detection of target bacteria present in seeds. Immunomagnetic separation (IMS) was performed prior to PCR assay to concentrate *Acidovorax avenae* subsp. *citrulli* present in watermelon seeds. A significant increase in sensitivity (100 fold) of detection by IMS-PCR was observed when compared to a direct PCR assay, the detection limit being 10 CFU/ml. As low as 0.1% of seed infection was revealed by IMS-PCR assay (Appendix 2) (Walcot and Gitaitis, 2000). The possibility of watermelon seeds getting infected from the inoculum from blossom was tested by IMS-PCR technique. *A. avenae* subsp. *citrulli* was detected in 44% seed lots assayed, despite lack of fruit symptoms. The blossoms may prove to be potential site of ingress for fruit and seed infestation by this pathogen (Walcott et al., 2003). Natural infestation of onion seeds by *Pantoea amantis*, causing center rot, was revealed by employing IMS-PCR assay with species-specific primers and polyclonal antibodies, the thresholds of detection being 10^1 to 10^3 CFU/ml of seed wash (Walcott et al., 2002).

Polymerase chain reaction (PCR) along with primer pairs developed from sequences of cloned random amplified polymorphic DNA (RAPD) fragments was applied for the detection of *Xanthomonas campestris* pv. *carotae (Xcc)*, causing bacterial leaf blight disease in carrot seeds, leaves, and stem tissues. DNA fragments of ~350 bp and ~900 bp (or ~2 kb) were amplified from genomic DNA of all known strains of *Xcc* by the primer pairs 3S and 9B.

However, the DNA of 13 other *Xanthomonas camperstris* pathovars or other bacterial species isolated from carrot tissues and seeds, was not amplified by these primers, indicating the specificity of the diagnostic assay. The 3S primer pair directed the amplification of the ~350 bp target fragment from all *Xcc*-like strains isolated from different substrates from California, Idaho, Oregon, Washington, and Canada. PCR assay using the 3S primer pair could be completed in a single day for the detection of *Xcc* in carrot seed lots with the contamination rates ranging from $2 \times 10\,(^{\wedge 2})$ to $2.3 \times 10\,(^{\wedge 8})$ CFU per gram of seed. The results of PCR assays correlated well with that of seed-wash dilution plating assays and it was more sensitive than ELISA-based assays (Meng et al., 2004).

2.1.2.3 Viral Pathogens

A. Immunodiagnostic Techniques: Development of molecular techniques for the detection of viruses transmitted through seeds was particularly essential, since the time-consuming and cumbersome biological methods had to be replaced by simple, sensitive, and rapid assay methods. The immunodiagnostic techniques were initially developed for the detection of viral pathogens and later adopted for the detection of fungal and bacterial pathogens. The presence of virus(es) in seeds may generally be detected by preparing the samples in two ways. The sample seeds may be ground and suspended in a buffer followed by centrifugation to remove the supernatant containing the viruses. The other method is to allow germination of test seeds and to prepare the extract of the seedlings alone or both seedlings and seed remnants using a suitable buffer. The latter method has the advantage in that the virus can multiply during germination of seeds and reach greater concentration, thus increasing the sensitivity of the immunoassay. Various immunodiagnostic techniques have been employed for the detection of viruses in the seeds of different crops, with different levels of sensitivity and specificity. The immunoassays should be highly sensitive so that the standards prescribed for seed-borne viruses can be satisfied. For example, in the United Kingdom, a seed lot of 5000 lettuce seeds should not have a single infected seed (Maude, 1996). Hence, a highly sensitive technique has to be used for the detection of viruses. The immunodiagnostic techniques based on the formation of precipitin lines, such as double diffusion, are somewhat less sensitive when compared to labeled antibody procedures, such as enzyme-linked immunosorbent assays (ELISA) and its variants.

An Ouchterlony double diffusion (ODD) test was conducted to detect the *Pea seed-borne mosaic virus* (PSbMV) by Zimmer (1979). The antigen and the specific antibody raised against the virus diffuse through the agar medium and visible precipitin lines are formed at the position where both reactants meet at optimal concentrations, indicating a positive reaction. The *Barley stripe mosaic virus* (BSMV) was detected by a radial immunodiffusion test by Slack and Shepherd (1975). The leaf segments of seedlings from barley seeds are planted in the agar medium containing the antibodies specific for BSMV. A ring of precipitin is formed around the leaf segment from infected seed. This test was reported to be more sensitive (10 fold) than the conventional double diffusion test. The latex agglutination test developed by Lunsgaard (1976) was, possibly, one of the early tests evaluated for the detection of BSMV by the International Seed Testing Association (ISTA). These immunoassays require large amounts of antiserum and the results can be obtained only after several days. They have been largely replaced by more sensitive techniques employing labeled antibodies such as ELISA.

Among the immunodiagnostic techniques currently in use, ELISA and its variations are the most widely applied for the detection of seed-borne viruses. Protocols for ELISA methods are available in several publications (Van Regenmorel, 1982; Fox, 1993; Narayanasamy, 2001). The development of the ELISA method by Clark and Adams (1977) is considered as a milestone in research on plant viruses, since it is possible to gather information on different aspects of microbial plant pathogens in general, and viral pathogens in particular. ELISA has been applied for the detection of viruses in seeds, propagative materials, and vectors that are involved in their transmission. The presence of *Squash mosaic virus* (SMV) was detected in extracts from seed coats, papery layers, and distal halves of embryos from individual cucurbit seeds (Nolan and Campbell, 1984). Among seed-borne viruses, *Lettuce mosaic virus* (LMV) has been studied intensively for the development of effective seed health system. The standard double antibody sandwich (DAS)-ELISA was demonstrated to detect one LMV-infected seed in a lot of 7000 healthy seeds. (Maury-Chovelon, 1984). Later, Dinant and Lot (1992) reported that ELISA test was sensitive to a tolerance level of zero infected seeds out of 30,000, revealing the suitability of ELISA test to satisfy standard for LMV control imposed in the United States.

ELISA has been demonstrated to be effective for the detection of several viruses present in the seeds, such as *Alfalfa mosaic virus* (Pesic and Hiruke, 1986), *Barley stripe mosaic virus* (Lister et al., 1981), *Bean common mosaic virus* (Jafarpour et al., 1979), *Broad bean stain virus* (Makkouk et al., 1987), *Cucumber mosaic virus* (Jones and Proudlove, 1991), *Cucumber green mottle virus* (Faris-Mukhayyish and Makkouk, 1983), *Melon necrotic spot virus* (Avegelis and Barba, 1986), *Pea seed-borne mosaic virus* (Hamilton and Nicholas, 1978; Ding et al., 1992), *Pea early browning virus* (van Vuurde and Maaqt, 1985), *Peanut mottle virus* (Bharathan et al., 1984), *Peanut stripe virus* Culver and Sherwood, 1988), *Prune dwarf virus* and *Prune necrotic ringspot virus* (Mink and Aichele, 1984), *Soybean mosaic virus* (Maury et al., 1985), *Squash mosaic virus* (Nolan and Campbell, 1984), and *Tobacco ringspot virus* (Lister, 1978). Sweet corn seedlings grown from 46,600 seeds were tested by ELISA for the presence of *High Plains virus* (HPV) and three plants tested positive for HPV, whose occurrence was first observed in 1993 (Forster et al., 2001). ELISA test was shown to be equally sensitive as the reverse transcription polymerase chain reaction (RT-PCR) for the detection of *Cowpea aphid-borne mosaic virus* (CAMV) in peanut seeds, when the tests were applied to large seed lots (Gillaspie Jr. et al., 2001).

Indirect ELISA format has been applied for the detection of many viruses in seeds. *Papaya lethal yellowing virus* (PLYV) was detected by using indirect ELISA format, in papaya seeds, soil and also water from the rhizosphere of infected plants (Cammarco et al., 1998). The presence of *Bean common mosaic virus* (BCMV) could be detected in individual bean seeds by using MABs in indirect ELISA format. It was shown that infection by BCMV could be assessed by testing flour samples from seeds, and it was possible to predict the incidence of BCMV in germplasm accessions (Klein et al., 1992). *Peanut clump virus* (PCV) infection in 16.5% of seeds was observed by Konate and Barro (1993). The extent of seed infection of *Indian peanut clump virus* (IPCV-H), as determined by ELISA, was correlated with the results obtained by carrying out a growing-on test, indicating the reliability of the immunoassay. Furthermore, all seeds from wheat plants infected by IPCV-H were ELISA positive, whereas none of the seeds of barley infected by this virus showed the presence of this virus, even though barley is a natural host of IPCV-H (Dalfosse et al., 1999).

Dot-immunobinding assay (DIBA) also known as dot-ELISA or dot blot assay, is similar to ELISA, the microtiter plate being replaced by nitrocellulose or nylon-based membrane. The cut surface of the presoaked test seed is brought in contact with the membrane. The free protein binding sites present in the membranes are then blocked by using bovine serum albumin (BSA) or nonfat milk powder, followed by application of virus-specific antiserum. Development of a colored dot indicates the positive reaction. The presence of *Barley stripe mosaic virus* and *Bean common mosaic virus* in single seed of French bean could be detected by DIBA technique (Lange and Heide, 1986). DIBA was reported to be slightly more sensitive than ELISA for the detection of *Pea seed-borne mosaic virus* (Lange, 1986). In the case of *Peanut mottle virus* (PMV), detection by DIBA was satisfactory only when monoclonal antibodies were employed, whereas ELISA provided the required level of sensitivity with both monoclonal and polyclonal antibodies (Sherwood et al., 1987). Use of plain paper in place of nitrocellulose membrane has been demonstrated to be equally effective for the detection of PSbMV, BSMV, BCMV, *Pea early browning virus*, and *Squash mosaic virus* making this assay very simple (Lange et al., 1991). DIBA has certain advantages over ELISA in that the membranes can be easily stored and transported during disease surveys and there is no need for sample preparation. However, ELISA formats have been the preferred procedures, employed very widely for the detection of seed-borne viruses. Enzyme-linked fluorescent assay (ELFA) has the potential to be used as routine test for the detection of LMV and SMV in seeds. ELFA can detect LMV in a single infected seeds present in lots of 500 seeds and it was found to be 10 to 25 fold more sensitive than DAS-ELISA (Dolares-Talens et al., 1989). For the detection of SMV, ELFA proved to be more sensitive than standard ELISA format (Hill and Durand, 1986). Another sensitive detection technique is radioimmunoassay (RIA), using radioactive gamma globulin. Although this technique could detect LMV in three infected seeds present in seed lots of 30,000, the need for using potentially dangerous isotopes (^{125}I) has restricted its large scale use (Ghabrial and Shepherd, 1982; Dinant and Lot, 1992).

Electron microscopy has been useful for spot checking the presence of viruses in infected tissues. With the development of immunosorbent electron microscopy (ISEM), the specificity of detection was improved significantly due to the use of specific PABs or MABs which react with the virus or strain concerned (Derrick, 1973; Roberts and Harrison, 1979). The virus particles in the virus preparation or seed extract are adsorbed onto electron microscope grids, and made specific for the target virus, followed by staining or metal shadowing. The adsorbed virus particles can be viewed under electron microscope. *Lettuce mosaic virus* (LMV) was detected in the pooled seed extract containing 1 part of infected seed extract and 100 parts of healthy seed extract. The other viruses detected by ISEM were *Barley stripe mosaic virus* (BSMV), *Tobacco ringspot virus* (TRSV), and *Soybean mosaic virus* (SoyMV), when there was low incidence of these viruses (Brlansky and Derrick, 1979).

Labeling the antibodies specific for the virus(es) to be detected, with gold particles facilitated the *in situ* observations on the viruses in different tissues/cells of infected plants and in vectors that transmit the viruses from plant to plant. *Alfalfa mosaic virus* (ALMV) antigen was detected in the cytoplasm and vacuoles of ovule integuments, microspores, mature pollen grains, and anther tapetum cells of infected alfalfa plants. Raft-like aggregates of virus particles and large crystalline bodies were also seen in the cytoplasm of pollen grains and anther tapetum cells (Pesic et al., 1988). By using immunogold labeling, *Cucumber mosaic virus* (CMV) could be detected in the cytoplasm of ovary wall cells, ovule integuments and nucellus, anther and seed coat cells, and also in fine-fibril-containing vesicles and electron dense inclusions of amorphous aggregates in the central vacuoles of these cells in the infected spinach seeds (Yang et al., 1997). These methods are not suitable for routine testing, because of the requirements of facilities and expertise. However, they can be used for confirmation of positive test results obtained from other assays such as enzyme-linked immunosorbent assay (ELISA), since the virus particles can be visualized.

B. Nucleic Acid-Based Techniques: Various nucleic acid-based assays have been employed to detect, differentiate, and quantify plant viruses in infected plants and propagative plant materials such as tubers, corms, bulbs, and cuttings. However, these methods have been applied to detect viruses in seeds in only a few cases. In the case of RNA plant viruses, reverse transcriptase is used to produce a complementary cDNA prior to PCR assay for the detection of the virus concerned in the seeds. *Pea seed-borne mosaic virus* (Kohnen et al., 1992) and *Cucumber mosaic virus* in lupin seeds (Wylie et al., 1993) were detected by RT-PCR assay. The anthers of stone fruit plants may be a source of *Plum pox virus* (PPV), as revealed by the results of RT-PCR assay on desiccated anthers stored at 0°C for 1 to 2 years. The results of Levy et al. (1995) pointed out the need for monitoring international movement of *Prunus* germplasm to prevent the spread of PPV to other nations. The presence PPV could be detected using RT-PCR assay in both seed coat and cotyledons of apricots, whereas ELISA could detect PPV in the seed coat only, indicating the greater sensitivity of RT-PCR assay in detecting the virus in tissue containing low virus titer (Pasquini et al., 1998). RT-PCR assay was able to detect more *Cowpea aphidborne virus* (CAMV) in peanut seeds than in the number of infected seedlings normally arising in the germination tests. Moreover, CAMV could be detected in samples consisting of one infected and 99 healthy leaves, indicating the sensitivity of the RT-PCR assay (Gillaspie Jr. et al., 2001).

The combination of an immunological reaction with RT-PCR results in a dramatic increase in the sensitivity of detection of plant viruses. The immunocapture (IC)-RT-PCR assay has been demonstrated to be 100 to 1000 times more sensitive than ELISA. In the case of LMV, the IC-RT-PCR procedure consists of trapping the virus particles on antiserum-coated wall of the tube, followed by washing to remove the inhibitory substances of plant origin (Vlugt et al., 1997). The IC-RT-PCR assay was applied to detect *Peanut stripe virus* (PStV) and *Peanut mottle virus* (PeMV) in extracts of small slices taken from test seeds distal to the radicle. This assay was more sensitive than ELISA and has the potential for large-scale testing of peanut germplasm (Gillaspie et al., 2000).

2.2 MICROBIAL PATHOGENS IN PROPAGATIVE PLANT MATERIALS

Microbial plant pathogens infect propagative plant materials such as tubers, corms, bulbs, and setts either through infected aerial plant parts or by the soilborne inoculum or inoculum carried by the irrigation water. Infected plant materials are the primary sources of inoculum for the incidence and spread of diseases in the subsequent crops. When they are stored for marketing to consumers, the microbial pathogens – fungi and bacteria – multiply and cause deterioration in the quality and reduce the market value to a great extent. Hence it is essential to detect and identify the pathogens present in these plant materials to prevent spread of the diseases and to provide commodities with acceptable quality to the consumers. Several vegetable crops, such as potatoes, onions, and yams, and other crops such as sugarcane are grown by planting propagative plant materials. Infection by fungal, bacterial, and viral pathogens can cause considerable reduction in quantity and quality of marketable produce. As in the case of true seeds, the pathogens present in propagative materials have been detected and identified by using both conventional and molecular methods.

2.2.1 Fungal Pathogens

Generally, the fungal pathogens cause visible symptoms on the infected propagative plant parts and it may be possible to identify the pathogens based on the symptoms. However, in the case of quiescent infections, the presence of fungal pathogens has to be detected by using conventional methods of isolation and studying the morphological characteristics of the pathogen(s) isolated in pure culture. As the conventional methods require a long time to yield results, rapid and reliable diagnostic methods are necessary to detect and identify the pathogens with certainty.

The propagules of fungal pathogens present in the soil or other complex substrates may form the sources of infection for tubers that may be carried over during storage and transport. *Pythium aphanidermatum* present in the soil causes leak disease and watery wound rot in potato tubers. A simple, rapid, reliable and quantitative technique involving the use of sterilized oat seeds was employed as bait to assess the pathogen population. This method could detect ca. 1 oospore per gram of dry soil (Priou and French, 1997).

Immunoassays have the potential to detect and quantify the soil-borne pathogens more precisely. ELISA formats have been shown to be useful in the detection of fungal pathogens infecting propagative plant materials. Polyclonal antibodies generated against purified mycelial proteins from *Verticillium dahliae* could be employed for the detection of the isolates infecting potato (Sundaram et al., 1991). The presence of *Spongospora subterranea* could be detected in potato tuber extract by using the PABs generated against the homogenate of spore balls (cystosori). The detection limit of ELISA was as little as 0.08 spore balls equivalent/ml (Harrison et al., 1993). The antisera raised against specific proteins of pathogen origin can be used to detect the fungal pathogens producing such protein. *Colletotrichum falcatum* causing red rot disease of sugarcane has a 101-kDa polypeptide that can be used to generate polyclonal antibodies (PABs) and these PABs can be used to detect the pathogen in root eyes, buds, leafscar, and pith region of the sugarcane stalk (Viswanathan et al., 1998).

Nucleic acid-based techniques have higher level of sensitivity and specificity compared with immunodiagnostic methods. By using appropriate DNA probes, fungi that cannot be detected and identified by other diagnostic procedures may be precisely and rapidly identified, in addition to establishing the relatedness of fungal pathogens. Polymerase chain reaction (PCR) assay has been demonstrated to be useful for the detection of different species of *Phytophthora*, including *P. infestans*. Primers designed from the sequence of the internal transcribed spacer (ITS) region specific to *P. infestans* could be employed to detect the pathogen in infected potato and tomato field samples (Trout et al., 1997). Furthermore, *P. infestans* and *P. erythroseptica*, causing pink rot of potato tubers could be detected and differentiated by employing primers designed from sequences of the ITS2 region of their respective DNA. It is possible to detect the presence of these pathogens in potato tubers as early as 72 h after inoculation, well before any visible symptoms could be observed (Tooley et al., 1998). *Phytophthora* spp. belonging to six taxonomic groups, including *P. infestans*, could be identified by adopting a PCR procedure based on amplification of 5.8 S rDNA gene and ITS4 and ITS5 primers developed by Liew et al. (1998). Potato tubers were inoculated with *P. infestans* which was detected in light sprouts and stems at different growth stages by PCR. The concentration of pathogen mycelium was below the detection limit during the course of plant growth, probably due to dilution effects. However, the symptoms induced later indicated the ability of the pathogen to cause symptom at very low concentration (Appel et al., 2001).

Colletotrichum coccodes causing black dot disease of potato was detected by employing species-specific nested primers (CcINFI/Cc2NRI). A rapid procedure for the direct extraction of DNA from potato tubers and soil was developed. The limit of sensitivity of detection of *C. coccodes* was 3.0 spores/g or the equivalent of 0.06 microsclerotia/g of soil. *C. coccodes* was detected in extracts of potato peel (Cullen et al., 2002). Another potato fungal pathogen, *Spongospora subterranea*, was detected in potato peel, tuber washings, and soil by employing specific primers (Sps1 and Sps 2) based on the sequences of the ITS region of rDNA of the target pathogen. These primers amplified a 391 bp product only from *S. subterranea*, but not from any other fungi associated with potato tubers, indicating the specificity of detection of the desired pathogen. The PCR assay had a detection of threshold for *S. subterranea* DNA equivalent to 25×10^5 cystosori or one zoospore per PCR. This assay procedure may be applied for disease risk assessments of seed potato stocks (Bell et al., 1999). A PCR-based protocol was developed for the detection of *Helminthosporium solani*, causing silver scurf disease responsible for significant losses in table stock, seed and, processing potatoes. The specific primers Hs1F1/Hs2R1 amplified a 447 bp product from 20 infected potato tissue samples and 54 single spore isolates of *H. solani*. The detection by PCR was rapid, providing results in 1 day and the results correlated well with those obtained from conventional methods, which required 4 to 5 weeks (Errampalli et al., 2001).

A solid-phase microextraction (SPME) fiber to trap volatiles from Russet Burbank potatoes inoculated with *Pythium ultimum*, *Phytophthora infestans*, or *Fusarium sambucinum* and gas chromatography with flame ionization detector (GC-FID) to fingerprint trapped volatiles has been used. The amount of volatiles produced, in general, depended on the severity of disease. In addition disease-specific volatiles were also detected. The chromatogram of *F. sambucinum* showed two unique peaks at retention time (RT), whereas *P. infestans* produced few peaks and the profile was quite similar to uninoculated control tubers. In contrast, *P. ultimum* produced many peaks and a distinct peak at RT = 1.71 min (Kushalappa and Lui, 2002).

2.2.2 Bacterial Pathogens

Bacterial pathogens may be present in all plant parts including seeds and vegetative propagules. Frequently, latent infections (expressing no visible symptoms) form important sources of inoculum for the diseases of several glasshouse and agricultural crops, such as tomatoes and potatoes. Development of rapid and accurate indexing procedures to detect latent infections forms a basic component of disease management systems, resulting in the reduction or elimination of inoculum and consequently disease incidence in subsequent crops. Both conventional and molecular techniques have been applied for the detection of bacterial pathogens in propagative plant materials. Classical agar plating technique has not been very successful for screening cassava planting material for contamination with bacterial pathogens due to nonavailability of a reliable semiselective agar medium. A new cefazolin trehalose agar (CTA) medium was developed, consisting of K_2HPO4 (3.0 g), NaH_2 PO4 (1.0 g), $MgSO_4$, $7H_2O$ (0.3 g), NH_4 Cl (1.0 g), D(+) trehalose (9.0 g), (+) glucose (1.0 g), yeast extract (1.0 g), cefazolin (0.025 g), lincomycin (0.0012 g), phosphomycin (0.0025 g), cycloheximide (0.25 g) and agar (14.0 g) in 1 L of distilled water. Isolation of *Xanthomonas campestris* pv. *manihotis* infecting cassava from contaminated planting materials and soils was at a much higher rate than on a starch-based semiselective medium (SXM). The CTA medium has the potential for use in identification of infected planting materials and their subsequent elimination, as well as in basic ecological studies of the pathogen (Fessehaie et al., 1999). Another new semi-selective medium (MMG), consisting of maltose (10 g), tryptone (5 g), K_2HPO_4 (2.75 g), trace elements (0.02 to 1.0 mg), methyl green (2 ml of 1% aqueous solution), amoxicillin (32 mg), cephalothin (32 mg), cycloheximide (50 mg), and agar (15 g) in 1 L of water, was useful in selective isolation of *Xanthomonas campestris* pv. *vitians* causing bacterial leaf spot of lettuce. MMG medium inhibited most background bacteria and permitted reliable identification of *X. campestris pv. vitians* (Toussaint et al., 2001).

Indexing propagative plant materials for latent bacterial infections by isolation methods often tend to be time consuming, laborious, and lacking in adequate sensitivity. Hence immunodiagnosis of bacterial diseases has been preferred approach, because of rapidity, reliability, and greater sensitivity of these techniques. Use of monospecific polyclonal antisera and monoclonal antibodies has enhanced the specificity of immunological detection of latent bacterial infections. Lipopolysaccharide is a structural component of the outer membrane of the Gram-negative bacterial pathogens and it is readily released into the surrounding medium. As the lipopolysaccharide forms one of the primary antigens, the antibodies generated against lipopolysaccharide can be used to detect the presence of the bacterium producing this compound. ELISA and immunofluorescence tests have been shown to be effective for the detection of latent infection.

Visual examination of seed potatoes, as used in the past, did not ensure selection of potatoes free of infection by *Erwinia* spp. By using DPEM medium for anaerobic amplification of *Erwinia chrysanthemi*, the ELISA test could be employed to detect the bacterial pathogen in seed potatoes. This protocol improved the sensitivity of the assay from 10^5 bacteria/ml to 10^3 bacteria/ml, resulting in the detection of latent infection in 10% of 133 seed lots that had been certified by visual examination. This procedure was recommended for large-scale application for detection as well as for prediction of disease outbreaks in Switzerland (Cazelles et al., 1995). By using an indirect ELISA and specific monoclonal antibodies, isolates of *E. carotovora* subsp. *atroseptica* were assigned to serotype I. The serotype I constituted 86% of the isolates from potato tubers. None of the isolates was found to be *E. chrysanthemi* in this study conducted in Poland (Sledz et al., 1999). *Clavibacter michiganensis* subsp. *sepedonicus* (*Cms*) causing bacterial ringrot of potato tubers was detected efficiently by ELISA, the sensitivity of which was equal to that of PCR assays (Slack et al., 1996). Factors such as inoculum dose (>10^9 CFU), cultivar, and interval after planting may affect the effectiveness of detection of *Cms*. The need to adjust the population of bacteria and optimal levels (>10^9 CFU) was indicated by the studies of De Boer and Hall (2000).

The possibility of detecting *Xanthomonas albilineans* in symptomatic stalks of sugar cane (99%) and asymptomatic stalks (14%) by ELISA format was reported by Comstock et al. (1997). Furthermore, the comparative usefulness of ELISA formats for the detection of sugarcane pathogens has been assessed. Indirect ELISA was found to be more sensitive in detecting *X. albilineans* than immunoblot assay. The pathogen concentration was greater in sugarcane internode tissues than in leaves (Viswanathan et al., 1998). Another sugarcane bacterial pathogen *Clavibacter xyli* subsp. *xyli* could be consistently detected in asymptomatic canes, thus providing the possibility for preventing infected setts being used for planting (Viswanathan, 1997).

The sensitivity of immunofluorescence (IF) tests may be influenced by factors such as sample concentration, volume of sample applied to the microscope slide, number of microscope fields observed, and the magnification of the sample viewed. *Clavibacter michiganensis* subsp *sepedonicus* (*Cms*) was detected in potato tissues by IF tests with varying sensitivities (De Boer et al., 1994). The IF test was useful in distinguishing three serovars among 215 strains of *X. albilineans* obtained from Australian, African, Asian, and North American countries (Rott et al., 1994). A protocol combining the use of rhodamine-labeled oligonucleotide probe and indirect IF test based on specific MAB detected with a FITC-labeled conjugate was useful in the accurate identification of *Cms* by microscopic examination (Li et al., 1997). The MABs prepared from phage display library, against the lipopolysaccharide of *R. solanacearum* (biovar 2, race 3) were used for the reliable detection of this pathogen causing potato brown rot disease (Griep et al., 1998). Sensitive and specific routine detection of *R. solanacearum* in symptomless potato tubers could be obtained by efficient enrichment followed by a reliable DAS-ELISA based on specific MAB 8B-IVIA which reacted with 168 strains of *R. solanacearum*. The MAB did not react with any of the 174 isolates of other pathogenic or unidentified bacteria isolated from potato tubers indicating the specificity of ELISA test. The detection limit of the immunoassay was 1 to 10 CFU of *R. solanacearum* per ml (Caruso et al., 2002).

Nucleic acid-based techniques are not only more rapid and sensitive, but also are capable of detecting bacteria that have not been grown successfully in culture. Different probes have been employed for the detection of *Cms* in symptomless potato tissues and it is known to persist in potato tissues for long periods without being detected, if cultural and immunological methods are applied (Verreault et al., 1988; Johansen et al., 1989; Mirza et al., 1993; Drennan et al., 1993). A 1.1 kb multicopy repeated sequence (Rs) from *Cms* was used as a probe in direct blotting of potato plant tissue (Drennan et al., 1993). A PCR method based on the amplification of a 258-bp fragment of *Cms* plasmid pCS1 was reported to be more sensitive than the probe-based methods, the detection limit being approximately 100 CFU/g of potato tissue (Schneider et al., 1993). By using radiolabeled probes obtained from three unique single copy DNA fragments (Cms 50, Cms 72 and Cms 85) isolated from CS strain, this bacterial pathogen was more efficiently detected by subtraction hybridization. All North American strains of *Cms*, including plasmidless and nonmucoid strains, could be specifically detected by employing these probes. The detection limit of a PCR employing a primer pair which amplified Cms 85 was 1000 CFU/ml (Mills et al., 1997).

Ralstonia solanacearum race 3 biovar 2 causing brownrot was detected in potato tissues by fluorescent *in situ* hybridization (FISH) technique, using specific probe RSLOB. Potato tissues showed a strong fluorescent signal (Wullings et al., 1998). Pectin methyl esterase (Pme) is essential for the pathogen development during pathogenesis. Primers designed from the *pme* gene sequence coding for Pme generated an 800-bp fragment involved in Pme production by *R. solanacearum*. When this sequence was employed as probe, PMR-coding plasmids of *R. solanacearum* genomic library were identified. This study indicated that Pme is necessary for the growth of but not for the virulence of *R. solanacearum* (Wullings Kersten et al., 1998).

Polymerase chain reaction (PCR) has been shown to be more sensitive than immunoassays in detecting bacterial pathogens in plant tissues. For the detection of *Clavibacter muchiganensis* subsp. *sepedonicus* (*Cms*), PCR primers (Sp1f and Sp5r) amplifying a 215-bp fragment of genomic DNA of *Cms* could be employed advantageously. There was no amplification of DNA from phenotypically and serologically related bacteria associated with potato stem or tubers, indicating the specificity of the PCR assay, which was shown to be more sensitive than ELISA and immunofluorescene (IF) tests using MABs. Tubers showing a negative reaction when tested by ELISA and IF tests were found to be infected by PCR assay (Li and De Boer, 1995; Li et al., 1997).

The BIO-PCR technique combining biological and enzymatic amplification of PCR targets was developed by Schaad et al. (1995) for the detection of *Pseudomonas syringae* pv. *phaseolicola* in bean seeds. Later the BIO-PCR assay was used in combination with an automated real-time fluorescence for the detection of *Clavibacter michiganensis* subsp. *sepedonicus* (*Cms*) in symptomless potato tubers. PCR primers and a fluorescent probe for use in PerkinElmer 7700 automated real-time PCR detection system (TaqMan®) were designed from a *Cms*-specific generic DNA fragment. The comparative efficacy of: (1) plating aliquots into agar media; (2) classical PCR; and (3) BIO-PCR was assessed by testing a total of 30 naturally rot-infected tubers. These methods detected *Cms in* 4, 8, and 26 tuber samples respectively. The BIO-PCR combined with TaqMan® automation-based detection was shown to be a rapid, reliable method of assaying large number of potato tubers extracts for *Cms*, which was difficult to detect because of low populations and its slow competitive growth on available media. In addition, there is no need for DNA extraction from the test organism before amplification and false positive results due to the presence of dead bacterial cells can be avoided by employing a BIO-PCR technique (Schaad et al., 1999). A new assay procedure involving the use of digoxigenin (dig)-labeled PCR, employing the primer pair MSIF1/CMSIR1, had high sensitivity on a nylon membrane and the signal was detected by a colorimetric assay using alkaline phosphatase. This assay could detect *Cms* in symptomless field-grown potato tubers (Lee et al., 2001).

The superiority of PCR assay over ELISA was also demonstrated in the case of another potato bacterial pathogen, *Erwinia carotovora* subsp. *atroseptica* (De Boer and Ward, 1995). The PCR assay, with a greater sensitivity of detection, has a crucial role in disease management programs, since infected seed potatoes form the primary sources of infection which have to be thoroughly eliminated to raise healthy potato crops. To achieve the objective of detecting several bacterial pathogens concurrently, a one-step PCR-based method was developed for the detection of five species of *E. carotovora*, including subsp. *carotovora* and *atroseptica* and all pathovars and biovars of *E. chrysanthemi*, in micropropagated potato plants. A DNA fragment of 119 bp was amplified from all strains of target bacterial species by the primers SR3F1 and SRIcR based on a conserved region of the 16S rRNA gene present in test bacterial species. The sensitivity of detection was significantly enhanced (about 200 fold) by incorporating an enrichment step prior to PCR amplification (Toth et al., 1999). A quantitative competitive PCR method for the detection of *E. carotovora* subsp. *atroseptica* (*Eca*) was developed for use in quarantine programs. This assay method detected (*Eca*) even in the presence of other *Erwinia* and saprophytes. Furthermore, all serogroups of *Eca* could also be detected. Based on the results, a key was formulated relating product ratios to the likelihood of blackleg disease incidence (Hyman et al., 2000).

Specific and unique regions in DNA in different bacterial pathogens may be identified and primers designed from the required sequences can be employed to identify the bacteria concerned. *Clavibacter xyli* subsp. *xyli* (*Cxx*) causing ratoon stunting disease of sugarcane could be detected in infected sugarcane tissues by using the oligonucleotide primers derived from the intergenic region between 16S and 23S rRNA genes. These primers amplified the specific DNA sequences in *Cxx*. The specificity of the test is revealed by the absence of amplification in closely related *C. xyli* subsp. *cyanodontis* (Fegan et al., 1998; Pan et al., 1998). The presence of *Erwinia amylovora*, causing fireblight disease of apple, was detected in asymptomatic plant material by using a nested PCR assay. The test was performed in a single closed tube and involved two consecutive PCRs. The first one was conducted at a higher annealing temperature that allowed amplification of only an external primer pair. The second, standard PCR was performed with the internal primer pair and directed the amplification of a specific DNA fragment from plasmid pEA 29. The specificity and sensitivity of this procedure were greater than those of a standard PCR procedure using a single primer pair. This method has the potential for detection of both endophytic and epiphytic populations of *E. amylovora* in epidemiological studies, as well as for routine use in quarantine surveys because of its high sensitivity, specificity, simplicity, and rapidity (Llop et al., 2000).

A rapid and cost-effective PCR-based technique was developed for the detection of *Streptomyces scabies* and *S. turgidiscabies* causing potato scab disease. A total of 1245 scab lesions from potato cvs. Matilda and Sabina was tested using species-specific primers. Both scab pathogens were detected in the field grown potato tubers and they were transmitted through infected tubers after storage for 24 weeks at 4°C (Lehtonen et al., 2004). Specific detection of *Ralstonia solanacearum* causing brown rot disease of potato tubers, was achieved by a new assay known as Ampli Det RNA technique. This technique is based on nucleic acid sequence based amplification (NASBA) of RNA sequences and homogeneous real-time detection of NASBA amplicons with a molecular beacon. The assay is performed in sealed tubes to reduce the risks of carry-over contamination. The Ampli Det RNA technique is rapid and reliable, providing the results in 90 min at a level of 10 cells per reaction, which is equivalent to ca. 10^4 cells/ml per sample. AmpliDet RNA assay could detect *R. solanacearum* in surface water at a level of 10 CFU/ml, after concentrating the bacteria present in 200 ml of surface water into 1 ml by centrifugation. This technique was found to be more sensitive and effective than an immunofluorescence (IF) cell staining method (Van der Wolf et al., 2004). Real-time PCR primers and probe and a highly sensitive BIO-PCR assay were developed for the specific detection of strains of race 3 biovar (bv) 2 of *R. solanacearum* in asymptomatic potato tubers. The primers and probes detected all 17 strains of bv. 2, including 12 from potato and five from geranium. Other strains of *R. solanacearum* and 13 bacterial species associated with potato did not react with these primers and probes, indicating high specificity of the assay. Asymptomatic infection of potato by *R. solanacearum* could be detected by this real-time BIO-PCR assay. In contrast, the standard real-time PCR could not do so, indicating the greater sensitivity and accuracy of the new technique (Ozakman and Schaad, 2004).

2.2.3 Viral Pathogens

Plant viruses are known to be easily transmitted along with vegetative propagative materials such as tubers, bulbs, corms, cuttings, or budwood for grafting, if such materials are taken from virus-infected mother plants. Hence, it is imperative that the health status of both the mother plants as well as the propagative plant materials is ascertained and freedom from virus infection is ensured prior to planting the propagules. Latent infection of viruses in plants has been observed frequently. Furthermore, the propagative plant materials may not exhibit distinct visible symptoms, as in the case of fungal and bacterial pathogens. Adequate care and proper sampling of plant materials will be essential to avoid the escape of infected plant materials. Indexing of propagative plant materials by applying appropriate sensitive and reliable test(s) is the most effective way to prevent unlimited and uncontrolled spread of viral pathogens. Viruses can be detected, based on the reactions of the assay or indicator host plant species or based on tests that can determine the variations in the intrinsic properties of coat protein and genomic nucleic acids.

As the assay techniques based on the biological properties are generally time consuming and expensive requiring large space in glasshouse, molecular techniques have been employed more commonly for indexing plant materials. Among serodiagnostic methods, DAS-ELISA formats have been very widely used for testing propagative plant materials. In the Netherlands, about 11×10^6 ELISA tests are performed annually on propagative materials, including about 5×10^6 tests for indexing of seed potatoes and about 6×10^6 tests for indexing ornamental plant materials, since the ELISA format is less expensive and easy to perform in large numbers by relatively untrained personnel (Clark and Adams, 1977; Huttinga, 1996).

It may not be possible to detect the viruses present as such and special treatments have to be used for reliable detection, even when sensitive assays are performed. In dormant potato tubers, *Potato virus Y* (PVY) and *Potato virus A* (PVA) cannot be detected by ELISA. Hence only after breaking of dormancy, can the tubers be indexed (Vetten et al., 1983; De Bokx, 1987). For the detection of *Iris severe mosaic virus*, just after lifting the bulbs a small piece of bulb tissue has to be removed and the wounded bulbs are stored at 17°C or 20°C for 3 weeks prior to testing by ELISA test (van der Vlugt,1993). The scales of lily are stored at 20°C for 2 to 3 weeks under fluorescent light (12–16 h/day) prior to detection of *Lily mottle potyvirus* (LMoV) by ELISA, whereas other viruses such as *Lily symptomless virus* (LSV), *Cucumber mosaic virus* (CMV), and *Lily X virus* can be detected readily without any prior treatment (Derks et al., 1997). Detection of *Potato moptop virus* (PMTV) by ELISA is facilitated by storing potato tubers at 20°C for 3 weeks (Sokmen et al., 1998). Storage of dahlia tubers for a minimum period of 4 weeks and taking mixed samples of three roots per tuber are the requirements for reliable detection of *Tomato spotted wilt virus* (TSWV) by ELISA, since the virus is not uniformly distributed in the tubers (Schadewijk, 1996). The sensitivity of ELISA was remarkably enhanced, when the alkaline phosphatase and its substrate were replaced by the inorganic pyrophosphatase (PPase) from *Escherichia coli* conjugated with antibodies and tetrazolium pyrophosphate as substrate for the detection of potato viruses X, Y, M, S, and leafroll (Mizenina et al., 1991).

The usefulness of ELISA formats for the detection of viruses in vegetatively propagated crops has been demonstrated. *Apple mosaic virus* (Clark et al., 1976) and *Citrus tristeza virus* (Bar-Joseph et al., 1979) were successfully detected by ELISA formats. Application of ELISA for the detection of *Citrus psorosis-associated viruses* led to recognition of a greater number of infected accessions in which biological indexing could not detect the virus infection, suggesting that ELISA may provide more reliable results (D'Onghia, et al., 1998). The extent of incidence of bacilliform virus in sugarcane germplasm was assessed by using ELISA in order to select resistant lines (Viswanathan and Premachandran, 1998). The *Dioscorea bacilliform virus* (DaBV) infecting yams was identified and its serological relationship with *Sugarcane bacilliform virus* and *Banana streak virus* was established by employing ELISA (Philips et al., 1999).

When new virus-like symptoms are observed, the identity of the virus(es) involved should be established rapidly, so that effective disease management system can be developed to contain the new disease. The occurrence of a new mutant strain of *Potato virus M*, designated potato virus M-ID (PVM-ID) was established by the application of ELISA (Cavileer et al., 1998). Based on the serological properties determined by the ELISA, a new strain of potyvirus infecting *Dioscorea alata* was identified (Odu et al., 1999). Sweet oranges suffer severely due to infection by a strain of *Citrus tristeza virus* (CTV). By employing different combinations of PABs and MABs in indirect DAS-ELISA format, the strain inducing stem pitting strain could be differentiated (Nikolaeva et al., 1998). A monoclonal antibody (MAB) highly specific for CTV was generated and it reacted only with citrus tissues infected by CTV but not with non-infected citrus plant extracts (Öztürk and Cirakolu (2003). Pallidosis is a disease of strawberry recorded more than four decades ago and yet its causative agent has not been conclusively established. Polyclonal antibodies were used to detect the putative causal agent in the petiole tissue blots. The nucleotide sequences of the regions encoding the major coat protein were used to produce primers. A sensitive RT-PCR assay provided evidence to consider the virus causing pallidosis as a member of Closteroviridea family, *Crinivirus* genus, and it is most closely related to *Cucumber yellows virus* and *Cucurbit yellow stunt disorder virus* (Tzanetakis and Halgren, 2004).

Serodiagnostic techniques, other than ELISA, have also been tested for their usefulness in the detection of plant viruses in propagative plant materials. The sensitivity of tissue blot immunoassay (TBIA) was determined in comparison to ELISA and dot immuno-binding assay (DIBA) for the detection of *Lily symptomless virus* (LSV) in lily bulb scales. TBIA was shown to be more efficient than the other two immunoassays, since TBIA detected LSV in more bulb scales (Gera et al., 1995). Direct TBIA could be effectively applied for the detection of *Carnation mosaic virus* (CarMV) in stems, leaves, and in completely opened flowers of carnation plants. TBIA is simple and as sensitive as ELISA or dot ELISA in the detection of CarMV (Zhang, 1999). DIBA was successfully applied for the detection of *Tulip breaking virus*-lily (TBC-L) and *Cucumber mosaic virus* (CMV) in addition to LSV in the scale segments of *Lilium* sp. (Nümi et al., 1999). These studies indicate the need for selecting a suitable immunoassay for the detection of virus(es) depending on the host–virus combination.

Different nucleic acid-based techniques have been reported to be effective in detecting plant viruses in propagative plant materials. By using digoxigenin (DIG)-labeled cRNA probe of approximately 2100 bp, *Potato leafroll virus* (PLRV) was readily detected in dormant potato tuber tissues and no cross-reaction with *Potato viruses X* and *Y* was observed, indicating the specificity of the hybridization technique, which was more sensitive (2000 fold) than ELISA (Loebenstein et al., 1997). The polymerase chain reaction (PCR) can be applied for the detection, quantification, and differentiation of viruses present in vegetative propagules. Single and mixed infections of potato tubers by PVY could be resolved by digesting the PCR products with appropriate restriction enzymes (Rosner and Maslenin, 1999). A simple, rapid, and inexpensive method of sampling large number of dormant tubers by using a common electric drill to simultaneously remove and macerate the tuber eye samples has been developed. PLRV could be detected by both PCR and ELISA techniques in dormant potato tubers from plants with primary infection (Souza-Dias et al., 1999). By employing a RT-PCR technique, the presence of PLRV was detected in dormant tubers from field grown plants and in vitro-propagated microtubers. On the other hand, ELISA could detect PLRV in microtubers, but not in tubers from field-grown plants (Spiegel and Martin, 1993). RT-PCR assay detected PLRV at higher rates than field inspection assessment. The sprouts from tubers planted in glasshouse were tested by ELISA and the results corroborated those of PCR test in 85% of the positive cases (Russo et al., 1999). By using paramagnetic beads coated with specific antiserum, PLRV was captured from potato tuber extracts, followed by addition of cell wall degrading enzymes, cellulases, and macrozyme to the tuber extract to improve the sensitivity of the assay performed in microplates. This procedure is rapid, reproducible, semiquantitative, and can be automated for large-scale use. The distinct advantage of this assay (IC-RT-PCR) is the significant reduction in inspection time from 5 days for conventional testing to just 1 day (Shoen et al., 1996). The IC-RT-PCR technique was applied to detect *Potato moptop virus* (PMTV) in potato tubers and the test was found to be at least 100 fold more sensitive than ELISA format (Rantanen et al., 1999). Another technique, named detection of immobilized amplified product in a one-phase system (DIAPOPS), was shown to be effective in specifically detecting all four strains of PVY (O, N, NTN, and C) in dormant potato tubers. But DIAPOPS could not detect these strains in some potato samples which were ELISA positive (Nielsen et al., 1998).

The potato tuber necrotic ringspot disease caused by a PVY strain (PVY^{NTN}) could be differentiated from PVY^{N} strain by employing IC-RT-PCR (Tomessoli et al., 1998). A PCR-based assay was developed exploiting a recombination site in the coat protein (CP) of PVY^{NTN} strain, allowing more reliable diagnosis of the strains PVY^{o}, PVY^{r}, and PVY. This assay was shown to be more reliable than most commonly employed RT-PCR and it was especially useful for the detection of PVY^{NTN} in symptomless tissues in disease surveys and seed health programs (Boonham et al., 2002). Three major recombinant junctions (RJs) were detected in the genome of PVY^{NTN} strain at sites HC/Pro.P3, NIa and the C-terminal region of CPgene. Specific primer pairs were designed to target the three RJs so that sense and antisense primers completely matched the nucleotide sequences at either side of the RJ. In a multiplex RT-PCR, when all three primer pairs were employed simultaneously, the three fragments (641, 448, and 290 bp) were amplified exclusively from the recombinant PVY^{NTN}, while only one fragment (641 bp) was amplified from PVY^{O} isolate. The specificity of the test was demonstrated, since there was no amplification from the non-recombinant PVY, including PVY^{O} and North American (NA)-PVY^{NTN} strains (Nie and Singh, 2003).

A modified RT-PCR protocol was applied for the detection *of Prunus necrotic ringspot virus* (PNRSV) in dormant bark tissues and this protocol was recommended for screening imported budwood materials in postentry quarantine programs and for production of virus-free planting materials (Spiegel et al., 1996). Plum pox virus (PPV) is another destructive and economically devastating viral pathogen infecting *Prunus* spp. Rapid and precise detection of PPV is essential for the containment efforts involving eradication of infected trees. Detection has been based on ELISA surveys but these are laborious and not sensitive to the required level. A real-time, fluorescent RT-PCR assay has been shown to be highly sensitive and reproducible, providing results rapidly. This assay was more sensitive than ELISA or conventional PCR procedure. PPV could be detected from multiple hosts and in various tissues such as leaf, stem, bud, and root, making it suitable for the detection of the virus in propagative plant materials (Schneider et al., 2004).

An immunocapture (IC) step was combined with RT-PCR technique to enhance the sensitivity of virus detection of propagative plant materials. It has been found to be difficult to diagnose reliably two important viruses, *Tobacco rattle virus* (TRV) and *Potato moptop virus* (PMTV), in tubers and leaves of infected potato plants. A multiplex assay for the detection of TRV and PMTV directly in potato tubers was developed by combining PCR with a tube fluorescent product detection (TaqMan®). The new protocol can be used in place of two separate tests – a TRV-RT-PCR and a PMTV-ELISA – currently used in single-tube multiplex format. This test could reliably detect over 40 different isolates of TRV and PMTV, in addition to being more sensitive (100 to 10,000 fold). Other advantages of this TaqMan® assay include reduction in contamination risk, elimination of the need for ethidium bromide staining, and removal of time and cost of gel running. This assay has the potential for routine testing, speeding up and streamlining the functions of diagnostic laboratory (Mumford et al., 2000). By using a RT-PCR assay targeting the coat protein gene in RNA3 of PMTV, the virus was detected in potato seed lots and ware potato during surveillance in United States and Canada. Of the 3221 samples tested, 4.3% proved positive for PMTV. Confirmatory tests using two primer sets targeting gene segments in RNA2 and RNA3 were carried out. Amplicons generated from RNA2 and RNA3 were identified by RFLP analysis. ELISA, ISEM, and infectivity tests on *Nicotiana debneyi* also provided supporting evidences for establishing the identity of the virus (Xu et al., 2004).

Introduction of geminiviruses infecting sweetpotato from other countries through germplasm has become common, after globalization following the GATT Agreement. Indexing the propagative materials by graft-inoculation onto the indicator host plant *Ipomoea setosa* is laborious and requires up to 8 weeks to provide results. A polymerase chain reaction (PCR) assay based on degenerate primers SPG 1/SPG G was found to be sensitive in detecting nine uncharacterized isolates of geminiviruses in sweetpotato from Asia and America, including the *Sweetpotato leaf curl virus* (SPLCV-Taiwan) from Taiwan. The PCR assay detected SPLCV-Taiwan in a highly diluted sample ($10^{\wedge -9}$) indicating its sensitivity and specificity. The degenerate primers had a broader detection range than virus-specific primers and could be employed to detect geminiviruses from in vitro plantlets (obtained through meristem culture), greenhouse-grown sweetpotato plants, and in many *Ipomoea* hosts. The results of PCR assay correlated well with those of indexing by grafting (Li et al., 2004).

A garlic virus-specific antiserum was developed against a mixture of flexuous rod-shaped viruses isolated from mosaic-diseased garlic plants. Seven viral coat protein (CP) bands from 38 to 32 kDa were identified using this antiserum in western blots. Based on western blot in conjunction with reverse transcription polymerase chain reaction (RT-PCR) assay, three of seven bands were considered to correspond to CPs of *Leek yellow stripe virus* (LYSV) (38 kDa) and two different *Onion yellow dwarf virus* (OYDV) strains (35.5 or 34 kDa), whereas the 35 kDa band corresponded to the CP of GV1-Carlavirus and the other four bands (36, 36, 33 and 32 kDa) were identified as the CPs of four miteborne viruses based on RT-PCR assay. Mixed infections of OYDV, LYSV, and miteborne viruses reduce the bulb weight substantially (Takaichi et al., 2001). To further improve the sensitivity of detection of OYDV and LYSV in garlic, a method of detection combining IC-RT-PCR/RT-PCR with the use of TaqMan® probes was developed. These viruses were detected using different fluorochromes in the probes which allowed unequivocal diagnosis of both viruses. Using another variation of this method for which the immunocapture step was not required, it was possible to apply TaqMan® RT-PCR assay directly on extracts of garlic cloves. These assay procedures showed an increase in sensitivity by 10^6 fold over ELISA. Both versions may be advantageously applied in virus-free certification schemes in garlic to achieve high-quality product (Lunello et al., 2004).

2.3 MICROBIAL PATHOGENS IN FRUITS

Microbial plant pathogens may infect fruits either prior to harvest under field conditions or after harvest during transit and storage. The symptoms of infection may be observed at different period after harvest depending on the host–pathogen combination, existing environmental condition, and the levels of resistance of the cultivar to the disease(s) in question. Many microbial pathogens may remain dormant for varying periods untill favorable conditions become available for their development, leading to visible symptoms from which the disease and the pathogen may be identified.

2.3.1 Fungal Pathogens

The latent, dormant, or quiescent infections (without visible symptom) are due to the parasitic relationship during which the pathogen remains in a quiescent stage until, under specific circumstances, it becomes active (Verhoeff, 1974). Quiescence may occur during any of the processes from fungal germination to colonization (Swinburne, 1983). The period of quiescence may depend on several factors, such as the pathogen's nutritional requirements, presence of preformed antifungal compounds, elicitation of defense-related compounds such as phytoalexin, and activation of factors enhancing pathogenicity (Prusky, 1996). The presence of microbial pathogens has to be rapidly recognized and precisely identified to apply suitable disease management practices.

Fungal pathogens are isolated from infected fruits showing symptoms by following standard procedures in artificial media such as potato dextrose agar. The morphological characteristics such as spore morphology, sporulation pattern, production of sporulating structures, and the characteristics of sexual spore forms are studied. Isolation and identification of fungal pathogens infecting fruits have been found to be still useful. The postharvest pathogen causing fruit rot in d'Anjou pears was identified as *Sphaeropsis pyriputrescens* and described as a new species. It is a low temperature pathogen, capable of growing from 0 to 25°C with the optimum between 15°C and 20°C. It cannot grow at temperatures above 30°C (Xiao and Rogers, 2004).

The fungal populations present in the atmosphere and the equipment in the packing houses were assessed by isolating them on media. The total populations of fungal pathogens, mostly belonging to the genera *Cladosporium* and *Penicillium*, were high over the entire sampling period with an average of 25 to 50 CFU per plate. Fungicide-resistant isolates of *P. digitatum* and *P. italicum* were detected (Palou et al., 2001). In addition, the ability to produce enzymes, toxins, and other metabolites required for successful establishment of the pathogen in the fruit tissues leading to characteristic symptoms of the disease(s) can be determined by using specific media containing necessary substrates/nutrients. Identification of fungal pathogens based on cultural characteristics requires a long time and the results obtained may show variations due to conditions under which the studies are conducted.

Chemicals have been utilized for the detection of some postharvest pathogens present in plant parts or in harvested produce. Treatment of grapevine berries with paraquat reveals the latent infection of *Botrytis cinerea* (Gindrat and Pezet, 1994). Likewise, latent infection of plums by *Monilina fructicola* could be detected by treatment with paraquat, paving the way for employing corrective measures (Northover and Cerkauskas, 1994). Application of paraquat (500 ppm a.i.) aided the recognition of latent infection of *Colletotrichum musae* in banana (Rajeswari et al., 1997). Banana fruits exposed to ethylene (1200 μl/l) for 24 h at 25°C and stored at 32°C for 5 days with CO_2 concentration at optimal levels exhibited the lesions caused by *C. musae* much earlier than the untreated control fruits. This method has the potential for quantification of inoculum level in the immature banana fruits (Bellaire et al., 2000). The efficacy of the herbicide paraquate and freezing was assessed for the detection of *Colletotrichum* spp. infecting strawberry. Apparently healthy petioles from field-grown plants and inoculated petioles from greenhouse grown plants were killed by freezing or treatment with 0.3% paraquat and incubated in petridish moist chambers. After 5 to 7 days of incubation, acervuli (fruit bodies) of *Colletotrichum* developed much more frequently on freeze- and paraquat-treated petioles than on control petioles which were not killed. Frequency of detection was similar in both treatments. *C. acutatum*, *C. gloeosporioides*, and *Glomerella cingulata* (teleomorph of *C. gloeosporioides*) were detected on petioles with no visible symptom. Freezing detection has the potential for wide for application as it is a viable and nonhazardous method (Mertely and Legard, 2004).

Fungal pathogens induce significant changes in infected fruits, as in the case of dried figs (*Ficus carica*). Fig fruits infected by *Aspergillus* spp. emit characteristic bright greenish yellow fluorescence (BGYF) under long-wave light. *A. flavus* and *A. parasiticus*, capable of producing the mycotoxin aflatoxin, and *A. tamari* and *A. albiacens* were associated with BGYF observed in naturally infected figs. It may be possible to eliminate the infected figs by observing the BGYF under some specific conditions prevailing in California (Doster and Michailides, 1998).

Molecular techniques, which are being employed with increasing frequency, provide reliable and reproducible results rapidly with greater accuracy compared to conventional methods involving isolation of the pathogens. Furthermore, establishing the identity of slow-growing fungal pathogens which may be overgrown by saprophytic fungi will be difficult by the conventional methods. On the other hand, molecular diagnostic methods depend on the characteristics of pathogen genomes and hence they can be applied for the identification of all fungal pathogens, including obligate pathogens which cannot be isolated on cell-free artificial media.

The desirability of using molecular diagnostic techniques to detect the presence of fungal pathogens in fruits has been indicated by several studies. Immunoassays have been demonstrated to be an effective tool for the early and rapid detection of fungal pathogens causing postharvest diseases. *Botrytis cinerea* was detected and quantified in pear stems after 6 and 8 months of cold storage by ELISA and isolation on selective medium. ELISA was more sensitive than the standard isolation method. Quantitative ELISAs indicated that over 200 μg of *B. cinerea* biomass per gram of stem tissue was present in visibly rotted fruits, while the stems from fruits without visible symptom had about 35 μg/g. The monoclonal antibody BC-12.CA4, with high specificity, could be used to determine the infection path of *B. cinerea* and for detection of latent infections (Meyer and Spotts, 2000). Antiserum raised against *Uncinula necator* infecting grape berries detected the antigens present on the conidia and hyphae. The antibody reacted with three antigens with MW of 21, 29, and >250 kDa on the conidia in an immunofluorescent assay and also in an ELISA (Markovic et al., 2002).

The immunological techniques can be applied to detect not only the fungal pathogens, but also the metabolites produced by them. Following infection of grape berries by *Botrytis cinerea* causing gray mold disease, invertase activity was stimulated. In the infected berries a new invertase species with a MW similar to that of Botrytis invertase (BIT) was detected. Western blotting with extracts of diseased berries revealed the presence of the invertase of pathogen origin. The anti-BIT-IgY antibodies (produced in chickens) were highly specific in their reaction with BIT, demonstrating the potential of BIT as a target molecule for immunological detection of *B. cinerea* (Ruiz and Ruffner, 2002).

Melanins derived from 1,8-dihydroxy naphthalene (DHN) have been shown to be essential for the pathogenicity and survival of fungal pathogens such as *Alternaria alternata*. The phage-display antibody (scFv) M1 bound specifically to 1.8 DHN in competitive inhibition enzyme-linked immunosorbent assays. There was no crossreaction with nine structurally related phenolic compounds, indicating the specificity of reaction. Immunogold labeling technique showed that melanin was located in the septa and outer (primary) walls of wild type *A. alternata* conidia. By employing M1 antibody, it is possible to detect melanized fungal pathogens in different tissues in host plant organs infected by them (Carzaniga et al., 2002).

PCR-based assays, as in the case of seeds and propagative plant parts, have been demonstrated to be more rapid, reliable, and specific for the detection and differentiation of fungal pathogens present in fruits. Sequences of specific regions, such as the internal transcribed spacers (ITS) of ribosomal DNA, may be amplified by PCR with universal primers and used to identify fungal species that are difficult to differentiate based on the morphological characters alone. The isolates (39) of *Colletotrichum gloeosporioides* (*Glomerella cingulata*) infecting different fruits were classified into 12 groups that could be linked to host plant species or geographical origin based on the results of the study using probes generated from the ITS region of rDNA of this fungal pathogen (Mills et al., 1992). By using the PCR primers specific to the 3′ regions of the intron present only in *Monilinia fructicola* infecting plum fruits, together with the small subunit (SSU) rDNA primer NS5, *M. fructicola* could be differentiated from related species *M. fructigena* and *M. laxa*. These primers amplified a 444-bp product only from *M. fructicola* and plum tissue infected with *M. fructicola*, but not from the other two fungal species, indicating the specificity and reliability of the assay (Fulton and Brown, 1997). The usefulness of PCR assay for the detection of *Venturia inaequalis* causing apple scab was reported by Schnabel et al., 1999).

A multiplex PCR assay based on species-specific primers was used to establish the etiology of bull's eye rot disease of pear occurring in the United States. *Neofabraea alba* was more frequently seen in the samples from Oregon and California, whereas another species, *N. perennans*, was predominantly found in samples from Washington. The presence of these fungal pathogens in small cankers and pruning stubs on pear trees was detected, indicating that these infected tissues might serve as sources of infection. Furthermore, bull's eye rot pathogens were detected in nine European pear cultivars, Asian pear, and quince by the PCR assay (Henriquez et al., 2004).

DNA sequence analysis of the ITS 1 and 2 regions of the nuclear rRNA gene provided the basis of separating isolates of *Alternaria infectori*, *A. arborescens*, and *A. tenuissima* considered to be involved in core rot of red apple cultivars in South Africa. The major pathogens associated with core rot disease of Top Red apples in South Africa belonged to the *A. tenuissima* species group (Serdani et al., 1998, 2002). Seven strains of *Fusarium* isolated from rotten banana fruits imported into Japan from Mexico were identified as *F. verticillioides* based on morphological and molecular characterization (Hirata et al., 2001).

Penicillium digitatum infecting citrus fruits develops resistance to fungicides acting as demethylation inhibitor (DMI). A PCR-based detection technique, involving detection of a tandem repeat of a transcriptional enhancer in the promoter region of *PdCYP51* which encodes the target enzyme of DMIs by PCR using conidia as template, has been developed. The presence of the tandem repeat results in overexpression of this gene, conferring DMI resistance to *P. digitatum*. The DMI-resistant isolates of *P. digitatum* could be detected within 5 to 6h by adopting this procedure (Hamamoto et al., 2001).

Penicillium spp. involved in blue mold disease of apple were recovered from rotten apple and pear fruits as well from water and floatation tanks in commercial apple juice facilities. The isolates of the *Penicillium* spp. were characterized by employing RFLP of the region including the ITS1 and ITS2 and the 5.8 S rRNA gene ribosomal DNA region and RAPD primers. RAPD analysis was found to be a rapid and reliable tool to identify and group the isolates into *P. expansum* and *P. solitum*. The involvement of *P. solitum* in blue mold disease was reported for the first time by Pianzzola et al. (2004).

Characterization by molecular methods and identification by morphological characters has been carried out with isolates of *Colletotrichum* spp. from almond, avocado, and strawberry. Taxon-specific primers analysis grouped the avocado isolates within the species *C. gloeosporioides* and the U.S. almond and Israeli strawberry isolates within *C. acutatum*. Sequence analysis of the complete ITS1-5.8S-ITS2 region revealed a similarity between 97.03 and 98.72% among *C. gloeosporioides* isolates from Israel, *C. acutatum* almond isolates from United States, and C. *acutatum* strawberry isolates from Israel. Although morphological characteristics showed that Israeli isolates from almond are distinct, this population was grouped within *C. acutatum* species based on molecular analysis (Freeman et al., 2000). In another study, isolates of *C. acutatum*,

C. fragariae and *C. gloeosporioides* infecting strawberry were analyzed by comparing sequences of 5.8S-ITS region. Specific primers were used to identify isolates belonging to the genus *Colletotrichum* and to distinguish isolates of *C. acutatum*. The specificity of the detection and identification of *C. acutatum* was demonstrated by the absence of amplification of DNA sequences from non-strawberry isolates of *Colletotrichum* (Martinez-Culebras et al., 2003). Quiescent infection of strawberries by *C. acutatum*, a quarantine pathogen, could be detected by using good quality nucleic acid, free of PCR inhibitors, prepared by a modified DNA extraction method. Species-specific primers were successfully employed in the PCR assay for the detection of *C. acutatum* in symptomatic and asymptomatic plant parts and in artificially and naturally infected strawberry tissues. Positive PCR results were obtained from ripe and unripe tissues and other vegetative plant parts (Parikka and Lemmetty, 2004). By employing specific primers for the amplification of desired sequences of the ITS4 region, the isolates of *Colletotrichum* infecting avocado, papaya, mango, and passion fruits were identified as *C. gloeosporioides*, whereas the natural guava isolate was identified as *C. actuatum* (Peres et al., 2002).

Fungal pathogens such as *Alternaria alternata* are known to produce host-specific toxins. Identification and differentiation of toxin-producing pathogenic isolates have been extremely difficult due to high levels of variability. The gene *AMT* that plays a crucial role in AM toxin biosynthesis was cloned and characterized. Using primers designed from the *AMT* gene, a PCR-based method was developed to specifically detect AM-toxin producing isolates of *A. alternate* apple pathotype (Johnson et al., 2000).

Detection of visible and nonvisible quiescent infections of fruits is essential to develop an effective management system to protect fruits against postharvest pathogens during storage. Early infections by *Monilinia fruicticola* causing brown rot disease of sweet cherry could be detected by employing two primer sets that were developed from DNA sequences of either ribosomal DNA (MF5 / ITS4 / ITS3) or a random amplified polymorphic DNA (RAPD) fragment (X-09 int F3/X-09R) that specifically amplified DNA from isolates of *M. fructicola* and *Monilinia* spp. respectively. These primer sets did not amplify the DNA from *Botrytis cinerea* and other fungi commonly associated with sweet cherry fruit, indicating the specificity of detection of the target pathogen. The brown rot pathogen could be detected from laboratory-inoculated sweet cherry fruit with early brown rot infections which showed no visible symptoms. Visible quiescent infections in field-collected fruits could also be recognized by these primers (Förster and Adaskaveg, 2000). A nested-PCR assay was developed for the detection of *M. fructicola* infecting stone fruits based on the sequences of microsatellite regions of the pathogen genome. The sequence of a species-specific DNA fragment amplified by microsatellite primer M13 fragment was utilized for developing primers. The external and internal primer pairs EM If F + Emf R and IMf F + IMfR amplified 571- and 468-bp fragments respectively from *M. fructicola*, but not from any other fungal species present in stone fruit orchards. The nested PCR assays were highly sensitive, since as little as 1 fg of DNA from only two conidia of *M. fructicola* could be detected. Furthermore, small numbers of conidia of *M. fructicola* trapped on spore traps could also be detected. The nested PCR is useful not only for the detection of *M. fructicola* in infected plant tissue, but also for assessing the pathogen population in the orchard for epidemiological studies (Ma et al., 2003).

In a later study, RAPD analysis was applied to monitor imported and exported fruits for the presence of quarantined *Monilinia fructigena*, *M. fructicola*, *M. laxa*, and *Monilia polystroma*. An *M. fructicola*-specific band, that was characterized by sequencing, was generated. Primers were designed to amplify the bands in the genomic region of *M. fructicola* and *M. laxa* and these bands were also sequenced. Based on all three sequences, a multiplex PCR method using a common reverse primer (MO 368-5) and three species-specific forward primers (MO 368-8R, MO 368-10R, and Laxa-R2) was formulated for differentiating the three *Monilinia* species. In this assay, a 402-bp PCR product for *M. fructigena*, a 535-bp product for *M. fructicola*, and a 351-bp product for *M. laxa* were produced. Furthermore, another specific 425-bp PCR product was amplified, enabling the precise identification of all isolates of recently characterized *Monilia polystroma*. This multiplex PCR assay could be used to amplify the PCR products of respective pathogens inoculated on apples, as the PCR band produced was specific to the species of *Monilinia* or *Monilia*, demonstrating the potential of the assay for effective monitoring for the presence of quarantine pathogens that have to be intercepted (Côté et al., 2004).

Botrytis cinerea causing grey mold in greenhouse tomatoes was detected by a dot blot technique. DNA from fungal cultures and infected plant tissue were amplified by using specific primers in PCR. Probes labeled with digoxigenin were employed for hybridization of amplified sequences. The probe Bot1 reacted positively with all isolates of *B. cinerea* and with fresh or frozen plant tissues infected by *B. cinerea* from research centers and commercial greenhouses. This assay procedure provided rapid and reliable results (Mathur and Utkhade, 2002). Random amplified polymorphic DNA (RAPD) and amplified-fragment length polymorphism (AFLP) techniques were applied to study the genetic relationship of *B. cinerea* populations. The genetic diversity of *B. cinerea* was found to be higher based on RAPD than on AFLP analysis, since RAPD generated more polymorphisms per loci and provided a better explanation of the genetic relationships between isolates (Moyano et al., 2003).

2.3.2 Bacterial Pathogens

The bacterial pathogens have been found to be associated with a few diseases of fruits and their primary location appears to be organs/tissues other than fruits in infected plants. The bacterial pathogens can be isolated using appropriate culture media from infected fruits. Physiological and biochemical tests have to be carried out for the identification of the bacteria in pure culture. These tests need a long time to provide the information to establish the identity of the bacterial pathogens. Frequently, saprophytic fungi and bacteria may be present along with bacterial pathogens causing general rotting to a different extent, depending on the storage conditions. This situation will make the isolation and characterization of bacterial pathogens difficult. Hence, the development of effective diagnostic methods based on immunological properties and genomic characteristics of the pathogenic bacteria becomes essential. A PCR-based protocol was developed to detect the presence of *Erwinia amylovora* causing fire blight disease of apple. Two oligonucleotide primers designed from the sequences of a unique DNA fragment of the plasmid pEA 29 were employed to detect the bacterial pathogen in leaves, axillary bud, and mature fruit calyx from infected apple trees (McManus and Jones, 1996). The PCR technique was used to monitor the population of *E. amylovora* in the calyxes of apple fruits under storage conditions (Hale and Taylor, 1999).

Sets of primers based on sequence differences in the ITS and on a sequence from the plasmid gene *pthA* involved in virulence of *Xanthomonas axonopodis* pv. *citri* (*Xac*) causing citrus bacterial canker (CBC) were employed for the identification of world-wide strains of the pathogen. Under specific conditions, pathotypes of *Xac* could be discerned and subgroups of pathotypes were identified. The origin of type A strains introduced into Florida could be inferred based on association of strains with certain geographic areas of the world (Cubero and Graham, 2002). The strains of CBC pathogen was rapidly and reliably differentiated by employing the real-time PCR assay using a portable, field-hardened RAPID machine and primers designed to detect all canker-causing strains. The assay provided results within a period of 4 h and had an apparent sensitivity of less than 10 CFU of target bacterial cells from single lesions. Detection of CBC pathogen in a herbarium sample from a 1912 canker was possible for the first time by using the real-time PCR protocol (Mavrodieva et al., 2004) (Fig. 2.1). Sweet orange trees affected by citrus variegated chlorosis (CVC) disease caused by *Xylella fastidiosa* produce severely undersized and hard fruits lacking juice and such fruits become unacceptable for either juice or fresh market. By employing PCR assay, *X. fastidiosa* was detected in all main fruit vascular bundles, as well as in the seeds, and samples of peduncle, peel, endocarp, septum, locular membrane, and central axis of infected fruits of three sweet orange cvs. Pera, Natal, and Valencia (Li et al., 2003) (Figs 2.2 and 2.3).

2.4 MICROBIAL PATHOGENS IN VEGETABLES

Vegetables are exposed to microbial pathogens which may induce symptoms either prior to harvest or during transit and storage, leading to a significant reduction in quality. Infection by fungal and bacterial pathogens remain latent in many vegetables. Early detection and rapid identification of the pathogen(s)

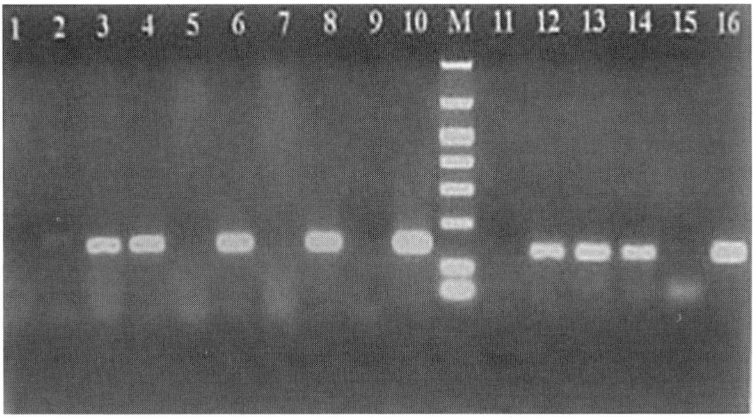

Figure 2.1 Real-time polymerase chain reaction (PCR) products of dried herbarium citrus leaf samples infected by *Xanthomonas citri* pv. citri using direct tissue extracts or DNA purified using Idaho Technology (IT) Kit from extracts, electrophoresed in 2% agarose gels and stained with ethidium bromide. Lanes 1 to 8: VM3 and 4 primer pair PCR products. Lane 1-A1 extract; 2-A1 DNA; 3-A2 extract; 4-A2 DNA; 5-F3 extract; 6-F3 DNA; 7-F4 extract; 8-F4 DNA; 9-water control; 10-10 ng of *Xanthomonas citri* 3213 DNA. Lane 11 to 14: Kingsley's primer pair products. Lane 11-A1 DNA; 12-A2 DNA; 13-F3 DNA; 14-F4 DNA; 15-water control; 16-10 ng of *X. citri* pv. *citri*A DNA. Lane M-Marker DNA (BioVentures Inc., Murfresboro, TN) of size: 1000, 700, 525, 400, 300, 200, 100 and 50 bp. (Courtesy of Mavrodieva et al., 2004; The American Phytopathological Society, St. Paul, MN, USA.)

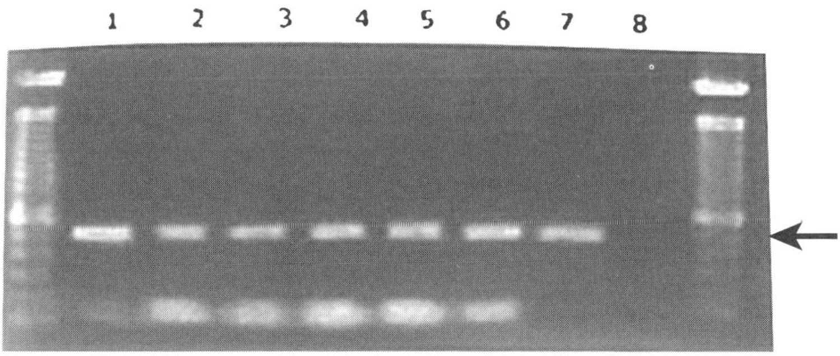

Figure 2.2 Detection of *Xylella fastidiosa* in sweet orange fruit tissues by polymerase chain reaction (PCR). Lane 1-peduncle; 2-exocarp; 3-mesocarp; 4-endocarp; 5-septa; 6-central axis; 7-*X. fastidiosa* strain 9a5c in culture; 8-central axis from healthy fruit (negative control). Lanes at both ends contain a 100-bp ladder as a size marker. Arrow indicates 472-bp PCR product. (Courtesy of Li et al., 2003; The American Phytopathological Society, St. Paul, MN, USA.)

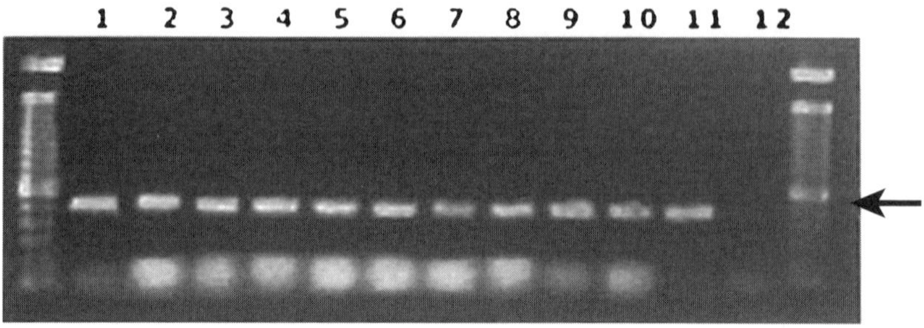

Figure 2.3 Detection of *Xylella fastidiosa* in seed parts and seedlings. Lanes-1 and 11-pure cultures of *X. fastidiosa* 9a5c; 2-seed coat; 3 and 4-embryos; 5 to 7-in vitro isolates from Valencia, Natal and Pera seedlings respectively; 8 to 10-seedlings of Pera, Natal, and Valencia; 12-water (control). Lanes at both ends contain a 100-bp DNA ladder as a size marker. Arrow indicates a 472-bp PCR reaction product. (Courtesy of Li et al., 2003; The American Phytopathological Society, St. Paul, MN, USA.)

involved become essential to eliminate the infected ones as well as to prevent the spread of disease during transit and storage. The molecular methods of disease diagnosis is preferable to conventional isolation methods because of their greater reliability, rapidity, and sensitivity.

2.4.1 Fungal Pathogens

The effectiveness of certain physical and chemical techniques in detecting infection of vegetables by fungal pathogens has been indicated by some studies. Prior to and after inoculation of tomatoes with conidial suspensions of *Fusarium oxysporum* and *Rhizopus stolonifer*, visible and near infrared (NIR) spectra were obtained. Spectral signatures in the frequency domain were analyzed using discriminate analysis and models capable of detecting spore-free (control) and inoculated tomatoes were obtained. The tomatoes with conidia of the fungal pathogens on their surface could be detected with high success rates and the infected ones could be differentiated accurately (Hahn, 2002). Ripening tomatoes stored at 20°C produce acetaldehyde, ethanol, ethylene, and carbon dioxide when they are inoculated with *B. cinerea*. In order to determine the possibility of using the production of these compounds as an early marker of infection in harvest fresh tomatoes, they were measured. Ethylene could be detected at more than 24 h before the first decay symptom was visible, whereas the production of other compounds were recorded much later. Hence it was suggested that ethylene production could be considered as an early indicator of infection of tomatoes by *B. cinerea* (Polevaya et al., 2002).

ELISA formats have been used for the detection of fungal pathogens infecting vegetables under field conditions and also in storage condition. *Pythium violae* causing cavities in field-grown carrots was detected by employing PABs in ELISA (Lyons and White, 1992). The sensitivity of detection can be enhanced by using monoclonal antibodies specific to certain epitopes on the fungal cell walls. *Botrytis cinerea* causes many postharvest diseases in fruits and vegetables. The presence of *B. cinerea* in infected vegetables could be detected by ELISA using specific MABs (Dewey, 1998). The possibility of detecting *Phoma betae* antigen at a wide range of concentrations in carrots and beet, and also in the seeds, with the aid of an amperometric immunoenzyme sensor was indicated by Khaldeeva et al. (2001).

Nucleic acid-based diagnostic methods have been developed for the specific detection of fungal pathogens infecting vegetables and other plants with export potential. *Botrytis cinerea* infects a wide range of crops. By employing the primer set PB 80-IF/ PB80-1R, which specifically amplified a 381-bp fragment from DNA of *B. elliptica*, both pathogens could be detected and differentiated. These primers did not amplify the DNAs from other pathogens tested. This highly specific and reliable detection method could be useful for regulatory, epidemiological, and also ecological studies (Chen et al., 1998).

For the detection of disease in potatoes, primers were designed based on sequences of the ITS1 region of rDNAs of *Pythium ultimum* and *P. aphanidermatum* associated with leak syndrome in potato tubers, and these pathogens could be detected and differentiated by employing the specific primers. This was necessary since both pathogens produce similar symptoms described for pink rot caused by *Phytophthora erythroseptica* (Triki et al., 2001). A rapid PCR assay was developed using the PINF and internal transcribed spacer (ITS) 5 primers whose sensitivity of detection of *Phytophthora infestans* in potato tubers was high and the pathogen DNA was detectable at concentration as low as 10 pg/ml. DNA from a single sporangium or oospore could be amplified by PCR after hexadecyltrimethyl ammonium bromide (CTAB) or NaOH lysis extraction methods. The PINF and ITS5 primers were shown to be a powerful tool for rapid and sensitive detection of zoospores, sporangia, and oospores of *P. infestans* and they could be easily employed to reduce the spread of the pathogen in potato tubers (Wangsomboondee and Ristaino, 2002).

RAPD-PCR analysis has been demonstrated to be useful for simultaneous identification of *Alternaria alternata*, *Fusarium semitectum*, *F. roseum*, and *Penicillium viridicatum* infecting melon fruits causing postharvest decay. Pure genomic DNA extracts or crude/unprocessed extracts from the diseased melon fruit peel tissues were used for detection of the pathogens. The results were stable, repeatable, and sensitive (Chen et al., 1999). A PCR-based assay was developed to detect *Alternaria radicina* in carrot seeds by using primers based on the sequence of a cloned RAPD fragment of *A. radicinia.* This method involved a 5-day incubation step in which the seed was maintained under high humidity conditions to increase fungal biomass. The detection limit of the assay was as low as 0.1% seed infection (Pryor and Gilbertson, 2001).

Onion neckrot caused by *Botrytis aclada* and other *Botrytis* spp. is a storage disease. It is very difficult to detect the infection at harvest by using conventional methods. Sequence characterized amplified region (SCAR) primers (BA2f/BA 1r) were designed based on a previously cloned and amplified DNA fragment of direct amplification of *Botrytis* spp. isolates associated with onion neck rot disease. The detection limit of the assay was 1 to 10 pg of purified fungal DNA. The presence of the pathogens could be detected in artificially inoculated onion bulb tissue and in mature onion leaves showing none of the symptoms of infection by *Botrytis* spp. This assay has the potential for application in ecological studies on this group of onion pathogens (Nielsen et al., 2002).

Isolation from immature (green) bell pepper showing severe anthracnose symptoms yielded two fungal isolates (AN 1 and AN 2). Based on morphological and cultural characteristics and polymerase chain reaction (PCR) with *Colletotrichum acutatum* species-specific primer (CaInt 2) and nucleotide sequencing, the fungal isolates were identified as *C. acutatum*. In addition to immature pepper, tomato and strawberry fruit were also susceptible to the AN 1 isolate. All bell pepper cultivars (22) evaluated under field conditions were found to be susceptible (Lewis Ivey et al., 2004).

2.4.2 Bacterial Pathogens

A semiselective diagnostic medium (T-5) in conjunction with low temperature incubation (5°C) was effective in the detection of *Pseudomonas viridiflora* causing bacterial streak and bulbrot of onions (Gaitaitis et al., 1997). The longer time required for isolation of the bacterial pathogens and biochemical tests are limitations of this test, as in the case of other conventional methods.

The production of volatile organic compounds (VOCs) by potato tubers inoculated with bacterial pathogens has been found to be a characteristic feature. Hence attempts have been made to assess the effectiveness of the VOCs as markers for the identification of the bacterial species. *Erwinia carotovora* subsp. *caratovora* (*Ecc*), causing soft rot disease in potato tubers when inoculated on tubers and stored at 10°C with 95% RH, induced the production of 22 volatiles unique to *Ecc* infections, including 10 alkanes, four alkenes, two aldehydes, one sulfide, one ketone, one alcohol, one aromatic, one acid, and one heterocyclic compound. Production of volatile nitrogen species from *Ecc*-infected tubers increased with time, but none was detected in the head space above control tubers, as determined by gas chromatography–mass spectrometry (GC-MS) technique, Although no specific volatile compound was useful as a marker, there was a significant increase in the volume of compounds evolved from infected tubers (Lacy Costello et al., 1999).

Immunodetection of bacterial pathogens in propagative materials and plants by applying different techniques has been effective in eliminating infected ones in quarantine and certification programs. An adaptation of the slide agglutination test using *Staphylococcus aureus* was employed in Portugal to rapidly detect the bacterial pathogen *Ralstonia solanacearum* infecting potato and tomato. The results showed that this slide agglutination technique may play a major role in the control program providing a sensitive tool for the rapid detection of *R. solanacearum* (Lyons et al., 2001).

Nucleic acid-based detection techniques have provided rapid and reliable results. The PCR-mediated amplification of DNA sequences located between specific interspersed repeated sequences (differently known as BOX, REP, and ERIC elements) in prokaryotic genomes is carried out in the repetitive sequence-based (rep)-PCR genomic finger printing technique. Highly specific DNA finger prints can be obtained by separating the amplified DNA sequences on agarose gel. The distinct advantage of this method is that there is no need for DNA extraction from samples and the test can be applied directly to cell suspensions prepared from infected tissues. By employing this technique, isolates of the pathovar *Xanthomonas campestris* pv. *vesicatoria* infecting pepper (chillies) could be differentiated (Louws et al., 1995). Another PCR-based technique, PCR-RFLP test based on a pectate lyase-encoding gene of *E. carotovora* subsp. *atroseptica* was applied for the detection of the bacterial pathogens in wash water, leaves, stem, and tuber peel extracts (Hélias et al., 1998).

The random amplified polymorphic DNA (RAPD) technique has been useful in identification and differentiation of related bacterial pathogens. *Erwinia* spp. are frequently associated with softrot diseases and hence rapid identification of the bacterial species causing softrot symptoms is necessary. By employing randomly chosen primers, *E. carotovora* subsp. *atroseptica* and *carotovora* could be differentiated by a RAPD-PCR technique (Mäki-Valkama and Karjalainen, 1994). In the new procedure, designated TP-RAPD, two primers from the sequences of 16S rDNA gene of the target bacterial species are used for amplification. With annealing at 45°C, the PCR product electrophoresed in agarose gels and produced band patterns that were different in different bacterial species and as well as in subspecies of *Clavibacter michiganensis* subsp. *sepedonicus* (*Cms*) causing potato ringrot disease. All strains of *Cms* formed the same TP-RAPD number pattern, indicating that this technique can be used for the rapid and reliable identification of strains of *Cms* (Rivas et al., 2002). Polymerase chain reaction (PCR) amplification of the *Cfl* gene from the gene cluster encoding the phytotoxin coronatine was used to identify coronatine-producing strains of *Pseudomonas syringae* pv. *maculicola* infecting leafy crucifers in Oklahoma. The expected 0.65 kb PCR product was detected in 19 strains of *P. syringae* pv. *maculicola* originating from diseased plants (Zhao et al., 2002).

2.4.3 Viral Pathogens

Plant viruses induce various characteristic symptoms on different plant parts in addition to a general reduction in growth and yield of plants. In certain host–virus combinations, symptoms of infection may be observed on seeds, propagative plant materials, fruits, and vegetables. External and internal necrosis of white cabbage during cold storage (0–2°C) was observed. In Bently and Lennox cabbage cultivars, when inoculated with a mixture of eight isolates of *Turnip mosaic virus* (TuMV), both external and internal necrosis appeared. A number of cabbages had latent virus infection without internal necrosis. Direct tissue blotting immunoassay (DTBIA) revealed the uneven distribution of the virus in cabbage heads (Krămer et al. 2000).

The relationship between viruses infecting two different fruit crop plants may be established by nucleic acid-based techniques. Molecular evidence of the relationship between the *Flat apple disease-associated virus* (FAV) and *Cherry raspleaf virus* (CRLV) was provided by reverse transcription polymerase chain reaction (RT-PCR) and tube capture (TC)/RT-PCR analyses. Comparison of amino acid residues derived from the 429-bp fragments of FAV and CRLV showed 95% identity. These assays were also used to confirm virus elimination in apple plants after heat therapy. Biological and serological evidence showed that FAV and CRLV are isolates of the same virus (James et al., 2001).

SUMMARY

Agricultural commodities are exposed to numerous microorganisms, both prior to and after harvest. Some of them cause spoilage during transit and storage. Significant reductions in the quantity and quality of harvested produce have been observed, because of the diseases induced by the microbial pathogens. Furthermore, some of the pathogens produce toxic metabolites in the infected produce leading to mycotoxicoses in animals and people when contaminated foods and feeds are consumed. It has, therefore, become essential to detect, identify, and quantify the microbial pathogen populations accurately and rapidly by employing techniques which are specific, sensitive, and rapid in providing results. Both conventional and modern molecular diagnostic techniques have been employed to detect the microbial pathogens causing postharvest diseases in seeds, fruits, and vegetables. Various techniques are discussed indicating their relative usefulness and limitations to facilitate the selection of suitable techniques for the different kinds of investigations.

APPENDIX 2

Detection of seed-borne bacterial pathogen by immunomagnetic separation and polymerase chain reaction (IMS-PCR) (Walcott and Gitaitis, 2000).

A. Production of Polyclonal Antibodies (PABs)

i. Heat-treat the whole bacterial cells [*Acidovorax avenae* subsp. *citrulli* (AAC)] at 100°C for 2 h and pellet the lysed bacterial cells by centrifugation (20,000 g for 15 min at 4°C).

ii. Collect the supernatant; adjust the pH to 7.0; add ammonium sulfate to saturation and incubate at 4°C overnight for completion of precipitation.

iii. Pellet the precipitate by centrifugation (20,000g for 15min at 4°C); resuspend the precipitate in sterile water and dialyze using MW 50,000 cut-off membrane against distilled water to remove excess ammonium sulfate present in the suspension.

iv. Adjust final antigen (bacterial cell surface protein) concentration to 300μg of protein/ml.

v. Immunize the rabbits with the bacterial antigen; purify the immunoglobulin G (IgG) fraction, using an Avid Chrom Protein Antibody Purification Kit (Sigma-Aldrich, St. Louis) and prepare the antibody reactant with a final concentration of 2mg/ml. This antiserum is designated anti-AAC.

B. Indirect Enzyme-Linked Immunosorbent Assay (ELISA)

i. Dispense approximately 5×10^6 CFU of test bacteria to each well of a polystyrene microtiter plate; add 10μl of coupling buffer consisting of $NaHCO_3$ (2.93g) and $NaCO_3$ (1.59g) in distilled water (800ml) and incubate for 30min at 37°C.

ii. Transfer 0.01M phosphate-buffered saline (PBS, 40μl) with 0.1% bovine serum albumin (PBS-BSA) to each well; gently agitate and incubate for 30min at room temperature.

iii. Wash the wells three times with wash buffer containing PBS plus 0.2% Tween 20 (PBST); add 20μl of diluted anti-AAC (1:1000); mix the reagents gently for 30s; incubate for 30min at room temperature and wash with wash buffer three times.

iv. Transfer 20μl of secondary antiserum [goat antirabbit, conjugated with alkaline phosphatase (ALP)] to each well; incubate for 30min and wash the wells with wash buffer three times.

v. Remove the excess fluid; add 20μl of p-nitrophenyl phosphate (pNPP) to each well and incubate the plates at room temperature until a yellow color develops in the positive control (approximately 5×10^8 CFU/ml of a known strain of AAC).

vi. Stop the reaction by adding 20μl of HCl (1.0N) and record the color intensity using a microplate reader (Bio-Tek Instruments Inc., Winooski, UT).

C. Coating of Immunomagnetic Beads (IMBs)

i. Use super-paramagnetic beads, precoated with sheep antirabbit antibodies (Dynabead M280 sheep antirabbit Dynal, Oslo, Norway); coat them with purified IgG fractions of anti-AAC (as per instructions of manufacturer) and agitate for 2 min.

ii. Transfer 500 μl (approximately 2×10^8 beads) to 4-ml vials; wash the IMBs with PBS four times while held by a magnetic particle concentrator (MPC; Dynal Oslo, Norway) and suspend IMBs in 3 ml of PBS.

iii. Add purified IgG protein (80 μg) and incubate on a tilt-shaker for 24 h at 4°C.

iv. Rinse the IMBs with 4 ml of PBS-BSA four times and resuspend in 3 ml of PBS-BSA to have a final concentration of approximately 1×10^7 beads/ml and store at 4°C.

v. To verify effective coating of IMBs, incubate 100 μl of processed IMB with 100 μl of goat anti-rabbit antibodies (1:1000 dilution) conjugated with ALP for 1 h; wash with PBS-BSA four times; add 100 μl of pNPP as substrate and incubate for 20 min.

vi. Remove the IMB with magnetic particle concentrator and determine OD of the buffer at 405 nm using an ELISA plate reader.

vii. OD values of coated beads twice that of negative control (noncoated beads) indicate successful coating of beads.

D. Threshold of Bacterial Cell Recovery

i. Place watermelon seeds (25 g) in sterile PBS; agitate in a rotary shaker at 150 rpm for 2 h; remove the seeds and sterilize the supernatant by autoclaving at 121°C and 15 psi of pressure for 15 min.

ii. Prepare bacterial dilutions by adding known concentrations of *A. avenae* subsp. *citrulli* (AAC) to glass vials containing either 2 ml of PBS or sterile PBS with seed debris.

iii. Assess the number of CFU per treatment by spread plating 100 μl of each suspension onto King's medium B (KMB; 10).

iv. To determine recovery efficiency, incubate 2 ml of each sample with 75 μl (7.5×10^5 beads) of anti-AAC-coated IMBs; gently mix and incubate for 15 min at room temperature.

v. Rinse the beads with PBS-BSA three times; resuspend in 50 μl of PBS-BSA and spread plate onto KMB.

vi. Incubate the plates at 28°C for 48 h and count AAC colonies.

vii. Repeat this procedure four times and carry out two-way analysis of variance (ANOVA) for comparing the effects of target cell population and seed-wash debris on AAC recovery by immunomagnetic separation (IMS).

E. Threshold and Detection by IMS-PCR

i. To assess the ability of IMS-PCR to detect AAC in the presence of inhibiting substances present in seeds, spike 2 ml of seed wash with

AAC to generate cell suspensions with varying bacterial cell concentration (1×10^4, 1×10^3, 1×10^2, and 10 and 0 CFU/ml).

ii. Perform PCR directly on 4 μl of cell lysates from each treatment using the primers WFB1 (sense) and 2 (antisense) designed from DNA sequence for the 16S ribosomal RNA (rRNA) gene from a strain of AAC (94–85).

iii. Perform IMS-PCR on the same cell suspensions by adding 100 μl of coated IMBs (1×10^6) and incubate at 15°C for 15 min with gentle agitation.

iv. Rinse the IMBs with PBS-BSA twice, once with sterile distilled water and resuspend in 15 μl of sterile distilled water.

v. Lyse the IMB-bound bacterial cells by boiling for 15 min and perform PCR using 10 μl of cell lysates and primers WFB1 and 2.

vi. Subject the PCR products (10 μl) to electrophoresis on 1% agarose gel followed by staining with ethidium bromide and visualization under UV light.

vii. Repeat the assay nine times.

F. Detection of AAC by IMS-PCR in Watermelon Seeds

i. Incubate seed lots (72 g) in 200 ml of bacterial suspension (1×10^8 CFU/ml) in PBS for 15 min; air dry and mix with healthy seeds (untreated) to generate seed lots with 0, 0.1, 1.0, 5.0, and 10.0% infestation.

ii. Incubate seed lots with PBST (200 ml) for 4 h with vigorous agitation on a wrist-action shaker; filter the suspension through a triple layer of cheese cloth; pellet 50 ml of seed wash by centrifugation (4000 g for 15 min) and resuspend the pellet in 5 ml of PBST; filter through Whatman No. 1 filter paper.

iii. Perform IMS-PCR with primers WFB 1 and 2 and repeat the assay five times.

iv. Perform indirect ELISA using 10 μl of seed wash from artificially infested seed lots for comparison of results with that of IMS-PCR assay.

REFERENCES

Agarwal, P.C., Mortensen, C.N. and Mathur, S.B. 1989. Seedborne diseases and seed health testing of rice. *Phytopathological papers*, **No.30**. Danish Government Institute of Seed Pathology for Developing Countries, Copenhagen and CAB International Mycological Institute, Kew, U.K.

Agarwal, V.K. and Sinclair, J.B. 1996. *Principles of Seed Pathology*, 2nd edition. CRC-Lewis Publishers, Boca Raton, USA.

Anil Kumar, Singh, A. and Garg, G.K. 1998. Development of seed immunoblot binding assay for the detection of Karnal bunt (*Tilletia indica*) of wheat. *Journal of Plant Biochemistry and Biotechnology*, **7**: 119–120.

Appel, R., Alder, N. and Habermeyer, J. 2001. A method for the artificial inoculation of potato tubers with *Phytophthora infestans* and polymerase chain reaction assay of latently infected sprouts and stems. *Journal of Phytopathology*, **149**: 297.

Audy, P., Braat, C.E., Saindon, G., Huang, H.C. and Laroche, A. 1996. A rapid and sensitive PCR-based assay for concurrent detection of bacteria causing common and haloblights in bean seed. *Phytopathology*, **86**: 361–366.

Avegelis, A. and Barba, M. 1986. Application of ELISA test for indexing melon necrotic spot virus in melon seeds. *Annali dell' Istituto Sperimentale per la Patologia* Vegetale, **11**: 107–110.

Banks, J.N., Cox, S.J., Clarke, J.H., Shamsi, R.H. and Northway, B.J. 1992. Towards the immunological detection of field and storage fungi. In: *Modern Methods in Food Mycology* (eds) R.A. Samson, A.D. Hocking, J.I. Pitt and A.D. King, pp. 247–252. Elsevier, Holland.

Ball, S. and Reeves, J. 1992. Application of rapid techniques to seed health testing – prospects and potential. In: *Techniques for the Rapid Detection of Plant Pathogens* (eds) J.M. Duncan and L. Torrance, pp. 193–207. Blackwell Scientific Publications, Oxford.

Bar-Joseph, M., Garnsey, S.M., Gonsalves, D., Moscovitz, M., Purcifull, D.E., Clark, M.F. and Loebenstein, G. 1979. The use of enzyme-linked immunosorbent assay for the detection of *Citrus tristeza virus*. *Phytopathology*, **69**: 190–194.

Bell, K.S., Roberts, J., Verrall, S., Cullen, D.W., Williams, N.A., Harrison, J.G. Toth, I.K., Cooke, D.E.L., Duncan, J.M. and Claxton, J.R. 1999. Detection and quantification of *Spongospora subterranea* f.sp. *subterranea* in soils and on tubers using specific PCR primers. *European Journal of Plant Pathology*, **105**: 905–915.

Bellaire, L. de, Chillet, M. and Mourichon, S. 2000. Elaboration of an early quantification of quiescent infections of *Colletotrichum musae* on bananas. *Plant Disease*, **84**: 128–133.

Benedict, A., Alvarez, A.M., Berestecky, J., Imanaka, W., Mizumoto, C., Pollard, L., Mew, T. and Gonzalez, C. 1989. Pathovar-specific monoclonal antibodies for *Xanthomonas campestris* pv. *oryzae*. *Phytopathology*, **79**: 322–328.

Bharathan, N., Reddy, D.V.R., Rajeshwari, R., Murthy, V.K., Rao, V.R. and Lister, R.M. 1984. Screening germplasm lines by enzyme-linked immunosorbent assay for seed transmission of peanut mottle virus. *Plant Disease*, **68**: 757–758.

Bluhm, B.H., Cousin, M.A. and Woloshuk, C.P. 2004. Multiplex real-time PCR detection of fumonisin-producing and trichothecene-producing groups of *Fusarium* species. *Journal of Food Protection*, **67**: 536–543.

Bonde, M.R., Peterson, G.L. and Matsumoto, T.T. 1989. The use of isozymes to identify teliospores of *Tilletia indica*. *Phytopathology*, **79**: 596–599.

Boonham, N., Walsh, K., Preston, S., North, J., Smith, P. and Barker, I. 2002. The detection of tuber necrotic isolates of *Potato virus Y* and the acute discrimination of PVY^{O}, PVY^{N} and PVY^{C} strains using RT-PCR. *Journal of Virological Methods*, **102**: 103–112.

Börjesson, T. and Johnson, L. 1998. Detection of common bunt (*Tilletia caries*) infestation in wheat with an electronic nose and human panel. *Journal of Plant Diseases and Protection*, **105**: 306–313.

Bragard, C. and Verhoyen, M. 1993. Monoclonal antibodies specific for *Xanthomonas campestris* bacteria pathogenic on wheat and other small grains in comparison with polyclonal antisera. *Journal of Phytopathology*, **139**: 217–228.

Brlansky, R.H. and Denrick, K.S. 1979. Detection of seedborne plant viruses using serologically-specific electron microscopy. *Phytopathology*, **69**: 96–100.

Cammarco, R.F.E.A., Lima, J.A.A. and Pio-Ribeiro, G. 1998. Transmission and presence in the soil of papaya lethal yellowing virus. *Fitopatologia Brasileira*, **23**: 453–458.

Candlish, A.A.G., Taylor, J.D. and Cameron, J. 1988. Immunological methods as applied to bacterial pea blight. *Proceedings of the Brighton Crop Protection Conference – Pests and Diseases*, **2**: 787–794.

Caruso, P., Gorris, M.T., Cambra, M., Palomo, J.L. Collar, J. and López, M.M. 2002. Enrichment double-antibody sandwich enzyme-linked immunosorbent assay that uses a specific monoclonal antibody for sensitive detection of *Ralstonia solanacearum* in asymptomatic potato tubers. *Applied and Environmental Microbiology*, **68**: 3634–3638.

Carzaniga, R., Fiocca, D., Bowyer, P. and ÓConnell, R.J. 2002. Localization of melanin in conidia of *Alternaria alternata* using phage antibodies. *Molecular Plant–Microbe Interactions*, **15**: 216–224.

Cavileer, T.D., Clarke, R.C., Corsini, D.L. and Berger, P.H. 1998. A new strain of potato carlavirus. *Plant Disease*, **91**: 98–102.

Cazelles, O., Ruchta, A. and Schwarzel, R. 1995. Bacterial soft rot of potato tubers: Progress in detection of latent infection of *Erwinia chrysanthemi* and forecasting field outbreaks. *Revue Suisse d' Agriculture*, **27**: 17–22.

Chang, G.H. and Yu, R.C. 1997. Rapid immunoassay of fungal mycelia in rice and corn. *Journal of Chinese Agricultural Chemical Society*, **35**: 533–539.

Chen, L.C., Chen, T.Z., Chen, H.L. and Yeh, H. 1998. Establishment of molecular markers for detection and diagnosis of *Botrytis cinerea* and *B. elliptica*. *Plant Pathology Bulletin*, **7**: 177–188.

Chen, S.W., Ma, H.Q. and Wang, H. 1999. Simultaneous identification of melon postharvest decay pathogens by the random amplified polymorphic DNA analysis with polymerase chain reaction. In: *Research Progress in Plant Protection and Plant Nutrition*, pp. 109–114, China Agriculture Press, Beijing, China.

Clark, M.F. and Adams, A.N. 1977. Characteristics of the microplate method of enzyme-linked immunosorbent assay. *Journal of General Virology*, **34**: 475–483.

Clark, M.F., Adams, A.N. and Barbara, D.J. 1976. The detection of plant viruses by enzyme-linked immunosorbent assay (ELISA). *Acta Horticulturae*, **67**: 43–49.

Comstock, J.C., Wang, Z.K. and Perdomo, R. 1997. The incidence of leaf scald and its effect on yield components: detection by serological and isolation techniques with particular reference to CP 180–1743 in Florida. *SugarCane*, **4**: 18–21.

Côté, M.J., Tardif, M.C. and Meldrum, A.J. 2004. Identification of *Monilinia fructigena*, *M. fructicola*, *M. laxa*, and *Monilia polystroma* on inoculated and natural infected fruit using multiplex PCR. *Plant Disease*, **88**: 1219–1225.

Cottyn, B., Bautista, A.T., Nelson, R.J., Leach, L.E., Swings, J. and Mew, T.W. 1994. Polymerase chain reaction amplification of DNA from bacterial pathogens of rice using specific oligonucleotide primers. *International Rice Research Notes*, **19**: 30–32.

Cubero, J. and Graham, J.H. 2002. Genetic relationship among world wide strains of *Xanthomonas* causing canker in *Citrus* species and design of new primers for their identification by PCR. *Applied and Environmental Microbiology*, **168**: 1257–1264.

Cullen, D.W., Lees, A.K., Toth, I.K. and Duncan, J.M. 2002. Detection of *Colletotrichum coccodes* from soil and potato tubers by conventional and quantitative realtime PCR. *Plant Pathology*, **51**: 1365–1369.

Culver, J.N. and Sherwood, J.L. 1988. Detection of peanut stripe virus in peanut seed by an indirect enzyme-linked immunosorbent assay using a monoclonal antibody. *Plant Disease*, **72**: 676–679.

Dalfosse, P., Reddy, A.S., Legréve, A., Devi, P.S., Devi, K.T., Maraite, H. and Reddy, D.V.R. 1999. Indian peanut clump virus (IPCV) infection on wheat and barley: symptoms, yield loss and transmission through seeds. *Plant Pathology*, **48**: 273–282.

De Boer, S.H. and Hall, J.W. 2000. Proficiency testing in a laboratory accreditation program for the bacterial ring rot pathogen of potato. *Plant Disease*, **84**: 649–653.

De Boer, S.H., Stead, D.E., Alivizatos, A.S., Janse, J.D., Van Vaerenbergh, J., De Haan, T.L. and Mawhinney, J. 1994. Evaluation of serological tests for detection of *Clavibacter michiganensis* subsp. *sepedonicus* in composite potato stem and tuber samples. *Plant Disease*, **74**: 1431–1434.

De Boer, S.H. and Ward, L.J. 1995. PCR detection of *Erwinia carotovora* subsp. *atroseptica* associated with potato tissue. *Phytopathology*, **85**: 854–858.

De Bokx, J.A. 1987. Characterization and identification of potato viruses and viroids: Biological properties. In: *Viruses of Potatoes and Seed-Potato Production* (eds) I.A. De Bokx and J.P.H. van der Want, pp. 58–82, Centre for Agricultural Publishing and Documentation, Wageningen.

De Parasis, J. and Roth, D.A. 1990. Nucleic acid probes for identification of genus-specific 16S rRNA sequences. *Phytopathology*, **80**: 618.

Derks, A.F.L.M., Lemmers, M.E.C. and Hollinger, T.C. 1997. Detectability of viruses in lily depends on virus, host and storage conditions. *Acta Horticulturae*, **430**: 633–640.

Derrick, K.S. 1973. Quantitative assay for plant viruses using serologically-specific electron microscopy. *Virology*, **56**: 652–653.

Dewey, F.M. 1998. Use of monoclonal antibodies to study plant invading fungi particularly *Botrytis cinerea* and *Septoria nodorum Beitraġe zur Züchtungs forschung-Bundesanstalt für Züchtungsforschung an Kulturpflangen*, **8**: 45–47.

Dewey, F., MacDonald, M. and Philipps, S. 1989. Development of monoclonal antibody-ELISA, DOT-BLOT and DIP-STICK immunoassays for *Humicola lanuginosa* in rice. *Journal of General Microbiology*, **135**: 361–374.

Dewey, J., MacDonald, M., Philipps, S. and Priestley, R. 1990. Development of monoclonal antibody ELISA and–DIP–STICK immunoassays for *Penicillium islandicum* in rice grains. *Journal of General Microbiology*, **136**: 753–760.

Dharam Singh and Maheswari, V.K. 2001. Influence of stack burn disease of paddy on seed health status. *Seed Research*, **29**: 205–209.

Dhingra, O.D. and Muchovej, J.J. 1980. Twin–stem abnormality disease of soybean seedlings caused by *Sclerotium* sp. *Plant Disease*, **64**: 176.

Dinant, S. and Lot, H. 1992. Lettuce mosaic virus: a review. *Plant Pathology*, **41**: 528–542.

Ding, X.S., Cockbain, A.J. and Govier, D.A. 1992. Improvements in the detection of *Pea seedborne mosaic virus* by ELISA. *Annals of Applied Biology*, **121**: 75–83.

Dolares-Talens, A.C., Hill, J.H. and Durand, D.P. 1989. Application of enzyme-linked fluorescent assay (ELFA) to detection of *Lettuce mosaic virus* in lettuce seeds. *Journal of Phytopathology*, **124**: 149–154.

D'Onghia, A.M., Djelouah, K., Alioto, D., Castellano, M.A. and Savino, V. 1998. ELISA correlates with biological indexing for detection of citrus psorosis-associated virus. *Journal of Plant Pathology*, **80**: 157–163.

Doster, M.A. and Michailides, T.J. 1998. Production of bright greenish yellow fluorescence in figs infected by *Aspergillus* species in California orchards. *Plant Disease*, **82**: 669–673.

Dowell, F.E., Boratynski, T.N., Ykema, R.E., Dowdy, A.K. and Staten, R.T. 2002. Use of optical sorting to detect wheat kernels infected with *Tilletia indica*. *Plant Disease*, **86**: 1011–1013.

Drennan, J.L., Westra, A.A.G., Slack, S.A., Delserone, L.M., Collmer, A., Gudmestad, N.C. and Oleson, A.E. 1993. Comparison of a DNA hybridization probe and ELISA for the detection of *Clavibacter michiganensis* subsp. *sepedonicus* in field grown potatoes. *Plant Disease*, **77**: 1243–1247.

Edwards, S.G., Pirgozliev, S.R., Hare, M.C. and Jenkinson, P. 2001. Quantification of trichothecene-producing *Fusarium* species in harvested grains by competitive PCR to determine efficacies of fungicides against fusarium head blight of winter wheat. *Applied and Environmental Microbiology*, **67**: 1575–1580.

Errampalli, D., Saunders, J. and Cullen, D.W. 2001. A PCR-based method for detection of potato pathogen-*Helminthosporium solani* in silver scurf infected tuber tissue and soils. *Journal of Microbiological Methods*, **44**: 59–68.

Faris-Mukhayyish, S. and Makkouk, K.M. 1983. Detection of four seedborne plant viruses by the enzyme-linked immunosorbent assay (ELISA). *Phytopathologische Zeitschrft*, **106**: 108–114.

Fegan, M., Croft, B.J., Teakle, D.S., Hayward, A.C. and Smith, G.R. 1998. Sensitive and specific detection of *Clavibacter xyli* subsp. *xyli* causal agent of ratoon stunting disease of sugarcane with a polymerase chain reaction-based assay. *Plant Pathology*, **47**: 495–504.

Feodorova, R.N. 1997. New and improved methods of detecting smuts in wheat and barley seeds. *Biuletyn Instyutu Hodowli i Aklimatyzacji Ros'lin*, **201**: 253–256.

Fessehaie, A., Wydra, K. and Rudolph, K. 1999. Development of a new selective medium for isolating *Xanthomonas campestris* pv. *manihotis* from plant material and soil. *Phytopathology*, **89**: 591–597.

Förster, H. and Adaskaveg, J.E. 2000. Early brown rot infections in sweet cherry fruit are detected by *Monilinia*-specific DNA primers. *Phytopathology*, **90**: 171–178.

Forster, R.L., Seifers, D.L., Strausbaugh, C.A. Jensen, S.G., Ball, E.M. and Harvey, T.L. 2001. Seed transmission of the High Plains virus in sweet corn. *Plant Disease*, **85**: 696–699.

Fox, R.T.V. 1993. *Principles of Diagnostic Techniques in Plant Pathology.* CAB International, Wallingford, Oxon, U.K.

Franken, A.A.J.M. 1992. Application of polyclonal and monoclonal antibodies for the detection of *Xanthomonas campestris* pv. *campestris* in crucifer seeds using immunofluorescence microscopy. *Netherlands Journal of Plant Pathology*, **98**: 81–94.

Frederick, R.D., Snyder, K.E., Tooley, P.W., Berthier-Schaad, Y., Peterson, G.L., Bonde, M.R., Schaad, N.W. and Knorr, D.A. 2000. Identification and differentiation of *Tilletia indica* and *T. walkeri* using the polymerase chain reaction. *Phytopathology*, **90**: 951–960.

Freeman, S., Minz, D., Jurkevitch, E., Maymon, M. and Shabi, E. 2000. Molecular analysis of *Colletotrichum* species from almond and other fruits. *Phytopathology*, **90**: 608–614.

Freigoun, S.O., El Faki, H.I., Gelie, B., Schirmer, M. and Lemattre, M. 1994. Phage sensitivity in relation to pathogenicity and virulence of the cotton bacterial blight pathogen of Sudan. *Plant Pathology*, **43**: 493–497.

Frommel, M.I. and Pazos, G. 1994. Detection of *Xanthomonas campestris* pv. *undulosa* in infested wheat seeds by combined liquid medium enrichment and ELISA. *Plant Pathology*, **43**: 589–596.

Fulton, C.E. and Brown, A.E. 1997. Use of SSU rDNA group I intron to distinguish *Monilinia fructicola* from *M. Laxa* and *M. fructigena. FEMS Microbiology Letters*, **157**: 307–312.

Gaitaitis, R., Summer, D., Gay, D., Smittle, D., McDonald, G., Maw, B., Johnson, W.C. III, Tollner, B. and Hung, Y. 1997. Bacterial streak and bulb rot of onion. I. A diagnostic medium for semiselective isolation and enumeration of *Pseudomonas viridiflava. Plant Disease*, **87**: 897–900.

Gera, A., Lawson, R.H. and Hsu, H.T. 1995. Identification and assay. In: *Viruses and Virus-like Diseases of Bulb and Flower Crops* (eds) G. Loebenstein, R.H. Lawson and A.A. Brunt, pp. 165–180. John Wiley and Sons, Chichester, U.K.

Ghabrial, S.A. and Shepherd, R.J. 1982. Radioimmunosorbent assay for the detection of *Lettuce mosaic virus* in lettuce seed. *Plant Disease*, **66**: 1037–1040.

Gilbertson, R.L., Maxwell, D.P., Hagedorn, D.J. and Leong, S.A. 1989. Development and application of a plasmid DNA probe for detection of bacteria causing common bacterial blight of bean. *Phytopathology*, **79**: 418.

Gillapsie, A.G. Jr., Pio-Ribeiro, G., Andrade, G.P. and Pappu, M.R. 2001. RT-PCR detection of seedborne *Cowpea aphidborne mosaic virus* in peanut. *Plant Disease*, **85**: 1181–1182.

Gillaspie, A.G. Jr., Pittman, R.N., Pinnow, D.L. and Cassidy, B.G. 2000. Sensitive method for testing peanut seed lots for *Peanut stripe and Peanut mottle viruses* by immunocapture-reverse transcription-polymerase chain reaction. *Plant Disease*, **84**: 559–561.

Gindrat, D. and Pezet, R. 1994. Paraquat, a tool for rapid detection of latent fungal infections and of endophytic fungi. *Journal of Phytopathology*, **141**: 86–98.

Gleason, M.L., Ghabrial, S.A. and Ferriss, R.S. 1987. Serological detection of *Phomopsis longicolla* in soybean seeds. *Phytopathology*, **77**: 371–375.

Glynn, N.C., Edwards, S.G., Hare, M.C., Burke, R. and Parry, D.W. 2000. Assessing the field performance of seed treatments for the control of *Fusarium* seedling blight of wheat using quantitative PCR. *Proceedings of BCPC Conference* Vol.1, pp. 479–482. British Crop Protection Council, U.K.

Glynn, N.C., Edwards, S.G., Hare, M.C., Parry, D.W. and Brande, F. 2001. Studies on the incidence and control of *Fusarium* seedling blight of wheat caused by *Microdochium nivale* var. *majus*, var. *nivale* and *Fusarium* spp. using PCR diagnostics. In: *Seed Treatment: Challenges and Opportunities* (ed.) A.J. Biddle, pp. 231–234, British Crop Protection Council, U.K.

Gnanamanickam, S.S., Shigaki, T., Medalla, E.S., Mew, T. and Alvarez, A.M. 1994. Problems in detection of *Xanthomonas oryzae* pv. *oryzae* in rice seeds and potential for improvement using monoclonal antibodies. *Plant Disease*, **78**: 173–178.

González-Jaén, M.T., Mirete, S., Patiño, B., López-Errasquín, E. and Vázquez. C. 2004. Genetic markers for the analysis of variability and for production of specific diagnostic sequences in fumonisin-producing strains of *Fusarium verticillioides*. *European Journal of Plant Pathology*, **110**: 525–532.

Griep, R.A., van Twisk, C., van Beckhoven, J.R.C.M., van der Wolf, J.M. and Schots, A. 1998. Development of specific recombinant monoclonal antibodies as against lipopolysaccharide of *Ralstonia solanacearum* race 3. *Phytopathology*, **88**: 795–803.

Grogan, R.G. 1983. *Lettuce mosaic virus* control by use of virus-indexed seed. *Seed Science and Technology*, **11**: 1043–1049.

Grogan, R.G. and Kimble, K.A. 1967. The role of seed contamination in the transmission of *Pseudomonas phaseolicola* in *Phaseolus vulgaris*. *Phytopathology*, **57**: 28–31.

Guillemette, T., Iacomi-Vasilescu, B. and USAMV. 2004. Conventional and real-time PCR-based assay for detecting pathogenic *Alternaria brassicae* in cruciferous seed. *Plant Disease*, **88**: 490–496.

Hahn, F. 2002. Fungal spore detection on tomatoes using spectral fourier signatures. *Biosystems Engineering*, **81**: 249–259.

Hale, C.N. and Taylor, R.K. 1999. Effect of cool storage on survival of *Erwinia amylovora* in apple calyxes. *Acta Horticulturae*, **489**: 139–143.

Halfon-Meiri, A. and Volcani, Z. 1977. A combined method for detecting *Colletotrichum gossypii* and *Xanthomonas malvacearum* in cotton seed. *Seed Science and Technology*, **5**: 129–139.

Hamamoto, H., Hasegawa, K., Nakaune, R., Lee, Y.J., Akutsu, K. and Hibi, T. 2001. PCR-based detection of sterol demethylation inhibitor-resistant strains of *Penicillium digitatum*. *Pest Management Science*, **57**: 839–843.

Hamilton, R.I. and Nichols, C. 1978. Serological methods for detection of *Pea seedborne virus* in leaves and seeds of *Pisum sativum*. *Phytopathology*, **68**: 539–543.

Harrison, T.G., Rees, E.A., Barker, H. and Lowe, R. 1993. Detection of spore balls of *Spongospora subterranea* on potato tubers by enzyme-linked immunosorbent assay. *Plant Pathology*, **42**: 181–186.

Helias, V., le Roux, A.C., Bertheau, Y., Andrivon, D., Gauthier, J.P. and Jouan, B. 1998. Characterization of *Erwinia carotovora* subspecies and detection of *Erwinia carotovora* subsp. *atroseptica* in potato plants, soil or water extracts with PCR-based methods. *European Journal of Plant Pathology*, **104**: 685–699.

Henriquez, J.L., Sugar, D. and Spotts, R.A. 2004. Etiology of bull's eye rot of pear caused by *Neofabraea* spp. in Oregon, Washington and California. *Plant Disease*, **88**: 1134–1138.

Hewett, P.D. 1977. Pretreatment in seed health testing: hypochlorite in the 2,4-D blotter for *Leptosphaeria maculans* (*Phoma lingam*). *Seed Science and Technology*, **5**: 599.

Hildebrand, M., Dickler, E. and Geider, K. 2000. Occurrence of *Erwinia amylovora* on insects in a fire blight orchard. *Journal of Phytopathology*, **148**: 251–256.

Hill, J.H. and Durand, D.P. 1986. *Soybean mosaic virus*. In: *Methods of Enzymatic Analysis* (eds) H.U. Bergmeyer, J. Bergmeyer and M. Grabl, Vol. XI, pp. 455–474, VCH Publishers, Weinheim.

Hirata, T., Kimishima, E., Aoki, T., Nirenberg, H.I. and ÓDonnell, K. 2001. Morphological and molecular characterization of *Fusarium verticillioides* from rotten banana imported into Japan. *Mycoscience*, **42**: 155–166.

Huang, B.C., Zhu, H., Hu, G.G., Li, Q.X., Gao, J.L. and Goto, M. 1993. Serological studies of *Xanthomonas campestris* pv. *oryzae* with polyclonal and monoclonal antibodies. *Annals of Phytopathological Society, Japan*, **59**: 123–127.

Huttinga, H. 1996. Sensitivity of indexing procedures for viruses and viroids. *Advances in Botanical Research*, **23**: 59–71.

Hyman, L.J., Birch, P.R.J., Dellagi, A., Avrova, A.O. and Toth, I.K. 2000. A competitive PCR-based method for detection and quantification of *Erwinia carotovora* subsp. *atroseptica* on potato tubers. *Letters in Applied Microbiology*, **30**: 330–335.

Iacomi-Vasilescu, B., Blancard, D., Guénard, M., Molinero-Demilly, V., Laurent, E. and Simoneau, P. 2002. Development of a PCR-based diagnostic assay for detecting pathogenic *Alternaria* species in cruciferous seeds. *Seed Science and Technology*, **30**: 87–95.

Jafarpour, B., Shepherd, R.J. and Grogan, R.G. 1979. Serologic detection of *Bean common mosaic* and *Lettuce mosaic viruses* in seed. *Phytopathology*, **69**: 1125–1129.

James, D., Howell, W.E. and Mink, G.I. 2001. Molecular evidence of the relationship between a virus associated with flat apple disease and *Cherry raspleaf virus* as determined by RT-PCR. *Plant Disease*, **85**: 47–52.

Johansen, I.E., Rasmussen, O.F. and Heide, M. 1989. Specific identification of *Clavibacter michiganense* subsp. *sepedonicus* by DNA-hybridization probes. *Phytopathology*, **79**: 1019–1023.

Johnson, R.D., Johnson, L., Kohmoto, K., Otani, H., Lane, C.R. and Kodama, M. 2000. Polymerase chain reaction-based method to specifically detect *Alternaria alternata* apple pathotype (*A. mali*) the causal agent of *Alternaria* blotch of apple. *Phytopathology*, **90**: 973–976.

Jones, R.A.C. and Proudlove, W. 1991. Further studies on *Cucumber mosaic virus* infection of narrow leafed lupin (*Lupinus angustefolius*): seedborne infection aphid transmission, spread and effects on grain yield. *Annals of Applied Biology*, **118**: 319–329.

Khaldeeva, E.V., Medyantseva, E.P., Glushko, N.I. and Budnikov, G.K. 2001. Amperometric immuno-enzyme sensor for evaluating the degree of infection of vegetable crops by phytopathogenic fungi. *Agrokhimiya*, **5**: 81–86.

Khanzada, A.K. and Mathur, S.B. 1988. Influence of extraction rate and concentration of stain on loose smut infection detection in wheat seed. *Pakistan Journal of Agricultural Research*, **9**: 218–222.

Khanzada, A.K., Rennie, W.J., Mathur, S.B. and Neergaard, P. 1980. Evaluation of two routine embryo test procedures for assessing the incidence of loose smut infection in seed samples of wheat (*Triticum aestivum*). *Seed Science and Technology*, **8**: 363.

Konate, G. and Barro, N. 1993. Dissemination and detection of *Peanut clump virus* in groundnut seeds. *Annals of Applied Biology*, **123**: 623–629.

Konstantinova, P., Bonants, P.J.M., Gent-Pelzer, M.P.E. van, Zouwen, P. van der and Bulk, R. van den. 2002. Development of specific primers for detection and identification of *Alternaria* spp. in carrot material by PCR and comparison with blotter and plating assays. *Mycological Research*, **106**: 23–33.

Klein, R.E., Wyatt, S.D., Keiser, W.J. and Mink, G.K. 1992. Comparative immunoassays of *Bean common mosaic virus* in individual bean (*Phaseolus vulgaris*) seed and bulked bean seed samples. *Plant Disease*, **76**: 57–59.

Knoll, S., Vogel, R.F. and Niessen, L. 2002. Identification of *Fusarium graminearum* in cereal samples by DNA detection Test Strips^{TM}. *Letters in Applied Microbiology*, **34**: 144–148.

Kohnen, P.D., Dougherty, W.G. and Hampton, R.O. 1992. Detection of *Pea seedborne potyvirus* by sequence-specific enzymatic amplification. *Journal of Virological Methods*, **37**: 253–258.

Krămer, R., Rabenstein, F., Schubert, J., Proll, E., Ehrig, F., Marthe, F. and Loptien, H. 2000. Necrosis in stored white cabbage infected with *Turnip mosaic virus* (TuMV). *Beiträge zur Züchtungsforschung-Bundesanstalt für Zuchtungsforschung an Kulturpflanzen*, **6**: 9–14.

Kritzman, G., Shwartz, H., Marcus, R., Manulis, S., Klietman, F., Oppenheim, D., Zilberstaine, M. and Shtienberg, D. 2003. Testing a rapid diagnostic medium for *Erwinia amylovora* and development of a procedure for sampling blossoms in pear orchards. *Phytopathology*, **93**: 931–940.

Kushalappa, A.C. and Lui, L.H. 2002. Volatile fingerprinting (SPME-GC-FID) to detect and discriminate diseases of potato tubers. *Plant Disease*, **86**: 131–137.

Lacy Costello, B.P.J. de, Evans, P., Ewen, R.J., Gunson, H.E., Ratcliff, N.M. and Spencer-Phillips, P.T.N. 1999. Identification of volatiles generated by potato tubers (*Solanum tuberosum* cv. Maris Piper) infected by *Erwinia carotovora*, *Bacillus polymyxa* and *Arthrobacter* sp. *Plant Pathology*, **48**: 345–351.

Lamka, G.L., Hill, J.H., McGee, D.C. and Braun, E.J. 1991. Development of an immunosorbent assay for seedborne *Erwinia stewartii* in corn seeds. *Phytopathology*, **81**: 839–846.

Lange, L. 1986. The practical application of new developments in test procedures for the detection of viruses in seed. *Developments in Applied Biology*, **1**: 269–281.

Lange, L. and Heide, M. 1986. Dot immunobinding (DIB) for detection of virus in seed. *Canadian Journal of Plant Pathology*, **8**: 373–379.

Lange, L., Jomanlor, A. and Heide, M. 1991. Testing seeds for viruses by dot immunobinding (DIB) directly on plain paper. *Tidiskrift for Planteaul*, **93**: 93–96.

Leach, J.E. and White, F.E. 1991. Molecular probes for disease diagnosis and monitoring. In: *Rice Biotechnology* (eds) G.S. Khush and G.H. Toenniessen, pp. 281–307. CAB International, U.K. and International Rice Research Institute, Philippines.

Lee, H.K., Tewari, J.P. 2001. A PCR-based assay to detect *Rhynchosporium secalis* in barley seed. *Plant Disease*, **85**: 220–225.

Lee, H.K., Tewari, J.F. and Turkington, T.K. 2001. Symptomless infection of barley seed by *Rhynchosporium secalis*. *Canadian Journal of Plant Pathology*, **23**: 315–317.

Lee, H.K., Tweari, J.P. and Turkington, T.K. 2002. Quantification of seedborne infection by *Rhynchosporium secalis* in barley using competitive PCR. *Plant Pathology*, **51**: 217–224.

Lee, I.M., Lukaesko, L.A. and Maroon, C.J.M. 2001. Comparison of dig-labeled PCR, nested PCR and ELISA for the detection of *Clavibacter michiganensis* subsp. *sepedonicus* in field-grown potatoes. *Plant Disease*, **85**: 261–266.

Lehtonen, M.J., Rantala, Kreuze, J.F., Bång, H., Kuisma, Koski, P., Virtanen, E., Vihlman, K. and Valkonen, J.P.T. 2004. Occurrence and survival of potato scab pathogens (*Streptomyces* species) on tuber lesions: quick diagnosis based on PCR-based assay. *Plant Pathology*, **53**: 280–287.

Levy, L., Hadidi, A., Kolber, M., Tokes, G. and Nemeth, M. 1995. 3' Non-coding region of RT-PCR detection and molecular hybridization of plum pox virus in anthers of infected stone fruit. *Acta Horticulturae*, **386**: 331–339.

Lewis Ivey, M.L., Nava-Diaz, C. and Miller, S.A. 2004. Identification and management of *Colletotrichum acutatum* on immature bell peppers. *Plant Disease*, **88**: 1198–1204.

Li, R., Salih, S. and Hurtt, S. 2004. Detection of geminiviruses in sweet potato by polymerase chain reaction. *Plant Disease*, **88**: 1347–1351.

Li, W.B., Pria, W.D. Jr., Lacava, P.M., Qin, X. and Hartung, J.S. 2003. Presence of *Xylella fastidiosa* in sweet orange fruit and seeds and its transmission to seedlings. *Phytopathology*, **93**: 953–958.

Li, X. and DeBoer, S.H. 1995. Selection of polymerase chain reaction primers from an RNA intergenic spacer region for specific detection of *Clavibacter michiganensis* subsp. *sepedonicus*. *Phytopathology*, **85**: 837–842.

Li, X., De Boer, S.H. and Ward, L.J. 1997. Improved microscopic identification of *Clavibacter michiganensis* subsp. *sepedonicus* by combining *in situ* hybridization with immunofluorescence. *Letters in Applied Microbiology*, **24**: 431–434.

Liew, E.C.Y., Maclean, D.J. and Irwin, J.A.G. 1998. Specific PCR-based detection of *Phytophthora medicaginis* using the intergenic spacer region of the ribosomal DNA. *Mycological Research*, **102**: 73–80.

Lister, R.M. 1978. Application of the enzyme-linked immunosorbent assay for detecting viruses in soybean seed and plants. *Phytopathology*, **68**: 1393–1400.

Lister, R.M., Caroll, T.W. and Zaske, S.K. 1981. Sensitive serological detection of *Barley stripe mosaic virus* in barley seed. *Plant Disease*, **65**: 809–814.

Llop, P., Bonaterra, A., Peñalver, J. and López, M.M. 2000. Development of a highly sensitive nested-PCR procedure using a single closed tube for detection of *Erwinia amylovora* in asymptomatic plant material. *Applied and Environmental Microbiology*, **66**: 2071–2078.

Loebenstein, G., Akad, F., Filatov, V., Sadvakasova, G., Manadilova, A., Bakelman, H., Teverovsky, E., Lachmann, O. and David, A. 1997. Improved detection of *Potato leafroll luteovirus* in leaves and tubers with a digoxigenin-labeled cRNA probe. *Plant Disease*, **81**: 489–491.

Louws, F.J., Fulbright, D.W., Stephens, C.T. and deBrujin, F.J. 1995. Differentiation of genomic structure of rep-PCR fingerprinting to rapidly classify *Xanthomonas campestris vesicatoria*. *Phytopathology*, **85**: 528–536.

Lunello, P., Mansilla, C., Concib, V. and Ponz, F. 2004. Ultra-sensitive detection of garlic potyviruses using a real-time fluorescent (TaqMan®) RT-PCR assay. *Journal of Virological Methods*, **118**: 15–21.

Lunsgaard, T. 1976. Routine seed health testing for stripe mosaic virus in barley seeds using the latex test. *Journal of Plant Diseases and Protection*, **83**: 278–283.

Luthra, J.C. 1953. Solar energy treatment of wheat loose smut *Ustilago tritici* (Pers.) Roster. *Indian Phytopathology*, **6**: 49–56.

Lyons, N., Cruz, L. and Santos, M.S. 2001. Rapid field detection of *Ralstonia solanacearum* in infected tomato and potato plants using the *Staphylococcus aureus* slide agglutination test. *Bulletin OEPP*, **31**: 91–93.

Lyons, N.F. and Taylor, J.D. 1990. Serological detection and identification of bacteria from plants by the conjugated *Staphylococcus aureus* slide agglutination test. *Plant Pathology*, **39**: 584–590.

Lyons, N.F. and White, J.G. 1992. Detection of *Pythium violae* and *Pythium sulcatum* in carrots with cavity spot using competition ELISA. *Annals of Applied Biology*, **120**: 235–244.

Ma, Z., Luo, Y. and Michailides, T.J. 2003. Nested PCR assays for detection of *Monilinia fructicola* in stone fruit orchards and *Botryosphaeria dothidea* from Pistachios in California. *Journal of Phytopathology*, **151**: 312–322.

Magan, N. and Evans, P. 2000. Volatiles as an indicator of fungal activity and differentiation between species, and the potential use of electronic nose technology for early detection of grain spoilage. *Journal of Stored Products Research*, **36**: 319–340.

Maguire, J.D. and Gabrielson, R.L. 1983. Testing techniques for *Phoma lingam. Seed Science and Technology*, **11**: 599–605.

Mäki-Valkama, T. and Karjalainen, R. 1994. Differentiation of *Erwinia carotovora* subsp. *atroseptica* and *carotovora* by RAPD-PCR. *Annals of Applied Biology*, **125**: 301–305.

Makkouk, K.M., Bos, L., Azam, O.I., Katul, L. and Rizkallah, A. 1987. *Broad bean stain virus*: identification detectability with ELISA in faba bean leaves and seeds, occurrence in West Asia and North Africa and possible wild hosts. *Netherlands Journal of Plant Pathology*, **93**: 97–106.

Malandrin, L. and Samson, R. 1998. Isozyme analysis for the identification of *Pseudomonas syringae* pv. strains. *Journal of Applied Microbiology*, **84**: 895–902.

Markovic, V.L., Stummer, B.E. and Hill, A.S. 2002. Immunodetection and characterization of antigens expressed by *Uncinula necator. Journal of Phytopathology*, **150**: 663–673.

Martin, H. and Woodcock, D. 1983. *The Scientific Principles of Crop Protection*. Edward Arnold, London.

Martinez-Culebras, P.V., Querol, A., Suarez-Fernandez, M.B., Garcia-Lopez, M.D. and Barrio, E. 2003. Phylogenetic relationships among *Colletotrichum* pathogens of strawberry and design of PCR primers for their identification. *Journal of Phytopathology*, **151**: 135–143.

Mathur, S. and Utkhede, R. 2002. Development of a dot blot technique for rapid identification of *Botrytis cinerea*, the causal organism of gray mould in greenhouse tomatoes. *Journal of Horticultural Science and Biotechnology*, **77**: 604–608.

Maude, R.B. 1996. *Seedborne Diseases and Their Control – Principles and Practice.* CAB International, Wallingford, Oxon, U.K.

Maury-Chovelon, V. 1984. *Recherches sur la Détection du Virus du la Mosaique de la Laitue par le Méthode Immunoenzymatique: adaptation au contrôle de l'état sanitaire des graines*. Doctoral Thesis, Montpellier, France.

Maury, Y., Duby, C., Bossennec, J.M. and Boudazin, G. 1985. Group analysis using ELISA: determination of the level of transmission of *Soybean mosaic virus* in soybean seed. *Agronomie*, **5**: 405–415.

Mavrodieva, V., Levy, L. and Gabriel, D.W. 2004. Improved sampling methods for real-time polymerase chain reaction diagnosis of citrus canker from field samples. *Phytopathology*, **94**: 61–68.

Mazarei, M. and Kerr, A. 1990. Distinguishing pathovars of *Pseudomonas syringae* on peas: nutritional, pathogenicity and serological tests. *Plant Pathology*, **39**: 278–285.

McAlpin, C.E., Wicklow, D.T. and Horn, B.W. 2002. DNA fingerprinting analysis of vegetative compatibility groups in *Aspergillus flavus* from peanut field in Georgia. *Plant Disease*, **86**: 254–258.

McDonald, J.G., Wong, E. and White, G.P. 2000. Differentiation of *Tilletia* species by rep-PCR genomic printing. *Plant Disease*, **84**: 1121–1125.

McManus, P.S. and Jones, A.C. 1996. Detection of *Erwinia amylovora* by nested PCR and PCR-dot blot and reverse blot hybridization. *Acta Horticulturae*, **411**: 87–90.

Meijerink, G. 1997. The international seed health initiative. In: *Seed Health Testing* (eds) D. Hutchins and J.C. Reeves, pp. 87–94, CAB International, U.K.

Meng, X.Q., Umesh, K.C., Davis, R.M. and Gilbertson, R.L. 2004. Development of PCR-based assays for detecting *Xanthomonas campestris* pv. *carotae*, the carrot leaf blight pathogen, from different substrates. *Plant Disease*, **88**: 1226–1234.

Mertely, J.C. and Legard, D.E. 2004. Detection, isolation and pathogenicity of *Colletotrichum* spp. from strawberry petioles. *Plant Disease*, **88**: 407–412.

Mew, T., Bride, J., Hibino, H., Bonman, J. and Merca, S. 1988. Rice pathogens of quarantine importance. *Proceedings of an International Workshop of Rice Seed Health*, pp. 101–115, International Rice Research Institute, Los Baños, Philippines.

Meyer, U.M. and Spotts, R.A. 2000. Detection and quantification of *Botrytis cinerea* by ELISA in pear stems during cold storage. *Plant Disease*, **84**: 1099–1103.

Mills, D., Russell, B.W. and Hanus, J.W. 1997. Specific detection of *Clavibacter michiganensis* subsp. *sepedonicus* by amplification of three unique DNA sequences isolated by subtraction hybridization. *Phytopathology*, **87**: 853–861.

Mills, P.R., Sreenivasaprasad, J. and Brown, A.E. 1992. Detection and differentiation of *Colletotrichum gloeosporioides* isolates using PCR. *FEMS Microbiology Letters*, **98**: 137–143.

Mink, G.I. and Aichele, M.D. 1984. Detection of *Prunus necrotic ringspot* and *Prune dwarf viruses* in *Prunus* seed and seedlings by enzyme-linked immunosorbent assay. *Plant Disease*, **68**: 378–381.

Mirza, M.S., Rademaker, J.L., Janse, J.D. and Akkermans, A.D.L. 1993. Specific 16S ribosomal RNA targeted oligonucleotide probe against *Clavibacter michiganensis* subsp. *sepedonicus. Canadian Journal of Microbiology*, **39**: 1029–1034.

Mizenina, O.A., Borisova, O.A., Novikov, V.K., Ertushenko, O.A., Baykov, A.A. and Atabekov, J.G. 1991. Inorganic pyrophosphatase from *E. coli* as a label for the detection of plant viruses by ELISA. *Journal of Phytopathology*, **133**: 278–288.

Mosqueda-Cano, G. and Herrera-Estrella, L. 1997. A simple and efficient PCR method for the specific detection of *Pseudomonas syringae* pv. *phaseolicola* in the bean seeds. *World Journal of Microbiology and Biotechnology*, **13**: 463–467.

Moyano, C., Alfonso, C., Gallego, J., Raposo, R. and Melgarejo, P. 2003. Comparison of RAPD and AFLP marker analysis as a means to study the genetic structure of *Botrytis cinerea* populations. *European Journal of Plant Pathology*, **109**: 515–522.

Mulé, G., Susca, A., Stea, G. and Moretti, A. 2004. A species-specific PCR assay based on the Calmodulin partial gene for identification of *Fusarium verticillioides, F. proliferatum* and *F. subglutinans. European Journal of Plant Pathology*, **110**: 495–502.

Mumford, R.A., Walsh, K., Barker, I. and Boonham, N. 2000. Detection of *Potato mop top virus* and *Tobacco rattle virus* using a multiplex real-time fluorescent reverse transcription-polymerase chain reaction assay. *Phytopathology*, **90**: 448–453.

Narayanasamy, P. 2001. *Plant Pathogen Detection and Disease Diagnosis, 2nd Edition.* Marcel Dekker, New York.

Narayanasamy, P. 2002. *Microbial Plant Pathogens and Crop Disease Management.* Science Publishers, Enfield, New Hampshire, USA.

Narayanasamy, P. 2005. *Immunology in Plant Health and its Impact on Food Safety.* Haworth Press, New York.

Neergaard, P. 1977. *Seed Pathology*, Vols 1 and 2. Macmillan, London, U.K.

Nicholson, P., Simpson, D.R., Wilson, A.H., Chandler, E. and Thomsett, M. 2004. Detection and differentiation of trichothecene and enniatin-producing *Fusarium* species on small-grain cereals. *European Journal of Plant Pathology*, **110**: 503–514.

Nie, X.Z. and Singh, R.P. 2003. Specific differentiation of recombinant PVYN: O and PVY^{NTN} isolates by multiplex RT-PCR. *Journal of Virological Methods*, **113**: 69–77.

Nielsen, K., Yohalem, D.S. and Jensen, D.F. 2002. PCR detection and RFLP differentiation of *Botrytis* spp. associated with neck rot of onion. *Plant Disease*, **86**: 682–686.

Nielsen, S.L., Husted, K. and Rasmussen, H.N. 1998. Detection of potato potyvirus Y in dormant tubers by a PCR-based method DIAPOPS (detection of immobilized amplified product in a one phase system). *DJF Rapport, Markburg*, **3**: 73–80.

Nikolaeva, O.V., Karasev, A.V., Garnsey, S.M. and Lee, R.F. 1998. Serological differentiation of the *Citrus tristeza virus* isolates causing stem pitting in sweet orange. *Plant Disease*, **82**: 381–386.

Nolan, P.A. and Campbell, R.N. 1984. *Squash mosaic virus* detection in individual seeds and seed lots of cucurbits by enzyme-linked immunosorbent assay. *Plant Disease*, **68**: 971–975.

Northover, J. and Cerkauskas, R.F. 1994. Detection and significance of symptomless latent infections of *Monilinia fructicola* in plums. *Canadian Journal of Plant Pathology*, **16**: 30–36.

Nümi, Y., Gondaira, T., Kutsuwada, Y. and Tsuji, H. 1999. Detection by ELISA and DIBA tests of *Lily symptomless virus* (LSV), *Tulip breaking virus*-lily (TBV-L) and *Cucumber mosaic virus* (CMV). *Journal of the Japanese Society for Horticultural Science*, **68**: 176–183.

Odu, B.O., Hughes, J. d' A., Shoyinka, S.A. and Dongo, L.N. 1999. Isolation, characterization and identification of a potyvirus from *Dioscorea alata* L. (water yam) in Nigeria. *Annals of Applied Biology*, **134**: 65–71.

Ou, S.H. 1985. *Rice Diseases*, 2nd edition. Commonwealth Mycological Institute, Surrey, U.K.

Ozakman, M. and Schaad, N.W. 2004. A real-time BIO-PCR assay for detection of *Ralstonia solanacearum* race 3, biovar 2, in asymptomatic potato tubers. *Canadian Journal of Plant Pathology*, **26**: 232–239.

Öztürk, S. and Cirakolu, B. 2003. Production of a monoclonal antibody specific for *Citrus tristeza virus*. *Food and Agricultural Immunology*, **15**: 65–73.

Palou, L., Usali, J., Pons, J., Cerda, M.C. and Viñas, I. 2001. Fungal populations in Tarrgona citrus packinghouses. *Investigación Agararia y producción Vegetales*, **16**: 445–462.

Pan, Y.B., Grisham, M.P., JBurner, D.M., Damam, K.E. Jr. and Wei, Q. 1998. A polymerase chain reaction protocol for the detection of *Clavibacter xyli* subsp. *xyli*, the causal bacterium of sugarcane ratoon stunting disease. *Plant Disease*, **82**: 285–290.

Parashar, R.D. and Leben, C. 1972. Detection of *Pseudomonas glycinea* from soybean seed lots. *Phytopathology*, **62**: 1075–1077.

Parikka, P. and Lemmetty, A. 2004. Tracing latent infection of *Colletotrichum acutatum* on strawberry by PCR. *European Journal of Plant Pathology*, **110**: 393–398.

Pasquini, C., Simeone, A.M., Conte, L. and Barba, M. 1998. Detection of *Plum pox virus* in apricot seeds. *Acta Virologica*, **42**: 260–263.

Patiño, B., Mirete, S., Gondález-Jaén, M.T., Mulé, G., Rodríquez, T. and Vázquez, C. 2004. PCR detection assay of fumonisin producing *Fusarium verticillioides* strains. *Journal of Food Protection*, **67**: 1278–1283.

Peres, N.A.R., Kuramae, E.E., Dias, M.S.C. and Souza, N.L. de. 2002. Identification and characterization of *Colletotrichum* spp. affecting fruit after harvest in Brazil. *Journal of Phytopathology*, **150**: 128–134.

Pesic, Z. and Hiruki, C. 1986. Differences in the incidence of *Alfalfa mosaic virus* in seed coat and embryo of alfalfa seed. *Canadian Journal of Plant Pathology*, **8**: 39–42.

Pesic, Z., Hiruki, C. and Chen, M.H. 1988. Detection of viral antigen by immunogold cytochemistry in ovules, pollen and anthers of alfalfa infected with *Alfalfa mosaic virus*. *Phytopathology*, **78**: 1027–1032.

Phillips, S., Briddon, R.W., Brunt, A.A. and Hull, R. 1999. The partial characterization of a badnavirus infecting the greater Asiatic or water yam (*Dioscorea alata*). *Journal of Phytopathology*, **147**: 265–269.

Pianzzola, M.J., Moscatelli, M. and Vero, S. 2004. Characterization of *Penicillium* isolates associated with blue mold in apple in Uruguay. *Plant Disease*, **88**: 23–28.

Polevaya, Y., Alkalai-Tuvia, S., Copel, A. and Fallik, E. 2002. Early detection of gray mould development in tomato after harvest. *Postharvest Biology and Technology*, **25**: 221–225.

Priou, S. and French, E.R. 1997. A simple baiting technique to detect and quantify *Pythium aphanidermatum* in soil. *Fitopatologia*, **32**: 187–193.

Prossen, D., Hatziloukas, E., Panopoulos, N.J. and Schaad, N.W. 1991. Direct detection of the halo blight pathogen *Pseudomonas syringae* pv. *phaseolicola* in bean seeds by DNA amplification. *Phytopathology*, **81**: 1159.

Prusky, D. 1996. Pathogen quiescence in postharvest diseases. *Annual Review of Phytopathology*, **34**: 413–434.

Pryor, B.M., Davis, R.M. and Gilbertson, R.L. 1994. Detection and eradication of *Alternaria radicina* in carrot seed. *Plant Disease*, **78**: 452–456.

Pryor, B.M. and Gilbertson, R.L. 2001. A PCR-based assay for detection of *Alternaria radicina* on carrot seed. *Plant Disease*, **85**: 18–23.

Rajeshwari, N., Shylaja, M.D., Krishnappa, M., Shetty, H.S., Mortensen, C.N. and Mathur, S.B. 1998. Development of ELISA for the detection of *Ralstonia solanacearum* in tomato: its application in seed health testing. *World Journal of Microbiology and Biotechnology*, **14**: 697–704.

Rajeswari, S., Palaniswami, A. and Rajappan, K. 1997. A chemodiagnostic method for the detection of symptomless latent infection of *Colletotrichum musae* in banana fruits. *Plant Disease Research*, **12**: 50–52.

Rantanen, T., Lehtinen, V. and Kurppa, A. 1999. Development of an immunocapture RT-PCR for the detection of *Potato moptop virus* in potato tubers. *Petria*, **9**: 149–152.

Rivas, R., Velázquez, E., Palomo, J.L., Mateos, P.F., García-Benavides, P. and Martinez-Molina, E. 2002. Rapid identification of *Clavibacter michiganensis* subsp. *sepedonicus* using two primers random amplified polymorphic DNA (TP-RAPD) finger prints. *European Journal of Plant Pathology*, **108**: 179–184.

Roberts, I.M. and Harrison, B.D. 1979. Detection of *Potato leaf roll* and *Potato mop top viruses* by immunosorbent electron microscopy. *Annals of Applied Biology*, **93**: 289–297.

Roebroeck, E.J.A., Groen, N.P.A. and Mes, J.J. 1990. Detection of latent *Fusarium oxysporum* in gladiolus corms. *Acta Horticulturae*, **266**: 468–476.

Rosner, A. and Maslenin, L. 1999. Differentiation of PVY^{NTN} by unique single-restriction cleavage of PCR products. *Potato Research*, **42**: 215–221.

Rott, P., Davis, M.J. and Baudin, P. 1994. Serological variability in *Xanthomonas albilineans*, causal agent of leaf scald disease of sugarcane. *Plant Pathology*, **43**: 344–349.

Ruiz, E. and Ruffner, H.P. 2002. Immunodetection of *Botrytis* specific invertase in infected grapes. *Journal of Phytopathology*, **150**: 76–85.

Russo, P., Miller, L., Singh, R.P. and Slack, S.A. 1999. Comparison of PLRV and PVY detection in potato seed samples tested by Florida winter field inspection and RT-PCR. *American Journal of Potato Research*, **76**: 313–316.

Says-Lesage, V., Meliala, C., Nicolas, P., Roeckel-Drevet, P., Tourvieille de Labrouhe, D., Archambault, D. and Billaud, F. 2001. Molecular test to show the presence of mildew (*Plasmopara halstedii*) in sunflower seeds. *OCL-Oléagineux, Corps Gras, Lipides*, **8**: 258–260.

Schaad, N.W. 1989. Detection and identification of bacteria. In: *Detection of Bacteria in Seed and Other Planting Material* (eds) A.W. Saettler, N.W. Schaad and D.A. Roth, pp. 9–16, APS Press, St. Paul, Minnesota, USA.

Schaad, N.W., Azad, H., Peet, R.C. and Panopoulos, N.J. 1989. Identification of *Pseudomonas syringae* pv. *phaseolicola* by a DNA hybridization probe. *Phytopathology*, **79**: 903.

Schaad, N.W., Bertheir-Schaad, Y., Sechler, A. and Knorr, D. 1999. Detection of *Clavibacter michiganensis* subsp. *sepedonicus* in potato tubers by BIO-PCR and an automated real-time fluorescence detection system. *Plant Disease*, **83**: 1095–1100.

Schaad, N.W., Cheong, S.S. Tamati, S., Hatziloukas, E. and Panopoulos, N.J. 1995. A combined biological and enzymatic amplification (BIO-PCR) technique to detect *Pseudomonas syringae* pv. *phaseolicola* in bean seed extracts. *Phytopathology*, **85**: 243–248.

Schadewijk, A.R. van. 1996. Detection of *Tomato spotted wilt virus* in dahlia. *Acta Horticulturae*, **432**: 384–391.

Schnabel, G., Schnabel, E.L. and Jones, A.L. 1999. Characterization of ribosomal DNA from *Venturia inaequalis* and its phylogenetic relationship to rDNA from other tree fruit *Venturia* species. *Phytopathology*, **89**: 100–108.

Schneider, A.L., Sherman, D.J., Stone, A.L., Damsteegt, V.D. and Frederick, R.D. 2004. Specific detection and quantification of *Plum pox virus* by real-time fluorescent reverse transcription PCR. *Journal of Virological Methods*, **120**: 97–105.

Schneider, J., Zhao, J.L. and Orser, C. 1993. Detection of *Clavibacter michiganensis* subsp. *sepedonicus* by DNA amplification. *FEMS Microbiology Letters*, **109**: 207–212.

Schnerr, H., Niessen, L. and Vogel, R.F. 2001. Real-time detection of the *tri5* gene in *Fusarium* species by Light Cycler™-PCR using SYBR® Green I for continuous fluorescence monitoring. *International Journal of Food Microbiology*, **71**: 53–61.

Serdani, M., Crous, P.W., Holz, G. and Petrini, O. 1998. Endophytic fungi associated with core rot of apples in South Africa, with specific reference to *Alternaria* species. *Sydowia*, **50**: 257–271.

Serdani, M., Kang, J.C., Andersen, B. and Crous, P.W. 2002. Characterization of *Alternaria* species-groups associated with core rot of apples in South Africa *Mycological Research*, **106**: 561–569.

Sheppard, J.W., Roth, D.A. and Saettler, A.W. 1989. Detection of *Xanthomonas* pv. *phaseoli* in bean. In: *Detection of Bacteria in Seed and Other Planting Material* (eds) A.W. Saettler, N.W. Schaad and D.A. Roth, pp. 17–29. APS Press, St. Paul, Minnesota, USA.

Sherwood, J.L., Sanborn, M.R. and Keyser, G.C. 1987. Production of monoclonal antibodies to *Peanut mottle virus* and their use in enzyme-linked immunosorbent assay and dot immunobinding assay. *Phytopathology*, **77**: 1158–1161.

Shoen, C.D., Knorr, D. and Leone, G. 1996. Detection of *Potato leafroll virus* in dormant potato tubers by immuno-capture and a fluorogenic 5' nuclease RT-PCR assay. *Phytopathology*, **86**: 993–999.

Slack, S.A., Drennan, J.L., Westra, A.A.G., Gudmestad, N.C. and Oleson, A.E. 1996. Comparison of PCR, ELISA and DNA-hybridization for the detection of *Clavibacter michiganensis* subsp. *sepedonicus* in field grown potatoes. *Plant Disease*, **60**: 319–331.

Slack, S.A. and Shepherd, R.J. 1975. Serological detection of seedborne *Barley stripe mosaic virus* by a simplified radial-diffusion technique. *Phytopathology*, **65**: 948–955.

Sledz, W., Jafra, S., Waleron, M., Malinowska, A. and Lojkowska, E. 1999. Diagnostics of pectinolytic *Erwinia* isolated from infected potato plants and tubers by PCR. *Petria*, **9**: 73–76.

Sokmen, M.A., Barker, H. and Torrance, L. 1998. Factors affecting the detection of *Potato moptop virus* in potato tubers and improvement of test procedures for more reliable assays. *Annals of Applied Botany*, **133**: 55–63.

Souza-Dias, J.A.C., de, Russo, P., Betti, J.A., Miller, L. and Slack, S.A. 1999. Simplified extraction method for ELISA and PCR detection of *Potato leafroll luteovirus* primary infection in dormant potato tubers. *American Journal of Potato Research*, **76**: 209–213.

Spiegel, S., Scott, S.N., Bowman-Vance, V., Tam, Y., Galiakparov, N.N. and Rosner, K. 1996. Improved detection of *Prunus necrotic ringspot virus* by the polymerase chain reaction. *European Journal of Plant Pathology*, **102**: 681–685.

Spiegel, S. and Martin, R.R. 1993. Improved detection of *Potato leafroll virus* in dormant potato tubers and microubers by polymerase chain reaction and ELISA. *Annals of Applied Biology*, **122**: 493–500.

Sundaram, S., Plasencia, S. and Banttari, E.E. 1991. Enzyme-linked immunosorbent assay for *Verticillium* spp. using antisera produced to *V. dahliae* from potato. *Phytopathology*, **81**: 1485–1489.

Swinburne, T.R. 1983. Quiescent infections in postharvest diseases. In: *Post-Harvest Pathology of Fruits and Vegetables* (ed.) C. Dennis, pp. 1–21, Academic Press, London.

Takaichi, M., Nagakubo, T. and Oeda, K. 2001. Mixed virus infections of garlic determined by a multivalent polyclonal antiserum and virus effects on disease symptoms. *Plant Disease*, **85**: 71–75.

Taylor, J.D. 1970. Bacteriophage and serological method for identification of *Pseudomonas phaseolicola* (Burkh.) Dowson. *Annals of Applied Biology*, **66**: 387–395.

Tomessoli, L., Lumia, V., Cerato, C. and Ghedini, R. 1998. Occurrence of potato tuber necrotic ringspot disease (PTNRD) in Italy. *Plant Disease*, **82**: 350.

Tooley, P.W., Carras, M.M. and Lambert, D.H. 1998. Application of a PCR-based test for detection of potato late blight and pinkroot in tubers. *American Journal of Potato Research*, **75**: 187–194.

Tooley, P.W., Goley, E.D. and Carras, M.M. 2002. AFLP comparisons among *Claviceps africana* isolates from the United States, Mexico, Africa, Australia, India and Japan. *Plant Disease*, **86**: 1247–1252.

Toth, I.K., Hyman, L.J. and Wood, J.R. 1999. A one step PCR-based method for the detection of economically important soft rot *Erwinia* species in micropropagated potato plants. *Journal of Applied Microbiology*, **87**: 1015–1021.

Toussaint, V., Morris, C.E. and Carisse, O. 2001. A new semi-selective medium for *Xanthomonas campestris* pv. *vitians*, the causal agent of bacterial leaf spot of lettuce. *Plant Disease*, **85**: 131–136.

Trans-Kersten, J., Guam, Y. and Allen, C. 1998. *Ralstonia solanacearum* pectin methyl esterase is required for growth on methylated pectin, but not for bacterial virulence. *Applied and Environmental Microbiology*, **64**: 4918–4923.

Triki, M.A., Pirou, S., El-Mahjoub, B. and Baudry, A. 2001. Leak syndrome of potato in Tunisia caused by *Pythium aphanidermatum* and *P. ultimum*. *Potato Research*, **44**: 221–231.

Trout, C.L., Ristaino, J.B., Madritch, M. and Wangsomboonde, T. 1997. Rapid detection of *Phytophthora infestans* in late blight-infected potato and tomato using PCR. *Plant Disease*, **81**: 1042–1048.

Trujillo, G.E. and Saettler, A.W. 1979. A combined semi-selective medium and serology test for the detection of *Xanthomonas* blight bacteria in bean seed. *Journal of Seed Science and Technology*, **4**: 35–41.

Tsushima, S., Narimatsu, C., Mizuno, A. and Kimura, R. 1994. Cloned DNA probes for detection of *Pseudomonas glumae* causing bacterial grain rot of rice. *Annals of Phytopathological Society, Japan*, **60**: 576–584.

Tzanetakis, I.E. and Halgren, A.B. 2004. Identification and detection of a virus associated with strawberry pallidosis disease. *Plant Disease*, **88**: 383–390.

Van der Vlugt, C.I.M., Derks, A.F.L.M., Bonekamp, P.M. and Goldbach, R.W. 1993. Improved detection of *Iris severe mosaic virus* in secondarily infected iris bulbs. *Annals of Applied Biology*, **122**: 279–288.

Van der Wolf, J.M., Van Beckhoven, J.R.C.M., De Haan, E.G., Van den Bovenkamp, G.W. and Leone, G.O.M. 2004. Specific detection of *Ralstonia solanacearum* 16S rRNA sequences by AmpliDet RNA. *European Journal of Plant Pathology*, **110**: 25–39.

Van Regenmortel, M.H.V. 1982. *Serology and Immunochemistry of Plant Viruses*. Academic Press, New York.

Van Vuurde, J.W.L. and Maat, D.Z. 1985. Enzyme-linked immunosorbent assay (ELISA) and disperse dye immunoassay (DIA): comparison of simultaneous and separate incubation of sample and conjugate for the routine detection of *Lettuce mosaic virus* and *Pea early browning virus* in seeds. *Netherlands Journal of Plant Pathology*, **91**: 3–13.

Van Vuurde, J.W.L. and Van den Bovenkamp, G.W. 1981. Routine methods for the detection of halo-blight (*Pseudomonas phaseolicola*) in beans. In: *Report of the 17th International Workshop on Seed Pathology*, p. 22. International Seed Testing Association, Zurich, Switzerland.

Verhoeff, K. 1974. Latent infections by fungi. *Annual Review of Phytopathology*, **12**: 99–110.

Verreault, H., Lafond, M., Asselin, A., Banville, G. and Bellemare, G. 1988. Characterization of two DNA clones specific for identification of *Corynebacterium sepedonicum*. *Canadian Journal of Microbiology*, **34**: 993–997.

Vetten, H.J., Ehlers, U. and Paul, H.L. 1983. Detection of *Potato virus Y* and *A* in tubers by enzyme-linked immunosorbent assay after natural and artificial break of dormancy. *Phytopathologische Zeitschrift*, **108**: 41–53.

Viswanathan, R. 1997. Detection of ratoon stunting disease (RSD) bacterium by ELISA. *Madras Agricultural Journal*, **84**: 374–377.

Viswanathan, R., Padmanaban, P., Mohanraj, D. and Nallathambi, P. 1998. Immunological techniques for the detection of leaf scald disease of sugarcane. *Indian Phytopathology*, **51**: 187–189.

Viswanathan, R. and Premachandran, M.N. 1998. Occurrence and distribution of sugarcane bacilliform virus in the sugarcane germplasm collection in India. *SugarCane*, **6**: 9–18.

Viswanathan, R., Samiyappan, R. and Padmanaban, P. 1998. Specific detection of *Colletotrichum falcatum* in sugarcane by serological techniques. *SugarCane*, **6**: 18–23.

Vlugt, R.A.A., Van der, Berendsen, M. and Koenraadt, 1997. Immunocapture reverse transcriptase PCR for detection of *Lettuce mosaic virus*. In: *Seed Health Testing: Progress towards the 21st century* (eds) J.D. Hutchins and J.C. Reeves, pp. 185–198, CAB International, Wallingford, Oxon, U.K.

Waalwijk, C., Heide, R. van der, de Vries, I., Lee, T., van der, Schoen, C., Coainville, G.C., Häuser-Hahn, I., Kastelein, P., Köhl, J., Demarquet, T. and Kema, G.H. 2004. Quantitative detection of *Fusarium* species in wheat using TaqMan. *European Journal of Plant Pathology*, **110**: 481–494.

Walcott, R.R. and Gitaitis, R.D. 2000. Detection of *Acidovorax avenae* subsp. *citrulli* in watermelon seed using immunomagentic separation and polymerase chain reaction. *Plant Disease*, **84**: 470–474.

Walcott, R.R., Gitaitis, R.D. and Castro, A.C. 2003. Role of blossom in watermelon seed infestation by *Acidovorax avenae* subsp. *citrulli. Phytopathology*, **93**: 528–534.

Walcott, R.R., Gitaitis, R.D., Castro, A.C., Sanders, F.H. Jr. and Diaz-Perez, J.C. 2002. Natural infestation of onion seed by *Pantoea anantis*, causal agent of center rot. *Plant Disease*, **86**: 106–111.

Wang, G.H., Nakajima, Y., Shaku, L. and Kumura, M. 2001. Effect of fungicide thiram on microbial communities on rice seed surface estimated by culture-dependent and culture-independent (PCR-RFLP) methods. *Soil Science and Plant Nutrition*, **47**: 303–314.

Wang, Y.C. and Yu, R.C. 1998. Detection of toxigenic *Aspergillus* spp. in rice and corn by ELISA. *Journal of Chinese Agricultural Chemical Society*, **36**: 512–520.

Wangsomboondee, T. and Ristaino, J.B. 2002. Optimization of sample size and DNA extraction methods to improve PCR detection of different propagules of *Phytophthora infestans. Plant Disease*, **86**: 247–253.

Watson, D.R.W. and Dye, D.W. 1971. Detection of bacterial disease in New Zealand garden pea stocks. *Plant Disease Reporter*, **55**: 517–521.

Wharton, A.L. 1967. Detection of infection of *Pseudomonas phaseolicola* (Burkh) Dowson in white-seeded dwarf bean seed stocks. *Annals of Applied Biology*, **60**: 305.

Wiwart, M. and Korona, A. 1998. Application of a colour image analysis of kernels in evaluation of the infection of triticale grown in different cultivation systems. *Plant Breeding and Seed Science*, **42**: 69–79.

Wu, F.M., Beuchat, L.R. Wells, J.G., Slutsker, L., Doyle, M.P. and Swaminathan, B. 2001. Factors influencing the detection and enumeration of *Escherichia coli* 0157: H7 on alfalfa seeds. *International Journal of Food Microbiology*, **71**: 93–99.

Wu, W.S. and Chen, T.W. 1999. Development of a new semiselective medium for detecting *Alternaria brassicola* in cruciferous seeds. *Seed Science and Technology*, **27**: 397–409.

Wullings, B.A., Beuningen, A.R. van, Janse, J.D. and Akkermans, A.D.L. 1998. Detection of *Ralstonia solanacearum* which causes brown rot of potato by fluorescent *in situ* hybridization with 23S rRNA-targeted probes. *Applied and Environmental Microbiology*, **64**: 4546–4554.

Wylie, S., Wilson, C.R., Jones, R.A.C. and Jones, M.G.K. 1993. A polymerase chain reaction assay for *Cucumber mosaic virus* in lupin seeds. *Australian Journal of Agricultural Research*, **44**: 41–51.

Xiao, C.L. and Rogers, J.D. 2004. A postharvest fruit rot in d' Anjou pears caused by *Sphaeropsis pyriputrescens* sp. nov. *Plant Disease*, **88**: 114–118.

Xu, H., DeHaan, T.L. and De Boer, S.H. 2004. Detection and confirmation of *Potato mop top virus* in potatoes produced in the United States and Canada. *Plant Disease*, **88**: 363–367.

Yang, Y., Kim, K.S. and Anderson, E.J. 1997. Seed transmission of *Cucumber mosaic virus* in spinach. *Phytopathology*, **87**: 924–931.

Yao, C.L., Fredriksen, R.A. and Magill, C.W. 1990. Seed transmission of sorghum downy mildew: detection by DNA hybridization. *Seed Science and Technology*, **18**: 201–207.

Yao, C.L., Magill, C.W. and Frederiksen, R.A. 1991. Use of an A-T rich DNA clone for identification and detection of *Peronosclerospora sorghi*. *Applied and Environmental Microbiology*, **57**: 2027.

Yashitola, J., Krishnaveni, D., Reddy, A.P.K. and Sonti, R.V. 1997. Genetic diversity within the population of *Xanthomonas oryzae* pv. *oryzae* in India. *Phytopathology*, **87**: 760–765.

Yu, F.H. and Chiu, F.S. 1998. Analysis of fumonisins and *Alternaria alteranta* toxin by liquid chromatography-enzyme-linked immunosorbent assay. *Journal of AOAC International*, **81**: 749–756.

Zhang, A.P. 1999. Direct tissue blotting ELISA for detection of *Carnation mosaic virus* in carnation. *Acta Agriculturae, Shangai*, **15**: 84–86.

Zhang, A.W., Hartman, G.L., Riccioni, L., Chen, W.D., Ma, R.Z. and Pedersen, W.L. 1997. Using PCR to distinguish *Diaporthe phaseolorum* and *Phomopsis longicolla* from other soybean fungal pathogens and to detect them in soybean tissues. *Plant Disease*, **81**: 1143–1149.

Zhao, Y., Damicone, J.P. and Bender, C.L. 2002. Detection, survival and sources of inoculum for bacterial diseases of leafy crucifers in Oklahoma. *Plant Disease*, **86**: 883–888.

Zhu, H., Gao, J.L., Li, Q.X. and Hu, G.X. 1988. Detection of *Xanthomonas oryzae* in rice by monoclonal antibodies. *Journal of Jiangsu Agricultural College*, **9**: 41–43.

Zimmer, R.C. 1979. Influence of agar on immuno-diffusion serology of *Pea seedborne mosaic virus*. *Plant Disease Reporter*, **63**: 278–282.

3

ECOLOGY OF POSTHARVEST MICROBIAL PATHOGENS

The seeds, fruits, and vegetables are exposed to numerous microorganisms, some of which are pathogenic, causing deterioration of the quality of harvested agricultural produce and resulting in a substantial reduction in their market value. The postharvest diseases are predominantly caused by fungal pathogens, whereas bacterial pathogens causing postharvest diseases are less numerous. The ecology of postharvest pathogens, influencing their survival, development, and overwintering, is discussed in this chapter.

3.1 SEED MICROFLORA

3.1.1 Mycoflora

Seed-borne pathogens may be defined as microorganisms carried with, on, or in seeds and transmitted by seeds, some of which may cause visible symptoms and damage to varying degrees (Baker 1972; Maude 1996). Seed-inhabiting fungi and storage fungi are grouped according to their ecological requirements, not on their taxonomic characteristics. The field fungi gain access to the seeds either during heading stage or after maturity but before harvest. Generally, these fungi will have done all the damage by harvest time and they cease to develop when the seeds are dried. The field fungi can do little or no damage during storage, since they require a moisture content in equilibrium with

Postharvest Pathogens and Disease Management, by P. Narayanasamy

relative humidities of 95 to 100%, which will provide free liquid water. Since such conditions are not available during storage, they become eliminated during storage. In contrast, storage fungi, which are unable to colonize actively metabolizing plant tissues, tend to invade dying, dead, or naturally dried seeds. These fungi are adapted to grow without free water, unlike field fungi. The storage fungi have a significant role in seed deterioration in storage, whereas the field fungi are responsible for various types of discoloration of seeds, poor filling, or reduction in seed size, 1000-grain weight, and other parameters required for getting vigorous and healthy seedlings or plants. The storage fungi associated with grains generally belong to two genera, *Aspergillus* and *Penicillium*, whereas different species of *Alternaria*, *Helminthosporium*, and *Cladosporium* infect grains under field conditions. The fungal pathogens causing seed-borne diseases and methods for detecting them are discussed in Chapter 2. Some of the mycoflora observed on seeds of economically important crops are presented in Table 3.1.

The development of fungi during storage has been studied in detail to assess the environmental requirements and the extent of deterioration in the quality of seeds to be used for human and animal consumption. The grains and other seeds are considered as durables. Various types of adverse effects are induced by storage fungi on durables, such as reduction in germination, discoloration, musty or sour odors, caking, nutritional alterations, and reduction in processing quality. Furthermore, contamination with mycotoxins produced by fungal pathogens is possibly the most serious adverse effect, resulting in dangerous health hazards to humans and animals.

3.1.1.1 Factors Influencing Development of Storage Fungi

A. Intrinsic Characteristics of Storage Fungi: The survival and development of fungi are dependent on the genetic constitution of the fungi. Based on the conidial morphology and spore-producing structures, five groups of storage fungi differing in the survival capacity can be recognized as suggested by Neergaard (1977): (1) hyaline fungi producing thin-walled conidia such as *Botrytis*, *Cercospora*, and *Fusarium*; (2) strongly pigmented fungi with thick-walled conidia such as *Drechslera* and *Alternaria*; (3) fungi producing acervuli or pycnidia which protect the conidia formed inside as in *Colletotrichum* and *Botryodiplodia*; (4) fungi producing deep-seated resting mycelium and heavily pigmented spores as in smut fungi; and (5) resting mycelia of internally seed-borne fungi.

The fungi belonging to first two groups can survive for short periods, whereas other groups may remain viable for longer periods depending on the storage conditions and the treatments applied to the stored grains.

B. Moisture Content: High moisture content generally favors the development of storage fungi. The water contained in cereal grains consists of bound and free water. Bound water is chemically absorbed to the substrate surface, initially in a monolayer of molecules, whereas free water is more weakly bound as more and more layers of water molecules accumulate on seed surfaces. The

TABLE 3.1 Seed Mycoflora of Some Economically Important Crops

Crop	Seed-inhabiting fungi	References*
Wheat	*Alteranria alternata*	Bhowmik, 1969
	Fusarium avenaceum	Mishra, 1973
	F. culmorum	Hope et al., 2000
	F. graminearum	Richardson, 1990
	F. moniliforme	Pidoplichko, 1970
	Microdochium nivale	Miller and Colhoun, 1969
	Septoria tritici	Cooke and Jones, 1970
	Tilletia carries	Swinburne, 1963
	T. controversa	Bechtel et al., 1999
	T. indica	Agarwal, 1977
	T. tritici	Swinburne, 1963
Rice	*Alternaria padwickii*	Mathur et al., 1972; Chidambaram et al., 1973
	Bipolaris oryzae	Mia and Safeeulla, 1998
	Cercospora oryzae	Cherewick, 1954
	Cladosporium sp.	Li et al., 2000
	Epicoccum nigrum	Tanaka and Tsuchiva, 1987
	Fusarium moniliforme	Anderson, 1948; Mia et al., 1986
	Microdochium oryzae	Manandhar, 1999
	Rhizopus spp.	Li et al., 2000
	Sarocladium oryzae	Shahjahan et al., 1977
	Tilletia barclayana	Chowdhury, 1951
Corn (Maize)	*Alternaria alternata*	Tanaka et al., 2001
	Aspergillus flavus	Payne et al., 1988
	A. ochraceus	Lee and Magan, 1999
	A. parasiticus	Zummo and Scott, 1990
	Bipolaris maydis	Tanaka et al., 2001
	Botryodiplodia theobromae	Kumar and Shetty, 1983
	Claviceps gigantea	Fuentes et al., 1962
	Epicoccum nigrum	Wright and Billeter, 1974
	Fusarium graminearum	Dungan and Koehler, 1944
	F. moniliforme	Marín et al., 2000; Tanaka et al., 2001
	Penicillium oxalicum	Halfon-Meiri and Solel, 1990
	Trichothecium roseum	Kumar and Shetty, 1985
	Ustilago zeae	Beke and Martin, 1958
Sorghum	*Aspergillus niger*	Reo et al., 2000
	Claviceps africana	Bhuiyan et al., 2002
	Claviceps purpurea	Futrell and Webster, 1965
	Colletotrichum capsici	Rao et al., 2000
	Colletotrichum graminicola	Basuchaudhary and Mathur, 1979
	Curvularia lunata	Singh and Agarwal, 1989
	Fusarium moniliforme	Wu and Mathur, 1987
	Sorisporium cruentum	Faris and Reed, 1925
	Rhizoctonia solani	Tanaka et al., 2001
	Rhizopus spp.	

TABLE 3.1 (*continued*)

Crop	Seed-inhabiting fungi	References*
Soybean	*Alternaria alternata*	Kunwar et al., 1986; Anuja Gupta and Aneja, 2001
	A. tenuissima	Shortt et al., 1982
	Aspergillus flavus	Umechuruba and Nwachukwu, 2002
	A. ruber	Dhingra et al., 1998
	Cercospora kikuchii	Ilyas et al., 1975
	Colletotrichum gloeosporioides	Roy, 1982
	Fusarium moniliforme	Solanke et al., 1998
	Glomerella cingulata	Ahn and Chung 1970
	Phomopsis longicolla	Hobbs et al., 1985.
Peanut (Groundnut)	*Aspergillus flavus*	Mehan et al., 1988
	Cylindrocladium crotalariae	Porter et al., 1991
	Fusarium solani	El-wakil and Ghonim, 2000
	Lasiodiplodia theobromae	Reddy et al., 1991
	Rhizoctonia bataticola	Amadioha, 2000
	R. solani	Ashworth and Langley, 1964
Tomato	*Alternaria alternata f. sp. lycopersici*	Franseschini et al., 1982
	Aspergillus spp.	Alimova et al., 2002
	Glomerella cingulata	Calo, 1957
	Penicillium spp.	Alimova et al., 2002
	Phythophthora infestans	Boyd, 1935
	Rhizoctonia solani	Baker, 1947
Brassica (Crucifers)	*Alternaria brassicae*	Maude and Humpherson-Jones, 1980
	A. brassicola	Knox-Davies, 1979
	Botrytis cinerea	Anon, 1982–1983.
Capsicum (Pepper)	*Alternaria alternata* *Aspergillus spp.*	Solanke et al., 2001
	Botryodiplodia palmarum	Maholay and Sohi, 1977
	Cercospora capsici	Doolittle, 1953
	Colletotrichum capsici	Grover and Bansal, 1970
	C. dematium	Gahukar et al., 1989
	Fusarium moniliforme	Solanke, et al., 2001
	Rhizoctonia solani	Baker, 1947

* Not necessarily the first report.

free water is more readily available for microbial growth and metabolism. Water availability is measured and expressed as water activity, which reflects the relationship between moisture in grains/foods and the ability of the fungi to grow on the stored materials. Water activity is defined as the ratio between the vapor pressure of water in a substrate (P) and the vapor pressure of pure water (Po) at the same temperature and pressure as expressed by the formula given below (Lacey et al., 1991).

$$a_w = \frac{P}{Po}$$

Water activity is found to be numerically equal to the equilibrium relative humidity (ERH) expressed as a decimal. The range of a_w allowing fungal growth is between 1.00 (pure water) and about 0.6 a_w. Tolerance of low a_w by various classes of fungi is well differentiated by their growth requirements. With freely available water (0.995 a_w), ecological similarity between *Aspergillus ochraceus*, *A. alternata*, *A. candidus*, and *A. flavus* in exploiting the same carbon sources at both 18°C and 30°C was evident (Lee and Magan, 1999).

Establishment, development, and growth of storage fungi during storage primarily depends on the moisture content of seeds, which may vary with moisture contents in equilibrium with RH of 65 to 90%. The development of all types of microbes is retarded when the moisture content drops below 13%. The limiting moisture contents of seeds may differ, based on host plant species and fungal species (Table 3.2).

C. Temperature: The development of storage fungi within seeds is affected by atmospheric, seed, and intergranular air temperature. Grain temperature is altered not only in response to changes in the ambient air temperature, but also in response to the activity of microbes and insects, resulting in a process

TABLE 3.2 Optimal Moisture/Water Activity (a_w) Levels for the Development of Storage Fungi

Crop	Fungal species	Moisture (%) or a_w	Reference
Barley	*Aspergillus spp.* *Penicillium spp.*	15–24%	Welling, 1969
Celery	*Septoria apicola*	7.3%	Roberts and Roberts 1972
Cotton	*Colletotrichum gossypii*	8.0%	Arndt, 1946
Maize	*Aspergillus flavus* *A. candidus* *A. orcharceus*	>18.5%	Lopez and Christensen, 1967
	F. moniliforme *F. proliferatum*	0.85–0.95 a_w	Marín et al., 2000
Soybean	*A. ruber*	>13.0%	Dhingra et al., 1998
	Cercospora kikuchii	>10.8%	Lehman, 1952
Sunflower	*A. glaucus* *A. restrictus*	>6.5%	Christensen, 1972
Wheat	*Fusarium culmorum*	0.995–0.97 a_w	Hope et al., 2000

known as spontaneous heating. Heating is due to the release of energy by respiration of seeds, microbes, and insects present in the stored seeds. Initially, some of the energy release may be due to seed respiration, but the microbial respiration soon dominates and grains are killed rapidly at temperatures of 40°C or above. The minimum, optimum, and maximum temperatures required for the growth of most storage fungi are 0 to 5°C, 30 to 33°C, and 50 to 55°C, respectively (Agarwal and Sinclair, 1996). Classification of microorganisms based on their ecological requirements, especially their water and temperature conditions, would be more meaningful. Most field fungi require at least 0.85 a_w for germination and 0.90 a_w for sporulation and growth. On the other hand, storage fungi germinate with 0.72 to 0.80 a_w and grow at temperatures ranging from 4 to 15.0°C to 30 to 55°C, reflecting the differences in the ecological requirements of these two groups of fungi (Lacey et al., 1991). Regular monitoring of temperature, relative humidity, and air flow at many points within the grain bulk helps to avoid the formation of hot spots. Generally, agricultural commodities are stored at temperatures favorable for fungal growth. Cold-tolerant fungi, such as *Cladosporium* spp. and *Penicillium* spp., develop rapidly at storage temperatures below 20°C, whereas at higher storage temperatures, under tropical conditions, *Aspergillus* spp. may be seen on stored produce very commonly. The parasitic potential of *A. flavus* infecting maize grain is remarkably increased at temperatures from 30 to 40°C (Payne et al., 1988). In general, survival of the storage fungi is progressively reduced with increase in storage period irrespective of storage temperature. Survival of *Microdochium oryzae* in infected rice seeds decreased over time when stored at 10°C (Manandhar, 1999).

D. Location of Storage Fungi in the Seeds: The longevity of storage fungi is markedly affected by the location of the fungi in or on the seeds. The organisms that are deep-seated are protected by the seed tissues for a longer time, whereas the organisms present on the seed surface are eliminated soon. There are very few seed-borne biotrophic fungal pathogens, such as downy mildew pathogens, present on mature seeds (Sackston, 1981). Differential survival of the two forms of inoculum, that is superficial conidia and internal mycelium, for short and longer periods has been demonstrated in the case of *Alternaria brassicola*. A linear relationship between the population of conidia produced on the seed surface of cabbage and the internal inoculum was observed by Maude and Humpherson-Jones (1980).

E. Preharvest Infestation/Infection: Seeds already infected under field conditions or prior to storage may deteriorate faster, since the storage fungi can continue to invade the seed tissues for longer periods when such seeds are stored at lower moisture- and temperature conditions, favoring rapid development of storage fungi. Seed spoilage under such conditions may be at a faster rate as in corn (maize) seeds infected by *Aspergillus flavus* (Qasem and Christensen, 1960).

F. Infestation of Seeds with Arthropods: Insects and mites by themselves can cause seed deterioration directly. Furthermore, they can spread the spores and other fungal propagules when they interact with storage fungi. In addition, the moisture content of seeds may increase due to the release of water from their digestive process (Christensen and Sauer, 1982). Insect respiration may also lead to a release of water and to moisture migration, both favoring fungal colonization of stored grains. The damage caused by the insects may provide avenues of entry for the fungi, while the fecal materials may form new substrates for fungal colonization. The rice weevil, *Sitophilus oryzae*, may be able to transmit *Aspergillus flavus* from infected to healthy rice seeds (Pande and Mehrotra, 1988). Infestation of maize kernel by *S. oryzae* and peanuts by *Rhyzopertha dominica* carrying seed-borne fungi accelerated seed deterioration (Mukherjee and Nandi, 1993). Several fungi provide food for insects and mites. Field fungi such as *Alternaria alternata* and *Cladosporium cladosporioides* and storage fungi such as *Aspergillus fumigatus* and *A. niger* support the development of insects and mites (Lacey et al., 1991).

3.1.2 Seed-inhabiting Bacteria

Seeds carry both pathogenic and saprophytic bacteria. Pathogenic bacteria enter seeds through wounds. Transmission of bacteria through seeds forms a principal mode of their survival, since they may remain dormant in or on seeds. The mucilage present around the bacterial cells protects them from desiccation and loss of viability during dry periods. The bacterial cells are frequently disseminated during the rainy season, when conditions favor splash spread. Splash spread generally occurs in the direction of the wind, as in the case of *Pseudomonas syringae* pv. *pisi* (Grondeau et al., 1991). The seed-borne bacterial pathogens may survive for longer in surface plant residues than when the residues are incorporated into the soil, indicating the limited possibilities for the survival of bacteria as soil inhabitants. *Xanthomonas campestris* pv. *malvacearum* causing cotton blackarm disease could overwinter in debris on the soil surface in Oklahoma, United States, for 230 days, whereas the infectivity of the bacterial pathogen in buried debris was lost in 40 to 100 days (Brinkerhoff and Fink, 1964). The bacterial species inhabiting seeds of some of economically important crops are presented in Table 3.3.

3.2 FRUIT MICROFLORA

Spoilage of fruits occurs because of two groups of factors: abiotic and biotic. The abiotic factors include various physiological changes induced after harvest and during transit and storage. Hydrolytic action of enzymes, oxidation of fats and putrification of proteins (enzymes) are the important physiological activities initiated after harvest leading to various degrees of spoilage. The biotic spoilage is due to the microbial action associated with fungi (yeasts and molds)

TABLE 3.3 Seed-Inhabiting Bacteria

Crop	Bacteria	References
Brassica (Crucifers)	*Pseudomonas syringae* pv. *maculicola*	Bakker, 1951
	Xanthomonas campestris pv. *campestris*	Shiomi, 1991
Capsicum (Pepper)	*Burkholderia solanacearum*	Moffett et al., 1981
	Xanthomonas campestris pv. *vesicatoria*	Crossan and Morehart, 1964
Maize	*Erwinia zeae*	Pauer, 1964
	E. stewartii	Invanoff, 1933
	Psedomonas syringae pv. *syringae*	
Peanut (groundnut)	*Burkholderia solanacearum*	Machmud and Middleton, 1991
Rice	*Pseudomonas fuscovaginae*	Ziegler and Alvarez, 1987
	P. glumae	Wakimoto et al., 1987
	Xanthomonas oryzae pv. *oryzae*	Fang et al., 1956
	X. oryzae pv. *oryzicola*	Shekhawat and Srivastava, 1972
Sorghum	*Pseudomonas syringae* pv. *syringae*	Tarr, 1962
Soybean	*Burkholderia solanacearum*	Krasnova, 1963
	Pseuomonas syringae pv. *glycinea*	Graham, 1953
	P. syringae pv. *tabaci*	
	Xanthomonas campestris pv. *glycines*	
Tomato	*Clavibater michiganensis* subsp. *michiganensis*	Bryan, 1930
	Pseudomonas corrugata	Kritzman, 1989
	P. syringae pv. *tomato*	Devash et al., 1980
Wheat	*Bacillus megaterium* pv. *ceralis*	Hosford Jr., 1982
	Clavibacter tritici	Mathur and Ahmad, 1964
	Pseudomonas fuscovaginae	Ziegler and Alvarez, 1987
	P. syringae pv. syringae	Otta, 1977

and bacteria. Two kinds of biotic spoilage may be recognized: (1) the spoilage may be due to microbial pathogens infecting the plants during flowering, fruit development, and carried over to storage; and (2) saprophytes may either invade wounded tissues or colonize tissues necrosed due to the action of metabolites produced by microbial pathogens.

3.2.1 Fungal Pathogens

The fungi are the primary causes of spoilage of fruits, due to the low pH of fruits and their different ecological requirements. The important ecological requirements for the development of the fungal pathogens are environmental conditions, host factors, and pathogen factors.

3.2.1.1 Environmental Conditions

A. Temperature: The growth of fungal pathogens is markedly influenced by the temperature to which the fruits are exposed. The temperature ranges optimal for the growth of the fungi have been shown to accelerate the development of postharvest diseases. Generally the incidence and development of the diseases is slow at 5°C and below. The fungal pathogens such as *Botrytis* spp., *Penicillium* spp., and *Rhizopus* spp. could develop only very slowly at low temperatures. The spore germination and growth of *R. stolonifer* are entirely inhibited at temperatures below 7.5°C (Dasgupta and Mandal, 1989). The influence of temperature on the development of different postharvest diseases caused by fungal pathogens has been studied by various workers. The optimum temperature for the growth of *Botrydiplodia theobromae* causing *Botryodiplodia* rot of mango was found to be 30°C, but severe infection of mangoes occurred at 35°C. The growth of the pathogen was either inhibited or markedly retarded at low temperatures. However, the normal growth was resumed when the infected mangoes were placed at ambient temperatures (Eckert and Somner, 1967). The isolates of *B. theobromae* from mango, guava, and sapota showed variations in the optimal temperatures required for spore germination and growth (Srivastava and Tandon, 1971), indicating the possible influence of host plant species. This possibility was indicated by Williamson and Tandon (1966) who reported that maximum rotting of banana was observed at 25 to 31°C, while at 7 to 9°C the rotting was minimal. Similar effects of temperature on growth of *B. theobromae* infecting mango and banana were noted by Palaniswami (1978). The temperature optima for *Monilinia fructigena* infecting apple and *Aspergillus niger* infecting mango were 23 to 25°C and 30°C, respectively (Jadeja, 2000; Xu et al., 2001).

The fungal pathogens requiring a higher minimum temperature for growth may be controlled by storage below this temperature, as in the case of *Gliocephalotrichum bulbitium*. This pathogen could not infect apples at temperatures below 10°C (Jamaludin et al., 1973). On the other hand, *Botrytis cinerea* could grow and infect strawberries at temperatures as low as 0°C (Dennis and Cohen, 1976). Hence, its development could be delayed by rapid cooling to temperatures below the optimum for growth of the pathogen. *Colletotrichum gloeosporioides*, causing mango anthracnose disease could successfully infect the fruits at temperatures ranging from 20 to 30°C. Isolates of *C. gloeosporioides* from different locations exhibited significant variations in the optimal temperature requirements for spore germination and formation of appressoria (Dodd et al., 1991). Apple blotch disease is a complex disease caused by three fungal pathogens, that is *Leptodontium elatius, Peltaster fructicola*, and *Geastrumia polystigmatis*, which grow on the apple fruits superficially without penetrating the cuticle (Johnson et al., 1996, 1997). The optimum temperature requirements of *P. fructicola* were slightly lower (12°C to 24°C) than that required for *L. elatius* (16° to 28°C) for mycelial growth. This difference was reflected in the ability these pathogens for surviving extended exposure to higher temperatures of ≥32°C. Mycelia of *L. elatius* could tolerate the higher temperature but not of *P. fructicola* (Johnson and Sutton, 2000). *Zygophiala jamaicensis*, causing another apple disease, flyspeck, could grow over a wide range of temperatures, from 5° to 28°C. The optimum temperatures for mycelial growth were 15° to 24°C (Williamson and Sutton, 2000). *Phytophthora* brown rot of citrus is caused primarily by *P. palmivora*, while *P. nicotianae* occurs as a secondary pathogen associated with the disease in Florida orchards. The optimum temperature range for fruit infections and brown rot development was found to be 27 to 30°C. Temperatures below 22°C did not favor the disease development. Sporangium production on the fruit surface occurred at 24°C, its production being reduced at higher or lower temperatures (Timmer et al., 2000). *Sphaeropsis pyriputerscens* causing a fruit rot in d'Anjou pears was able to grow at temperatures from 0 to 25°C, with optimum growth between 15 and 20°C, little or no growth at 30°C, and no growth at 35°C. This pathogen was essentially a low-temperature species (Xiao and Rogers, 2004).

B. Relative Humidity (RH): Increase in RH may result in the appearance of free water at or near saturation levels (approaching 100% RH). In contrast, as the temperature increases the RH generally drops under natural conditions. The capacity of air to hold moisture varies with changes in temperature. Hence the interaction between temperature and moisture (RH) may be a critical factor affecting the spore germination and growth of fungal pathogens infecting fruits. Initiation of infection of banana by *Colletotrichum musae* causing anthracnose disease and by *Fusarium moniliforme* and *F. roseum* causing postharvest rots was reported to be favored by high RH (Khanna and Chandra, 1995). The development of *Mycosphaerella* rot of papaya was rapid at high RH conditions (Chan and Alvarez, 1979). The incidence of *Colletotrichum gloeosporioides* (*Glomerella cingulata*) causing mango anthracnose disease was greater at temperatures ranging from 28° to 34°C and 70 to 87% RH, whereas the incidence of stem-end rot caused by *Botryodiplodia theobromae* increased at temperatures ranging from 32° to 36°C and 84 to 100% RH (Banik et al., 1998). Free water or RH above 95% is required for germination of conidia of *C. gloeosporioides* and for appressoria formation as well, although conidia could survive for 1 to 2 weeks at RH as low as 62% and then germinate when placed at 100% RH (Fitzell et al., 1984; Dodd et al., 1991). The studies to determine the RH requirements of *Peltaster fructicola* and *Leptodontium elatius* causing sooty blotch complex of apple showed that both fungal pathogens could grow at 88% RH and their mycelial growth was inhibited at less than 95% RH. The germination of conidia of *P. fructicola* was maximum at 24°C and 97% RH as compared to 32°C and 99% RH required for *L. elatius* (Williamson and Sutton, 2000). The conidia of *Monilinia fructigena*, causing brown rot of apple could germinate only at near saturation humidity levels (≥97% RH) (Xu et al., 2001).

3.2.1.2 Host Factors The levels of susceptibility/resistance to a pathogen depends basically on the genetic constitution of the plant species, although the environmental factors may have influence on disease incidence by predisposing the host plants in a particular location. The plants have certain natural barriers for preventing the penetration of host surface and the presence of preformed antimicrobial substances that may restrict the development of microbial pathogens. Morphological and anatomical barriers contribute to passive resistance to the invasion of fungal pathogens, as in the grape–*Botrytis cinerea* pathosystem where the thickness of grapes berry skin (Karadimtcheva, 1981), the number of skin cell layers (Sarig et al., 1998), and the thickness of berry cuticle and wax content (Commenil et al., 1997; Riberau-Gayon et al., 2000) contribute to resistance.

A study involving 42 genetically diverse cultivars and selections of grapevine with various levels of resistance to *B. cienrea* showed that the number of pores present in berries was negatively correlated with resistance. Highly resistant cultivars had few or no pores in the berry surface. In addition, the number and thickness of epidermal and hypodermal cell layers and cuticle and wax contents were found to be positively correlated with resistance to *B. cinerea* (Mlikota Gabler et al., 2003). In addition, host defense-related compounds such as phytoalexins and pathogenesis-related (PR) proteins, produced in response to infection, may also slow down the invasion of the microbial pathogens. Thus, the interaction between the pathogen and host tissue may result in the development of symptoms of infection in susceptible plants, whereas the response of resistant plants may vary from the development of hypersensitive lesions to complete absence of any visible symptoms. The susceptible plants may support the production of varying amounts of pathogen populations, depending on the extent of nutrient availability from infected plant tissues and environmental conditions.

The fungal pathogens may have a restricted or wide host range capable of infecting a large number of plant species, as in *Botrytis cinerea*. This pathogen can infect many fruit and vegetable crops. Hence the infected plant organs, such as seeds, leaves, shoots, flowers, and fruits, may provide inoculum for further spread. Furthermore, *B. cinerea* may exist as a saprophyte in strawberry plantations and produce spores on dead leaves and plant debris lying on the soil surface (Jarvis, 1962). The presence of air-borne conidia of *B. cinerea* was found prior to and during flowering and fruiting season (Jordon, 1978). Necrotic stamens and sepals infected by *B. cinerea* were associated with marketable fruits having latent infections, indicating that the infected floral parts might be the sources of infection for harvested strawberry fruits (Dennis, 1983).

Infected plant organs of one crop may serve as sources of infection for another crop. *B. cinerea* may sporulate on infected grapes berries on the vineyard floor, or on fruit wounds caused by rodents or birds in California. Furthermore, this pathogen can sporulate on senescing leaves of weeds (grasses and broad leaf species). The inoculum from these infected plant tissues may be the sources of infection for kiwifruit (*Actinidia deliciosa*), with a sweet tropical taste, which is being grown in several countries. The pollen grains of grapevine could enhance the aggressiveness of *B. cinerea* by stimulating the germination of conidia, which may be carried by honey bees visiting both grapevine and other susceptible crops (Chou and Preece, 1968; Rose, 1996).

Colletotrichum gloeosporioides causing mango anthracnose disease, produces conidia abundantly on the lesions present on leaves, twigs, panicles, and mummified fruits (Fitzell et al., 1984). Secondary infection may be due to rain-splashed conidia that can reach developing fruits causing preharvest loss (Gantotti and Davis, 1993). Late infections on developing fruits may remain quiescent until the onset of ripening after harvest. The lesions begin to develop when the climacteric period commences and no fruit-to-fruit infection occurs at the postharvest phase (Arauz, 2000). *C. gloeosporioides* can infect other host plant species such as avocado, papaya, banana, coffee, and citrus. The isolates of *C. gloeosporioides* from these crops could infect mango fruits, indicating that these infected crops may act as sources of infection for mangoes (Freeman and Shabi, 1996; Freeman et al., 1998).

Nutrient availability is a critical factor influencing spore germination, penetration into, and colonization of, host tissues by the fungal pathogens. The spores of *Penicillium digitatum* causing green mold of citrus fruits were stimulated by the ascorbic acid and some terpenes present in the fruits skin. These compounds released the nutrients required for the sustained and vigorous growth of the pathogen during pathogenesis (Eckert, 1978). The studies to determine surface colonization of, and penetration into, grape berries by *Botrytis cinerea* using fluorescence microscopy, revealed that germination of conidia was delayed on immature berries. Dieback of conidia and germlings was observed at a significantly higher level on immature than on mature berries. Induction of stilbene and suberin generally occurred in the early stages of berry development. The nonavailability of nutrients and induction of antimicrobial compounds in immature berries might have played a role in the failure of infection by *B. cinerea* (Coertze et al., 2001).

The physiological changes occurring in harvested fruits may contribute to the success or failure of infection by fungal pathogens. Secretion of pectate lyase (PL), required for the growth of *Colletotrichum gloeosporioides* infecting avocado fruits, occurred at pH > 5.8 of the enzyme-inducing medium. Under in vivo conditions, the pH of unripe pericarp of freshly harvested avocado cv. Fuerte fruits, resistant to infection by *C. gloeosporioides*, was 5.2 (below the optimum pH). In contrast, the pericarp of ripe fruits, which were susceptible to infection by this pathogen, showed an increase to pH 6.3 which was favorable for the secretion of PL, whose macerating activity led to release of nutrients required for the growth of *C. gloeosporioides* (Yakoby et al., 2000). Various physiological activities operating in unripe fruits thought to check or inhibit the development of fungal pathogens, resulting in the phenomenon of quiescence, are discussed in the next section (3.2.1.3).

3.2.1.3 Pathogen Factors Successful postharvest fungal pathogens should be aggressive or highly virulent so that it can infect a wide range of plant species and they should be able to survive adverse environmental conditions by producing structures such as sclerotia, chlamydospores, and sexual spores. The ability to survive as a saprophyte in the absence of a living host plant is another important factor for perpetuation. Abundant production of spores or other forms of inoculum may be useful for rapid development of the diseases. Plant diseases may be classified into two categories, depending on the ability of the pathogens to cause one or more secondary infections in a single season. For example, *Colletotrichum gloeosporioides*, causing mango anthracnose disease, forms conidia from lesions on leaves, panicles, and mummified fruits and these conidia when spread by rain splash can cause secondary infections on the leaves and fruits. At this phase, the disease is polycyclic in nature. On the other hand, fruit infection in the field prior to harvest remains quiescent and, when lesions develop, there is no spread of the disease to other fruits under storage conditions. Thus at the postharvest phase, the disease is a monocyclic disease, since there is no secondary infection from the inoculum on infected fruits (Arauz, 2000).

The population diversity within isolates of *Glomerella cingulata*, causing apple bitter rot disease present in orchards of cv. Granny Smith and Gala, was studied. The presence of *G. cingulata* was predominantly correlated with bitter rot disease. Three vegetative compatibility groups (VCGs), VCG-1, VCG-2, and VCG-6 of *G. cingulata* were differentiated. Differences in frequencies among the different morphological types found within three orchards remained similar throughout the season and from year to year. The results suggested that the frequencies of occurrence of *G. cingulata* in relation to *Colletotrichum gloeosporioides* and *C. acutatum* remained unaltered after these pathogens had established in the orchards (González and Sutton, 2004).

Fungal pathogens such as *Botrytis cinerea*, *Colletotrichum gloeosporioides*, and *Botryodiplodia theobromae* have wide host ranges that include fruit crops. The infected plants of one species can provide inoculum for infection of another plant species and such pathogens may survive in one or more host plants during adverse conditions. On the other hand, pathogens confined to one particular plant species will have a low degree of survivability. The fungal pathogens may spread from infected plants to healthy plants through seeds, wind, rain/water, rain splashes, and soil. The pathogens that can spread through more than one mode may spread at a faster rate. Wind-borne inoculum may reach distant places where susceptible host plants are available.

Botrytis cinerea, present in dead leaves and debris, sporulates and airborne conidia are always present at all periods during flowering and fruiting in strawberry (Jarvis, 1962; Jordon, 1978). *B. cinerea*, present in infected grape berries, senescing leaves of infected weed hosts, and fruit remnants of kiwifruits, could sporulate abundantly. Furthermore, the overwintering sclerotia and mycelium also produce enormous amounts of conidia in winter and spring (Michailides and Elmer, 2000). The populations of *B. cinerea* and *Peni-*

cillium spp. on pear fruit surface had a bearing on the disease development during storage. Significant positive correlation between fruit decay and fruit surface populations of both pathogens were observed. The inoculum level in fruit surfaces may be used as a predictor of fruit decay (Lennox et al., 2003). The fruit rot intensity of kiwifruits was proportional to inoculum concentration. These infected plant materials form the sources of infection for kiwifruits (Bautista-Baños et al., 2001). *Peltaster fructicola*, a component of apple sooty blotch disease, overwintering in apple twigs and fruits produces conidia which are spread by wind. Secondary infections may occur through conidia produced on apple fruit (Williamson and Sutton, 2000). The relationship between infection of pear stems and calyxes by *B. cinerea* and the incidence of gray mold disease in storage was investigated. Calyxes were susceptible to infection soon after full bloom. Calyx-end gray mold was detected at low levels in peaches stored for up to 8 months. Preharvest calyx infection levels could not be used as an indicator of gray mold disease incidence in fruits and air storage (Lennox and Spotts, 2004).

The phenomenon of quiescence has been observed in infections of fruits and vegetables by various fungal pathogens. During the early phases of pathogenesis, the fungal pathogens may remain quiescent (dormant) at different stages such as spore germination, germ tube formation, appressorium formation, and colonization. When conditions favorable for the development of the pathogen become available, the pathogens develop further, leading to expression of visible symptoms. The period of quiescence is enforced during the various processes of fungal infection and it appears to be a tactic adopted by the fungal pathogens to overcome the possible defense mechanisms that may inhibit or prevent fungal infections. When the toxic, preformed antimicrobial compounds are converted into nontoxic ones as the fruits mature, followed by the conversion of sugars to forms that will favor the growth of the pathogens, they bounce back and grow to reach the levels necessary for production of symptoms. The pathogen quiescence is discussed in detail in Chapter 8.

3.2.2 Bacterial Pathogens

Bacterial pathogens causing fruit diseases are far fewer compared with fungal pathogens. *Xanthomonas campestris* pv. *viticola* inducing cankerous lesions on grape bunches and berries could develop at temperatures between 5° and 35°C with an optimal temperature of 30°C for maximum growth. Relative humidities above 80% were found to be favorable for bacterial development (Chand, 1996). *Xanthomonas campestris* pv. *mangiferae* causing cankers on mango fruits required temperatures ranging from 28 to 32°C and RH of 70% for its development (Liao, 1975). For the development of *Xanthomonas axonopodis* pv. *citri*, temperatures ranging from 20° to 30°C with well-distributed rains (high RH) were conducive. Presence of free water on the plant surface for about 20 min appeared to be essential for successful infection of citrus fruits by the bacterial pathogen (Ramakrishnan, 1954). The bacterial pathogens may overwinter in the plant tissues and resume the activities when conditions are favorable. They are spread by rain splashes, irrigation water, and in some cases by arthropod vectors. *Erwinia amylovora*, causative agent of fire blight disease of apple, was detected by using a technique, marketed by BIO-RAD, on 15 out of 348 insects collected which have different habitats. They belong to four different orders, namely Homoptera, Coleoptera, Hymenoptera, and Diptera. The bacteria could survive at least 5 days on/in the green lace wing (*Chrysoperla carnea*) and for 12 days on aphids (*Aphis carnea*) (Hildebrand et al., 2000).

3.3 VEGETABLE MICROFLORA

Vegetables, after harvest, have active metabolic activities and the quality of vegetables does not improve during postharvest period, irrespective of the conditions of storage. Several factors adversely affect the quality of vegetable and they include: (1) respiratory and metabolic activities leading to reduction in nutritive value and sensory quality; (2) loss of turgidity and withering due to transpiration; (3) growth phenomena such as sprouting, rooting, and color changes; (4) mechanical damage and injuries due to handling, packing, and transport; and (5) physiological disorders and pathogen invasions and saprophytic association. Of these factors, the postharvest diseases caused by microbial pathogens are the most important, the magnitude of losses being greater than the losses incurred in the field due to the added costs of harvesting, transport, and storage (Dennis, 1987; Gorini, 1987). Many factors, such as high water content, nutrient composition, and pH, predispose the vegetables to attack by fungi and bacteria. In addition, vegetables exhibit increasing susceptibility as they ripen. The vegetables, however, may defend themselves against microbial infection either through the presence of preformed antimicrobial substances or formation of antimicrobial compounds such as phytoalexins in response to infection by microbial pathogens. The vegetables may be grouped as root vegetables (potato, sweetpotato, and onion) and fruit vegetables (tomato, capsicum, and cucurbits) which are infected by microbial pathogens both in pre- and postharvest stages.

3.3.1 Fungal Pathogens

3.3.1.1 Environmental Conditions Among the fungi causing postharvest diseases, those belonging to the genera *Aspergillus, Botrytis, Colletotrichum, Fusarium, Penicillium, Phytophthora, Pythium, Rhizopus*, and *Sclerotinia* are important. The system of classifying the fungal pathogens causing postharvest diseases as field fungi and storage fungi seems to be inappropriate. For example, *Ceratocystis fimbriata*, causing black rot disease, invades sweetpotato tubers in the field. But the symptoms of infection develop only during storage (Bulgarelli and Brackett, 1991). The fungi not only affect quality of fresh vegetables, but make the infected vegetables harmful because of the secretion of toxins in the infected plant tissues.

Potato tubers are infected by the inoculum of *Phytophthora infestans*, causing late blight disease, from infected haulm. Rainfall and soil moisture influence tuber infection to a great extent (Logan, 1983). Temperatures ranging from 4° to 27°C at 100% RH favored tuber infection, depending on the levels of resistance/susceptibility of potato cultivars (Lozoya-Saldaña and Hernandez-Vilchis, 2001). The spores of *P. infestans* survived 0 to 16 days in water suspensions under nonshaded conditions and 2 to 20 days under shaded conditions. When soil was added to the water suspension the spores survived longer periods. Exposure to direct sunlight reduced the spore survival markedly (Porter and Johnson, 2004). Pink rot, caused by *Phytophthora erythroseptica*, rapidly developed under temperatures between 15° and 25°C with moist conditions. In the potato piles, pockets of wet, rotting tubers may be infected, as the pathogen penetrates directly through the eyes or lenticels in the tubers (Logan, 1983; Salas et al., 2000). Leak, another fungal disease, caused by *Pythium* spp. exhibits symptoms quite similar to that of pink rot disease. Characteristic oozing of liquid from infected tubers is seen. The liquid that leaks carries the fungal spores, spreading the disease to healthy tubers in contact with infected tubers. High moisture and the higher temperatures developed due to heat of fermentation favor the development and spread of the disease. *Helminthosporium solani*, causing potato silver scurf disease, accounts for serious losses in the United States and Europe and infects rapidly when pulp temperatures remain warm with RH above 95% along with recirculation of the internal air. Secondary infection, to a limited extent, may be possible through contact with healthy tubers at about 5 to 10°C (Shetty et al., 1996).

Sweetpotato is affected by a number of postharvest diseases/disorders including physiologically induced problems due to chilling (<10°C). Rhizopus soft rot caused by *Rhizopus stolonifer* is one of the most important diseases under warm conditions. *R. arrhizus* and other *Rhizopus spp.* may also cause severe infections. The pathogen is ubiquitous and invades the roots through wounds. Release of nutrients from wounded/crushed tuber is essential for infection and further development of the pathogen. Exposure to sunlight or chilling predisposes the sweetpotatoes to soft rot disease caused by *Rhizopus* spp. (Clark, 1992). *Botryodiplodia theobromae*, with a wide host range, infects sweetpotato causing Java black rot disease. This pathogen rapidly develops at temperatures ranging from 20° to 30°C (Clark, 1992; Pati et al., 2001). The development of *Aspergillus niger*, causing black mold rot, was favored by temperatures above 35°C (Mandal, 1981).

Tomato fruits are infected by several fungi both at pre- and postharvest stages. The environmental requirements for pathogen development and disease progression may vary depending on the pathogen species and the geographical and storage conditions (Table 3.4). Tomato fruits are attacked by several fungal pathogens such as *Alternaria alternata*, *Aspergillus flavus*, *A.*

TABLE 3.4 Environmental Requirements for Development of Tomato Pathogens

Pathogen	Environmental conditions		References
	Temperature (0°C)	RH (%)	
Alternaria alternata	21–30	100	Pearson and Hall, 1975
	26–30		Mehta et al., 1975
Aspergillus flavus	35		Akhtar et al., 1999
Aspergillus niger	38		Dasgupta and Mandal, 1989
Botrytis cinerea	20–25	100	Ferrer and Owen, 1959; Chastagner and Ogawa, 1978
Fusarium moniliforme (Gibberella fujikuroi)	25		Silveira et al., 2000
F. verticillioides	25	>95	Silveira, et al., 2001
Geotrichum candidum	30	>95	Butler, 1969
	25		Silveira et al., 2001
Phytophthora infestans	18–21	100	Bashi et al., 1982
P. parasitica	>27		Obrero and Aragaki, 1965
Rhizopus stolonifer	5–25	80	Akhtar et al., 1999; Silveira et al., 2000

niger, *Botrytis cinerea*, *Colletotrichum gloeosporioides*, *Fusarium* spp., *Geotrichum candidum*, *Mucor circinelloides*, *Phytophthora infestans*, *Pythium* spp., *Rhizopus stolonifer*, *Sclerotium rolfsii*, and *Sclerotinia sclerotiorum*, inflicting differing degrees of losses depending on the susceptibility of tomato cultivars.

Seed infection has been shown to be responsible for the appearance of onion neck rot disease caused by *Botrytis allii* in storage. Direct relationships were observed between seed-borne infection and the percentage of neck rot-infected stored bulbs and between the percentage of plants infected under field conditions and the incidence of neck rot disease under storage conditions (Maude and Presley, 1977a, b). No visible symptoms could be observed on onion leaves. Hence, the seed to plant transfer of *B. allii* generally was not recognized earlier. The onion debris, and sclerotia present on the soil surface and cull piles of affected onions may act as potential sources of infection for spring-sown onion crops (Maude, 1983). In the case of *Aspergillus niger*, causing black mold disease, the principal source of infection was soil-borne inoculum which contaminated outer scales of onion bulbs and was carried into storage. The amount of rain received prior to harvest and high RH during storage influenced the development of the disease (Miller and Dillon, 1979). Infection was at a faster rate at temperatures from 28° to 32°C (Maude, 1983).

Mechanical damages or injuries not only reduce the market value of vegetables directly, but also serve as avenues of entry for the fungal pathogens leading to additional losses, because of the diseases they induce. Saprophytic fungi such as *Fusarium* spp. and *Geotrichum* spp. may cause much spoilage due to the conditions conducive for opportunistic invasion. Vegetables invaded by fungal pathogens present the risk of becoming toxic because of the mycotoxins secreted by some of the fungal pathogens. The importance of mechanical injury in providing infection sites for *Botrytis cinerea* infecting cabbage was revealed by the studies of Lockhart (1976). The accelerated spread of *Botrytis* infection following midterm trimming of stored winter white cabbage was reported by Shipway (1981). *B. cinerea* could cause secondary infection by invading lesions caused by other fungal pathogens such as *Alternaria* spp. or *Mycosphaerella brassicola* (Geeson and Browne, 1978) and necrotic leaf lesions caused by *Turnip mosaic virus* (Walkey and Webb, 1978), demonstrating the potential of *B. cinerea* both as pathogen and as opportunistic invader.

Chill injury may predispose vegetables to infection by fungal pathogens. The frequency of rots in tomatoes caused by *Botrytis cinerea* increased with increase in the period of exposure to chill injury. Ripe tomatoes were found to be more susceptible than quarter-ripe fruit to increased rotting after storage at 2°C. The pathogen population has a decisive influence on the rate and extent of spoilage of tomato fruits. The number of tomato fruits infected at 20°C after an initial storage at 2°C for 7 days, was significantly greater for fruit sprayed with a high concentration (10^4 spores/ml) of *B. cinerea* when compared with uninoculated controls. The rate of germination and production of appressoria at 20°C increased after initial storage at 2°C. The percentage of germ tubes forming appressoria by *B. cinerea* was higher on the surface of ripe tomato fruit than on quarter-ripe fruit indicating the greater susceptibility of ripe fruit to infection by *B. cinerea* (Dennis and Davis, 1980; Dennis, 1983). On immature fruits, *B. cinerea* could cause ghost-spotting symptom, since no development occurred after the initial germination of conidia and penetration of the cuticle of immature fruits. The fungus could not be isolated from the ghost-spots from the immature fruits stored at 2 to 20°C (Verhoeff, 1970; Verhoeff and Liem, 1975). Exposure of tomatoes to lower temperatures appeared to induce susceptibility of tomato fruits to infection by *Alternaria* spp. Tomatoes that were chilled developed rots caused by *Alternaria* spp. in proportion to the duration of exposure to 0°C or to the decrease in temperature below 10°C. The tomato fruits at ambient temperatures were not infected by this pathogen (Dennis et al., 1979).

3.3.1.2 Host Factors The susceptibility or resistance to different pathogens of a vegetable plant species varies considerably. The genetic constitution of the plant species exerts decisive influence on the development of resistance to postharvest diseases. Potato cultivars carrying R genes exhibit resistance to the late blight pathogen, *Phytophthora infestans*. The potato tissues of resistant cultivars respond to infection by producing hypersensitive response (HR) leading to the death (necrosis) of infected tissues, when inoculated with incompatible races of *P. infestans* (Varns et al., 1997).

Factors contributing to disease resistance may be of two types, that is constitutive and induced. Morphological or anatomical barriers and presence of antifungal compounds may contribute to passive resistance, whereas production of antimicrobial compounds such as phytoalexins and pathogenesis-related (PR-) proteins may represent active resistance which develops following inoculation/infection. The development of fungal pathogens on the surface of tomato and capsicum fruits may be inhibited in the early stages of ripening. The period of quiescence indicates the time between the pathogens remaining dormant in unripe fruits and resuming their pathogenic activities, when the physiological status of the fruits become favorable for their development. Quiescence of *Colletotrichum piperatum* in *Capsicum* and *Botrytis cinerea* in ghost-spot in tomato fruits has been observed (Verhoeff, 1970; Grover, 1971).

The presence of antimicrobial substances considered to offer resistance to infection by microbial pathogens was demonstrated by the classical work of Walker and Stahmann (1955). An onion cultivar with red scales was found to be resistant to the smudge disease caused by *Colletotrichum circinans*. The resistance was attributed to the presence of the phenolic compounds, catechol and protocatechuic acid. These compounds were inhibitory to another onion pathogen *Lasiodiplodia theobromae*, but not to *Aspergillus niger*, which can attack both red and white onion cultivars. The role of phytoalexins in the development of resistance to plant diseases has yet to be clearly established. The production of various kinds of phytoalexins, such as rishitin in potatoes and ipomoeamarone in sweetpotatoes, has been demonstrated following inoculation with fungal pathogens *Phytophthora infestans* and *Ceratocystis fimbriata*, respectively. Variations in the accumulation of ipomoeamarone in response to infection by postharvest pathogens were observed. Ipomoeamarone concentrations were higher in restricted lesions than in expanding lesions, indicating the possible inhibitory effect of this compound on pathogen development (Clark, 1992).

The cultivation practices, method of harvesting, extent of mechanical damage or injury caused, and type of storage also may influence the rate of disease incidence and spread. The pathogens such as *Botrytis cinerea* and *Sclerotinia sclerotiorum* infecting carrot rarely infected the roots in the field, but they could spread from the lamina into the base of the older petioles which were not removed at the time of harvest. These pathogens, present in the foliage debris adhering to the roots, were found to be important sources of primary infection (Goodliffe and Heale, 1975; Geary, 1978). The timing of harvest is considered to be an important factor affecting the susceptibility of carrot to postharvest pathogens. Carrot crops of different ages may show variations in the degree of susceptibility to storage diseases. The concentrations of glucose and sucrose at harvest seem to have a bearing on disease incidence (Lewis and Garrod, 1983). The unripe tomato fruits are less susceptible to *B. cinerea* than ripe fruits, possibly because of the presence of the toxic glycoalkaloid, tomatine in unripe fruits (Verhoeff and Liem, 1975).

3.3.1.3 Pathogen Factors The pathogen, to be successful, should be virulent and able to overcome host defenses and initiate various processes such as triggering pathogenic factors resulting in secretion of required enzymes and other toxic metabolites. Maceration of host tissues and release of nutrients are essential to sustain the pathogen growth and reproduction. The pathogens may have a wide host range or may be confined to a limited number of host plant species depending on their pathogenic potential and ability to produce different forms or races capable of infecting various plant species. The fungal pathogens, depending on the availability of their ecological requirements, develop and cause both quantitative and qualitative losses. Generally they are favored by warm temperatures and high humidity prevailing in storage. The ability to infect seeds, and other plant parts on which the fungal pathogens can overwinter, may aid in the survival of the pathogen, as in the case of *Botrytis allii* causing onion neck rot disease. Infected onion seeds, dead onion tissues supporting the saprophytic growth of *B. allii* and the black sclerotia, weather-resistant overwintering structures, play important role in the incidence and spread of the disease (Tichelaar, 1967; Maude and Presley, 1977). *Colletotrichum piperatum* causing fruit rot of *Capsicum frutescens* also has the ability to survive in infected seeds and plant debris (Grover and Bansal, 1968).

Ceratocystis fimbriata causing sweetpotato black rot disease can survive as mycelium and spores in soil-borne debris from diseased plants, and also in stored roots, for about 2 years. Infected transplants and seed stock are responsible for long distance spread of this pathogen. Chlamydospores and the perithecia (sexual spore-bearing structures) resistant to adverse conditions can aid in the overwintering and survival of the pathogen (Sherf and Macnab, 1986). The fungal pathogens such as *Alternaria alternata* and *Botrytis cinerea*, with wide host range, can survive on several fruit and vegetable crops and inoculum may be available in all seasons as different crops may be grown in different seasons. Recent evidence indicates that *B. cinerea* could generate reactive oxygen species (ROS) and this could contribute to the virulence of this fungal pathogen on beans. Mutants (Dbcgod1) defective for a putative glucose oxidase gene (*bcgod1*) displayed normal virulence, whereas Dbcsod1 mutants generated by deletion of a Cu-Zn-superoxide dimutase gene (*bcsod1*) had reduced virulence. The results indicated that the Cu-Zn-SOD- activity is an important single virulence factor in bean–*B. cinerea* pathosystem (Rolke et al., 2004).

3.3.2 Bacterial Pathogens

The spoilage of fruits is caused primarily by fungal pathogens, due to their low pH, whereas both bacteria and fungi are responsible for vegetable spoilage. The bacteria belonging to the genera *Clavibacter, Erwinia, Pseudomonas*, and *Xanthomonas* induce postharvest diseases of various vegetables. Furthermore, bacterial species capable of causing human ailments, such as *Bacillus cereus, Clostridium botulinum, Escherichia coli, Salmonella typhimurium*, and *Vibrio cholerae* are associated with vegetables and food products. Bacterial pathogens commonly cause 'soft rot' type of diseases. They induce maceration of parenchyma tissues resulting in areas of rotting that may spread rapidly if conditions are conducive, ultimately leading to collapse of tissues entirely. Often growth of secondary, non-rotting bacteria develop contributing to emanation of offensive odor from affected tissues.

The bacterial pathogens generally invade through wounds and natural openings on the vegetables. The incidence of soft rot-causing *Erwinia* spp. is favored by mechanical damages or wounds. *E. carotovora* subsp. *carotovora* (*Ecc*) and *E. carotovora* subsp. *atroseptica* (*Eca*) are commonly found on the surface of potatoes due to contamination occurring after decay of the mother tuber. They can survive in the lenticles and to a lesser extent on the tuber surface for about 6 to 7 months during storage (Pérombelon, 1974; Pérombelon and Kelman, 1980). The erwineas may be disseminated through soil water resulting in transfer of the pathogens from decaying mother tubers to daughter tubers on the same or adjacent plants. Transmission through farm implements, such as cutting knives, graders, tractors, and harvesting machinery, may also be possible (Lund, 1983).

3.3.2.1 Environmental Conditions *Pseudomonas cepacia*, causing sour skin disease of onion, requires free moisture and high temperature for infection, and rain storms and flooding may accelerate disease incidence and spread. Pathogen development is apparently enhanced at acidic pH which supports the activity of macerating enzymes such as polygalacturonase, produced by *P. cepaciae* (Kawamoto and Lorbeer, 1974; Ulrich, 1975). *P. gladioli* pv. *alliicola*, another onion pathogen causing bulbrot, inflicts greater damage at higher temperature (25°C) and lower pH levels (4.9–5.1) (Hildebrand, 1971; Taylor et al., 1980). *Erwinia carotovora* subsp. *carotovora*, causative agent of bacterial soft rot disease of onion, grows at temperature ranges of 2 to 37°C with the optimum of 24° to 33°C, if optimum moisture is available. This bacterium can enter only through wounds. Onion maggots (*Hylemya antigua*) carrying the bacteria facilitate the entry of the bacterial pathogen which persists in the intestinal tract of maggot larvae and adult flies (Sherf and Macnab, 1986).

The bacterial pathogen *Clavibacter michiganensis* subsp. *sepedonicus* (*Cms*) causing ringrot disease of potato may be disseminated via surface water and potatoes may be infected through the inoculum in irrigation water. *Cms* could survive in nonsterile surface water at 10°C for a maximum period of 7 days, during which period the pathogen can be transported over long distances, indicating the possibility of infection of potato crops and tubers later by *Cms*. Nevertheless, the inoculum may be strongly diluted (Wolf and Beckhoven, 2004).

Erwinia carotovora is known to cause various diseases of vegetables, both in the field and in storage or transit all over the world. *E. carotovora* subsp. *atroseptica* (*Eca*) is generally restricted to temperate regions, causing blackleg (stem infection) disease in the field or tuber soft rot in stored potatoes. On the other hand, *E. carotovora* subsp. *carotovora* (*Ecc*) occurs in all countries world wide, causing blackleg and aerial stem rot diseases of potato (Molina and

Harrison, 1980; Pérombelon and Kelman, 1987). Temperature, humidity, and aeration exert marked influence on the incidence and severity of diseases caused by *Eca* and *Ecc*. The temperature has the greatest effect on the ability of *Eca* and *Ecc* to infect potatoes. The optimal temperature for infection by *Eca* was found to be between 15° and 20°C, while *Ecc* could infect at higher temperatures around 25°C (Pérombelon and Salmond, 1995). In a later study, the thermodependence of growth and enzymatic activities of *Eca* and *Ecc* required for infection was studied. The specific growth rate and the pectate lyase and protease activities of eight strains of *Ecc* and *Eca* were determined in vitro. These two bacterial pathogens responded distinctly to exposure to various temperature regimes. However, the protease activity of both *Eca* and *Ecc* appeared to be similarly thermoregulated. The in vitro assessment was in agreement with ecological data, demonstrating that infection by *Eca* was favored by lower temperatures (20°C), whereas successful infection by *Ecc* occurred at higher temperatures (Smadja et al., 2004).

Xanthomonas campestris pv. *vesicatoria* causing bacterial spot on fruits and leaves of capsicum (pepper) is both internally and externally seed-borne. Infection by the bacterial pathogen is favored by high relative humidity with free moisture for long periods. The optimum temperature for disease development is about 20°C (Diab et al., 1982). Watermelon fruit blotch caused by *Acidovorax avenae* subsp. *citrulli* did not appear to be transmitted to healthy fruit in contact with diseased fruit for short period (1 week), but the frequency of transmission was increased significantly when stored at 11°C (Rushing et al., 1999). *Streptomyces scabies* and *S. turgidiscabies* survived in scab lesions in tubers of two potato cvs. Matilda and Sabina stored at 4°C for 24 weeks (Lehtonen et al., 2004).

3.3.2.2 Host Factors The levels of susceptibility/resistance to bacterial diseases are dependent on the genetic constitution of cultivars. However, the expression of symptoms may be influenced by the environmental conditions. *Erwinia chrysanthemi* pv. *chrysanthemi* causes bacterial soft rot in sweetpotatoes. Infected sweetpotatoes may remain symptomless until temperatures above 32°C are reached. Severe rotting occurs if such roots are stored (Schaad and Brenner, 1977). Availability of both nutrients and environmental conditions, favoring succulent growth of tomatoes predisposes them to infection by *Clavibacter michiganensis* subsp. *michiganensis* causing bacterial canker of tomato fruits (McKeen, 1973). Tomato seeds carrying *Xanthomonas campestris* pv. *vesicatoria* form the primary sources of infection and other host plant species, such as black night shade, may also provide inoculum. Young leaves and fruits are more susceptible than older tissues (Bashan et al., 1982). The rhizosphere of cruciferous crops such as lettuce and carrot may support the soft rot erwinias infecting potatoes and other vegetables (Kikumoto, 1974; Mew et al., 1976; Burr and Schroth, 1977; de Mendonca and Stanghellini, 1979). The susceptibility of potatoes to *E. carotovora* was reported to be conditioned largely by the depletion of oxygen in the tuber. The presence of a film of water on the surface of tubers for several hours, exposure to high temperatures, or storage in closed containers resulted in rapid depletion of oxygen, thereby increasing the susceptibility of potatoes (Lund, 1979). Likewise, the susceptibility of *Capsicum* (pepper) fruits to *E. carotovora* subsp. *carotovora* was increased by high moisture content, high nitrogen fertilization, and hydrocooling (Coplin, 1980).

3.3.2.3 Pathogen Factors The pathogens with wide host range have greater degree of survivability than the pathogens with restricted host range. *Erwinia carotovora* subsp. *carotovora* (*Ecc*) is known to cause soft rots of a wide variety of vegetables both in the pre- and postharvest stages. *Ecc* causes the blackleg disease under field condition, and it is recognized as the major postharvest pathogen of potato, accounting for serious losses in temperate regions (Pérombelon and Kelman, 1980). The ability to secrete required amounts of pectate lyases (Pels) and hydrolases is essential for the development of bacterial pathogens. The mutants of *E. chrysanthemi* defective in *PelI* and *PelL* genes had reduced virulence on potato tubers because of their inability to produce enzymes sufficient for tissue maceration to cause soft rot disease (Jafra et al., 1999), indicating that induction of symptoms is strongly correlated with the massive secretion of cell wall-degrading enzymes.

The internal tissue of apparently healthy vegetables may show the presence of soft rot bacteria including *Erwinia* spp. Meneley and Stanghellini (1974) reported that 56% of field-grown cucumbers contained soft rot bacteria and the soft rot symptoms could be observed only when they were incubated at 37°C. This observation reveals the ability of soft rot bacteria to escape detection by visual examination.

Potato ring rot disease caused by *Clavibacter michiganensis* subsp. *sepedonicus* cannot be detected in potato tubers, without cutting open the tuber, until the disease development is in advanced stage. Such a situation facilitates the spread of the disease during storage. Infection of potato tubers by soft rot pathogen *Erwinia carotovora* has been shown to result in the production of specific volatile compounds. The amount of volatiles produced was dependent on the initial amount of bacterial population and the growth of bacteria (Lyew et al., 2001).

E. carotovora subsp. *carotovora* (*Ecc*) not only has a wide range of hosts but also has the ability to survive in soil and plant debris. The ommipresence of *Ecc* on carrot roots is indicative of its ability multiply within the soil. Furthermore, any type of wound caused due to harvest bruises, freezing injury, fungus invasion, and insect wounds may provide points of entry for *Ecc*. Flies, especially *Hylemya cillicrura platura* and *H. brassicae*, carrying the bacteria in their intestinal tracts act as vectors of *Ecc* (Sherf and Macnab, 1986). *Xanthomonas campestris* pv. *vesicatoria* (*Xcv*) could overwinter in pepper leaves buried 30 cm below the soil surface but not on the leaves on soil surface. *Xcv* present in vascular tissues appears to be protected by the plant tissues (Lewis and Brown, 1961).

3.3.3 Viral Pathogens

The adverse effects of virus infection are more conspicuous on vegetables than on fruits, leading to a marked reduction both in quantity and market quality. Of the several viruses infecting tomato, infection of fruits by *Tomato mosaic virus* alone (ToMV), *Tobacco mosaic virus* (TMV) in combination with *Potato virus X* (PVX), and *Tomato spotted wilt virus* cause appreciable losses. ToMV is both internally and externally seed-borne. The presence of ToMV on the seed coat and endosperm has been detected (Broadbent, 1965). As the ToMV is highly stable, the virus present in plant debris can retain its infectivity for about 2 years and for about 6 months in dry soil. In addition, the infected weeds form effective reservoirs for further disease spread. Since there is no demonstrable transmission of ToMV by any vector, the disease is spread mainly through leaf contact and handling (Broadbent et al., 1965). The complex due to dual infection by ToMV and PVX produces more severe symptoms on tomato plants and fruits. Exposure of infected plants to temperatures >26°C results in masking of symptoms and only a faint mosaic pattern is induced (Phillip et al., 1967). Tomato is infected by another important disease, caused by *Tomato spotted wilt virus* (TSWV). In contrast to ToMV and PVX, TSWV is highly unstable and it is inactivated even at low temperatures (45–50°C). The virus is transmitted by larvae and adults of the thrips (*Thrips tabaci, Frankliniella schultzei, F. occidentalis*, and *F. fusca*). The virus has a very wide host range which includes 163 species in 34 families of plants (Best, 1968; Francki and Hatta, 1981).

Plum pox virus (PPV) causes one of the most devastating diseases of stone fruits (*Prunus* spp.). Disease symptoms and effects on fruit production depend on the host species. PPV is transmitted by several aphid species, among which *Aphis fabae, A spiraecola, Myzus persicae*, and *Brachycaudus persicae* are more efficient transmitters of PPV. The possibility of virus-infected peach fruit being the source for long-distance dispersal has been reported. The presence of PPV in peach fruit with no visible symptom was checked using an enzyme-linked immunosorbent assay (ELISA). The infected fruits, after storage for 2 to 4 weeks at 4°C, were used as the virus sources for the four aphid species to acquire and transmit PPV to healthy peach seedlings (Table 3.5). The results suggest that PPV-infected fruits may be another mode of PPV movement, allowing the virus to bypass natural barriers and invade new geographical locations (Gildow et al., 2004).

All viruses infecting potato plants are transmitted through the tubers and cause different degrees of deformation and tissue discoloration. Tubers infected by *Potato leafroll virus* (PLRV) exhibit a brown ring in the vascular system and germination of tubers is poor. Potato viruses are transmitted by different aphid species. *Potato virus Y* (PVY) has a wide host range and occurs in the form of strains. *Potato virus X* (PVX) is highly infectious and stable, and it can be transmitted through leaf contact and knives. During environmental conditions that favor the multiplication and movement of vector insects, the

TABLE 3.5 Transmission of *Plum pox virus* (PENN-7 Strain) by Aphid Species From Infected Peach Fruit to Healthy Peach Seedlings

Aphid species	Test 1	Test 2	Total	Percent transmission
Myzus persicae	5/10*	5/10	10/20	50.0
Aphis spiraecola	5/10	2/10	7/20	35.0
Aphis fabae	0/10	0/10	0/20	0.0
Brachycaudus persicae	0/10	0/10	0/20	0.0

* Number of peach seedlings infected/total inoculated.
Infection of seedlings was verified by ELISA initially and later by real-time polymerase chain reaction (PCR).
Source: Gildow et al., 2004.

disease spread will be rapid. The levels of resistance/susceptibility of potato cultivars will be a crucial factor influencing the extent of adverse effects of viruses on tuber development.

The feathery mottle complex observed on sweetpotato was considered to be due to the combined infection of different viruses. The feathery mottle, russet crack, and internal cork symptoms are due to different strains of *Sweetpotato feathery mottle virus* (SPFMV) and the variations in symptoms may be due to differential varietal response (Hildebrand, 1960; Moyer et al., 1980). Propagative planting materials, *Ipomoea* spp., and weeds form sources of virus infection and disease spread (Esbenshade and Moyer, 1982). Different aphid species transmit the virus in a non-persistent manner. The disease incidence in North Carolina was found to be positively related to aphid populations, which are correlated with winter temperatures (Nielsen, 1968; Cadena-Hinojosa and Campbell, 1981).

SUMMARY

Postharvest diseases of agricultural produce are primarily due to fungal and bacterial pathogens. A few viruses may also infect fruits and vegetables directly, although the viruses may be responsible for a general reduction in growth and yielding capacity of plants. Various ecological factors may influence the development, survival, and overwintering of microbial pathogens affecting the incidence and spread of diseases. The knowledge on the sources of infection and modes of disease spread may provide basic information for the development of effective management systems. The optimum environmental conditions required for pathogen growth and reproduction are discussed to focus attention on the possibilities of controlling such factors during storage, so that the development of postharvest pathogens is effectively contained.

REFERENCES

Agarwal, V.K. and Sinclair, J.B. 1996. *Principles of Seed Pathology*, 2nd Edition, CRC Press, Boca Raton, FL, USA.SW.

Agarwal, V.K., Verma, H.S. and Khetarpal, R.K. 1977. Occurrence of partial bunt on triticale. *FAO Plant Protection Bulletin*, **25**: 210.

Ahn, J.K. and Chung, B.K. 1970. Effects of seed disinfectants for controlling the soybean anthracnose. *Journal of Plant Protection, Korea*, **9**: 21.

Akhtar, K.P., Shakir, A.S. and Sahi, S.T. 1999. Physiological studies of fungi causing postharvest losses of tomato fruits. *Pakistan Journal of Phytopathology*, **11**: 25–29.

Alimova, F.K., Zakharova, N.G., Asghar, M.G., Fattahkova, A.N. and Garusov, A.V. 2002. The quarantine control and the methods of disinfection of tomato and cucumber seeds in different regions of Russian Federation. *Pakistan Journal of Scientific and Industrial Research*, **45**: 191–196.

Amadioha, A.C. 2000. The production and activity of amylase by *Rhizoctonia bataticola. Archives of Phytopathology and Plant Protection*, **33**: 1–9.

Andersen, A.L. 1948. The development of *Gibberella zeae* head blight of wheat. *Phytopathology*, **38**: 595.

Anonymous, 1982–1983. Brassica diseases, vegetable crop production and storage. *Annual Report of Edinburgh School of Agriculture* , 147 pp.

Anuja Gupta and Aneja, K.R. 2001. Mycofloral spectrum during storage and its effect on seed viability of soybean (*Glycine max* (L.) Merrill] seeds under ambient conditions. *Proceedings of the National Academy of Science, India, Sect. B.Biological Sciences*, **71**: 245–253.

Arauz, L.F. 2000. Mango anthracnose: Economic impact and current options for integrated management. *Plant Disease*, **84**: 600–611.

Arndt, C.H. 1946. Effect of storage conditions on survival of *Colletotrichum gossypii. Phytopathology*, **36**: 24–29.

Ashworth, L.J. and Langley, B.C. 1964. The relationship of pod damage to kernel damage by molds in Spanish peanut. *Plant Disease Reporter*, **48**: 875.

Baker, K.F. 1947. Seed transmission of *Rhizoctonia solani* in relation to control of seedling damping-off. *Phytopathology*, **37**: 912.

Baker, K.F. 1972. Seed Pathology. In: *Seed Biology*, Vol.2 (ed.) T.T. Kozlowski, pp. 317–416, Academic Press, New York.

Bakker, M. 1951. Bacterial spot disease of cauliflower and other *Brassica* species caused by *Pseudomonas maculicola* (McCulloch) Stevens. *Tijdschr. Plantenziekten*, **57**: 75.

Banik, A.K., Kaiser, S.A.K.M. and Dhua, R.S. 1998. Influence of temperature and humidity on the growth and development of *Colletotrichum gloeosporioides* and *Diplodia natalensis*, causing postharvest fruit rot of mango. *Advances in Plant Sciences*, **11**: 51–57.

Bashan, Y., Okon, Y. and Henis, Y. 1982. Long term survival of *Pseudomonas syringae* pv. *tomato* and *Xanthomonas campestris* pv. *vesicatoria. Phytopathology*, **72**: 1143–1144.

Bashi, E., Ben-Joseph, Y. and Roten, J. 1982. Inoculum potential of *Phytophthora infestans* and development of potato late blight epidemics. *Phytopathology*, **72**: 1043–1047.

Basuchaudhary, K.C. and Mathur, S.B. 1979. Infection of sorghum seeds by *Colletotrichum graminicola.* I. Survey, location in seed and transmission of the pathogen. *Seed Science and Technology*, **7**: 87.

Bautista-Baños, S., Barrera-Necha, L.L., Long, P.G., Ganesh, S. and Cheah, L.H. 2001. Inoculum variables affecting pathogenicity of *Botrytis cinerea* infection of kiwifruit. *Revista Mexicana de Fitopatologia*, **19**: 161–167.

Bechtel, D.B., Wilson, J.D., Eustace, W.D., Behnke, K.C., Whitaker, T., Peterson, G.L. and Sauer, D.B. 1999. Fate of dwarf bunt fungus teliospores during milling of wheat into flour. *Cereal Chemistry*, **76**: 270–275.

Beke, F. and Martin, J. 1958. The influence of blister stunt of maize on yield. *Mag. Mezog*, **13**: 17.

Best, R.J. 1968. *Tomato spotted wilt virus*. *Advances in Virus Research*, **13**: 65–146.

Bhowmik, T.P. 1969. *Alternaria* seed infection of wheat. *Plant Disease Reporter*, **53**: 77.

Bhuiyan, S.A., Ryley, M.J., Galea, V.J., Tay, D. and Lisle, A.T. 2002. Survival of conidia of sorghum ergot (caused by *Claviceps africana*) on panicles, seed and soil in Australia. *Australian Plant Pathology*, **31**: 137–141.

Boyd, O.C. 1935. Evidence of the seedborne nature of late blight (*Phytophthora infestans*) of tomatoes. *Phytopathology*, **25**: 7.

Brinkerhoff, L.A. and Fink, G.B. 1964. Survival and infectivity of *Xanthomonas malvacearum* in cotton plant debris and soil. *Phytopathology*, **54**: 1198–1201.

Broadbent, L. 1965. The epidemiology of tomato mosaic XI. Seed transmission of TMV. *Annals of Applied Biology*, **56**: 177–205.

Broadbent, L., Read, W.H. and Last, F.T. 1965. The epidemiology of *Tomato mosaic virus*. X. Persistence of TMV infected debris in soil and the effects of soil partial sterilization. *Annals of Applied Biology*, **55**: 471–483.

Bryan, M.K. 1930. Studies on bacterial canker of tomato. *Journal of Agricultural Research*, **41**: 825.

Bulgarelli, M.A. and Brackett, R.E. 1991. The importance of fungi in vegetables. In: *Handbook of Applied Mycology, Vol.3: Foods and Feeds* (eds) D.K. Arora, K.G. Mukerji and E.H. Marth, pp. 179–199, Marcel Dekker, Inc., New York.

Burr, T.J. and Schroth, M.N. 1977. Occurrence of softrot *Erwinia* spp. in soil and plant material. *Phytopathology*, **67**: 1382–1387.

Butler, E.E. 1969. Pathogenicity and taxonomy of *Geotrichum candidum*. *Phytopathology*, **50**: 665–672.

Cadena-Hinojosa, M.A. and Campbell, R.N. 1981. Characterization of isolates of four aphid-transmitted sweetpotato viruses. *Phytopathology*, **71**: 302–305.

Calo, R.P. 1957. Anthracnose of tomato. *Philippine Agriculture*, **41**: 273.

Chan, K.F. and Alvarez, A.M. 1979. Role of *Mycosphaerella* ascospores in stem-end rot of papaya fruit. *Plant Pathology*, **69**: 500–503.

Chand, R. 1996. Grapevine bacterial canker and its management. In: *Advances in Diseases of Fruit Crops in India* (ed.) S.J. Singh, pp. 161–172, Kalyani Publishers, Ludhiana, India.

Chastagner, G.A. and Ogwa, J.M. 1978. Dispersal of conidia of *Botrytis cinerea* in tomato fields. *Phytopathology*, **68**: 1171–1176.

Cherewick, W.J. 1954. Studies on seedborne microflora and the effect of seed treatment of rice. *Malaysia Agricultural Journal*, **37**: 169.

Chidambaram, P., Mathur, S.B. and Neergaard, P. 1973. Identification of seedborne *Drechslera* species. *Friesia*, **10**: 165.

Chou, M.C. and Preece, T.F. 1968. The effect of pollen grains on infections caused by *Botrytis cinerea* Fr. *Annals of Applied Biology*, **62**: 11–22.

Chowdhury, S. 1951. Studies on the bunt of rice (*Oryzae sativa* L.). *Indian Phytopathology*, **4**: 25.

Christensen, C.M. 1972. Moisture content of sunflower seeds in relation to invasion by storage fungi. *Plant Disease Reporter*, **56**: 173.

Christensen, C.M. and Sauer, D.B. 1982. Microflora. In: *Storage of Cereal Grains and Their Products* (ed.) C.M. Christensen, pp. 219. American Association of Cereal Chemists, St. Paul, MN, USA.

Clark, C.A. 1992. Postharvest diseases of sweetpotatoes and their control. *Postharvest News and Information*, **3**: 75N–79N.

Coertze, S., Holz, G. and Sadie, A. 2001. Germination and establishment of infection on grape berries by single airborne conidia of *Botrytis cinerea*. *Plant Disease*, **85**: 668–677.

Commenil, P., Brunet, L. and Audran, J.C. 1997. The development of the grape berry cuticle in relation to susceptibility to bunch rot disease. *Journal of Experimental Botany*, **48**: 1599–1607.

Cooke, B.M. and Jones, D.G. 1970. The epidemiology of *Septoria tritici* and *S. nodorum* II. Comparative studies of head infection by *Septoria tritici* and *S. nodorum* on spring wheat. *Transactions of British Mycological Society*, **54**: 395.

Coplin, D.L. 1980. *Erwinia carotovora* var. *crotovora* on bell peppers in Ohio. *Plant Disese Reporter*, **64**: 191–194.

Crossan, D.F. and Morehart, A.L. 1964. Isolation of *Xanthomonas vesicatoria* from tissues of *Capsicum annuum*. *Phytopathology*, **54**: 358.

Dasgupta, M.K. and Mandal, N.C. 1989. *Postharvest Pathology of Perishables*. Oxford and IBH Publishing Co. Pvt. Ltd., New Delhi., India.

de Mendonca, M. and Stanghellini, M.E. 1979. Endemic and soilborne nature of *Erwinia carotovora* var. *atroseptica*, a pathogen of mature sugar beets. *Phytopathology*, **69**: 1096–1099.

Dennis, C. 1983. Soft fruits. In: *Post-Harvest Pathology of Fruits and Vegetables* (ed.) Dennis, pp. 23–42, Academic Press, London.

Dennis, C. 1983. Salad Crops. In: *Post-Harvest Pathology of Fruits and Vegetables* (ed.) C. Dennis, pp. 157–177, Academic Press, London.

Dennis, C. 1987. Fungi. In: *Postharvest Physiology of Vegetbles* (ed.) J. Weichmann, pp. 377–411, Marcell Dekker, Inc., New York.

Dennis, C., Browne, K.M. and Adamicki, F. 1979. Controlled atmosphere storage of tomatoes. *Acta Horticulturae*, **93**: 75–83.

Dennis, C. and Cohen, E. 1976. The effect of temperature on strains of soft fruit spoilage fungi. *Annals of Applied Biology*, **82**: 51–56.

Dennis, C. and Davis, R.P. 1980. Fungal spoilage of stored tomatoes. *Biennial Report of Field Research Institute, 1977 and 1978*, 44–45.

Devash, Y., Okon, Y. and Henis, Y. 1980. Survival of *Pseudomonas tomato* in soil and seeds. *Phytopathologische Zeitschrift*, **99**: 175.

Dhingra, O.D., Jham, G. and Napoleao, I.T. 1998. Ergosterol accumulation and oil quality changes in stored soybean invaded by *Aspergillus rubber* (*A. glaucus* group). *Mycopathologia*, **143**: 85–91.

Diab, S.Y., Bashan, Y., Okon, Y. and Henis, Y. 1982. Effects of relative humidity on bacterial scab caused by *Xanthomonas campestris* pv. *vesicatoria* on pepper. *Phytopathology*, **72**: 1257–1260.

Dodd, J.C., Estrada, A.B., Matcham, J., Jeffries, P. and Jeger, M.J. 1991. The effect of climatic factors on *Colletotrichum gloeosporioides*, causal agent of mango anthracnose in the Philippines. *Plant Pathology*, **40**: 568–575.

Doolittle, S.P. 1953. Diseases of peppers. In: *Year book of' Agriculture*, U.S. Department of Agriculture, Washington, D.C.

Dungan, G.H. and Koehler, B. 1944. Age of seed corn in relation to seed infection and yielding capacity. *Journal of American Society of Agronomy*, **36**: 436.

Eckert, J.W. 1978. Pathological diseases of fresh fruits and vegetables. In: *Postharvest Biology and Biotechnolgy* (eds) H.O. Hutlin and M. Milner, pp. 161–209, Food and Nutrition Press, USA.

Eckert, J.W. and Sommer, N.F. 1967. Control of diseases of fruits and vegetables by postharvest treatment. *Annual Review of Phytopathology*, **5**: 391–392.

El-Wakil, A.A. and Ghonim, M.I. 2000. Survey of seedborne mycoflora of peanut and their control. *Egyptian Journal of Agricultural Research*, **78**: 47–61.

Esbenshade, P.R. and Moyer, J.W. 1982. Indexing system for *Sweetpotato feathery mottle virus* in sweetpotato using enzyme-linked immunosorbent assay. *Plant Disease*, **66**: 911–913.

Fang, C.T., Liu, C.F. and Chu, C.L. 1956. A preliminary study on the disease cycle of the bacterial leaf blight of rice. *Acta Phytopathologica Sinica*, **2**: 173.

Faris, J.A. and Reed, G.M. 1925. Modes of infection of sorghum by loose kernel smut. *Mycologia*, **17**: 51.

Ferrer, J.B. and Owen, J.H. 1959. *Botrytis cinerea*, the cause of ghost-spot disease of tomato. *Phytopathology*, **49**: 411–417.

Fitzell, R.D., Peak, C.M. and Darnell, R.E. 1984. A model of estimating infection levels of anthracnose disease of mango. *Annals of Applied Biology*, **104**: 451–458.

Franseschini, A., Corda, P. and Carta, C. 1982. Observations on pathogenesis and epidemiology of tomato wilt due to *Fusarium oxysporum* f.sp. *lycopersici*, Grogan, Kimble and Misaghi. *Studi Sassaresi*, **29**: 61.

Francki, R.I.B. and Hatta, T. 1981. *Tomato spotted wilt virus*. In: *Hand book of Plant Virus Infection and Comparative Diagnosis* (ed.) E. Kurstak pp. 492–512, Elsevier/North-Holland Biomedical Press, Amsterdam.

Freeman, S. and Shabi, E. 1996. Cross-infection of subtropical and temperate fruits by *Colletotrichum* species from various hosts. *Physiological and Molecular Plant Pathology*, **49**: 395–404.

Freeman, S., Katan, T. and Shabi, E. 1998. Characterization of *Colletotrichum* species responsible for anthracnose diseases of various fruits. *Plant Disease*, **82**: 596–605.

Fuentes, S., De Lourdes de la Isla, M., Ullstrup, A.J. and Rodriguez, A.E. 1962. Ergot of maize in Mexico. *Phytopathology*, **52**: 733.

Futrell, M.C. and Webster, O.J. 1965. Ergot infection and sterility in grain sorghum. *Plant Disease Reporter*, **49**: 680.

Gahukar, K.B., Raut, J.G. and Deshmukh, R.N. 1989. Seed infection in relation to pericarp infection of *Colletotrichum dematium* in Chilli. *PKV Research Journal*, **13**: 52.

Gantotti, B.V. and Davis, M.J. 1993. Pectic zymogram analysis for characterizing genetic diversity of the mango anthracnose pathogen. *Acta Horticulturae*, **341**: 353–359.

Geary, J.R. 1978. Host–parasite interactions between the cultivated carrot (*Daucus carota* L.) and *Sclerotinia sclerotiorum* (Lib.) De Bary. Doctoral Thesis, University of East Anglia, U.K.

Geeson, J.D. and Browne, K.M. 1978. Careful harvest is vital for white cabbage storage success. *Grower*, **89**: 27–31.

Gildow, F., Damsteegt, V., Stone, A., Schneider, W., Luster, D. and Levy, L. 2004. Plum pox in North America: Identification of aphid vectors and a potential role for fruit in virus spread. *Phytopathology*, **94**: 868–874.

González, E. and Sutton, T.B. 2004. Population diversity within isolates of *Colletotrichum* spp. causing Glomerella leaf spot and bitter rot of apples in three orchards in North Carolina. *Plant Disease*, **88**: 1335–1340.

Goodliffe, J.P. and Heale, J.B. 1975. Incipient infections caused by *Botrytis cinerea* in carrots entering storage. *Annals of Applied Biology*, **80**: 243–246.

Gorini, F. 1987. Market preparation methods and shelf life. In: *Postharvest Physiology of Vegetables* (ed.) J. Weichmann, pp. 489–495, Marcel Dekker, Inc., New York.

Graham, J.H. 1953. Overwintering of three bacterial pathogens of soybean. *Phytopathology*, **43**: 189.

Grondeau, C., Samson, R. and Poutier, F. 1991. Spread of *Pseudomonas syringae* pv. *pisi*: can prevailing winds help bacterias progression? *Proceedings of the 4th International Working Group on Pseudomonas syringae pathovars*, 211–213.

Grover, R.K. 1971. Participation of host exudates chemicals in appressorium formation of *Colletotrichum piperatum*. In: *Ecology of Leaf surface Microorganisms* (ed.) T.F. Preece and C.H. Dickinson, pp. 509–518. Academic Press, London.

Grover, R.K. and Bansal, R.D. 1968. Occurrence and overwintering of *Colletotrichum piperatum* on *Capsicum frutescens* in India. *Indian Phytopathology*, **21**: 116–118.

Grover, R.K. and Bansal, R.D. 1970. Seedborne nature of *Colletotrichum capsici* in chilli seeds and its control by seed dressing fungicides. *Indian Phytopathology*, **23**: 664.

Halfon-Meiri, A. and Solel, Z. 1990. Factors affecting seedling blight of sweet corn caused by seedborne *Penicillium oxalicum*. *Plant Disease*, **74**: 36.

Hildebrand, D.C. 1971. Pectate and pectin gels for differentiation of *Pseudomonas* sp. and other bacterial plant pathogens. *Phytopathology*, **61**: 1430–1436.

Hildebrand, E.M. 1960. The feathery mottle virus complex of sweetpotato. *Phytopathology*, **50**: 751–757.

Hildebrand, M., Dickler, E. and Geider, K. 2000 Occurrence of *Erwinia amylovora* on insects in a fire blight orchard. *Journal of Phytopathology*, **148**: 251–256.

Hobbs, T.W., Schmitthenner, A.F. and Kutter, G.A. 1985. A new *Phomopsis* species from soybean. *Mycologia*, **77**: 535.

Hope, R.J., Colleate, A., Baxter, E.S. and Magan, N. 2000. Interactions between environmental stress and fungicides affect growth and mycotoxin production by *Fusarium culmorum* isolates from wheat grain. *BCPC conference: Pests and Diseases*, vol. 3, 889–894, U.K.

Hosford, R.J. Jr. 1982. White blotch incited in wheat by *Bacillus megaterium* pv. *cerealis. Phytopathology*, **72**: 1453.

Ilyas, M.B., Dhingra, O.D., Ellis, M.A. and Sinclair, J.B. 1975. Location of mycelium of *Diaporthe phaseolorum* var. *sojae* and *Cercospora kikuchii* in infected soybean seed. *Plant Disease Reporter*, **59**: 17.

Ivanoff, S.S. 1933. Stewart's wilt disease of corn, with emphasis on the life history of *Phytomonas stewartii* in relation to pathogenesis. *Journal of Agricultural Research*, **47**: 749.

Jadeja, K.B. 2000. Ecological association of two postharvest pathogens on mango fruits. *Journal of Mycology and Plant Pathology*, **30**: 406–407.

Jafra, S., Figura, L., Hugouvieux-Cotte-Pattat, N. and Lojkowska, E. 1999. Expression of *Erwinia chrysanthemi* pectinase genes *pelI*, *PelL* and *PelZ* during infection of potato tubers. *Molecular Plant–Microbe Interactions*, **12**: 845–851.

Jamaludin, Tandon, M.P. and Tandon, R.N. 1973. Studies on fruit rot of apple (*Malus sylvestris*) caused by *Gliocephalotrichum bulbilium* Ellis. *Proceedings of National Academy of Science, India*, **B,43**: 154–158.

Jarvis, W.R. 1962. The infection of strawberry and raspberry fruits by *Botrytis cinerea* Fr. *Annals of Applied Biology*, **50**: 569–575.

Johnson, E.M. and Sutton, T.B. 2000. Response of two fungi in the apple sooty blotch complex to temperature and relative humidity. *Phytopathology*, **90**: 362–367.

Johnson, E.M., Sutton, T.B. and Hodges, C.S. 1996. *Peltaster fructicola*: A new species in the complex of fungi causing apple sooty blotch disease. *Mycologia*, **88**: 114–120.

Johnson, E.M, Sutton, T.B. and Hodges, C.S. 1997. Etiology of apple sooty blotch disease in North Carolina. *Phytopathology*, **87**: 88–95.

Jordan, V.W.L. 1978. Epidemiology and control of fruit rot of *Botrytis cinerea* on strawberry. *Pflanzen schutz–Nachrisht, Bayer*, **31**: 1–10.

Karadimtcheva, B. 1981. Characteristics of the anatomical structure of the grape skin in relation to resistance to grey mold. *Gradinar Lozar, Nauka*, **18**: 94–99.

Kawamoto, S.O. and Lorbeer, J.W. 1974. Infection of onion leaves by *Pseudomonas cepacia. Phytopathology*, **64**: 1440–1445.

Khanna, K.K. and Chandra, S. 1995. Studies on storage diseases of fruits and vegetables. II. Factors affecting banana fruit rot caused by *Fusarium moniliforme* and *F. roseum. Proceedings of National Academy of Sciences, India B*, **45**: 105–108.

Kikumoto, T. 1974. Ecological studies of the soft rot bacteria of vegetables (13). The role of Chinese cabbage culture in seasonal trends in the populations of the soft rot bacteria in soil. *Bulletin of Institute of Agricultural Research, Tohoku University, Japan*, **25**: 125–137.

Knox-Davies, P.S. 1979. Relationship between *Alternaria brassicola* and *Brassica* seeds. *Transactions of British Mycological Society*, **73**: 235.

Krasnova, M.V. 1963. The effect of some physical factors on the causal agents of bactericosis in soybean seeds. *Journal of Microbiology, Kiev*, **25**: 50.

Kritzman, G. 1989. Tomato seed assay for seedborne bacteria. *Hassadeh*, **69**: 614.

Kunwar, I.K., Manandhar, J.B. and Sinclair, J.B. 1986. Histopathology of soybean seeds infected with *Alternaria alternata. Phytopathology*, **76**: 543.

Kumar, V. and Shetty, H.S. 1983. Transmission of *Botryodiplodia theobromae* in maize (*Zea mays*). *Seed Science and Technology*, **11**: 781.

Kumar, V. and Shetty, H.S. 1985. An ear and kernel rot of maize caused by *Trichothecium roseum*. *Current Science*, **54**: 486.

Lacey, J., Ramakrishna, N. and Hamer, A. 1991. Grain fungi. In: *Handbook of Applied Mycology–Foods and Feeds*, Vol. 3, pp. 121–177, Marcel Dekker, Inc., New York.

Lee, H.B. and Magan, N. 1999. Environmental factors and nutritional utilization patterns affect niche overlap indices between *Aspergillus ochraceus* and other spoilage fungi. *Letters in Applied Microbiology*, **28**: 300–304.

Lehman, S.G. 1952. Survival of purple stain fungus in soybean seeds. *Phytopathology*, **42**: 285.

Lehtonen, M.J., Rantala, H., Kreuze, J.F., Bång, H., Kuisa, J., Koski, P., Vitanen, E., Vihlman, K. and Valkonen, J.P.T. 2004. Occurrence and survival of potato scab pathogens (*Streptomyces species*) on tuber lesions: quick diagnosis based on a PCR based assay. *Plant Pathology*, **53**: 280–287.

Lennox, C.L. and Spotts, R.A. 2004. Timing of preharvest infection of pear fruit by *Botrytis cinerea* and the relationship to postharvest decay. *Plant Disease*, **88**: 468–473.

Lemox, C.L., Spotts, R.A., and Cervantes, L.A. 2003. Populations of *Botrytis cinerea* and *Penicillium* spp. on pear fruit and in orchards and packinghouses, and their relationship to postharvest decay. *Plant Disease*, **87**: 639–644.

Lewis, G.D. and Brown, D.H. 1961. Studies on the overwintering of *Xanthomonas vesicatoria* in New Jersey. *Phytopathology*, **51**: 599.

Lewis, B.G. and Garrod, B. 1983. Carrots. In: *Post-Harvest Pathology of Fruits and Vegetables* (ed.) C. Dennis, pp. 103–124, Academic Press, London, U.K.

Liao, C.H. 1975. Studies on mango fruit spot. II. Pathogenicity. *Bulletin of Taiwan Agricultural Research Institute*, **No.32**, 62–66.

Liu, X.L., Li, J.Q. Zhu, C.Y., Zhang, G.Z., Zhang, L., Luo, J., Li, X.L., Zhu, J.J. and Fang, S.L. 2000. Testing of seedborne fungi of major rice varieties from China and disinfection effect of several fungicides to seed pathogens. *Journal of China Agricultural University*, **5**: 42–47.

Lockhart, C.L. 1976. Reducing disease losses in stored cabbage. *Canadian Agriculture*, **21**: 9–10.

Logan, C. 1983. Potatoes. In: *Post-Harvest Pathology of Fruits and Vegetables* (ed.) C. Dennis, pp. 179–217, Academic Press, London.

Lopez, L.C. and Christensen, C.M. 1967. Effect of moisture content and temperature on invasion of stored corn by *Aspergillus flavus. Phytopathology*, **57**: 588.

Lozoya-Saldaña, H. and Hernandez-Vilchis, A. 2001. Postharvest decay of potato tubers in moist chamber and in two types of soils. *Revista Mexicana de Fitopatologia*, **19**: 140–146.

Lund, B.B. 1983. Bacterial spoilage. In: *Post-Harvest Pathology of Fruits and Vegetables* (ed.) C. Dennis, pp. 219–257, Academic Press, London.

Lyew, D., Gariépy, Y., Raghavan, G.S.V. and Kushalappa, A.C. 2001. Changes in volatile production during an infection of potatoes of *Erwinia carotovora*. *Food Research International*, **34**: 807–813.

Machmud, M. and Middleton, K.J. 1991. Transmission of *Pseudomonas solanacearum* through groundnut seeds. *Bacterial Wilt News*, **7**: 4.

Maholay, M.N. and Sohi, H.S. 1977. Seedborne infection of *Capsicum* by *Botryodiplodia palmarum* (Cook) Petrak and Syd. and Syd. and its control. *Indian Journal of Horticulture*, **34**: 316.

Manandhar, J.B. 1999. Isolation of *Microdochium oryzae* and *Pinatuba oryzae* from rice seeds and their survival on stored seeds. *European Journal of Plant Pathology*, **105**: 139–145.

Mandal, N.C. 1981. *Postharvest diseases of fruits and vegetables.* Doctoral Thesis, Visva-Bharati Unviersity, India.

Marín, S., Magan, N., Abellana, M., Canela, R., Ramos, A.J. and Sanchis, V. 2000. Selective effect of propionates and water activity on maize mycoflora and impact on fumonisin B_1 accumulation. *Journal of Stored Products Research*, **36**: 203–214.

Mathur, R.S. and Ahmad, Z.U. 1964. Longevity of *Corynebacterium tritici* causing tundu disease of wheat. *Proceedings National Academy of Sciences, India, Section, B.* **34**: 335.

Mathur, S.B., Mallya, J.I. and Neergaard, P. 1972. Seedborne infection by *Trichoconis padwicikii* in rice: distribution and damage to seeds and seedlings. *Proceedings of International Seeds Testing Association*, **37**: 803.

Maude, R.B. 1983. Onions. In: *Post-Harvest Pathology of Fruits and Vegetables* (ed.) *C. Dennis*, pp. 73–101, Academic Press, London.

Maude, R.B. 1996. *Seedborne Diseases and Their control–Principles and Practice* CAB International, Wallingford, Oxon, U.K.

Maude, R.B. and Humpherson-Jones, J.M. 1980. Studies on the seedborne phases of dark leaf spot (*Alternaria brassicola*) and grey leaf spot (*Alternaria brassicae*) of brassicas. *Annals of Applied Biology*, **95**: 311–319.

Maude, R.B. and Presley, A.H. 1977a. Neck rot (*Botrytis allii*) of bulb onions. I. Seed-borne infection and its relationship to the disease in onion crop. *Annals of Applied Biology*, **86**: 163–180.

Maude, R.B. and Presley, A.H. 1977b. Neck rot (*Botrytis allii*) of bulb onions. II. Seed-borne infection in relation to the disease in store and the effect of seed treatment. *Annals of Applied Biology*, **86**: 181–188.

McKeen, C.D. 1973. Occurrence, epidemiology and control of bacterial canker of tomato in southwestern Ontario. *Canadian Plant Disease Survey*, **53**: 127–130.

Mehan, V.K., Rao, R.C.N., McDonald, D. and Williams, J.H. 1988. Management of drought stress to improve field screening of peanuts for resistance to *Aspergillus flavus*. *Phytopathology*, **78**: 659.

Meneley, J.C. and Stanghellini, M.E. 1974. Detection of enteric bacteria within locular tissue of healthy cucumbers. *Journal of Food Science*, **39**: 1276–1278.

Mew, T.W., Ho, W.C. and Chu, L. 1976. Infectivity and survival of softrot bacteria in Chinese cabbage. *Phytopathology*, **66**: 1325–1327.

Mia, M.A.T. and Safeeulla, K.M. 1998. Survival of seedborne inoculum of *Bipolaris oryzae*, the causal agent of brown spot disease of rice. *Seed Research*, **26**: 78–32.

Mia, M.A.T., Safeeulla, K.M. and Shetty, H.S. 1986. Seedborne nature of *Gerlachia oryzae*, the incitant of leaf scald of rice in Karnataka. *Indian Phytopathology*, **39**: 92.

Michailides, T.J. and Elmer, P.A.G. 2000. Botrytis gray mold of kiwifi fruit caused by *Botrytis cinerea* in the United States and New Zealand. *Plant Disease*, **84**: 208–223.

Miller, C.S. and Colhoun, J. 1969. Fusarium diseases of cereals. VI. Epidemiology of *Fusarium nivale* on wheat. *Transactions of British Mycological Society*, **52**: 195.

Miller, M.E. and Dillon, R.C. Jr. 1979. Survey of bulb diseases on onions in South Texas. *Journal of Rio Grande Valley Horticulture Society*, **33**: 25–28.

Mishra, C.B.P. 1973. Studies on *Fusarium* species on wheat caryopses and proof of their pathogenicity as foot disese agents. *Arch. Phytopathologische Pflanzenschultz*, **9**: 123.

Mlikota Gabler, F., Smilanick, J.L., Mansour, M., Ramming, D.W. and Mackey, B.E. 2003. Correlations of morphological, anatomical and chemical features of grape berries with resistance to *Botrytis cinerea*. *Phytopathology*, **93**: 1263–1273.

Moffett, M.L., Wood, B.A. and Hayword, A.C. 1981. Seed and soil sources of inoculum for the colonization of the foliage of solanaceous hosts by *Pseudomonas solanacearum*. *Annals of Applied Biology*, **98**: 403.

Molina, J.J. and Harrison, M.D. 1980. The role of *Erwinia carotovora* in the epidemiology of potato blackleg. II. The effect of soil temperature on disease severity. *American Potato Journal*, **57**: 351–363.

Moyer, J.B., Cali, B.B., Kennedy, G.G. and Abou-Ghadir, M.F. 1980. Identification of two *Sweetpotato feathery mottle virus* strains in North Carolina. *Plant Disease*, **64**: 762–764.

Mukherjee, P.S. and Nandi, B. 1993. Insect–fungus associations influencing seed deterioration in storage. *Journal of Mycopathology Research*, **31**: 87.

Neergaard, P. 1977. *Seed Pathology*, Vols1 and 2, Macmillan, London, U.K.

Nielsen, L.W. 1968. Summer spread of *Sweetpotato internal cork virus* in North Carolina in relation to minimum winter temperature. *Phytopathology*, **58**: 1717–1718.

Obrero, F.C. and Aragaki, M. 1965. Some factors influencing infection and disease development of *Phytophthora parasitica* Dast. on tomato. *Plant Disease Reporter*, **49**: 327–333.

Otta, J.D. 1977. Occurrence and characteristics of isolates of *Pseudomonas syringae* on winter wheat. *Phytopathology*, **67**: 22.

Palaniswami, A. 1978. *Studies on fruit rot disease of mango and banana caused by Botryodiplodia theobromae* Pat. Doctoral thesis, Tamil Nadu Agricultural University, Coimbatore, India.

Pande, N. and Mehrotra, B.S. 1988. Rice weevil (*Sitophilus oryzae* Linn.) vector of toxigenic fungi. *National Academy Science Letters, India*, **11**: 3.

Pati, S.P., Maheswari, S.K. and Ray, R.C. 2001. Effect of culture media, temperature, pH, carbon and nitrogen sources on growth of *Botryodiplodia theobromae* causing Java black rot of sweetpotato tubers. *Journal of Mycopathological Research*, **39**: 15–19.

Pauer, G.D. 1964. *Erwinia carotovora* f.sp. *zeae* the bacterium causing stalk rot of maize in the Republic of South Africa. *South African Journal of Agricultural Sciences*, **7**: 581.

Payne, G.A., Thompson, D.L., Lillehoj, E.B., Zuber, M.S. and Adkins, C.R. 1988. Effect of temperature on the preharvest infection of maize kernels by *Aspergillus flavus*. *Phytopathology*, **78**: 1376.

Pearson, R.D. and Hall, D.H. 1975. Factors affecting the occurrence and severity of black mold in ripe tomato fruit caused by *Alternaria alternata*. *Phytopathology*, **65**: 1352–1359.

Pérombelon, M.C.M. 1974. The role of seed tuber in the contamination by *Erwinia carotovora* of potato crops in Scotland. *Potato Research*, **17**: 187–199.

Pérombelon, M.C.M. and Kelman, A. 1980. Ecology of the soft rot Erwinias. *Annual Review of Phytopathology*, **18**: 361–387.

Pérombelon, M.C.M. and Kelman, A. 1987. Blackleg and other diseases caused by soft rot erwinias: Proposal for revision of terminology. *Plant Disease*, **71**: 283–285.

Pérombelon, M.C.M. and Salmond, G.P.C. 1995. Bacterial soft rots. In: *Pathogenesis and Host Specificity in Plant Disease* (eds) U.S. Singh and K. Kohmoto, vol.1., pp. 1–20, Pergamon Press Ltd., Oxford, U.K.

Phillip, M.S.S., Homma, S. and Murakishi, H. 1967. The effect of mixed infections with *Tobacco mosaic virus* and *Potato virus* X on flowering and yield of field-grown tomatoes. *Michigan Quarterly Bulletin*, **50**: 25–30.

Pidoplichko, V.M. 1970. Phytotoxicity of *Fusarium* spp. causal agents of root rot of winter wheat. *Mykrobiol. Zh.* **32**: 700.

Porter, D.M., Wright, F.S., Taber, R.A. and Smith, D.H. 1991. Colonization of peanut seed by *Cylindrocladium crotalariae. Phytopathology*, **81**: 896.

Porter, L.D. and Johnson, D.A. 2004. Ecosurvival of *Phytophthora infestans* in surface water. *Phytopathology*, **94**: 380–387.

Qasem, S.A. and Christensen, C.M. 1960. Effect of moisture content and temperature on invasion in stored corn by *Aspergillus flavus. Phytopathology*, **50**: 703.

Ramakrishnan, T.S. 1954. *Common Diseases of Citrus*. Government of Madras Publication, Chennai, India.

Rao, P.H.V., Giridhar, P. and Reddy, S.M. 2000. Fungal succession on sorghum seed harvested at different stages of maturity. *Indian Phytopathology*, **53**: 102–104.

Reddy, G.R., Reddy, A.G.R. and Rao, K.C. 1991. Effect of certain seedborne fungi on seed germination and seedling vigor of groundnut. *Journal Research, APAU*, **19**: 77.

Riberau-Gayon, P., Dubordieu, D., Doneche, B. and Lonvand, A. 2000. *Handbook of Enology, Vol.1. The Microbiology of Wine and Vinifications.* John Wiley and Sons, New York.

Richardson, M.J. 1990. *An Annotated List of Seedborne Diseases*, 3rd Edition. Commonwealth Mycological Institute, Kew, Surrey, U.K.

Roberts, E.H. and Roberts, D.L. 1972. Moisture content of seeds. In: *Viability of Seeds* (ed.) E.H. Roberts, pp. 424–437, Chapman and Hall, London.

Rolke, Y., Liu, S., Quilde, T., Williamson, B., Schouten, A., Weltring, R.M., Siewers, V., Tenberge, K.P., Tudzynski, B. and Tudzynski, P. 2004. Functional analysis of H_2O_2-generating system in *Botrytis cinerea*, the major Cu-Zn-superoxide dimutase (BCSOD1) contributes to virulence on Frenchbean, whereas glucose oxidase is dispensable. *Molecular Plant Pathology*, **5**: 17–27.

Rose, A. 1996. Vectoring of *Botrytis cinerea* (Persoon: Fries) to kiwifruit (*Actidinidia deliciosa*) flowers by honey bees (*Apis melifera* Linnaeus). M.S. Thesis, Lincoln University, Canterbury, New Zealand.

Roy, K.W. 1982. Seedling diseases caused in soybean by species of *Colletotrichum* and *Glomerella*. *Phytopatholgy*, **72**: 1093.

Rushing, J.W., Keinath, A.P. and Cook, W.P. 1999. Postharvest development and transmission of watermelon fruit blotch. *HortTechnology*, **9**: 217–219.

Sackston, W.E. 1981. Downy mildew of sunflower. In: *The Downy Mildews* (ed.) D.M. Spencer, pp. 545–575, Academic Press, London, U.K.

Salas, B., Stack, R.W., Secor, G.A. and Gudmestad, N.C. 2000. The effect of wounding temperature and inoculum on the development of pink rot of potatoes caused by *Phytophthora erythroseptica*, *Plant Disease*, **84**: 1327–1333.

Sarig, P., Zutkhi, Y., Monjauze, A., Lisker, N. and Ben-Arie, R. 1998. Natural and induced resistance of table grapes to bunch rots. *Acta Horticulturae*, **464**: 65–70.

Schaad, N.W. and Brenner, D. 1977. A bacterial wilt and root rot of sweetpotato caused by *Erwinia chrysanthemi*. *Phytopathology*, **67**: 302–308.

Shahjahan, A.K.M., Harahap, Z. and Rush, M.C. 1977. Sheath rot of rice caused by *Acrocylindrium oryzae* in Louisiana. *Plant Disease Reporter*, **61**: 307.

Shekhawat, G.S. and Srivastava, D.N. 1972. Mode of seed infection and transmission of bacterial leaf streak of rice. *Annals of Phytopathological Society, Japan*, **38**: 4.

Sherf, A.F. and Macnab, A.A. 1986. *Vegetable Diseases and Their Control*. 2nd Edition, John Wiley and Sons, New York.

Shetty, K.K., Frazier, M.J., Gale E. Kleinkopf, G.E. and Nolte, P. (1996). *Managing Silver Scurf*. Available at http://www.uidaho.edu/ag/plantdisease/sscurf.htm.

Shiomi, T. 1991. Dry heat sterilization for seed stain of cabbage black rot disease. *Agriculture and Horticulture*, **66**: 1177.

Shipway, M.R. 1981. Refrigerated storage of winter white cabbage. *Annual Report of Kirton Experiment Horticulture Station*, 1980, 44–50.

Shortt, B.J., Sinclair, J.B., Helm, C.G., Jeffords, M.R. and Kogan, M. 1982. Soybean seed quality losses associated with bean leaf beetles and *Alternaria tenuissima*. *Phytopathology*, **72**: 615.

Silveira, N.S.S., Michereff, S.J., Mariano, R. de L.R., Noronha, M.A. and Pedrosa, R.A. 2000. Virulence and dispersion gradient of fungi causing postharvest rots in tomato fruits. *Summa Phytopathologica*, **26**: 422–428.

Silveira, N.S.S., Michereff, S.I., Mariano, R.L.R. Tavares, L.A. and Maia, L.C. 2001. Influence of temperature, wetness period and fungal inoculum concentration of incidence of postharvest tomato fruit rots. *Fitopatologia Brasileira*, **26**: 33–38.

Singh, D.P. and Agarwal, V.K. 1989. Effect of different degrees of grain mold infection on yield and quality of sorghum seed. *Indian Journal of Plant Pathology*, **7**: 103.

Smadja, B., Latour, X., Trigui, S., Burini, J.F., Chevalier and Orange, N. 2004. Thermodependence of growth and enzymatic activities implicated in pathogenicity of two *Erwinia carotovora* subspecies (*Pectobacterium* spp.). *Canadian Journal of Microbiology*, **50**: 19–27.

Solanke, R.B., Deosarkar, D.B. and Jawale, L.N. 2001. Seedborne fungi of chilli and response of *Fusarium moniliforme* to various seed dressers. *Journal of Maharashtra Agricultural Universities*, **26**: 187–188.

Solanke, R.B., Jawale, L.N., Hussaini, M.M. Bonde, V.J. and Sudewad, S.M. 1998. Effect of seed treatment on seed health of soybean under storage. *Journal of Maharashtra Agricultural Universities*, **23**: 176–178.

Srivastava, M.P. and Tandon, R.N. 1971. Postharvest diseases of banana in India. *Indian Phytopathology*, **24**: 115–118.

Swinburne, T.R. 1963. Infection by *Tilletia caries* (DC.) Tul., the causal organism of bunt. *Transactions of British Mycological Society*, **46**: 145.

Tanaka, F. and Tsuchiva, S. 1987. Factors affecting the incidence of pink coloring of rice grains, caused by *Epicoccum purpurescens*. I Conidial dispersion and infection to the glume and kernels. *Bulletin of Hokkaido Prefectural Agricultural Experiment Station*, **56**: 51.

Tanaka, M.A. de, S., Maeda, J.A. and Plazas, I.H. de A.Z. 2001. Fungi associated to corn seeds under storage conditions. *Scientia Agricola*, **58**: 501–508.

Tarr, S.A.J. 1962. *Diseases of Sorghum, Sudan Grass and Broom Corn.* Commonwealth Mycological Institute, Kew, Surrey, U.K.

Taylor, J.D. 1972. Field studies on halo-blight of beans (*Pseudomonas phaseolicola*) and its control by foliar sprays. *Annals of Applied Biology*, **70**: 191–197.

Taylor, J.D., Dudley, C.L. and Littlejohn, I.H. 1980. Bacterial rots of stored onion. *Report of National Vegetable Research Station*, 1979, 76–77.

Tichelaar, G.M. 1967. Studies on the biology of *Botrytis allii* and *Allium cepa. Netherlands Journal of Plant Pathology*, **73**: 157–160.

Timmer, L.W., Zitko, Gottwald, T.R. and Graham, J.H. 2000. *Phytophthora* brown rot of citrus: Temperature and moisture effects on infection, sporangium production and dispersal. *Plant Disease*, **84**: 157–163.

Ulrich, J.M. 1975. Pectic enzymes of *Pseudomonas cepacia* and penetration of polygalacturonase into cells. *Physiological Plant Pathology*, **5**: 37–44.

Umechuruba, C.I. and Nwachukwu, E.O. 2002. Long-term survival and pathogenicity of seedborne fungi on soybeans stored at 5°C. *Tropical Science*, **42**: 42–45.

Varns, J.L., Currier, W.W. and Kuć, J. 1971. Specificity of rishitin and phytuberin accumulation by potato. *Phytopathology*, **61**: 968–971.

Verhoeff, K. 1970. Spotting of tomato fruits caused by *Botrytis cinerea. Netherlands Journal of Plant Pathology*, **76**: 219–226.

Verhoeff, K. and Liem, J.I., 1975. Toxicity of tomatine to *Botrytis cinerea* in relation to latency. *Phytopathologische Zeitschrift*, **82**: 333–338.

Wakimoto, S., Akai, M. and Tsuchiya, K. 1987. Serological specificity of *Pseudomonas glumae*, the pathogenic bacterium of grain rot disease of rice. *Annals of Phytopathological Society, Japan*, **54**: 633.

Walker, J.C. and Stahmann, W.A. 1955. Chemical nature of disease resistance. *Annual Review of Plant Physiology*, **6**: 351–366.

Walkey, D.G.A. and Webb, M.J.W. 1978. Internal necrosis of stored white cabbage caused by *Turnip mosaic virus*. *Annals of Applied Biology*, **89**: 435–441.

Welling, B. 1969. Fungus flora and germination of barley. *Tidsskrift for Planteavl*, **73**: 291.

Williamson, D. and Tandon, R.N. 1966. Some pathological studies on *Botryodiplodia theobromae* Pat. causing banana rot. *Mycopathologia Mycologia Applicata*, **29**: 245–253.

Williamson, S.M. and Sutton, T.B. 2000. Sooty blotch and flyspeck of apple: etiology, biology and control. *Plant Disease*, **84**: 714–724.

Wolf, J.M. and Beckhoven, J.R.C.M. 2004. Factors affecting survival of *Calvibacter michiganensis* subsp. *sepedonicus* in water. *Journal of Phytopathology*, **152**: 145–152.

Wright, W.R. and Billeter, B.A. 1974. Red kernel disease of sweet corn on the retail market. *Plant Disease Reporter*, **58**: 1065.

Wu, W.S. and Mathur, S.B. 1987. Evaluation of methods for detecting *Fusarium moniliforme* in sorghum seeds. *Seed Science and Technology*, **15**: 821.

Xiao, C.L. and Rogers, J.D. 2004. A postharvest fruit rot in d'Anjou pears caused by *Sphaeropsis pyriputrescens. Plant Disease*, **88**: 114–118.

Xu, X.M., Guerin, L. and Robinson, J.D. 2001. Effects of temperature and relative humidity on conidial germination and viability, colonization and sporulation of *Monilinia fructigena. Plant Pathology*, **50**: 561–568.

Yakoby, N., Kobiler, I., Dinoor, A. and Prusky, D. 2000. pH regulation of pectate lyase secretion modulates the attack of *Colletotrichum gloeosporioides* on avocado fruits. *Applied and Environmental Microbiology*, **66**: 1026–1030.

Ziegler, R.S. and Alvarez, E. 1987. Bacterial sheath brown rot of rice caused by *Pseudomonas fuscovaginae* in Latin America. *Plant Disease*, **71**: 592.

Zummo, N. and Scott, G.E. 1990. Relative aggressiveness of *Aspergillus flavus* and *A. parasiticus* on maize in Mississippi. *Plant Disease*, **74**: 978.

4

DISEASE DEVELOPMENT AND SYMPTOM EXPRESSION

Postharvest diseases of durables and perishables caused by fungal pathogens are more numerous, whereas bacterial pathogens induce a limited number of diseases. The phenomenon of pathogenesis encompasses distinct phases of disease development. The phases of disease development are well defined in some pathosystems, while they are indistinct and overlap in others. Nevertheless, these phases in the infection process, such as recognition of host plant species, adhesion of pathogen propagules to plant surface, germination of spores, formation of germ tube and appressorium, penetration into host plant, colonization, and expression of symptoms, may be significantly influenced by environmental conditions and host plant resistance. The nature of interactions between host plant and pathogen is dependent on the virulence, tolerance to adverse conditions, and ability to overcome the resistance of the host plant due to the presence of preformed antimicrobial compounds and production of defense-related compounds following initiation of infection process. The responses of host plant tissues to postharvest pathogens are discussed in this chapter.

4.1 SEED INFECTION

4.1.1 Fungal Pathogens

Fungal pathogens may get established on the seed surface or within any part of the seed. They invade the seed tissues systemically, either through the vas-

Postharvest Pathogens and Disease Management, by P. Narayanasamy

cular system or plasmodesmatal connections or directly through floral parts by penetrating the ovary wall, seed coat, or natural openings. The presence of oospores of *Phytophthora capsici* in the embryos of squash seeds has been observed; the pathogen was able to penetrate the cuticle directly followed by invasion of the embryo, passing through the integument stratum (Melo et al., 1992). *P. infestans* was shown to take a similar route while infecting tomato seeds. Abundant oospores were seen in the vascular tissues, pericarp, columella, and placenta. Occasionally, oospores were enclosed in between epidermal hairs of the seed coat and in a few seeds the oospores were detected inside the embryo (Rubin et al., 2001).

4.1.1.1 Entry through Floral Parts/Fruits The fungal pathogens infecting different tissues of mother plants may reach the seeds, passing through vascular elements, as in the case of fungi causing smut, downy mildews, wilts, and root diseases. *Tilletia tritici*, causing bunt disease of wheat, and *Sphacelotheca sorghi*, causing grain smut of sorghum, invade young coleoptiles/radicles, producing systemic mycelium which advances along with developing tissues of the shoots of infected plants. The pathogens invade the floral tissues, replacing ovaries with teliospores enclosed in fungal tissues and pericarp. The oospores of *Sclerospora graminicola*, causing downy mildew and green ear disease of pearl millet, lie in the soil along with infected plant debris, and, on germination, penetrate the radicle tissues and keep pace with developing apical growing tissues, and later invade the floral primordia, converting the affected floral parts into green leafy shoots in place of grains. *Fusarium moniliforme* and other *Fusarium* spp. may invade the seeds via the xylem of the mother plant. *F. oxysporum* f.sp. *lagenarium* infecting bottle gourd enters the seeds through the vascular bundles of fruits (Kuniyasu and Kishi, 1977). The sunflower downy mildew pathogen (*Plasmopara halstedii*) infects the captiulum and invades the seeds (Cohen and Sackston, 1974; Sackston, 1981).

Infection of seeds may occur by penetration through stigma. The wheat loose smut pathogen *Ustilago segetum tritici* releases wind-blown teliospores from smutted ears of infected wheat plants to alight on floral stigmas of healthy plants. The basidiospores produced from the germinating teliospores penetrate and invade stigmatic tissues. The hyphae develop between the cells down the stylar canal and enter the young seed through the micropyle or through the inner integument to the nucellus and then reach the endosperm, embryo, and its growing point (Lang, 1917; Gaümann, 1950).

Fusarium graminearum causes *Fusarium* head blight (FHB), also known as scab disease, in wheat and barley. Ascopores and conidia infect spikelets involving the entire reproductive head with a pink to reddish mycelial growth resulting in floral sterility and shrunken kernels, in addition to contamination of grains with mycotoxin (trichothecenes) produced by this pathogen. The trichothecenes have been demonstrated to have a role in pathogenesis. A mutant of *F. graminearum*, when inoculated into spikelets of spring wheat lines (*Triticum aestivum*), did not spread into the rachis since it could not produce trichothecene. In contrast, the wild parent type capable of producing trichothecene, rapidly spread in the spikes, confirming that trichothecenes are a principal determinant of the aggressiveness of *F. graminearum* (Eudes et al., 2001). The green fluorescent protein gene (*gfp*) was used to transform *F. graminearum* and the transformant was inoculated on intact barley spikes to study the infection pattern. The brush hairs (ovary epithelial hairs) were colonized within 7 h after inoculation. The aleurone and starchy endosperm were not infected even after 16 days, although the pericarp was intensively colonized by the pathogen. The activities of amylase did not show any change following infection. Use of the *gfp*-expressing strain of *F. graminearum* has provided the possibility of following infection routes involving different tissues of the seeds by making light and electron microscopic observations. This study has opened up the possibility of viewing single hyphae (because of the fluorescence) and providing insights into pathogenesis in various pathosystems (Skadsen and Hohn, 2004). *Fusarium culmorum*, another FHB pathogen, also infects when the wheat flowering ears are sprayed with the conidial suspension. High moisture (provided by mist irrigation) resulted in a higher mean disease severity, but in a overall lower toxin contamination, as compared to the nonirrigated treatments. There was a positive correlation between visual symptoms and deoxynivalenol (DON) content of grains (Lemmens et al., 2004). The presence of different mycotoxins in seeds of various crops and the changes induced by the mycotoxigenic fungi are discussed in Section 4.2.1.3.

Botrytis cinerea (*Sclerotinia fuckeliana*-teleomorph) has a wide host range and infects the petals of flowers of many fruit and vegetable crops. The pathogen may become quiescent in the infected stamens and styles. The senescing floral parts provide nutrients for the saprophytic existence of the pathogen which attacks the developing fruits later. In some cases, such as peas and bean pods, seeds may also be infected (Jarvis, 1980). More aggressive necrotrophic pathogens, such as *Ascochyta pisi* infecting peas and *Alternaria brassicola* infecting brassicas, infect the floral parts directly, deriving nutrition from the pollen for rapid germination of spores and further growth of hyphae (Channon, 1970). *Botrytis allii* causing onion neck rot, directly infects floral parts such as anthers, stigmas, and pedicels of individual flowers, causing flower abortion in severe cases. The fungal penetration into the seeds appears to be superficial (Presly, 1977; Maude and Presly, 1977).

The seed coat may be penetrated by the fungal hyphae present in the fruits, as in the case of anthracnose disease of watermelons caused by *Colletotrichum lagenarium*. The pathogen was able to colonize rind splits and invade the epidermal layer of palisade cells and sclerenchyma layers of the seed coat. Later the parenchyma and other seed coat tissues were colonized successively (Rankin, 1954). Direct penetration by *Alternaria alternata* into soybean through cuticle or hilum tissues, or indirectly through the micropyle or pits where seed coat was thin, was observed by Kunwar et al. (1986), whereas penetration by *Phomopsis longicolla* occurred through the funiculus and hilum or directly through the seed coat (Roy and Abney, 1988). Pepper (chili) seeds may be invaded by *Colletotrichum capsici* by the mycelium present in the inner surface of diseased fruit, which grows through the seed coat and later colonizes the outer layer of the endosperm. Another route of entry is through the mycelium present in the placenta remnant and invading the outer layer of the endosperm (Sangchote and Juanbhanich, 1984). Tomato fruits were inoculated with *Botrytis cinerea* to investigate localization and changes in the activity β-glucosidase. Disorders in embryogenesis were accompanied by decreased activity of β-glucosidase activity in all ovules. In the seeds of harvested tomato distant from the invaded area, there was no change in the enzymatic activity (Georgieva et al., 2001).

4.1.1.2 Entry through Natural Openings and Injuries Some fungal pathogens may enter seeds through natural openings such as the hilum and micropyle or through injuries caused during thrashing. Entry of *Phoma betae* into sugar beet seeds occurred through the basal pore in the pericarp, as a slit is formed by the peripheral zone between the seed cap and pericarp and the apical pore in the seed cap (El-Naashaar and Bugbee, 1981). Likewise, *Cercospora sojinae'* was able to penetrate the soybean seeds indirectly through pores and the parenchymatous seed coat tissues and cracks, as shown by histopathological observations (Singh and Sinclair, 1985).

Some fungi are able to penetrate undamaged grain but damage to kernels facilitates the entry of fungi directly into internal tissues of seeds. *Aspergillus* spp. and *Penicillium* spp. may penetrate through the aleurone layer of wheat grains to the subaleurone and endosperm cells (Chelkowski and Cierniewski, 1983). Damage to kernels is reported to increase their susceptibility to fungal invasion, in addition to the rapid development of visible molding. The extent of colonization of barley grains by *Penicillium* spp. increased with increase in damage to grains. On the other hand, colonization of seeds by *Aspergillus* spp. was not affected by the degree of damage to seeds (Welling, 1968). The interaction between peanut (groundnut) and *Aspergillus parasiticus* was studied histologically in seeds inoculated without wounding. An *A. parasiticus* strain was transformed with a β-glucuronidase (GUS) reporter gene under the control of *nor-1* promoter from the aflatoxin biosynthetic pathway. This strain was used to follow infection and aflatoxin (AF) production during colonization of undamaged, drought-stressed peanuts. *A. parasiticus* colonized all tissues of the peanut pod and appeared to gain ingress through the corky layer of the pericarp. Both intra- and intercellular colonization were seen. Fungal colonization of cotyledons resulted in visible depletion of storage bodies within cells. GUS activity was found in hyphae infecting pericarp, embryo, and cotyledons, indicating expression of AF biosynthetic genes in these tissues, but GUS activity was not detected in the hyphae colonizing the testa (Xu et al., 2000).

4.1.2 Bacterial Pathogens

4.1.2.1 Entry through Floral/Fruit Tissues *Pseudomonas syringae* pv. *lachrymans* was able to survive in the vascular elements of cucumber (*Cucumis* *sativus*) and reach the seeds through the base of the stem or flower. The seeds are considered symptomless carriers (Kritzman and Zutra, 1983). Similar route seems to be followed by *Xanthomonas campestris* pv. *phaseoli* to infect the seeds of bean which exhibit a yellowish discoloration of the hilum following infection (Burkholder, 1921; Lelliot, 1988). Infection of flowers of cabbage by *X. campestris* pv. *campestris*, causing black rot disease, leads to the invasion of xylem of the pedicels and suture of vein pods. The bacterium gains access to the funicle and occasionally the xylem of the seed coat (Cook et al., 1952). Floral and pod infection preceded the invasion of pea seeds by *P. syringae* pv. *pisi*, bacterium entering through the funiculus, micropyle, and into the seed coat in succession (Skoric, 1927).

The bacterial pathogens *X. campestris* pv. *vesicatoria* in pepper, *Clavibacter michiganensis* subsp. *michiganensis* and subsp. *sepedonicus* in tomato have been reported to invade seeds through the vascular system of infected mother plants (Bryan, 1930; Larson, 1944; Crossan and Morehart, 1964). Citrus seeds have very prominent vascular bundles which are attached through ovular and seed bundles to the xylem system of the fruit. *Xylella fastidiosa* causing citrus variegated chlorosis (CVC) disease was detected in all main fruit vascular bundles as well as in the seeds and in dissected seed parts by using a polymerase chain reaction (PCR) assay, indicating the transfer of *X. fastidiosa* from infected fruits to the seeds through vascular bundles (Li et al., 2003). Seed infection may occur when the flowers are affected, as in pepper (chilli) by *X. campestris* pv. *vesicatoria*. The bacteria infects flowers, resulting in bacterial proliferation in warts on fruits. The bacterial slime might cover the entire fruit surface (Bashan and Okon, 1986).

4.1.2.2 Entry through Natural Openings and Injuries Generally, bacterial pathogens gain entry into plants through natural openings and injures on the plant surface, since they have no mechanism to penetrate through intact plant surface, as the fungal pathogens do. Bacteria may enter through hilum which is highly absorptive. In bean seeds, the micropyle was shown to be the point of entry of *X. campestris* pv. *phaseoli* (Baker, 1972). Likewise, cucumber seeds were also infected by *P. syringae* pv. *lachrymans* present in the funicle and entering through the micropyle (Naumann, 1963). Infection of bean seeds by *P. syringae* pv. *phaseolicola* was favored by cracks or injuries in the seed coat (Baker, 1972).

4.1.3 Viral Pathogens

Most of the viruses transmitted through true seeds have been detected in the embryos, which are infected through systemic invasion of the ovule directly from the infected mother plant. In addition, the viruses may be introduced into the embryo sac through infected pollen or, occasionally, by direct penetration of the embryo during development (Bennett, 1969). Seed infection forms the major source of virus carry-over from one generation to the next.

Hence, in crops such as lettuce, specific monitoring programs are in operation to take steps to contain seed infection within tolerance limits. Seed transmission is at higher rates if infection takes place before flowering or before fertilization of ovules. Infection of embryos by viruses can occur if the viruses are able to reach the embryo before callose wall formation. The plasmodesmata seem to be the avenue of virus movement through meristem tissues to reach the ovules. The callose layers do not have plasmodesmata and prevent the movement of viruses from mother plants, as in barley infected by *Barley stripe mosaic virus* (BSMV) (Bennett, 1969; Carroll, 1981). Direct invasion of immature embryos by *Pea seed-borne mosaic virus* (PSbMV) has been reported by Wang and Maule (1992). Many viruses have been reported to successfully infect embryos resulting in seed transmission: *Bean common mosaic virus* (BCMV) in bean (Medina and Grogan, 1961); *Cowpea aphid-borne mosaic virus* (CAMV) in cowpea (Tsuchizaki et al., 1970); *Lettuce mosaic virus* (LMV) in lettuce (Ryder, 1964); *Prunus necrotic ringspot virus* (PNRV) in *Cucurbita maxima* (Das et al., 1961); and *Tomato ringspot virus* (TRSV) in raspberry (Braun and Keplinger, 1973).

Transmission of viruses through pollen may lead to seed infection, though the mother plant is not infected. The presence of BSMV in the pollen tube between the integument and ovary wall was observed during embryogenesis. BSMV virions were located in the embryo integument and ovary wall (Brlansky et al., 1986). *Prunus necrotic ringspot virus* (PNRV) was pollen-borne in cherry and seeds in healthy plants were infected when pollinated by virus-contaminated pollen. PNRV was detected in hand-collected sweet cherry pollen by using enzyme-linked immunosorbent assay (ELISA) (George and Davidson, 1963; Hamilton et al., 1984). *Prune dwarf virus* (PDV) particles could be observed in the cytoplasm of pollen grains from sweet cherry plants grown from PDV-infected seeds (Kelley and Cameron, 1986). The *Cherry leaf roll virus* (CLRV) introduced through pollen grains was able to multiply in the embryo as revealed by immunoassay (Cooper et al., 1984). Fresh, unwashed pollen of several cucurbits showed the presence of *Squash mosaic virus* antigen (Nolan and Campbell, 1984). Replication of *Apple mosaic virus* within germinating pollen was reported by Gotlieb (1974).

4.2 SEED DETERIORATION

Seed infection depending on the pathogen population, storage conditions, and storage period may cause different effects on seed quality or suitability of grains to be used as foods or feeds.

4.2.1 Fungal Pathogens

4.2.1.1 Effect on Seed Germination The quality of seeds is judged by the purity without any admixture of other seeds, percentage of germination, and vigor of seedlings growing from the seed lots. The storage fungi form one of the major causes of reduction in germination of seeds. The extent of reduction is influenced by moisture content, temperature, and period of storage. Most storage fungi invade seed embryos preferentially, resulting in significant or even total loss of germination. Adverse effects on the seed germination caused by storage fungi vary depending on the crop and composition of fungal population: see examples for rice (Singha, 2000), corn (Tanaka, 2001), soybean (Bringel et al., 2001; Anuja Gupta and Aneja, 2001; Umechuruba and Nwachukwu et al., 2002), melon (Bankole et al., 1999), and pepper (chilli) (Asalmol et al., 2001).

Generally, seeds stored with high moisture and at high temperatures are invaded by storage fungi and suffer significant reduction in germinability fairly rapidly. Thus storage fungi may be the primary causes of losses in germinability. The conditions favorable for fungal proliferation may lead to a rapid decrease in seed viability. A linear relationship between initial water content and percentage germination of seeds was observed (Hill and Lacey, 1983). Wheat stored at 16% moisture deteriorated at a much slower rate than maize stored under similar conditions, indicating the differential response of crops to attack by storage fungi (Wilcke et al., 1999). The germination of oil-type safflower (*Carthamus tinctorius*) seeds significantly decreased with increasing moisture content and length of storage when infected by *Penicillium chrysogenum* (Hasan, 2000).

4.2.1.2 Discoloration of Seeds and Abnormalities Both field and storage fungi may cause discoloration of seeds. Infection of rice seeds by *Drechslera oryzae* and *Curvularia lunata* results in considerable seed discoloration and reduction in germinability. *Cladosporium cladosporioides* has been reported to be responsible for discoloration of soybean seeds in addition to losses in quality and nutrient status (Krishnamurthy and Raveesha, 2001). Different species of *Aspergillus* are able to invade the embryos and grow profusely on the seed surface and pods, as in peanut and cereal grains. The seeds of wheat affected by *A. restrictus*, *A. candidus*, and *A. glaucus* turn jet black. Development of caking from abundant mycelium into solid masses and emanation of strong musty odors may follow at later stages (Christensen, 1991). The seeds may exhibit different abnormalities in shape due to shriveling and poor filling, because of infection by field fungi. This may result in reduced 1000-grain weight, which is one of the important yield attributes.

4.2.1.3 Seed Quality

A. Carbohydrate Degradation: Cereal grains consist of about 75% carbohydrates, which include both functional and storage saccharides, mostly as polysaccharides (Aspinall and Greenwood, 1962). Starch is hydrolyzed during respiration through the action of both amylase and amylopectin. Amylase is present in seeds as well as in invading fungi. A proportional increase in amylase activity following fungal invasion was detectable, in addition to increases in reducing sugar contents (Hummel et al., 1954; Farag et al., 1985). Due to the activities of maltase, glucose is formed from maltose and dextrin is converted into glucose by other enzymes. Glucose thus formed is utilized by the fungi as an energy source or for production of ribose sugar for nucleic acid synthesis (Vonk and Western, 1984). During respiration, oxygen and starch are converted into carbon dioxide and water, releasing energy for use in different physiological processes in seeds. A storability test, involving determination of dry matter loss over time by measuring carbon dioxide produced through respiration by storage fungi growing on maize kernels, was conducted. No significant differences in storability or microbial infection levels could be noted between freshly harvested samples and samples that were stored wet or dry at −10°C for up to 8 months (Wilcke et al., 2001). The rate of respiration is stimulated as the storage temperature increases, leading to loss of seed weight and quality. Accelerated respiration contributes to appreciable increase in temperature. As the cereal and legumes seeds have low heat conductivity, heat damage may also occur due to fungal activity.

B. Protein Degradation: Total protein content is an important component of dry matter in wheat and other cereals, legumes, and vegetables. Dry matter loss may be indicative of protein degradation, leading to a reduction in storability of seeds (Wilcke et al., 1999, 2001). The amount and quality of protein are heritable attributes and contribute to the physical and structural characteristics of baked wheat products. Some proteins determine the nutritive value of the products, while some seed proteins play a role in the development of resistance to plant pathogens (Narayanasamy, 2005). Oils, proteins, and vitamins are stored as nutritional reserves in the endosperm or cotyledons. The seed proteins are stored primarily within protein bodies. Proteins are hydrolyzed, during germination, to polypeptides which are further broken down to amino acids by proteases and peptides and then translocated to the developing embryo for further development. Following invasion of seeds by fungal pathogens, which also produce protease, the degradation of proteins is accelerated to meet the requirements of pathogen development. In peanuts, infection leads to decomposition of proteins to low molecular weight components, depletion of some enzymes, and stimulation of other enzymes (Mali et al., 1986). The increase in enzymatic activity following fungal invasion may result in higher levels of total proteins (Farag et al., 1985; Hasan, 2000). In contrast, the total protein content was significantly reduced in soybeans infected by *Diaporthe phaseolorum* var. *sojae* (Fábrega et al., 2000).

C. Lipid Degradation: Lipids in seeds are mainly triglycerides that are used as energy reserves during seed germination. They may be degraded endogenously and also through pathogen activity, by both oxidation and hydrolysis, involving lipoxygenases and lipases (St. Angelo et al., 1979; St. Angelo and Ory, 1983). Lipid degradation is indicated by increased fatty acid contents, the amount depending on the form of damage and also by the nature of fungi involved (Hummel et al., 1954; Goodman and Christensen, 1952). Heat damage can adversely affect the oil quality in oilseed crops such as sunflower (Robertson et al., 1985). Quantitative evaluation of oil quality, based on parameters such as percent free fatty acids (FFA), may be desirable. Infestation of oil-type safflower seeds by *Penicillium chrysogenum* was positively correlated with the increase in FFA, under storage at 10°C. Similar increases in FFA contents were associated with the infestation of safflower seeds at 25°C by *Aspergillus flavus* and *A. niger*. Seed deterioration was accompanied by a corresponding rise in FFAs, free amino acids, and soluble proteins. Furthermore, the high levels of activities of amylase, cellulase, lipase, and protease detected in these fungi might have had a role in seed deterioration. The growth and lipase activity of *A. flavus* were favored by the carcumin and turmerol oils of safflower (Hasan, 2000). Soybean seeds inoculated with *Diaporthe phaseolorum* var. *sojae* had higher oil and FFA contents (Fàbrega et al., 2000). Increase in free fatty acids contents was observed in seeds of safflower invaded by *Penicillium chrysogenum, Aspergillus flavus* and *A. niger*, whereas no differences in soybean seeds infected by *Diaporthe phaseolorum* var. *sojae* could be detected (Hasan, 2000; Fábrega et al., 2000).

D. Contamination of Seeds with Mycotoxins: Among the fungi colonizing seeds, species of *Aspergillus*, *Fusarium*, *Alternaria*, and *Penicillium* have been shown to produce mycotoxins, affecting the quality of foods and feeds. In addition, the mycotoxins are known to induce mycotoxicoses in humans and animals following consumption of contaminated foods and feeds. *Aspergillus flavus* and *A. parasiticus* produce aflatoxins (AF) in seeds of many crops such as: corn (maize) (Anderson et al., 1975; Côrtes et al., 2000); peanut (groundnut) (Diener et al., 1987); mustard (Bilgrami et al., 1992); and rice and wheat (Tsai and Yu, 1999). Another mycotoxin, fumonisins, produced by *Fusarium moniliforme* (Wilson et al., 1999) and by *F. proliferatum* (Proctor et al., 1999) in corn has been detected. Several fungi (*Fusarium* spp., *Cephalosporium* sp., *Trichothecium* sp., *Phomopsis* sp.) produce a group of toxic compounds designated trichothecenes, which include deoxynivalenol (DON), nivalenol (NIV), and T-2 toxin in cereals. Wheat and corn grains invaded by *F. graminearum* contain DON, NIV, and their ester zearalenone (ZEA) (Li et al., 1999; Ngoko et al., 2001). High concentrations of DON, 15-acetyl DON, and zearalenone were detected in barley seeds inoculated with *F. graminearum*, whereas barley seeds inoculated with *F. poae* had low concentrations of DON.

The seed infection by *F. graminearum* drastically reduced kernel plumpness and seed germination, in addition to the reduction in the quality of malt. Furthermore, very pronounced effects on wort-soluble nitrogen, free aminonitrogen, and wort color were also evident, indicating possibly greater proteolysis in infected kernels (Schwarz et al., 2001). The relationship between lipid metabolism and development of *Aspergillus parasitcus* was studied by generating an *A. parasiticus* 12-desaturase mutant which was unable to convert oleic acid to linoleic acid. This mutant was improved in polyunsaturated fatty acid biosynthesis and it had delayed spore germination, a two-fold reduction in growth, a reduced level of conidiation, and complete loss of sclerotial development, compared with a wild-type parent. Colonization of peanut and corn seed by the mutant was also reduced, indicating the crucial role of lipid metabolism in the development of *A. parasiticus* producing aflatoxin (Wilson et al., 2004). Ochratoxins, produced by *Aspergillus* spp. and *Penicillium* spp., and *Alternaria* toxins, produced by *Alternaria alternata*, are the other mycotoxins present in grains, fruits, and vegetables (Narayanasamy, 2005).

4.2.2 Bacterial Pathogens

The effects of bacterial pathogens on seeds have been studied in a few cases. The seeds of bean (*Phaseolus vulgaris*) infected by common bacterial blight (CBB) caused by *Xanthomonas campestris* pv. *phaseoli* were grouped into three categories: seeds with severe, slight to moderate symptoms, and symptomless. The failure of seedling emergence was high in the case of seeds with severe symptoms of CBB disease (Tefera, 2001). An unusual effect of the inoculation of the bacterial pathogen *X. axonopodis* pv. *vignicola* on seeds of cowpea (*Vigna unguiculata*) has been reported by Gour et al. (2001). The inoculated seed lots had significantly high percentage of germination compared with controls. The maximum increase in germination was observed in highly susceptible lines. In the case of Valencia sweet orange fruits, seeds infected by *Xylella fastidiosa* causing citrus variegated chlorosis disease, weighed less and germinated at lower rates compared with seeds from normal fruits. No adverse effect on the embryo formation in infected seed could be seen (Li et al., 2003). *Burkholderia glumae*, causing rice seedling rot and grain rot, produces the toxin toxoflavin consisting of two acidic proteins. The nontoxigenic (TOX^-) Tn5 mutants did not produce a yellow pigment and they were avirulent to rice plants. The acidic proteins TRP-1 and TRP-2 were produced only by the wild type. The anti-TRP-1 antibody positively reacted with a single protein band of about 30 kDa only on blots from sodium dodecylsulfate (SDS)-polyacrylamide gel electrophoresis (PAGE) separations of *B. glumae* (Suzuki et al., 1998).

4.2.3 Viral Pathogens

The effects of virus infection may be recognized, in some cases, by the formation of small, ill-filled, and shriveled seeds, and such seeds may be separated from healthy seeds by various methods. A general reduction in 1000-seed weight and grain yield may be observed. The quality of malting cultivars of barley infected by *Barley yellow dwarf virus* (BYDV) was determined. While the yields of three malting cultivars were reduced, by 32.5 to 38.0% in 1989 and 8.5 to 19.8% in 1990, higher proportions of thin seeds due to BYDV infection contributed to the effects on several quality parameters. Increases in wort proteins (2.5 to 14.5%) and diastatic power (3.8 to 12.6%) and decreases in malt extract (1.1 to 5.6%) were observed (Edwards et al., 2001).

4.3 INFECTION OF FRUITS AND VEGETABLES

Harvesting of fruits and vegetables initiates senescence in physiologically mature, or even in immature, storage tissues. Ripening of fruits is a specialized and unique form of senescence. The loss of preformed barriers to pathogens, as well as a reduction in the capacity to mount an active defense, are the important consequences of senescence, resulting in enhanced susceptibility to the pathogens in the infected fruits and vegetables. Several postharvest pathogens infect fruits early in their development and these infections remain quiescent until ripening commences (Swinburne, 1983; Prusky, 1996). Verhoeff (1974) considered the latent, dormant, or quiescent parasitic relationship as a condition in which the pathogen spends long periods during the life of the host in a quiescent stage until, under specific circumstances, it becomes active. The period of quiescence may occur during any phases of disease development from spore germination to colonization of host tissues by the pathogen (Swinburne, 1983).

A simple procedure to quantify quiescent infections of *Colletotrichum musae* in banana was developed by Bellaire et al. (2000). This method is based on the treatment of banana fruits with 1200 µl of ethylene/L of air for 24 h at 25°C. The fruits are placed at 32°C for 5 days, maintaining the ethylene concentration, so that conditions are optimal for the development of symptoms of infection. The technique was useful for routine testing of immature fruits at age 5 to 6 weeks after inflorescence emergence. Presence of high CO_2 concentration had an inhibitory effect on development of lesions on banana fruits. Latent infections in peach and sweet cherry fruits by *Monilinia fructicola* and *Botrytis cinerea* could be detected by treating the immature fruits with paraquat. *M. fructicola* was more frequently isolated than *B. cinerea* (Emery et al., 2000).

The early events during development of quiescent infection by *Colletotrichum gloeosporioides* in unripe avocado fruit were traced. Avocado pericarp tissue was inoculated with *C. gloeosporioides* and avocado cell cultures were treated with the cell wall elicitor of *C. gloeosporioides*. Both treatments increased the production of reactive oxygen species (ROS), which was detected within minutes in treated cell suspensions. On the other hand, ROS was detected only after 2 h in inoculated pericarp tissue. The production of ROS in uninduced pericarp tissue of freshly harvested, unripe, resistant fruit was twice as high as in ripe susceptible fruit. ROS production was further accelerated in resistant fruit when challenged with inoculation. It is suggested that ROS production might be induced by fungal infection of unripe fruits and, consequently, might modulate resistance, leading to the inhibition of fungal development and quiescence (Beno-Moualem and Prusky, 2000).

The period of quiescence represents a dynamic equilibrium between host, pathogen, and environment. Physiological and biochemical responses of the host after harvest may trigger changes on that equilibrium to activate the pathogen. On the other hand, the pathogenicity factors may be activated as the conditions become conducive, leading to active parasitic development and ending the period of quiescence during which the pathogen has low metabolic functions. The studies on pathogen–host interaction have elucidated the regulation of natural defense mechanisms and indicated the possibility of maintaining the natural defense mechanisms in harvested fruits and vegetables. Furthermore, such studies have provided the basis for the development of new strategies to reduce postharvest losses due to microbial pathogens. In each host–pathogen interaction, several factors may be operating simultaneously or sequentially resulting in success or failure of infection.

4.3.1 Process of Infection by Fungal Pathogens

4.3.1.1 Spore Germination Postharvest diseases caused by fungal pathogens occur more frequently than the diseases caused by bacterial pathogens. The infection process commences with host recognition by the pathogen as soon as propagules arrive at the fruit surface during the growing period or at later stages. Host recognition is undoubtedly regulated by signals arising from the host. Various chemical signals from the host have been identified as being stimulatory to germination of the spores produced by fungal pathogens. The conidia of *Colletotrichum piperatum* may remain quiescent on the surface of green unripe pepper fruits. However, the conidia germinated much more freely on the leachates of red fruits of *Capsicum frutescens* than on green fruits (Grover, 1971). It appears that presence of sucrose in leachates of red fruits may be a stimulatory factor. In contrast, greater germination and appressorium formation on unripe than on ripe fruit was reported in the case of *C. capsici* and *Glomerella cingulata* (Adikaram, 1981). Neither stimulation nor inhibition of conidial germination of *C. musae* due to stage of fruit ripening was evident on banana fruits. Formation of appressoria was, however, found to be stimulated on the surface of unripe banana fruits (Swinburne, 1976). The anthranilic acid present in leachates of banana fruit was rapidly degraded by the germinating conidia of *C. musae* to form 2,3-dihydroxy benzoic acid (DHBA), which possibly stimulated spore germination (Harper and Swinburne, 1979). DHBA was considered to have a role in the active transport of iron in many microorganisms (Harper et al., 1980).

Dothiorella dominicana infects floral parts of mango and develops endophytically in the fruit pedicel. The pathogen remains quiescent until the maturity of fruits, when the fruit was infected through the stem end (Johnson et al., 1991).

Botrytis cinerea can infect the floral parts of several fruit crops, such as strawberries, raspberries, and grapes, and remain dormant in the infected tissues until it is able to invade the fruits during growth or ripening (Jarvis, 1977; Elad and Evensen, 1995). Inflorescences of field-grown grapevines cv. Gamay were inoculated with *B. cinerea* conidial suspension or dried conidia at different stages during bloom in moist weather. About 10% of conidia germinated within 72 h and latent infections were two or three times more than in uninoculated controls in pea size (7-mm diameter) berries. Latent *B. cinerea* was also present in the style and in mature berries (Keller et al., 2003). Aqueous suspension of conidia of *B. cinerea* were spray-inoculated on specific floral organs, including the calyptra, stigma, and receptacle area, of glasshouse-grown grapevines cv. Cabernet Savignon. By using light, scanning electron microscopy (SEM), and transmission electron microscopy (TEM) the initial stages of pathogenesis were studied. After germination, the conidia attached themselves to host surface within 48h after inoculation. A high proportion of conidia accumulated in a channel-like gap between the ovary and calyx that extended in a narrowing fashion into the flower interior where the ovary was in contact with receptacle. A few conidia attached to the style did not appear to grow towards the ovaries. On the other hand, abundant hyphal growth could be observed in the receptacle area irrespective of the site of inoculation. A sheath of extracellular material could be seen around conidia and germ tubes in the open area at 48h after inoculation (Fig. 4.1A). Later, hyphae of *B. cinerea* growing toward the internal tissues between receptacle and style was observed (Fig. 4.1B). Colonization by the fungal hyphae of calyx tips, corresponding to fused sepals, occurred. Development of intercellular intramural hyphae in the host cell walls was also revealed by the SEM and TEM observations (Appendix 4) (Viret et al., 2004). Necrosis of calyx tips and the presence of mycelium in the gap between the ovary and calyx could be seen at 72h after inoculation, indicating that calyx tips might be colonized by *B. cinerea* preferentially (Fig. 4.1C). The infected calyptras that remained attached in the flower cluster might be a potential source of inoculum. The results suggested that the receptacle area may be the preferred site of infection by *B. cinerea* (Viret et al., 2004).

Infection of grapes by different densities of airborne conidia of *B. cinerea* on table grapes stored at –0.5°C was investigated. Fluorescence microscopy of skin segments showed that conidia were consistently deposited individually, and not in pairs or groups, on berry surfaces. Lesion numbers tended to increase exponentially at higher doses (3.24 to 3.88 conidia/mm^2 of berry surface). Individual conidia induced no disease symptoms on fresh berries (Coertze and Holz, 1999). Germination of the solitary conidia and appressorium formation in *B. cinerea* were observed by fluorescence microscopy after incubation for varying periods (3 to 96h) at ±93% RH. Microscopic observations revealed delayed germination on immature berries, whereas germination proceeded at a high rate on mature berries. Growth of the pathogen was invariably restricted on moist berries and penetration of berries by *B. cinerea* was always direct (Coertze et al., 2001).

Botrytis cinerea causes postharvest decay of apples. Conidia of *B. cinerea* on polycarbonate membranes were incubated on filter paper disks saturated with water and exposed for 24h at 22°C to 0–16µl of ethyl, butyl, or hexyl acetates injected into the head spaces of 500-ml glass jars. Conidia were stimulated to germinate by 4µl of butyl acetate and 8µl of hexylacetate. *B. cinerea* infection of apple wounds was increased by butyl acetate. The presence of

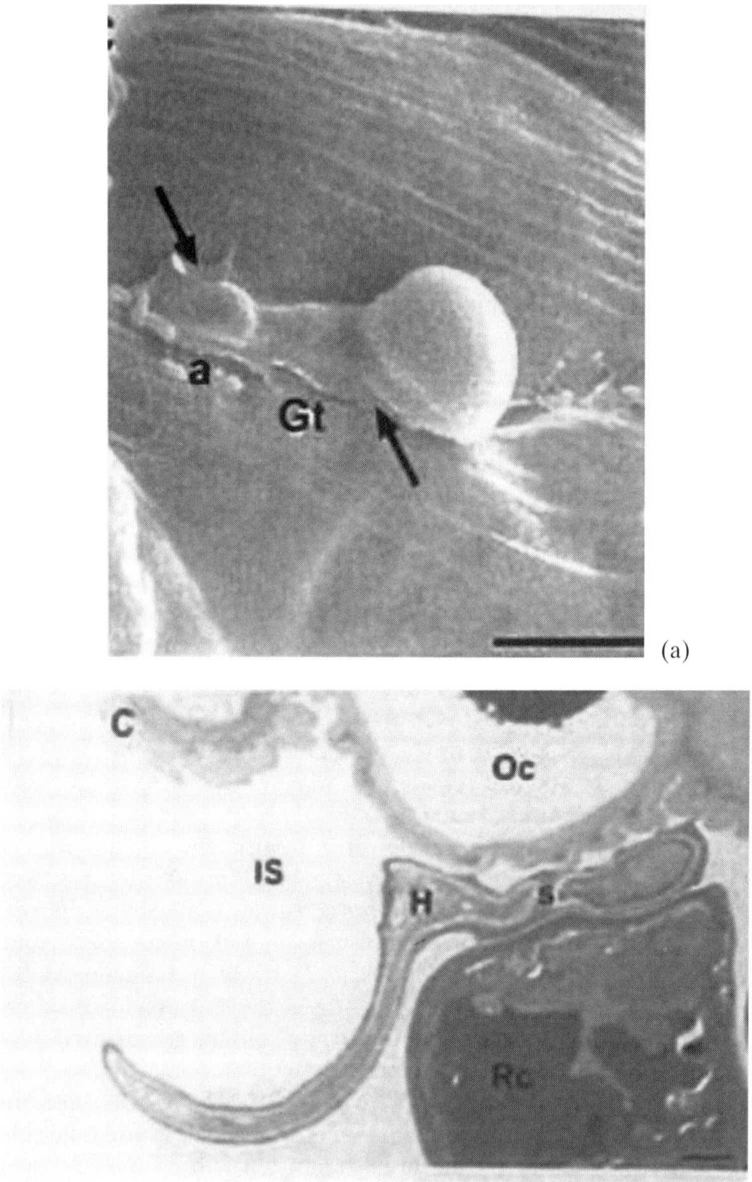

(a)

(b)

Figure 4.1 Electron micrographs of grape flowers inoculated with *Botrytis cinerea* on the stigma at full bloom. (a) Scanning electron micrograph showing formation of appressoria-like structure (a) at the terminal end of germ tube (Gt) attached by an extracellular sheath on the host surface (arrows) at 48 h postinocuation. (b) Transmission electron micrograph of a longitudinal section of a hypha (H) with septum (S) growing in the intercellular space (IS) between receptacle cells (Rc) and ovary cells (Oc) at 72 h postinoculation.

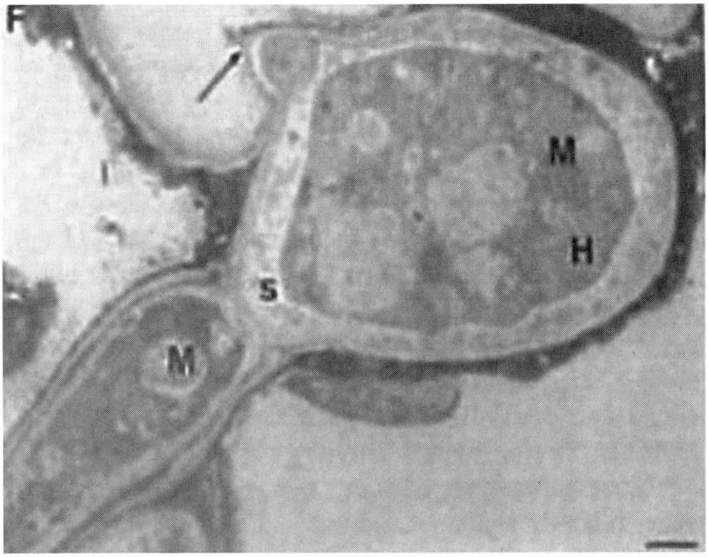

(c)

Figure 4.1 (*continued*) (c) Transmission electron micrograph of a longitudinal section of an intercellular hypha showing a septum (S) and mitochondria (M) along with an emerging side hypha (arrow) in the intercellular space of the calyx tip. (Courtesy of Viret et al., 2004; The American Phytopathological Society, St. Paul, MN, USA.)

butyl and hexyl acetate could be detected in the head space of apples cv. Golden Delicious by GLC. The acetate esters produced by apples stimulated germination of *B. cinerea* (Filonow, 1999). Further investigation to assess the role of butyl acetate, a volatile aroma and flavor compound produced by apples, showed that the conidia of *B. cinerea* (three strains) had greater adhesion to, and greater germination on, polycarbonate membrane filters on water inside sealed glass jars that were injected with butyl acetate than the unexposed controls. Conidial germination was highly correlated with conidial adhesion. Butyl acetate did not appear to be a food source for *B. cinerea* conidia (Filonow, 2001).

The quantitative relationship between inoculum dose of *B. cinerea* and *Penicillium expansum* and infection of pear fruit was studied using dry conidia applied to wet or dry pears, cv. D'Anjou, in a settling tower. Gray mold incidence on wet fruit increased as conidial concentration of *B. cinerea* was increased from 0 to 8.6 spores/L of air. Likewise, blue mold disease caused by *P. expansum* increased with greater concentration of conidia (0.1 to 803.5 spores/L of air) (Spotts and Cervantes, 2001). The *Botrytis* storage rot infection in kiwifruits was positively correlated with inoculum concentration of *B. cinerea.* The grape and kiwifruit isolates were able to induce higher disease severity than blueberry and strawberry isolates, indicating the variations in pathogenic potential of isolates, irrespective of the environmental conditions

(Bautista-Baños et al., 2001). The role of latent infection of *Monilinia fructicola* of immature peach fruit in the carryover to preharvest period (fruit rot phase) was studied. The incidence of latent infection was correlated significantly with both the incidence of blossom blight earlier in the season and the incidence of fruit rot at harvest. The results suggested that latent infections might serve as a source of inoculum for subsequent fruit rot in peach orchards in Georgia (Emery et al., 2000). A study to determine the factors affecting latent infection and sporulation of *Monilinia fructicola* on prune fruit revealed that inoculum concentration and wetness duration significantly influenced secondary infection. Increased inoculum concentration and wetness duration increased the percentage of fruit with latent infections, while increased temperature reduced the percentage of fruit with latent infections (Luo et al., 2001). Latency and development of *B. cinerea* was studied in two grape varieties, Gamay (susceptible) and Gamaret (resistant). The percentages of latent infection of berries were the same for both resistant and susceptible varieties. However, significant difference could be observed in the severity of visible gray mold, which was much less in the resistant compared with the susceptible variety. The results indicated that the development of the pathogen in the susceptible berries was not affected as in the resistant variety (Pezet et al., 2003).

4.3.1.2 Entry into Host Tissue Under conducive conditions, spores germinate to produce a germ tube, which terminates as a thick-walled structure known as an appressorium. Postharvest pathogens belonging to the genera *Gloeosporium* and *Colletotrichum* produce deeply pigmented, characteristic appressoria which aid in penetration into intact plant surfaces. The appressoria are closely attached to the outer layers of the cuticle, sometimes by producing mucilaginous secretions, as in *Colletotrichum musae* (Emmet and Parbery, 1975) (Fig. 4.2). Infection hyphae, produced from appressoria formed by *C. capsici* infecting pepper, penetrate cuticles and initially grow beneath the cuticle in the anti- and periclinal walls of epidermal cells, causing extensive wall degradation (Pring et al., 1995). Both physical and enzymatic processes may be involved in direct penetration.

Colletotrichum gloeosporioides infecting avocados form thin infection pegs from appressoria soon after inoculation of immature, unripe fruit and may be detached from appressoria during deposition of wax, as revealed by electron microscopy (Coates et al., 1993). The fungus remains quiescent until harvest, when antifungal dienes in the skin of avocado fruit breakdown due to degradation by lipoxygenase activity (Prusky et al., 1990; Adikaram et al., 1992). Formation of appressoria may be induced when the conidia land on hydrophobic surfaces such as fruit wax, suggesting that surface waxes may serve as signals for the initiation of host–pathogen interaction (Prusky et al., 1982; Podilia et al., 1993).

Certain compounds present in the cuticle may stimulate spore germination and elongation of germ tube of fungal pathogens. The differential stimulatory effect of surface waxes from avocado on germination and formation of appres-

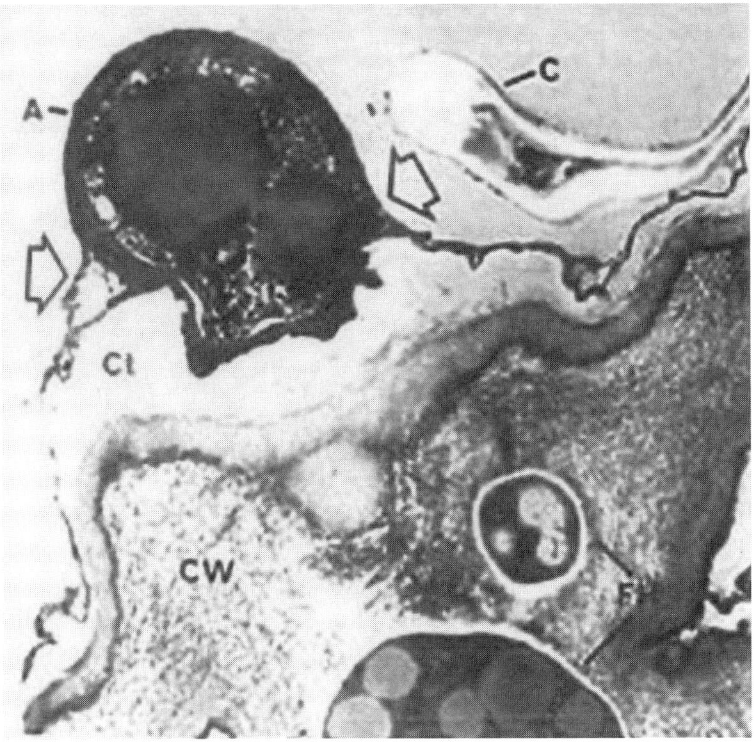

Figure 4.2 Formation of pigmented, thick-walled appressorium (A) from the germinated conidium (C) of *Colletotrichum musae* partially embedded in the host cuticle (Ct) and adhering to the surface with mucilage-like deposits (arrows). Fungal hyphae (FH) in the host cell wall (CW) may be seen in the transmission electron micrograph. [Courtesy of T.R. Swinburne and T.W. Fraser, 1983; (Academic Press, London), Elsevier, Oxford, United Kingdom. Specimen prepared by T.R. Swinburne; sectioned and photographed by T.W. Fraser.]

soria from the conidia of *Colletotrichum gloeosporioides*, and the lack of such effect for waxes from nonhost species, was demonstrated. In addition, the avocado wax could not induce appressorium formation in other *Colletotrichum* spp. that did not infect avocado, indicating the host-specific action of waxes in the plant–pathogen interaction (Podila et al., 1993). The gene *Cap50*, cloned from *Colletotrichum*, was expressed uniquely during appressorium formation induced by host cuticular waxes. In tomato, the transcripts for this gene were detected at the infection front, when infected by *C. gloeosporioides*. In contrast, mutants that did not express the gene had decreased virulence on tomato and avocado fruits (Hwang et al., 1995).

Cutin offers the first barrier to infection presented by intact plant tissues. Insertion of a cutinase gene derived from *Fusarium solani* f.sp. *pisi* (infecting peas) into *Mycosphaerella* sp. (normally a wound pathogen) resulted in direct penetration of papaya fruit by the transformant through the nonwounded cuticle (Dickman et al., 1989). However, mutants of *B. cinerea* deficient in cutinase A were able to infect gerbera flowers and tomato (van Kan et al., 1997). Cutinase production by *Monilinia fructicola*, infecting peaches, was inhibited by chlorogenic and cafeic acids. The resistance of peach fruit to *M. fructicola* was reported to be positively correlated with levels of these two phenolic acids, indicating the possible role of cutin in the early stages of pathogenesis (Bostock et al.,1999). These studies indicated that cutinase may not be required for penetration of the cuticle by all pathogens.

A pepper esterase gene (*PepEST*), expressed during the incompatible interaction between pepper (*Capsicum annuum*) and *Colletotrichum gloeosporioides*, was cloned. Glutathione-S-transferase-tagged recombinant PepEST protein expressed in *Escherichia coli* appears to regulate appressorium formation by modulating the cAMP-dependent signaling pathway in this pathogen. The PepEST esterase activity may inhibit appressorium formation of *C. gloeosporioides*, which may result in protection of the unripe fruit against the fungus (Kim et al., 2001).

Further studies demonstrated that contact with a hard surface induced appressorium formation in *C. gloeosporioides*, whereas hydrophobicity of the contact surface did not affect infection-related differentiation. Exogenous addition of Ca^{2+}, regardless of concentration, augmented conidial germination, whereas appressorial differentiation was observed at higher concentration. It was suggested that biochemical processes controlled by the calcium/calmodulin signaling system are involved in the induction of penetration morphogenesis in *C. gloeosporioides* infecting red pepper (*Capsicum annum*) (Uhm et al., 2003).

Colletotrichum lagenarium forms darkly melanized appressoria to penetrate the host epidermis. The *CMK1* gene of *C. lagenarium* regulates conidial germination, appressorium production, and infection hyphae. The mitogen-activated protein kinase (MAPK) gene *CMK1* is involved in different steps of infection process, including appressorium formation. A second MAPK gene, *MAF1*, was functionally characterized. The *maf1* gene replacement mutant had significantly reduced conidiation and fungal pathogenicity. *MAF1* seems to be required for the early differentiation phase of appressorium formation, whereas *CMK1* is involved in the maturation of appressoria (Kojima et al., 2002). The *Ste 12* homologue *CST1* gene from *C. legenarium* was isolated and characterized. The *cst1* Delta strains produced conidia that could germinate and form melanized appressoria which, however, did not produce infection hyphae. The results suggested that *CST1* was essential for appressorial penetration. Abundant lipid droplets were present in the *cst1* Delta strains, but they were rapidly degraded during the formation of appressoria. This degradation of lipids may, possibly, be related to the failure of formation of infection hyphae and penetration into the host tissue by the *cst1* Delta strain (Tsuji et al., 2003).

In another study, pathogenicity of *C. lagenarium* was shown to be dependent on *RPK1*, encoding the regulatory subunit of PKA. Three growth-suppressor mutants were isolated from the *rpk1* mutant and these *rpk1* suppressor mutants were nonpathogenic and showed amino acid changes in the PKA catalytic subunit Cpk1 (a protein involved in spore germination through cAMP signaling). Knockout mutants of CPK1 and the adenylate cyclase gene *CAC1* were generated to assess the roles of the cyclic-AMP (cAMP) signaling. The mutants germinated poorly, indicating possible involvement of cAMP signaling in spore germination in *C. lagenarium*. The germinating conidia of the *cpk1* and *cac1* mutants formed nonfunctional appressoria which contained a larger number of lipid bodies compared with the wild type, as revealed by cytological analyses. In contrast, lipid levels of the *rpk1* mutant were lower, suggesting the requirement of cAMP-mediated regulation of lipid metabolism for appressorium functionality. The *Cmk1* mitogen-activated protein kinase (MAPK) is known to regulate conidial germination, appressorial formation, and growth of infection hyphae of *C. lagenarium*, as described above. The results suggest that cAMP signaling controls multiple steps of fungal infection in cooperative regulation with Cmk MAPK in *C. lagenarium* (Yamaguchi et al., 2004).

Ultrastructural studies using scanning electron microscopy showed that infection hyphae of *Colletotrichum acutatum*, infecting guava fruit, grew directly from conidia and penetrated the cuticle directly or entered the fruit wall through stoma (Das and Bora, 1998). *C. acutatum* can infect citrus petals and later cause premature fruit drop and formation of persistent calyces. The effect of infection on hormonal and growth regulator metabolism and differential gene expression, in affected flowers and young fruit, was assessed. Dramatic increase in ethylene evolution (three fold) and rapid accumulation of indole-3-acetic acid (IAA) (140 fold) were evident, while no significant change could be recognized in the levels of abscisic acid (ABA). A two-fold increase in the concentrations of salicylic acid (SA) in affected petals, but not in pistils, was seen. The genes encoding ACC oxidase or ACC synthase and 12-oxo-phytodienoic acid (12-oxo PDA) reductase were highly expressed in affected citrus flowers. Up regulation of genes encoding auxin-related proteins was also noted. The results indicate that symptom development and premature fruit drop might be due to an imbalance of IAA, ethylene, and SA in citrus flowers infected by *C. acutatum* (Lahey et al., 2004). On the other hand, conidia of *B. cinerea* were unable to infect uninjured apple skin, regardless of inoculum density and presence of nutrients. However, after the removal of surface wax by washing apples with chloroform, *B. cinerea* could infect them. The requirement of a MAP kinase for infection by *B. cinerea* was tested by using a polymerase chain reaction (PCR)-based approach to isolate MAP kinase homologues from *B. cinerea*. The *Botrytis* MAP kinase required for pathogenesis *(BMP) MAP* kinase gene is highly homologous to the *Magnaporthe grisea PMK1*. The mutants generated by *bmp1* gene replacement produced normal conidia and mycelia, but with reduced growth rate. The *bmp1* mutants were nonpathogenic and reintroduction of the wild type *BMP1* allele into the *bmp1* mutant restored normal growth and pathogenicity. Furthermore, *bmp1* mutants also seemed to be defective in infecting through wounds. The results indicated that *BMP1* was essential for *B. cinerea* to infect plants (Li et al., 2000).

The signal transduction pathways of *B. cinerea* involved in host infection was studied by isolating two genes *bcg1* and *bcg2*, encoding alpha subunits of heterotrimeric GTP-binding proteins using heterologous hybridization and PCR techniques. RT-PCR assays demonstrated the expression of both genes at very early stages of infection. Gene replacement experiments showed that *bcg1* null mutants differed in colony characters from wild type strain, with no secretion of proteases and consequent reduced pathogenicity on bean and tomato; the infection process terminated after formation of a primary lesion. On the other hand, *bcg2* mutants had only slightly reduced pathogenicity and retained the wild-type colony characters. Full recovery of colony morphology, protease secretion, and pathogenicity on both host plants occurred following complementation of *bcg1* mutants with a wild-type gene copy. The results suggest an essential role of BCG1 in different signaling pathways (Gronover et al., 2001).

Many pathogens causing postharvest diseases gain entry into the fruits through injuries incurred during harvesting and handling operations. Such pathogens do not usually penetrate intact fruit and the process of colonization follows germination almost immediately. *B. cinerea* usually does not penetrate intact fruit surfaces. Grape berries were inoculated with different densities of air-borne conidia of *B. cinerea* and incubated for 24 h. Removal of the pathogen from the surface of fresh berries by ethanol and subsequent incubation of excised skin segments showed that less than 2% of skin segments had been penetrated, regardless of conidial density or wetness regime. Further increasing densities of conidia neither increased surface colonization nor skin penetration by *B. cinerea* (Coertze and Holz, 1999).

Preharvest cuticular fractures in sweet cherry have been shown to facilitate invasion by *B. cinerea* and *Monilinia laxa*. Five categories of cuticular fracturing were recognized (from 1 = no visible fractures to 5 = severe fracturing). A significant linear relation between the category of fracturing and the percentage of fruits infected by both pathogens was observed (Børve et al., 2000). Compression bruising of strawberry fruits altered their volatile profile for a period after the injury, increasing the production of volatile compounds and more common aroma compounds such as ethyl butanoate. Conidia of *B. cinerea* placed in close proximity to a bruise site on a ripe fruit and to gently shaken fruit exhibited higher rates of germ tube elongation compared to spores placed in proximity to a nonwounded surface. As no significant accumulation of CO_2 and ethylene from bruising could be detected, the wound volatile compounds may have a functional role in promoting fungal growth after a wound event (Archbold et al., 2002). The stimulatory effect of ethyl, butyl, and hexyl acetates injected into the head spaces on the germination of conidia of *B. cinerea* infecting apples was earlier reported (Filonow, 1999).

Some fungal pathogens may penetrate through natural openings present in fruits and vegetables. *Phytophthora infestans*, causing late blight disease of potato, may be able to penetrate both directly and also through stomata in the leaves. Lenticels, formed by loosely packed cells, are present in the periderm of potato tubers. *P. infestans* and *Streptomyces scabies* (causing common scab disease) gain entry into the tuber tissues through lenticels. Potato blackleg pathogen *Erwinia carotovora* subsp. *atroseptica* also finds its way into the tuber through lenticels. *Phytophthora erythroseptica*, causing pink rot disease of potato, invades the tubers directly through the eyes. Potato silver scurf pathogen *Helminthosporium solani* can penetrate the periderm of potato or gain entry through natural openings on the surface of the tuber (Rich, 1983). A later study, using TEM and SEM, showed that the pathogen entered potato tubers primarily via hyphae, although germ tubes could also penetrate the tubers directly. An extracellular sheath was observed around the hyphae growing over the surface of tubers and the host cell wall appeared lyzed at the point of penetration. Periderm penetration by the fungus may be due to mechanical and enzymatic action (Martinez et al., 2004).

4.3.1.3 Colonization of Host Tissues After successful entry into the plant, the fungal pathogens have to disrupt the cell wall in order to colonize the host tissue. A cocktail of enzymes and/or toxins are secreted by pathogens to digest the host cell wall. Pectinases, cellulases, proteases, and xylanases of pathogen origin are required for pathogenic and saprophytic development of the fungi. Pathogenic fungi have multiple genes governing the secretion of extracellular hydrolytic enzymes required for degrading the physical barriers presented by host plants to restrict invasion by pathogens.

Colletotrichum gloeosporioides, infecting avocado fruit, produced pectate lyase (PL), the maximum concentration of PL becoming detectable at the leading edge of lesions. The antibodies specific to PL inhibited the development of *C. gloeosporioides*, indicating that PL is required for the pathogenicity of this fungus (Wattad et al., 1997). The importance of PL for pathogenicity was further demonstrated by transforming *C. magna*, a weak pathogen of watermelon, with the *pel* gene from *C. gloeosporioides*. The transformed isolates of *C. magna* showed additive macerating capacities on avocado pericarp (Yakoby et al., 2000a). PL is considered as a pathogenicity factor, required for penetration and colonization of host plant tissues by fungal pathogens. Pectate lyase secretion by *C. gloeosporioides* occurred at a pH range of 5.8 to 6.5 in pectolytic enzyme-inducing medium (PEIM). PL gene (*pel*) transcript production began at pH 5.0 and increased up to pH 5.7. The pH of unripe pericarp of freshly harvested avocado fruits resistant to *C. gloeosporioides* was 5.2. But the pH of ripe fruits, when decay symptoms were exhibited, increased to 6.3. The pH of the pericarp of resistant cultivars Ardit and Ettinger was less than 5.5 and did not increase during ripening. These findings suggest that the host pericarp pH regulates the secretion of PL and may affect the pathogenicity of *C. gloeosporioides* (Yakoby et al., 2000b).

Disruption of the *pelB* gene in *C. gloeosporioides* resulted in a 36 to 45% reduction in estimated decay diameter on avocado fruit. In addition, *pelB* mutants induced significantly higher phenylalanine ammonia lyase (PAL) activity as well as antifungal diene, indicating a higher defense response to mutants. PLB appears to be an important factor because of its ability to be a virulence factor and its role in induction of resistance (Yakoby et al., 2001). Accumulation of ammonia and associated tissue alkalinization predisposes fruit to infection by *C. gloeosporioides*. PL and other extracellular proteins are secreted and accumulate as external pH increases from 4.0 to 6.0. Both inorganic and organic N sources increased secretion of PL. The transcript level of *pelB* increased in parallel as a function of pH. The ambient pH and the source of N were indicated to be independent regulatory factors for processes linked to PL secretion and virulence of *C. gloeosporioides* (Prusky et al., 2003).

Polygalacturonase is another enzyme involved in the colonization of plant tissues by fungal pathogens. The requirement and production of endopolygalacturonase (endo-PG) for both saprophytic growth of, and infection by, *C. lindemuthianum* has been demonstrated. Secretion of an endo-PG by *C. lindemuthianum* was controlled by *CLPG1* and *CLPG2* genes. By using specific antibodies, the protein encoded by *CLPG1* was detected in planta, and this protein degraded the cell wall extensively (Centis et al., 1997). The endo-PG encoded by the gene *Bcpg1* was essentially required for the primary colonization of tomato tissues by *Botrytis cinerea*. Elimination of polygalacturonase1 by partial gene replacement led to a significant reduction in development of lesions on tomato fruit, indicating PG1 was essential for full virulence of *B. cinerea* (Have et al., 1998). Purified PG1 from *Phomopsis cucurbitae*, causing musk melon fruit decay, effectively macerated mature fruit tissue, suggesting that this isoenzyme might be involved in pathogenesis (Zhang et al., 1999).

The expression of all *Bcpg* genes (1–6) in tomato, apple, and *Cucurbita pepo* fruits was studied. The patterns of gene expression indicated that *B. cinerea* had a flexible enzymatic pectate degradation machinery (Have, 2001). Likewise, *Alternaria citri*, causing citrus blackrot, secreted an endo-PG that was required for tissue maceration, whereas *A. alternata* colonizes citrus tissues by producing the host-selective toxin. Although the endo-PG produced by *A. alternata* had similar properties to that of the endo-PG produced by *A. citri*, it was not involved in pathogenesis. This study showed that the cell wall-degrading enzymes produced by these two citrus pathogens may play different roles in the disease development and pathogen growth (Isshiki et al., 2001). By employing specific antisera generated against four exo-polygalacturonases (exo-PGs) secreted by *B. cinerea*, the expression of exo-PGs was analyzed by immuno-histochemical investigations; this indicated an important role for the exo-PGs at the early stages of pathogenesis in cucumber (Rha et al., 2001).

The PG activity of the citrus race of *Geotrichum candidum*, causing sour rot disease, was much greater than that of the noncitrus race isolates in culture medium and inoculated lemon peel. A significant correlation between PG activity and pathogenicity was observed. Expression of the *S31pg1* gene of the citrus race could be detected in mycelium grown in liquid cultures of citrus race 31 and also in inoculated lemon peel. In contrast, no transcripts of *S63 pg1* gene of the noncitrus race were detectable, both in culture and inoculated lemon peel. These results indicated that PG may play an important role in the development of sour rot symptoms and may be involved in the difference of pathogenicity between the two races of *G. candidum* (Nakamura et al., 2001).

Cell wall degradation and pectin dissolution were determined in mandarin fruit inoculated with *Penicillium digitatum* and *P. italicum* causing blue and green molds. PG, α-and β-galactosidase, α-glucosidase, and β-arabinosidase showed relatively higher activities in fruits inoculated with *P. digitatum* than in fruits inoculated with *P. italicum*. β-galactosidase and α-arabinosidase were active, in addition to PG, in hydrolyzing pectin chains in this pathosystem (Hwang et al., 2002).

The transient correlation between the expression of tomato PG and susceptibility of fruits to fungal infection was demonstrated. The level of PG expressed by tomato lines was reflected in the severity of postharvest diseases caused by *Alternaria alternata* and *Rhizopus stolonifer*. The results suggested that the developmentally regulated plant genes functioned as biochemical determinants of susceptibility of these diseases. In addition, toxin production by *A. alternata* f. sp. *lycopersici* (AAL) on resistant tomato lines was induced specifically by PG-solubilized pectic polysaccharides from tomato acting as signal molecules. The AAL toxin production was significantly increased by the PG-solubilized signals (Prusky, 1996). In the case of *Colletotrichum acutatum*, seven isozymes of PG are secreted. The genes (*capg1* and *capg2*) encoding two of these PGs have been cloned and sequenced. These two genes were differentially regulated with respect to pH and carbon source of the medium. The production of PG and expression of *capg1* were reduced in medium containing apple cell walls from fruit resistant to *C. gloeosporioides*. The expression pattern of *cpg1* was found to be positively correlated with total PG activity, as observed in several experiments (McEvoy et al., 2003).

Diplodia viticola, causing bunch rot of grapes, produced arabinosidase, xylanase, and β-1, 3-glucanase, which have been reported to be associated with rotting symptoms. Xylanase degrades xylan into a variety of xylan oligosaccharides (Strobel, 1963). Stimulation of invertase (β-fructofuranosidase) activity was observed. In berries infected by *B. cinerea*, a new invertase similar to that of *Botrytis* invertase (BIT) was detected. The antibodies specific against this new invertase were employed for the detection of the native BIT in the extracts of diseased berries (Ruiz and Ruffner, 2002). In the case of *Phytophthora capsici* infecting pepper (chilli), immunogold labeling technique revealed specific labeling of chitinase on the cell wall of the pathogen. The labeling was uniformly distributed over the entire fungal wall in intimate contact with host cell wall (Lee et al., 2000).

Apple fruits inoculated with four wild strains of *B. cinerea* showed two types of symptoms, that is firm or soft rot. Different specific polygalacturonase (PG) isoenzymes appeared in soft-rot affected apples. No PG activity was found in apples infected by firm-rotting wild type strains or transformants, whereas two new PG isoenzymes were expressed in soft-rotting strains. No association could be discerned between the production of these PG isozymes and colonization level. No variation in pectin methyl esterase (PME) isozyme patterns in vitro could be related with the culture substrate on apple or rotting type (Reignault et al., 2000). The pectinolytic enzyme pectin methylesterase (PME) hydrolyses pectin in methanol and PG. A *BcPme1* cDNA from *Botrytis cinerea*, with 1041 bp encoding a 346-amino acid protein of 37 kDa T4, was identified. A mutant in *Bcpme1* was generated by gene disruption. This mutant had reduced growth on pectin medium. Two isozymes of PI 7.4 and 7.1 were detected in the wild strain, whereas the mutant had only the PI 7.1 isozyme. The PME activity of the mutant was reduced by 75% and the mutant was less virulent on apple fruits, grapevine, and *Arabidopsis* leaves. The results indicated that *B. cinerea* has more than one PME-encoding gene and that BCPME1 is an important determinant of virulence of *B. cinerea* (Valette-Collet et al., 2003). The *Dbcsod* 1 mutants of *B. cinerea* generated by deletion of a Cu-Zn superoxide dimutase gene (*bcsod1*) had significantly reduced virulence on French beans, indicating that the Cu-Zu SOD-activity is an important single virulence factor in French bean–*B. cinerea* pathosystem (Rolke et al., 2004).

B. cinerea secreted aspartic proteinase (AP) activity in axenic cultures. The fluid extracted from *B. cinerea*-infected tissues of apple, pepper, tomato, and zucchini also contained AP activity. Five genes of *B. cinerea*, *Bcap1–5* encoding AP were cloned and the encoded proteins had novel characteristics. All the *Bcap* genes were expressed in liquid cultures. Transcript levels of *Bcap1*, *Bcap2*, *Bcap3*, and *Bcap4* were subject to glucose and peptone repression. The presence of all *Bcap* transcripts could be detected in infected plant tissues, indicating that at least part of the AP activity in planta had pathogen origin (ten Have et al., 2004).

Production of toxins by *B. cinerea* has been demonstrated. The strains of the pathogen showed variations in the nature and amount of toxins produced, depending on their virulence. A correlation between toxin production and virulence of the pathogen isolates was observed. Toxin production by less virulent isolates was at a low level and they produced botrydial or its derivatives only. In contrast, more virulent isolates formed high concentrations of two classes of toxins, that is botryane skeleton and botcinolide derivatives. Synergism in the activities of several toxins involved in the phytotoxicity induced by *B. cinerea* was also indicated by the studies of Reino et al. (2004).

Expression of pathogen and host genes during the susceptible interaction was investigated in *Penicillium expansum*–apple fruit (cv. Golden Delicious) pathosystem. Northern blot analysis showed that 18 genes were highly expressed during the apple–fungus interaction. Among the differentially expressed genes, one gene of fungal origin encoding an unknown protein and two apple genes were identified. The apple genes were homologous to a β-glucosidase and a phosphatase 2C. These genes exclusively expressed during the infection process. Several up-regulated pathogen genes appeared to mediate adaptive responses to the environment (Sánchez-Torres and González-Candelas, 2003).

Senescence is accelerated by ethylene and its biosynthesis in infected tissues is enhanced. As a consequence, decay is generally increased (Boller, 1991). Evolution of ethylene following infection of plant tissues has been reported in some pathosystems. Ethylene production was greatly enhanced in sweet-potato root tissues infected by *Caratocystis fimbriata* causing black rot disease. The rate of ethylene production in the surface layers (0–0.5 mm) in response to fungal invasion was increased (Okumura et al., 1999; Yoshioka et al., 2001). The appressorial dormancy on the fruit surface is terminated by ethylene produced by the host specifically at ripening, as it is able to act as a signal for the induction of appressorium formation. Ethylene at concentrations much lower than that produced during fruit ripening could induce both spore germination and appressorium formation in *Colletotrichum gloeosporioides* and *C. musae* (Flaishman and Kolattukudy, 1994). Spores of *C. gloeosporioides* placed on oranges which did not produce significant amounts of ethylene after harvest did not produce multiple appressoria (Brown, 1975, 1977). Likewise, no germination of spores of *C. gloeosporioides* on transgenic tomatoes (nonhost) could be observed. But on exposure to exogenous ethylene, the spores did germinate, produced multiple appressoria and ultimately produced lesions (Flaishman and Kolattukudy, 1994). The ethylene-induced PG activity in tomato fruit may be related to decreased resistance to decay (Labavitch, 1998). The activities of PG as well as abscission enzymes and cellulase in oranges were increased by treatment with ethylene, resulting in increased infection by *Diplodia natalensis* causing stem-end rot disease, indicating a possible role for these enzymes in pathogenesis (Brown and Burns, 1998). In citrus, effects of ethylene treatment have been assessed. Severity of decay caused by wound pathogens decreased by ethylene treatments, while the incidence of stem-end rot caused by quiescent infection was increased. Reduction in green mold caused by *P. digitatum* by exposure to ethylene at 30°C with RH of 90 to 96% for 2 to 3 days was reported by Brown (1973). Furthermore, the growth of *P. italicum* was reduced by treating orange fruits to 1000 ppm ethylene at 20°C or 2°C prior to inoculation, as indicated by diameter of lesions and glucosamine content (El-Kazzaz et al., 1983). Ethylene seemed to induce defense responses in treated produce.

Ethylene has been shown to regulate genes in both climacteric and nonclimateric fruit, resulting in enhanced or reduced expression depending on the gene concerned (Alonso et al., 1995; Lelievre et al., 1997). The genes encoding chitinase, endochitinase, β-1,3-glucanases, and thaumatin-like antifungal proteins were expressed actively, as reflected by the presence of abundant transcripts in ripening banana fruit (Medina-Suarez et al., 1997; Clendennen and May 1997). Treatment of grapefruit with 1-MCP, an irreversible inhibitor of ethylene action, increased ethylene production. However, the rate of disease caused by *Penicillium digitatum* was not altered. Infection of grapefruit flavedo led to accumulation of ACC synthase, glucanase, and PAL transcripts. No change in the levels of chitinase could be detected (Mullins et al., 2000).

The fungal pathogens produce proteolytic enzymes at the initial stages of pathogenesis. *Sclerotinia sclerotiorum* has a proteolytic system mainly constituted by acidic proteases, but a neutral protease (serine protease) was also detected in the early stages. The proteolytic enzymes secreted by *S. sclerotiorum* reflected the environmental niche created during mycelial growth (Billon-Grand et al., 2002). This pathogen has been shown to have a phytoalexin detoxification mechanism. Camalexin and 6-methoxycamalexin were detoxified via 6-hydoroxycamalexin to 6-*O*-β-D-glucopyranosylcamalexin. Furthermore, brassinin, 1-methoxybrassinin, and cyclobrassinin present in cruciferous plants were metabolized by *S. sclerotiorum* into their corresponding glucosyl derivatives displaying no detectable antifungal activity. Inducible brassinin glucosyltransferase (BGT) activity was detected in crude cell-free extracts of *S. sclerotiorum*. Camalexin and brassinin analogs, methyltryptamine dithiocarbamate and methyl 1-methyltryptamine dithiocarbamate, induced BGT activity in *S. sclerotiorum*. It appears that *S. sclerotiorum*, in its continuous adaptation and coevolution with brassinin-producing plants, might have acquired different glucosyltransferase(s) that can disarm some of the most active plant chemicals involved in defense reactions against microbial pathogens (Pedras et al., 2004).

Cell wall-degrading enzymes are required for breaching the cell wall barrier, facilitating the pathogen development. Furthermore, they are also involved in the generation of signals that form the components of the recognition process, aiding the recognition of pathogens by host plant species. Recognition of the presence of the pathogen is a prerequisite for the activation of the plant's defense system. The plants respond to the products generated by the activities of cell wall-degrading enzymes (hydrolases) produced by incompatible pathogens, but not to the hydrolases as such. Polygalacturonase (PG)-inhibiting proteins (PGIP) may be present in infected and uninfected plant tissues (Abu Goukh and Labavitch, 1983). PGIP purified from pear inhibited the growth of *B. cinerea* and other fungi tested, but it did not inhibit endogenous PG activity in pear fruit (Abu-Goukh et al., 1983a). The susceptibility to disease development after fruit ripening was enhanced by a decrease in the contents of PG inhibitor in Bartlett pear fruit (Abu-Goukh et al., 1983b). The PG inhibitor activity varied in different pear tissues. The PGIP activity in pear fruit was 200- and 1400-fold greater than in flowers and leaves respectively. It was suggested that the PGIP promoter in pear may affect a fruit-specific expression of the gene leading to inhibition of *B. cinerea* (Stotz et al., 1993).

Preformed antimicrobial compounds present in the developing fruits have a crucial role in disease development/resistance. The antimicrobial compounds may inhibit the spore germination and/or appressorial development that precede colonization of tissues in fruits and vegetables. A cDNA encoding PGIP from mature apple fruit was cloned and characterized. The PGIP transcripts levels showed variations in fruits with different maturities, suggesting that the gene is developmentally regulated. Very high levels of PGIP transcripts occurred in areas showing decay and tissues adjacent to the inoculation sites of *Penicillium expansum* and *Botrytis cinerea*. Although PGIP expression following wounding was noted, the extent of expression was much less compared with inoculated tissues. The results suggested that apple PGIP might have different roles during fruit development and stress response (Yao et al., 1999).

Unripe avocado fruits contain two major antifungal compounds, 1-acetoxy-2-hydroxy-4-oxo-heneicosa-12,15-diene (Prusky et al., 1982) and 1-acetoxy-2,4-dihydroxy-n-heptadeca-16-ene (Prusky et al., 1991). The diene was more fungitoxic than the monoene. The cultivars that decayed faster showed faster reduction in both diene and monoene during ripening. The establishment and development of *Colletotrichum gloeosporioides* infecting avocado appear to depend on a threshold concentration of the antifungal diene present in unripe fruit (Adikaram et al., 1992). It can be expected that the mechanism controlling reduction in the contents of diene may predispose fruit to activation of quiescent infection by *C. gloeosporioides*. The diene forms a substrate for oxidation by lipoxygenase which degrades diene and induces susceptibility to *C. gloeosporioides* in ripening avocado fruit (Prusky, 1988; Prusky et al., 1983). The preformed (Z,Z)-1-acetoxy-2- hydroxy-4-oxo-heneicosa-12,15-diene (AFD) affected the quiescence of *C. gloeosporioides* in unripe avocado fruit. It was hypothesized that one of the genes encoding D12 fatty acid desaturase (*avfad 12*) may take part in the biosynthesis of AFD. Hence, its expression pattern and enzymatic activity were determined in relation to the content of AFD. Young fruits and leaves of avocado which contained the highest concentrations of AFD, showed high levels of expression of *avfad 12*, whereas mature leaves and fruits contained no transcripts from *avfad 12* gene family and lower contents of AFD (Beno-Moualem et al., 2004).

The presence of a mixture of antifungal compounds, that is 5–12-cis-heptadecenyl resorcinol and 5-pentadecenyl resorcimol, was detected in unripe mango fruit that was resistant to *Alternaria alteranta*. The concentration of these 5-substituted resorcinols were reduced in susceptible cultivars at a faster rate than in resistant cultivars. The reduction in antifungal compounds may facilitate the colonization of mango fruit tissues by *A. alternata* (Droby et al., 1986, 1987). The antifungal compounds present in mango peel may act on the germinated hyphae of the pathogen (Prusky, 1996). The presence of a preformed monoterpene aldehyde, citral, in the young, mature green lemon correlated with resistance to infection by *Penicillium* spp. Decreases in citral contents in older, yellow lemon fruit paves way for the rapid development of decay by this pathogen (Rodov et al., 1994). In grape berries infected with *Phomopsis viticola* and *Uncinula nectator*, the expression of osmotin and thaumatin-like proteins, which belong to the group of pathogenesis-related

(PR) proteins, was induced. These two proteins were identified by immunological techniques and N-terminal sequencing and they showed strong antifungal activity against these two pathogens, in addition to *Botrytis cinerea* (Monteiro et al., 2003).

The preformed antifungal compounds and the elicitors of resistance formed following initiation of infection by fungal pathogens may determine the progress of disease development. If the pathogen is able to breach host barriers and overcome the effects of different defense-related compounds, it can grow adequately to induce specific visible symptoms and changes in the quality of fruits and vegetables. The changes in the level of aspartic protease (AP) in response to infection of potato tubers by *Phytophthora infestans* and wounding were studied. The intracellular washing fluids (IWFs) from tuber disks of two potato cultivars differing in their susceptibility were examined for the AP content. Induction of AP was greater and faster in the resistant cultivar than in susceptible one, and in wounded tuber. AP inhibited the germination of cysts of *P. infestans* and conidia of *Fusarium solani.* These results suggested a role for AP in the defense response of potato to *P. infestans* (Guevara et al., 2002). *P. infestans* has a unique elicitor-like gene family. The mating-induced gene *M81* elicits a defense reaction in the host plant. *M81C* and *M81D* mRNA accumulated only during zoosporogenesis, whereas *M81E* was expressed only in the hyphae. The protein products of these genes were predicted to be extracellular or cellular transglutaminases. The genes with an elicitor or proline- and threonine-rich repeat, and both elicitor and repeat domains, have been shown to be widely distributed in *P. infestans.* Studies on the products of these genes have been useful to understand the natural functions of elicitors in pathogen biology (Fabritius and Judelson, 2003).

The heterotrimeric G-proteins have been shown to be essential signaling components involved in development and pathogenicity of fungal pathogens. The G-protein subunit PiGPA1 in *P. infestans* controls zoospore motility and it is required for its virulence. Hypovirulent *P. infestans* strains were generated by silencing the *Pigpa1* gene. The gene expression profiles of the hypovirulent strains were analyzed, using an optimized cDNA-AFLP protocol. The expression profiles in sporangia and mycelium of the wild-type strain were compared with that of the hypovirulent strains. This comparison revealed a substantial number of mycelium- or sporangia-specific transcript derived fragments (TDFs) (Dong et al., 2004). *Phoma exigua* var. *foveata* causes dry rot (gangrene) disease in potato tubers. The activity of aminopeptidase and esterase was detected in tissues adjacent to infected, and also in apparently healthy, tissues by using a nitrocellulose blotting method. On the other hand, activity of glucuronidase and succinic and glucose-6-phosphate dehydrogenases (G-6-PDH) was confined to tissues adjacent to the rotted tissue. Maximum activity of glycosidases was observed in rotten tissue, since the pathogen could actively produce them in the rotten tissue filled with pathogen mycelium. These enzymes may be involved in alteration of cell metabolism and destruction of diseased tuber tissue (Giebel and Dopieraa, 2004).

4.3.1.4 Symptom Expression In susceptible cultivars, when the conditions are conducive, the compatible pathogen develops by drawing nutrition from the host plant, producing asexual and sexual spores/propagules for its perpetuation. In doing so, various characteristic symptoms are induced due to the interaction with host tissues. The enzymes, toxins, or other metabolic products of the pathogens have been shown to be responsible for many of the symptoms of diseases.

A. Anthracnose Disease: Different species of *Colletotrichum* cause the anthracnose diseases in several fruits and vegetables (Table 4.1). The symptoms of infection by *Colletotrichum gloeosporioides* in leaves, inflorescence, and fruits are characterized by the appearance of brown to black lesions. On fruits, lesions of different sizes coalesce to form large irregular areas. The discolored areas enlarge, typically in a tear-stain pattern, developing from the basal toward the distal end of the fruit. Generally the lesions may be restricted to the peel, but may invade the pulp in the case of severe infections. The asexual spore-bearing structures, acervuli, producing abundant orange to salmon pink masses of conidia, are formed. Early infection of young fruits may result in mummification of affected fruits. This pathogen produces different kinds of compounds during germination of spores, development of appressorium, and colonization of host tissues, resulting in breach of host barriers and structural integrity. Fruit crops such as avocado, banana, grapes, mango, and papaya suffer significantly due to anthracnose diseases (Fig. 4.3A and B).

On beans, anthracnose symptoms appear initially on pods as small rusty spots and, as the pods mature, the lesions are characterized by cankerous edges with a slightly raised black ring and a cinnamon-colored periphery. Salmon-colored mass of conidia ooze out from central zone of the lesions. The infected seeds exhibit yellowish to brown sunken cankers. Cucurbits, pepper, tomato, and watermelon also are infected by anthracnose disease (Fig. 4.4A and B).

B. Mold Disease: Mold diseases are caused by fungal pathogens belonging to the genera *Botrytis*, *Botryotinia*, and *Penicillium*. Fruit crops such as apple, citrus, grapes, pear, and strawberry are frequently affected by these diseases. *Botrytis cinerea* causing gray mold disease has a wide host range, encompassing several fruit, vegetable, and flower crops. The affected areas appear darker than the uninfected healthy portions of fruits. Whitish to gray mycelium and abundant spores are formed from the infected tissues. The internal tissues may turn dark green and water-soaked. Sclerotia may be formed on the infected fruits. In the case of kiwifruit, the adjacent fruits may be connected by the spreading mycelium forming "nests" of disease. The relationship between levels of infection by *B. cinerea* of pear stems and calyxes in the orchard during the growing season and the development of gray mold in storage was studied. Calyxes were susceptible to infection soon after bloom. However, inoculation of calyxes in April to May did not have any bearing on the levels of infection

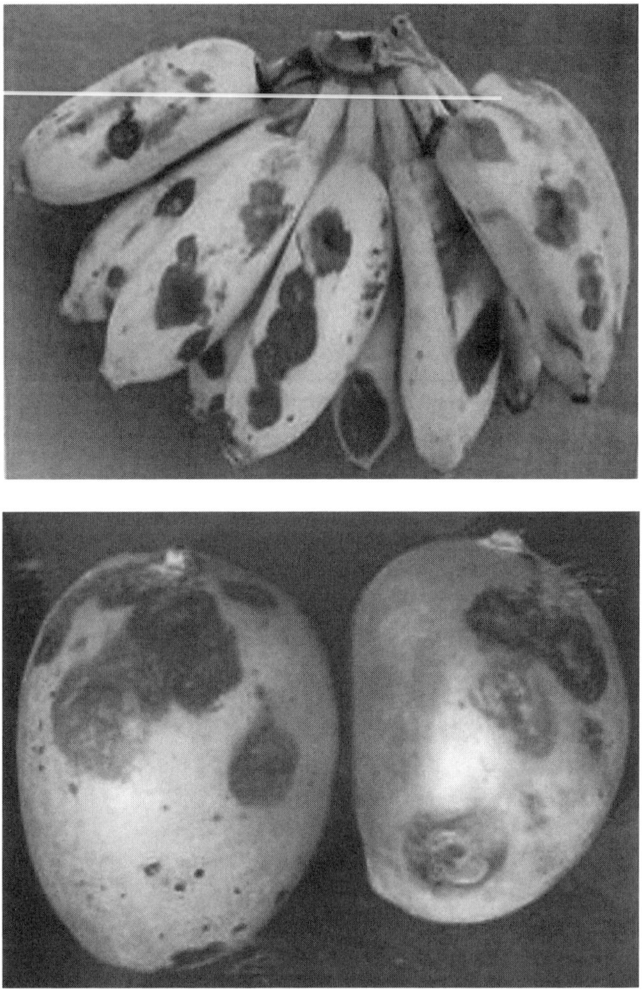

(a)

(b)

Figure 4.3 (a) Symptoms of anthracnose disease of banana. (Courtesy of S. Rajeswari and A. Palaniswami, Tamil Nadu Agricultural University, Coimbatore, India.) (b) Symptoms of anthracnose disease of mango. (Courtesy of T.K. Ashokkumar and A. Palaniswami, Tamil Nadu Agricultural University, Coimbatore, India.)

observed in storage, indicating that preharvest level of calyx infection cannot be used as an indicator of calyx gray mold in storage (Lennox and Spotts, 2004).

In pepper, the affected surface areas turn olive green with distinct margins and become slightly sunken and soft. A mass of gray powdery mold is formed on affected areas when the skin is broken. Tomato fruits infected by *B. cinerea* present initially with water-soaked and soft areas that become grayish or yel-

TABLE 4.1 Types of Symptoms on Fruits and Vegetables Caused by Fungal Pathogens

Type of symptom	Crop	Casual organism
1. Anthracnose	Avocado	*Colletotrichum acutatum*
		Colletotrichum gloeosporioides
		Glomerella cingulata
	Banana	*Colletortichum musae*
	Beans	*Colletotrichum lindemuthianum*
	Cucurbits	*Colletotrichum orbiculare*
	Grapes	*Elsinoe ampelina*
	Mango	*Colletotrichum gloeosporioides*
	Papaya	*Glomerella cingulata*
	Pepper (capsicum)	*Colletotrichum capsici*
	Pulses	*Colletotrichum lindemuthianum*
	Tomato	*Colletotrichum* spp.
2. Molds	Apple (Blue)	*Penicillium expansum*
	Aubergine (Gray)	*Botryotinia fuckeliana*
	Citrus (Blue)	*Penicillium italicum*
	Citrus (Green)	*Penicillium digitatum*
	Grapes (Gray)	*Penicillium digitatum*
	Kiwifruit (Gray)	*Botrytis cinerea*
	Pepper (Gray)	*Botrytis* spp.
	Pear (Blue)	*Penicillium* spp.
	Potato (Blue)	*Penicillium* spp.
	Strawberry (Gray)	*Botryotinia fuckeliana*
	Tomato (Grey)	*Botryotinia fuckeliana*
3. Rots	Apple – Bitter rot	*Glomerella cingulata*
	– Brown rot	*Monilinia fructigena*
	Avocado – Rhizopus	*Rhizopus stolonifer*
	– Stem-end rot	*Botryodiplodia theobromae*
		Phomopsis spp.
		Dothiorella gregaria
	Banana – Crown rot	*Botryodiplodia theobromae*
		Colletotrichum musae
		Fusarium palidoroseum
		Verticillium theobromae
	Citrus	
	– Alternaria black rot	*Alternaria citri*
	– Brown rot	*Phytophthora citrophthora*
	– Stem-end rot	*Diaporthe citri*
		Botryodiplodia theobromae
	Crucifers – Soft rot	*Rhizopus* spp.
	– Watery soft rot	*Sclerotinia sclerotiorum*
	Cucurbits – Black rot	*Didymella bryoniae*
	– Charcoal rot	*Macophomina phaseolina*
	– Fusarium rot	*Fusarium* spp.
	– Soft rot	*Rhizopus* spp.
	– Stem-end rot	*Botryodiplodia theobromae*

TABLE 4.1 (*continued*)

Type of symptom	Crop	Casual organism
	Grapes – Alternaria rot	*Alternaria* spp.
	– Black rot	*Guignardia bidwellii*
	– Bunch rot	*Botrytis cinerea*
	– Rhizopus rot	*Rhizopus stolonifer*
	Mango – Black mold rot	*Aspergillis niger*
	– Stem-end rot	*Botryoplodia theobromae*
	Onion – Black rot	*Aspergillus niger*
	– Neck rot	*Botrytis allii*
	– White rot	*Sclerotium cepivorum*
	Papaya – Blackrot	*Phoma caricae – papaya*
	– Ripe fruit/stem – end rot	*Botryodiplodia theobromae*
	Pepper – Alternaria rot	*Alternaria* spp.
	– Rhizopus rot	*Rhizopus* spp.
	Pear – Alternaria rot	*Alternaria alternata*
	– Bitter rot	*Glomerella cingulata*
	– Fruit rot	*Sphaeropsis pyriputrescens*
	– Late storage rot	*Nectria galligena*
	Pineapple – Brown rot	*Gibberella fujikuroi*
	– Fruitlet core rot	*Penicillium funiculosum*
	Potato – Dry rot	*Fusarium* spp.
	– Pink rot	*Phytophthora erythroseptica*
	– Tuber rot	*Corticium rolfsii*
	Pulses – Pod rot	*Thanatephorus cucumeris*
	– Watery soft rot	*Sclerotinia* spp.
	Stone fruit – Brown rot	*Sclerotinia* spp.
	– Rhizopus rot	*Rhizopus stolonifer* *R. nigricans*
	Strawberry – Rhizopus rot	*Rhizopus* spp.
	Sweetpotato – black rot	*Ceratocystis fimbriata*
	– blue – mold rot	*Penicillium* spp.
	– charcoal rot	*Macrophomina phseolina*
	– Dry rot	*Diaporthe batatis*
	– Java black rot	*Botryodiplodia theobromae*
	– Soft rot	*Rhizopus* spp.
	– Surface/end rot	*Fusarium* spp.
	Tomato – Alternaria rot	*Alternaria alternata*
	– Cladosproium rot	*Cladosporium herbarum*
	– Fruit rot	*Colletotrichum capsici* *Didymella lyopersici*
	– Phoma rot	*Phoma destructiva*
	– Rhizopus rot	*Rhizopus stolonifer*
	– Sclerotium rot	*Corticium rolfsii*
	– Soil rot	*Thanatephorus cucumeris*
	– Watery rot	*Geotrichum candidum*

TABLE 4.1 (*continued*)

Type of symptom	Crop	Casual organism
4. Blight	Potato – Late blight Tomato – Early blight – Late blight	*Phytophthora infestans* *Alternaria solani* *Phytophthora infestans*
5. Melanose	Citrus – Melanose	*Diaporthe citri*
6. Scab	Citrus	*Elsinoe fawcettii* *E. australis*
7. Scurf	Potato – Black surf – Silver scurf	*Thanatephorus cucumeris* *Helminthosporium solani*
8. Smudge	Onion – Smudge	*Colletotrichum circinans*
9. Smut	Onion – smut Potato – Smut	*Urocystis cepulae* *Thecaphora solani*
10. Spots	Citrus – Septoria spot Crucifers – Ring spot Pulses – Pod spot	*Septoria depressa* *Mycosphaerella brassicola* *Aschochyta* spp.
11. Wart	Potato – Wart	*Synchytrium endobioticum*

Source: Waller, 2002.

lowish green with lighter margins. Later, darker grayish growth of the pathogen is seen on the fruit surface. Ghost spots develop on young green fruits when the pathogen is killed after penetration of the fruit epidermis, due to exposure to warm sunny weather (Fig. 4.5A and B).

C. Rots: Fungal pathogens cause various types of rotting of infected fruits and vegetables, making them unfit for human consumption. Furthermore, the fungi, *Alternaria alternata*, *Penicillium expansum*, and *Stemphylium vericarium* infecting pear fruits produce the mycotoxin patulin in the infected tissues. The concentrations of patulin have been found to surpass the accepted maximum residue limit (MRL), even in apparently sound tissues. This report appears to be the first evidence indicating patulin diffusion to different depths of pear fruits (Ladou et al., 2001).

Colletotrichum gloeosporioides (Glomerella cingulata), causing bitter rot disease of apple and pear, can infect several other fruit crops, such as banana and mango. Apple fruits are infected under field conditions, but symptoms develop when the fruits begin to mature or ripen. The lesions enlarge under humid conditions and turn dark brown, developing sunken or saucer-shaped depressions. Fruiting structures (acervuli) appear near the center of the lesions, producing large number of spores in a creamy mass, salmon pink in color. They are frequently arranged in concentric circles. The rotten flesh beneath the surface of lesion is watery. *Monilinia fructigina* causing brown rot

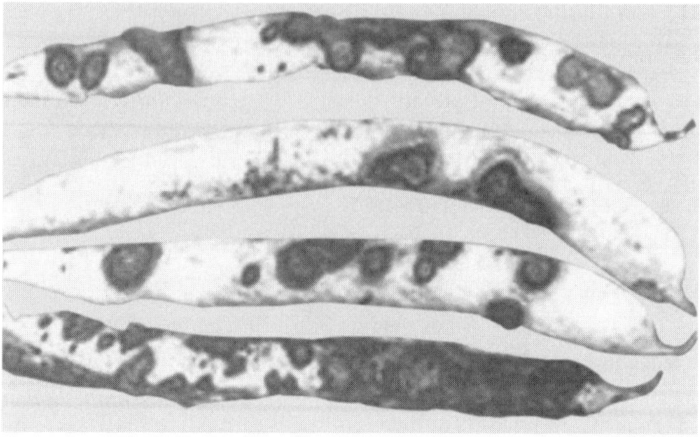

(a)

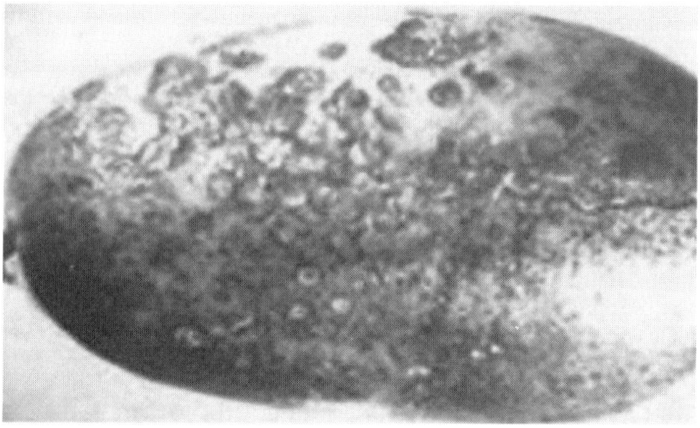

(b)

Figure 4.4 (a) Symptoms of anthracnose disease of beans. (Courtesy of Sherf and Macnab, 1986: John Wiley & Sons, Hoboken, NJ, USA.) (b) Symptoms of anthracnose disease of watermelon. (Courtesy of New South Wales Department of Agriculture, Rydalmere, Australia and Sherf and Macnab, 1986: John Wiley & Sons, Hoboken, NJ, USA.)

is a wound pathogen and infection may occur at both pre- and postharvest phases. The lesions on the fruits are firm, irregular in shape, and enlarge rapidly under conducive conditions, producing masses of brown oval conidia. Profuse white mycelium may be produced under moist conditions. The disease can spread by contact between sound and infected fruits under storage, increasing the disease incidence to high proportions (Edney, 1983). *C. gloeosporioides* infection of papaya fruit did not alter firmness, malic acid, and ascorbic acid concentrations and total soluble solids, but the appearance of peel and flesh was significantly affected (Ramos et al., 2001). A new postharvest fruit rot in d'Anjou pears caused by *Sphaeropsis pyriputrescens*, causing stem-end rot, calyx-end rot and wound-associated rot, has been observed. The decayed area on the fruit was firm or spongy and turned brown (Xiao and Rogers, 2004).

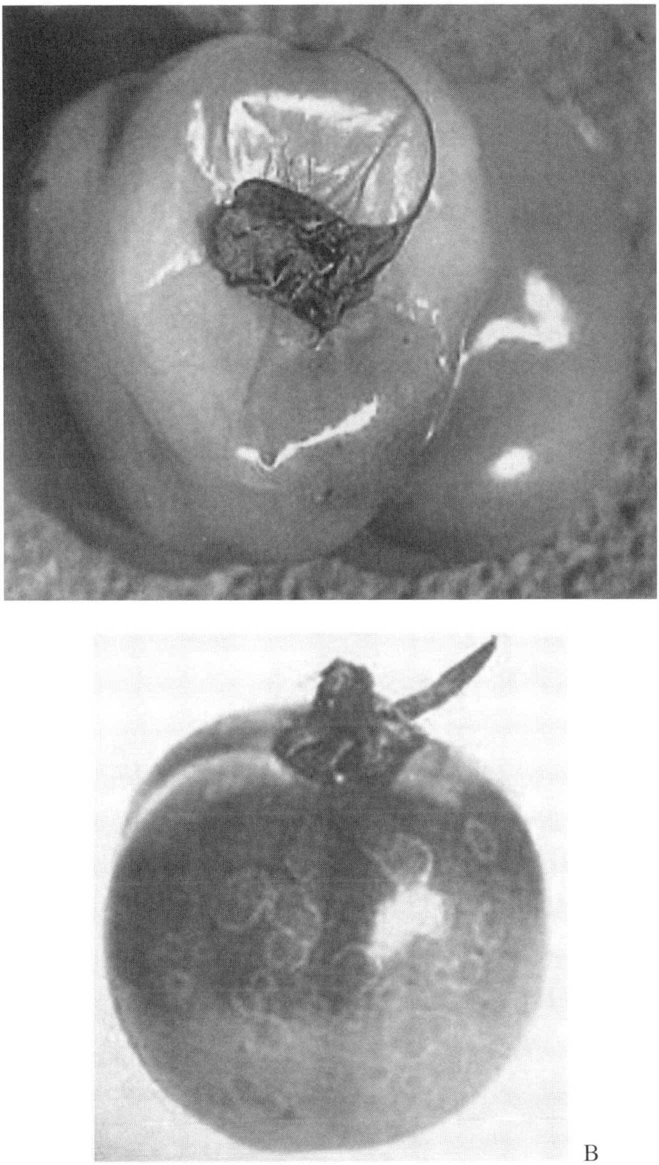

A

B

Figure 4.5 (a) Symptoms of gray mold disease of pepper (chilli). (Courtesy of Black et al., 1991; Asian Vegetable Research and Development Center, Taipei.) (b) Symptoms of gray mold (ghost spot) disease of tomato. (Courtesy of New South Wales Department of Agriculture, Rydalmere, Australia and Sherf and Macnab, 1986: John Wiley & Sons, Hoboken, NJ, USA.)

The stem-end rots caused by *Botrydiplodia theobromae* (*Diplodia natalensis*) and *Phomopsis citri* (*Diaporthe citri*) are important postharvest diseases of citrus. Softening of the rind around the button followed by brown discoloration of affected areas are the principal symptoms induced by *B. theobromae*. Development of decay at both ends of the fruit appears to be a characteristic feature of this disease. In the case of *P. citri* infection, the decay of softened tissues is rapid and the affected tissues turn tan or brown in color. It may be difficult to differentiate these two diseases based on the symptoms alone, necessitating the accurate identification of the pathogen involved. When ethylene was used at concentrations above the optimal level for degreening, a significant increase in disease incidence was observed. The increased activity of the abscission enzymes, polygalacturonase (PG) and cellulase (CX) was also associated with increased disease incidence (Brown and Burns, 1998). *Lasidiplodia theobromae* causes characteristic black spots on infected banana fruits. When the culture filtrate of *L. theobromae* was injected into the banana fruit peel, typical black spots were formed. The toxic compound was isolated both from the culture filtrate and infected banana fruit peel and identified as (3S.4R)-3-carboxy-2-methylene-heptan-4-olide. The results indicate that the role of the pathogen toxin in the production of disease symptom and its possible use for screening the banana cultivars for their resistance to *L. theobromae* (He et al., 2004)

Alternaria citri, causing Alternaria black rot disease in many citrus cultivars, infects at the stem-end of the fruit, resulting in black discoloration and maceration of the fruit core during storage (Brown and Eckert, 2000). *A. citri* produces extracellular enzymes that can degrade cell wall polymers during infection. The role of cell wall-degrading enzymes and their genes have been studied in pathogenicity, including penetration, maceration, nutrient acquisition, plant defense induction, and symptom expression. The endopolygalacturonases from *A. citri* (endoPG 3.2.1.15) that can degrade pectic polysaccharides in plant cell walls were purified and the gene *Acpg1* from *A. citri* and *Aapg1* from *A. alternata* causing brown rot in citrus were cloned (Isshiki et al., 1997; Isshiki et al., 2001). The endoPG-disrupted mutants of *A. citri* had reduced ability to macerate plant tissue and cause black rot symptoms. The plasmid pTEFEGFP carrying a green fluorescent protein (GFP) gene was introduced into wild type *A. citri* and its endoPG-disrupted mutant (M60). Green fluorescence was detected in spores, germ tubes, appressoria, and infection hyphae of transformants G1 (derived from wild type) and GM4 (derived from M60). The hyphae of G1, but not GM4 could vertically penetrate the peel, but the hyphae of both G1 and GM4 spread equally in juice sac area of citrus fruit. EndoPG activity was induced by the pectin present in the peel, but repressed completely in the juice sac area, possibly due to carbon catabolite repression by sugars in the juice (Isshiki et al., 2003).

Grapevine bunch rot disease caused by *Botrytis cinerea* is characterized by the development of soft brown rot in immature berries in bunches. During favorable environmental conditions, gray to buff-colored spore masses are formed. During cold-storage, small water-soaked spots appear on mature berries which soften and turn brown. As the rot spreads, gray-buff-colored tufts of fungal growth emerge from splits in the skin of infected berries. In compact bunches, the rot may spread rapidly from berry to berry until entire bunches are rotted with a matted gray velvet-like appearance. Under dry conditions, berries dry up and become raisin-like, but remain attached to bunches. This disease is reported to adversely affect the must and wine quality appreciably, in addition to reduction in yield. Wines produced with infected grapes had consistently lower color capacity, a yellowish tint, and had lower anthocyanin and total phenol contents. The reduction in sugar content as reflected by density and Brix was the important effect of bunch rot disease (Meneguzzo et al., 1999; Pszczolkowski et al., 2001). The foaming properties of Champagne wines are important in terms of product attractiveness for the consumer. Dramatic effects of Champagne foam characteristics were observed when grape berries were severely infected by *B. cinerea*. Foamability suffered considerably (60–65%) when *B. cinerea* infection was about 20% compared with control (Marchel et al., 2001). The effects of infection of grapevine by powdery mildew disease caused by *Uncinula necator* on yield, juice, and wine quality were quantified for the cv. Cabernet Sauvignon (Csa). An average weight reduction (by 12 to 20%) and increase in total acidity following infection was evident. In Csa wines, the concentration of 3-mercapto hexanol, a component of varietal aroma was decreased by the powdery mildew disease. The quality of wines from berries with more than 30% infection was very poor (Calonnec et al., 2004). Blackrot is another important disease caused by *Guignardia bidwellii*. Berry infection is initiated by ascospores released from infected plant parts. Small, circular reddish brown lesions are formed on the berries and they develop a black border. Small black pimple-like pycnidia are produced in the reddish brown part of the lesions. On green mature fruits, whitish circular spots appear at the point of infection. As the disease progresses rapidly, the berries begin to shrivel and turn into hard bluish raisin.

Penicillium expansum, *P. digitatum*, and *P. italicum* cause postharvest diseases in a wide range of deciduous and tropical fruits. They cause blue and green mold diseases in apples, citrus, and pear fruits. Infected apple tissues become pale in color and watery in texture. With high humidity, blue-green cushion of fruiting structures bearing millions of conidia make their appearance on the fruit skin. The conidia may initiate infection on sound fruits by contact with infected fruits (Fig. 4.6 A and B). These pathogens elaborate significant amounts of hydrolytic enzymes, including polygalacturonases (PGs) occurring predominantly in many pathosystems. PG has been shown to be important for virulence of several fungal pathogen. However, in some cases, disruption of cell wall-degrading enzymes caused only partial or no reduction in pathogenicity, indicating that some of the enzymes may be required for pathogen growth but not for pathogenicity (Narayanasamy, 2002). The pathogens may be able to enhance their virulence by locally modulating the host's ambient pH as in the case of *Colletotrichum* spp. The genes encoding

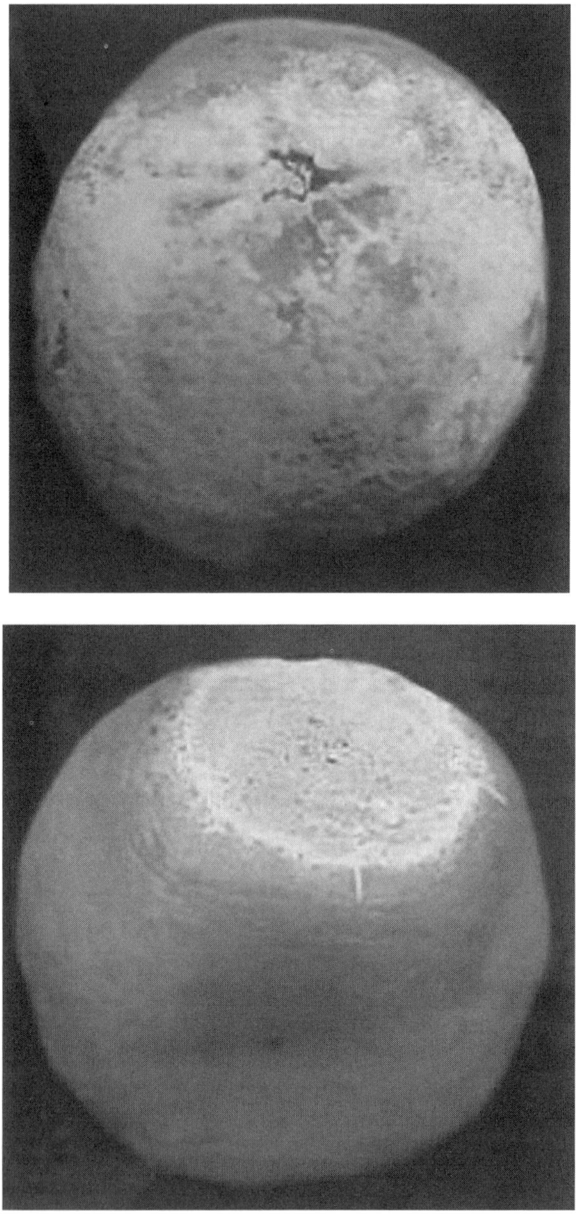

A

B

Figure 4.6 (a) Symptoms of green mold disease of mandarins. (b) Symptoms of blue mold disease of mandarins. (Courtesy of T. Ashokkumar and A. Palaniswami, Tamil Nadu Agricultural University, Coimbatore, India.)

cell wall-degrading extracellular enzymes may be expressed and their products secreted under the optimal pH conditions for their activity (Prusky et al., 2001).

The postharvest pathogens *P. expansum*, *P. digitatum*, and *P. italicum* acidify the ambient environments of apple and citrus fruits during decay (rot) development via two mechanisms, that is by producing organic acids, mainly citric and gluconic acids, and/or by utilizing the $NH4^+$ associated with H^+ efflux. *P. expansum* and *P. digitatum* produced significant amounts of citric and gluconic acids in decayed tissue and the pH of host tissue was reduced by 0.5 to 10 units (Fig. 4.7; Table 4.2). Reduction in ambient pH was directly correlated with ammonium depletion from the growth medium or from the fruit tissue. Acidification seemed to be a pathogenicity enhancing factor, since the transcripts encoding endoPG gene *pepg1* of *P. expansum* accumulated under acidic conditions from pH 3.5 to 5.0. The results suggested that ambient pH was a regulatory one for processes linked to pathogenicity of postharvest pathogens and specific genes may be expressed as a result of the modified host pH created by pathogens (Prusky et al., 2004).

Potato tubers are infected by fungal pathogens resulting in various types of rots. The characteristic symptoms of dry rot disease caused by *Fusarium solani* and *F. roseum* include formation of sunken areas of brown, firm rot, frequently involving large portions of tubers. The surface of the infected areas become sunken, wrinkled, and often with blue or white protuberances. In contrast, prink rot (watery rot) caused by *Phytophthora erythroseptica* induces oozing of liquid from infected areas of the tubers and the tissues become flaccid and rubbery. A sharp line demarcating healthy and infected tissues may be seen in the cut tubers. The affected tissue, when exposed, turns pink, gradually becoming black after some time. Invasion of infected tubers by soft-rot bacteria is usually seen.

In tomato, *Alternaria alternata* causes black rot disease which induces small brown to black spots (2–3 cm in diameter) with a sunken central portion as the spots enlarge. The spots coalesce, turning into ivory-black areas which have inner soft tissues with a dark brown to black cylindrical dry core. A dense velvety olive green to black conidial mass on gray superficial moldy fungal growth may be observed. Watery (sour) rot disease due to *Geotrichum candidum* is responsible for huge losses of the produce. Wounds and cracks on the ripe fruits form the entry points and water-soaked areas rapidly develop. The affected tissues become soft, exhibiting superficial white scum-like growth emitting a characteristic sour odor and considerable amount of juice oozes out through diseased areas. The symptoms of infection of tomato fruits by *Phoma destructiva* appear usually at the stem scar. Brown to black spots develop on green fruits. As fruits ripen, the spots enlarge rapidly and become leathery. The rot spots have sunken, dark central areas with light colored margins. Dark pycnidia (fruiting bodies) develop in the dark areas of the spots. *Rhizopus stolonifer* may infect tomatoes at any time during transit and storage and enter the fruits through wounds or cracks. The lesions are generally large and indis-

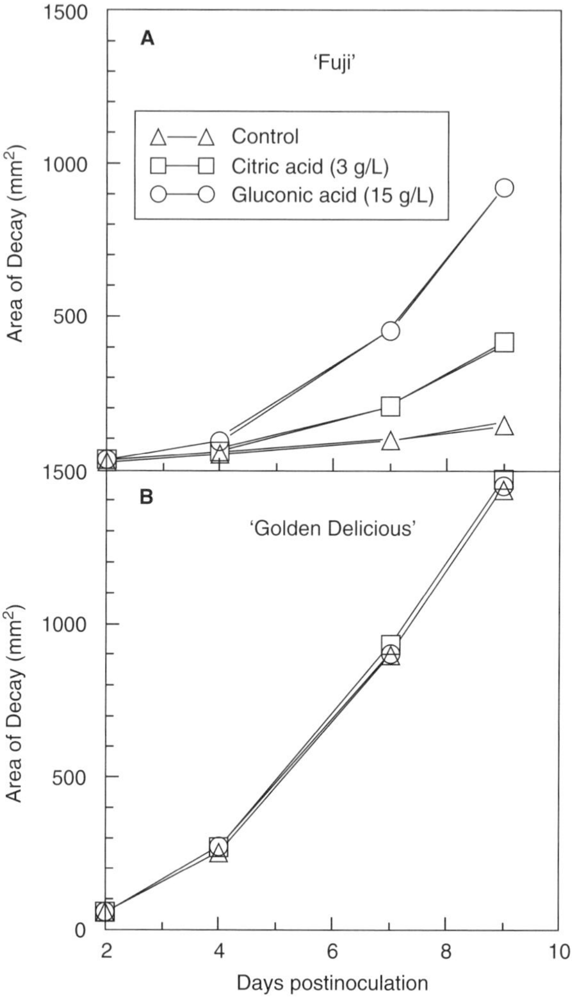

Figure 4.7 Influence of citric and gluconic acid treatment on the virulence of *Penicillium expansum* in apples. (Courtesy of Prusky et al., 2004; The American Phytopathological Society, St. Paul, MN, USA.)

tinct, affected area becoming puffy. As water-soaked areas becoming visible, rot develops deeply and rapidly involving the entire fruit. The fruit tissue ruptures, ultimately collapsing into a wrinkled mass of soft tissue covered by gray-black fungal growth. Emission of fermented odor is a characteristic feature of this disease.

Sweetpotatoes are subjected to various rots caused by fungal pathogens. *Ceratocystis fimbriata* causing black rot disease reduces both yield and quality

TABLE 4.2 Levels of Organic Acids Present in Healthy and Decayed 'Oro Blanco' Grapefruit and 'Golden Delicious' Apple Inoculated with *Penicillium Digitatum* and *P. Expansum* Respectively

Tissue	Concentration of organic acids (μg/fresh weight)			
	Citric	Fumaric	Ascorbic	Gluconic
Grapefruit				
Healthy	5979 ± 234	0	42 ± 41	6320 ± 2470
Diseased	9097 ± 362	0	473 ± 520	15,832 ± 2980
Apple				
Healthy	591 ± 280	0	4 ± 30	0
Diseased	970 ± 256	6 ± 1	0	1586 ± 165

All measurements were repeated four times from the diseased lesions of different fruits.
Source: Prusky et al., 2004.

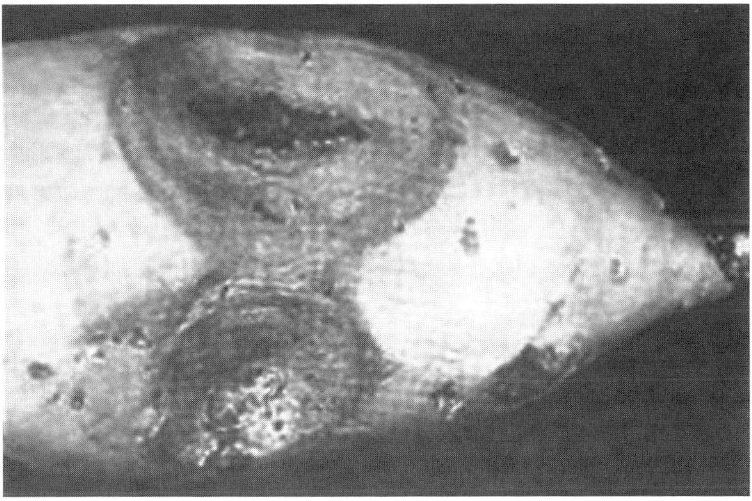

Figure 4.8 Symptoms of black rot disease of sweetpotato. (Courtesy of Sherf and Macnab, 1986; John Wiley & Sons, Hoboken, NJ, USA.)

of sweetpotatoes, giving a bitter taste. Small, circular, slightly sunken, dark brown spots are formed initially and the affected areas enlarge and turn greenish black when wet and oil grayish black when dry. The rot generally remains firm and shallow. The tissue near discolored areas may have a bitter taste and ultimately the entire root may rot (Fig. 4.8). Java rot caused by *Botryodiplodia theobromae* is one of the most important disease causing losses both in the field as well in storage. The disease may progress from one or both ends of the root. The affected tissues turn yellowish first, then reddish brown, and finally black as the decay advances. The affected areas are firm and moist. The infected root decays entirely within a period of 2 weeks and they become dry, mummified, and extremely hard. Decay in storage is usually restricted to the tip of the root (1 to 2 cm) and the tissue at the center of the lesion is usually black and hard. A gradual decline in the contents of starch and ascorbic acid in diseased tubers was observed. But no change in total sugar and proline contents could be detected following infection by *B. theobromae* (Pati and Ray, 2000). Charcoal rot disease caused by *Macrophomina phaseolina* is initiated as a reddish brown, firm, moist rot commencing just beneath the sweetpotato skin. As the decay progresses, the central tissues are invaded, causing further rot. The affected tissues may be differentiated by the leading edge with reddish brown area and a zone of black actively decaying black tissues. The entire root may be affected resulting in the development of dry, hard, and mummified root.

D. Blights: Potato tuber infection by *Phytophthora infestans* causing late blight disease results in small, slightly sunken, dark spots. Under warm and moist environments, the affected areas shrink rapidly and become darker. A foul-smelling rot is frequently seen, when secondary infection by other rot-causing organisms occurs (Fig. 4.9). Early blight disease caused by *Alternaria solani* in tomato starts as brown spots on pedicels and calyx attached to blossoms or fruits. On the fruit, characteristic brown to black spots appear at the stem-end and rot radiates out from the area of attachment between the calyx and the fruit. The affected areas become firm and develop depressions and distinct concentric rings of discolored portions.

E. Melanose: The fungal pathogen *Diaporthe citrii* infects leaves and twigs, in addition to citrus fruits, in all citrus growing areas. The other infected plant organs are the sources of infection for the fruits. Small, raised, brown spots are

Figure 4.9 Symptoms of tuber rot disease of potato. [Courtesy of Rich, 1983; (Academic Press, New York, USA) Elsevier, Oxford, United Kingdom.]

produced on the fruits. As the infected areas enlarge they turn rough, giving an unsightly appearance and reducing the marketability of the produce.

F. Scab: The citrus scab disease caused by *Elsinoe fawcettii* and *E. australis* affect leaves and twigs which provide inoculum for fruit infection. Raised reddish or yellowish irregular spots appear. They turn scabby or corky, soon coalescing to cover larger areas of the fruits. The fruit surface becomes rough and distorted. The scab stroma on young lemon fruits are initially pink to light brown. As more hyperplastic cells are formed, the spots turn grayish brown and more corky.

G. Scurf: *Rhizoctonia solani* (*Thanatephorus cucumeris*) causing black scurf disease, grows superficially on tuber surface and dark chocolate-colored hard fungal growth containing numerous sclerotia, is produced. Although the internal tissues are not infected, the market value of the tuber is greatly reduced. In contrast, *Helminthosporium solani* causing silver scurf disease, can penetrate into the tuber through the periderm or through natural openings on the surface of the tuber. Primary infection is generally restricted to tuber skin at the stem-end as small gray and silver sheen. Infection of red skinned potatoes may lead to partial or complete discoloration of the skin. In severe cases the affected areas shrivel, resulting in shrinkage of tissues of the tubers, as a major portion of tuber surface shows black circular lesions which coalesce disfiguring the tubers. The spores produced in these lesions infect tubers during storage (Fig. 4.10).

H. Smudge: Onion smudge disease caused by *Colletotrichum circinans* infects onions in the field prior to harvest and continues to develop during the storage. Small dark green or black spots are seen on the outer scales of bulbs. The fruiting structures (acervuli) are formed as black dots in concentric circles. Sometimes the black discoloration follows the veins of scales or may cross transversely. In the case of severe infection, the pathogen may penetrate into the outer dry scales reaching the living tissue resulting in the collapse of fleshy scales. Generally red skinned onions show resistance to the disease (Fig. 4.11).

I. Smut: The fungal pathogens *Urocystis cepulae* and *U. colchici* cause clearly visible streaks within the tissues of leaf, leaf sheaths, and bulbs of onion. Dark brown, powdery mass of spores are formed along these streaks. The smut pustules may provide points of entry for other organisms causing soft rot symptoms. Potato smut disease causes serious yield losses (≈80%) in the Andean region of South America. The development of galls specifically located on the lower stems, stolons, and tubers is the primary symptoms of the disease. On tubers, galls (1 mm to 4 cm or more in diameter) enclosing oval to irregular locular sori of variable sizes containing a reddish-dark, granular-powdery mass of teliospores are produced (Mordue, 1988). The quality of tubers is markedly reduced in addition to quantitative loss due to the disease. The pathogen

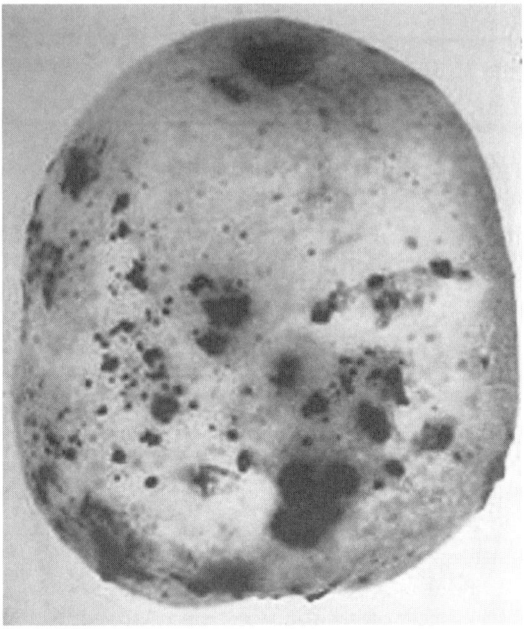

Figure 4.10 Symptoms of black scurf disease of potato. [Courtesy of Rich, 1983; (Academic Press, New York, USA) Elsevier, Oxford, United Kingdom.]

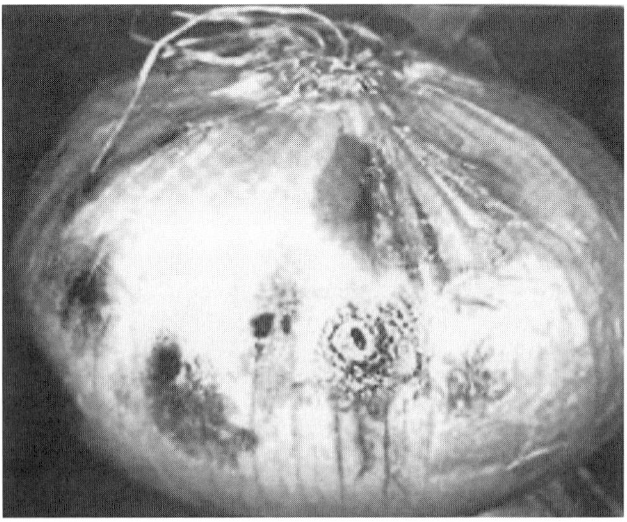

Figure 4.11 Symptoms of smudge disease of onion. (Courtesy of Sherf and Macnab, 1986; John Wiley & Sons, Hoboken, NJ, USA.)

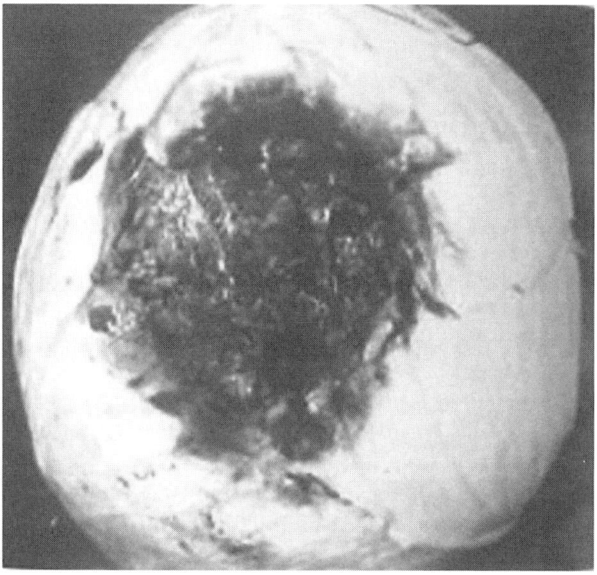

Figure 4.12 Symptoms of Alternaria spot disease of cabbage. [Courtesy of Geeson, 1986; (Academic Press) Elsevier, Oxford, United Kingdom.]

Thecaphora solani has recently been cultured in artificial medium for the first time and its identity was established by DNA fingerprinting and partial sequencing of the large subunit (LSU) rDNA region of the pathogen genome (Andrade et al., 2004).

J. Spots: Leaf spot caused by *Alternaria brassicae* and *A. brassicola* and ringspot due to *Mycospherella brassicola* are commonly observed on cabbage and cruciferous vegetables (Fig. 4.12). The disease due to *Alternaria* spp. begins as small, circular, yellow areas which later enlarge in concentric circle, assuming a black, sooty color. In cases of severe infection, numerous spots appear, coalescing to cover large areas of the leaves. This may necessitate heavy trimming before cabbage is placed in storage. Infection by *M. brassicola* is characterized by the appearance of snow-white specks arranged in concentric rings. The spore-bearing pycnidia produce spore masses which exude in tendrils and appear raised until they become wet and spread out to form a white film. The affected petioles become hard and brown, often splitting longitudinally, resulting in twisting and distortion of leaves.

K. Wart: *Synchytrium endobioticum* causing potato wart diseases induces the infected cells of tubers to proliferate (hyperplasia), resulting in the formation of protuberances of different sizes. The warts are greenish-white initially, turning dark brown or even black later as they grow in size, sometimes becoming larger than the tuber itself. The wart consists of distorted and proliferated, branched structures of altered tuber tissue, and they become a mass of putrefied tissue when invaded by secondary saprophytic microbes.

4.3.2 Process of Infection by Bacterial Pathogens

The postharvest spoilage by bacterial pathogens occurs more frequently in vegetables than in fruits. Cucumbers, peppers, and tomatoes are more frequently infected by bacterial pathogens compared with fruit such as apples, mangoes, and bananas, probably because of the higher pH values within the fruity vegetables (4.5–7.0), whereas the low pH (usually <4.5) of most fruits may inhibit the development of bacterial pathogens.

4.3.2.1 Entry into Host Tissues The bacterial pathogens of fruits and vegetables do not penetrate the intact plant surface. But they can enter through natural openings such as stomata, hydathodes, and lenticels (Billing, 1982). Bacteria do not form specialized structures to initiate the infection process, as in the case of fungal pathogens. Stomata present in cucumber provide the points of entry for *Pseudomonas lachrymans* (Wiles and Walker, 1951). There appears to be no special mechanism of entry through stomata. The bacteria seem to gain entry by a simple, passive process of moving through the film of water contaminated with bacteria. The conformation of the cuticle around the stomata in citrus fruits appears to be a critical factor in preventing or allowing the passage of water droplets contaminated with *Xanthomonas axonopodis* pv. *citri* to initiate the process of infection. The connective tissue at the stem end was reported to be an entry point for the bacteria present in the sepals of young tomato fruit (Samish and Etinger-Tulczynska, 1963). Lenticels formed by loosely packed cells present in the peridium of potato tubers provide entry points for *Streptomyces scabies* causing common scab disease and *Erwnia carotovora* subsp. *atroseptica* causing black leg disease. The soft rot pathogen *Erwinia carotovora* subsp. *carotovora* enters the tubers through enlarged lenticels which result from excess soil moisture and poor aeration in the field.

The bacterial pathogens are primarily wound invaders. Injuries or wounds caused during harvesting, handling, and storage, as well as damage caused by various agents such as fungi, nematodes, animals, birds, rain, and hail, may allow entry of bacterial pathogens. The bacterial pathogens, such as *Erwinia carotovora* subsp. *carotovora* in potato and *E. chrysanthemi* in sweetpotato, gain entry into the tubers through wounds. The damage due to frost may also help the entry of bacterial pathogens as in the case of potato soft rot disease. After inoculation of cabbage with *X. campestris* (NRRL B-1459) chromosomally tagged with the green fluorescent protein (*gfp*), the infection process was studied visually. Using a fluorescence microscope, the migration and distribution of *gfp*-labeled cells could be easily visualized from the commencement of bacterial pathogen cells in cabbage tissues to the appearance of visible necrosis (So et al., 2002). Some isolates of *Pseudomonas fluorescens* have inducible pectinolytic activity and can cause soft rot disease in refrigerated plant products. *P. fluorescens* strain 123 isolated from celery (*Apium graveolens* cv. Dulce) caused soft rot of cut vegetables such as broccoli, cauliflower, and lettuce, since this strain is able to develop well at 4°C, the temperature at which, refrigerated cut vegetables are generally shipped. *P. fluorescens* strain 123, when subjected to mini-*Tn5* mutagenesis, formed six mutant phenotypes with altered pathogenicity and inability to produce a surfactant involved in dispersing *P. fluorescens* cells in a surface aqueous environment. The mutant *EG3*, with reduced pathogenicity on broccoli and cauliflower, reacquired its pathogenicity when complemented with a 7.1-kb segment of *P. fluorescens* genomic DNA. As the mutants lost the aqueous surface-tension activity, it is suggested that a biosurfactant may contribute to pathogenicity, enhancing invasion of host plant tissues by the bacterial pathogen (Hernandez-Anguiano et al., 2004).

4.3.2.2 Colonization of Host Tissues Bacterial pathogens cannot grow through compact plant tissues as such. The pectic enzymes produced by the bacteria accelerate breakdown of pectic substances, such as calcium pectate required for keeping the plant cells intact. *Erwinia* spp., causing soft rot diseases, macerate the plant tissues by producing pectinesterase and lyases, and hydrolases. Cellular breakdown and subsequent release of nutrients from plant cells facilitate intercellular colonization. Several bacterial pathogens are known to secrete large amounts of extracellular polysaccharides (EPS) both in culture and in planta during pathogenesis. The production of polysaccharides in culture has been related to the virulence of the bacterial species/strains. *E. chrysanthemi* producing pectate lyase and pectate hydrolase was able to macerate the plant tissue, whereas a mutant strain lacking the ability to secrete pectate lyase was unable to macerate the plant tissue (Chatterjee and Starr, 1977). The pectate lyases (Pels) produced by *E. chrysanthemi* designated PelA, PelB, PelC, and PelE are important virulence factors involved in degradation of plant cell walls (Preston et al., 1992). The cell wall-degrading enzymes form the principal virulence factors of *E. carotovora* subsp. *carotovora* and their synthesis is coordinately regulated by a complex network (Eriksson et al., 1998).

Virulence of *Erwinia carotovora* subsp. *carotovora* (Ecc) causing soft rot disease depends on the production of main virulence determinants, controlled by a complex regulatory network. Transcriptional activation of the extracellular enzyme genes of *Ecc* is regulated by one of the global regulators, the response regulator *ExpA*, a GacA homolog. A central role of the RsmA–rsmB regulatory system during pathogenesis in integrating signals from the *ExpA* (GacA) and *KdgR* global regulators of extracellular enzyme production in *Ecc* has been indicated by (Hyytiäiens et al. 2001). In order to identify and characterize Nip (necrosis-inducing virulence proteins) of *E. carotovora* subsp. *carotovora* (*Ecc*), different isolates of the pathogen were analyzed by Southern blot using a harpin-encoding gene *hrpN* as probe. Several isolates of *Ecc* did not have the *hrpN* gene. Regulation of virulence determinants in one strain SCC 3193 was extensively characterized and it was fully virulent on potato and in *Arabidopsis thaliana*. The mutant RpoS of SCC 3193, capable of secreting high amounts of proteins, caused necrotic lesions resembling a hypersensitive response. The protein inducing cell death was purified and sequenced, followed by cloning of the corresponding gene. The Nip exhibited homology to necrosis- and ethylene-inducing elicitors of fungi and Oomycetes. The mutant (Nip^-) had reduced virulence on potato tuber, but not on potato stem tissue or other host plant species tested (Mattinen et al., 2004). The production of pectinase is controlled by many regulatory factors. The major regulatory proteins KdgR, CRP, Pir, and PecS, mainly from *E. chrysanthemi*, have been characterized. Homologs of KdgR and Crp, but not Pir and Pecs, were detected by Southern blot analysis in *Ecc* when plasmids containing the regulatory genes from *E. chrysanthemi* were introduced into *Ecc*. Pel production was controlled as predicted from their roles in *E. chrysanthemi*, except for PecS. The results suggest that KdgR and CRP homologues of *Ecc* may regulate Pel and Peh production as in *E. chrysanthemi* (Matsumoto et al., 2003b). In another study, a new regulatory factor involved in the virulence was identified. The (Delta) *cytR* mutant produced low concentrations of polygalacturonase (Peh) and had no motility. The virulence of the mutant was reduced in potato and Chinese cabbage compared to wild-type strain. The results suggested that the CytR homolog of *Ecc* positively controlled Peh production and synthesis of flagella in addition to having an important role in its pathogenicity (Matsumoto et al., 2003a).

Ralstonia solanacearum produces three extracellular polygalacturonases (PGs), PehA, PehB, and PehC, which hydrolyze pectin's polygalacturonic acid backbone. However, each enzyme releases different reaction products. PehA and PehB contribute significantly to the virulence of this bacterial pathogen, possibly by invasion and colonization of susceptible tissues. PehC may degrade polygalacturonide elicitors of host defense, resulting in protection of the pathogen against antimicrobial responses of the host plant. It is considered that galacturonic acid released by pathogen PGs from plant cell walls provide nourishment to the pathogen during pathogenesis. The *exuT* gene, a galacturonate transporter gene, is cotranscribed with *pehC*. The *exuT* mutant produced all three isozymes of PG, but could not take up PG degradation products. Tomato was susceptible to the *exuT* mutant to the same level as the wild type. The results suggested that successful establishment of the bacterial pathogen is not primarily due to the metabolism of polygalacturonic acid (González and Allen, 2003).

The genes activated by the regulatory gene *hrp* (hypersensitive response and pathogenicity) have been shown to be important for the ability of *R. solanacearum* strain GMI 1000 to successfully infect potato and other crop plants. *HrpB* controls the expression of the type III secretion system (TTSS) and pathogenicity effectors transiting through this pathway. Direct biochemical evidence showed that five *R. solanacearum* effector proteins were translocated into plant host cells through TTSS. These novel TTSS effectors, RipA and RipG, belong to multigenic families, RipG defining a novel class of proteins with leucine-rich repeats (Cunnac et al., 2004). In another study, *hrpB*-regulated genes in *R. solanacearum* were screened. By employing a transposon-based system, 30 novel *hpx* (*hrpB*-dependent expression) genes outside the *hrp* gene cluster were isolated. A plant-inducible promoter (PIP) box-like motif was present in the promoter regions of most of the *hpx* genes. Seven *hpx* genes encoded homologs of known type III effectors and type III-related proteins present in other plant pathogens. Furthermore, two novel *hrpB*-regulated genes, *hpaZ* and *hpaB* downstream of *hrpY* in the *hrp* cluster, were also identified. The *hpaB* gene, but not *hpaZ*, was essential for both the pathogenicity of and ability to induce hypersensitive reaction on plants by *R. solanacearum* (Mukaihara et al., 2004).

The pathogenicity of, and elicitation of resistance by, bacterial pathogens such as *Xanthomonas campestris* pv. *vesicatoria* may be governed by *hrp* genes. Secretion of pathogenicity factors as well as avirulence proteins, which induce resistance in incompatible interactions, are controlled by *hrp* genes (Bonas et al., 2000). Production of a heat-stable, protease-sensitive elicitor produced by virulent strains of *Clavibacter michiganensis* subsp. *sepedonicus*, causing potato ring rot disease, was demonstrated by Nissimen et al. (1997). Translocation of bacterial effector proteins by the type III secretion (TTS) system determines the pathogenicity of *X. campestris* pv. *vesicatoria* (*Xcv*). A genome-wide regulon of putative virulence genes that are coexpressed with TTS system has been identified using transcriptome analysis. The genes *xopC* and *xopJ* encoding outer proteins (Xops) were secreted by the TTS system. The translocation into the host plant cell of these proteins was demonstrated using the BvrBs3 effector domain as reporter (Noël et al., 2003). In a further study, the *hpaB* (*hpa, hrp*-associated) gene, which encodes a pathogenicity factor with typical features of a TTS chaperon, was characterized. *HpaB* was required for efficient secretion of at least five effector proteins, but not for the secretion of noneffectors, such as XopA and the TTS translocation protein HrpF. HpaB interacted with two unrelated effector proteins, AvrBs1 and AvrBs3, but not with XopA. The results indicated that HpaB might play a pivotal role in the exit control of the TTS system (Büttner et al., 2004).

Pseudomonas syringae pv. *tomato* DC 3000 (*Pst*) is believed to depend on an Hrp secretion system (type III) for its ability to develop in planta. Multiple effector proteins are injected by this secretion system into plant cells. ORF5 in the conserved effector locus of *Pst* Hrp pathogenicity island has been reported to encode an Hrp-secreted protein and the presence of a similarly secreted homolog encoded in an effector-rich pathogenicity island has also been detected elsewhere in the pathogen genome. These putative effector genes were designated *hop Pto A1* and *hop Pto A2* respectively. The strains such as DC 3000 seem to have two copies of the gene *hop Pto A1*, which is widespread among pathovars of *P. syringae*. The results suggest that these effector genes may contribute redundantly to the formation of colonies in infected plant tissues (Badel et al., 2002). Constitutive expression of a neutral leucine aminopeptidase (LAP-N) and an acid LAP (LAP-A) during floral development and in leaves of tomato in response to infection by *Pst* has been observed. LAP-A expression was monitored after infection with coronatine-producing antideficient *Pst* strains. LPA-A RNA and its activity could be detected only in tissues infected by the coronatine-producing *Pst* strain (Pautot et al., 2001).

The pathogenicity of *Pst* DC 3000 depends on both the type III secretion system, which delivers virulence effector proteins into the host cells, and the phytotoxin coronatine (COR), which is considered to mimic the action of jasmonic acid (JA). Sensitivity to JA seems to be a critical factor in the response of tomato genotypes to *Pst* DC 3000. A JA-insensitive mutant (jai 1) of tomato was unresponsive to COR and highly resistant to *Pst* DC 3000, while the tomato genotypes defective in JA biosynthesis were as susceptible to the pathogen as the wild-type (WT) tomato plants. COR was considered to act through *Jai 1* to induce the massive expression of JA and wound response genes, as deduced from the analysis of host gene expression cDNA microarrays. Under the physiological condition of the host cells, the type III secretion system and COR repressed the expression of PR genes in *Pst* DC 3000-infected WT plants. A high level of PR-gene expression and reduced expression of JA / wound response genes are associated with the resistance of jai1 plants to *Pst* DC 3000. COR promoted bacterial virulence by activating the JA signaling pathway of the host. The type III secretion may probably modify the host defense by targeting the JA signaling pathway in susceptible tomato plants (Zhao et al., 2003).

Analysis of the whole genome sequence of *Pst* DC 3000 identified a locus, designated *HopPtoF* which was homologous to the avirulence gene locus *avrPphF* in *P. syringae* pv. *phaseolicola*. The *HopPtoF* locus had two genes, *ShcF(Pto)* and *HopF (Pto)*. The gene *Shc F (Pto)* interacted with and stabilized the HopF (Pto) protein in the pathogen cells (Shan et al., 2004). The compatible and incompatible interaction of *Pst* DC 3000 with tomato and *Nicotiana spp.* respectively requires Hrp type III secretion system (TTSS) which injects Hop (Hrp outer protein) effectors into plant cells. HopPtoN was translocated into tomato cells via Hrp TTSS. Large numbers of necrotic 'speck' lesions (eight fold) were produced by a *hopPtoN* mutant compared to *Pst* DC 3000, although both the mutant and the wild type (WT) strain developed similarly on tomato leaves. The mutant of hopPtoN induced more cell death in the nonhost (*N. tabacum*) leaves. On the other hand, a DC 3000 strain overexpressing HopPtoN induced less necrosis, in addition to lower levels of electrolyte leakage, in comparison with the WT strain. HopPtoN exhibited cysteine protease activity in vitro. The results show that HopPtoN is a TTSS effector that can suppress plant cell death events in both compatible and incompatible interactions (López-Solanilla et al., 2004).

Pseudomonas syringae pv. *phaseolicola* produces a non-host-specific toxin, phaseolotoxin, which is toxic to beans. However, the pathogen remains unaffected by the toxin because of its ability to synthesize ornithine carbamoyl-transferase (ROCT), which protects the bacterial pathogen. ROCT encoded by *argK* was expressed coordinately with phaseolotoxin synthesis at 18°C. The thermoregulation of *argK* was lost at 28°C, apparently due to accumulation of carbamoyl phosphate, one of the substrates of phaseolotoxin-sensitive OCTase. Carbamoylphosphate, with a structure similar to that of the inorganic moiety of phaseolotoxin, was able to induce *argK* expression in M9 medium at 28°C. The results indicated that *argK* expression was regulated directly by a compound resembling phaseolotoxin and indirectly by temperature variation (Lopéz-Lopéz et al., 2004).

The isolates of *P. syringae* pv. *phaseolicola* can be separated into two genetic lineages based on their ability to produce the toxin (ToX^+) or not (ToX^-). The ToX^- isolates occurring in Spain seemed to lack the entire DNA region for the biosynthesis of phaseolotoxin (*argK-tox* gene cluster), as revealed by PCR amplification and DNA hybridization using DNA sequences specific for separated genes of this cluster. Furthermore, TOX^+ and TOX^- isolates exhibited genomic divergence that included differences in enterobacterial repetitive intergenic consensus (ERIC)-PCR and arbitrarily primed-PCR profiles. A pathogenicity island required for pathogenicity of *P. syringae* pv. *phaseolicola* on beans seemed to be conserved among TOX^+ but not among TOX^- isolates, which also lacked the characteristic large plasmid carrying the pathogenicity island. It is suggested that TOX^+ and TOX^- isolates may be grouped into two distinct genetic lineages, designated Pph1 and Pph2 respectively (Oguiza et al., 2004).

4.3.2.3 Symptom Expression Bacterial pathogens induce different kinds of rots, which are the result of disintegration of cellular integrity and collapse of cells releasing nutrients required for the development of the invading bacteria. These pathogens are also responsible for other types symptoms, such as canker and scab in some pathosystems (Table 4.3).

Soft rots caused by *Erwinia* spp. are characterized by the presence of small water-soaked areas which become soft after some time. Under conducive conditions, maceration of parenchyma progresses rapidly resulting in rotting of the tissues, which spreads causing collapse of the tissue entirely. Offensive odor emanating from the affected tissues may be due to invasion by secondary non-rotting bacteria. The damaged portions of cauliflower curd can become infected, whereas infection of cabbage and lettuce may start from injured stems. Lenticels or wounds in potato tubers provide entry points for this bacterial pathogen. Moist conditions and factors that may lead to anaerobic environments favor faster rotting of the tubers (Fig. 4.13).

E. carotovora subsp. *carotovora* causes serious losses of onion in storage. The bacteria may enter the neck tissue as the plants reach maturity and invade the scales causing water-soaked areas which turn pale yellow to light brown color. As the rotting of tissues progresses, the fleshy scales become soft. If the affected bulb is pressed, a watery, foul-smelling fluid oozes out from the neck

TABLE 4.3 Types of Symptoms on Fruits and Vegetables Caused by Bacterial Pathogens

Type of symptom	Crop/disease	Causal organism
1. Rots	Crucifers	
	– Bacterial soft rot	*Erwinia carotovora* subsp. *carotovora*
	– Black rot	*Xanthomonas campestris* pv. *campestris*
	Onion	
	– Bacterial soft rot	*Erwinia carotovora* subsp. *carotovora*
	Pepper (capsicum)	
	– Bacterial soft rot	*Erwinia carotovora* subsp. *carotovora*
	Potato	
	– Brown rot	*Ralstonia solanacearum*
	– Bacterial ring rot	*Clavibacter michiganensis* subsp. *sepedonicus*
	– Soft rot	*Erwinia carotovora* subsp. *carotovora*
	Pulses	
	– Bacterial soft rot	*Erwinia carotovora* subsp. *carotovora*
	Tomato	
	– Bacterial soft rot	*Erwinia carotovora* subsp. *carotovora*
2. Canker	Tomato	
	– Bacterial canker	*Clavibacter michiganensis* subsp. *michiganensis*
3. Scab	Potato	
	– Common scab	*Streptomyces scabies*
4. Spot	Stone fruit	
	– Bacterial spot	*Xanthomonas pruni*

Source: Waller, 2002.

of diseased bulbs. In pepper, small soft spot may be seen near the stem. Sunken areas are formed, initially from the peduncle or around a skin puncture, and spread to cover the entire fruit, which soon becomes a soft slimy mass. The entire fruit seems to be a sack of liquid enclosed in the fruit skin. The contents flow out as the fruit skin breaks and finally dries up.

Brown rot disease caused by *Ralstonia solanacearum* induces wilting of infected plants, while infected tubers exhibit a characteristic brown ring due to the discoloration of the vascular bundles. The skin and eyes, in severe cases of infection, turn black. Grayish-white bacterial ooze comes out of the vascular bundles, when the tuber is cut open or squeezed. Ring rot disease of potato caused by *Clavibacter michiganensis* subsp *sepedonicus* is characterized by the presence of a yellowish color of the vascular ring just beneath the skin. The infection in the early stages cannot be recognized, unless the tuber is cut open to find a brown ring of infected vascular bundles. Secondary infections by other bacteria may result in rapid rotting of the tubers. Potato common scab disease caused by *Streptomyces scabies* may induce shallow and deep scab in tubers.

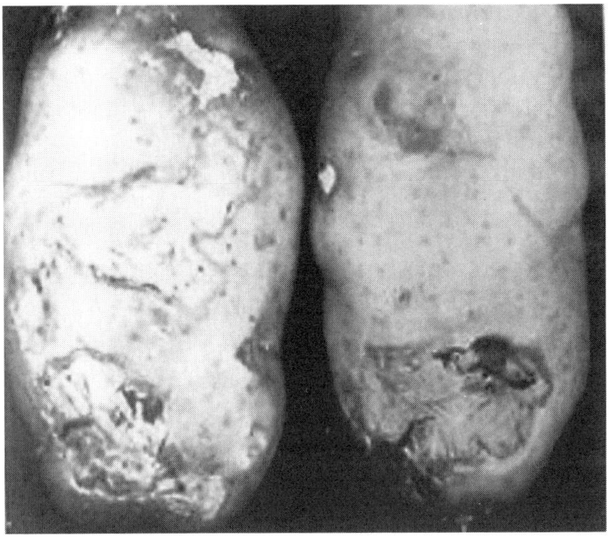

Figure 4.13 Symptoms of soft rot disease of potato. [Coutesy of Rich, 1983; (Academic Press, New York) Elsevier, Oxford, London]

Shallow scab starts as superficial rough areas consisting of corky tissues formed because of abnormal cell division occurring in the periderm of the tuber. When deeper tissues are infected, the lesions are darker developing pits consisting of corky tissues.

The symptoms of infection by *Clavibacter michiganesis* pv. *michiganensis* causing bacterial canker disease are seen on parts of tomato. Systemic infection spreads from leaves, petioles, and stem to fruits. Infected young fruits become stunted and distorted and show the presence of yellow-colored vascular bundles. Secondary infection of fruits may be recognized by the formation of small, round, white, superficial, and slightly raised spots. The central area of the spots ruptures and develops a rough yellow to brown surface. The spots may be surrounded by a white halo. *Xanthomonas campestris* pv. *campestris* causing black rot disease of cabbage and other cruciferous crops is the most serious disease present in all regions of the world. Numerous brown spots appear on the infected leaves, which may turn yellow and finally drop to the ground. A black ring is seen when the affected stem or petiole is cut cross-wise due to the invasion of vascular bundles by the bacteria. Soft rot bacteria may enter black rot lesions, reach the heads and render them worthless. *Xanthomonas pruni* causes bacterial spots on stone fruits. The spots are initially small, circular, and brown in color. As the disease progresses, the spots become darker and depressed, often with water-soaked margins. Later, pitting and cracking or enlarging of fruits occur, rendering the affected fruits unmarketable.

Edible ginger, a popular spice crop raised from a rhizome, is grown primarily for fresh market. The affected rhizomes may show decay or remain symptomless. A facultative, anaerobic, rod-shaped bacterium was found to be consistently associated with decay of ginger rhizome. Based on biochemical tests and 16S rDNA sequence analysis, the bacterium was identified as *Enterobacter cloacae*. Ginger strains of *E. cloacae* caused basal stem and root rot, with foliar chlorosis and necrosis in tissue-cultured ginger plantlets. *E. cloacae* induced, on inoculation, internal yellowing of ripe papaya fruit and internal rot of onion bulbs. The bacterium from all infected hosts was reisolated and identified as *E. cloacae*. This study suggests that *E. cloacae* may exist as an endophyte of ginger rhizomes and reports, for the first time, that *E. cloacae* causes ginger rhizome rot disease (Nishijima et al., 2004).

4.3.3 Process of Infection by Viral Pathogens

4.3.3.1 Replication of Viral Pathogens Plant viruses differ widely from other microbial pathogens. They have simple constitution, lacking all organelles present in cells of other organisms. They possess essentially a protein coat that encloses the viral genomic RNA or DNA molecule(s). Consequently they follow a unique set of replication steps that are not observed in any other organism. The viral coat protein (CP) and genomic nucleic acids are synthesized at different sites using different amino acids and nucleotides available in the host cell and finally assembled together to produce progeny virions. The required information for the synthesis of coat protein, virus-associated nonstructural protein, and movement protein (MP) that facilitates cell-to-cell movement of virus is contained in viral nucleic acid.

The movement protein (MP) and coat protein (CP) are required for the cell-to-cell movement of *Cucumoviruses*. The CMV-MP was able to function with the CP of *Tomato aspermy virus* (TAV) (chimera RT). In contrast, the TAV-MP was unable to promote the cell-to-cell movement in the presence of CMV-CP (chimera TR) (Salánki et al., 2004). Deletion of the C-terminal 33 amino acids of the CMV 3a MP (in the mutant designated 3a) led to CP-independent cell-to-cell movement, but not long-distance movement (Kim et al., 2004). The systemic movement of *Cucumber green mottle mosaic virus* (CGMMV) in cucumber plants appears to occur more efficiently via phloem, although CGMMV can also move systemically via xylem. In stem internodes, CGMMV was first localized in the companion cells of the external phloem and subsequently in all tissues except the medulla. These observations suggested leakage of the virus from, and reloading into, the transport phloem during systemic movement. CGMMV could be simultaneously detected in the xylem and phloem of systemically infected sink leaves (Moreno et al., 2004).

The viral pathogens are incapable of entering the susceptible hosts through natural openings. But some viruses, such as *Potato virus X*, are able to infect healthy plants when the leaves of infected and healthy plants come in contact or through equipments used for cutting propagative materials. Insects, eriophyid mites, nematodes, and fungi can successfully inoculate healthy plants when they probe healthy plants, using contaminated stylets or through saliva carrying the virions. It is essential that the viruses should be directly placed into a susceptible cell to initiate the process of infection.

4.3.3.2 Symptom Expression As viruses do not have any detectable physiological functions such as respiration, alterations observed in the virus-infected plants are entirely due to changes in the physiology of the host plants induced by virus infection. The changes in color, shape, and size observed in fruits and vegetables are the resultant of abnormal physiological functions of diseased plants. A general reduction in fruit size and number of fruits formed may be observed, due to the systemic infection of plants by viruses.

In grapevine infected by *Grapevine leafroll virus*, fruit clusters are smaller but the berry number is increased. The pigmentation in the berry skin is reduced and ripening of berries is considerably delayed. *Papaya ringspot virus* induces characteristic symptoms on fruits. The infected papaya fruits exhibit yellow spots developing yellow rings with solid green centers. Later, blistering and deformation of fruits are also seen. The flavor and aroma of affected fruits show considerable difference from that of healthy fruits. On the other hand, infection of papaya by leafcurl diseases may result in total loss of flowering or the affected plants may produce a few small and abnormal fruits.

The potato crop is affected by several viruses causing varying degrees of reduction in yield, because of decrease in number and size of tubers formed in infected plants. In addition to the general effect on yield, some viruses induce specific symptoms in tubers. In potato plants infected by *Potato leafroll virus* (PLRV), necrosis of primary phloem is a characteristic, diagnostic symptom. Because of phloem necrosis, the photosynthates and other food materials are prevented from translocation to tubers resulting in the poor development of tubers. Necrosis may extend to tubers and a brown ring may be seen in the vascular system between the cortex and core.

The fleshy roots of sweetpotato infected by *Sweetpotato feathery mottle virus* are much smaller and exhibit internal cork, characterized by dark brown to black corky spots. These corky spots consist of a central area of collapsed dead cells surrounded by a distinct cork layer of several cells thick. During storage at 21 to 27°C the development of cork symptom is favored. Cavities on the surface may be formed at positions with shallow internal lesions, due to the death of tissues near the surface of the growing root. In addition, russet-crack symptoms, consisting of dark lesions with fine cracks in the skin and outer cortical areas of the fleshy root, may also be noted in some cultivars.

Cucumber fruits infected by *Cucumber mosaic virus* (CMV) develop yellowish green mottle, initially at the stem end of young fruit and later covering the fruit completely. The entire fruit may show light yellowish green patches intermingled with darker green areas. The greenhouse-grown fruits may have raised dark green areas giving the appearance of warts. CMV causes characteristic concentric rings or solid circular spots on pepper (chilli) fruits. The spots may be yellowish first, later becoming brown when the affected tissues turn necrotic. The infected fruits may be somewhat flattened. Tomato fruits,

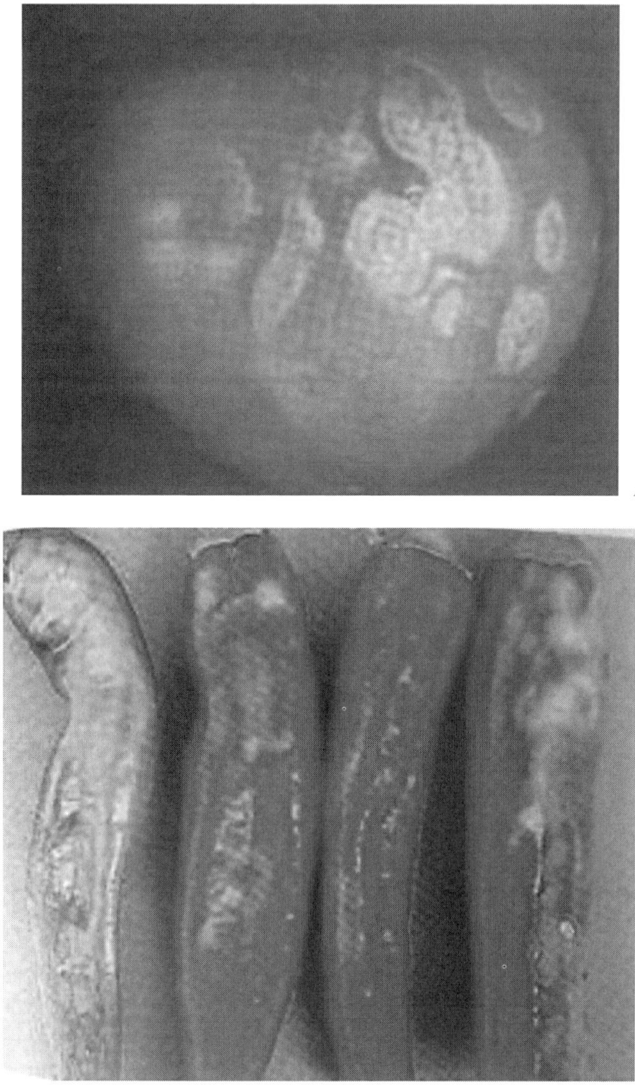

A

B

Figure 4.14 Symptoms of tomato spotted wilt disease. (a) Tomato; (b) pepper (chilli). [Courtesy of (a) Sherf and Macnab, 1986; John Wiley & Sons, Hoboken, NJ, USA, and (b) Asian Vegetable Research and Development Center, Taipei.]

following infection by *Tomato spotted wilt virus* (TSWV), show the presence of yellowish spots in the early stage of fruit development. These spots develop distinct concentric zones of shades of yellow or brown alternating with green and later with pink or red rings. The development of concentric rings on mature red fruit reduces the market value of the produce (Fig. 4.14).

Lettuce plants growing from seeds infected by *Lettuce mosaic virus* (LMV) may fail to head or produce small, loose heads of poor quality. Leaves show various intensities of mottling depending on the susceptibility of the cultivars. Mild to severe brown discoloration may be observed in susceptible varieties. The leaf margins become ruffled and distorted at later stages. The quality of the salad crop is drastically affected.

SUMMARY

Microbial pathogens – fungi, bacteria and viruses – cause diseases at both pre- and postharvest stages in durables and perishables. The phenomenon of pathogenesis shows variations depending on the nature of the pathogen and host plant species. The development of disease during different phases in infected seeds, fruits, and vegetables is traced by elucidative evidences obtained from studies on various pathosystems. The usefulness of molecular techniques in studying the interaction between host plant and the pathogen is highlighted. The symptom expression, as the culmination of successful pathogenesis leading to formation of various symptoms characteristic of diseases caused by fungal, bacterial, and viral pathogens, is discussed.

APPENDIX 4: ELECTRON MICROSCOPE STUDIES OF SITES OF INFECTION BY FUNGAL PATHOGEN (*BOTRYTIS CINEREA*) (VIRET ET AL., 2004)

A. Inoculation of Inflorescences of Grapevine Plants

i. Thin out the flowers to about 50 flowers/inflorescence; inoculate the flowers at full bloom with 0.5 µl drops of conidial suspension of *B. cinerea* (1.4×10^6 / ml) placed on either the calyptra, stigma, or receptacle area of each individual flower and maintain three replications for each position in three separate plants.
ii. Incubate the inflorescence or whole plants (inoculated and controls) by enclosing them in plastic bags for 24 h to maintain 100% RH; remove the bags and collect three replicates of six flowers for each treatment at 24, 48, 72, and 96 h after inoculation for scanning electron microscopy (SEM) and transmission electron microscopy (TEM).
iii. Take ten additional flowers for each treatment for the isolation of the pathogen on potato dextrose agar (PDA) to confirm the infection of flowers by the pathogen.

B. Scanning Electron Microscopy (SEM)

i. Fix the whole inoculated flowers by immersing them in 5% glutaraldehyde in 0.03 M Pipes buffer (pH 6.8 to 7.0) for 2 h at room temperature under vacuum for easy infiltration of the fixative and rinse the fixed tissues in buffer (three times for 5 min each).

ii. Dehydrate the tissues in a graded ethanol series (10% steps) for 15 min in each step and finally rinse in absolute dry ethanol (three times for 30 min each).

iii. Dry the tissues to critical point; mount them on aluminum stubs; sputter with gold and observe under scanning electron microscope at 5 to 10 kV.

C. Transmission Electron Microscopy (TEM)

i. Fix the samples as in step B (i) with inclusion of caffeine (1.0%) to the buffer to stabilize phenolic compounds and rinse the tissues in buffer, three times for 5 min. each.

ii. Postfix the tissues in 1% osmium tetraoxide in buffer for 2 h at room temperature and rinse the tissues with distilled water three times for 5 min each.

iii. Dehydrate in graded ethanol series as in step B (ii) and wash in absolute acetone for 30 min.

iv. Infiltrate 25, 50, or 75% firm-grade Spurr's resin in dry acetone into the tissue followed by infiltration in 100% resin (once for 30 min) and finally again overnight on an inclined rotator at 4°C.

v. Polymerize the embedded tissues at 60 to 70°C overnight.

vi. Cut sections (80 to 90 nm thick) using an ultramicrotome with a diamond knife; transfer sections to filmed copper grids (200 to 300 mesh); stain the sections with uranyl acetate for 10 min; thoroughly rinse in 50% methanol followed by lead citrate; rinse for 10 min and finally in distilled water.

vii. Observe under transmission electron microscope.

REFERENCES

Abu-Goukh, A.A. and Labavitch, H.M. 1983. The *in vivo* role of Bartlett pear fruit polygalacturonase inhibitors. *Physiological Plant Pathology*, **23**: 123–135.

Abu-Goukh, A.A., Greve, L.C. and Labavitch, J.M. 1983a. Purification and partial characterization of 'Bartlett' pear polygalacturonase inhibitors. *Physiological Plant Pathology*, **23**: 111–122.

Abu-Goukh, A.A., Strand, L.L. and Labavitch, J.M. 1983b. Development-related changes in decay susceptibility and polygalacturonase inhibitor content of 'Bartlett' pear fruit. *Physiological Plant Pathology*, **23**: 101–109.

Adikaram, N.K.B. 1981. *A study of latent infections of Capsicum* spp. *by Colletotrichum capsici* and *Glomerella cingulata*. Doctoral Thesis, Queens University, Belfast, U.K.

Adikaram, N.K.B., Edwing, D.F., Karunaratne, A.M. and Wijeratne, W.M.K. 1992. Antifungal compounds from immature avocado fruit peel. *Phytochemistry*, **31**: 93–96.

Alonso, J.M., Chamarro, J. and Gramell, A. 1995. Evidence for the involvement of ethylene in the expression of specific RNAs during maturation of the orange, a non-climacteric fruit. *Plant Molecular Biology*, **29**: 385–390.

Anderson, H.N., Nehring, E.W. and Wisher, W.R. 1975. Aflatoxin contamination of corn in the field. *Journal of Agricultural and Food Chemistry*, **23**: 775, 767–769.

Andrade, O., Muñoz, G., Galdames, R., Durán, P. and Honorato, R. 2004. Characterization, *in vitro* culture and molecular analysis of *Thecaphora solani*, the causal agent of potato smut. *Phytopathology*, **94**: 875–882.

Anuja Gupta and Aneja, K.R. 2001. Mycofloral spectrum during storage and its effect on seed viability of soybean (*Glycine max* (L.) Merrill) seeds under ambient conditions. *Proceedings of the National Academy of Science, India, Section B, Biological Sciences*, **71**: 245–253.

Archbold, D.D., Hamilton-Kemp, T.R. and Fallik, E. 2002. Postharvest injury, strawberry volatiles and *Botrytis* development. *Acata Horticulturae*, **567**: 767–769.

Asalmol, M.N., Kale, V.P. and Ingle, S.T. 2001. Seedborne fungi of chilli-incidence and effect on seed germination. *Seed Research*, **29**: 76–79.

Aspinall, G.C. and Greenwood, C.T. 1962. Aspects of the chemistry of cereal polysaccharides. *Journal of Institute of Brewery*, **68**: 167–178.

Badel, J.L., Charkowski, A.O., Deng, W.L. and Collmer, A. 2002. A gene in the *Pseudomonas syringae* pv. *tomato* Hrp pathogenicity island conserved effector locus *hopPtoA1* contributes to efficient formation of bacterial colonies *in planta* and is duplicated elsewhere in the genome. *Molecular Plant–Microbe Interactions*, **15**: 1014–1024.

Baker, K.F. 1972. Seed Pathology. In: *Seed Biology*, Vol.2 (ed.) T.T. Kozlowski pp. 317. Academic Press, New York.

Bankole, S.A., Ikotun, B. and Ekpo, E.J.A. 1999. Fungal deterioration of melon seeds stored in jute sacks and polyethylene bags in Ago-Iwoye, Southwestern, Nigeria. *Mycopathologia*, **146**: 135–146.

Bashan, Y. and Okon, Y. 1986. Internal and external infections of fruits and seeds of pepper by *Xanthomonas campestris* pv. *vesicatoria. Canadian Journal of Botany*, **64**: 2865.

Bautista-Baños, S., Barrera-Necha, L.L., Long, P.G., Ganesh, S. and Cheah, L.H. 2001. Inoculum variables affecting pathogenicity of *Botrytis cinerea* infection of kiwifruit. *Revitsa Mexicana de Fitopatologia*, **19**: 161–167.

Bellaire, L. de, Chillet, M. and Mourichon, X. 2000. Elaboration of an early quantification method of quiescent infections of *Colletotrichum musae* on bananas. *Plant Disease*, **89**: 128–133.

Bennett, C.W. 1969. Seed transmission of plant viruses. *Advances in Virus Research*, **14**: 221.

Beno-Moualem, D. and Prusky, D. 2000. Early events during quiescent infection development by *Colletotrichum gloeosporioides* in unripe avocado fruits. *Phytopathology*, **90**: 553–559.

Beno-Moualem, D., Lichter, A., Wang, X., Leikin-Frenkel, A., Kobiler, I. and Prusky, D. 2004. Expression of 12 fatty acid desaturase during the induced accumulation of the antifungal diene in avocado fruits. *Molecular Plant Pathology*, **5**: 575–585.

Bilgrami, K.S., Choudhary, A.K. and Ranjan, K.S. 1992. Aflatoxin contamination in field mustard (*Brassica juncea*) cultivars. *Mycotoxin Research*, **8**: 21.

Billing, E. 1982. Entry and establishment of pathogenic bacteria in plant tissues. In: *Bacteria and Plants* (eds) M.E. Rhodes-Roberts and F.A. Skinner, pp. 51–70, Society for Applied Bacteriology Symposium Series, No. 10, Academic Press, London and New York.

Billon-Grand, G., Pousserean, N. and Févre, M. 2002. The extracellular proteases secreted *in vitro* and *in planta* by the phytopathogenic fungus *Sclerotinia sclerotiorum*. *Journal of Phytopathology*, **150**: 507–511.

Black, L.L., Green, S.K., Hartman, G.L. and Poulos, J.M. 1991. *Pepper Diseases: A Field Guide.* Asian Vegetable Research Center, AVRDC Publication No. 91, p. 437.

Boller, T. 1991. Ethylene in pathogenesis and disease resistance. In: *The Plant Hormone Ethylene* (eds) A.K. Mattoo and J.C. Suttle, CRC Press, Boca Raton, FL.

Bonas, V., van den Ackerveken, G., Büttner, D., Hahn, K., Marios, E., Nennstiel, D., Noel, L., Rossier, O. and Szurek, B. 2000. How the bacterial plant pathogen *Xanthomonas campestris* pv. *vesicatoria* conquers the host? *Molecular Plant Pathology*, **1**: 73–76.

Børve, J., Sekse, L. and Stensvand, A. 2000. Cuticular fractures promote postharvest fruit rot in sweet cherries. *Plant Disease*, **84**: 1180–1184.

Bostock, R.M., Wilcox, S.M., Wang, G. and Adaskaveg, J.E. 1999. Suppression of *Monilinia fructicola* cutinase production by peach fruit surface phenolic acids. *Physiological and Molecular Plant Pathology*, **54**: 37–50.

Braun, A.J. and Keplinger, J.A. 1973. Seed transmission of *Tomato ringspot virus* in raspberry. *Plant Disease Reporter*, **57**: 431.

Bringel, J.M.M., Moraes, M.H.D. de, Menten, J.O.M. and Bedendo, I.P. 2001. Health and physiological quality of soybean seeds produced in the southern region of Maranhão/Brazil. *Summa Phytopathologica*, **27**: 438–441.

Brlansky, R.H., Carroll, T.W. and Zaske, S.K. 1986. Some ultra-structure aspects of the pollen transmission of *Barley stripe mosaic virus* in barley. *Canadian Journal of Botany*, **64**: 853.

Brown, G.E. 1973. Development of green mold in degreened oranges. *Phytopathology*, **63**: 1104–1107.

Brown, G.E. 1975. Factors affecting postharvest development of *Colletotrichum gloeosporioides* in citrus fruits. *Phytopathology*, **65**: 404–409.

Brown, G.E. 1977. Ultrastructure of penetration of ethylene-degreened Robinson tangerines by *Colletotrichum gloeosporioides* after ethylene treatment. *Phytopathology*, **67**: 700–706.

Brown, G.E. and Burns, J.K. 1998. Enhanced activity of abscission enzymes predisposes oranges to invasion by *Diplodia natalensis* during ethylene degreening. *Postharvest Biology and Technology*, **14**: 217–227.

Brown, G.E. and Eckert, J.W. 2000. Alternaria rot. In: *Compendium of Citrus Diseases*. 2nd edition (eds) L.W. Timmer, S.M. Garnsey and J.H. Graham, pp. 37, American Phytopathological Society, St. Paul, MN.

Bryan, M.K. 1930. Studies on bacterial canker of tomato. *Journal of Agricultural Research*, **41**: 825.

Burkholder, W.H. 1921. The bacterial blight of the bean: a systemic disease. *Phytopathology*, **11**: 61.

Bryk, H. 1999. The study on the infection of apple fruits by *Botrytis cinerea* Pers. after harvest. *Acta Agrobotanica*, **52**: 19–29.

Büttner, D., Gürlebeck, D., Noel, L.D. and Bonas, U. 2004. HpaB from *Xanthomonas campestris* pv. *vesicatoria* acts as an exit control protein in type III-dependent protein secretion. *Molecular Microbiology*, **57**: 755–768.

Calonnec, A., Cartolaro, P., Poupot, C., Dubourdieu, D. and Darriet, P. 2004. Effects of *Uncinula necator* on the yield and quality of grapes (*Vitis vinifera*) and wine. *Plant Pathology*, **53**: 434–445.

Carroll, T.W. 1981. Seedborne viruses: virus–host interactions. In: *Plant Disease and Vectors* (eds) K. Maramorosch and K.F. Harris pp. 293, Academic Press, New York.

Centis, S., Guillas, I., Séjalon, N., Esequerre-Tugaye, M.T. and Dumas, B. 1997. Endopolygalacturonase genes from *Colletotrichum lindemuthianum*: cloning of *CLPG 2* and comparison of its expression to that of *CLPG1* during saprophytic and parasitic growth of the fungus. *Molecular Plant–Microbe Interactions*, **10**: 769–775.

Channon, A.G. 1970. Tests of fungicides against *Alternaria brassicola* and studies on some factors enhancing attack by the fungus. *Annals of Applied Biology*, **65**: 481–487.

Chatterjee, A.K. and Starr, M.P. 1977. Donor strains of the soft rot bacterium *Erwinia chrysanthemi* and conjugational transfer of the pectolytic capacity. *Journal of Bacteriology*, **132**: 862–869.

Chelkowski, J. and Cierniewski, A. 1983. Mycotoxins in cereal grains Part X. Invasion of fungal mycelium into wheat and barley kernels. *Nahrung*, **6**: 533–536.

Christensen, C.M. 1991. Fungi and seed quality. In: *Handbook of Applied Mycology Vol.3: Foods and Feeds* (eds) D.K. Arora, K.G. Mukerji and E.H. Marth, pp. 99–119, Marcel Dekker, Inc., New York.

Clendennen, S.K. and May, G.D. 1997. Different gene expression in ripening banana fruit. *Plant Physiology*, **115**: 463–469.

Coates, L.M., Muirhead, I.F., Irwing, J.A.G. and Gowanlock, D.H. 1993. Initial infection processes by *Colletotrichum gloeosporioides* on avocado fruit. *Mycological Research*, **97**: 1363–1370.

Coertze, S. and Holz, G. 1999. Surface colonization, penetration and lesion formation on grapes inoculated fresh or after cold storage with single airborne conidia of *Botrytis cinerea*. *Plant Disease*, **83**: 917–924.

Coertze, S., Holz, G. and Sadie, A. 2001. Germination and establishment of infection on grape berries by single airborne conidia of *Botrytis cinerea*. *Plant Disease*, **85**: 668–677.

Cohen, Y. and Sackston, W.E. 1974. Seed infection and latent infection of sunflowers by *Plasmopara halstedii*. *Canadian Journal of Botany*, **52**: 231–238.

Cook, A.A., Larson, R.H. and Walker, J.C. 1952. Relation of the black rot pathogen to cabbage seed. *Phytopathology*, **42**: 316–320.

Cooper, J.I., Massalski, P.R. and Edwards, M. 1984. *Cherry leaf roll virus* in the female gametophyte and seed of birch and its relevance to vertical virus transmission. *Annals of Applied Biology*, **105**: 55.

Côrtes, N. de A., Cassetari Neto, D. and Correa, B. 2000. Occurrence of aflatoxins in maize produced by traditional cultivation methods in agricultural communities in the state of Marto Grosso. *Higiene Alimentar*, **14**: 16–26.

Crossan, D.F. and Morehart, A.L. 1964. Isolation of *Xanthomonas vesicatoria* from tissues of *Capsicum annuum*. *Phytopathology*, **54**: 358.

Cunnac, S., Occhialini, A., Barberis, P., Boucher, C. and Genin, S. 2004. Inventory and functional analysis of the large *Hrp* regulon in *Ralstonia solanacearum*: identification of novel effector proteins translocated to plant host cells through the type III secretion system. *Molecular Microbiology*, **53**: 115–128.

Das, C.R., Milbrath, J.A. and Swenson, K.G. 1961. Seed and pollen transmission of *Prunus ringspot virus* in Buttercup squash. *Phytopathology*, **51**: 64.

Das, M. and Bora, K.N. 1998. Ultrastructural studies on infection processes by *Colletotrichum acutatum* on guava fruit. *Indian Phytopathology*, **51**: 353–356.

Dickman, M.B., Podila, G.K. and Kolattukudy, P.E. 1989. Insertion of cutinase gene into a wound pathogen enables it to infect intact host. *Nature*, **342**: 446–448.

Diener, U.L., Cole, R.J., Sanders, T.H., Payne, G.A., Lee, L.S. and Klich, M.A. 1987. Epidemiology of aflatoxin formation by *Aspergillus flavus*. *Annual Review of Phytopathology*, **25**: 249.

Dong, W., Latijnhouwers, M., Jiang, R.H.Y., Meijer, H.J.G. and Govers, F. 2004. Downstream targets of the *Phytophthora infestans* Ga subunit PiGPA1 revealed by cDNA–AFLP. *Molecular Plant Pathology*, **5**: 483–494.

Droby, S., Prusky, D., Jacoby, B. and Goldman, A. 1986. Presence of an antifungal compound in the peel of mango fruits and their relation to latent infections of *Alternaria alternata*. *Physiological and Molecular Plant Pathology*, **29**: 173–183.

Droby, S., Prusky, D., Jacoby, B. and Goldman, A. 1987. Induction of antifungal resorcinols in flesh of unripe mango fruits and its relation to latent infection by *Alternaria alternata*. *Physiological and Molecular Plant Pathology*, **30**: 285–292.

Edney, K.L. 1983. Top fruit. In: *Post-Harvest Pathology of Fruits and Vegetables* (ed.) C. Dennis, pp. 43–71, Academic Press, London.

Edwards, M.C., Fetch, T.G. Jr., Schwarz, P.B. and Steffenson, B.J. 2001. Effect of *Barley yellow dwarf virus* infection on yield and malting quality of barley. *Plant Disease*, **85**: 202–207.

Elad, Y. and Evensen, K. 1995. Physiological aspects of resistance to *Botrytis cinerea*. *Phytopathology*, **85**: 637–643.

El-Kazzaz, M.K., Chordas, A. and Kader, A.A. 1983. Physiological and compositional changes in orange fruits in relation to modification of their susceptibility to *Penicillium italicum* by ethylene treatment. *Journal of American Society for Horticultural Science*, **108**: 618–621.

El-Naashaar, H. and Bugbee, W.M. 1981. Scanning electron microscopic examination of sugar beet flowers and fruits infected with *Phoma betae*. *Journal of American Society of Sugarbeet Technologists*, **21**: 11.

Emery, K.M., Michailides, T.J. and Scherm, H. 2000. Incidence of latent infection of immature peach fruit by *Monilinia fructicola* and relationship to brown rot in Georgia. *Plant Disease*, **84**: 853–857.

Emmett, R.W. and Parbery, D.G. 1975. Appressoria. *Annual Review of Phytopathology*, **13**: 147–167.

Eriksson, A.R., Andersson, R.A., Pirhonen, M. and Palva, E.T. 1998. Two component regulators involved in the global control of virulence of *Erwinia carotovora* subsp. *carotovora*. *Molecular Plant–Microbe Interactions*, **11**: 743–752.

Eudes, F., Comeau, A., Rioux, S. and Collin, J. 2001. Impact of trichothecenes on *Fusarium* head blight (*Fusarium graminearum*) development in spring wheat (*Triticum aestivum*). *Canadian Journal of Plant Pathology*, **23**: 318–322.

Fábrega, A.R., Lamarque, A.L., Giorda, L.M., Guzman, C.A. and Maestri, D.M. 2000. Effect of *Diaporthe phaseolorum* var. *sojae* infection on seed quality of soybean genotypes. *Fitopatologia Brasileira*, **25**: 624–627.

Fabritius, A.L. and Judelson, H.S. 2003. A mating-induced protein of *Phytophthora infestans* is a member of a family of elicitors with divergent structures and stage-specific patterns of expression. *Molecular Plant–Microbe Interactions*, **16**: 926–935.

Farag, R.S., Mohsen, S.M., Khalil, F.A. and Basyony, A.E. 1985. Effect of fungal infection on the chemical composition of wheat, soybean and sesame seeds. *Grasas Aceites*, **36**: 357–361.

Filonow, A.B. 1999. Yeasts reduce the stimulatory effect of acetate esters from apple on the germination of *Botrytis cinerea*. *Journal of Chemical Ecology*, **25**: 1555–1565.

Filonow, A.B. 2001. Butylacetate and yeasts interact in adhesion and germination of *Botrytis cinerea* conidia *in vitro* and fungal decay of Golden Delicious apples. *Journal of Chemical Ecology*, **27**: 831–844.

Flaishman, M.A. and Kolattukudy, P.E. 1994. Timing of fungal invasion using hosts ripening hormone as a signal. *Proceedings of the National Academy of Sciences*, USA, **91**: 6579–6583.

Gaümann, E. 1950. *Principles of Plant Infection*. Crosby Lockwood and Son, London, U.K.

Geeson, J.D. 1983. Brassicas. In: *Post-Harvest Pathology of Fruits and Vegetables* (ed.) C. Dennis, pp. 125–156, Academic Press, London.

George, J.A. and Davidson, T.R. 1963. Pollen transmission of necrotic ringspot and sour cherry yellows viruses from tree to tree. *Canadian Journal of Plant Science*, **43**: 276.

Georgieva, I.D., Edreva, A. and Rodeva, R. 2001. Response of β-glucosidase to fungal infections in seed, ovary and fruit. *Biologia Plantarum*, **44**: 573–578.

Giebel, J. and Dopieraa, U. 2004. Pathogenesis of potato gangrene caused by *Phoma exigua* var. *foveata*: II. Activities of some hydrolases and dehydrogenases. *Journal of Phytopathology*, **152**: 399–403.

González, E.T. and Allen, T. 2003. Characterization of a *Ralstonia solanacearum* operon required for polygalacturonate degradation and uptake of galacturonic acid *Molecular Plant–Microbe Interactions*, **16**: 536–544.

Goodman, J.J. and Christensen, C.M. 1952. Grain storage studies. XI. Lipolytic activity of fungi isolated from stored corn. *Cereal Chemistry*, **29**: 299–308.

Gotlieb, A.R. 1974. *Isolation, characterization and transmission of Apple mosaic virus in white birch and isolation and characterization of latent virus in yellow birch*. Doctoral Thesis, University of Wisconsin, USA.

Gour, H.N., Jaydeo, A. and Ranjan, N. 2001. Effect of bacterial inoculum (*Xanthomonas axonopodis* pv. *vignicola*) on seed germination, fresh and dry weight of cowpea [*Vigna unguiculata* (L.) Walp] seedlings. *Journal of Mycology and Plant Pathology*, **31**: 46–49.

Gronover, C.S., Kasulke, D., Tudzynski, P. and Tudzynski, B. 2001. The role of G protein alpha subunits in the infection process of the gray mold fungus *Botrytis cinerea*. *Molecular Plant–Microbe Interactions*, **14**: 1293–1302.

Grover, R.K. 1971. Participation of host exuded chemicals in appressorium formation of *Colletotrichum piperatum*. In: *Ecology of Leaf Surface Microorganisms* (eds) T.F. Preece and C.H. Dickinson, pp. 509–518, Academic Press, London.

Guevara, M.G., Olivia, C.R., Huarte, M. and Daleo, G.R. 2002. An aspartic-protease with antimicrobial activity is induced after infection and wounding in intercellular fluids of potato tubers. *European Journal of Plant Pathology*, **108**: 131–137.

Hamilton, R.I., Nichols, C. and Valentine, B. 1984. Survey for *Prunus necrotic ringspot* and other viruses contaminating the exine of pollen collected by bees. *Canadian Journal of Plant Pathology*, **6**: 196.

Harper, D.B. and Swinburne, T.R. 1979. 2,3-Dihydroxy-benzoic acid and related compounds as stimulants of germination of conidia of *Colleotrichum musae* (Ber. and curt) Arx. *Physiological Plant Pathology*, **14**: 363–370.

Harper, D.B., Swinburne, T.R., Moore, S.K., Brown A.E. and Graham, H. 1980. A role for iron in germination of conidia of *Colleotrichum musae*. *Journal of General Microbiology*, **121**: 169–174.

Hasan, H.A.H. 2000. Fungal association and deterioration of oil-type safflower (*Carthamus tinctorius*) seed during storage. *Czech Mycology*, **52**: 125–137.

Have, A. ten, Breuil, W.O., Wubben, J.P., Visser, J. and Kan, J.A.L. van. 2001. *Botrytis cinerea* endopolygalacturonase genes are differentially expressed in various plant tissues. *Fungal Genetics and Biology*, **33**: 97–105.

Have, A. ten, Dekkers, E., Kay, J., Phylip, L.H. and van Kan, J.A.L. 2004. An aspartic proteinase gene family in the filamentous fungus *Botrytis cinerea* contains members with novel features. *Microbiology*, **150**: 2475–2489.

Have, A. ten, Mulder, W., Visser, J. and van Kan. J.A.L. 1998. The endopolygalacturonase gene *Bcpg1* is required for full virulence of *Botrytis cinerea*. *Molecular Plant–Microbe Interactions*, **11**: 1009–1016.

He, G., Matsuura, H. and Yoshihara, T. 2004. Isolation of an α-methylene-butyrolactone derivative, a toxin from the plant pathogen, *Lasiodiplodia theobromae*. *Phytochemistry*, **65**: 2803–2807.

Hernandez-Anguiano, A.M., Suslow, T.V., Leloup, L. and Kado, C.I. 2004. Biosurfactants produced by *Pseudomonas fluorescens* and soft rotting of harvested florests of broccoli and cauliflower. *Plant Pathology*, **53**: 596–601.

Hill, R.A. and Lacey, J. 1983. Factors determining the microflora of stored barley grain. *Annals of Applied Biology*, **102**: 467–483.

Hummel, B.C.W., Cuendet, L.S., Christensen, C.M. and Geddes, W.F. 1954. Grain storage studies. XIII. Comparative changes in respiration, viability and chemical composition of mould-free and mould-contaminated wheat upon storage. *Cereal Chemistry*, **31**: 143–149.

Hwang, B.H., Kim, J., Yang, Y.J. and Kim, K. 2002. Cell wall degradation from mandarin fruit by *Penicillium* spp. during postharvest storage. *Journal of the Korean Society for Horticultural Science*, **43**: 195–200.

Hwang, C.S., Flaishman, M.A. and Kolattukudy, P.E. 1995. Cloning of a gene expressed during appressorium formation by *Colletotrichum gloeosporioides* and a marked decrease in virulence by disruption of this gene. *Plant Cell*, **7**: 183–193.

Hyytiäinen, H., Montesano, M. and Palva, E.T. 2001. Global regulators ExpA (GacA) and Kdg R modulate extracellular enzyme gene expression through RsmA-rsm B system in *Erwinia carotovora* subsp. *carotovora. Molecular Plant–Microbe Interactions*, **14**: 931–938.

Isshiki, A., Akimitsu, K., Nishio, K., Tsukamoto, M. and Yamamoto, H. 1997. Purification and characterization of an endopolygalacturonase from the rough lemon pathotype of *Alternaria alternata*, the cause of brown spot disease. *Physiological and Molecular Plant Pathology*, **51**: 155–167.

Isshiki, A., Akimitsu, K., Yamamoto, M. and Yamamoto, H. 2001. Endopolygalacturonase is essential for citrus black rot caused by *Alternaria citri*, but not for brown spot caused by *A. Alternata. Molecular Plant–Microbe Interactions*, **14**: 749–757.

Isshiki, A., Ohtani, K., Kyo, M., Yamamoto, H. and Akimitsu, K. 2003. Green fluorescent detection of fungal colonization and endopolygalacturonase gene expression in interaction of *Alternaria citri* with citrus. *Phytopathology*, **93**: 768–773.

Jarvis, W.R. 1977. *Botryotinia* and *Botrytis* species: taxonomy, physiology and pathogenicity. *Monograph No.15*, Canadian Department of Agriculture, Ottawa, Canada.

Jarvis, W.R. 1980. Epidemiology. In: *The Biology of Botrytis* (eds) J.R. Coley-Smith, K. Verhoeff and W.R. Jarvis, pp. 219–245, Academic Press, London, U.K.

Johnson, G.I., Mead, A.J., Cooke, A.W. and Dean, R.I. 1991. Mango stem-end rot pathogens – infection levels between flowering and harvest. *Annals of Applied Biology*, **119**: 465–473.

Keller, M., Viret, O. and Cole, F.M. 2003. *Botrytis cinerea* infection in grape flowers: Defense reaction, latency and disese expression. *Phytopathology*, **93**: 316–322.

Kelley, R.D. and Cameron, H.R. 1986. Location of *Prune dwarf* and *Prunus necrotic ringspot viruses* associated with sweet cherry pollen and seed. *Phytopathology*, **76**: 317.

Kim, S.H., Kalinina, N.O., Andreev, I., Ryabov, E.V., Fitzerald, A.G., Taliansky, M.E. and Palukaitis, P. 2004. C-terminal 33 amino acids of the *Cucumber mosaic virus* 3a protein affect virus movement RNA binding and inhibition of infection and translocation. *Journal of General Virology*, **85**: 221–230.

Kim, Y.S., Lee, H.H., Ko, M.K., Song, C.E., Bae, C.Y. Lee, Y.H. and Oh, B.J. 2001. Inhibition of fungal appressorium formation by pepper (*Capsicum annuum*) esterase. *Molecular Plant–Microbe Interactions*, **14**: 80–85.

Kojima, K., Kikuchi, T., Takano, Y., Oshiro, E. and Okuno, T. 2002. The mitogen-activated protein kinase gene MAF1 is essential for the early differentiation phase of appressorium formation in *Colletotrichum lagenarium. Molecular Plant–Microbe Interactions*, **15**: 1268–1276.

Krishnamurthy, Y.L. and Raveesha, K.A. 2001. Seedborne nature and location of *Cladosporium cladosporioides* in soybean. *Plant Disease Research*, **16**: 203–209.

Kritzman, G. and Zutra, D. 1983. Systemic movement of *Pseudomonas syringae* pv. *lachrymans* in stem, leaves, fruits and seeds of cucumber. *Canadian Journal of Plant Pathology*, **5**: 273.

Kuniyasu, K. and Kishi, K. 1977. Seed transmission of *Fusarium* wilt of bottle gourd, *Lagenaria siceraria* Standle, used as rootstock of watermelon. II. The seed infection course from infected stem of bottle gourd to the fruit and seed. *Annals of Phytopathological Society, Japan*, **43**: 192.

Kunwar, J.K., Manandhar, J.B. and Sinclair, J.B. 1986. Histopathology of soybean seeds infected with *Alternaria alternata*. *Phytopathology*, **76**: 543.

Labavitch, J.M. 1998. Fruit ripening and defense against pathogens – Loss of resistance or gain of susceptibility? In: *Disease Resistance in Fruit* (eds.) G.I. Johnson, E. Highley and D.C. Joyce, pp. 53–59, ACIAR Proceedings Series, 80.

Lahey, K.A., Yuan, R., Burns, J.K., Ueng, P.P., Timmer, L.W. and Chung, K.R., 2004. Induction of phytohormones and differential gene expression in citrus flowers infected by the fungus *Colletotrichum acutatum*. *Molecular Plant–Microbe Interactions*, **17**: 1394–1401.

Laidou, I.A., Thanassoulopoulos, C.C. and Liakopoulou-Kyriakides, M. 2001. Diffusion of patulin in the flesh of pears inoculated with four postharvest pathogens. *Journal of Phytopathology*, **149**: 457–461.

Lang, W. 1917. Zur Angsteckung der Gerste durch *Ustilago nuda*. *Bericht der Deutschen botanischen Gesellscaft*, **35**: 4–20.

Larson, R.H. 1944. The ring rot bacterium in relation to tomato and eggplant. *Journal of Agricultural Research*, **69**: 309.

Lee, Y.K., Hippe-Sanwald, S., Jung, H.W., Hong, J.K., Hause, B. and Hwang, B.K. 2000. *In situ* localization of chitinase mRNA and protein in compatible and incompatible interactions of pepper stems with *Phytophthora capsici*. *Physiological and Molecular Plant Pathology*, **57**: 111–121.

Lelievre, J.M., Latche, A., Jones, B., Bouzayen, M. and Pech, J.C. 1997. Ethylene and fruit ripening. *Physiologia Plantarum*, **101**: 727–739.

Lelliot, R.A. 1988. *Xanthomonas campestris* pv. *phaseoli* (Smith) Dye. In: *European Handbook of Plant Diseases* (eds) I.M. Smith, J. Dunez, D.H. Phillips, R.A. Lelliot and S.A. Archer, pp. 165–166, Blackwell Scientific Publications, Oxford, U.K.

Lemmens, M., Buerstmayr, H., Krska, R., Schuhmacher, R., Gransgruber, H. and Ruckenbauer, P. 2004. The effect of inoculation treatment and long-term application of moisture on *Fusarium* head blight symptoms and deoxynivalenol contamination in wheat grains. *European Journal of Plant Pathology*, **110**: 299–308.

Lennox, C.L. and Spotts, R.A. 2004. Timing of preharvest infection of pear fruit by *Botrytis cinerea* and the relationship to postharvest decay. *Plant Disease*, **88**: 468–473.

Li, F.Q., Luo, X.Y. and Yoshizawa, T. 1999. Mycotoxins (trichothecenes, zearalenone and fumonisins) in cereals associated with human red mold intoxications stored since 1989 and 1991 in China. *Natural Toxins*, **7**: 93–97.

Li, W.B., Pria, W.D. Jr., Lacava, P.M., Qin, X. and Hartung, J.S. 2003. Presence of *Xylella fastidiosa* in sweet orange fruit and seeds and its transmission to seedlings. *Phytopathology*, **93**: 953–958.

Li, Z., Campbell, M., Murphy, J., Lam, S. and Xu, J.R. 2000. The *BMP1* gene is essential for pathogenicity in the gray mold fungus *Botrytis cinerea*. *Molecular Plant–Microbe Interactions*, **13**: 724–732.

Lopéz-Lopéz, K., Hernández-Flores, J.L., Cruz-Aquilar, M. and Alvarez-Morales, A. 2004. In: *Pseudomonas syringae* pv. *phaseolicola* expression of *argK* gene, encoding the phaseolotoxin-resistant ornithine carbamoyltransferase, is regulated indirectly by temperature and directly by a precursor resembling carbamoylphosphate. *Journal of Bacteriology*, **186**: 146–153.

López-Solanilla, E., Bronstein, P.A., Schneider, A.R. and Collmer, A. 2004. HopPto N is a *Pseudomonas syringae* Hrp (type III secretion system) cysteine protease effector that suppresses pathogen-induced necrosis associated with both compatible and incompatible plant interactions. *Molecular Microbiology*, **57**: 353–367.

Luo, Y., Ma, Z. and Michailides, T.J. 2001. Analysis of factors affecting latent infection and sporulation of *Monilinia fructicola* on prune fruit. *Plant Disease*, **85**: 999–1003.

Mali, O.P., Pateria, H.M. and Chauhan, S.K. 1986. Mycoflora and aflatoxin in wet harvest sorghum. *Indian Phytopathology*, **39**: 409–413.

Marchel, R., Tabary, L., Valade, M., Moncomble, D., Viaux, L., Robillard, B. and Jeandet, P. 2001. Effects of *Botrytis cinerea* infection on champagne wine foaming properties. *Journal Science of Food and Agriculture*, **81**: 1371–1378.

Martinez, C., Rioux, D. and Tweddell, R.J. 2004. Ultrastructure of the infection process of potato tuber by *Helminthosporium solani*, causal agent of potato silver scurf. *Mycological Research*, **108**: 828–836.

Matsumoto, H., Jitareerat, P., Baba, Y. and Tsujumu, S. 2003a. Comparative study of regulatory mechanisms of pectinase production by *Erwinia carotovora* subsp. *carotovora* and *E. chrysanthemi*. *Molecular Plant–Microbe Interactions*, **16**: 226–237.

Matsumoto, H., Muroi, H., Umehara, M., Yoshitake, Y. and Tsuyumu, S. 2003b. Peh production, flagellum synthesis and virulence reduced in *Erwinia carotovora* subsp. *carotovora*. *Molecular Plant–Microbe Interactions*, **16**: 389–397.

Mattinen, L., Tshuikina, M., Mäe, A. and Pirhonen, M. 2004. Identification and characterization of Nip, necrosis-inducing virulence protein of *Erwinia carotovora* subsp. *carotovora*. *Molecular Plant–Microbe Interactions*, **17**: 1366–1375.

Maude, R.B. and Presly, A.H. 1977. Neck rot (*Botrytis allii*) of bulb onions. 1. Seed-borne infection and its relationship to the disease in the onion crop. *Annals of Applied Biology*, **86**: 163–180.

McEvoy, J.L., Park, E. and Conway, W.S. 2003. Virulence factors involved in postharvest apple decay by fungal pathogens. *Phytopathology*, **93**: S 105 (Abst.).

Medina, A.C. and Grogan, R.G. 1961. Seed transmission of bean mosaic viruses. *Phytopathology*, **51**: 452.

Medina-Suarez, R., Manning, K., Fletcher, J., Aked, J., Bird, C.R. and Seymour, G.B. 1997. Gene expression in the pulp of ripening bananas. *Plant Physiology*, **115**: 453–461.

Melo, M.B. de, Matsuoka, K. and Ansari, C.V. 1992. Histopathological examination of natural and induced infections by *Phytophthora capsici* on squash seeds. *Fitopatologia Brasileira*, **17**: 96.

Meneguzzo, J., Rizzon, L.A., Miele, A. and Ayub, M.A.Z. 1999. The effect of bunch rot on the composition of the must of *Vitis vinifera* cv. Gewurztraminer. *Fitopatologia Brasileira*, **24**: 45–48.

Monteiro, S., Barakat, M., Picarra-Pereira, M.A., Teixeira, A.R. and Ferreira, R.B. 2003. Osmotin and thaumatin from grape: A putative general defense mechanism against pathogenic fungi. *Phytopathology*, **93**: 1505–1512.

Mordue, J.E.M. 1988. *Thecaphora solani*. CMI descriptions of pathogenic fungi and bacteria No.966. *Mycophytopathologia*, **103**: 177–178.

Moreno, I.M., Thompson, J.R. and Garcí-Arenal, F. 2004. Analysis of the systemic colonization of cucumber plants by *Cucumber green mottle mosaic virus*. *Journal of General Virology*, **85**: 749–759.

Mukaihara, T., Tamura, N., Murata, Y. and Iwabuchi, M. 2004. Genetic screening of *Hrp* type III-related pathogenicity genes controlled by the *HrpB* transcriptional activator in *Ralstonia solanacearum*. *Molecular Microbiology*, **54**: 863–875.

Mullins, E.D., McCollum, T.G. and McDonald, R.E. 2000. Consequences on ethylene metabolism of inactivating the ethylene receptor sites in diseased nonclimacteric fruit. *Postharvest Biology and Technology*, **19**: 155–164.

Nakamura, M., Suprapta, D.N., Iwai, H. and Arai, K. 2001. Comparison of endopolygalacturonase activities of citrus and noncitrus races of *Geotrichum candidum* and cloning and expression of the corresponding genes. *Molecular Plant Pathology*, **2**: 265–274.

Narayanasamy, P. 2002. *Microbial Plant Pathogens and Crop Disease Management.* Science Publishers, Enfield, Newhampshire, USA.

Narayanasamy, P. 2005. *Immunology in Plant Health and its Impact on Food Safety.* The Haworth Press, Inc., New York.

Naumann, K. 1963. On the occurrence of bacteria in cucumber seeds from fruits infected with *Pseudomonas lachrymans*. *Phytopathologische Zeitschrift*, **48**: 258.

Ngoko, Z., Marasas, W.F.O., Rheeder, J.P., Shephard, G.S., Wingfield, M.J. and Cardwell, K.F. 2001. Fungal infection and mycotoxin contamination of maize in the humid forest and western highlands of Cameroon. *Phytoparasitica*, **29**: 352–360.

Nishijima, K.A., Alvarez, A.M., Hepperly, P.R., Shintaku, M.H., Keith, L.M., Sato, D.M., Bushe, B.C., Armstrong, J.W. and Zee, F.T. 2004. Association of *Enterobacter cloacae* with rhizome rot of edible ginger in Hawaii. *Plant Disease*, **88**: 1318–1327.

Nissimen, R., Lai, F.M., Laine, M.J., Bauer, P.J., Reilley, A.A., Li, X., De Boer, S.H., Ishimaru, C.A. and Metzler, M.C. 1997. *Clavibacter michiganensis* subsp. *sepedonicus* elicits a hypersensitive response in tobacco and secretes hypersensitivity-inducing proteins. *Phytopathology*, **87**: 678–684.

Nöel, L., Thieme, F., Gäbler, J., Büttner, D. and Bonas, U. 2003. XopC and XopJ, two novel type III effector proteins from *Xanthomonas campestris* pv. *vesicatoria*. *Journal of Bacteriology*, **185**: 7092–7102.

Nolan, P.A. and Campbell, R.N. 1984. *Squash mosaic virus* detection in individual seeds and seed lots of cucurbits by enzyme-linked immunosorbent assay. *Plant Disease*, **68**: 971.

Oguiza, J.A., Rico, A., Rivas, L.A., Sutra, L., Vivian, A. and Murillo, J. 2004. *Pseudomonas syringae*, pv. *phaseolicola* can be separated into two genetic lineages distinguished by the possession of the phaseolotoxin biosynthetic cluster. *Microbiology*, **150**: 473–482.

Okumura, K., Hyodo, H., Kato, M., Ikoma, Y. and Yano, M. 1999. Ethylene biosynthesis in sweetpotato root tissue infected by black rot fungus (*Ceratocystis fimbriata*). *Postharvest Biology and Technology*, **17**: 117–125.

Pati, S.P. and Ray, R.C. 2000. Physiology of post-harvest spoilage of sweetpotato (*Ipomoea batatas* L.) *Journal of Mycopathological Research*, **28**: 59–64.

Pautot, V., Holzer, F.M., Chaufaux, J. and Walling, L. 2001. The induction of tomato leucine aminopeptidase genes (LapA) after *Pseudomonas syringae* pv. *tomato* infection is primarily a wound response triggered by coronatine. *Molecular Plant–Microbe Interactions*, **14**: 214–224.

Pedras, M.S.C., Ahiahonu, P.W.K. and Hossai, M. 2004. Detoxification of the cruciferous phytoalexin brassinin in *Sclerotinia sclerotiorum* requires an inducible glucosyltransferase. *Phytochemistry*, **65**: 2685–2694.

Pezet, R., Viret, O., Perret, C. and Tabacchi, R. 2003. Latency of *Botrytis cinerea* Pers. Fr. and biochemical studies during growth and ripening of two grape berry cultivars respectively susceptibility and resistant to gray mold. *Journal of Phytopathology*, **152**: 208–214.

Podila, G.K., Rogers, L.M. and Kolatukkudy, P.E. 1993. Chemical signals from avocado surface wax trigger germination and appressorium formation in *Colletotrichum gloeosporioides*. *Plant Physiology*, **103**: 267–272.

Presly, A.H. 1977. *Botrytis species on overwintered salad onions.* Doctoral Thesis, University of Birmingham, Birmingham, U.K.

Preston, J.F., Rice, J.D., Ingram, L.O. and Keen, N.T. 1992. Differential depolymerization mechanisms of pectate lyases secreted by *Erwinia chrysanthemi* EC 16. *Journal Bacteriology*, **174**: 2039–2042.

Pring, R.J., Nash, C., Zakaria, M. and Bailey, J.A. 1995. Infection process and host range of *Colleotrichum capsici. Physiological and Molecular Plant Pathology*, **46**: 137–152.

Proctor, R.H., Desjardins, A.E., Plattner, R.D. and Hohn, T.M. 1999. A polyketide synthase gene required for biosynthesis of fumonisin mycotoxins in *Gibberella fujikuroi* mating population A. *Fungal Genetics and Biology*, **27**: 100–112.

Prusky, D. 1998. The use of antioxidants to delay the onset of anthracnose and stem- end rot in avocado fruits after harvest. *Plant Disease*, **72**: 381–384.

Prusky, D. 1996. Pathogen quiescence in postharvest diseases. *Annual Review of Phytopathology*, **34**: 413–434.

Prusky, D., Drori, N., Kramer-Haimovich, H. and Rollins, J. 2003. External pH and nitrogen source affect secretion of pectate lyase by *C. gloeosporioides. Phytopathology*, **93**: S 105 (Abst.).

Prusky, D., Karni, L., Kobiler, I. and Plumbley, R.A. 1990. Induction of the antifungal diene in unripe avocado fruits: effect of inoculation with *Colletotrichum gloeosporioides*. *Physiological and Molecular Plant Pathology*, **37**: 425–435.

Prusky, D., Keen, N.T. and Eaks, I. 1983. Further evidence for the involvement of a preformed antifungal compound in the latency of *Colletotrichum gloeosporioides* on unripe avocado fruits. *Physiological Plant Pathology*, **22**: 189–198.

Prusky, D., Keen, N.T., Sims, J.J. and Midland, S.L. 1982. Possible involvement of an antifungal disease in the latency of *Colletorichum gloeosporioides* on unripe avocado fruits. *Phytopathology*, **72**: 1578–1582.

Prusky, D., Kobiler, I., Flaishman, Y., Sims, J.J., Midland, S.L. and Keen, N.T. 1991. Identification of an antifungal compound in unripe avocado fruits and its possible involvement in the quiescent infection of *Colletotrichum gloeosporioides. Journal of Phytopathology*, **132**: 319–327.

Prusky, D., McEvoy, J.L., Leverentz, B. and Conway, W.S. 2001. Local modulation of host pH by *Colletotrichum* species as a mechanism to increase virulence. *Molecular Plant–Microbe Interactions*, **14**: 1105–1113.

Prusky, D., McEvoy, J.L., Saftner, R., Conway, W.S. and Jones, R. 2004. Relationship between host acidification and virulence of *Penicillium* spp. on apple and citrus fruit. *Phytopathology*, **94**: 44–51.

Prusky, D., Plumbley, R.A. and Kobiler, I. 1991. The relationship between antifungal diene levels and fungal inhibition during quiescent infection of unripe avocado fruits by *Colletotrichum gloeosporioides. Plant Pathology*, **40**: 45–52.

Pszczolkowski, P., Latorre, B.A. and Ceppi di Lecco, C. 2001. Mold on late harvested Cabernet Sauvignon grapes and their effect on must and wine quality. *Ciencia e Investigación Agraria*, **28**: 157–163.

Ramos, A.M., Nieto-Angel, D., Dominguez-Alavarez, J.L. and Delgadillo-Sánchez, F. 2001. Quality and tolerance in papaya fruits (*Carica papaya*, L.) to the inoculation of the fungus *Colletotrichum gloeosporioides* Penz. in postharvest. *Revista Chapingo, Serie Horticultura*, **7**: 119–130.

Rankin, H.W. 1954. Effectiveness of seed treatment for controlling anthracnose and gummy stem blight of watermelon. *Phytopathology*, **44**: 675.

Reignault, P., Kunez, C., Delage, N., Moreau, E., Vedel, R., Aamada, W., Bompeix, G. and Boccara, M. 2000. Host-and symptom-specific pectinase isozymes produced by *Botrytis cinerea. Mycological Research*, **104**: 421–428.

Reino, J.L., Hernández-Galán, R., Durán-Patrón, R. and Collado, I.G. 2004. Virulence-toxin production relationship in isolates of plant pathogenic fungus *Botrytis cinerea. Journal of Phytopathology*, **152**: 563–566.

Rha, E., Park, H.J., Kim, M.O., Chung, Y.R., Lee, C.W. and Kim, J.W. 2001. Expression of exopolygalacturonases in *Botrytis cinerea. FEMS Microbiology Letters*, **201**: 105–109.

Rich, A.E. 1983. *Potato Diseases*, Academic Press, New York, USA.

Robertson, J.A., Roberts, R.G. and Champman, G.W. Jr. 1985. An evaluation of 'heat damage' and fungi in relation to sunflower seed quality. *Phytopathology*, **75**: 142.

Rodov, V., Ben-Yehoshua, S., Fang, D., Kim, J.J. and Ashkenazi, R. 1995. Preformed anti-fungal compounds of lemon fruit: citral and its relation to disease resistance. *Journal of Agricultural and Food Chemistry*, **43**: 1057–1061.

Rolke, Y., Liu, S., Quidde, T., Williamson, B., Schouten, A., Weltring, K.M., Siewers, V., Tenberge, K.B., Tudzynski, B. and Tudzynski, P. 2004. Functional analysis of H_2O_2-generating system in *Botrytis cinerea* the major Cu-Zn- superoxide dimutase (BCSOD1) contributes to virulence on French bean, whereas glucose oxidase is dispensable. *Molecular Plant Pathology*, **5**: 17–27.

Roy, K.W. and Abney, T.S. 1988. Colonization of pods and infection of seeds by *Phomopsis longicolla* in susceptible and resistant soybean lines inoculated in the greenhouse. *Canadian Journal of Plant Pathology*, **10**: 317.

Rubin, E., Baider, A. and Cohen, Y. 2001. *Phytophthora infestans* produces oospores in fruits and seeds of tomato. *Phytopathology*, **91**: 1074–1080.

Ruiz, E. and Ruffner, H.P. 2002. Immunodetection of *Botrytis*-specific invertase in infected grapes. *Journal of Phytopathology*, **150**: 76–85.

Ryder, E.J. 1964. Transmission of common *Lettuce mosaic virus* through the genetics of the lettuce plant. *Plant Disease Reporter*, **48**: 522.

Sackston, W.E. 1981. Downy mildew of sunflower. In: *The Downy Mildews* (ed.) D.M. Spencer, pp. 545–575. Academic Press, London, U.K.

Salánki, K., Gellért, A., Huppert, E., Náray- Szabó,G. and Balázs, E. 2004. Compatibility of the movement protein and coat protein of cucumoviruses is required for cell-to-cell movement. *Journal of General Virology*, **85**: 1039–1048.

Samish, S. and Elinger-Tulczynska, R. 1963. Distribution of bacteria within the tissue of healthy tomatoes. *Applied Microbiology*, **11**: 7–10.

Sánchez-Torres, P. and González-Candelas, L. 2003. Isolation and characterization of genes differentially expressed during the interaction between apple fruit and *Penicillium expansum. Molecular Plant Pathology*, **4**: 447–457.

Sangchote, S. and Juangbhanich, P. 1984. Seed transmission of *Colletotrichum capsici* on pepper (*Capsicum* spp.) *Kasetsart Journal of Natural Science* **18**: 7.

Schwarz, P.B., Schwarz, J.G., Zhou, A., Prom, L.K. and Steffenson, B.J. 2001 Effect of *Fusarium graminearum* and *F. poae* infection on barley and malt quality. *Monaltsschrift für Brauwissenchaft*, **54**: 55–63.

Shan, L., Oh, H., Chen, J., Guo, M., Zhou, J., Alfano, J.R., Collmer, A., Jia, X. and Tang, X. 2004. The *HopPto* locus of *Pseudomonas syringae* pv. *tomato* DC 2000. encodes a type III chaperone and a cognate effector. *Molecular Plant–Microbe Interactions*, **17**: 447–455.

Sherf, A.F. and Macnab, A.A. 1986. *Vegetable Diseases and Their Control.* John Wiley and Sons, New York.

Singh, T. and Sinclair, J.B. 1985. Histopathology of Cercospora sojina in soybean seeds. Phytopathology, 75: 185.

Singha, K.D. 2000. Mycoflora on discoloured grains of hybrid rice and its effect on germination in Assam. *Annals of Agri Bio Research*, **5**: 155–157.

Skadsen, R.W. and Hohn, T.M. 2004. Use of *Fusarium graminearum* transformed with *gfp* to follow infection patterns in barley and *Arabidopsis*. *Physiological and Molecular Plant Pathology*, **64**: 45–53.

Skoric, V. 1927. Bacterial blight of pea: overwintering, dissemination and pathological histology. *Phytopathology*, **17**: 611.

So, J.S., Lim, H.T., Oh, E.T., Heo, T.R., Koh, S.C., Leung, K.T., Lee, H. and Trevors, I.T. 2002. Visualizing the infection process of *Xanthomonas campestris* in cabbage using green fluorescent protein. *World Journal of Microbiology and Biotechnology*, **18**: 17–21.

Spotts, R.A. and Cervantes, L.A. 2001. Disease incidence-inoculum dose relationships for *Botrytis cinerea* and *Penicillium expansum* and decay of pear fruit using dry, airborne conidia. *Plant Disease*, **85**: 755–759.

St. Angelo, A.J., Kuc, J.C. and Ory, R.L. 1979. Role of lipoxygenase and lipid oxidation in quality of oil seeds. *Journal of Agriculture and Food Chemistry*, **27**: 229–234.

St. Angelo, A.J. and Ory, R.L. 1983. Lipid degradation during seed deterioration. *Phytopathology*, **73**: 315–317.

Stotz, H.U., Powell, A.L.T., Damon, S.E., Greve, C.L., Benett, A.B. and Labavitch, J.M. 1993. Molecular characterization of a polygalacturonase inhibitor from *Pyrus communis* L. cv. Bartlett. *Plant Physiology*, **102**: 133–138.

Strobel, G.A. 1963. A xylanase system produced by *Diplodia viticola. Phytopathology*, **52**: 592–596.

Suzuki, F., Zhu, Y.F., Sawada, H. and Matsuda, I. 1998. Identification of proteins involved in toxin production by *Pseudomonas glumae*. *Annals of the Phytopathological Society, Japan*, **64**: 75–79.

Swinburne, T.R. 1976. Stimulants of germination and appressoria formation by *Colletotrichum musae* (Berk & Ccurt.) Arx. In banana leachate. *Phytopathologische Zeitschrift*, **87**: 74–90.

Swinburne, T.R. 1983. Quiescent infections in postharvest diseases. In: *Post-Harvest Pathology of Fruits and Vegetables* (ed.) C. Dennis, pp. 1–21, Academic Press, London.

Tanaka, M.A.S. 2001. Survival of *Fusarium moniliforme* in corn seeds stored under two different conditions. *Fitopatologia Brasileira*, **26**: 60–64.

Tefera, T. 2001. Seed transmission of bean common bacterial blight and its influence on yield. *Tropical Science*, **41**: 185–191.

Tsai, G.J. and Yu, S.C. 1999. Detecting *Aspergillus parasiticus* in cereals by an enzyme-linked immunosorbent assay. *International Journal of Food Microbiology*, **50**: 181–189.

Tsuchizaki, T., Yora, K. and Asuyama, H. 1970. Seed transmission of viruses in cowpea and Azuki bean plants. II. Relations between seed transmission and gamete infection. *Annals of Phytopathological Society, Japan*, **36**: 237.

Tsuji, G., Fujii, S., Tsuge, S., Shiraishi, T. and Kubo, Y. 2003. The *Colletotrichum lagenarium* Ste 12-like gene CST1 is essential for appressorium penetration. *Molecular Plant–Microbe Interactions.* **16**: 315–325.

Uhm, K.H., Ahn, I.P., Kim, S. and Lee, Y.H. 2003. Calcium/calmodulin-dependent signaling for prepenetration development in *Colletotrichum gloeosporioides. Phytopathology*, **93**: 82–87.

Umechuruba, C.I. and Nuachukwu, E.O. 2002. Long-term survival and pathogenicity of seedborne fungi on soybeans stored at 5°C. *Tropical Science*, **42**: 42–45.

Valette-Collet, O., Cimerman, A., Reignault, P., Levis, C. and Boccara, M. 2003. Disruption of *Botrytis cinerea* pectin methylesterase gene *Bcpme1* reduces virulence on several host plants. *Molecular Plant–Microbe Interactions*, **16**: 360–367.

Van Kan, J.A.L., van't Kloster, J.W., Wagemakers, C.A.M., Dees, D.C.T. and van der Vlugt-Bergmans, C.J.B. 1997. Cutinase A of *Botrytis cinerea* is expressed, but not essential during penetration of gerbera and tomato. *Molecular Plant–Microbe Interactions*, **10**: 30–38.

Verhoef, K. 1974. Latent infections by fungi. *Annual Review of Phytopathology*, **12**: 99–110.

Viret, O., Keller, M., Jaudzems, V.G. and Cole, F.M. 2004. *Botrytis cinerea* infection of grapes flowers: light and electron microscopical studies of infection sites. *Phytopathology*, **94**: 850–857.

Vonk, H.J. and Western, J.R.H. 1984. *Comparative Biochemistry and Physiology of Enzymatic Digestion*, Academic Press, London, U.K.

Waller, J.M. 2002. Postharvest diseases. In: *Plant Pathologists Pocketbook* (eds) J.M. Waller, J.M. Lenné and S.J. Waller, pp. 39–54, CAB International, Wallingford, U.K.

Wang, D. and Maule, A.J. 1992. Early embryo invasion as a determinant in pea of the seed transmission of *Pea seedborne mosaic virus*. *Journal of General Virology*, **73**: 1615.

Wattad, C.D., Kobiler, D., Dinoor, A. and Prusky, D. 1997. Pectate lyase of *Colletotrichum gloeosporioides* attacking avocado fruits: cDNA cloning and involvement in pathogenicity. *Physiological and Molecular Plant Pathology*, **50**: 197–212.

Welling, B. 1968. The influence of thresh damage on microflora and germination of barley during storage. *Tidsskrift for Planteavl*, **72**: 513–519.

Wilcke, W.F., Ileleji, K.E., Gupta, P., Morey, R.V. and Meronuck, R.A. 2001. Comparison of sample storage and conditioning methods for corn storability tests. *Transactions of the ASAE*, **44**: 369–376.

Wilcke, W.F., Ng, H.F., Meronuck, R.A., Morey, R.V., Gupta, P. and Lang, J.P. 1999. Storability of wheat infected with *Fusarium* head blight. *Transactions of the ASAE*, **42**: 733–742.

Wiles, A.B. and Walker, J.C. 1951. The relation of *Pseudomonas lachrymans* to cucumber fruits and seeds. *Phytopathology*, **41**: 1059–1064.

Wilson, D.M., Sydenham, E.W., Lombaert, G.A., Trucksess, M.W., Abramson, D. and Bennett, G.A. 1998. Mycotoxins analytical techniques. In: *Mycotoxins in Agriculture and Food Safety* (eds) K.K. Sinha and D. Bhatnagar, pp. 135–181. Marcel Dekker, New York.

Wilson, R.A., Calvo, A.M., Chang, P.K. and Keller, N.P. 2004. Characterization of the *Aspergillus-parasiticus* 12 desaturase gene: a role for lipid metabolism in the *Aspergillus*-seed interaction. *Microbiology*, **150**: 2881–2888.

Xiao, C.L. and Rogers, J.D. 2004. A postharvest fruit rot in d'Anjou pears caused by *Sphaeropsis pyriputrescens* sp. nov. *Plant Disease*, **88**: 114–118.

Xu, H., Annis, S., Linz, J. and Trail, F. 2000. Infection and colonization of peanut pods by *Aspergillus parasiticus* and the expression of the aflatoxin biosynthetic gene *nor-1* in infection hyphae. *Physiological and Molecular Plant Pathology*, **56**: 185–196.

Yakoby, N., Beno-Moualem, D., Keen, N.T., Dinoor, A., Pines, O. and Prusky, D. 2001. *Colletotrichum gloeosporioides pel E* is an important virulence factor in avocado fruit–fungus interactions. *Molecular Plant–Microbe Interactions*, **14**: 988–995.

Yakoby, N., Freeman, S., Dinoor, A., Keen, N.T. and Prusky, D. 2000a. Expression of pectate lyase from *Colletotrichum gloeosporioides* in *C. magna* promotes pathogenicity. *Molecular Plant–Microbe Interactions*, **13**: 887–891.

Yakoby, N., Kobiler, I., Dinoor, A. and Prusky, D. 2000b. pH regulation of pectate lyase secretion modulates the attack of *Colletotrichum gloeosporioides* in avocado fruits. *Applied and Environmental Microbiology*, **66**: 1026–1030.

Yamaguchi, J., Takayanagi, N., Komeda, K., Takano, Y. and Okuno, T. 2004. cAMP-PKA signaling regulates multiple steps of fungal infection cooperatively with Cmk1 MAP kinase in *Colletotrichum lagenarium*. *Molecular Plant–Microbe Interactions*, **17**: 1355–1365.

Yao, C.L., Conway, W.S., Ren, R.H., Smith, D., Ross, G.S. and Sams, C.E. 1999. Gene encoding polygalacturonase inhibitor in apple fruit is developmentally regulated and activated by wounding and fungal infection. *Plant Molecular Biology*, **39**: 1231–1241.

Yoshioka, S., Okumura, K. and Hyodo, H. 2001. Ethylene biosynthesis in sweet potato root tissue infected by *Ceratocystis fimbriata*. *Acta Horticulturae*, **553**, 139–141.

Zhang, J.X., Bruton, B.D. and Biles, C.L. 1999. Purification and characterization of a prominent polygalacturonase isozyme produced by *Phomopsis cucurbitae* in decayed muskmelon fruit. *Mycological Research*, **103**: 21–27.

Zhao, Y., Thilmony, R., Bender, C.L., Schaller, A., He, S.H. and Howe, G.A. 2003. Virulence systems of *Pseudomonas syringae* pv. *tomato* promote bacterial speck disease in tomato by targeting the jasmonte signaling pathway. *Plant Journal*, **36**: 485–499.

PART II

FACTORS INFLUENCING DISEASE DEVELOPMENT IN STORAGE

5

INFLUENCE OF CULTIVATION PRACTICES AND HARVESTING METHODS

Infection of fruits and vegetables by the microbial pathogens causing postharvest diseases frequently occurs under field conditions prior to harvest. Obviously, the cultivation practices that predispose the plants, as well as the produce, can be expected to contribute to the level of disease incidence occurring at the postharvest stage. Furthermore, much of the primary inoculum of many postharvest diseases is carried from the field. Studies on the relationship between cultivation practices and incidence of diseases in storage have stressed the need to pay attention during all stages of crop growth to reduce the losses due to postharvest diseases.

5.1 CULTIVATION PRACTICES

5.1.1 Sources of Inoculum

The seeds and seed materials infected by microbial pathogens form the primary sources of inoculum. The plants growing from such infected planting materials carry the pathogens systemically and the pathogens may remain quiescent, expressing symptoms later during storage. The crucial role of infected onion seeds in the incidence of neck rot disease caused by *Botrytis allii* was established by the positive relationship between the percentage of infected seeds and the percentage of infected stored bulbs (Maude and Presley, 1977).

Postharvest Pathogens and Disease Management, by P. Narayanasamy

The importance of seed health testing to reduce the incidence of *Leptosphaeria maculans* in brassica, *Clavibacter michiganensis* subsp. *michiganensis* in tomato, and *Lettuce mosaic virus* (LMV) in lettuce has been clearly demonstrated (Maddox, 1998). Tolerance limits for important seed-borne pathogens have been worked out. The excellent control achieved in California for more than 20 years in the case of lettuce mosaic disease by adopting a tolerance of 0 infected seeds in 30,000 seeds is an eye opener for all involved in the effective management of postharvest diseases. The outbreaks of a relatively new economically important disease, watermelon bacterial fruit blotch (BFB) caused by *Acidovorax avenae* subsp. *citrulli* (*Aac*), indicate that assessment of seed health by employing sensitive techniques is essential to obtain results rapidly (Minsavage et al., 1995).

A rapid polymerase chain reaction (PCR) assay to detect *Phytophthora infestans*, causing late blight disease in seed potato tubers, has been developed. By using the PINF and internal transcribed spacer (ITS) 5 primers, the pathogen DNA from zoospores or oospores could be detected at concentrations as low as 10pg/ml. These two primers could be employed for rapid and sensitive detection of *P. infestans* in potato tubers and hence to reduce the spread of the pathogen through potato tubers (Wangsomboondee and Ristaino, 2002). It is very difficult to detect *Botrytis aclada* and other *Botrytis* spp. causing onion neck rot disease at the time of harvest by using conventional isolation methods. By employing sequence-characterized amplified region primers (BA2f/BA1r), it was possible to identify and distinguish five groups of *B. acalada* types AI and AII (*B. allii*), *B. byssoidea*, *B. squamosa*, and *B. cinerea*. *B. aclada* could be detected in artificially inoculated onion bulb tissue and in mature onion leaves showing no visible symptoms (Nielsen et al., 2002). Various diagnostic techniques for the detection and identification of pathogens present in seeds and propagative plant materials have been described in Chapter 2. It will be possible to greatly reduce the infected planting materials by using one or more suitable detection methods and to prevent introduction of pathogen(s) into the field or storage facilities.

It is well known that plant viruses can cause both quantitative and qualitative reduction in the yields of fruit and vegetable crops, although the effects of virus infection in harvested produce have not been clearly determined. *Plum pox virus* (PPV), causing a destructive disease of stone-fruit crops, is transmitted by various aphid species among which *Aphis fabae*, *A. spiraecola*, *Brachycaudus persicae*, and *Myzus persicae* are more efficient. Not only the infected plants and seedlings can be sources of virus infection; the infected peach fruits, which may not clearly exhibit symptoms, can also serve as sources of infection. When the aphids were given a 3-day acquisition access period, PPV was transmitted to 35–50% of healthy peach seedlings, suggesting that PPV-infected fruit has the potential to be sources of infection for long-distance dispersal (Gildow et al., 2004). The effect of rouging PPV-infected peach trees on the spread and persistence of severe strain PPV-M in 19 peach orchard blocks in Southern France was assessed. During a 7 to 10-year period, disease incidence was reduced in all blocks with an annual incidence varying from 2 to 6%. However, new PPV-M infection was seen because of exogenous sources of inoculum, symptom expression in trees with latent infection, and failure to recognize infection by visual inspection. It is suggested that reliable detection techniques should be followed for effective removal of all potential sources of infection within, and in the vicinity of, orchards to contain the PPV incidence (Dallot et al., 2004).

5.1.2 Crop Sanitation

Sanitation involving the collection and incineration of fallen leaves, twigs, and immature fruits has been found to be effective in reducing disease incidence. Elimination of dried panicles and mummified mango fruits reduced inoculum of *Colletotrichum gloeosporioides* causing anthracnose disease (Lim and Khoo, 1985). A similar reduction in incidence of both mango anthracnose and bacterial black spot (caused by *Xanthomonas campestris* pv. *mangiferae-indicae*) was reported by Fitzell and Peak (1986). Removal of all nonbearing flower panicles and all dead leaves and branches accounted for a significant reduction in the incidence and severity of both anthracnose and soft brown rot (caused by *Nattrassia mangiferae*) in mango compared with trees left without this treatment (Saaiman, 1997). *Botrytis cinerea* sporulated on infected grape berries on the vineyard floor, senescing leaves of weeds (grasses and broad-leaf species), necrotic kiwifruit leaf and fruit remnants, and senescing leaves of nearby crops such as *Citrus* sp. Proper disposal of these infected plant materials may lower disease incidence in both grapes and kiwifruits (Michailides and Elmer, 2000).

Soil-borne sources of infection have been shown to be important in the incidence of diseases such as onion neck rot caused by *B. allii*. Infected onion debris present in the soil are potential overwintering sources of *B. allii* from which crops raised in following spring might become infected (Maude, 1983). In the case of carrot, *B. cinerea* and *Sclerotinia sclerotiorum* were present among foliage debris adhering to the roots during harvest and the infected debris was considered to be an important source of primary infection (Goodliffe and Heale, 1975; Geary, 1978). During early growth stages of kiwifruit (first female flowers opening), the overwintering mycelium of *B. cinerea* present on prunings on the ground provided the majority of inoculum for infection. The sclerotia present on the prunings produced conidiophores bearing conidia (Elmer et al., 1993).

The survivability of strawberry isolates of *Colletotrichum gloeosporioides* and *C. acutatum* was studied under both laboratory and field conditions. At 11% moisture content of soil, a reduction in population occurred, reaching 95% of the initial levels in 114 to 124 days and in 72 to 74 days in the case of *C. gloeosporioides* and *C. acutatum* respectively, indicating the differential sensitivity of these two pathogens (Freeman et al., 2002). Removal and destruction of potato haulms, infected tubers, and rouging wilted and volunteer potato plants have been recommended to reduce the sources of inoculum and to avoid contamination of neighboring healthy tubers by *Ralstonia solanacearum* causing potato brown rot disease. The possibility that *R. solanacearum* race 3, biovar infecting geranium may move into potato fields, where it could cause both direct economic damage and quarantine problems, was indicated by Williamson et al. (2002). Symptomless tomato and pepper plants collected from fields with a history of tomato bacterial wilt disease contained *R. solanacearum* as revealed by immunocapture-PCR assay, suggesting that weeds and latent hosts might have a role in the survival of *R. solanacearum* between cropping cycles (Dittapongpitch and Surat, 2003).

The survival and infectivity of oospores in soils naturally infested with *Phytophthora infestans* were determined during winter fallow in central Mexico. Oospore population, viability, and infectivity differed in soils from different locations. However, viability and infectivity of oospores decreased following intercropping. The number of stem lesions and initial disease severity in potato were significantly greater in soils with moderate oospore population (20–39/g of soil) than in soils with low oospore concentration (0–19/g of soil). Thus the oospores surviving in the soil could be a source of primary inoculum for infection of both potato plants and tubers (Fernández-Pavía et al., 2004).

Xanthomonas campestris armoraciae and *Pseudomonas syringae* pv. *maculicola* infecting leafy crucifers could be recovered from infected plant debris left on the soil surface for up to 5 months. *X. campestris* pv. *campestris* and *P. syringae* pv. *maculicola* were isolated from surface plant debris and regrowth from crop stubble left after harvest in the field (Zhao et al., 2002).

5.1.3 Crop Sequences

Crop sequences, the order in which crops are grown over a period of time, have considerable influence on the incidence of field crop diseases. Some pathogens, which infect various plant parts/ organs under field conditions, can remain quiescent until they find favorable conditions in storage. Monocropping or monoculture may lead to a build up of inoculum levels, especially in the case of seed-borne and soil-borne pathogens. The incidence of basal rot disease of onion caused by *Fusarium oxysporum* f.sp. *cepae* was at high levels in many countries where onions were grown for many years continuously. Likewise, onion neck rot disease due to *Botrytis cinerea* was perpetuated by the accumulation of inoculum in soil when onions were grown without rotation of crops (Maude, 1983). A combination of collecting haulm debris after harvest and burning immediately and a 4- or 5-year rotation with nonhost crops reduced the inoculum of *Phytophthora infestans* and *P. erythroseptica* in soil (Logan, 1983). Rotation with nonhosts, such as cereals and graminaceous pastures, decreases the soil inoculum of *Ralstonia solanacearum* infecting potatoes and consequently tuber infection is reduced. Beneficial effect of crop rotation to reduce sweetpotato black rot caused by *Ceratocystis fimbriata* is also known.

The incidence and severity of *Mycocentrospora acerina*, causing licorice disease on carrot roots, were assessed during a period of 10 years when carrot was grown as a monoculture in Norway. The extent of leaf infection and storage period correlated with the severity of decay in storage (Hermansen et al., 2000). In a study to determine the influence of crop rotation on storage diseases caused by *M. acerina* and *Sclerotinia sclerotiorum* (Sclerotinia rot), it was observed that disease incidence was influenced by the frequency of carrot cultivation in the previous 4 to 5 years. The disease incidence was 24% in fields where carrot was not grown in the previous 5 years, whereas the incidence doubled (49%) in fields with a history of carrot cultivation. An increase in storage diseases was observed, especially licorice rot, when carrots had been raised for more than 1 year in the previous 5 years (Suojala and Tahvonen, 1999).

The effect of the previous crop in the crop sequence on the incidence of tomato bacterial wilt caused by *Burkholderia solanacearum* was assessed. The population of the bacteria was reduced after cowpea and rice, and a similar effect was seen on disease incidence when these crops were raised prior to tomato (Michel et al., 1996). The *Phytophthora* blight of red pepper (*Capsicum annuum*) caused by *Phytophthora capsici* was effectively reduced by raising sesame (*Sesamum indicum*) or peanut (groundnut) as intercrops in Korea (Kim, 1989). In soils treated with extract of soils on which sesame or peanut had been cultivated, the incidence of blight was markedly reduced (Lee et al., 1990, 1991).

The effects of previous crop residues on *Fusarium* head blight (FHB) of wheat were determined. FHB incidence and severity were maximum when wheat followed corn and least when wheat followed soybeans. Yields of wheat were approximately 15% lower in plots where wheat followed corn or wheat than in wheat following soybeans. The mycotoxin (deoxynivalenol, DON) content of harvested grain was significantly correlated with FBH incidence and severity. The DON level in wheat following soybeans was 25% lower than in wheat following wheat and 50% of the level in wheat following corn. Selection of cultural practices that reduce inoculum-borne residues may be effective in reducing FHB incidence (Dill-Macky and Jones, 2000). Inoculation of the seeds of wheat, barley, oat, rye, triticale, flax, bean, lentil, and chickpea with a floret infected by *Fusarium graminearum* reduced seedling emergence and caused root rot. As the temperature increased from 10 to 30°C, seedling emergence and establishment were reduced and root rot severity increased. The results showed that if these crops are grown in rotation they may serve as alternative hosts to *F. graminearum* causing FBH disease (Chongo et al., 2001).

Corn and cotton seeds may be infected by *Aspergillus flavus*, resulting in aflatoxin contamination. As corn–cotton rotations are followed commonly in South Texas, the effect of long-term residues of corn cobs on the soil surface on *A. flavus* infection was determined. Corn cobs colonized by *A. flavus*, either prior to harvest or while in soil, formed potential sources of inoculum for corn and cotton. Corn cobs from the previous season contained on average over

190-times more *A. flavus* propagules than soil from the same field. Further, even 2-year old corn cobs retained high pathogen populations (45 fold) compared with soil. The results revealed the requirement for removal and proper disposal of all sources of inoculum to effectively reduce the incidence of disease and subsequent mycotoxin contamination (Jaime-Garcia and Cotty, 2004).

The effect of crop rotation was studied on the survival of *Phytophthroa capsici* infecting pepper (chilli) at a naturally infested site in Michigan. Amplified fragment length polymorphism (AFLP) marker profiles were determined to establish the identity of the isolates of *P. capsici* recovered from the infested soil. The results showed that *P. capsici* persisted as resistant oospores for the 2 years, when cucumbers and tomatoes were rotated with pepper, and hence no significant reduction in disease incidence was seen following crop rotation (Lamour and Hausbeck, 2003).

The influence of green manures (buckwheat and canola) applied in conjunction with three crop sequences (alfalfa–potato, corn–potato, and potato–potato) on the incidence of potato scab disease was assessed in a 2-year field trial. Potatoes in fields planted to corn or alfalfa had significantly lower scab ratings compared to fields with the potato–potato sequence. Consequently, increases in yields was also realized. Streptomycetes in soils from green manure-treated plots tended to have greater in vitro pathogen-inhibitory activity. The results indicate that green manure application may contribute to active management of scab disease due to enhancement of inhibitory activity of Streptomycete community (Wiggins and Kinkel, 2005).

5.1.4 Nutrient Management

Among the various cultivation practices, crop nutrition has a significant influence on the growth and yield of crop plants as well as the development of diseases. Both major and minor nutrients may influence the susceptibility/resistance levels of crops to postharvest diseases, as in the case of field diseases. The principal sources of plant nutrients are the organic matter and inorganic fertilizers.

5.1.4.1 Organic Matter Application of organic matter enhances soil fertility and consequently the availability of nutrients to plants improves. Soil structure, water retention capacity, and aeration in soils are favorably altered by incorporation of green manures, farm yard manures, various types of composts, and other organic matter. A distinct advantage of the addition of organic matter is the activation of many microbes antagonistic to soil-borne pathogens and also the microbes that are involved in the decomposition of plant debris carrying pathogens that may infect fruits and vegetables. Organic matter incorporated into the soil promote biological destruction of pathogen inoculum through the germination-lysis mechanism and intensification of microbiological activity resulting in the decay of pathogen propagules. Organic matter may also be placed or spread on the soil surface as mulch.

Application of composted cattle manure at a rate equivalent to the recommended inorganic fertilizer rate, significantly reduced the intensity of diseases caused by *Colletotrichum piperatum* and *Erwinia carotovora* in pepper (Huelsman and Edwards, 1998). Kiwifruit crop grown using organic inputs had a lower incidence of gray mold in cold storage compared with fruit from conventional KiwiGreen systems (Michailides and Elmer, 2000). The incidence of anthracnose fruit rot of tomato caused by *Colletotrichum coccodes* was reduced in organic tomato plots amended with a high rate of compost cannery wastes compared with control plots when disease incidence was high. Marketable yield increased by 33% in compost-added organic plots. Plots amended with a high compost rate had more ripe fruits than untreated control (Abbasi et al., 2002).

5.1.4.2 Inorganic Fertilizers Application of inorganic fertilizers becomes necessary when adequate organic matter is not available to meet the crop requirements for satisfactory growth of plants to realize expected yield levels. Balanced nutrition is required to maintain both the quality and quantity of agricultural commodities.

A. Nitrogen: Nitrogen is known to favor vegetative growth of plants and also to increase the susceptibility of crop plants to several diseases caused by microbial pathogens. Nitrogen is generally applied to soils as either ammonium ($NH4^+$) ions and/or as nitrate ($NO3^-$) ions. Susceptibility of apples to rotting was increased by high nitrogen and decreased by high potash contents (Gregory and Horne, 1928). Later Muskett et al. (1938) observed that addition of nitrogen increased the nitrogen content of apple fruit and facilitated radial spread of rotting by *Cytosporina ludibunda*. Montgomery and Wilkinson (1962) recorded consistent adverse effects of applied nitrogen by increasing losses due to *Gloeosporium album*. It is possible that the influence of nitrogen may be indirect in affecting the vigor of plants, resulting in an imbalance in the K/Ca ratio of the fruit, as suggested by Sharples (1968). Berry clusters in grapevines supplied with 35% N urea and 65% N as Ca $(NO_3)_2$ had less water loss after 56 days of storage than clusters of vines supplied with 100% N as urea or 70% urea + 30% Ca $(NO_3)_2$ through fertigation (that is fertilizers added to the irrigation water). The storage decay caused by *Alternaria alternata*, *Cladosporium herbarum*, *Peniciliium* sp. *Rhizopus* sp., and *Aspergillus niger* was the least in the former treatment than in other treatments (Choudhury et al., 1999). The effect of N application on the incidence of kiwifruit rot caused by *B. cinerea* was studied. The rot in cold storage significantly increased due to excessive nitrogen application both in the year of application and in the following year (Pertot and Perin, 1999).

Excessive doses of fertilizer application have been found to favor development of diseases infecting vegetables. Higher doses of nitrogen increased the susceptibility of potato tubers to *Phytophthora infestans* (Heitefuss, 1989). The incidence and intensity of diseases infecting potato tubers during storage were assessed. Mineral fertilization (at higher rates) was responsible for higher incidence of diseases such as black scurf caused by *Rhizoctonia solani*, late blight caused by *Phytophthora infestans*, common scab due to *Streptomyces scabies*, and soft rot induced by *Erwinia carotovora* subsp. *carotovora* (Czajka et al., 1999). The effect of N fertilization on the development of *Fusarium* head blight (FHB) in wheat and consequent deoxynivalenol (DON) contamination in kernels was assessed. The FHB intensity and DON contamination in the grain were progressively increased with increasing N doses from 0 to 80 kg/ha. Hence, the amount of N fertilizer applied with a view to increasing yield has to be carefully determined due to the disease risk involved (Lemmens et al., 2004).

B. Calcium: Among the minerals required for plant growth and reproduction, calcium (Ca) has a marked influence on plant health. Ca deficiency itself may cause physiological disorders, such as apple bitter rot, and also increases susceptibility to pathogenic diseases such as rotting caused by *Gloeosporium* sp. The susceptibility was associated with low Ca and high nitrogen content. Electron microscopy of stored apple fruit with low and high calcium content at the time of harvest revealed that the disorganization of cell walls was very severe in long-stored, low-calcium fruits, especially in the central parts of the cell walls, suggesting that the disorganization might have originated in the middle lamella (Edney, 1983). A negative correlation between Ca content of apple tissue and intensity of decay caused by *Penicillium expansum* was observed. Apple with a maximum Ca content had less than 50% decay compared to controls with low Ca content (Conway and Sams, 1983). Fertilization with $Ca(NO_3)_2$ or $Ca SO_4$ was found to increase Ca contents of potato tubers. The percent surface area of potato tuber decay caused by the bacterial pathogen *Erwinia carotovora* pv. *atroseptica* was negatively correlated with the tuber Ca content. The calcium application improved the structural integrity of the cell wall (McGuire and Kelman, 1984).

Calcium may be applied to apples as foliar sprays as many as 12 times prior to harvest to increase the calcium contents of apple fruits in New Zealand. As an alternative to several foliar sprays of Ca, a combination mulch (black polythene topped with saw dust to hold it in place) to afford mechanical protection and to provide thermal insulation was used to cover the soil surface after sprinkling coarse ground gypsum (1000 or 4000 kg/ha). The mulch increased root length density and roots came closer to the surface of the soil, which had higher Ca, K, and Mg. The cation exchange capacity of the soil was also increased. The incidence and severity of bitter pit due to Ca deficiency was significantly reduced. Mulching seems to be a sustainable and low input management option that enhances apple fruit quality (Lang et al., 2001).

C. Other Minerals: Phosphate fertilizers favor plant growth and also enhance resistance to crop diseases. Higher levels of phosphorus have been reported to lower the incidence of potato common scab disease caused by *Streptomyces* *scabies* (Davis et al., 1976), potato late blight disease (Heitefuss, 1989), and cowpea anthracnose disease (Adebitan, 1996). A balanced fertilizer (composed of 12:12:18 N, P, K) reduced the susceptibility of potato to dry rot disease caused by *Fusarium solani* var. *coeruleum*, while additional nitrogen supply enhanced tuber susceptibility to disease (Boyd, 1967), indicating the possible influence of other minerals (P and K) in the incidence of storage diseases. Potassium and sodium salts suppressed sporulation and lesion formation by *Helminthosporium solani* causing silver scurf disease of potato tubers (Olivier et al., 1999).

5.1.5 Tillage

Tillage is essential to preserve the soil status in an optimum condition for plant growth. The soil is turned, mixed, loosened, or compacted by using various implements. Retention of water by soils, aeration, and soil temperature may be altered to different levels. These conditions may significantly influence the survival and dissemination of plant pathogens present in the soil or plant debris. Tillage provides a general beneficial effect due to the destruction of volunteer plants and weeds which may be the sources of inoculum for the microbial pathogens of fruits and vegetables. Deep ploughing may bury pathogens present on the soil surface or in the infected plant debris. In addition, inactivation of pathogens surviving in deeper soil layers is likely because of their exposure to solar radiation (solarization) after ploughing.

Deep ploughing has been found to be useful for the management of bean white mold disease caused by *Sclerotinia sclerotiorum*, as the sclerotia buried deep in the soil could not germinate (Abawi and Grogan, 1979). Reduction of the sources of inoculum for succeeding apple and grapevine crops was possible by deep ploughing plant debris containing resting spores which could liberate airborne spores of *Venturia inaequalis* and *Plasmopara viticola* causing scab and downy mildew diseases, respectively, on apple and grapevines. A study to compare the influence of moldboard plow, chisel plow, and no till showed that the incidence and severity of *Fusarium* head blight (FHB) of wheat were lower in moldboard-plowed plots than in either plowed or no-till plots and grain yield was greater than the other two treatments. It was suggested that the changes toward conservation tillage or reduced-till systems may contribute to FHB epidemics (Dill-Macky and Jones, 2000).

5.1.6 Effects of Rootstocks

It may be possible to reduce disease incidence by using appropriate rootstocks. The effect of 12 different rootstocks, on which Sampion apple trees were grafted, on the incidence of diseases was assessed. Apple fruits from trees on three of the 12 rootstocks tested had lowest physiological disorders, though no significant difference in the incidence of fungal diseases was noted (Kurlus and Ugolik, 1999). On the other hand, the incidence and severity of avocado anthracnose disease caused by *Colletotrichum gloeosporioides* was significantly less on 'Hass' grafted to 'Velvick' Guatemalan seedling rootstock compared with 'Duke 6' Mexican seedling rootstock. The susceptibility to anthracnose was negatively related to concentrations of antifungal dienes in leaves and mineral nutrients in leaves and fruits. Leaf diene concentration were up to 1.5-fold higher in 'Hass' trees on the 'Velvick' than the 'Duke 6' rootstock. The Velvick–Hass combination contained significantly lower leaf N content and a significantly greater Mn level in fruit flesh and significantly lower and higher leaf N / Ca and Ca+ Mg / K ratios respectively. A significant correlation (r = 0.82) between anthracnose severity and skin N/Ca ratio was also evident (Willingham et al., 2001).

5.1.7 Production System and Plant Density

The effect of production systems on the incidence and intensity of diseases affecting grapevines was determined. The mean incidence of *Botrytis* fruit rot (BFR) was 88–94% lower in grapevines grown in tunnels than in field-grown strawberry indicating that BFR could be effectively managed by using a tunnel cultivated system without application of fungicides (7-day captan schedule). Shorter periods of leaf wetness and higher temperatures that existed in tunnels might have contributed to a lower incidence of BFR (Xiao et al., 2001).

The plant density, as varied by spacing plants within and between rows, has a significant effect on the amount of foliage canopy. Consequently, the microclimate around the crop is distinctly altered. The severity of potato silver scurf caused by *Helminthosporium solani* and black scurf caused by *Rhizoctonia solani* increased with increasing planting density (Firman and Allen, 1995). The influence of within-row plant spacing on the incidence of BFR and marketable yield of annual strawberry was evaluated. Spacing exerted a dramatic influence on the cumulative and weekly *Botrytis* incidence for the late and whole season periods during the 1998–1999 season. Narrower spacings had higher disease incidence than wider spacing. Whole season marketable yields were higher at the narrower spacing, but with higher incidence of *Botrytis* (Legard et al., 2000).

5.1.8 Irrigation Systems

Irrigation can, possibly, be expected to exert greater influence on the development of both the host plant and microbial plant pathogens than other cultivation practices. Irrigation is necessary to maintain the required soil moisture levels to facilitate absorption of plant nutrients from the soil and for maintaining turgidity of tissues of fruits and vegetables. The time and frequency of irrigation as well as the kinds of irrigation, such as flooding, furrow irrigation, and sprinkling system, have a marked effect on the pathogen population build up and consequent incidence of diseases. Both flooding and furrow irrigation systems may spread the pathogen present in plant debris lying in the soil. Sprinkling systems require lower volumes of water to irrigate the same planted area compared with other irrigation system; however, this system may be responsible for enormous splash dissemination of pathogen propagules.

By adopting furrow irrigation and reducing the frequency, the progress of root and fruit rots caused by *Phytophthora capsici* in squash was significantly reduced. The disease onset was delayed and final disease severity was also reduced (Café-Filho et al., 1995). The effects of microsprinkler and flood irrigation on pear growth and postharvest quality of fruits were evaluated. The cumulative fruit diameter, weight at harvest, yield of fruits, and rate of decrease of fruit firmness during storage at 0°C did not show any difference due to the irrigation systems. However, the diameter of decay lesions due to inoculation with *Penicillium expansum* after storage for 2 months was larger in fruits from microsprinkler irrigation than in fruits from flood irrigation in the first year, but no differences were seen in the following year (Chen et al., 2001).

5.1.9 Organic and Inorganic Mulches

The potential of both organic and inorganic mulches in reducing the incidence of diseases was assessed under field conditions. The effect of application of straw mulch on the incidence of *Potato virus Y* (PVY) and the infestation of aphids, which are the vector of the virus, was determined. Straw mulching reduced the aphid infestation and PVY incidence in tubers in all the 3 years of experimentation. Straw mulch was most effective when the vector pressure and plant susceptibility to virus infection were high during the early phase of plant growth. When mulching was combined with presprouting of tubers, a synergistic (complementary) effect on reduction of PVY incidence was observed (Saucke and Döring, 2004).

Plastic foils have been used to suppress the microbial pathogens infecting high-value crops. Soil solarization involves the use of thin polyethylene sheets to cover the soil, after irrigation, during summer months or seasons with high temperatures. Solarization selectively destroys microbial pathogens, while high biological activity in the soil is maintained. Soil solarization using photoselective, low-density polyethylene film for a period of 32 to 49 days prior to planting tomatoes significantly decreased the density of *Burkholderia solanacearum* up to depths of 15 cm. The temperature in the soil at this depth was 7.1°C (Chellemi et al., 1994). Use of green-pigmented polyethylene sheet in the greenhouse reduced the conidial load of *Botrytis cinerea* and gray mold by 35 to 75% (Elad, 1997). Reflecting film mulching combined with rain-shelter treatment appreciably reduced the incidence of *Monilinia fructicola* and *Gloeosporium laeticolor* (*Glomerella cingulata*) on peach. Fruits from this combination of treatments matured early and were of better quality compared with fruits from control trees (Hyun et al., 1995). Soil mulching with polyethylene film reduced tomato bacterial canker disease caused by *Clavibacter michiganensis* subsp. *michiganensis* (Antoniou et al., 1995). The incidence of tuber blight of potato caused by *Phytophthora infestans* in plots mulched with black film averaged 32% compared with 56% plots without mulch (Glass et al., 2001). The use of plastic mulches provided greater and longer protection to summer squash (*Cucurbita pepo*) cv. Dividend compared to cv. Multipik. High total marketable yields could be obtained earlier by using either black or white mulch, due to effective control of *Watermelon mosaic virus* (WMV). However, row covers reduced the disease incidence and severity to a greater extent than mulches (Walter, 2003).

5.1.10 Organic Farming

Organic farming relies on holistic management systems for the production of agricultural and horticultural crops by the use of natural, recyclable organic materials and avoiding the use of synthetic external inputs such as chemical fertilizers and plant protection chemicals. There is growing realization of the harmful effect of the various agrochemicals utilized in crop cultivation with the aim of increasing production levels. Some studies of nonchemical and biological methods for the management of diseases of fruits and vegetables caused by microbial pathogens have demonstrated their effectiveness. In New Zealand, kiwifruits raised organically had a lower incidence of gray mold caused by *Botrytis cinerea* compared with fruits from conventional Kiwi-Green systems. Both incidence and size of *B. cinerea* populations were significantly less, even in transitional organic orchards (Michailides and Elmer, 2000).

The prevalence of food-borne human pathogens, such as *Escherichia coli* 0157: H7, *E. coli* and *Salmonella* spp., in fruits and vegetables produced in organic and conventional farms was determined. A total of 476 and 129 produce samples (tomatoes, leafy greens, lettuce, green peppers, cabbage, cucumbers, broccoli, strawberries, and apples) from 32 organic and eight conventional farms, respectively, were examined. All organic farms had applied aged or composted animal manure. The conventional and organic produce showed 1.6 and 9.7% infestation of *E. coli* respectively. Organic lettuce showed the maximum (22.4%) incidence of *E. coli* compared with other produce types. Serotype *E. coli* 0157: H7 was not found in any produce samples, whereas one organic lettuce and one organic green pepper had *Salmonella* spp. at detectable levels (Mukherjee et al., 2004). This report appears to be the first microbiological assessment of organic fruits and vegetables at farms.

Biological wastes as agricultural amendments have been reported to have both benefits and risks to plant and human health. Properly composted wastes have suppressive effects on plant pathogens in addition to the supply of nutrients to plants. In recent years, production of compost tea from compost, water and nutrient additives has been taken up by several agencies in order to transfer and multiply the microbial populations from compost into an aqueous phase that can be applied to plant surfaces and soils in ways not logistically or economically feasible with compost. This practice has resulted in both crop disease control and potential risks to increase residual populations of human pathogenic bacteria from the starting materials (Scheuerell, 2004). Compost tea is prohibited for application on crops within 90/120 days prior to harvest. A Compost Tea Task Force was formed to examine human bacterial pathogen risk from compost tea application especially from manure-based composts and to develop guidelines and recommendations (Diver, 2004).

5.2 HARVEST-RELATED OPERATIONS

5.2.1 Operations Prior to Harvest

Destruction of potato haulms 2 to 3 weeks prior to harvesting tubers has been considered to be effective for the management of tuber rot caused by *Phytophthora infestans*, since the infected foliage as a source of inoculum may be eliminated. The severity of tuber blight on surfaces and internal tissue was highly correlated. When the tubers were harvested within 1 to 4 days after haulm kill, the percentage of tuber blight was lowest. The severity of tuber blight at harvest was positively correlated with its severity after 6 months in storage (Miller et al., 2002). In contrast, the interval between haulm destruction (using diquat dibromide) and harvest had less effect on the incidence of dry rot in daughter tubers caused by *Fusarium solani* var. *coeruleum* (Carnegie et al., 2001). In the case of silver scurf disease of potato caused by *Helminthosporium solani*, early removal of haulms and harvesting about 2 weeks after removal of haulms in dry weather were useful in reducing disease incidence in storage (Stachewicz, 1999). Staggered harvesting of potato tubers (two or four rows at a time) reduced the spread of *Erwinia carotovora*. Tuber drying in the field before storage also reduced the incidence of *Rhizoctonia solani* causing black scurf disease in storage (Bouman, 1998). Delaying the harvest decreased storage decay of carrots caused by *Mycocentrospora acerina* and *Botrytis cinerea*. Continuous frost at the very end of growth impaired storage quality (Suojala, 1999).

5.2.2 Methods of Harvesting

In the crop production system various operations are undertaken with the ultimate objective of obtaining commodities that will have all the features expected by the consumer. Harvesting is the most important operation for the grower, since this operation will determine his level of profit for all his efforts, right from sowing or planting, and several factors have to be considered at this stage. The fruits and vegetables should be harvested at optimum maturity without injury or damage. Proper methods of harvesting, mode of packaging, and transportation should be selected to avoid bruising due to impact (due to shedding from trees/plants), vibration (due to loose packing), or compression (due to overfilling of containers).

The time of harvesting, when the produce reaches harvest maturity, is important to obtain fruits and vegetables of high quality. Maturation is the stage of development when the produce reaches physiological and horticultural maturity. Harvest maturity indices are determined by applying computational, physical, chemical, and physiological methods. The parameters on which harvest maturity indices are determined are described in publications on postharvest technology.

Hand or mechanical harvesting methods may be adopted depending on the quantity of the produce. Harvesting methods should minimize the wounds/injuries to fruits and vegetables, since they form the major avenues of pathogen entry into the harvested produce. From this point of view, hand harvesting may be preferred. However, cost of any operation will be the deciding factor to choose the right method under the conditions existing in a country or location. Mechanical harvesting may be used to handle large quantities of commodities rapidly. However, the injuries caused may be more, resulting in greater losses due to higher incidence of postharvest diseases. Adequate care should be taken to avoid injuries at all stages of harvest, packing, and transport prior to storage to reduce infection by postharvest pathogens (Sharma and Singh, 2000).

The feasibility of employing an electronic nose (E-nose) to predict the optimal harvest date for apples has been assessed. The volatiles of apple cvs. Jonagold and Braeburn during preclimacteric stage for two consecutive harvest years were determined. Among the parameters used to construct calibration models, the number of days before commercial harvest had a validation correlation of 0.91 for Jonagold and 0.84 for Braeburn apples. The quality characteristics such as soluble solids, acidity, starch, and firmness could also be predicted reasonably correctly. However, interpretation and use of models for predicting the optimal harvest date demand adequate care and expertise (Saevels et al., 2003).

SUMMARY

The influence of cultivation practices on the incidence and intensity of postharvest diseases in fruits and vegetables has been studied in certain pathosystems. Reduction in sources of infection, crop sanitation, crop sequences, application of organic manures and mulches, and irrigation systems have been shown to have marked influence on the extent of storage decay caused by microbial pathogens. The need for avoiding/ reducing injuries to fruits and vegetables during harvesting and transport to contain storage diseases is highlighted.

REFERENCES

Abawi, G.S. and Grogan, R.G. 1979. Epidemiology of diseases caused by *Sclerotinia* species. *Phytopathology*, **69**: 899–904.

Abbasi, P.A., Al-Dahmani, J., Sahin, F., Hoitnik, H.A.J. and Miller, S.A. 2002. Effect of compost amendments on disease severity and yield of tomato in conventional and organic production system. *Plant Disease*, **86**: 151–161.

Adaskaveg, J.E., Förster, H. and Thompson, D.F. 2000. Identification and etiology of visible quiescent infections of *Monilinia fructicola* and *Botrytis cinerea* in sweet cherry fruit. *Plant Disease*, **84**: 328–333.

Adebitan, S.A. 1996. Effects of phosphorus and weed interference on anthracnose of cowpea of Nigeria. *Fitopatologia Brasileira*, **21**: 173–179.

Antoniou, P.P., Tjmos, E.C., Andreou, M.T. and Panago-Poulos, C.G. 1995. Effectiveness, modes of action and commercial application of soil solarization for control of *Clavibacter michiganensis* subsp. *michiganensis* of tomatoes. *Acta Horticulturae*, **382**: 119–128.

Bouman, A. 1998. Staged harvesting and improvement in quality. *Kartoffelbau*, **49**: 228–230.

Boyd, A.E.W. 1967. The effects of the length of the growth period and of nutrition upon potato tuber susceptibility to dry rot (*Fusarium coeruleum*). *Annals of Applied Biology*, **60**: 231–240.

Café-Fiho, A.C., Duniway, J.M. and Davis, R.M. 1995. Effects of the frequency of furrow irrigation on root and fruit rots of squash caused by *Phytophthora capsici*. *Plant Disease*, **79**: 44–48.

Carnegie, S.F., Cameron, A.M. and Haddon, P. 2001. The effect of date of haulm destruction and harvest on the development of dry rot caused by *Fusarium solani* var. *coeruleum* on potato tubers. *Annals of Applied Biology*, **139**: 209–216.

Chellemi, D.O., Olson, S.M. and Mitchell, D.J. 1994. Effects of soil solarization and fumigation on survival of soil-borne pathogens of tomato in Northern Florida. *Plant Disease*, **78**: 1167–1172.

Chen, C., Roseberg, R.I., Sugar, D. and Selker, J.S. 2001. Effect of microsprinkler irrigation on pear fruit growth and postharvest quality. *HortTechnology*, **11**: 56–61.

Chongo, G., Gossen, B.D., Kutcher, H.R., Gilbert, J., Turkington, T.K., Fernandez, M.R. and McLaren, D. 2001. Reaction of seedling roots of 14 crop species to *Fusarium graminearum* from wheat heads. *Canadian Journal of Plant Pathology*, **24**: 132–137.

Choudhury, M.M., Lina, M.A. de, Soares, J.M. and Faria, C.M.B. de 1999. Effect of nitro gen sources and calcium application on postharvest quality of cv. Italia grapes. *Revista Brasileira de Fruticultura*, **21**: 322–326.

Conway, W.S. and Sams, C.E. 1983. Calcium infiltration of Golden Delicious apples and its effect on decay. *Phytopathology*, **73**: 1068–1071.

Czajka, W., Cwaljna, Czajka, M. and Fabisiewicz, M. 1999. Infestation of potato tubers by pathogens in relation to mineral fertilization. *Progress in Plant Protection*, **39**: 852–855.

Dallot, S., Gottwald, T., Laonne, G. and Quiot, J.B. 2004. Factors affecting the spread of *Plum pox virus* strain M in peach orchards subjected to rouging in France. *Phytopathology*, **94**: 1390–1398.

Davis, J.R., Dole, R.E. and Callihan, R.H. 1976. Fertilizer effects on common scab of potato and the relation of calcium and phosphate-phosphorus. *Phytopathology*, **66**: 1236–1241.

Dill-Macky, R. and Jones, R.K. 2000. The effect of previous crop residues and tillage on Fusarium head blight of wheat. *Plant Disease*, **84**: 71–76.

Dittapongpitch, V. and Surat, S. 2003. Detection of *Ralstonia solanacearum* in soil and weeds from commercial tomato fields using immunocapture and polymerase chain reaction. *Journal of Phytopathology*, **151**: 239–246.

Diver, S. 2004. Compost teas: A tool for rhizosphere + phyllosphere agriculture. *Phytopathology*, **94**: S 137 (Abst).

Edney, K.L. 1983. Top fruit. In: *Post-Harvest Pathology of Fruits and Vegetables* (ed.) C. Dennis, pp. 43–71, Academic Press, London.

Elad, Y. 1997. Effect of filtration of solar light on the production of conidia by field isolates of *Botrytis cinerea* and on several diseases of greenhouse grown vegetables. *Crop Protection*, **16**: 635–642.

Elmer, P.A.G., Boyd-Wilson, K., Cook, D., Gaunt, R.E., Frampton, C. and Pyke, N. 1993. Sources of *Botrytis cinerea* inoculum in kiwifruit orchards and the relationship between total inoculum potential, external fruit contamination and stem-end rot of kiwifruit. *Sixth International Plant Pathology Congress, Montreal, Canada*, p. 110.

Fernández-Pavía, S.P., Grünwald, N.J., Díaz-Valasis, M., Cadena-Hinojosa, M. and Fry, W.E. 2004. Soilborne oospores of *Phytophthora infestans* in Central Mexico survive winter fallow and infect potato plants in the field. *Plant Disease*, **88**: 29–33.

Firman, D.M. and Allen, E.J. 1995. Effects of seed size, planting density and planting pattern on the severity of silver scurf (*Helminthosporium solani*) and black scurf (*Rhioctonia solani*) diseases of potato. *Annals of Applied Biology*, **127**: 73–85.

Fitzell, R.D. and Peak, C.M. 1986. Management strategies for control of anthracnose and bacterial black spot in Northern New South Wales. *Proceedings of First Australian Mango Research Workshop, Melbourne, Australia*, pp. 253–257.

Freeman, S., Shaler, Z. and Katan, J. 2002. Survival in soil of *Colletotrichum acutatum* and *C. gloeosporioides* pathogenic on strawberry. *Plant Disease*, **86**: 965–970.

Geary, J.R. 1978. *Host–Parasite Interactions Between the Cultivated Carrot (Daucus carota L.) and Sclerotinia sclerotiorum (Lib) De Barr*. Doctoral Thesis, University of East Anglia, U.K.

Gildow, F., Damsteegt, V., Stone, A., Schneider, W., Luster, D. and Levy, L. 2004. Plum pox in North America: Identification of aphid vectors and a potential role for fruit in virus spread. *Phytopathology*, **94**: 868–874.

Glass, J.R., Johnson, K.B. and Powelson, M.L. 2001. Assessment of barriers to prevent the development of potato tuber blight caused by *Phytophthora infestans*, *Plant Disease*, **85**: 521–528.

Goodliffe, J.P. and Heale, J.B. 1975. Incipient infections caused by *Botrytis cinerea* in carrots entering storage. *Annals of Applied Biology*, **80**: 243–246.

Gregory, F.G. and Horne, A.S. 1928. A quantitative study of the course of fungal invasion of the apple fruit and its bearing on the nature of disease resistance. Part I. A statistical method of studying fungal invasion. Part II. The application of the statistical method to certain specific problems. *Proceedings of Royal Society, London, Series B*, **102**: 427–466.

Heitefuss, R. 1989. *Crop and Plant Protection: The Practical Foundations*, Ellis Horwood Limited, Chichester, U.K.

Hermansen, A., Amundsen, T., Taksdal, G., Dragland, S., Synnevag, G., Flønes, M. and Sundheim, L. 2000. Variations in infection by *Mycocentrospora acerina* in carrot monoculture plots at four sites during 1984–1995. *Acta Agriculturae Scandinavica, Section B, Soil and Plant Science*, **49**: 248–257.

Huelsman, M.F. and Edwards, C.A. 1998. Management of diseases in cucumbers (*Cucumis sativus*) and peppers (*Capsicum annum*) by using compost as fertility inputs. *Brighton Crop Protection Conference: Pests and Diseases*, **3**: 881–886.

Hyun, C.D., Soo, K.J., Tak, Y.J., Yong, C.S. and Sull, C.B. 1995. Effect of rain shelter and reflecting film mulching on fruit quality and diseases infection on peach. *RDA Journal of Agricultural Science and Horticulture*, **37**: 456–460.

Jaime-Garcia, R. and Cotty, P.J. 2004. *Aspergillus flavus* in soils and corncobs in South Texas: Implication for management of aflatoxins in corn-cotton rotations. *Plant Disease*, **18**: 1366–1371.

Kim, C.H. 1989. *Phytophthora* blight and other diseases of red pepper in Korea-Disease and Pest problems from continuous cropping. II. Soilborne diseases. *FFTC Extension Bulletin*, **302**: 10–17.

Kurlus, R. and Ugolik, M. 1999. The effect of rootstocks on the storage value of Sampion apples. *Prace z Zakresu Nauk Rolniczych*, **87**: 87–94.

Lamour, K.H. and Hausbeck, M.K. 2003. Effect of crop rotation on the survival of *Phytophthora capsici* in Michigan. *Plant Disease*, **87**: 841–845.

Lang, A., Behboudian, M.H., Kidd, J. and Brown, H. 2001. Mulch enhances apple fruit storage quality. *Acta Horticulturae*, **557**: 433–439.

Lee, H.U., Kim, C.H. and Lee, F.J. 1990. Effect of pre- and mixed cropping with nonhost plants on incidence of *Phytophthora* blight of red pepper. *Korean Journal of Plant Pathology*, **6**: 440–446.

Lee, H.U., Kim, C.H. and Nam, K.W. 1991. Suppression of *Phytophthora* blight incidence of red pepper by cropping system. *Korean Journal of Plant Pathology*, **7**: 140–146.

Legard, D.E., Xiao, C.L., Mertely, J.C. and Chandler, C.K. 2000. Effects of plant spacing and cultivar on the incidence of *Botrytis* fruit rot in annual strawberry. *Plant Disease*, **84**: 531–538.

Lemmens, M., Lew, H., Ruckenbauer, P. and Hain, 2004. The effect of nitrogen fertilization of *Fusarium* head blight development and deoxynivalenol contamination in wheat. *Journal of Phytopathology*, **152**: 1–8.

Lim, T.K. and Khoo, K.C. 1985. *Diseases and Disorders of Mango in Malaysia*. Tropical Press, Kuala Lumpur, Malaysia.

Logan, C. 1983. Potatoes. In: *Post-Harvest Pathology of Fruits and Vegetables* (ed.) C. Dennis, pp. 179–217, Academic Press, London.

Maddox, D.A. 1998. Implications of new technologies for seed health testing and the worldwide movement of seed. *Seed Science Research*, **8**: 277–284.

Maude, R.B. 1983. Onions. In: *Post-Harvest Pathology of Fruits and Vegetables* (ed.) C. Dennis, pp. 73–101, Academic Press, London.

Maude, R.B. and Presley, A.H. 1977. Neck rot (*Botrytis allii*) of bulb onions. II. Seed-borne infection in relationship to the disease in store and the effect of seed treatment. *Annals of Applied Biology*, **86**: 181–188.

McGuire, R.G. and Kelman, A. 1984. Reduced severity of *Erwinia* soft rot in potato tubers with increased calcium content. *Phytopathology*, **74**: 1250–1256.

Michailides, T.J. and Elmer, P.A.G. 2000. Botrytis gray mold of kiwifruit caused by *Botrytis cinerea* in the United States and New Zealand. *Plant Disease*, **84**: 208–223.

Michel, V.V., Hartman, G.L. and Midmore, D.J. 1996. Effect of previous crop on soil population of *Burkholderia solanacearum* bacterial wilt and yield of tomatoes in Taiwan. *Plant Disease*, **80**: 1367–1372.

Miller, J.S., Cummings, T.F., Mikitzel, L.J. and Johnson, D.A. 2002. Influence of timing of harvest in relation to haulm killing and planting date on potato tuber rot caused by *Phytophthora infestans*. *Plant Disease*, **86**: 264–268.

Minsavage, G.V., Hoover, R.J., Kucharek, T.A. and Stall, R.E. 1995. Detection of watermelon fruit blotch pathogen on seeds with the polymerase chain reaction. *Phytopathology*, **85**: 1162.

Montgomery, H.B.S. and Wilkinson, B.G. 1962. Storage experiments with Cox's Orange Pippin apples from a manurial trial. *Journal of Horticultural Science*, **37**: 150–158.

Mukherjee, A., Speh, D., Dyck, E. and Diez-Gonzalez, F. 2004. Preharvest evaluation of coliforms, *Escherichia coli, Salmonella and Escherichia coli* 0157: H7 in organic and conventional produce grown by Minnesota farmers. *Journal of Food Protection*, **67**: 874–900.

Muskett, A.E., Horne, A.S. and Colhoun, I. 1938. The effect of manuring upon apple fruits. *Annals of Applied Biology*, **25**: 50–67.

Nielsen, K., Yohalem, D.S. and Jensen, D.F. 2002. CR detection and RFLP differentiation of *Botrytis* species associated with neck rot of onion. *Plant Disease*, **86**: 682–686.

Olivier, C., Mac Neil, C.R. and Loria, R. 1999. Application of organic and inorganic salts to field-grown potato tubers can suppress silver scurf during potato storage. *Plant Disease*, **83**: 814–818.

Pertot, I. and Perin, L. 1999. Influence of N-fertilization on rot caused by *Botrytis cinerea* on kiwifruit in cold storage. *Notiziario ERSA*, **12**: 39–41.

Saaiman, W.C. 1997. Orchard sanitation as a means of controlling postharvest diseases. *Yearbook–South African Mango Growers Association*, **17**: 73–74.

Saevels, S., Lammertyn, J., Berna, A.Z., Veraverbeke, E.A., Natale, C.D. and Nicolai, B.M. 2003. Electronic nose as a non-destructive tool to evaluate the optimal harvest date of apples. *Postharvest Biology and Technology*, **30**: 3–14.

Saucke, H. and Döring, T.F. 2004. *Potato virus Y* reduction by straw mulch in organic potatoes. *Annals of Applied Biology*, **144**: 347–355.

Scheuerell, S. 2004. Going from compost to compost tea: Weighing plant health benefits against human pathogen uncertainties. *Phytopathology*, **94**: S 127 (Abst).

Sharma, R.M. and Singh, R.R. 2000. Harvesting, postharvest handling and physiology of fruits and vegetables. In: *Postharvest Technology of Fruits and Vegetables* (*eds.*) L.R. Verma and V.K. Joshi, pp. 94–147. Indus Publishing Co., New Delhi, India.

Sharples, R.O. 1968. The structure and composition of apples in relation to storage quality. *Report of East Malling Research*, 1967, pp. 185–189.

Stachewicz, H. 1999. Possibilities of controlling silver scurf. *Kartoffelbau*, **50**: 74–77.

Suojala, T. 1999. Effect of harvest time on the storage performance of carrot. *Journal of Horticultural Science and Biotechnology*, **74**: 484–492.

Suojala, T. and Tahvonen, R. 1999. Effect of crop rotation on storage diseases of carrot. In: *Agri-Food Quality II: Quality Management of Fruits and Vegetables – From Field to Table* (eds) M. Hägg, R. Ahvenainen, A.M. Evers and K. Tiilikkala, pp. 76–78, Royal Society of Chemistry, Finland.

Walters, S.A. 2003. Suppression of *Water melon mosaic virus* in summer squash with plastic mulches and rowcovers. *HortTechnology*, **13**: 352–357.

Wangsomboondee, T. and Ristaino, J.B. 2002. Optimization of sample size and DNA extraction methods to improve PCR detection of different propagules of *Phytophthora infestans*. *Plant Disease*, **86**: 247–253.

Wiggins, B.E. and Kinkel, L.L. 2005. Green manures and crop sequences influence potato diseases and pathogen inhibitory activity in indigenous Streptomycetes. *Phytopathology*, **95**: 178–185.

Williamson, L., Nakaho, K., Hudelson, B. and Allen, C. 2002. *Ralstonia solanacearum* Race 3, biovar 2 strains isolated from geranium are pathogenic on potato. *Plant Disease*, **86**: 987–991.

Willingham, S.L., Pegg, K.G., Cooke, A.E., Coates, L.M., Langdon, P.W.B. and Dean, J.R. 2001. Rootstock influences postharvest anthracnose development in 'Hass' avocado. *Australian Journal of Agricultural Research*, **52**: 1017–1022.

Xiao, C.L., Chandler, C.K., Price, J.F., Duval, J.R., Mertely, J.C. and Legard, D.E. 2001. Comparison of epidemics of Botrytis fruit rot and powdery mildew of strawberry in large plastic tunnel and field production system. *Plant Disease*, **85**: 901–909.

Zhao, Y., Damicone, J.P. and Bender, C.L. 2002. Detection, survival and sources of inoculum of bacterial diseases of leafy crucifers in Oklahoma. *Plant Disease*, **86**: 883–888.

6

INFLUENCE OF POSTHARVEST HANDLING AND STORAGE CONDITIONS

Proper postharvest management is essential to provide commodities of high quality to the consumer. Losses in both quantity and quality of durable and perishable commodities, occurring between harvesting and final utilization, are substantial. These losses may be greater than the gains attained through improvements in crop production in the field. There is an imperative need to understand the influence of the various methods of handling, transportation, packing, and storage, so that effective corrective measures can be taken to prevent the enormous losses due to postharvest diseases caused by microbial pathogens.

After harvest, seeds, fruits, and vegetables continue to have physiological functions, such as respiration and enzymatic activities, and as senescence sets in their natural resistance to postharvest diseases is reduced. It is well established that the quality of harvested commodities cannot be improved. But it may be possible to retain the quality by adopting proper postharvest handling methods and providing suitable storage conditions to slow down the ripening process by reducing the metabolic activity of the harvested produce.

6.1 POSTHARVEST HANDLING

6.1.1 Seeds

Profound changes occur in the seed ecosystem, from the extremes of environments to which the earheads or seed-bearing organs are exposed to the

Postharvest Pathogens and Disease Management, by P. Narayanasamy

comparatively equable environments of seed storage. These changes may influence the distribution and development of microbial pathogens colonizing the seeds.

Seeds are dried to decrease the water content so that the mycoflora may not be able to find favorable conditions for development. Grain driers are used for drying the seeds. Three kinds of drying have been adopted. Continuous-flow drying involves exposure of thin layers of grains to high temperatures (which vary depending on factors such as the crop, its water content, and its end use) for relatively short periods of time (Anon, 1982). After continuous flow drying, the seeds are allowed to cool by intermittent aeration to prevent moisture migration. In the batch drying procedure, discrete batches of seeds are dried to completion using temperatures at 6 to 12°C above ambient air temperatures. In the low temperature and ambient air in-store drying system, the stored grains or seeds are ventilated with air at near-ambient temperatures so that water and respiratory heat are removed before grain discoloration occurs. This system is the least expensive and most commonly used drying method. The extent of heating is determined by the temperature and relative humidity of the ambient air at the time of drying (Nellist, 1988).

Seeds are cleaned by using mechanical separators or winnowing machines. The fungal sclerotia of ergot pathogens may be removed by immersing the contaminated seed lots in salt solutions. The healthy seeds at the bottom are removed, washed, and dried to reduce the moisture content to the required level.

6.1.2 Fruits and Vegetables

6.1.2.1 Postharvest Cooling Postharvest cooling eliminates field heat rapidly from freshly harvested fruits and vegetables before shipment, storage, or processing. Proper postharvest cooling and handling may ensure that the quality of the commodities is maintained at a desirable level until they reach the consumer. Proper postharvest cooling is effective in: (1) suppressing enzymatic degradation and respiratory activity; (2) reducing water loss; (3) inhibiting the development of microbial pathogens causing postharvest diseases; and (4) decreasing production of ethylene or minimizing the reaction of commodities to ethylene. The choice of cooling method depends on the nature of the commodity, packaging requirements, and economic constraints.

A. Room Cooling: The produce is placed in an insulated room equipped with refrigeration units to chill the air. This method is suitable for most of commodities. However, it is slow, taking a longer time to attain the desired level of cooling. This system can be relatively energy efficient if it is properly designed.

B. Forced-air Cooling: This method is used in conjunction with a cooling room and it offers beneficial effects for many produce. The cooling rate can be increased by providing additional fans to draw cool air through the packages of produce. This method is usually 75 to 90% faster than room cooling.

C. Hydrocooling: Commodities that are not sensitive to wetting can be subjected to hydrocooling, which involves directing a flow of chilled water over the produce resulting in rapid removal of heat. The chilled water removes heat about 15 times faster than air. Hydrocooling is much less energy efficient compared with room (20–40%) and forced-air cooling (70–80%). Tomatoes hydrocooled with a cell suspension of *Erwinia carotovora* subsp. *carotovora* containing 50 to 200 mg of free chlorine per liter (at pH 7.0 and 10°C) remained decay-free during a 10-day storage at 20°C. But infection by *Rhizopus stolonifer* occurred sporadically under similar conditions. Hydrocooling seems to be a viable method for rapid cooling of tomatoes. The treated tomatoes ripened in the same way as the control with no evidence of phytotoxicity (Vigneault et al., 2000).

D. Top or Liquid Icing: In the top icing method, crushed ice is placed by hand or machine in a container over the top of the produce. In the case of liquid icing, a slurry of water and ice is injected into the produce packages through vents or hand holds without removing their tops. Icing is especially useful for dense packages that cannot be cooled using forced air.

E. Vacuum Cooling: This method is particularly effective for commodities that have a high ratio of surface area to volume, such as leafy greens and lettuce, since it is very difficult to cool them with forced air or hydrocooling. The air is evacuated from large containers holding the produce. The temperature is lowered as the water is evaporated rapidly. Care should be taken to maintain the moisture at the optimum level.

F. Evaporative Cooling: In this system the temperature is reduced by misting or wetting the produce in the presence of a stream of dry air, while high relative humidity is maintained. Evaporative cooling is most effective when the relative humidity of the air is below 65%.

The postharvest cooling requirements of different commodities are presented in Table 6.1. The requirements of some selected fruits and vegetables are detailed below:

a. Apples: Production of quality apples depends primarily on the harvesting of mature but not yet fully ripe or overripe fruits. Firmness and soluble solids content of apple fruits are dependable indicators of maturity to determine picking time. Apples have to be handled very carefully, since they are easily bruised and damaged. Damage from rough handling is likely to accelerate deterioration as the injuries form the points of entry for the microbial pathogens. Apples placed in bulk boxes should not be allowed to remain for long periods in direct sunlight before cooling is started. The boxes should be kept clean and in good conditions and should not be overfilled. Adequate air

TABLE 6.1 Recommended Postharvest Cooling System and Optimum Temperature Range for Different Commodities

Commodity	Temperature range (°C)	Type of postharvest cooling system			
		Room cooling	Forced air cooling	Hydrocooling	Icing
Apples	–1.0 to 4.5	*	**	**	–
Beans (butter)	2.8 to 5.0	*	**	**	–
Cabbage	0.0	**	**	–	–
Cucumber	7.0 to 10.0	–	**	**	–
Onions (Green)	0.0	–	–	**	**
Pepper	7.0 to 10.0	**	**	–	–
Potatoes	3.0 to 4.5	**	**	–	–
Strawberries	0.0	**	**	–	–
Sweetpotatoes	13.0	**	–	–	–
Tomatoes	7.0 to 10.0	**	**	–	–
Watermelon	10.0 to 15.5	**	–	–	–

** Preferred; * acceptable.
Source: Boyette et al., 2003a, b, c.

circulation and drainage should be provided by keeping open 5 to 8% of lateral surfaces and 3 to 5% of the bottom.

The objective of postharvest cooling is to reduce the respiration and to increase storage life of apples. It is essential to remove the field heat as soon as possible after harvest. The apples may be either hydrocooled or forced-air cooled without removing them from bulk boxes. Forced-air cooling is faster (four to 10 times) than room cooling. However, care should be taken to cool until the apple pulp temperature reaches 3.5 to 4.5°C and to maintain the relative humidity at the optimum level of 90 to 95% (Boyette et al., 2003d).

b. Kiwifruit: In New Zealand, kiwifruits are placed in wooden bins in a suitable containment facility for a minimum period of 48h before cooling to reduce the potential for the incidence of gray mold disease caused by *Botrytis cinerea*. In California, the kiwifruits are subjected to a 48h delayed cooling at 15°C and high air velocity (ethylene free). This procedure arrests the development of decay caused by *B. cinerea*. Provision of 48-h delayed cooling at 15°C, high (2m/s) air velocity and 95% relative humidity was the most effective in reducing gray mold disease (Crisosto et al., 1997).

c. Tomatoes: Greenhouse- and field-grown tomatoes are one of the leading commodities in the Unites States in terms of value and volume. The appearance, quality, and flavor are the characteristics expected by consumers. Proper postharvest handling and storage methods are essential to maintain acceptable quality and longer shelf life of tomatoes. Tomatoes for the fresh market are hand-harvested, either with or without a harvesting aid. Tomatoes harvested at the mature green stage will ripen into a product similar to vine-ripened fruit. The right stage of maturity of tomato fruit may be determined by slicing the

TABLE 6.2 Color Classification of Tomatoes

Color	Description
Green	Fruit surface with light to dark green color
Breakers	Color breaking from green to tannish yellow with pink or red skin covering not more than 10% of fruit surface
Turning	Fruit surface (10–30%) turning from green to tannish yellow, pink, red, or a combination of these colors
Pink	Fruit surface (30–60%) becoming pinkish red or red in color
Light red	Fruit surface (60–90%) turning pinkish red or red
Red	Entire fruit surface or more than 90% showing red color

Source: Boyette et al., 2003a.

tomato with a sharp knife. If the seeds are cut, it indicates immaturity and such fruits will not ripen properly. A fruit maturity color chart may be provided to harvesting crew (Table 6.2). All diseased and misshapen fruits should be removed to avoid the spread and build up of diseases and insect pests. Clean hands and personal sanitary habits of staff are required and suitable instructions to workers have to be given. Frequently checks of harvesting pails for trash and poor quality tomatoes may be useful to maintain high quality and consequent market value of the produce. Tomatoes should never be packed tightly into harvesting containers or exposed to sunlight for long periods.

Passing tomatoes along a roller conveyer that slowly turns each tomato is the best method of inspecting for damaged, misshapen, or cracked fruit. The good quality indicators are firm flesh, shiny skin, and uniform color. Tomatoes should be washed sufficiently to remove dust and foreign material by spraying a small amount of chlorinated water as they move over a set of soft brush rolls. The moisture present on the fruit surface can be removed by sponge-rubber doughnut rolls alone or in combination with an air-blast drier. Tomatoes with surface injury should be removed carefully to prevent entry and spread of postharvest diseases such as gray mold, Alternaria rot, Rhizopus rot, and sour rot.

The tomatoes are subjected to postharvest cooling to remove excessive field heat in order to maintain the quality to the desired level and to increase the shelf life appreciably. Furthermore, it is also possible to reduce the effects of dehydration and minimize decay by applying suitable cooling system. Room cooling system may be adopted after the containers are loosely stacked allowing space between the containers for sufficient air circulation (Boyette et al., 2003a).

d. Peppers: Peppers are commonly harvested by hand into pails and then transferred to bulk bins. During picking and handling, appreciable mechanical damage may occur, resulting in reduction in market value for the produce. Hence quality rather than quantity of produce harvested should receive greater attention. After harvest, peppers suffer due to water loss, sun scald, and heat damage if they are exposed to direct sun light for more than an hour. Hence, the peppers should be kept in the shade immediately and cooled as soon as possible. To overcome these problems, field packing has been suggested. Field packing involves packing limited quantities and hence it is possible to reduce mechanical damage, in addition to lower packing costs. But field packing requires intensive training and careful supervision of pickers. A light waxing has been found to increase storage life and reduce damage by scuffing and abrasion. The peppers of high quality, uniform maturity, and desirable color, shape, and size have to be selected and packed. Forced-air or room cooling is the preferred system for peppers (Boyette et al., 2003e).

e. Onions: Onions have a period of active growth which is followed by a period of dormancy. As for the crops discussed earlier, the determination of harvesting time when onions reach maturity is very important. Onions have to be harvested when approximately 10 to 20% of tops fall over and begin to dry. A shallow undercutting 2 to 4 cm below the bulbs is performed to hasten the change from growth to dormancy. Field drying of onions may be possible, depending on the weather conditions existing during harvesting. Warm, humid, or rainy weather may favor the incidence of diseases. Hand harvesting may be preferable to mechanical harvesters because of the risk of damage. Furthermore, hand harvesting seems to be the cost-effective alternative and, in addition, offers gentler handling and field grading compared to mechanical harvesters.

During harvesting, onion tops should not be cut less than 2 cm from the bulbs and roots should be trimmed leaving at least 0.5 cm segments attached to the bulb. Onions are packed in mesh bags in the field, thoroughly dried, and removed from the bags for grading. Then they are gathered in the same bags and sent to markets. Proper drying is essential so that the onions are in a state of dormancy. The neck of onions, if dried well, should not slide back and forth when squeezed between the thumb and fore finger. *Botrytis allii* causing neck rot disease frequently infects onions by entering through incompletely dried neck wounds. Artificial drying, using heated air, has to be done if natural drying is not sufficient. Heated air (about 38°C) may be used for drying. During drying the weight of onions may be reduced by 5 to 8%. The tops and roots can be removed in the grading and packing line, after drying. Onions should be transported and stored separately from other fruits and vegetables which may acquire the odor of onion. Furthermore, onions may absorb moisture readily from other fresh vegetables (Boyette et al., 2003b).

6.1.2.2 Washing, Cleaning, and Trimming Fruits and vegetables may require washing and cleaning to remove the adhering soil particles and plant debris which carry microbial pathogens capable of causing diseases under storage conditions. Chlorinated water is used commonly to wash the fruits and vegetables. Leafy vegetables such as cabbage need to be trimmed to remove leaves infected by pathogens.

TABLE 6.3 Classification of Tomatoes Based on Fruit Diameter

Size group	Diameter (cm)	
	Minimum	Maximum
Extra small	4.672	5.397
Small	5.397	5.794
Medium	5.794	6.429
Large	6.429	7.302
Extra large	7.302	8.811

Source: Boyette et al., 2003a.

6.1.2.3 Sorting and Grading Abnormal and pathogen-infected, insect-infested, or damaged fruits and vegetables may be removed by hand. Based on the market requirement/preference, healthy commodities are separated into different grades on the basis of color, shape, and size or weight. The culled ones may be sent for processing. Tomatoes are graded by size on one or more sizing belts (Table 6.3).

6.1.2.4 Curing and Waxing The water loss during storage from hardy vegetables, such as onions, garlic, potato, and sweetpotato, has to be reduced by curing. Depending on the nature of the commodity the curing method may vary. Onion and garlic bulbs are cured by spreading them under shade until their green tops, outer skins, and roots are dried well. When the kiwifruits are held at ambient temperatures between harvest and cold storage for up to 4 days, gray mold incidence caused by *Botrytis cinerea* was significantly reduced and no adverse effect on fruit quality could be observed. A marked reduction in infection by *B. cinerea* in fruits exposed to greater than 95% humidity during curing was observed (Long and Bautista-Baños, 1994; Bautista-Baños et al., 1997; Michailides and Elmer, 2000).

A coating, consisting of a wax and/or other chemicals, may be applied to the surface of fruit. Waxing reduces the respiration and transpiration rates. Fungicides, growth regulators, or preservatives may be incorporated to reduce spoilage by pathogens or to inhibit sprouting. Waxing may also be useful to reduce moisture losses and shriveling of stored fruits (Kasmir, 1985; Anon, 1989; Agrahari et al., 1998).

6.1.2.5 Packaging An appropriate method of packaging is essential to retain freshness and to prevent mechanical injuries, moisture loss, and other physiological process hastening ripening/senescence during transportation. The quality of fruits cannot be improved by packaging. However, it is possible to preserve and protect the commodity from further deterioration and hazards during transportation. The traditional packaging using gunny bags, grasses, and leaves has been replaced by packaging using wooden boxes, baskets woven with bamboo, jute bags, plastic punnettes, and corrugated fiber board (CFB) boxes (Dordi, 1994). Tomatoes may be packed in a variety of containers in baskets or cartons. Corrugated cartons may be used for packing peppers. Onions are generally packed in mesh bags. Fiber board cartons have been used for packing onions in recent years.

6.2 STORAGE CONDITIONS

6.2.1 Seeds

6.2.1.1 Development of Mycoflora After harvest viable seeds respire, as they do when they are attached to the mother plant. Starch in the seed is converted into CO_2, water, and energy during respiration. With an increase in storage temperature, respiration rate is increased, resulting in loss of nutrients present in the seed. Seed moisture is the most important factor favoring the development of microbial pathogens. The accelerated fungal development may also increase temperature, leading to heat damage to seeds. The effects of heat damage are revealed by the discoloration of sunflower seed meats (Anon, 1979). The fungal infection of wheat grains with different levels of moisture (15, 18, 21, and 24%), temperatures (4°, 15°, 25°, and 40°C), and of mechanical damage (0, 15, and 30%) was assessed. Fungal infection increased very slowly with low moisture (15%), low temperature (4°C), and low damage (0%), whereas it increased rapidly under high moisture, temperature, and damage levels. The mycoflora consisted of *Penicillium* spp., *Aspergillus candidus*, *A. niger*, and *A. flavus* (Al-Yahya, 1999).

Colonization by fungi of maize grain conditioned at constant temperatures and humidities was studied. Colonization of kernels by *Eurotium chevalieri* reached the maximum percentage (50–96%) at a moisture level of 17.5%, irrespective of the various temperatures (10° to 40°C) at which the seeds were conditioned. Grain samples with greater than 33% infection by *E. chevalieri* showed reduced infection by *Gibberella fujikuroi* and *Aspergillus zeae*. The individual patterns of fungal colonization during grain conditioning appeared to be a function of the survival rates for fungal colonists and their potential replacement by *E. chevalieri* (Wicklow et al., 1998). The influence of storage time or method on the storability of maize was assessed by determining dry matter loss over time and by measuring CO_2 produced through respiration of grain storage fungi developing on maize kernels. There was no significant difference in storability or microbial infection levels between freshly harvested samples and samples that had been stored wet (23.5% moisture, wet basis) or dry at –10°C for up to 8 months (Wilcke et al., 2001).

The effect of storage of maize seed for 12 months in a cold chamber (14°C) on the final incidence of *Fusarium moniliforme* (*G. zeae*) was compared with that under an uncontrolled environment. The incidence of *F. moniliforme* decreased progressively with increasing period of storage. However, there was no influence of temperature, as the initial and final incidences of *F. monili-*

forme did not show significant differences with storage in a cold chamber or under an uncontrolled environment (Tanaka, 2001).

The effects of interacting conditions of water activity (a_w, 0.99–0.85), temperature (15 and 25°C), and duration (40 days) on growth and production of deoxynivalenol (DON) and nivalenol (NIV) by *Fusarium culmorum* on a wheat-based medium were determined. DON and NIV reached the maximum concentrations at 0.995 a_w and 0.981a_w at 25°C, respectively, after an incubation period of 40 days. The interactions between a_w, temperature and incubation time, and $a_w \times$ temperature and temperature $\times$ incubation period resulted in significant variations in DON /NIV production (Hope and Magan, 2003).

The effect of storage of melon seeds in jute or polyethylene bags on the extent of deterioration by fungi was assessed. The germination percentage was reduced from 96.3 to 28.7% in jute bags and to 45.3% in polyethylene bags, while the incidence of visible moldiness increased from 1.8% before storage to 8.9 and 4.8% in jute and polyethylene bags respectively. An increase in incidence of *Aspergillus*, *Penicillium*, and *Rhizopus* with prolonged storage was observed (Bankole et al., 1999).

In modified atmosphere storage, the intergranular atmosphere in stored grain can be manipulated artificially by introducing gases such as CO_2 or nitrogen (N_2) (Shejbal, 1980). Atmospheres containing 99.7% N_2 + 0.3% O_2, 61.7% CO_2 + 8.7% O_2 + 26.6% N_2, and 13.5% CO_2 + 0.5% O_2 + 84.8% N_2 could only delay deterioration of maize grain by *Aspergillus flavus* and *Fusarium moniliforme*, but did not arrest their growth. *A. flavus*-inoculated grains showed mold growth after 4 weeks in an atmosphere consisting of 61.7% CO_2 + 8.7% O_2 + 29.6% N_2 (Wilson et al., 1975). The tolerance of fungal pathogens to CO_2 varies considerably and large concentrations are required for inhibition of spore germination and growth of fungi such as *Penicillium aurantiogriseum* (Lillehoj et al., 1972).

6.2.1.2 Production of Mycotoxins Microbial pathogens, primarily fungi, infect seeds resulting in poor quality of seeds. They also produce toxic metabolites, known as mycotoxins, contaminating seeds. Such contaminated seeds are harmful, causing serious ailments in humans and animals consuming the contaminated foods and feeds. Different species of *Aspergillus*, *Fusarium*, *Penicillium*, and *Alternaria* produce different kinds of mycotoxins in seeds in the field and during transit and storage. The growth of, and mycotoxin production by, toxigenic fungi are influenced by the presence of other contaminant molds and environmental factors, as in the case of *F. culmorum, Aspergillus ochraceus*, and *Penicillium verrucosum* (Magan et al., 2003).

A. Aflatoxin (AF): Aflatoxins are produced by *Aspergillus flavus* and *A. parasiticus* in several crops, such as maize (Anderson et al., 1975; Côrtes et al., 2000), peanut (groundnut) (Diener et al., 1987; Xu et al., 2000; Victoria Novas and Cabral, 2002), rice and wheat (Tsai and Yu, 1999), cotton (Ashworth Jr. et al., 1969; Hasan, 2001), mustard (Bilgrami et al., 1992), soybean (Swelim et al., 2001), and Sudanese guar (El-Nagerabi and Elshafie, 2001). Aflatoxin (AF) B_1 and B_2 are produced by *A. flavus*, while G_1, G_2, and M_1 are produced in addition to B_1 and B_2 by *A. parasiticus*. Among them, AFB_1 is the most important aflatoxin in agricultural produce – accumulation of high levels of AFB_1 is reported to be an important risk factor for the development of liver cancer (Wogan, 1991).

Aspergillus parasiticus can colonize all tissues of peanut and seems to gain ingress through the corky layer of the pericarp. Both inter- and intracellular colonization was seen and depletion of storage bodies from infected cotyledons was evident. Expression of the aflatoxin biosynthetic gene *nor-1* in the embryo and cotyledons was indicated (Xu et al., 2000). Vegetative compatibility (VC) of *A. flavus* isolates from peanut seed was determined. Five isolates simultaneously produced aflatoxins B, G, and cyclopiazonic acid (CPA) and hence they were considered atypical. The isolates (25) were grouped into 13 VC groups (VCGs). All isolates within the same VCG were characterized by their ability to produce or not to produce sclerotia. The atypical isolates formed a single and exclusive VCG (Victoria Novas and Cabral, 2002). On the other hand, some isolates of *A. flavus* from soybean seeds produced specific types of aflatoxins (B_1, B_2, G_1, and G_2), while others did not produce any of the aflatoxins.

Aflatoxin B_1 levels in peanut extract and peanut paste were determined by high performance liquid chromatography (HPLC) and enzyme-linked immunosorbent assay (ELISA). The results from both assays on 28 peanut samples artificially contaminated with *A. flavus* and *A. parasiticus* spores were highly correlated ($r = 0.977$, $P < 0.0001$). The detection limits and matrix influence associated with the ELISA protocol developed in this study were more sensitive than those reported for other ELISA formats because of the specific antiserum used (Asis et al., 2002). Some isolates of *Penicillium funiculosum* produced only AFB_2, while some isolates produced G_1 and G_2 or G_1 only (Swelim et al., 2001). The fungi *A. niger*, *A. flavus*, and *Penicillium corylophilum* produced aflatoxin in different parts of cotton boll such as fiber, valves, and seeds. Application of pencycuron to cotton seed favored the growth of these fungi at 1 g/kg, but inhibited their growth at higher concentrations (3 and 5 g/kg). *A. niger* utilized pencycuron as nitrogen source (Hasan, 2001).

A. flavus, causing an important ear rot disease of corn, produces aflatoxins which are known to be carcinogenic. The α-amylase of *A. flavus* promotes aflatoxin production in the endosperm of infected maize kernels. A 36-kDa α-amylase inhibitor from *Lablab purpureus* (AILP) inhibited conidial germination and hyphal growth of *A. flavus* (Fakhoury and Woloshuk, 2001). Maize kernels contain a ribosome-inactivating protein (RIP1). Treatment of conidia of *A. flavus* with RIP1 resulted in increased hyphal branching. However, in the case of *A. nidulans*, a nonpathogen of maize, RIP1 was observed to cause lysis, a response that was apparently avoided by *A. flavus* (Nielsen et al., 2001). The maize inbred Tex6 shows resistance to colonization and accumulation of aflatoxin produced by *A. flavus* and a protein present in aqueous extract of mature Tex6 seeds inhibited the growth of *A. flavus*. This inhibitory protein was identified as an endochitinase that also had exochitinase activity. It inhibited the growth of *A. flavus* by 50% at a concentration of 20 μg/ml. This chitinase appears to make a major contribution to the antifungal activity in this maize genotype (Moore et al., 2004). These studies show that the aflatoxin production depends on the fungal species, crop genotype, storage conditions, and chemical applied.

The relationship between lipase gene expression and aflatoxin production by *A. flavus* and *A. parasiticus* has been studied. The gene *lipA* encodes a lipase involved in the breakdown of lipids from aflatoxin-producing *A. flavus* and *A. parasiticus* and two nonaflatoxigenic *A. flavus* isolates (wool-1 and wool-2). The expression of the lipase gene was observed only in growth media supplemented with 0.5% soyabean or peanut oil or in lipid-rich medium such as coconut medium. The expression of the lipase gene under substrate-induced conditions correlated well with production of aflatoxin in *A. flavus* and *A. parasiticus* (Yu et al., 2003). Production of aflatoxin by *A. flavus* in infected cotton seeds was shown to be governed by the polyketide synthase gene (*pksA*). The strain AF36 belongs to a vegetative compatibility group (VCG) that includes several hundred isolates of *A. flavus* that do not produce detectable quantities of aflatoxin. The *pksA* gene in AF36 required for aflatoxin biosynthesis showed a nucleotide polymorphism near the beginning of the coding sequence. Such a nucleotide change introduced a premature stop codon into the coding sequence resulting in prevention of enzyme production and aflatoxin accumulation (Ehrlich and Cotty, 2004).

The gene *VeA* in *A. prasiticus* infecting peanut kernels is required for the expression of *aflR* and *aflJ* which regulated the activation of an aflatoxin gene cluster. Deletion of *VeA* from *A. parasiticus* led to the blockage of sclerotial formation in addition to the production of aflatoxin intermediates, both on culture medium and peanut seeds (Calvo et al. 2004).

Aspergillus spp. produce aflatoxins, ochratoxin A, and patulin which contaminate various agricultural commodities. Phylogenetic analysis of sequences of the ribosomal RNA gene cluster has been shown to be of use in clarifying taxonomic relationships among toxigenic *Aspergilli* causing pre- and postharvest contamination of plant products. Phylogenitically unrelated *Aspergillus* spp. produce the same mycotoxins, indicating that ability to produce mycotoxins might have altered during the evolution of the genus *Aspergillus*. Biosynthetic gene-based probes have to be employed for the molecular detection and differentiation of the mycotoxin-producing *Aspergillus* spp. (Varga et al., 2004).

The effect of metal ions, such as Zn^{2+}, Cu^{2+}, and Fe^{2+}, either individually or in combination, applied to *A. flavus* in submerged cultures, on growth, total RNA, aflatoxin pathway gene expression, and production of aflatoxin and its precursor O-methyl sterigmatocystin (OMST), was assessed. Expression of the aflatoxin pathway gene (*omtA*) was evaluated by quantifying mRNA levels by reverse transcription-polymerase chain reaction (RT-PCR) assay. The ions induced changes in total RNA, mRNA levels, biosynthesis of aflatoxin, and

OMST production in addition to enhanced fungal growth (Cuero et al., 2003). The influence of antioxidants on aflatoxin production by *A. flavus* (section Flavi) was assessed. Among the antioxidants tested, propyl paraben (PP) and butylated hydroxyanisole (BHA) inhibited aflatoxin B_1 production, in addition to a reduction in the percentage of spore germination, growth rate, and 0.937 a_w (water activity) (Nesci et al., 2003).

B. Fumonisins (FB): The fungal species complex consisting of *Fusarium moniliforme* (*Gibberella fujikuroi*) and other *Fusarium* spp. has world-wide occurrence causing great concern, as they produce the mycotoxins, fumonisins, contaminating corn (maize). The putative polyketide synthase gene (*FUM5*) involved in fumonisin biosynthesis has been characterized (Proctor et al., 1999). *F. proliferatum*, causing corn ear rot disease, produces fumonisin B_1 (FB_1), fumonisin B_2 (FB_2), beauvericin, and fusaproliferin. The mycotoxin contamination increased with increase in ear rot index (Pascale et al., 1999). The presence of fumonisins in corn under natural conditions has been reported from several countries, including the United States (Plattner et al., 1990), Poland (Lew et al., 1991), South Africa (Sydenham et al., 1990), Taiwan (Tseng et al., 1995), China (Li et al., 1999), and Spain (Velluti et al., 2001). Corn-based foods and feeds for human and animal consumption were tested for the presence of FB_1 by HPLC; FB_1 was detected in 23% and 86% of the samples meant for human and animal consumption respectively (Velluti et al., 2001). The effectiveness of direct competitive ELISA was evaluated for determining total fumonisins (B_1, B_2, and B_3) in corn. Methanol–water (7:3) was used to extract the fumonisins in test samples, followed by filtration and suitable dilution. Naturally and artificially contaminated corn samples were tested by 13 collaborators in the United States. Average recoveries of total fumonisins were 120, 100, and 90% respectively for samples spiked with 1.0, 3.0, and 5.0 mg of total fumonisins (B_1, B_2, and B_3), indicating acceptable levels of intra- and interlab precision (Bird et al., 2002).

Fumonisin production by *F. verticillioides* in corn kernels was detected by reflectance and transmittance visible and near-infrared spectroscopy. The color and chemical constituents of infected kernels form the basis for the detection of fumonisin in single kernels. It is possible to employ this detection technique for the rapid detection of fumonisin in corn, as single corn kernels can be screened nondestructively (Dowell et al., 2002). Mutants generated by disrupting *FUM1*, the gene encoding a polyketide synthase required for fumonisin biosynthesis, were able to cause ear rot disease in maize, as occurred with the wild strain of *F. verticillioides*. This evidence confirmed that fumonisins are not required to cause ear rot in maize (Desjardins et al., 2002). Production of fumonisin in naturally contaminated corn and starch samples was demonstrated by employing a direct competitive ELISA as a postcolumn monitoring system after liquid chromatography (LC), following infection by *Alternaria alternata*. Both FB_1 and *Alternaria alternata* toxins (AAL) were detected in the culture extracts (Yu and Chu, 1998).

The effect of genetically modified Bt-corn hybrids on deoxynivalenol (DON) content in grain at harvest was assessed during 1997 to 1999. Bt-corn hybrids were developed by incorporating the *Bt* gene from *Bacillus thuringiensis* for managing corn borer, which is considered to predispose plants to infection by *Fusarium* spp. When there was a high intensity of stalk injury by *Ostrinia nubialis*, the use of Bt-hybrids was effective in reducing DON concentrations by an average of 59% compared with the Bt-isoline. But when the intensity of *O. nubialis* injury was less, there was no variation in the DON concentrations of Bt and non-Bt lines (Schaafsma et al., 2002). An attempt was made to identify important weather variables and their timing for predicting DON contents in mature grain at wheat heading, during 1996 to 2000 in southern Ontario, Canada. In the first period, 4 to 7 days before heading, DON generally increased with the number of days of less than 10°C. In the second period, 3 to 6 days after heading, DON content increased with the number of days of rain more than 3 mm and decreased with days exceeding 32°C. In the third period, 7 to 10 days after heading, DON content increased with number of days with more than 3 mm of rain (Hooker et al., 2002).

Although the US Food and Drug Administration has recommended that cleaned and dry milled corn meant for human consumption should not contain more than 2 to 4 µg/g of fumonisin (total homologues of FB_1, FB_2 and FB_3) (Anon, 2001), a substantial portion of the corn crop in the United States would have a fumonisin concentration above the prescribed limit. Further, fumonisin was detected at levels of concern from asymptomatic corn kernels that did not exhibit symptoms of ear and kernel rot (Clements, 2002). Phylogenetic analysis of *F. verticillioides* (*Gibberellla moniliforme*, *G. fujikuroi* mating population A) showed that *F. verticillioides* occurred as a major fumonisin-producing population with a wide geographical distribution and host preferences (to cereals) and also as a minor fumonisin non-producing population with restricted host preference (banana) had low variability (Mirete et al., 2004). Because of the nonavailability of maize genotypes highly resistant to infection by *F. verticilloides* or to fumonisin contamination, it has become necessary to find alternative methods for prevention of fumonisin contamination. In a study to determine the effect of harvest time on fumonisin contamination, it was shown that, under years conducive for fumonisin contamination, early harvest (greater than 25% grain moisture) might reduce the level of contamination (Bush et al., 2004).

C. Trichothecenes: These mycotoxins are produced by different species of *Fusarium* and also by *Cephalosporium* sp., *Phomopsis* sp., and *Trichothecium* sp. Samples of wheat and corn are commonly contaminated by deoxynivalenol (DON), nivalenol (NIV), and T-2 toxin. High concentrations of DON accumulated in the wheat spikelets inoculated with *Fusarium graminearum*, causing Fusarium head blight (FHB) disease (Savard et al., 2000). Contamination of maize kernels with DON increased in proportion to the incidence of *F. graminearum* (Ngoko, et al., 2001). High concentrations of DON, 15-acetyl DON, and zearalenone in barley seeds inoculated with *F. graminearum* were detected, whereas barley seeds inoculated with *F. poae* had low concentrations of DON alone, revealing the varying capacities of *Fusarium* spp. in elaborating the trichothecene compounds in plant tissues (Schwarz et al., 2001). Disruption of the trichodiene synthase gene (*Tri5*) of *F. graminearum* resulted in the failure of production of DON and the spread of the defective strain from inoculated wheat spikelets to other spikelets was inhibited, indicating a significant role for DON in the progression of disease in infected ears (Bai et al., 2002).

The levels of resistance/susceptibility of 15 wheat cultivars were assessed by artificially inoculating them with *F. graminearum*; they were grouped into partially resistant and highly susceptible genotypes. Kernels of highly susceptible cultivars contained higher levels of mycotoxins and ergosterol (representing the amount of fungal mycelium). In some cultivars, there was no relationship between fungal mycelium and level of DON, suggesting that these wheat cultivars possess resistance to accumulation of DON (Muthoni et al., 2002). The role of DON in disease development has been examined in some detail. Aggressiveness of isolates of *F. graminearum* and *F. culmorum* was found to be closely associated with DON production (Gang et al., 1998; Miedaner et al., 2000; Mesterhazy, 2002).

A rapid, surface plasmon resonance (SPR)-based indirect inhibitive immunoassay was developed for the quantification of DON. The recovery of DON in spiked wheat was 104 ± 15%. The analysis of a sample could be completed in 15 min including sample preparation (10 min) and quantification (5 min). Wheat samples with different levels of DON contamination were analyzed by this technique (Schnerr et al., 2002). A soluble mixture of extracellular fungal products designated exoantigens (ExAgs) was used to develop an immunoassay for quantifying fungal biomass present within host tissue. A linear correlation between ExAgs of *F. graminearum* detected by an indirect ELISA and DON content ($r = 0.76$–0.80) was observed (Abramson et al., 1998). Fungal colonization was shown to be a predictive and sensitive indicator of presence of DON (Lamper et al., 2000). Non-trichothecene-producing isolates of *F. graminearum*, generated by gene disruption, were less efficient in colonizing wheat ears compared to wild type isolates (Eudes et al., 2001). However, investigations on field isolates indicated that fungal biomass in vivo was not proportional to the development of disease symptoms and DON production. Hence, it is essential to determine DON production relative to the amount of fungal biomass in host tissue. Hence, a segregating population of *Gibberrella zeae* (anamorph *F. graminearum*) (50 progenies) were inoculated onto a susceptible winter wheat cultivar. Fungal colonization, as determined as *Fusarium* (ExAgs) content, and DON production were analyzed with two ELISA variants. There was a high correlation between DON production and *Fusarium* ExAg content across environments ($r = 0.8$, $P = 0.01$). Genotypic variation was found to be significant for all traits including aggressiveness and DON production (Appendix 6) (Cumagun et al., 2004). *Fusarium culmorum* also produces DON in barley. There was a significant correlation between

DON concentration and disease index, indicating that production of DON influenced the aggressiveness of the isolates of *F. culmorum* towards barley (Hestbjerg et al., 2002). A Vogel's medium amended with ammonium sulfate, which favored growth and DON production and expression of genes involved in trichothecene biosynthesis (*Tri* genes), has been developed by Covarelli et al. (2004).

By using the real-time PCR assay, the major species of *Fusarium* involved in the FHB complex could be monitored and quantified. Wheat ears and harvested grains collected from 40 wheat crops in the Netherlands showed that *F. graminearum* was the major species. Similar predominance of *F. graminearum* in 29 wheat crops in France was noted. The concentration of DON correlated equally well with the incidence of *F. culmorum* and *F. graminearum* in the grain samples ($r^2 = 0.8232$) as well as with the total DNA of both these species ($r^2 = 0.8259$) (Waalwijk et al., 2004).

The mycotoxin-producing abilities and molecular variability of 37 *F. culmorum* isolates along with isolates representing related species were assessed by thin layer chromatography (TLC) and by PCR using primer pairs specific for the *Tri7* and *Tri13* genes. The chemotype I included 30 isolates capable of producing DON and 3-acetyl-DON, whereas the chemotype II had seven isolates that produced NIV and/or fusarenone X. Production of NIV was correlated with the presence of a functional *Tri7* gene (Toth et al., 2004). A diverse collection of 158 barley lines was artificially inoculated with spore suspension of *F. graminearum*. The FHB index was worked out based on the product of disease incidence and severity and this formed the basis for the identification of barley lines resistant to FHB (McCallum et al., 2004).

Fifteen *Fusarium* species belonging to the *Gibberella fujikuroi* (Gf) complex and 12 other species were examined for fumonisin production and the presence of fumonisin biosynthetic genes (*FUM*). Fumonisin production could be detected only in *F. fujikuroi, F. globosum, F. proliferatum, F. nygamai, F. oxysporum*, and *F. verticillioides*. The *FUM* genes were also detected in these five species belonging to Gf complex and also in *F. anthophilum* included in a third clade of the Gf complex. Fumonisin production and presence of *FUM* genes were observed only in *F. oxysporum* outside the Gf complex. The results indicated the discontinuous distribution of *FUM* genes in the Gf complex resulting in the differences in the abilities of closely related *Fusarium* spp. to produce fumonisins (Procter et al., 2004). Evaluation of 203 strains of *Fusarium verticillioides* (*G. fujikuroi* mating population A) and 79 strains of *F. proliferatum* (*G. fujikuroi* mating population D) for the production of fumonisins, fusaproliferin, and beauvericin showed that 193 and 65 strains respectively produced fumonisins. On the other hand, *F. subglutinans* (*G. fujikuroi* mating population E) strains produced low levels of, or no, fumonisin. However, the fusaproliferin production was more consistent in *F. subglutinans* strains. Coproduction of fumonisin, fusaproliferin, and beauvericin among the strains in *G. fujikuroi* D and E was also seen, suggesting the need for assessing the interactions between these conditions from the toxicological point of view

(Reynoso et al., 2004). A trichothecene-producing wild strain (Tri5$^+$) of *F. graminearum* was generally more aggressive than the nonproducing (mutated) strain (Tri 5$^-$) on wheat. However, this aggressiveness was markedly influenced by the host plant species. The difference in aggressiveness was less pronounced in rye, which was highly resistant. In barley with moderate type II resistance, the pathogen was found to move externally from one floret to another within the dense spike without penetrating the rachis. The results suggested that trichothecenes are a major determinant of fungal spread and disease development in Triticeae (Langevin et al., 2004).

D. Ochratoxins: Species of *Aspergillus* and *Penicillium* produce ochratoxins. Three isolates of *A. ochraceus* from maize samples were found to elaborate high concentrations of ochratoxin B (Shivendra Kumar and Roy, 2000). High concentration of ochratoxins were detected in wheat, wheat bran (feed), and oats (seed) (Vrabcheva et al., 2000). The ochratoxins are considered as likely causes of a nephrotoxic disease with associated tumors in certain European populations (Richard, 2003). *A. ochraceus* and *P. verrucosum* producing ochratoxins in corn, barley, and wheat were detected in storage. The occurrence and the production of ochratoxins seemed to be associated with water activity of the substrate and temperature. The presence of ochratoxin A-producing toxigenic *A. ochraceus* and *A. niger* in coffee beans was detected. The levels of contamination were 80 to 92% for parchment and dry cherry coffee and 15 to 34% for green coffee (Suárez-Quiroz et al., 2004). The European legislative limit for orchratoxin A (OTA) in cereal grain is 5 mg of OTA/kg grain in relation to *P. verrucosum*. By inoculating cereal samples with spores of *P. verrucosum* and storing for up to 9 months at 10 to 25°C and water activities of 0.77 to 0.95, a logistic regression analysis was performed. The risk of exceeding 5 mg OTA/kg of grain increased with increasing levels of *P. verrucosum*. A threshold of 100 CFU was suggested to predict whether or not the legislative limit would be exceeded (Lindblad et al., 2004).

6.2.2 Fruits

Fruits and vegetables are kept in storage to extend their shelf life and to suit the market requirements. Various kinds of storage techniques (such as ground, ambient, refrigerated, air-cooled, zero energy, modified atmospheric, hypobaric, and controlled atmosphere storage), based on the nature of produce and period of storage required, have been used. It is essential to adopt a suitable storage method to avoid losses. In India, the cumulative annual loss was estimated to be as high as Rs.300 to 400 millions because of improper postharvest operations and storage methods (Swaminathan, 1981). The process of ripening or senescence can be delayed or extended to prevent the development of most microbial pathogens. The fruits are divided into two groups: climacteric fruits and nonclimacteric fruits. The climacteric fruits require a ripening period before reaching a desirable stage of edibility, whereas the nonclimacteric fruits ripe on the plants. Apple, banana, kiwifruit, mango, papaya, and tomato are climacteric fruits, whereas citrus, grapes, pomegranate, and strawberry belong to the non-climacteric group.

The evolution of organic acids such as quinic, malic, shikimic, acetic, citric, and succinic acids in apples and pears stored at 0 to 3°C for 17 weeks and kiwifruits stored for 8 weeks at 0 to 3°C was determined by HPLC. Production of organic acids progressively decreased with increase in the period of cold storage (Rico et al., 1999). The relative amounts of phenolics such as (+)catechin, phloretin glycosides, and quercetin glycosides in peels of apples cv. Golden Delicious stored at low temperature were higher than in the flesh. Quercetin glycosides could be detected only in the skin. Chlorogenic acid and (−)epicatechin contents were not altered by low temperature storage. Chlorogenic acid inhibited spore germination and mycelial growth of *Phlyctaena vagabunda* (*Pezicula alba*) causing postharvest rot. The results suggested that phloridzin and chlorogenic acid in combination with polyphenoloxidase activity might arrest *P. vagabunda* in quiescent infections associated with immature and ripening apple fruit (Lattanzio et al., 2001).

The conidia of *Monilinia fructigena*, causing apple brown rot disease, were viable for up to 20 days (longest assessment time) depending on storage temperature (10 or 20°C) and relative humidity (RH) (45 or 85%). The conidia were more sensitive to temperature than to RH. Storage of apples at 10°C and 85% RH for up to 20 days seemed to have no effect on conidial viability. It appears that the environmental conditions existing during the main growing seasons in temperate countries are unlikely to be limiting factors for the development of brown rot disease of apples (Xu et al., 2001). A new postharvest fungal pathogen *Sphaeropsis pyriputrescens* causing stem-end rot, calyx-end rot, and wound-associated rot of d'Anjou pears was isolated from stored pears. The pathogen developed well at a range of temperatures from 0 to 25°C with an optimum between 15 and 20°C (Xiao and Rogers, 2004).

Controlled atmosphere (CA) prolongs marketable life by reducing the oxygen concentration and increasing the carbon dioxide concentration in the storage atmosphere. An atmosphere containing 5% CO_2 and 3% O_2 at a temperature of 0°C has been found to be suitable for many apple varieties for storage for a period of about 10 months (Boyette et al., 2003d). Cool storage of mature, export-quality apples cv. Gala in either laboratory (0 ± 0.5°C) or commercial packhouse (2 ± 0.5°C) for a period of 25 days reduced the survival of *Erwinia amylovora* in calyxes of both inoculated and naturally infected fruit. Populations of *E. amylovora* did not increase to levels detectable by PCR in fruits which were cool-stored and then incubated at room temperature (about 20°C) for 4 days to simulate possible retail conditions. The cool-stored apples were considered unlikely to be vectors of *E. amylovora* (Hale and Taylor, 1999). In a further study, the survival of *E. amylovora* was assessed during storage at 2°C. The populations of *E. amylovora* on tissue in apple calyx decreased during cold storage, whereas in nutrient media the pathogen population increased at low temperatures. Cold storage of apples, even with low levels of *E. amylovora*, may enable the export of apples with a negligible risk of introducing the disease into countries where fireblight does not occur (Taylor and Hale, 2003).

During a 150-day period of cold storage, the time of expression of calyx-end rot caused by *Botrytis cinerea* in pears cv. Packham's Triumph, harvested from a high (H1) or low (H2) disease incidence forecasted orchards, was determined. The disease incidence in pears from H1 orchards with greater firmness (8.2 kg) was significantly lower compared with less firm fruit (6.8 kg). The initial maturity had a marked influence on the level and time of expression of calyx-end rot disease in pears stored at –0.5°C. The type of packaging (wood or pulpboard containers) did not have an effect on disease development (Lolas et al., 1998). In peeled Gialla Cactus pear stored at 4°C, fungal growth was seen after 11 days compared with fruits stored at 15°C, which showed fungal growth after 4 days, indicating that peeled cactus pears could be stored safely for 8 days at 4°C (Piga et al., 2000).

The internal gas atmosphere of Conference pears held at 20°C was markedly altered by application of a 1.5% preparation of the sucrose-ester coating, TALPro-long. The O_2 concentration dropped from 14% to less than 2% and over a period of 18 h, whereas CO_2 levels increased from 6% to a peak value between 20 and 24%, and ethylene production was also markedly reduced. In pears inoculated with *Monilinia fructigena* prior to coating, rate of lesion development was significantly reduced compared with controls. The ability of TAL Pro-long treatment to inhibit the lesion development may possibly be due to a physiological response of the fruits to the gas environments, but not due to a direct consequence of elevated levels of CO_2 and lower concentrations of O_2 on the metabolism of the fungal pathogens (Bancroft, 2000).

The effect of heat treatments (ranging from 4 to 46°C) on the quality of eight apple cultivars and two pear cultivars was assessed. Heating enhanced fruit firmness and color and suppressed decay by microorganisms. The Brix ratio (the sugar to alcohol conversion ratio measured at 20°C) was increased in all cultivars except one (Granny Smith), while ripening was delayed in apples, but there was no change in pears (Neven et al., 2000). Application of hot water rinsing and brushing (HWRB) treatment at 55°C for 15 s to apples significantly reduced decay development in fruits inoculated with *Penicillium expansum*, after 4 weeks at 20°C. This HWRB treatment was also effective in reducing disease development in naturally infected fruit after prolonged storage of 4 months at 1°C plus 10 days at 20°C. The HWRB treatment inhibited ripening processes, as evidenced by low respiratory rate and ethylene production in addition to delayed color development. The application of this prestorage treatment has greater potential for commercial exploitation than dry heat treatment (Fallik et al., 2001).

The effects of prestorage heat, controlled-atmosphere (CA) storage, and pre- and poststorage treatments with an ethylene action inhibitor, 1-methylcyclopropene (1-MCP), on decay due to *Penicillium expansum, Botrytis cinerea*, and *Colletotrichum acutatum* (*Glomerella acutata*-teleomorph) in wound-inoculated Golden Delicious apples, were assessed. Prestorage 1-MCP,

TABLE 6.4 Effects of Methylcyclopropene (MCP), Heat, and Controlled Atmosphere (CA) Storage Treatments on Apples Wound-Inoculated with Postharvest Pathogens and Stored for 5 Months at 0°C and 7 Days at 20°C

Treatment	Lesion size (mm)*		
	B. cinerea	*P. expansum*	*C. acutatum*
Prestorage treatments of preclimacteric fruit			
Control	43.0a **	40.2a	35.4a
CA	26.8c	25.8c	27.4c
2MCP	28.4c	33.0b	27.8c
Heat	38.5b	37.1ab	31.4b
MCP + Heat	36.3b	35.0b	32.0b
Poststorage treatments of preclimacteric fruit			
Control	44.1a	40.4a	37.1a
MCP	42.7a	39.9a	35.3a

* Mean of 240 lesion measurements from three 10-fruit replications.
** Means followed by the same letter in the same column are not significantly different by Tukey's HSD ($\alpha = 0.05$).
Source: Saftner et al., 2003.

heat (up to 5 months at 0°C and 7 days at 20°C), 1-MCP plus heat treatments, and CA (1.5 kPa at 20°C, CO_2 at 0°C) decreased decay severity caused by all three pathogens (Table 6.4). Both prestorage 1-MCP treatment and CA-storage delayed ripening as indicated by better retention of green peel color, titratable acidity, and flesh firmness. The results indicated that 1-MCP could be an effective alternative to CA for reducing decay severity and for production of good quality of apples during postharvest storage of Golden Delicious apples (Saftner et al., 2003).

Shattering of grape berries occurring frequently during storage may be of two types, that is physiological and decay shattering. Among the several fungi, including species of *Alternaria, Botrytis, Cladosporium, Pestalotioposis, Rhizopus*, and *Stemphylium*, isolated from shattered grape berries, inoculation with *B. cinerea* mainly reproduced the decay shattering with an extensive growth of the fungus on inoculated berries. In association with ethylene treatment, berry shattering was as high as 60% during 24 days of incubation at 5°C, indicating that fungal infection induced both physiological and decay shattering of grape berries (Xu et al., 1999). The effect of harvest maturity of Thompson Seedless table grapes, subjected to different RH, wetness periods, and storage temperatures, on infection by *Aspergillus niger* and *Rhizopus stolonifer* was assessed. Susceptibility to infection by both pathogens was increased with increase in harvest maturity. The wetness period showed a positive correlation with decay potential of the pathogens, while no consistent influence of RH was evident. Development of both pathogens was favored by storage at 7.5°C for

14 days and inhibition of fungal growth was observed at –0.5°C and 3°C (Witbooi et al., 2000). Grapes treated with 2 g/kg potassium metasulfate, packed in corrugated paper boxes, and stored in ambient conditions (at 30.2–34.7°C and 33.33–51.1% RH) were free from molds (*Aspergillus, Penicillium, Rhizopus, Botrytis*, and *Fusarium* spp.) for up to 9 days after storage. Cold storage (at 6.5 ± 0.5°C and 69.5% RH) increased the period of mold-free days to 21 when combined with potassium metasulfate treatment (Mangasuli et al., 2000).

Assessment of fungal populations in the atmosphere and on the surfaces of equipments and facilities in citrus packing houses in Spain was undertaken during 1995 to 1997 seasons. The fungal population consisted primarily of species of *Cladosporium* and *Penicillium* (average of 25–50 CFU/plate). Initially, *Cladosporium* was predominant, but the frequency of *Penicillium* increased later, especially on the packing lines and cold storage rooms. *Rhizopus* was present in higher populations on the surfaces of bins and packing lines. The detection of fungicide-resistant strains of *P. italicum* and *P. digitatum* was indicative of the need for alternative disease management strategies (Palou et al., 2001). *Phyllosticta citricarpa*, causing citrus blackspot disease, was exposed to –0.5°C and 25°C under artificial and natural conditions. The conidia could not infect unwounded citrus fruit (*Citrus sinensis* cv. Valencia), but the pathogen could be isolated infrequently from artificially inoculated, wounded fruit kept at 25°C. Conidial viability was reduced to zero on black spot-infected fruit stored for 3 weeks at 25°C (Korf et al., 2001).

Cold storage and controlled atmosphere storage have been shown to maintain resistance of mangoes to decay by anthracnose disease caused by *Colletotrichum gloeosporioides* by delaying the ripening process. However, there are two limitations to this strategy for disease management – mangoes are sensitive to chilling and are injured at temperatures lower than 10 to 13°C, depending on the cultivar and duration of exposure, and disease develops normally once the cold-stored fruit is returned to ambient conditions. Antifungal compounds present in the immature mango are lost by oxidative processes as the fruit ripens. Treatment with antioxidant compounds butylated hydroxy anisole (BHA) resulted in reduced severity of anthracnose (Arauz, 2000). Hot water treatment at 46.5°C for 45 min in combination with 2 days of intermittent warming (34°C) during 12 days of cool storage (13°C) resulted in a significantly low incidence and severity of external injury and diseases, softer fruits, higher Brix, and better general appearance (Nyanjage et al., 1998).

Gray mold disease of kiwifruit caused by *Botrytis cinerea* may be effectively contained by exposing the fruit to greater than 95% humidity during curing, a process followed for holding the fruit at ambient temperatures prior to cold storage. The potential for gray mold development is significantly reduced by the curing process. A 48-h delayed cooling at 15°C and high air velocity (ethylene free) inhibited decay of wound-inoculated kiwifruit and also slowed naturally occurring gray mold. Application of 48-h delayed cooling at 15°C, high (2 m/s) air velocity, and relative humidity of 95% was the most effective in reducing gray mold incidence (Crisosto et al., 1997). The development of gray mold was increased by forced-air cooling at 0°C. Controlled atmosphere of 2.5% oxygen and 5% CO_2 delayed the fruit ripening process appreciably, thereby reducing the production of ethylene. *B. cinerea* infection promoted ethylene production to some extent in cold storage. When such infected fruits were stored along with healthy fruits, gray mold incidence was increased significantly after 45 days of cold storage (Michailides and Elmer, 2000). The effects of three ranges of RH (40–59%, 65–80%, and 92–97%) during a storage period of one week at 0°C, on the physiological changes and infection levels by *Botrytis cinerea* in kiwifruit were determined. Weight loss registered an increase with decrease in RH, whereas fruit firmness was significantly higher at 92 to 97% RH than at lower RH ranges, irrespective of time of harvest. Infection levels decreased with increase in RH range in artificially inoculated fruits (Bautista-Baños et al., 2000).

Colletotrichum gloeosporioides causing anthracnose disease of papaya, when inoculated on papaya fruits and stored at 18 ± 3°C, did not influence fruit firmness and contents of malic acid, ascorbic acid, and total soluble solids, in addition to fruit color. The results suggested that under low temperature storage, the pathogen did not affect most of the desirable characteristics of papaya fruits which differed due to varietal tolerance to *C. gloeosporioides* (Ramos et al., 2001).

6.2.2.1 Production of Mycotoxins As in the case of seeds, the fungal pathogens, in addition to causing decay and spoilage of fruit in storage, produce harmful mycotoxins, some of which have been proved to be carcinogenic. Species of *Penicillium* and *Aspergillus*, especially *P. expansum*, have been demonstrated to produce mycotoxins, designated patulin, in infected fruits. The occurrence of patulin as a natural contaminant of apple juice is a world-wide problem. Various analytical methods, such as HPLC with UV or, preferably, photodiode assay gas chromatography and TLC, have been employed for the detection of patulin (Shephard and Leggott, 2000). The effect of modified atmospheres and packaging materials on the growth of *P. expansum* and patulin production in apple cv. Granny Smith was determined. Patulin production in apples packaged in polyethylene was almost entirely inhibited by all three gas combinations, that is 58% CO_2/42% N_2, 48% CO_2/52 N_2, and 88% CO_2/12% N_2. Gas chromatographic determination of the polyethylene-packaged samples before and after the incubation period (14 days at 25°C) revealed that CO_2 levels declined and N_2 levels increased for all of the atmospheres tested. The results demonstrated that polyethylene is an excellent packaging material for the storage of apples and it inhibited development of *P. expansum* and limited the production of patulin to less than 3.2 μg/ml, regardless of gaseous environment (Moodley et al., 2002).

The simultaneous occurrence of patulin and citrinin in seven different varieties of apples with small rotten areas was detected by a rapid, multidetection TLC technique. The percentage of contamination only with patulin was higher

(68.6%) than that with citrin alone (3.9%). Patulin and cirinin were present together in 19.6% of fruits tested. The highest mean patulin content was 80.5 mg/kg for the Richard apple. All apple varieties had low citrinin contamination, ranging from 0.32 to 0.92 mg/kg. There appears to be a potential risk of human exposure to patulin through consumption of juices and jams manufactured with apples with small rotten areas (Martins et al., 2002). Pear fruits cv. Abatefetel were inoculated with *Penicillium expansum, Aspergillus flavus, Alternaria alternata*, and *Stemphylium vesicarium*. The presence of patulin in fruits inoculated with all fungi, except *A. flavus*, was detected in the flesh of pears with lesions of 10 mm diameter. Patulin concentrations were higher than the permitted maximum residue limit in all sections of fruits, even in apparently sound tissues. This seems to be the first report of diffusion of the mycotoxin into different depths of pear fruits (Laidou et al., 2001).

The dried raisin fruits were examined for the presence of toxicogenic fungi. *Aspergillus flavus, A. fumigatus, A. niger, Penicillium chrysogenum, P. oxalicum*, and *Rhizopus stolonifer* were frequently present in the samples tested. Chromatographic estimation showed that aflatoxin B_1 was present in two samples at concentrations of 300 and 220 µg/kg fruit, whereas the third sample was contaminated with ochratoxin A at 250 µg/kg dried fruit. The fungal isolates (66 out of 90) produced at least one of the following mycotoxins: alterinariol, aflatoxins B_1, B_2, G_1, and G_2 fumagillin, ochratoxin A, penicillic acid, sterigmatocystin, citrinin, and terrin, as detected by chromatographic analysis (Youssef et al., 2000). *A. niger* var. *niger* and *A. niger* var. *awamori* were isolated in higher frequency than *A. carbonarius* from black dried vine fruits. Ochratoxin A (OA) was detected in 74% of dried vine fruits samples. In *Aspergillus* section Nigri, 28% of strains tested were ochratoxigenic, whereas *A. carbonarius* contained a higher proportion (82.6%) of strains capable of producing OA. The results provide evidence suggesting that dried vine fruits could be an important source of OA (Magnoli et al., 2004). In a survey conducted by the UK Food Standards Agency, during 2000 to 2001, OA was detected in one sample of pistachio nuts and one sample of dried figs above the limit of 10 µg/kg of dried vine fruit, the standard set in the European Union and United Kingdom (Anon, 2002). *Aspergillus ochraceus* and *A. alliaceus* were isolated from tree nut orchards, nuts, and figs in California, USA. *A. alliaceus* a little known species, was found to be the important ochratoxin-producing fungus and it might be responsible for the ochratoxin contamination occasionally observed in figs (Bayman et al., 2002).

6.2.3 Vegetables

6.2.3.1 Development of Pathogens Vegetables may be stored for varying periods depending on the market requirements. Potatoes were inoculated with two biotypes of *Phytophthora infestans* and stored at 3, 7, 10, and 15°C. The average reflective intensity (ARI) of diseased tissue from cut surfaces of sample tubers was determined by image analysis; the mean ARI of samples indicated the extent of tuber infection and infection rate. Average tuber tissue infection and infection rate were the least at 3°C. Increase in temperature above 3°C increased the mean ARI, depending on the cultivar and biotype of *P. infestans*. Rate of blight infection in tuber tissue showed increases from –0.2 ARI/day at 10°C (Kirk et al., 2001). *P. erythroseptica*, causing pink rot disease, could grow from rotting, wound-inoculated tubers to the surfaces of adjacent tubers in which oospores were formed, during storage at 9°C under high humidity. Thus, these tubers may spread the disease to daughter tubers in previously uninfected fields (Cunliffe et al., 1977). Pink rot development occurred very rapidly when the temperature and humidity increased due to poor aeration in storage (Logan, 1983).

Development of black eye disease caused by *Fusarium solani* f.sp. *eumartii* in potato tubers with symptoms (apparent infection) and without symptoms (latent infection), stored at 4°C or at room temperature (27 ± 2°C), was studied. Storage of infected tubers at 4°C did not reduce the rate of disease development. Tubers with visible symptoms rotted much earlier than the tubers with latent infection (with no visible symptoms). Tubers stored at room temperature rotted more rapidly (Lima and Lopes, 1998). During storage *Helminthosporium solani* causing silver scurf disease of potato, developed well when free moisture was available. Free moisture is formed by condensation on the ceiling and walls due to poor insulation or to improper management of relative humidity and ventilation systems. The difference in temperature between the top and bottom of the potato pile should not be allowed to exceed 1°C (Shetty et al., 2003). The development of black dot disease caused by *Colletotrichum coccodes* in potato tubers during storage was studied by inoculating 13 isolates from different geographical origins. The minitubers were successfully infected at 5, 10 and 15°C, although symptoms developed more slowly at low temperatures. Latent infection of up to 21% was detected in tubers and the pathogen was able to develop at low temperatures in controlled conditions (Glais-Varlet et al., 2004).

Soft rot disease caused by *Erwinia* spp. in potatoes stored in a netted wooden box under natural conditions developed at different rates in various cultivars depending on their level of susceptibility to the disease (Rasul et al., 1999). Infection and expansion of lesions in potato cv. Russet Burbank tubers inoculated with *Erwinia carotovora* subsp. *carotovora* and stored at 10 to 25°C for 3 to 48 h of wetness was assessed. Infection was maximum with 12 h of wetness at 20°C. Lesion expansion was faster at 16°C after about 60 days of storage (Kushalappa and Zulfiqar, 2001).

Streptomyces scabies causing potato scab disease also infects other roots crops such as carrot, radish, sugarbeet, and parsnip (Stevenson et al., 2001). Another important pathogen involved in potato scab disease, occurring in Japan and northern Scandinavia was reported to be due to *S. turgidiscabies* (Takeuchi et al., 1996; Lindholm et al., 1997). Indentificalion of *Streptomyces* spp. from potato tubers using conventional methods is time consuming and laborious because of the slow growth rate of the pathogen and high diversity of *Streptomyces* spp. inhabiting scab lesions (Lindholm et al., 1997). By employing species-specific primers in a PCR-based assay, *S. scabies* and *S. turgidiscabies* could be reliably detected and identified, in addition to the third pathogen, *S. aureofaciens*, associated with scab disease. *S. scabies* and *S. turgidiscabies* were detected in the same fields, tubers, and scab lesions from freshly harvested or stored potato tubers incubated at room temperature (18 to 21°C) under humid conditions for a few days. The incidence of *S. scabies* was relatively high in freshly harvested tubers but the percentage of tubers with *S. scabies* was much lower than those with *S. turgidiscabies* following storage. Both scab pathogens were transmitted through seed tubers of Matilda and Sabina potatoes after storage at 4°C for 24 weeks (Lehtonen et al., 2004).

Ceratocystis fimbriata, causing black rot disease of sweet potato, produces abundant spores during storage, contaminating washing machines, crates, and structures, and also the hands of workers. Under such conditions entire lots may be infected, since the pathogen can spread rapidly to roots surrounding a rotting sweetpotato. Java black rot disease caused by *Botryodiplodia theobromae* is favored by warm temperatures between 20° and 30°C and by a wide range of relative humidities. Susceptibility to Java black rot disease was increased with increase in storage time or when chilled roots are returned to higher temperatures. Accumulation of phenol at high levels (two to 13 times) occurred in sweetpotato following inoculation with *B. theobromae* (Mohapatra et al., 2001). Rhizopus soft rot disease caused by *Rhizopus stolonifer* is favored by relative humidities in the range 75 to 85%. Chilling and heat damage predispose the tubers to infection. Susceptibility of stored sweetpotato roots of cultivars Beauregard and Hernandez to *R. stolonifer* was assessed at an interval of 4 to 6 weeks over a storage period of 335 days. The roots were totally resistant to infection at 30 to 60 days after harvest. The disease incidence increased later and reached the maximum level during 100 to 175 days. The susceptibility thereafter declined to levels comparable to that of freshly harvested roots. The period of enhanced susceptibility was longer in Hernandez than in Beauregard (Holmes and Stange, 2002).

Tomatoes are highly sensitive to chilling and the optimum storage temperature varies with the maturity of the tomato fruit. The influence of temperatures (5, 15, 25, and 35°C), wetness periods (0, 3, 6, 12, 24, 48, and 72h) and inoculum concentrations (10^1, 10^2, 10^6, and 10^8 spores/ml) on the intensity of postharvest rots caused by *Fusarium verticillioides, Geotrichum candidum*, and *Rhizopus stolonifer* was determined. Incubation of fruits at 25°C resulted in maximum disease incidence. *F. verticillioides* did not infect at 5 or 35°C, indicating high and low temperatures were not suitable for infection. High incidence of *R. stolonifer* was observed at 5, 15, and 25°C, reaching 97.5 to 100%, and this pathogen could develop well at the wide range of temperatures tested. Wetness period had no effect on disease incidence, as all the pathogens could develop at different periods of wetness, though longer periods of free water increased disease incidence. *R. stolonifer* caused 100% infection even in the absence of free water. As expected, increase in spore concentration resulted in proportional increase in disease incidence, implying the importance of spore load in the storage facilities to the disease incidence (Silveira et al., 2001).

Proper temperature control is a critical requirement for maintaining the quality and extending shelf-life of tomatoes. The optimum temperature for ripening mature green tomatoes is from 18 to 24°C. Mature green tomatoes stored at temperatures below 10°C were found to be susceptible to decay by *Alternaria alteranta*. Ripe tomatoes (light red) can be stored at lower temperatures (about 4°C) than mature green tomatoes. However, longer storage at low temperature may lead to loss of color, firmness, shelf-life, and, more importantly, taste (Boyette et al., 2003a).

Carrot roots are frequently invaded by postharvest pathogens through wounds. The exposure of carrots to high temperature and humidity for a short period after harvest to "harden" the periderm and to heal wounds has been shown to be effective in containing the invasion by postharvest pathogens. This treatment promoted the wound healing process, involving lignification and suberization, in addition to development of callus in treated carrot roots (Garrod et al., 1982). Infection of wounds by *Mycocentrospora acerina* was dramatically reduced when carrot roots were exposed to high temperature and humidity near saturation (Lewis and Garrod, 1982). The level of resistance of wounds to pathogen invasion is considered to be an important factor in determining the storage potential of carrots harvested at different stages of growth. The young roots inoculated with *M. acerina* permitted an increase in lesion numbers more slowly compared with more mature roots. These differences were discernible for at least 5 months of storage. Enhancement of susceptibility of older roots was attributed to a reduction in healing potential of wounds (Lewis et al., 1981).

Peppers are sensitive to chill injury. Temperatures below 4.5°C may result in softening, pitting, and predisposition to decay. Controlled atmospheres low in oxygen and high in CO_2 may retard color change due to ripening and to help maintain quality during storage and transit. Incidence of soft rot disease caused by *Fusarium equiseti* and *F. semitectum* ranged from 2.8 to 9.6% at locations where the temperature was high (more than 35°C) during storage. Storing bell pepper at 10 to 15°C may be effective in avoiding soft rot disease (Shukla and Sharma 2000).

Onions are generally packed in mesh bags which offer little protection against damage due to mishandling. Onions may be stored at a temperature range of 0 to 2°C and RH of 65 to 75% for several months. At temperatures below 0°C, onions may freeze, become soft, and rapidly decay. Storage losses due to temperatures have been shown to be appreciable due to sprouting and other changes that may lead to unacceptable market quality (Table 6.5). In the case of garlic, the quality depends on the conditions of storage and initial colonization of garlic bulbs by fungal pathogens under field conditions. *Penicillium* spp. was isolated from the bulbs stored at 2 to 8°, 4°, and 10 to 24°C for 3, 6, and 9 months. *P. viridicatum* had initiated colonization of the underground

TABLE 6.5 Effect of Storage Temperatures on Market Losses of Onions

Storage period	Percentage of loss at different storage temperatures		
	21°C (70°F)	4.5°C (40°F)	1°C (34°F)
2 weeks	8.9	5.2	5.3
1 month	10.2	7.0	6.1
3 months	25.2	15.9	10.6
6 months	61.8	32.7	14.0

Source: Boyette et al., 2003b.

TABLE 6.6 Bacterial Species Producing Toxic Metabolites

Bacteria	Toxic metabolites
Clostridium botulinum	Exotoxin
Escherichia coli	
– enteropathogenic (EPEC) – enterotoxigenic (ETEC) – enteroinvasive (EIEC) – enteroaggragative (EAEC) – enterohemorrhagic (EHEC)	Different kinds heat stable (100°C for 15 min) and heat labile toxins; *Shigella* enterotoxin-2 or EIEC enterotoxin; enterotoxin (EAST-1); 4 kDa, 108 kDa and 12 kDa proteins Verotoxin
Staphylococcus aureus	Enterotoxin

Source: Pandey et al., 1999.

parts of garlic in the field. Most of the garlic ecotypes grown in Poland were susceptible to decay by *Penicillium* spp. (Machowicz-Stefaniak et al., 1998).

6.2.4 Food-borne Human Pathogens

Fruits and vegetables may harbor not only microbial pathogens that cause spoilage of commodities during harvest, transit, and storage, but also pathogens that can induce several ailments in humans and animals, when contaminated commodities are consumed. Food poisoning is the illness caused by the toxins produced by microorganisms present in the food. Several species of bacteria, belonging to the genera *Bacillus, Clostridium, Escherichia, Listeria, Salmonella*, and *Staphylococcus*, have been shown to be the cause of food poisoning. The toxins or the toxic metabolites cause diarrhea, botulism, fever, and vomiting in the affected individuals (Table 6.6). The species of microorganisms associated with spoilage vary depending on the nature of fruits or vegetables, environmental conditions, storage temperature, relative humidity, and gases and their proportions in the storage. *Listeria monocytogenes* has been considered to be associated with a number of outbreaks in the United States of a food-borne illness known as listeriosis. This bacterium could survive and increase in apple fruit slices when fruits were stored at 20 or 10°C, but they were unable to multiply at 5°C. The fruit decay (brown rot) caused by *Glomerella cingulata* was enhanced following inoculation with *L. monocytogenes*. The results indicated that safety of fresh-cut apples depended essentially on proper storage temperature and for maintenance of disease-free conditions of fruits (Conway et al., 2000). A recent Salmonellosis outbreak was epidemiologically associated with consumption of fresh mangoes imported from Brazil. A commercial deinfestation method used to eliminate dephritid fly larvae from mangoes and the subsequent product cooling procedures were studied for their influence in the internalization of *Salmonella enteritidis*. The heat treatment with water (47°C for 90 min followed by immersion in water at 21°C for 10 min) might promote internalization of the human bacterial pathogen if heat-disinfested mangoes were cooled using contaminated water (Penteado et al., 2004)

Rapid and accurate identification of the microorganisms associated with fruits and vegetables is the basic requirement for the prevention of food poisoning and other associated problems. Standard enrichment techniques were employed to isolate and identify enteric pathogens such as *Salmonella, Campylobacter*, *E. coli, Listeria*, and *Aeromonas* spp. from commercially available organic vegetables. The most commonly observed pathogen was *A. schubertii*, accounting for a maximum of 21% of all samples. *A. hydrophila* (5.8%), *A. trota* (5.8%), *A. caviae* (3.5%), and *A. veronii* biovar *veronii* (2.3%) were also present in the organic vegetables. The absence of accepted enteric pathogens in organic vegetables despite the presence of *Aeromonas* spp. provides a significant support for the use of organic vegetables (McMahon and Wilson, 2001). An enrichment protocol in conjunction with an immunoprecipitation kit VIP (Biocontrol, Bellevue, WA) was demonstrated to be effective for the detection of *E. coli* in sprouted seed products (Weagant and Bound, 2001). Since *E. coli* is entrapped in protected areas of the seed coat of alfalfa, it was difficult to isolate and identify the bacteria. Soaking seeds at 37°C for 1 h increased the *E. coli* population and pummeling seeds for 1 min, an enrichment step in modified tryptic soy broth, followed by the use of immunomagnetic beads for the separation of *E. coli* were effective in the detection of this dreadful bacteria (Wu et al., 2001). The presence of human pathogens *E. coli* and *Salmonella enterica* was detected and identified by fluorescence microscopy technique (Charkowski et al., 2002).

Nucleic acid-based techniques are considered to be more sensitive, rapid, and reliable compared with other diagnostic methods. The human pathogen *Listeria monocytogenes* was detected in vegetables such as tomatoes, cabbage, and coriander leaves by employing the *hlyA* gene (required for intracellular survival in the host) in a colony hybridization assay (Anu et al., 2002). A quantitative real-time polymerase chain reaction (PCR) detection method was employed for the specific detection of *L. monocytogenes* in cabbage. The DNA of the pathogen was purified by an organic solvent extraction technique and analyzed by real-time PCR, which provided results within 8 h and was relatively inexpensive, showing good potential for routine analytical application (Hough et al., 2002). The efficacy of an immunochromatographic method and the TaqMan® technique was compared for the detection of *E. coli*. TaqMan® PCR assay was more sensitive than the immunochromatographic method, since it detected the bacteria in a greater number of samples. However, the immunoassay was easier and more rapid for detecting *E. coli* (Fratamico and Bagi, 2001).

Many outbreaks of an illness linked to the consumption of imported raspberries contaminated with *Cyclospora* have been reported since 1995 in the United States and Canada. Washing of raspberries appeared to be not adequate for removing the coccidial contamination (Lee and Lee, 2001). Cantaloupe melon has been linked with outbreaks of *Salmonella* infection (salmonellosis) based on the theory that bacterial surface charge and hydrophobicity may affect bacterial attachment and complicate bacterial detachment from cantaloupe surfaces. Whole cantaloupes were inoculated by submerging them in suspensions of *Salmonella*, *E. coli*, and *L. monocytogenes* for 10 min followed by drying for 1 h in a biosafety cabinet and storing for 7 days at 4°C. After washing the cantaloupes, bacteria still attached to the melon surface, as well as those in wash water, were enumerated. *Salmonella* was more strongly attached at 0, 3 and 7 days compared with other bacterial species. When present in a cocktail (mixture of genera or strains), the attachment of *Salmonella* strain was the strongest. However, *E. coli* exhibited more extensive attachment than that of *L. monocytogenes* on days 0, 3 and 7. The strength of the bacterial attachment to cantaloupe surfaces showed a linear relationship with bacterial cell surface hydrophobicity, negative charge, and positive charge (Ukuku and Fett, 2002). Unpasteurized apple juice and cider have been implicated in outbreaks of *Escherichia coli* 0157: H7 and *Salmonella* infections (Kenney and Beuchat, 2002).

Consumption of several types of raw fruits and vegetables contaminated with *Salmonella* before harvest seemed to be associated with salmonellosis. Water and soil as reservoirs of *Salmonella* for contamination of mature green tomatoes were examined. *Salmonella* could survive in inoculated, moist soil for 45 days and its population on tomatoes in contact with soil increased during storage for 4 days at 20°C and were maintained at a higher level during storage for an additional 10 days. Populations of *Salmonella* was assessed by PCR fingerprinting. The serotype *Salmonella* Montevideo was the most persistent on tomatoes in contact with inoculated soil and on spot-inoculated tomatoes, when compared with the other four serotypes (Michigan, Poona, Hartford, and Enteritidus). The results suggest that preharvest contact of produce with contaminated water or soil may intensify problems associated with the postharvest removal of pathogens or their accessibility to treatment with sanitizers (Guo et al., 2002).

The harvested fruits and vegetables may carry viruses that infect humans (Madden, 1992). Lettuce has been implicated as a vehicle for transmission of food-borne viruses, especially hepatitis A virus when stored at 4°C and room temperature under normal air and different modified atmosphere-packaged conditions (Bidawid et al., 2001).

SUMMARY

Postharvest management aims to preserve and maintain high quality of the harvested produce both in the field immediately after harvest and in storage, as well as in transit. Proper methods have to be adopted for postharvest handling of commodities to avoid damage. Various methods have been found to be useful for handling seeds, fruits, and vegetables. Postharvest cooling effectively eliminates field heat rapidly from harvested fruits and vegetables prior to storage. The requirements for precooling different fruits and vegetables vary considerably. It is essential to provide optimal temperatures for the maintenance of market quality. The storage conditions markedly influence the development of microorganisms causing postharvest diseases, in addition to production of mycotoxins capable of inducing mycotoxicoses in humans and animals, when the contaminated foods and feeds are consumed. Furthermore, fruits and vegetables may act as vehicles of transmission of some human pathogens. The imperative need to be aware of such problems associated with improper handling and inadequate vigilance to avoid incidence of postharvest diseases is highlighted.

APPENDIX 6: QUANTIFICATION OF FUNGAL PATHOGEN BIOMASS (CUMAGUN ET AL., 2004)

A. Preparation of Samples (Antigen Extract)

i. Grind the samples of grain flour (0.1 g) in 2 ml of extraction buffer (phosphate-buffered saline (PBS) without Tween, containing ethylenediaminetetraacetic acid disodium salt (EDTA)) in a mortar and pestle and incubate overnight at 4°C in a refrigerator.
ii. Maintain two replications for each sample.

B. Plate-trapped Antigen (PTA)-ELISA

i. Prepare polyclonal antisera using the surface washings and/or mycelial homogenates of cultures of *Fusarium culmorum* (Fc) and *F. graminearum* (Fg) for immunizing rabbits.
ii. Select the polyclonal antibodies (PABs) that can react with all *Fusarium* species infecting cereal grains, without showing cross-reactions with other fungal species outside the genus *Fusarium*.

iii. Dispense antigen extract (100 µl) to each well in NUNC Polysorb ELISA plates; incubate for 2 h at 37°C; pour out the plates without washing the wells; add blocking solution (200 µl) (containing 1% nonfat dry milk powder in PBS) per well and incubate for 1 h at 37°C.
iv. Wash the wells three times with PBS-Tween 20.
v. Transfer to each well 100 µl of IgG (selected antiserum containing PABs) in blocking solution; incubate for 1 h at 37°C and wash the wells four times as in step (iv).
vi. Add to each well 100 µl of alkaline phosphatase-conjugated goat antirabbit IgG (H+L) (Dianova, Hamburg, Germany) diluted to 1:2000 in 0.05 M Tris-HCl buffer, pH 8.0 containing 1% nonfat dry milk powder; incubate for 1 h at 37°C and wash the wells four times as in step (iv).
vii. Add to each well 200 µl of substrate (p-nitrophenyl phosphate, 1 mg/ml in substrate buffer, pH 9.6) for 1 h at 20°C and read the absorbance at 405 nm in ELISA reader (Tecan "Rainbow" ELISA reader, Tecan SLT Laboratory Instruments, Crailsheim, Germany).

REFERENCES

Abramson, D., Gan, Z., Clear, R.M., Gilbert, J. and Marquardt, R.R., 1998. Relationship among deoxynivalenol, ergosterol and *Fusarium* exoantigens in Canadian hard and soft wheat. *International Journal of Food Microbiology*, **45**: 217–224.

Agrahari, P.R., Sharma, R.M. and Tripathi, V.K. 1998. Use of waxes to increase shelf life of fruits. *Farmers and Parliament*, **33**: 13.

Al-Yahya, S.A. 1999. Change of fungal infection during wheat storage at different conditions. *Arab Universities Journal of Agricultural Sciences*, **7**: 531–545.

Anderson, H.W., Nehring, E.W. and Wisher, W.R. 1975. Aflatoxin contamination of corn in the field. *Journal of Agricultural and Food Chemistry*, **23**: 775.

Anon, 1979. *Report of Grain Inspection Division*, Minnesota Department of Agriculture, USA.

Anon, 1982. *Grain Drying and Storage No.1. Grain Storage Methods*. MAF Booklet No. 2417.

Anon, 1989. Prevention of postharvest food losses. In: *Fruits, Vegetables and Root crops – A Training Manual*, No.17/2 Food and Agriculture Organization, Rome.

Anon, 2001. *Guidance for Industry: Fumonisin Levels in Human Foods and Animal Feeds*. Online publication. USFDA Center for Food Safety and Applied Nutrition and the Center for Veterinary Medicine. Docket No.00D-1277. Available at www.cfsan.ida.gov/~dms/fumongu2.html.

Anon, 2002. Survey of Nuts, Nut Products and Dried Tree Fruits for Mycotoxins. Food Survey Information sheet, No.21/02, 50 pp. Food Standards Agency, London, UK.

Anu K., Ussuf, K.K. and Bongirwar, D.R. 2002. Detection of *Listeria monocytogenes* in foods: comparison of colony hybridization technique with conventional method. *Journal of Food Science and Technology* (Mysore), **39**: 214–218.

Arauz, L.F. 2000. Mango anthracnose: Economic impact and current options for integrated management. *Plant Disease*, **84**: 600–611.

Ashworth, Jr. L.J., McMeans, J.L. and Brown, M. 1969. Infection of cotton by *Aspergillus flavus*: Time of infection and influence of fibre moisture. *Phytopathology*, **59**: 383.

Asis, R., Di Paola, R.D. and Aldao, M.A.J. 2002. Determination of aflatoxin B_1 in highly contaminated peanut samples using HPLC and ELISA. *Food and Agricultural Immunology*, **14**: 201–208.

Bai, G.H., Desjardins, A.E. and Plattner, R.D. 2002. Deoxynivalenol-non producing *Fusarium graminearum* causes initial infection but does not cause disease spread in wheat spikes. *Mycopathologia*, **153**: 91–98.

Bancfroft, R.D. 2000. The influence of a surface coating on the internal gas atmosphere and rate of postharvest rot development in Conference pears. *Acta Horticulturae*, **518**: 79–86.

Bankole, S.A., Ikotun, B. and Ekpo, E.J.A. 1999. Fungal deterioration of melon seeds stored in jute sacks and polyethylene bags in Ago-Iwoye, Southwestern Nigeria. *Mycopathologia*, **146**: 135–146.

Bautista-Buñas, S., Long, P.G. and Ganesh, S. 1997. Curing of kiwifruit for control of postharvest infection by *Botrytis cinerea. Biological Technology*, **12**: 137–145.

Bautista-Baños, S., Long, P.G. and Ganesh, S. 2000. The role of relative humidity during early storage on *Botrytis cinerea* incidence of kiwifruit. *Revista Mexicana de Fitopatologia*, **18**: 92–96.

Bayman, P., Baker, J.L., Doster, M.A., Michailides, T.J. and Mahoney, N.E. 2002. Ochratoxin production by the *Aspergillus ochraceus* group and *Aspergillus alliaceus. Applied and Environmental Microbiology*, **68**: 2326–2329.

Bidawid, S., Farber, J.M. and Sattar, S.A. 2001. Survival of hepatitis A virus on modified atmosphere-packaged (MAP) lettuce. *Food Microbiology*, **18**: 95–102.

Bilgrami, K.S., Choudhary, A.K. and Ranjan, K.S. 1992. Aflatoxin in field mustard (*Brassica juncea*) cultivars. *Mycotoxin Research*, **8**: 21.

Bird, C.B., Malone, B., Rice, L.G., Ross, P.F., Eppley, R. and Abouzied, M.M. 2002. Determination of total fumonisins in corn by competitive enzyme-linked immunosorbent assay. Collaborative study. *Journal of AOAC, International*, **85**: 404–410.

Boyette, M.D., Sanders, D.C. and Estes, E.A. 2003a. *Postharvest Cooling and Handling of Field and Greenhouse-grown Tomatoes*. North Carolina State University, Raleigh and USDA Publication, 14 pp.

Boyette, M.D., Sanders, D.C. and Estes, E.A. 2003b. *Postharvest Cooling and Handling of Onions*. North Carolina State University, Raleigh and USDA Publication, 10 pp.

Boyette, M.D., Wilson, L.G. and Estes, E.A. 2003c. *Introduction to Proper Postharvest Cooling and Handling Methods*. North Carolina State University Hand book No.66, 7 pp.

Boyette, M.D., Wilson, L.G. and Estes, E.A. 2003d. *Postharvest Cooling and Handling of Apples*. North Carolina State University, Raleigh and USDA Publication, AG–413–1, 10 pp.

Boyette, M.D., Wilson, L.G. and Estes, E.A. 2003e. *Postharvest Cooling and Handling of Peppers*. North Carolina State University, Raleigh and USDA Publication, AG–413–3, 5 pp.

Bush, B.J., Carson, M.L., Cubeta, M.A., Hagler, W.M. and Payne, G.A. 2004. Infection and fumonisin production by *Fusarium verticillioides* in developing maize kernels. *Phytopathology*, **94**: 88–93.

Calvo, A.M., Bok, J., Brooks, W. and Keller, N.P. 2004. *veA* is required for toxin and sclerotial production in *Aspergillus parasiticus*. *Applied and Environmental Microbology*, **70**: 4733–4739.

Charkowski, A.O., Barak, J.D., Sarreal, C.Z. and Mandrell, R.E. 2002. Differences in growth of *Salmonella enterica* and *Escherichia coli* 0157: H7 on alfalfa sprouts. *Applied and Environmental Microbiology*, **68**: 3114–3120.

Clements, M.J. 2002. *Resistance to Fumonisins and Fusarium Ear Rot Corn*. Ph.D. Thesis, Department of Crop Sciences, University of Illinois, Urbana, USA.

Conway, W.S., Leverentz, B., Saftner, R.A., Janisiewicz, W.J., Sams, C.E. and Leblanc, E. 2000. Surival and growth of *Listeria monocytogenes* on fresh-cut apple slices and its interaction with *Glomerella cingulata* and *Penicillium expansum*. *Plant Disease*, **84**: 177–181.

Côrtes, N. de A., Cassetari Neto, D. and Correa, B. 2000. Occurrence of aflatoxins in maize produced by traditional cultivation methods in agricultural communities in the State of Marto Grosso. *Higiene Alimentar*, **14**: 16–26.

Covarelli, L., Turner, A.S. and Nicholson, P., 2004. Repression of deoxynivalenol accumulation and expression of *Tri* genes in *Fusarium culmorum* by fungicides *in vitro*. *Plant Pathology*, **53**: 22–28.

Crisosto, C.H., Michailides, T.J. Garner, D. and Crisosto, G. 1997. Improving postharvest temperature management of kiwifruit. *Annual Report of Kiwifruit Commission, Sacramento.*

Cuero, R., Ouellet, T., Yu, J. and Mogongwa, N. 2003. Metal enhancement of fungal growth, gene expression and aflatoxin synthesis by *Aspergillus flavus*: RT-PCR characterization. *Journal of Applied Microbiology*, **94**: 953–961.

Cumagun, C.R., Rabenstein, F. and Miedaner, T. 2004. Genetic variation and covariation for aggressiveness, deoxynivalenol production and fungal colonization among progeny of *Gibberella zeae* in wheat. *Plant Pathology*, **53**: 446–453.

Cunliffe, C., Londsdale, D. and Epton, H.A.S. 1977. Transmission of *Phytophthora erythroseptica* on stored potatoes. *Transactions of British Mycological Society*, **69**: 27–30.

Desjardins, A.E., Munkvold, G.P., Plattner, R.D. and Proctor, R.H. 2002. *FUM1* – a gene required for fumonisin biosynthesis but not for maize ear rot and ear infection by *Gibberella moniliformis* in field tests. *Molecular Plant–Microbe Interactions*, **15**: 1157–1164.

Diener, U.L., Cole, R.J., Sanders, T.H., Payne, G.A., Lee, L.S. and Klich, M.A. 1987. Epidemiology of aflatoxin formation by *Asperigllus flavus*. *Annual Review of Phytopathology*, **25**: 249.

Dordi, M.C. 1994. Packaging of horticultural produce. *Indian Horticulture*, **39**: 16.

Dowell, F.E., Pearson, T., Maghirang, E.B., Xie, F. and Wicklow, D.T. 2002. Reflectance and transmittance spectroscopy applied to detecting fumonisin in single corn kernels infected with *Fusarium verticillioides*. *Cereal Chemistry*, **79**: 222–226.

Ehrlich, K.C. and Cotty, P.J. 2004. An isolate of *Aspergillus flavus* used to reduce aflatoxin contamination in cotton-seed has a defective polyketide synthase gene. *Applied Microbiology and Biotechnology*, **65**: 473–478.

El-Nagerabi, S.A.F. and Elshafie, A.E. 2001. Determination of seedborne fungi and aflatoxins in Sudanese guar seeds. *Tropical Science*, **41**: 31–35.

Eudes, F., Comeau, A., Rioux, S. and Collin, J. 2001. Impact of trichothecenes on *Fusarium* head blight (*Fusarium graminearum*) development in spring wheat (*Triticum aestivum*). *Canadian Journal of Plant Pathology*, **23**: 318–322.

Fakhoury, A.M. and Woloshuk, C.P. 2001. Inhibition of growth of *Aspergillus flavus* and fungal alpha-amylases by a lectin-like protein from *Lablab purpureus. Molecular Plant–Microbe Interactions*, **14**: 955–961.

Fallik, E., Tuvia-Alkalai, S., Feng, X. and Lurie, S. 2001. Ripening characterization and decay development of stored apples after a short pre-storage hot water rinsing and brushing. *Innovative Food Science and Emerging Technologies*, **2**: 127–132.

Fratamico, P.M. and Bagi, L.K. 2001. Comparison of an immunochromato-graphic method and the TaqMan® *E. coli* 0157: H7 assay for detection of *Escherichia coli* 0157: H7 in alfalfa sprout spent irrigation water and in sprouts after blanching. *Journal of Industrial Microbiology and Biotechnology*, **27**: 129–134.

Gang, G., Miedaner, T., Schumacher, U., Schollenberger, M. and Geiger, H.H. 1998. Deoxynivalenol and nivalenol production by *Fusarium culmorum* isolates differing in aggressiveness toward winter rye. *Phytopathology*, **88**: 879–884.

Garrod, B., Lewis, B.G., Brittain, M.J. and Daves, W.P. 1982. Studies on the contribution of lignin and suberin to the impedance of wounded carrot root tissue to fungal invasion. *New Phytologist*, **90**: 99–108.

Glais-Varlet, I., Bouchek-Mechiche, K. and Andrivon, D. 2004. Growth *in vitro* and infectivity of *Colletotrichum coccodes* on potato tubers at different temperatures. *Plant Pathology*, **53**: 398–404.

Guo, X., Chen, J., Brackett, R.E. and Beuchat, L.R. 2002. Survival of *Salmonella* on tomatoes stored at high relative humidity, in soil and on tomatoes in contact with soil. *Journal of Food Protection*, **65**: 274–279.

Hale, C.N. and Taylor, R.K. 1999. Effect of cool storage on survival of *Erwinia amylovora* in apple calyxes. *Acta Horticulturae*, **489**: 139–143.

Hasan, H.A.H. 2001. Role of pencycuron in aflatoxin production and cotton seed protection. *Journal of Natural Toxins*, **10**: 127–136.

Hestbjerg, H., Felding, G. and Elmholt, S. 2002. *Fusarium culmorum* infection of barley seedlings: correlation between aggressiveness and deoxynivalenol content. *Journal of Phytopathology*, **150**: 308–312.

Holmes, G.J. and Stange, R.R. 2002. Influence of wound type and storage duration on susceptibility of sweet potatoes to *Rhizopus* soft rot. *Plant Disease*, **86**: 345–348.

Hooker, D.C., Schaafsma, A.W. and Tamburic-Ilincic, L. 2002. Weather variables pre- and post-heading to predict deoxynivalenol content in winter wheat. *Plant Disease*, **86**: 611–619.

Hope, R. and Magan, N. 2003. Two-dimensional environmental profiles of growth, deoxynivalenol and nivalenol production by *Fusarium culmorum* on a wheat-based substrate. *Letters in Applied Microbiology*, **37**: 70–74.

Hough, A.J., Harbison, S.A., Savill, M.G. Melton, L.D. and Fletcher, G. 2002. Rapid enumeration of *Listeria monocytogenes* in artificially contaminated cabbage using real-time polymerase chain reaction. *Journal of Food Protection*, **65**: 1329–1332.

Kasmir, R.F. 1985. Postharvest handling systems: Leafy, root and stem vegetables. In: *Postharvest Technology of Horticultural Crops* (eds) A.A. Kader *et al.*, p. 131, Cooperative Extension University of California, DANR, Berkeley, USA.

Kenney, S.J. and Beuchat, L.R. 2002. Comparison of aqueous cleaners for effectiveness in removing *Escherichia coli* 0157: H7 and *Salmonella muenchan* from the surface of apples. *International Journal of Food Microbiology*, **74**: 47–55.

Kirk, W.W., Niemira, B.A. and Stein, J.M. 2001. Influence of storage temperature on rate of potato tuber tissue infection caused by *Phytophthora infestans* (Mont.) de Bary estimated by digital image analysis. *Potato Research*, **44**: 86–96.

Korf, H.J.G., Schutte, G.C. and Kotzé, J.M. 2001. Effect of packhouse procedures on the viability of *Phyllosticta citricarpa*, anamorph of the citrus black spot pathogen. *African Plant Protection*, **7**: 103–109.

Kushalappa, A.C. and Zulfiqar, M. 2001. Effect of wet incubation time and temperature on infection and of storage time and temperature on soft rot lesion expansion in potatoes inoculated with *Erwinia carotovora* subsp. *carotovora*. *Potato Research*, **44**: 233–242.

Laidou, I.A., Thanassoulopoulos, C.C. and Liakopouloukyriakides, M. 2001. Diffusion of patulin in the flesh of pears inoculated with four postharvest pathogens. *Journal of Phytopathology*, **149**: 457–461.

Lamper, C., Teren, J., Bartok, T., Komoroczy, R., Mesterhazy, A. and Sagi, F. 2000. Predicting DON contamination in *Fusarium*-infected wheat grains via determination of ergosterol content. *Cereal Research Communications*, **28**: 337–344.

Langevin, F., Eudes, F. and Comeau, A. 2004. Effect of trichothecenes produced by *Fusarium graminearum* during *Fusarium* head blight development of six cereal species. *European Journal of Plant Pathology*, **110**: 735–746.

Lattanzio, V., Venere, D. di, Linsalata, V., Bertolini, P., Ippolito, A. and Salerono, M. 2001. Low temperature metabolism of apple phenolics and quiescence of *Phlyctaena vagabunda*. *Journal of Agricultural and Food Chemistry*, **49**: 5817–5821.

Lee, M.B. and Lee, E.H. 2001. Coccidial contamination of raspberries: mock contamination with *Eimeria acervulina* as a model for decontamination treatment studies. *Journal of Food Protection*, **64**: 1854–1857.

Lehtonen, M.J., Rantala, H., Kreuze, J.F., Bang, H., Kuisma, L., Koski, P., Virtanen, E., Vihlman, K. and Valkonen, J.P.T. 2004. Occurrence and survival of potato scab pathogens (*Streptomyces* species) on tuber lesions: quick diagnosis based on a PCR-based assay. *Plant Pathology*, **53**: 280–287.

Lew, H., Alder, A. and Edinger, W. 1991. Moniliformin and the European cornborer (*Ostrinia nubilalis*). *Mycotoxin Research*, **7**: 71.

Lewis, B.G., Davies, W.P. and Garrod, B. 1981. Wound healing in carrot roots in relation to infection by *Mycocentrospora acerina*. *Annals of Applied Biology*, **99**: 35–42.

Lewis, B.G. and Garrod, B. 1983. Carrots. In: *Post-Harvest Pathology of Fruits and Vegetables* (ed.) C. Dennis, pp. 103–124, Academic Press, London.

Li, F.Q., Luo, X.Y. and Yoshizawa, T. 1999. Mycotoxins (trichothecenes, zearalenone and fumonisins) in cereals associated with human red mold intoxications stored since 1989 and 1991 in China. *Natural Toxins*, **7**: 93–97.

Lillehoj, E.B., Milburn, M.S. and Ciegler, A. 1972. Control of *Penicillium martensii* development by atmospheric gases and temperature. *Applied Microbiology*, **24**: 198–201.

Lima, M.F. and Lopes, C.A. 1998. Development of black eye on potato tubers. *Horticultura Brasileira*, **16**: 38–44.

Lindblad, M., Johnsson, P., Jonsson, N., Lindquist, R. and Oslen, M. 2004. Predicting noncompliant levels of ochratoxin A in cereal grain from *Penicillium verrucosum* counts. *Journal of Applied Microbiology*, **97**: 609–616.

Lindholm, P., Kortemaa, H., Kokkola, M., Haahtela, K., Salkinoja-Salonen, M. and Valkonen, J.P.T. 1997. *Streptomyces* spp., isolated from potato scab lesions under Nordic conditions in Finland. *Plant Disease*, **81**: 1317–1322.

Ling, X., Kusakari, S.I., Toyoda, H. and Ouchi, S. 1999. Role of fungi in the shattering of grape berries during storage. *Bulletin of the Institute for Comprehensive Agricultural Sciences, Kinki University*, No.7, pp. 97–106.

Logan, C. 1983. Potatoes. In: *Post-Harvest Pathology of Fruits and Vegetables* (ed.) C. Dennis, pp. 179–217. Academic Press, London, U.K.

Lolas, M., Moggia, C. and Chacón, F. 1998. Harvest maturity, packaging and storage time related to calyx-end rot expression on Packham's Triumph pears under refrigerated storage. *Acta Horticulturae*, **475**: 459–469.

Long, P.G. and Bautista-Baños, S. 1994. Curing and infection of kiwifruit by *Botrytis cinerea*. In: *Proceedings of National Conference, New Zealand Marketing Board*, pp. 31–34, New Zealand.

Machowicz-Stefaniak, Z., Dyduch, J., Zalewska, E. and Nurzyńska-Wierdak, R. 1998. Fungi damaging garlic bulbs (*Allium sativum* L.) during storage. *Annales Universitatis Mariae Curie – Sklodowska, Sectio EEE, Horticultura*, **6**: 27–37.

Madden, J.M. 1992. Microbial pathogens in fresh produce – the regulatory perspective. *Journal of Food Protection*, **55**: 821.

Magan, N., Hope, R., Cairns, V. and Aldred, D. 2003. Postharvest fungal ecology: Impact of fungal growth and mycotoxin accumulation in stored grain. *European Journal of Plant Pathology*, **109**: 723–730.

Magnoli, C., Astoreca, A., Ponsone, L., Combina, M., Palacio, G., Rosa, C.A.R. and Dalcero, A.M. 2004. Survey of mycoflora and ochratoxin A in dried vine fruits from Argentina markets. *Letters in Applied Microbiology*, **39**: 326–331.

Mangasuli, N.S., Reddy, B.S. and Patil, D.R. 2000. Effect of packages and chemicals on mould growth and shelf-life of Thompson Seedless grapes under different storage conditions. *Karnataka Journal of Agricultural Sciences*, **13**: 220–223.

Martins, M.L., Gimeno, A., Martins, H.M. and Bernardo, F. 2002. Co-occurrence of patulin and citrinin in Portuguese apples with rotten spots. *Food Additives and Contaminants*. **19**: 568–574.

McCallum, B.D., Tekawz, A. and Gilbert, J. 2004. Reaction of diverse collection of barley lines to *Fusarium* head blight. *Plant Disease*, **88**: 167–174.

McMahon, M.A.S. and Wilson, I.G. 2001. The occurrence of enteric pathogens and *Aeromonas* species in organic vegetables. *International Journal of Food Microbiology*, **70**: 155–162.

Mesterhazy, A. 2002. Role of deoxynivalenol in aggressiveness of *Fusarium graminearum* and *F. culmorum* and in resistance to *Fusarium* head blight. *European Journal of Plant Pathology*, **108**: 675–684.

Michailides, T.J. and Elmer, P.A.G. 2000. Botrytis gray mold of kiwifruit caused by *Botrytis cinerea* in the United States and New Zealand. *Plant Disease*, **84**: 208–223.

Miedaner, T., Reinbrecht, C. and Schilling, A.G. 2000. Association among aggressiveness, fungal colonization and mycotoxin production of 26 isolates of *Fusarium graminearum* in winter rye head blight. *Journal of Plant Diseases and Plant Protection*, **107**: 124–134.

Mirete, S., Vázquez, C., Mulé, G., Jurado, M. and González-Jaén, M.T. 2004. Differentiation of *Fusarium verticillioides* from banana fruits by IGS and EF-1 sequence analyses. *European Journal of Plant Pathology*, **110**: 515–523.

Mitchell, F.G. 1985. Cooling horticultural commodities. In: *Postharvest Technology of Horticultural Crops* (eds) A.A. Kader et al. pp. 35, Cooperative Extension, University of California, DANR, Berkeley. USA.

Mohapatra, N.P., Pati, S.P. and Ray, R.C. 2001. Spoilage of sweetpotato tubers in tropics. V. Java blackrot by *Botryodiplodia theobromae* Pat: phenol accumulation in culture and in tubers during spoilage. *Journal of Mycopathological Research*, **39**: 21–27.

Moodley, R.S., Govinden, R. and Odhav, B. 2002. The effect of modified atmospheres and packaging on patulin production in apples. *Journal of Food Protection*, **65**: 867–871.

Moore, K.G., Price, M.S., Boston, R.S., Weissinger, A.K. and Payne, G.A. 2004. A chitinase from Tex6 maize kernels inhibits growth of *Aspergillus flavus*. *Phytopathology*, **94**: 82–87.

Muthoni, J.W., Oerke, E.C., Dehne, H.W. and Mutitu, E.W. 2002. Susceptibility of Kenyan wheat varieties to head blight fungal invasion and deoxynivalenol accumulation in heads inoculated with *Fusarium graminearum*. *Journal of Phytopathology*, **150**: 30–36.

Nellist, M.E. 1988. Near-ambient grain drying. *Agricultural Engineer*, **5**: 93–101.

Nesci, A., Rodriguez, M. and Etcheverry, M. 2003. Control of *Aspergillus* growth and aflatoxin production using antioxidants at different conditions of water activity and pH. *Journal of Applied Microbiology*, **95**: 279–287.

Neven, L.G., Drake, S.R. and Ferguson, J.H. 2000. Effects of the rate of heating on apple and pear fruit quality. *Journal of Food Quality*, **23**: 317–325.

Ngoko, Z., Marasas, W.F.O., Rheeder, J.P., Shephard, G.S., Wingfield, M.J. and Cardwell, K.F. 2001. Fungal infection and mycotoxin contamination of maize in the humid forest and western highlands of Cameroon. *Phytoparasitica*, **29**: 352–360.

Nielsen, K., Payne, G.A. and Boston, R.S. 2001. Maize ribosome-inactivating protein inhibits normal development of *Aspergillus nidulans* and *A. flavus*. *Molecular Plant–Microbe Interactions*, **14**: 164–172.

Nyanjage, M.O., Wainwright, H. and Bishop, C.F.H. 1998. The effects of hot-water treatments in combination with cooling and or storage on the physiology and disease of mango fruits (*Mangifera indica* Linn). *Journal of Horticultural Science and Biotechnology*, **73**: 589–597.

Palou, L., Usall, J., Pons, J., Cerda, M.C. and Viñas, I. 2001. Fungal populations in Tarragona citrus packinghouses. *Investigación Agraria, Producción y Protection Vegetables*, **16**: 447–462.

Pandey, A., Joshi, V.K., Nigam, P. and Soccol, C.R. 1999. Enterobacteriaceae, coliform and *E. coli*. In: *Encyclopedia of Food Microbiology* (eds) R. Robinson, C. Batt and P. Patel, p. 604, Academic Press, London.

Pascale, M., Visconti, A., Avantaggiato, G., Prończuk, M. and Chelkowski, J. 1999. Mycotoxin contamination of maize hybrids after infection with *Fusarium proliferatum*. *Journal of the Science of Food and Agriculture*, **79**: 2094–2098.

Penteado, A.L., Eblen, S. and Miller, A.J. 2004. Evidence of *Salmonella* internalization into fresh mangoes during simulated postharvest insect disinfestations procedures. *Journal of Food Protection*, **67**: 181–184.

Piga, A., D'Aquino, S., Agabbio, M., Emonti, G. and Farris, G.A. 2000. Influence of storage temperature on shelf-life of minimally processed cactus pear fruits. *Lebensmittel-Wissenschaft und Technologie*, **33**: 15–20.

Plattner, R.D., Neffed, W.P., Bacon, C.W. et al., 1990. A method of detection of fumonisins in corn sample associated with field cases of equine leuko- encephalomalacia. *Mycologia*, **82**: 898.

Proctor, R.H., Desjardins, A.E., Plattner, R.D. and Hohn, T.M. 1999. A polyketide synthase gene required for biosynthesis of fumonisin mycotoxins in *Gilberella fujikuroi* mating population A. *Fungal Genetics and Biology*, **27**: 100–112.

Procter, R.H., Plattner, R.D., Brown, D.W., Seo, J.A. and Lee, Y.W. 2004. Discontinuous distribution of fumonisin biosynthetic genes in *Gibberella fujikuroi* species complex. *Mycological Research*, **108**: 815–822.

Ramos, A.M., Nieto-Ángel, D., Domínguez-Álvarez, J.L. and Delgadillo-Sanchez, F. 2001. Quality and tolerance in papaya fruits (*Carcia papaya,* L.) to the inoculation of fungus *Colleotrichum gloeosporioides* Penz. In postharvest. *Revista Chapingo, Serie Horticultura*, **7**: 119–130.

Rasul, M.G., Islam, M.S. and Sheikh, M.H.R. 1999. Storability of different potato varieties under natural conditions. *Bangladesh Journal of Scientific and Industrial Research*, **34**: 86–90.

Reynoso, M., Torres, A.M. and Chulze, S.N. 2004. Fusaproliferin, beauvericin and fumonisin production by different mating populations among the *Gibberella fujikuroi* complex isolated from maize. *Mycological Research*, **108**: 154–160.

Richard, J.L. 2003. Ochratoxins – risks associated with their occurrence in small grains. *Phytopathology*, **93**: 599.

Rico, C., Centelles, E. and Pujolá, M. 1999. Application of HPLC for the determination of volatile organic acids: study on their evolution in cold-stored fruits. *Alimentaria*, **302**: 101–103.

Saevels, S., Lammertyn, J., Berna, A.Z., Veraverbeke, E.A., Natale, C.D. and Nicolai, B.M. 2003. Electronic nose as a nondestructive tool to evaluate the optimal harvest date of apples. *Postharvest Biology and Technology*, **30**: 3–14.

Saftner, R.A., Abbott, J.A. and Conway, W.S. 2003. Effects of 1-methylcyclopropene and heat treatments of ripening and postharvest decay in Golden Delicious apples. *Journal of American Society of Horticultural Science*, **128**: 120–127.

Savard, M.E., Sinha, R.C., Seaman, W.L. and Fedak, G. 2000. Sequential distribution of the mycotoxin deoxynivalenol in wheat spikes after inoculation with *Fusarium graminearum. Canadian Journal of Plant Pathology*, **22**: 280–285.

Schaafsma, A.W., Hooker, D.C., Bante, T.S. and Illincic-Tamburic, L. 2002. Effect of Bt-Corn hybrids on deoxynivalenol content in grain at harvest. *Plant Disease*, **86**: 1123–1126.

Schnerr, H., Vogel, R. and Niessen, L. 2002. A biosensor, based immunoassay for rapid screening deoxynivalenol contamination in wheat. *Food and Agricultural Immunology*, **14**: 313–321.

Schwarz, P.B., Schwarz, J.G., Zhou, A., Prom, L.K. and Steffenson, B.J. 2001. Effect of *Fusarium graminearum* and *F. poae* infection on barley and malt quality. *Monatsschrift für Brauwissenchaft*, **54**: 55–63.

Shejbal, J. 1980. *Controlled Atmosphere Storage of Grains*. Elsevier, Amsterdam.

Shephard, G.S. and Leggott, N.L. 2000. Chromatographic determination of the mycotoxin patulin in fruit and fruit juices. *Journal of Chromatography*, **802**: 17–22.

Shetty, K.K., Frazier, M.J., Kleinkopf, G.E. and Nolte, P. 2003. Managing silver scurf In: *Potato Disease Problems*, University of Idaho, USA

Shivedra Kumar and Roy, A.K. 2000. Occurrence of ochratoxin in maize grains. A report. *National Academy of Science Letters*, **23**: 101–103.

Shukla, A. and Sharma, R.L. 2000. Incidence of soft rot of bell pepper in Himachal Pradesh. *Journal of Mycology and Plant Pathology*, **30**: 107–109.

Silveira, N.S.S., Michereff, S.J., Mariano, R.L.R., Tavares, L.A. and Maia, L.C. 2001. Influence of temperature, wetness period and fungal inoculum concentration on incidence of postharvest tomato fruit rots. *Fitopatologia Brasileira*, **26**: 33–38.

Stevenson, W.R., Loria, R., Franc, G.D. and Weingartner, D.P. (eds) 2001. *Compendium of Potato Diseases*, 2nd edition. American Phytopathological Society, St. Paul, MN.

Suárez-Quiroz, M., González-Rios, O., Barel, M., Guyot, B., Schorr-Galindo, S. and Guiraud, J. 2004. Study of ochratoxin-A. producing strains in coffee processing strains in coffee processing. *International Journal of Food Science and Technology*, **39**: 501–507.

Swaminathan, M.S. 1981. *Chairman's Report of the Group on Perishable Agricultural Commodities*. Ministry of Agriculture, Department of Cooperation, Government of India, New Delhi.

Swelim, M.A., Abdallah, S.A., Basyony, A., El-Sayed, T.A., Amer, M.M. and Ahmed, S.S. 2001. Aflatoxin production by fungi and chemical analysis of apparently healthy-looking field-grown soybean seeds. *African Journal of Mycology and Biotechnology*, **9**: 43–61.

Sydenham, E.W., Thiel, P.G., Marasas, W.F.O., Shepherd, G.S., Van Schalkwyk, D.J. and Koch, K.R. 1990. Natural occurrence of some *Fusarium* mycotoxins in corn from low and high esophageal cancer prevalence areas of the Transkei, South Africa. *Journal of Agriculture and Food Chemistry*, **38**: 1900.

Takeuchi, T., Sawada, H., Tanaka, F. and Matusuda, I. 1996. Phylogenetic analysis of *Streptomyces* spp. causing potato scab based on 16S rRNA sequences. *International Journal of Systematic Bacteriology*, **46**: 476–479.

Tanaka, M.A.S. 2001. Survival of *Fusarium moniliforme* in corn seeds stored under two different conditions. *Fitopatologia Brasileira*, **26**: 60–64.

Taylor, R.K. and Hale, C.N. 2003. Cold storage affects survival and growth of *Erwinia amylovora*. *Letters in Applied Microbiology*, **37**: 340–343.

Tóth, B., Mesterházy, Á., Nicholson, P., Teren, J. and Varga, J. 2004. Mycotoxin production and molecular variability of European and American isolates of *Fusarium culmorum* . *European Journal of Plant Pathology*, **110**: 587–599.

Tsai, G.J. and Yu, S.C. 1999. Detecting *Aspergillus parasiticus* in cereals by an enzyme-linked immunosorbent assay. *International Journal of Food Microbiology*, **50**: 181–189.

Tseng, T.C., Lee, K.L., Deng, C.Y., Liu, C.Y. and Huang, J.W. 1995. Production of fumonisins by *Fusarium* species of Taiwan. *Mycopathologia*, **130**: 117.

Ukuku, D.O. and Fett, W.F. 2002. Relationship of cell surface charge and hydrophobicity to strength of attachment of bacteria to cantaloupe rind. *Journal of Food Protection*, **65**: 1093–1099.

Varga, J., Juhász, A., Kevei, F. and Kozakiewicz, Z. 2004. Diversity of agriculturally important *Aspergillus* species. *European Journal of Plant Pathology*, **110**: 627–640.

Velluti, A., Marín, S., Sanchis, V. and Ramos, A.J. 2001. Occurrence of fumonisin B_1 in Spanish corn-based foods for animal and human consumption. *Food Science and Technology International*, **7**: 433–437.

Victoria Novas, M. and Cabral, D. 2002. Association of mycotoxin and sclerotia production with compatibility groups in *Aspergillus flavus* from peanut in Argentina. *Plant Disease*, **86**: 215–219.

Vigneault, C., Bartz, J.A. and Sargent, S.A. 2000. Postharvest decay risk associated with hydrocooling tomatoes. *Plant Disease*, **84**: 1314–1318.

Vrabcheva, T., Usleber, E., Dietrich, R., and Martlabauer, E. 2000. Cooccurrence of ochratoxin A and citrinin in cereals from Bulgarian villages with a history of Balken endemic nephropathy. *Journal of Agricultural and Food Science*, **48**: 2483–2488.

Waalwijk, C., Heide R. van der, de Vries, I., Lee, T. van der, Schoen, C., Corainville, G.C., Häuser-Hahn, I., Kastelein, P., Köhl, J., Lonnet, P., Demarquet, T. and Kema, G.H.J. 2004. Quantitative detection of *Fusarium* species in wheat using TaqMan. *European Journal of Plant Pathology*, **110**: 481–494.

Weagant, S.D. and Bound, A.J. 2001. Evaluation of techniques for enrichment and isolation of *Escherichia coli* 0157: H7 from artificially contaminated sprouts. *International Journal of Food Microbiology*, **71**: 87–92.

Wicklow, D.T., Weaver, D.K. and Throne, J.E. 1998. Fungal colonists of maize grain conditioned at constant temperatures and humidities. *Journal of Stored Products Research*, **34**: 355–361.

Wilcke, W.F., Ileleji, K.E., Gupta, P., Morey, R.V. and Meronuck, R.A. 2001. Comparison of sample storage and conditioning methods for corn storability tests. *Transactions of the ASAE*, **44**: 369–376.

Wilson, D.M., Huang, L.H. and Jay, E. 1975. Survival of *Aspergillus flavus* and *Fusarium moniliforme* in high moisture corn stored under modified atmospheres. *Applied Microbiology*, **30**: 592–595.

Witbooi, W.R., Taylor, M.A. and Fourie, J.F. 2000. *In vitro* studies on the effect of harvest maturity, relative humidity, wetness period and cold-storage temperature on the infection potential of *Aspergillus niger* and *Rhizopus stolonifer* on Thompson Seedless grapes. *Deciduous Fruit Grower*, **50**: S1–S14.

Wogan, G.N. 1991. Aflatoxin as risk factors for primary hepato-cellular carcinoma in human. In: *Mycotoxins, Cancer and Health* (eds) G.A. Bray and D.H. Ryan, p. 18, Louisiana University Press, Baton Rouge, USA.

Wu, F.M., Beuchat, L.R., Wells, J.G., Slutsker, L., Doyle, M.P. and Swaminathan, B. 2001. Factors influencing the detection and enumeration of *Escherichia coli* 0157: H7 on alfalfa seeds. *International Journal of Food Microbiology*, **71**: 93–99.

Xiao, C.L. and Rogers, J.D. 2004. A postharvest fruit rot in d'Anjou pears caused by *Sphaeropsis pyriputrescens* sp. nov. *Plant Disease*, **88**: 114–118.

Xy, H., Annis, S., Linz, J. and Trail, F. 2000. Infection and colonization of peanut pods by *Aspergillus parasiticus* and the expression of the aflatoxin biosynthetic gene *nor-1*, in infection hyphae. *Physiological and Molecular Plant Pathology*, **56**: 185–196.

Xu, L., Kusakari, S.I., Toyoda, H. and Ouchi, S. 1999. Role of fungi in the shattering of grape berries during storage. *Bulletin of the Institute for Comprehensive Agricultural Sciences, Kinki University*, No.7, pp. 97–106.

Xu, X.M., Guerin, L. and Robinson, J.D. 2001. Effects of temperature and relative humidity on conidial germination and viability. Colonization and sporulation of *Monilinia fructigena. Plant Pathology*, **50**: 561–568.

Youssef, M.S., Abo-Dahab, N.F. and Abou-Seidah, A.A. 2000. Mycobiota and mycotoxin contamination of dried raisins in Egypt. *African Journal of Mycology and Biotechnology*, **8**: 69–86.

Yu, F.Y. and Chu, F.S. 1998. Analysis of fumonisins and *Alternaria alternata* toxin by liquid chromatography-enzyme-linked immunosorbent assay. *Journal of AOAC International*, **81**: 749–756.

Yu, J., Mohawed, S.M., Bhatnagar, D. and Cleveland, T.E. 2003. Substrate-induced lipase gene expression and aflatoxin production in *Aspergillus parasiticus* and *Aspergillus flavus. Journal of Applied Microbiology*, **95**: 1334–1342.

PART III

PRINCIPLES AND PRACTICES OF POSTHARVEST DISEASE MANAGEMENT

7

PREVENTIVE AND PHYSICAL METHODS

Some of the strategies adopted for management of field crop diseases are also applicable for the management of postharvest diseases, as many pathogens are carried from the field into the storage after harvest. It is, therefore, essential to follow cultivation practices that may reduce the incidence of postharvest diseases of fruits and vegetables (Chapter 6). As a basic step in disease management systems, exclusion of microbial pathogens has to be strictly enforced to reduce the inoculum levels that may reach the harvested produce later.

7.1 SEEDS AND PROPAGATIVE MATERIALS

7.1.1 Preventive Methods Applicable During Preharvest Stage

The infected seeds and propagative planting materials, such as tubers and bulbs, form the primary sources of infection. The fungal pathogens from infected seeds may spread to healthy seeds and they may also be introduced into new locations where they may not be present or currently not important. The seeds may be indexed to assess the pathogen populations by employing various diagnostic methods (Chapter 2). It is possible to eliminate seeds infected by fungal, bacterial, and viral pathogens. The tolerance limits have been prescribed by the International Seed Testing Association (ISTA). Now, most countries enforce zero tolerance to prevent the introduction of new

Postharvest Pathogens and Disease Management, by P. Narayanasamy

TABLE 7.1 Standards for Seed-Borne Pathogens Adopted in United States and United Kingdom

Crop	Pathogen	Seed category	Standard*
Brassicas	*Phoma lingam*	Basic	0 in 1000 seeds
Red beet	*Phoma betae*	Basic	0 in 200 seeds
Celery	*Septoria apicola*	Basic and certified	0 in 400 seeds
	Phoma apicola	Basic and certified	0 in 400 seeds
Peas	*Ascochyta spp.*	Basic and certified	0 in 200 seeds 20 seeds in 200 seeds
	Colletotrichum lindemuthianum	Basic	0 in 600 seeds
Phaseolus bean	*Pseudomonas syringae* pv. *phaseolicola*	Basic	0 in 5000 seeds
Vicia bean	*Ascochyta fabae*	Basic	0 in 600 seeds
Lettuce	*Lettuce mosaic virus (*LMV*)*	Basic and certified	0 in 5000 seeds 0 in 30,000 seeds**
Cabbage	*Xanthomonas campestris* pv. *campestris*	Certified	One in 1000 seeds**

* Maximum permitted; ** standard followed in United States.
Source: Rennie, 1993; Maddox, 1998.

pathogens into a country or area (Table 7.1). The movement of germplasm and other genetic material has been rapid in recent years, posing a considerable threat to plant health and necessitating the imposition of quarantine measures. The possibility of the spread of diseases through seeds, such as celery leaf spot (*Septoria apicola*), bean halo blight (*Pseudomonas syringae* pv. *phaseolicola*), carrot leaf blight (*Alternaria dauci*), onion neckrot (*Botrytis allii*), *Lettuce mosaic virus* (LMV), *Soybean mosaic virus* (SMV), and *Bean common mosaic virus* (BCMV), has been well recognized. The need for the production of disease-free seeds to prevent the introduction of microbial pathogens into other countries has been emphasized (Agarwal and Sinclair, 1996; Maude, 1996; Narayanasamy, 2002).

By prescribing stringent tolerance levels, it has been demonstrated that LMV disease could be controlled in California with a set tolerance of zero infected seed in 30,000 seeds (Maddox, 1998). Likewise, for bacterial diseases of vegetables, samples of 30,000 seeds were tested as recommended by NCR-100 (a consortium of plant pathologists) as in the case of cabbage black rot disease caused by *Xanthomonas campestris* pv. *campestris* (*Xcc*). By testing 30,000 seeds, the confidence level of detecting the targeted pathogen at a 0.1% contamination level (one seed in 10,000 seeds) was achieved. Crucifer seed lots infected with *Xcc* at one seed in 10,000 (0.01%) did not result in black rot disease incidence, whereas with three seeds in 10,000 seeds (0.03%), occurrence of the disease was observed in South Carolina (Schaad et al., 1980). These findings show that adequate care has to be taken in the selection and use of disease-free seeds to have healthy and marketable agricultural commodities.

Crop inspection is carried out to determine the extent and intensity of infection by seed-borne pathogens. Techniques for crop inspection for quantitative assessment of infection are employed to decide whether the seed crop can be accepted or not. Infected plants are rogued to eliminate the pathogen sources. The presence of any bacterial blight infection (due to *Xanthomonas campestris* pv. *phaseoli*) was considered to be sufficient to warrant the immediate destruction of the crop by ploughing under to eliminate seed infection (Copeland et al., 1975). The International Seed Health Initiative (ISHI), founded in 1993, is an international consortium of seed industry and seed health testing plant pathologists. The ISHI aims to develop efficient, reliable seed health testing protocols in a timely manner, to assure that seed lots are sufficiently healthy for world-wide movement and to have a means of quickly testing new technologies for incorporation into seed health testing procedures (Maddox, 1998).

Use of disease-free seeds not only helps reduce the infection in the crop grown in the next season, but this strategy will sustain the attempt to have better produce with desirable market quality. The seed-borne diseases of bean, cherry, citrus, lettuce, pea, *Prunus*, strawberry, and sugarbeet could be effectively tackled by using disease-free seeds (Mandahar, 1981). The practical utility of this approach is dramatically revealed by locating much of the U.S. seed industry in the semiarid western States such as Oregon, California and Idaho where transmission of seed-borne pathogens is at a low level, and there is a consequent reduction in losses (Wallen, 1964).

Harvested seed lots may contain inert materials, plant debris carrying microbial pathogens, and dormant/fungal propagules that aid in overwintering, as well as discolored and abnormal seeds due to infection. Such seed lots have to be cleaned mechanically before storage, either by hand picking or by separators. Immersion of seeds in a salt solution (10–20%) is useful to remove the ill-filled seeds, ergot sclerotia, and other fungal propagules that may float on the solution. Lettuce seeds infected by *Lettuce mosaic virus* (LMV) are lighter in weight than healthy seeds. The infected seeds may be separated by using a vertical air stream (Ryder and Johnson, 1974). Removal of small seeds or seeds with growth cracks resulted in a reduction in incidence of *Barley stripe mosaic* virus (BSMV) and *Pea seed-borne mosaic virus* (PSbMV) (Phatak and Summanwar, 1967; Stevenson and Hagedorn, 1970). Washing and sieving of whole wheat samples through a 60-μm nylon sieve, followed by pelleting by centrifugation decreased the number of teliospores of *Tilletia controversa*. Large number of spores were found in materials that were sieved or aspirated from the grain. The results indicated that mechanical cleaning removed a high percentage, but not all, spores (Bechtel et al., 1999).

7.1.2 Preventive Methods Applicable During Storage

The effect of type of storage bags on deterioration of melon seeds by fungal pathogens was studied. Melon seeds were stored in jute and polyethylene bags under ambient conditions for 12 months. The moisture content, incidence of visible moldiness, and germinability of seeds were determined at monthly intervals. The incidence of visible moldiness increased to 5.5% and 10.7% in the seeds stored in polyethylene and jute bags, respectively, from an initial 2.1%. The decrease in germination from 96.3% occurred more slowly in polyethylene bag-stored seeds (45.3%) compared with jute bag-stored seeds (28.7%). This study showed that storage of melon seeds in polyethylene bags is preferable (Bankole et al., 1999). In the case of soybean, polylined bags harbored a low incidence of mycoflora (37.6%) as compared to cloth bag packaging (62.4%). The difference in the seed mycoflora population during storage was attributed to the variations in temperature and relative humidity existing in the storage facilities (Anuja and Aneja, 2001). The critical moisture contents (% dry basis) is determined by assessing the moisture contents to be maintained at 25°C that will not allow fungal growth. The critical moisture contents were 13% for rice grains and 12.9% for glutinous rice grains. It is essential to maintain a moisture content lower than 13% for ordinary rice grains and 12.9% for glutinous rice grains for long-term storage (Abdullah et al., 2000).

The effect of storage of maize under uncontrolled (ambient) and cold conditions on the physiological quality of seeds and survival of fungal pathogens was assessed. In general, the survival of fungi decreased with storage period under uncontrolled atmosphere compared to cold environment (14°C and 40% RH). However, the effect was different for some fungi. Incidence of *Aspergillus* and *Penicillium* increased with storage under ambient conditions, while the opposite effect was observed in the case of *F. moniliforme*. Although the cold condition exerted a favorable effect on seed viability, it reduced the sanitary quality of seeds because of longer period of survival of fungal pathogens (Tanaka et al., 2001). Depending on the nature of the host plant and fungal flora associated with seeds, the storage conditions have to be suitably altered to contain the fungal populations.

7.1.3 Physical Methods

Application of heat in different forms in order to eliminate seed-borne pathogens has been demonstrated to be effective. Jensen (1888) showed that hot water treatment of seeds could be adopted for the control of the smut diseases of barley and oats. The seeds were immersed in water at 20 to 30°C to stimulate the dormant mycelium of *Ustilago segetum* var. *tritici* and *U. nuda*, causing loose smut in wheat and barley, respectively. Then the seeds were immersed in hot water, at 50 to 52°C, to kill the developing mycelium (Appel and Riehm, 1911). This procedure was modified to utilize the solar heat available in tropical countries to kill the pathogen present in wheat seeds (Luthra,

1953). The seeds are soaked in cold water for about 4h and spread in a thin layer under sun for 4h (temperatures about 50°C). The comparative effectiveness of electrons and hot water treatment for the control of bunt disease of wheat caused by *Tilletia caries* was reported by Winter et al. (1998). No adverse effect on seed germination due to electron treatment could be seen, while hot water treatment (52°C for 10min) reduced seed germination. Maize seeds were first pretreated in water at 18 to 22°C for 4h and at 60°C for 5min to eliminate *Fusarium moniliforme* (Daniels, 1983). Hot water treatment at 60°C for 10min was the most effective in suppressing the colonization of rice seeds by *Drechslera oryzae* (Krishnamurthy et al., 2001).

Soaking onion seeds in hot water (50°C for 20min) was found to be more effective than fungicidal application in reducing the seed-borne infection by *Alternaria porri* and *Stemphylium vesicarium*, though seed germination and seedling emergence were reduced (Aveling et al., 1993). Hot air treatment was applied for the control of *Colletotrichum capsici* infecting pepper (chili) seeds. Exposure of seeds to hot air at 85°C for 5 days reduced seed-borne infection with some reduction in seed germination (Kobayashi, 1990). *Phytophthora infestans* carried in discolored seeds from infected tomato fruits was eliminated by drying seeds in an oven at 29.5 to 37.5°C for 6h (Vartanian and Endo, 1985). Likewise, tomato seeds were freed of *Tomato mosaic virus* by dry heat treatment at 78°C for 2 days. No adverse effect on seed germination was seen due to heat treatment, which was more effective than soaking the seeds in trisodium phosphate (Green et al., 1987). A new high-precision method, involving the use of warm air which is in moisture equilibrium with the grain to be treated, has been developed. Sanitation of wheat grain using warm air resulted in control of seed-borne fungal pathogens such as *Tilletia caries* and *Microdochium nivale* comparable to the chemical seed treatment. No adverse effect on plant development, following warm air treatment, was observed. The method is suitable for treatments of seeds with thick layers at short durations, and has potential for large scale treatments at a low cost (Forsberg, 2001). Heat treatment of wheat and barley seeds with high level of infection by *Fusarium graminearum* (84 and 23% respectively) was evaluated. The pathogen was eliminated from Canada Western red spring wheat seeds treated at 60°C for 15 days, or at 70°C for 5 days, or at 80°C for 2 days. Likewise, barley seeds were freed of the pathogen after treatment at 60°C for 21 days, or at 70°C for 9 days, or at 80°C for 5 days. Germination rates in most samples were not affected. Hence, it is suggested that thermotherapy can be applied to control national and international movement of the pathogen (Clear et al., 2003).

The sterilization effect of infrared irradiation on wheat seeds contaminated with microorganisms has been determined. The bacterial counts were effectively reduced from 5.7×10^4 CFU/g to 0.73×10^3 CFU by irradiation at 2.0kW range for 60sec. Intermittent irradiation was more effective in maintaining the internal quality of wheat. Reduction in bacterial counts was proportional to the irradiation dose (Hamanaka et al., 2000). Exposure of cabbage seeds to ultraviolet (UV-C) light at 3.6kJ/m was effective in reducing black rot disease caused by *Xanthomonas campestris* pv. *campestris* (*Xcc*) and the density of the bacterial pathogen population. Furthermore, the cabbage plants growing from treated seeds had the most desirable color, highest weight, largest head diameter, and delayed maturity (Brown et al., 2001).

7.2 FRUITS AND VEGETABLES

7.2.1 Mechanical Methods

It is well established that damages and injuries occurring at different stages, that is harvesting, handling, transport, packaging, and storing, are the important avenues for the entry of fungal and bacterial pathogens. Even a single bad bruise on an apple, due to improper handling, may result in an increase in the rate of water loss from the damaged apple by as much as 400%, and consequent desiccation and reduction in market quality of the commodity. By careful handling and employing improved packaging techniques, losses in marketable agricultural commodities can be effectively reduced to a great extent.

The containers should provide good ventilation and have adequate cushioning materials. Proper design of the container can help avoid losses, as, for example, in the case of tomatoes. In Ghana, the traditional inverted cone-shaped baskets were replaced by baskets with broad bases and narrow necks, and this resulted in a reduction in disease incidence. Polyethylene packaging may reduce desiccation and injury appreciably (Waller, 2002). Careful inspection and the application of appropriate diagnostic techniques to rapidly and precisely identify and quantify infection by postharvest pathogens will be very useful to cull and dispose off the infected fruits and vegetables.

The application of waxes to fruit surfaces is done to extend their postharvest shelf life. Waxing reduces the respiration and transpiration rates. Various advantages of waxing, such as improvement in fruit appearance, reduction in spoilage due to chilling injury, creation of a diffusion barrier resulting in reduced availability of oxygen to fruit tissues and consequent reduced respiration rate, reduction in loss of moisture, and protection against infection by microbial pathogens, have been reported. However, the fact that waxing does not improve the quality of commodities that have not been handled properly has to be borne in mind. Fungicides, growth regulators, and preservatives have been incorporated to prevent sprouting and microbial spoilage. The major disadvantage of using wax for coating the fruits is the development of off flavor, if it is not applied properly (Sharma and Singh, 2000). The sanitizing effects of waxes applied on orange fruit surfaces was assessed. Effective bactericidal activity on *Escherichia coli* of combined alkali and heat treatments was observed on orange fruit surfaces. At pH 11, dipping at, or above, 50°C for 2 min or longer provided similar bactericidal effects. Wax treatments were found to be effective at the nonstem scar area (Pao et al., 1999). Such practices can be expected to have a significant effect in restricting the spread of the diseases to healthy commodities.

7.2.2 Physical Methods

Primarily, two of physical agents, irradiation and heat treatments, have been observed to provide appreciable protection against the microbial pathogens causing postharvest diseases of fruits and vegetables.

7.2.2.1 Radiation Under suitable conditions, radiation may reduce populations of, or eliminate, microbial pathogens, and retard physiological processes such as ripening or senescence and sprouting. The radiation systems, though expensive, may be integrated with other storage and handling methods. Gamma rays (Cobalt-60 or Caesium-137), fast electrons (linear accelerators), and ultraviolet light (UV) have been used as radiation sources. UV light has been employed to treat several fruits and vegetables as a disease management strategy.

Three ranges of wave lengths of UV light, present in sunlight, have been designated UV-A (310–390 nm), UV-B (280–320 nm), and UV-C (190–280 nm). All ranges of UV light can damage plant DNA and alter physiological processes, depending on the duration of exposure (Luckey, 1980; Stapleton, 1992). Low doses of UV-C light and gamma radiation treatments reduced storage rots of onions (Lu et al.,1987) and sweetpotatoes (Stevens et al., 1990). Later, a wide array of fruits and vegetables have been reported to respond positively to treatment with UV light and the treated commodities exhibited resistance to the respective pathogens (Table 7.2; Fig 7.1).

TABLE 7.2 Management of Postharvest Diseases by UV-C Light Treatment

Crop	Pathogen	Reference
Apple	*Penicillium expansum*	Capdeville et al., 2002
Peaches	*Botrytis cinerea* *Monilinia fructicola*	Crisosto et al., 1998
	Monilinia fructicola	Lu et al., 1993
Citrus	*Penicillium digitatum*	Wilson et al.,1994
Lemon	*Penicillium digitatum*	Ben-Yehoshua et al., 1992; Kim and Ben-Yehoshua 2001
Satsuma mandarins	*Penicillium digitatum*	Yildiz et al., 2001
Grapefruit	*Penicillium digitatum*	Droby et al., 1993; Porat et al., 1999
Grapes	*Botrytis cinerea*	Cia et al., 2000
Strawberry	*Botrytis cinerea* *Monilinia fructigena*	Nigro et al., 2000; Marquenie et al., 2000, 2002a, b.
Tomatoes	*Rhizopus stolonifer*	Liu et al., 1993
	Alternaria alternata	Rong and Feng, 2001
Cabbage	*Xanthomonas campestris* pv. *campestris*	Brown et al., 2001
Carrot	*Botrytis cinerea*	Mercier et al., 1993; Mercier et al., 2000
Beans	*Colletotrichum lindemuthianum*	Andebrhan and Wood, 1980

A

B

Figure 7.1 (a) Effect of UV-C treatment (7.5 kj/m^2) on the incidence of natural rot infection in Golden Delicious apples stored at 24°C for 30 days. Left-untreated (control); right-treated. (b) Effect of UV-C treatment (0.88 kj/m^2) on the incidence of natural infection in bell pepper cv. Bell Boy stored at 13°C for 18 days. Left-untreated (control); right-treated. (Courtesy of Wilson et al., 1994; The American Phytopathological Society, St. Paul, MN, USA.)

The optimum dose of UV-C light for different fruits and vegetables varies depending on the nature of the commodities. The time required for development of maximum protection after UV-C treatment also varies depending on the commodities. Maximum protection for peaches against *Monilinia fructicola* and for tomatoes against *Rhizopus stolonifer* developed at 48 to 72 h after UV-C treatment (Lu et al., 1991; Liu et al., 1993). On the other hand, sweet-potatoes required 1 to 7 days after UV-C treatment for the development of resistance against Java black rot disease caused by *Diplodia gossypina* (Stevens et al., 1990). Prestorage treatment of carrots with UV-C light reduced infection by storage pathogens considerably (Mercier et al., 1993). Following treatment with UV-C light at 0.5 kJ/m^2 decay due to *Penicillium digitatum* in grapefruits Marsh Seedless and Star Ruby and orange cultivars Washington Navel, Biondo Commune, Tarocco, and Valencia Late was significantly reduced. UV-irradiation significantly inhibited the growth of *P. digitatum* causing green mold disease at the fruit wound sites and enhanced the resistance of treated fruits (Porat et al., 1999). The cultivars of orange responded to UV-C light differently based on treatment dose and harvest date, indicating their differential response to UV-C treatment. UV-C treatment with 3.0 kJ/m^2 of Washington Navel and Biondo Commune oranges improved the control of decay compared with a lower dose (0.5 kJ/m^2) (D'hallewin et al., 1999). However, for Star Ruby grapefruit, UV-C irradiation at 0.5 kJ/m^2 was effective in reducing decay and increasing the dose above 0.5 kJ/m^2 did not further improve decay control (D'hallewin et al., 2000).

Postharvest decay of strawberries caused by *Botrytis cinerea* was significantly reduced by UV-C treatment (0.5 and 1.00 kJ/m^2) followed by storage at $20 \pm 1°$ or 3°C. The results indicated that reduction in disease incidence may be due to a direct germicidal effect of UV-C light by disinfecting the external contamination in addition to the induction of resistance by activating host defense systems (Nigro et al., 2000). The growth of *B. cinerea* was severely reduced after exposure to 0.05 J/cm^2. *Monilinia fructigena*, another postharvest pathogen of strawberry fruit, was entirely inhibited by treatment with UV-C at 0.50 J/cm^2. When UV-C treatment was combined with heat treatment (45°C for 15 min), the intensity of both treatments could be reduced, indicating the synergistic action of these treatments (Marquenie et al., 2000, 2002a). In the case of *Botrytis cinerea*, exposing strawberry fruits to UV-C light at >0.01 J/cm^2 resulted in significant reduction in pathogen growth. Temperature treatment exerted no effect on the development of *B. cinerea,* but the highest temperature (48°C) tested, damaged the berry surface and reduced fruit firmness (Marquenie et al., 2002b).

Treatment of carrots with UV-C radiation inhibited the development of *B. cinerea* (Mercier et al., 2000). Application UV-C (254 nm) hormesis on tomato fruits stimulated beneficial responses leading to reduction in disease development initiated by *Rhizopus stolonifer* causing soft rot. The delayed ripening of tomatoes treated with UV-C was considered to be partly due to the high level of putrecine and supermine polyamines as well as to accumulation of tomatine (Stevens et al., 1998). Further, mature green tomato fruits irradiated with UV-C light (24–36 kJ/m^2) exhibited delayed ethylene production and reduced respiration rate. Infection by *Alternaria alternata* was reduced in addition to retarded ripening and color development in tomato fruits (Rong and Feng, 2001). Successful treatment of cabbage seed with UV-C light is described in Section 7.1.3.

The efficacy of gamma radiation on the development of fungal flora in dry onion bulbs has been evaluated. The average storage moisture maintained at room temperature and 4°C were 75% and 85%, respectively. There was a positive correlation between the reduction in initial fungal flora and dose of irradiation. The combination of chilling and ionization was the most effective in reducing fungal flora at a 0.31 kGy dose (Benkeblia and Selselat-Attou, 1997). Bunches of grapes cv. Italia were irradiated at different doses of gamma rays and stored at 0°C and 95% RH for 30 days and then transferred to 25°C and 85% RH to evaluate the effect of treatment on the development of gray mold disease caused by *Botrytis cinerea*. Rate and severity of infection decreased at a dose of 1.5 or 2.0 kGy on inoculated fruit. The spread of the disease to adjacent and distant fruits also was significantly reduced. Irradiation treatment did not adversely affect the color and composition of fruits. Favorable effects following irradiation were a low rates of berry splitting and natural shattering but there were greater losses of fruit weight and firmness (Cia et al., 2000).

Short wave infrared radiation (IR) has been recognized as a highly efficient form of heating. The efficacy of IR was compared with the commercially employed hot water treatment for the management of postharvest diseases of mango. Exposure of mango fruits to IR for 3 min was found to be as effective as the hot water dip at 50°C for 5 min followed by a 20-s ambient temperature and prochloraz dip in controlling anthracnose disease caused by *Colletotrichum gloeosporioides* and soft brown rot caused by *Nattrassia mangiferae* on seven mango cultivars. The fruit quality was retained as in the case of hot water-treated mangoes. The rapidity and lower treatment cost are additional benefits of IR treatment (Saaiman, 2003).

The possibility of using microwaves for the control of postharvest diseases of peach cv. J.H. Hale was explored. Peach fruits inoculated with *Botrytis cinerea* were treated with microwaves for 2 min. Percentage of infection and lesion diameter were significantly ($P \leq 0.05$) reduced compared with control fruits. The storage experiments demonstrated the effectiveness of use of microwave power as a prestorage treatment in controlling natural infection by fungal pathogen. Furthermore, the fact that the microwave treatment neither caused surface damage to fruits nor impaired fruit quality, is an additional advantage for the wider application of this approach to the management of postharvest diseases of various crops (Karabulut and Baykal, 2002).

The presence of human pathogens, such as *Escherichia coli* 0157:H7, has been detected in fruits and vegetables. Two strains of *E. coli*, two bacterial species, three yeasts, and two molds randomly selected from a collection of microorganisms found on apples or apple cider were exposed to doses between 0.0 and 2.3 kGy electron beam irradiation. The D values for *E. coli* 0157:H7 strains ranged between 0.25 and 0.34 kGy. (The killing effect of irradiation on microorganisms is expressed as D (destruct)-values. One D-value is the amount of irradiation required to kill 90% of that organism. For example, it may require 0.3 kilo Gray (dose of irradiation) to kill 90% of *E. coli* strain 0157. So the D-value for this *E. coli* strain is 0.3 kGy.) It was calculated that irradiation at 0.247 kGy might result in a 5-log reduction of *E. coli* 0.157:H7 in apple cider, while naturally occurring yeast might not be adversely affected by irradiation at this dose (Wang et al., 2004).

7.2.2.2 Heat Treatment Treatment of fresh fruits and vegetables by employing different form of heat offers a pesticide-free method for the management of postharvest diseases caused by microbial pathogens. Postharvest decay control through the use of heat differs essentially from other uses of heat on agricultural commodities, such as curing to promote wound healing or heating to eliminate nematodes, insects, or viral pathogens. Postharvest heat treatment for disease management is restricted to about 3 to 5 min, whereas for other uses, heat treatments have to be applied for longer periods. Many pathogens are present on the surface or within a few outer cell layers of the infected produce and many of them have thermal sensitivity much lower than the fruits and vegetables. Hence, treatment of *Monilinia fructicola* for 5 to 10 min at 40°C has been found to be effective (Barkai-Golan and Philips, 1991). The location of the target pathogen in or on the fruit and the thermal sensitivity of the target organism and the fruit are the major factors determining the effectiveness of heat treatment (Coates and Johnson, 1993).

Heat treatment may alter various physiologic functions of both the pathogens and the host tissues. Protein denaturation, lipid liberation, destruction of hormones, depletion of food reserves, and metabolic derangement, with or without accumulation of toxic intermediates, are some of the changes induced in pathogens. In the host tissues, process of ripening, fruit color, electrolyte leakage, sugar metabolism, ethylene and ethanol production, pectic enzyme activity, and induction of resistance are the alterations observed. Heat treatment may involve the use of hot air, water, or steam. Generally, moist heat is more effective than dry heat. The incidence and/or intensity of postharvest diseases of several fruit and vegetables have been reduced substantially by the application of heat treatment (Table 7.3). The temperature is frequently adjusted to a level that is not injurious to the commodity to be treated. Hence, the temperature has to be carefully measured and controlled. The initial temperature and size of the produce and the position of the produce within the treatment chamber have a marked influence on the effectiveness of treatment.

Several improvements have been made over the years to increase the thermotolerance of the commodities. Preconditioning the fruits prior to heat treatment may be useful in avoiding injury to the fruit. Lemons held at 15.5°C for 2 to 8 days exhibited greater tolerance of heat treatment (Houck, 1967). During hot air treatment, condensation of water should be avoided by proper

TABLE 7.3 Management of Postharvest Diseases by Temperature Treatment

Commodity	Pathogen	Heating method	Temperature 0°C/Time (min)	Reference
Apple	*Gloeosporium* sp.	HW	45/10	Edney and Burchill, 1967
	Penicillium expansum	HA	45/15	
Banana	*Colletotrichum musae* *Fusarium proliferatum*	HW	45–47.5/10–30	López-Cabrera and Marrero-Domínguez, 1998
Bean	*Pythium butleri* *Sclerotinia sclerotiorum*	HW	52/0.5	Wells and Cooley, 1973
Grapefruit	*Penicillium digitatum*	HW	50/5	Spalding and Reeder, 1972
Lemon	*Penicillium digitatum*	HW	52/5–10	Houck, 1967
		HW	52–53/2	Nafussi et al., 2001
Mango	*Colletotrichum gloeosporioides*	HW	51/15	Pennock and Maldonaldo, 1962
		HW	55/5	Smoot and Segall, 1963
		HW	52/5	Spalding and Reeder, 1972
Melon	*Fusarium* spp.	HW	57/0.5	Wells and Stewart, 1968
	Various fungi	HW	52/2	Teitel et al., 1989
Orange	*Phomopsis citri* *Penicillium digitatum*	HW	53/5	Smoot and Melvin, 1963, 1965
Papaya	*Colletotrichum* sp.	HW	43–49/20	Akamine and Arisumi, 1953
Pear	*Botryosphaeria berengeriana*	HA	30/35	Teitel et al., 1989
		HW	50/5–7	Seo et al., 1998
Pepper (*Capsicum*)	*Alternaria alternata* *Botrytis cinerea*	HW	50–53/2	Lurie et al., 1998
		HA	38/48–72 h	
Strawberry	*Botrytis cinerea*	HA	44/40–60	Couey and Follstad, 1966
	Rhizopus stolonifer	HA		

HA = Hot air; HW = hot water.

moisture control. By employing wax with or without fungicides, the enhanced water loss following heat treatment may be reduced (Wells, 1972). Wrapping the commodities with plastic films before or after heat treatment is reported to prevent water loss from heat-treated produce (Anthony et al., 1989; Teitel et al., 1989).

Fungal pathogens differ considerably in their sensitivity to temperatures. Furthermore, within a species, the asexual spores, mycelium, sexual spores, and overwintering structures such as sclerotia show variations in their sensitivity to heat (Quimio and Quimio, 1974; Michailides and Ogawa, 1989). The mycelial growth, conidial germination, and pathogenicity of *Colletotrichum musae* from banana and *C. gloeosporioides* from mango were reduced by treatment at 55 and 60°C for 20 min. *C. musae* was more sensitive to heat than *C. gloeosporioides* (Liu et al., 1999). The mango stem-end rot disease occurring in Queensland, Australia is caused by *Dothiorella dominicana, Lasiodiplodia theobromae,* and *Phomopsis mangiferae.* Vapor heat treatment of mango fruits cv. Kensington Pride (fruit seed surface temperature of 46.5°C) for 10 min controlled the stem-end rot caused by *D. dominicana* (Coates et al., 2003).

It is essential to determine the in vitro heat tolerance of different kinds of spore forms or infection structures of postharvest pathogens of concern to develop effective heat treatment systems. *Helminthosporium solani* causes silver scurf disease in Irish potato tubers. The growth and proliferation of undisturbed colonies of the pathogen could be eliminated in vitro by radiant heat treatment above 50°C and lasting 5 min. In contrast, detached conidia were not affected by a 3-min treatment at 70°C. These conidia can pose problems if potato tubers are heat-treated. It is suggested that removal of conidia by washing the tubers in hot water followed by heat treatment lethal to the remaining mycelium may be effective in reducing fungal inoculum on seed potatoes (Johnson et al., 2003).

The efficacy of vapor heat treatment, with a view to replacing the use of SO_2 for the control of *Botrytis cinerea* infecting table grapes, was assessed. Vapor heat treatment at 52.5°C for 21 or 24 min and at 55°C for 18 or 21 min provided most effective protection against gray mold infection. These treatments were effective against natural and artificial infections when bunches were stored at 0.5 ± 1°C after treatment. Artificially inoculated bunches showed less than 1% infection in treated bunches as against about 50% infection in control bunches at 10 weeks after treatment. In the case of naturally infected bunches treated and stored for 10 weeks at 0.5 ± 1°C followed by 1 week shelf life at 20°C, only 0.7 to 1.5% infection was recorded in heat-treated bunches. On the other hand, 1.7% infection in bunches stored with a release SO_2 pad was observed and control bunches had more than 30% infection (Lydakis and Aked, 2003).

The fact that treatments may eliminate not only pathogens, but may reduce population of beneficial antagonists present on the fruit surface has to be borne in mind. Such a condition may favor rapid colonization by opportunistic pathogens. In addition, physiological disorders, such as bitter pit in apple and pear, may be exacerbated by the heat treatment (Neven et al., 2000). Nevertheless, postharvest heat treatment has been shown to be a potential nonchemical disease management strategy acting by directly inhibiting pathogen growth, activating the natural resistance of the host, and slowing down the ripening process.

The effect of curing treatments at 33° or 30°C for 24 to 72h applied to oranges for the control of green and blue mold diseases was assessed. Mould development in artificially inoculated oranges was prevented by curing at 33°C for 65h, while a reduction of more than 90% in naturally infected fruits was observed, when stored at 20°C for 7 days. This curing treatment was also effective in reducing green mold, but not blue mold in oranges stored at 4°C for 2 months plus shelf life at 20°C. As there was no effect on juice yield and soluble solids in cured oranges, curing treatment may be considered as an alternative strategy to chemical control (Plaza et al., 2003).

Various methods for the of application of heat have been evaluated for their efficacy in reducing the incidence and intensity of postharvest diseases. The effectiveness of a prestorage dry heat treatment and hot water dip in reducing storage rots of *Capsicum* bell peppers and tomatoes caused by *Alternaria alternata* and *Botrytis cinerea* was assessed. Treatment with hot air at 38°C for 48 to 72h or hot water at 50 to 53°C for 2 to 3min, resulted in reduction in the pathogenicity and development of these pathogens in inoculated pepper. In addition, the heat-treated fruits did not develop chilling injury under cold storage. The dry heat treatment was not only effective against fungal pathogens, but also against insect pests such as *Ceratitis capitata*, in addition to exerting a favorable effect on fruit physiology (Lurie et al., 1998).

The effects of hot water dips (HWD) on mycelial growth and conidial germination of *Colletotrichum musae* and *Fusarium proliferatum* involved in banana crown rot disease were determined. The development of these pathogens was effectively arrested by treatment at 45 to 47.5°C for 15 to 30min. Temperatures above 50°C induced peel darkening and incomplete soluble solids accumulation. Apparently inhibition of spore germination seems to be the principal mode of action of HWD (López-Cabrera and Marrero-Domínguez, 1998). Anthracnose diseases caused by *Colletotrichum musae* and *C. gloeosporioides* in banana and mango, respectively, were effectively controlled by HWD. A temperature range of 47 to 52°C for 10 to 20min inhibited *C. musae* as well as *Botryodiplodia theobromae* and *Fusarium* spp. causing finger rot disease of banana. Wrapping fruits in a plastic bag prior to HWD reduced the temperature requirement to 47 to 49°C (indirect HWD) as against 50 to 52°C for direct HWD. Mangoes treated in hot water at 40°C for 10 to 20min prior to a 54 to 56° HWD responded better than mangoes dipped in hot water at 51 to 54°C for 10min. HWD inhibited ripening in banana but not in mangoes, and no adverse effect due to HWD could be discernible in either bananas or mangoes (Acedo Jr. et al., 2001). Decay of lemon fruit caused by *Penicillium digitatum* was significantly prevented by HWD for 2min at 52 to 53°C. There was a direct transient inhibitory effect on the pathogen by arresting its growth for 24 to 48h prior to triggering the defense responses of the host tissue to pathogen infection (Nafussi et al., 2001). The effect of warm air (35°C at 80 ± 1 RH), warm water (25, 45, 55°C), with 20°C at atmospheric RH as control, on wound healing on sweet lime (*Citrus limettoides*) fruits was assessed. After 10 weeks of storage no decay could be seen in fruits with superficial wounds (2 mm) and inoculated before or after warm air treatment or in deep wound (5 mm)-inoculated fruits before warm air treatment. Warm water dip at 25 and 45°C for 5 min also effectively reduced decay due to *Penicillium italicum* during storage. Substitution of heat treatments for chemical use seems to be a feasible approach to control decay of sweet lime (Mahmoodabadi et al., 2000).

The procedure termed phased hot water treatment involves the use two temperature regimes, that is 36.5°C for 60 min plus 46.5°C for 43 min. The phased hot water treatment and 1 day of intermittent warming at 34°C resulted in significant reduction in incidence of mango anthracnose and also the severity of internal and external injury. The off flavors from the fruit were at a lower level compared with fruit treated with hot water at 46.5°C for 45 min and untreated control (Nyanjage et al., 1998). Strawberries were exposed to heat treatment in an air oven at 45°C for 3 h and stored at 0°C for periods up to 14 days and later placed at 20°C for 24 to 96 h. The fruits heat-treated and stored for 7 or 14 days at 0°C and placed at 20°C for 48 h had lower mold population than controls. The heat-treated fruits were firmer, had lower acidity, and total sugars content was not altered (Vincente et al., 2002).

The hot water brushing (HWB) method was evaluated for its efficacy against *Penicillium digitatum* causing green mold decay in citrus fruits. The method involves rinsing with hot water and brushing fruits at 24 h after artificial inoculation with *P. digitatum* spore suspension. HWB treatment at 56, 59, and 62°C for 20 s reduced decay development in the infected wounds to 20, 5, and less than 1%, respectively, of that in untreated control fruits. The epiphytic microbial counts (CFU) present on the treated fruit surface were 24, 12, and less than 1%, respectively, of the fruits rinsed and brushed with tap water. HWB treatment at 56°C for 20 s provided additional benefit of smoothening the fruit epicuticular waxes and consequently covering and sealing stomata and cracks on the fruit surface which are known to be potential sites of invasion by the pathogen. Organically grown citrus fruits also could be effectively protected by HWB at 56°C for 20 s which did not cause any surface damage, in addition to retaining all internal quality parameters as revealed scanning electron microscopic observations (Porat et al., 2000a). In a further study, the HWB technique comprising of rinsing hot water on the fruits as they move along a belt of brush rollers was tested on grapefruit cv. Star Ruby, which was wound-inoculated with *P. digitatum*. A 20-s HWB treatment at 59°C or 62°C reduced decay of fruit by 52 and 70%, respectively, compared with untreated control fruits. Furthermore, HWB treatments significantly reduced chilling injury and the percentages of fruits displaying chilling injury symptoms were reduced by 42 and 58%, respectively, after 6 weeks storage at 2°C and additional week at 20°C. Cleaning of fruits and improvement in general appearance of treated fruits are the additional advantages offered by HWB technique (Porat et al., 2000b).

Alternaria alternata is one of the principal postharvest pathogens infecting mangoes. A simple postharvest treatment of mangoes, involving a hot water (50–60°C) spray and simultaneously brushing the fruits (HWB), was developed. The decay due to *A. alternata* in several mango cultivars stored at 14°C for 3 weeks was effectively reduced to a level that could be achieved by application of prochloraz. However, the fungicide was more effective when mangoes were stored for longer periods of time (Prusky et al., 2003).

A brief postharvest hot water drench was developed for the control of green mold caused by *Penicillium digitatum* and sour rot caused by *Geotrichum citriaurantii* in lemon cv. Eureka and orange cv. Valencia. The treatment composed of a brief (15 or 30s) high-volume, low-pressure, hot water drench at 20.0, 48.9, 54.4, 60.0, or 62.8°C applied over rotating brushes. It was compared to immersion of fruit in 3% sodium carbonate at 35.0°C for 30s, a common commercial practice followed in California. With increase in temperature and treatment duration, the effectiveness of treatment increased. At 62.8°C for 30s the treated fruits had 14.5% and 9.4% infection, respectively, on lemon and orange as against 98% infection in untreated fruits. The hot water drench treatment had no effect on sour rot disease and it was less effective compared to treatment with sodium carbonate (Smilanick et al., 2003).

The usefulness of a rapid method for simultaneously rinsing and disinfecting freshly harvested Galia melon (*Cucumis melo* cv. *reticulatus*) fruit employing a hot-water rinse and brushes (HWRB) was assessed. A 15-second treatment at 59 ± 1°C was found to be optimum for reducing decay by *Alternaria alternata* and *Fusarium solani,* retaining fruit quality after prolonged storage. The epiphytic microbial population was significantly reduced by HRWB technique which removed soil, dust, and fungal spores from the fruit surface and partially or entirely sealed natural openings in the fruit epidermis as indicated by observations under scanning electron microscope (Fallik et al., 2000). The efficacy of HRWB treatment to reduce the decay of apples caused by *Penicillium expansum* was assessed in comparison with a dry heat treatment at 38°C for 96h. HWRB treatment at 55°C for 15s significantly reduced ($P = 0.05$) decay development in inoculated fruit after 4 weeks at 20°C or in naturally infected fruit after an extended storage for 4 months at 1°C plus 10 days at 20°C. Delayed ripening, low respiration rate and ethylene production, slow development of color, and maintenance of fruit quality are additional desirable effects of HWRB treatment (Fallik et al., 2001).

A short prestorage hot water rinsing and brushing (HWRB) method was shown to be effective for reducing infection by *Botrytis cinerea* in tomatoes. Fruit of the pink tomato cv. 189 were placed at 5 or 12°C for 15 days, plus 3 days at 22°C. Application of HWRB treatment at 52°C for 15s or dipping the fruit at 52°C for 1min (hot water dip, HWD) significantly reduced decay development. Furthermore, chilling injury symptoms were entirely inhibited after storage. In addition, HWRB treatment enhanced the resistance of tomato fruits to artificially inoculated *B. cinerea* and also their storability was extended to over 3 weeks at 5°C by minimizing chilling injury (Fallik et al., 2002).

The effectiveness of steam treatment for the reduction of decay of carrot caused by *Alternaria* spp. and *Sclerotinia sclerotiorum* was demonstrated. Winter carrots were exposed to steam for 3s prior to packaging. After 60 days of storage at 0.5°C plus an additional week at shelf conditions (20°C), treated carrot exhibited 2% decay as against 23% decay in untreated carrots. Artificial inoculation with *Alternaria alternata*, *A. radicina*, and *S. sclerotiorum* followed by steam treatment resulted in 5% decay, whereas the untreated carrots had 65% decay (Afek et al., 1999).

Comparative efficacy of heat treatments and other disease management strategies has been assessed. The effectiveness of immersion of oranges in water at up to 75°C for 150s, 2 to 4% sodium carbonate at 20°, or 45°C for 60 or 150s, or in 1 to 4% sodium bicarbonate at room temperature for 150s followed by storage at 20°C for 7 days, was determined for the control of blue mold disease of oranges caused by *Penicillium italicum*. Sodium carbonate and sodium bicarbonate were more effective than hot water treatment. Sodium carbonate solution (3 or 4%) at 45°C for 150s reduced decay by more than 90%. Sodium bicarbonate at room temperature, sodium carbonate at 45°C, and hot water at 45°C reduced blue mold disease incidence on artificially inoculated oranges to 6, 14, and 27%, respectively, after 3 weeks of storage at 3°C. In addition, the incidence of green mold caused by *P. digitatum* was reduced to 6, 1, and 12%, respectively, whereas the untreated fruits were entirely (100%) infected by green mold disease (Palou et al., 2001c). Hot water dipping at 45° or 50°C for 60 or 150s did not provide satisfactory control of both blue and green molds of Clementine mandarins. On the other hand, sodium carbonate solutions (2 or 3%) enhanced decay control compared to water alone at all temperatures (20, 40, 45, or 50°C) and for all immersion periods (60 or 150s). Treatment with sodium carbonate (3%) at 50°C for 150s entirely eliminated the infection by both blue and green mold diseases without any observable injury to the fruit (Palou et al., 2002).

The efficacy of different heat treatment methods, such as: (1) hot water treatment (HWT; 53°C for 10min); and (2) hot water brush (HWB; 60°C for 20 to 35s) and brief exposure to hot water dip (HWD; 60° for 20 to 35s) for the control of postharvest diseases of mangoes was assessed. Fruits were exposed to vapor heat treatment and ripened at 25°C. Incidence of anthracnose and stem-end rot diseases was reduced by all heat treatments. Hot water treatment followed by vapor heat treatment induced faster peel color change compared with other treatments. No significant change in physico-chemical characteristics of the fruits was noticed due to heat treatments (Esguerra et al., 2004). In another study, hot water brushing that a 15 to 20s combined water spray and fruit brushing was found to be effective in reducing incidence of postharvest pathogens. Furthermore, this technology improved keeping quality of mango fruits in addition to reducing the concentration of fungicides required to protect the mango fruits against infection by microbial pathogens (Prusky et al., 2004).

A combination of one or more disease management strategies may prove to be more effective. Various heat treatments, such as curing at 36°C for 72 h, hot water dip at 52°C for 2 min, or hot water drench brushing at 52, 56, or 60°C for 10 s, were combined with individual fruit sealing, packaging in polyethylene liners, or waxing, which included the addition of thiabendazole (TBZ) and 2,4-D isopropyl ester. Hot water dip or curing controlled the development of postharvest pathogens *Penicillium* spp. in 'Oroblanco' fruit (*Citrus grandis* x *C. paradisi*). The polyethylene liner packaging increased the risk of disease development in the absence of decay-reducing measures (Rodov et al., 2000). Melon cv. Galia fruits were treated with hot water at 55°C for 90 s and then packed into perforated polyethylene bags for cold storage at 2°C and 85 to 90% RH. Hot water treated, unpacked fruits had less infection by *Alternaria* spp., *Penicillium* spp., *Cladosporium* spp., *Mucor* spp., *Botrytis cinerea,* and *Aspergillus* spp. compared with hot water treated and packed, and control fruits without heat treatment. Treated fruits packed in polyethylene bags showed less weight loss than unpacked fruit (Halloran et al., 1999). The enhanced water loss that frequently follows hot water treatment can be reduced by applying wax after treatment. Nectarine fruits wrapped in plastic bags exposed to hot air at 52°C for 15 min, 90% RH were protected from water loss or gain and also from recontamination and discoloration, resulting in improvement of fruit quality (Barkai-Golan and Philips, 1991). The effect of combined treatments of shrink wrapping, curing, quarantine heat disinfestations, and modified atmosphere packaging (MAP) on the storage life of Washington Navel and Valencia oranges was determined. Shrink wrapping and curing significantly reduced development of *Penicillium* spp. MAP increased stem-end rot caused by *Phomopsis citri* in Washington Navel orange, but no effect on Valencia orange was observed. The combined treatment adversely affected flavor, peel color, and increased the ethanol content in both cultivars. These quality parameters were not altered in the absence of MAP treatment (Shahbake, 1999).

The aim of heat treatments is to kill the postharvest pathogens causing decay of fruit surface, while maintaining fruit quality during prolonged storage and marketing. These treatments can be applied easily, requiring short operating time, as they are efficient in heat transfer. When compared with a commercial vapor heat treatment system, the typical hot water system is economical and cost-effective. The time and temperature applied depend on the cultivar, fruit maturity, fruit size, and conditions during the growing season. Inhibition of ripening, reduction in extent of decay, and induction of resistance in certain harvested produce are the principal benefits of prestorage hot water treatments (Fallik, 2004). The adverse effects of heat treatments are enhanced water loss, discoloration, increased susceptibility to contaminating microorganisms, and reduction in shelf or storage life. In addition, the lack of residual protection against recontamination by postharvest pathogens and injury to host tissue are the other important limitations of heat treatments. However, when they are used in tandem with chemicals they may prove to be highly effective against the postharvest pathogens of fruits and vegetables.

7.2.3 Modification of Storage Atmosphere

The contents of oxygen (O_2) and carbon dioxide (CO_2) of the storage atmosphere may be modified to reduce deterioration of fruits and vegetables during storage (see also Section 6.2.2). The controlled atmosphere (CA) storage method adopted primarily for some cultivars of apples is formulated by reducing oxygen levels and increasing CO_2 levels. The proportion of O_2 and CO_2 is altered either by introducing CO_2 into a closed, refrigerated system or by allowing the respiratory CO_2 to accumulate with a simultaneous reduction in O_2 concentration. A very high concentration of CO_2 is required to inhibit the development of pathogens and it is very difficult to control decay after infection has occurred. Retardation of respiration, quality changes, tissue breakdown, and color changes are some of the reported effects of CA storage. Excess of CO_2 may result in skin injury and internal disorders, while very low O_2 may be responsible for off flavors and alcohol injury due to fermentation (Waller, 2002).

The CA storage used to maintain postharvest flesh firmness favored the spread of *Botrytis cinerea*, causing gray mold of kiwifruit. To overcome this problem, curing of kiwifruits (at ambient temperature) was carried out in the refrigerated room by gradually reducing the temperature from 10 to 0°C over a period of about 10 days. Postponing the provision of CA conditions by 30 to 50 days after harvest, helped avoid the negative impact of CA storage (i.e. increasing incidence of gray mold disease) with no adverse effects on fruit firmness (Tonini et al., 1999). The levels of CO_2, O_2, and ethylene have to be carefully monitored daily and controlled within narrow limits. Concentrations of CO_2 greater than 10% may injure apples, while O_2 concentrations of less than 1% can induce storage disorders. For many apple cultivars, an atmosphere of 5% CO_2 and 3% O_2 at a temperature of 0°C (32°F) has been recommended (Boyette et al., 2003).

In contrast, no adverse effect of high CO_2 was observed in the case of sweet cherries. With increase in concentrations of CO_2, growth of *Monilinia fructicola* was inhibited both in vitro and in vivo. The lesion size was significantly reduced with 15 to 25% CO_2 and lesion formation was entirely prevented with 30% CO_2 at 25°C. At low temperature (0°C) loss of pathogenicity of *M. fructicola*, in addition to inhibition of growth, was observed. There was neither any injury nor off-flavor at 0°C after 18 days of storage at any CO_2 concentration tested (Tian et al., 2001).

The development of blue mold disease of sour cherry caused by *Penicillium expansum* was contained by cold storage. The percentage of healthy fruits after storage at 2°C in 10% CO_2 + 3% O_2 air for 24 days and after subsequent storage for 3 days at 20°C was greater than in the fruits stored at 0°C in air (Nabialek et al., 1999). In the case of decay of strawberry caused by *Botrytis cinerea*, 15 kPa CO_2 in air and in combination with 40 kPa O_2 suppressed the mycelial growth most effectively when stored at 5°C for 7 days. Increasing O_2 concentration to 100 kPa reduced decay even more effectively, both in vitro and in vivo. However, this treatment increased production of fermentative metabolites that reduced organoleptic acceptance of fruits, making this treatment a doubtful alternative to other methods for the control of decay of strawberry (Wszelaki and Mitcham, 2000). The effect of high concentrations of CO_2 (10, 15, and 20%) was determined in the case of leaf yellowing and rot caused by *Alternaria brassicola* in cauliflower. The development of disease was delayed by CO_2 treatment. However, injury inside the stem and production of off flavors reduced the level of acceptance of the CO_2 treatment as a feasible alternative (Menniti and Casalini, 2000).

The efficacy of CO_2 treatment was compared with that of sulfur dioxide (SO_2) in reducing fruit decay and maintaining fruit quality of grape cv. Red Globe. Both CO_2 and SO_2 application significantly reduced decay percentage when the fruits were stored at 0°C and 90% RH. Reduction in percentage total weight loss and shatter and improvement in berry firmness and berry adherence strength over control were additional advantages of the treatments (Ahmed and El-Rayes, 2001). The efficacy of CO_2-enriched atmospheres on decay control of organically-grown 'Thompson Seedless' and Red Globe table grapes was assessed during storage at 0°C. Grapes were inoculated with conidial suspension of *B. cinerea* (10^3 conidia/ml). CA with CO_2 at 15% or higher reduced disease incidence significantly as in the case of treatment using metabisulfite pads. Stalks and pedicel were greener when metabisulfite pads were also used (Retamales et al., 2003).

Another method for CA storage that has been studied for the control of postharvest diseases is corona discharge, which can increase ozone and negative ions. A commercially available generation system consists of single corona generating plate sandwiched between two aluminum sheets. Corona was generated by a pulsing a voltage of 2.0 kV of alternating polarity of 101 Hz, while the second sheet was earthed. A control system cycled on and off to maintain average ozone levels of 115, 208, or 530 ppb in combination with 19, 49, and 72 × 10^2 negative air ions per ml. The efficacy of corona discharge in combination with ozone, on postharvest decay caused by *Penicillium* spp., *Sclerotinia sclerotiorum,* and *Botrytis cinerea* in onions, grapes, lemon, carrots, celery (*Apium graveolens*), and apples, was assessed. Among the pathogens tested, *Penicillium* spp. were more sensitive than *S. sclerotiorum* and *B. cinerea* (Hildebrand et al., 2001).

Exposure to ozone has been demonstrated to be effective for the control of postharvest diseases. Continuous exposure to ozone at 0.3 ppm (v/v) inhibited aerial mycelial growth and sporulation of *Monilinia fructicola*, *Botrytis cinerea*, *Mucor piriformis,* and *Penicillium expansum* inoculated in 'Elegant Lady' peaches, stored for 4 weeks at 5°C and 90% RH. However, only the incidence of brown rot caused by *Monilinia fructicola* was significantly reduced. Gray mold nesting due to *B. cinerea* in Thompson Seedless table grape was completely inhibited under 0.3 ppm ozone in fruits stored for 7 weeks at 5°C. Ozone or ambient atmosphere treatments did not cause phytotoxic injuries of fruit tissues. The incidence of citrus green and blue molds due to *P. digitatum* and *P. italicum,* respectively, on oranges was delayed by 1 week under 0.3 ppm ozone. A synergistic effect between ozone treatment and low temperature was observed. Sporulation by the mold fungi was reduced or prevented by ozone exposure. This effect is particularly useful in the case of fungicide resistant strains that frequently develop during storage, resulting in possible extension of useful period of postharvest fungicides (Palou et al., 2001a, b, 2002).

Ion generators can be employed to produce positive and negative air ions or these ions may be generated as by-products of other industrial processes. Air ions are short-lived, lasting only minutes before their charge is lost. Negative air ions (NAI) have been found to be useful in reducing the populations of airborne molds and contaminants at concentrations ranging from 5×10^4 to 5×10^6 ions/cm^3. A synergistic effect of combination of ozone with NAI was demonstrated for suppressing microbial development and decay of fresh agricultural commodities. The viability of cells of *Erwinia carotovora* exposed to 10^6 NAI/cm^3 and/or 50 pp ozone at 10°C for 6 h was assessed. The combined treatment reduced the viability of bacteria to 4% as compared to 70% of the bacterial cells exposed to ozone alone. Likewise, when carrots were exposed to 0.1 to 0.5 ppm ozone in combination with 2×10^4 to 7×10^4 NAI/cm^3, the growth rates of *Botrytis cinerea* and *Sclerotinia sclerotiorum* were reduced by 69 and 57% respectively. In contrast, the growth rates of these pathogens were reduced by 12 and 16% by treatment with ozone. NAI can be advantageously exploited for the management of other postharvest pathogens by making suitable changes in their concentrations (Forney et al., 2001).

The effect of a combination of ozone and corona discharge on the control of decay in stored onions was determined. The corona system was operated at 350 Hz and 2.5 kV and the ozone concentration was controlled at set points of 50 ppb during the day and 250 ppb during the night. Corona discharge-treated onions exhibited low incidence of molds (two times less) and less weight loss compared to untreated onions during a storage period of 2 or 4 weeks. The initial concentration of airborne conidia from the fungal pathogens was significantly reduced in the storage room after corona discharge treatment, leading to the reduced levels of disease incidence in onions (Fan et al., 2001).

SUMMARY

It is well known that prevention is better than cure. This strategy has been followed for the management of all diseases affecting plants and their produce. Exclusion of microbial pathogens at all stages after harvest is the basic and essential step to prevent access to the harvested produce. Elimination of infected, and apparently infected, commodities by employing sensitive and rapid detection techniques, may reduce the spread of the pathogens during storage. Use of appropriate packaging and efficient handling and transport to reduce injuries to the produce will contribute to the maintenance of quality and market value of the produce. Various treatments with different physical agents such as UV-C light, heat in different forms, and modification of storage atmosphere have been demonstrated to reduce decay of produce to varying degrees. A combination of one or more physical methods has been reported to have a synergistic effect in reducing incidence of postharvest diseases. The suitability of the different methods for the management of the postharvest diseases caused by microbial pathogens is discussed.

REFERENCES

Abdullah, N., Nawawi, A. and Othman, I. 2000. Fungal spoilage of starch-based foods in relation to its water activity (a_w). *Journal of Stored Products Research*, **36**: 47–54.

Acedo, A.L. Jr., Benetiz, M.M. and Milan, L.A. 2001. Heat treatment effects on ripening and disease infection of Philippines bananas and mangoes. *Acta Horticulturae*, **553**: 417–419.

Afek, U., Orenstein, J. and Nuriel, E. 1999. Steam treatment to prevent carrot decay during storage. *Crop Protection*, **18**: 639–642.

Agarwal, V.K. and Sinclair, J.B. 1996. *Principles of Seed Pathology*, 2nd edition. CRC Lewis Publishers, Boca Raton, USA.

Ahmed, D.M. and El-Rayes, D.A. 2001. Carbon dioxide treatment as a potential alternative to sulfur dioxide to control fruit decay in Red Globe table grape. *Assiut Journal of Agricultural Sciences*, **32**: 199–213.

Akamine, E.K. and Arisume, T. 1953. Control of postharvest storage decay of fruits of papaya (*Carica papaya* L.) with special reference to the effect of hot water. *Proceedings of the American Society for Horticultural Science*, **61**: 270–274.

Andebrhan, T. and Wood, R.K.S. 1980. The effect of ultraviolet radiation on the reaction of *Phaseolus vulgaris* to species of *Colletotrichum. Physiological Plant Pathology*, **17**: 105–110.

Anthony, B.R., Phillips, D.J., Badr, S. and Aharoni, Y. 1989. Decay control and quality maintenance after moist air heat treatment of plastic wrapped nectarines. *Journal of American Society of Horticultural Science*, **114**: 946–949.

Anuja, G. and Aneja, K.R. 2001. Mycoflora spectrum during storage and its effect on seed viability of soybean [*Glycine max* (L.) Merrill] seeds under ambient conditions. *Proceedings of the National Academy of Sciences, India, Section B. Biological Sciences*, **71**: 245–253.

Appel, O. and Riehm, E. 1911. [As quoted by M. Martin and D. Woodcock, 1983. *The Scientific Principles of Crop Protection*, 7th edition. Edward Arnold, London.]

Aveling, T.A.S., Snyman, H.G. and Maude, S.P. 1993. Evaluation of seed treatments for reducing *Alternaria porri* and *Stemphylium vesicarium* on onion seed. *Plant Disease*, **77**: 1009.

Bankole, S.A., Ikotun, B. and Ekpo, E.J.A. 1999. Fungal deterioration of melon seeds stored in jute sacks and polyethylene bags in Agro-Iwoye, southwestern Nigeria. *Mycopathologia*, **146**: 135–146.

Barkai-Golan, R. and Phillips, D.J. 1991. Postharvest heat treatment of fresh fruits and vegetables for decay control. *Plant Disease*, **75**: 1085–1089.

Bechtel, D.B., Wilson, J.D., Eustace, W.D., Behnke, K.C., Whitaker, T., Peterson, G.L. and Sauer, D.B. 1999. Fate of dwarf bunt fungus teliospores during milking of wheat into flour. *Cereal Chemistry*, **76**: 270–275.

Benkeblia, N. and Selselet-Attou, G. 1997. Evolution of fungal flora of irradiated dry onions (*Allium cepa* L.) during storage. *Microbiologie, Aliments, Nutrition*, **15**: 71–77.

Ben-Yehoshua, S., Shapiro, B., Kim, J.J., Sharoni, J., Carmeli, S. and Kashman, Y. 1988. Resistance of citrus fruit to pathogens and its enhancement by curing. In: *Proceedings of the Sixth International Citrus Congress* (eds) R. Goren and K. Mendel, pp. 1371–1379. Tel Aviv, Israel.

Boyette, M.D., Wilson, L.G. and Estes, E.A. 2003. *Postharvest Cooling and Handling of Apples*. Agricultural Extension Service Publication AG–413–1, North Carolina State University, USDA, USA.

Brown, J.E., Lu, T.Y., Stevens, C., Khan, V.A., Lu, J.Y., Wilson, C.L., Collins, D.J., Wilson, M.A., Igwegbe, E.C.K., Chalutz, E. and Droby, S. 2001. The effect of low dose ultraviolet light-C seed treatment on induced resistance in cabbage to black rot (*Xanthomonas campestris* pv. *campestris*). *Crop Protection*, **20**: 873–883.

Capdeville, G.de, Wilson, C.L., Beer, S.V. and Aist, J.R. 2002. Alternative disease control agents induce resistance to blue mold in harvested 'Red Delicious' apple fruit. *Phytopathology*, **92**: 900–908.

Cia, P., Benato, E.A., Anjos, V.D. and Vietes, R.L. 2000. Effect of irradiation on storage of grapes cv. Italia. *Revista Brasileira de Fruiticultura*, **22**: 62–67.

Clear, R.M., Patrick, S.K., Turkington, T.K. and Wallis, R. 2003. Effect of heat treatment on seedborne *Fusarium graminearum* and other cereal pathogens. *Canadian Journal of Plant Pathology*, **25**: 489–498.

Coates, L.M., Cooke, A.W. and Dean, J.R. 2003. The response of mango stem-end rot pathogens to heat. *V International Mango Symposium, Acta Horticlturae*, **455**.

Coates, L.M. and Johnson, G.I. 1993. Effective disease control in heat disinfested fruit. *Postharvest News and Information*, **4**: 35N–40N.

Copeland, L.O., Adams, M.W. and Bell, D.C. 1975. An improved bean seed programme for maintaining disease-free seed of field beans. *Seed Science and Technology*, **3**: 719–724.

Couey, H.M. and Follstad, M.N. 1966. Heat pasteurization for control of postharvest decay in fresh strawberries. *Phytopathology*, **56**: 1345–1347.

Crisosto, C.H., Seguel, X. and Michailides, J. 1998. Comparing pulstated ultraviolet light and postharvest fungicide for peach fruit decay control. *Acta Horticlturae*, **465**: 471–479.

Daniels, B.A. 1983. Elimination of *Fusarium moniliforme* from corn seeds. *Plant Disease*, **67**: 609.

D'hallewin, G., Schirra, M., Pala, M. and Ben-Yehoshua, S. 2000. Ultraviolet C irradiation at $0.5\,kJm^{-2}$ reduces decay without causing damage or affecting postharvest quality of star Ruby grapefruit (*Citrus paradisi* Macf.). *Journal of Agricultural and Food Chemistry*, **48**: 4571–4575.

D'hallewin, G., Schirra, M., Manueddu, E., Piga, A. and Ben-Yehoshua, S. 1999. Scoparone and scopoletin accumulation and ultraviolet-C induced resistance to postharvest decay in oranges as influenced by harvest date. *Journal of the American Society for Horticultural Science*, **124**: 702–707.

Droby, S., Chalutz, E., Horev, B., Cohen, L., Gaba, V., Wilson, C.L. and Wisniewski, M.E. 1993. Factors affecting UV-induced resistance in grapefruit against green mold decay caused by *Penicillium digitatum*. *Plant Pathology*, **42**: 418–424.

Edney, K.L. and Burchill, R.T. 1967. The use of heat to control the rotting of Cox's Orange Pippin apples by *Gloeosporium* sp. *Annals of Applied Biology*, **59**: 389–400.

Esguerra, E.B., Valerio-Traya, R.F. and Lizada, M.C.C. 2004. Efficacy of different heat treatment procedures in controlling diseases of mango fruits. *Acta Horticulturae*, **645**.

Fallik, E. 2004. Prestorage hot water treatments (immersion, rinsing and brushing). *Postharvest Biology and Technology*, **32**: 125–134.

Fallik, E., Aharoni, Y., Copel, A., Rodov, V., Tuvia-Alkalai, S., Horev, B., Yekutieli, O., Wiseblum, A. and Regev, R. 2000. Reduction of postharvest losses of Galia melon by a short hot-water rinse. *Plant Pathology*, **49**: 333–338.

Fallik, E., Ilic, Z., Alkalai-Tuvia, S., Copel, A. and Polevaya, Y. 2002. A short hot water rinsing and brushing reduces chilling injury and enhances resistance against *Botrytis cinerea* in harvested tomato. *Advances in Horticultural Science*, **16**: 3–6.

Fallik, E., Tuvia-Alkalai, S., Feng, X. and Lurie, S. 2001. Ripening characterization and decay development of stored apples after a short pre-storage hot water rinsing and brushing. *Innovative Food Science and Emerging Technologies*, **2**: 127–132.

Fan, L.H., Song, J., Hildebran, P.D. and Forney, C.F. 2001. Corona discharge reduces mold on commercially stored onions. *Acta Horticulturae*, **553**: 427–428.

Forney, F.C., Fan, L., Hildebrand, P.D. and Song, J. 2001. Do negative air ions reduce decay of fresh fruits and vegetables. *Acta Horticulturae*, **553**: 421–424.

Forsberg, G. 2001. Heat sanitation of cereal seeds with a new efficient, cheap and environmentally friendly method. In: *Seed Treatment: Challenges and Opportunities* (ed.) A.J. Biddle, pp. 69–72, British Crop Protection Council, Farnham, U.K.

Green, S.K., Hwang, L.L. and Kuo, Y.J. 1987. Epidemiology of *Tomato mosaic virus* in Taiwan and identification of strains. *Journal of Plant Diseases and Protection*, **94**: 386.

Halloran, N., Kasim, M.U., Cagiran, R. and Karakaya, A. 1999. The effect of postharvest treatments on storage duration of cantaloupes. *Acta Horticulturae*, **492**: 207–212.

Hamanaka, D., Dokan, S., Yasunaga, E., Kuroki, S., Uchino, T. and Akimoto, K. 2000. The sterilization effects of infrared ray on the agricultural products spoilage microorganisms. *Proceedings of ASAE Annual International Meeting*, Milwaukee, Wisconsin, USA, pp. 1–9, American Society of Agricultural Engineers, USA.

Hildebrand, P.D., Song, J., Forney, C.F., Renderos, W.E. and Ryan, D.A.J. 2001. Effects of corona discharge on decay of fruits and vegetables. *Acta Horticulturae*, **553**: 425–426.

Houck, L.G. 1967. Hot water treatments for control of *Penicillium digitatum* green mold of Eureka lemons. *Phytopathology*, **57**: 99.

Jensen, J.L. 1888. The propagation and prevention of smut in oats and barley. *Journal of Royal Society, England*, **24**: 397–415.

Johnson, T.P.C., Grout, B.W.W., Bishop, C.F.H. and Perera, A. 2003. The use of radiant heat to reduce the inoculum level of silverscurf (*Helminthosporium solani*) on potato tubers (*Solanum tuberosum*) before storage. *Acta Horticulturae*, **619**.

Karabulut, O.A. and Baykal, N. 2002. Evaluation of the use of microwave power for the control of postharvest diseases of peaches. *Postharvest Biology and Technology*, **26**: 237–240.

Kim, J.J. and Ben-Yehoshua, S. 2001. Histological detection of phytoalexin scoparone from heat-treated and UV-irradiated lemon fruits after inoculation with *Penicillium digitatum*. *Plant Pathology Journal*, **17**: 271–275.

Kobayashi, T. 1990. Impact of seedborne pathogen on quarantine in Japan. *Seed Science and Technology*, **18**: 427.

Krishnamurthy, C.D., Lokesh, S. and Shetty, H.S. 2001. Occurrence, transmission and remedial aspects of *Drechslera oryzae* in paddy (*Oryza sativa* L.). *Seed Research*, **29**: 63–70.

Liu, J., Stevens, C., Khan, V.A., Lu, J.Y., Wilson, C.L., Adeyeye, O., Kabwe, M.K., Pusey, P.L., Chalutz, E., Sultana, T. and Droby, S. 1993. Application of ultra violet-C light on storage rots and ripening of tomatoes. *Journal of Food Protection*, **56**: 868–872.

Liu, X.J., Huang, S.M., Huang, Z.H., Yang, Y., Hu, M.J. and Wang, X.S. 1999. Effect of heat treatment on growth and pathogenicity of two species of *Colletotrichum corda* on latent infection. *Acta Phytopathologica Sinica*, **29**: 91–95.

López-Cabrera, J.J. and Marrero-Domínguez, A. 1998. Use of hot water dips to control the incidence of banana crown rot. *Acta Horticulturae*, **490**: 563–569.

Lu, J.Y., Lukombo, S.M., Stevens, C., Khan, V.A., Wilson, C.L., Pusey, P.L. and Chalutz, E. 1993. Low dose UV and gamma radiation on storage rot and physicochemical changes in peaches. *Journal of Food Quality*, **16**: 301–309.

Lu, J.Y., Stevens, C., Yakubu, P. and Loretan, P.A. 1987. Gamma electron beam and UV radiation on control of storage rots and quality of Walla Walla onions. *Journal of Food Processing and Preservation*, **12**: 53–62.

Lu, J.Y., Stevens, C., Khan, V.A., Kabwe, M.K. and Wilson, C.L. 1991. The effect of ultraviolet irradiation on shelf life and ripening of peaches and apples. *Journal of Food Quality*, **14**: 299–305.

Luckey, T.D. 1980. *Hormesis with Ionizing Radiation*. CRC Press, Boca Raton, FL, USA.

Lurie, S., Klein, J.D., Fallik, E. and Varjas, L. 1998. Heat treatment to reduce fungal rots, insect pests and to extend storage. *Acta Horticulturae*, **464**: 309–313.

Luthra, J.C. 1953. Solar energy treatment of wheat loose smut *Ustilago tritici* (Pers.) Rostr. *Indian Phytopathology*, **6**: 49–56.

Lydakis, D. and Aked, J. 2003. Vapor heat treatment of Sultanina table grapes. I. Control of *Botrytis cinerea*. *Postharvest Biology and Technology*, **27**: 109–116.

Maddox, D.A. 1998. Implications of new technologies for seed health testing and the worldwide movement of seed. *Seed Science Research*, **8**: 277–284.

Mahmoodabadi, K., Rahemi, M. and Banihashemi, Z. 2000. Postharvest curing of sweet lime by heat treatments to reduce fruit decay by *Penicillium italicum*. *Iranian Journal of Plant Pathology*, **36**: 245–259.

Mandahar, C.L. 1981. Virus transmission through seed and pollen. In: *Plant Diseases and Vectors: Ecology and Epidemiology* (eds) K. Maramorosch and K.F. Harris, pp. 241–292, Academic Press, New York.

Marquenie, D., Lammertyn, I., Geeraerd, A.H., Soontjens, C., Impe, J.F. van, Nicolai, B.M. and Michiels, C.W. 2002a. Inactivation of conidia of *Botrytis cinerea* and *Monilinia fructigena* using UV-C and heat treatment. *International Journal of Food Microbiology*, **74**: 27–35.

Marquenie, D., Michiels, C., Schenk, A., Impe, J. and van Nicolai, B. 2000. Prolonged storage of strawberries by the use of physical methods. *Fruitteeltnieuws*, **13**: 12–13.

Marquenie, D., Schenk, A., Nicolai, B., Michiels, C., Soontjens, C. and Impe J. van 2002b. Use of UV-C and heat treatment to reduce storage rot of strawberry. *Acta Horticulturae*, **567**: 779–782.

Maude, R.B. 1996. *Seedborne Diseases and Their Control Principles and Practice*. CAB International, Wallingford, Oxon, UK.

Menniti, A.M. and Casalini, L. 2000. Prevention of postharvest diseases on cauliflower. *Colture Protette*, **29**: 67–71.

Mercier, J., Arul, J., Ponnampalam, R. and Boulet, M. 1993. Induction of 6-methoxymellein and resistance to storage pathogens in carrot slices by UV-C. *Journal of Phytopathology*, **137**: 44–55.

Mercier, J., Roussel, D., Charles, M.T. and Arul, T. 2000. Systemic and local responses associated with UV-C and pathogen-induced resistance to *Botrytis cinerea* in stored carrot. *Phytopathology*, **90**: 981–986.

Michailides, T.J. and Ogawa, J.M. 1989. Effects of high temperatures on the survival and pathogenicity of propagules of *Mucor piriformis. Phytopathology*, **79**: 547–554.

Nabialek, A., Ben, J.M. and Cisek, B. 1999. Effect of various temperatures and CO_2 and D_2 concentrations on the control of blue mold rot development in 'Shatten morelle' sour cherry. *Acta Horticulturae*, **485**: 281–285.

Nafussi, B., Ben-Yehoshua, S., Rodov, V., Peretz, J., Ozer, B.K. and D'hallewin, G. 2001. Mode of action of hot water dip in reducing decay of lemon fruit. *Journal of Agricultural and Food Chemistry*, **49**: 107–113.

Narayanasamy, P. 2002. *Microbial Plant Pathogens and Crop Disease Management.* Science Publishers, Enfield, Newhampshire, USA.

Neven, L.G., Drake, S.R. and Ferguson, H.J. 2000. Effects of the rate of heating on apple and pear fruit quality. *Journal of Fruit Quality*, **23**: 317–325.

Nigro, F., Ippolito, A., Lattanzio, V., Venere, D.D. and Salerno, M. 2000. Effect of ultraviolet C light on postharvest decay of strawberry. *Journal of Plant Pathology*, **82**: 29–37.

Nyanjage, M.O., Wainwright, H. and Bishop, C.F.H. 1998. The effects of hotwater treatments in combination with cooling and/or storage on the physiology and disease of mango fruits (*Mangifera indica* Linn.) *Journal of Horticultural Science and Biotechnology*, **73**: 589–597.

Palou, L., Crisosto, C.H., Smilanick, J.L., Adaskaveg, J.E. and Zoffoli, J.P. 2001a. Evaluation of the effect of ozone exposure on decay development and fruit physiological behavior. *Acta Horticulturae*, **553**: 429–430.

Palou, L., Crisoto, C.H., Smilanick, J.L., Adaskaveg, J.E. and Zoffoli, J.P. 2002. Effects of continuous 0.3 ppm ozone exposure on decay development and physiological responses of peaches and table grapes in cold storage. *Postharvest Biology and Technology*, **24**: 39–48.

Palou, L., Smilanick, J.L., Crisosto, C.H. and Mansour, M. 2001b. Effect of gaseous ozone exposure on the development of green and blue molds on cold stored citrus fruit. *Plant Disease*, **85**: 632–638.

Palou, L., Smilanick, J.L., Usall, J. and Viñas, I. 2001c. Control of postharvest blue and green molds of oranges by hot water sodium carbonate and sodium bicarbonate. *Plant Disease*, **85**: 371–376.

Palou, L., Usall, J., Muñoz, J.A., Smilanick, J.L. and Viñas, I. 2002. Hotwater, sodium carbonate and sodium bicarbonate for the control of postharvest green and blue molds of Clementine mandarins. *Postharvest Biology and Technology*, **24**: 93–96.

Pao, S., Davis, C.L., Kelsey, D.F. and Petracek, P.D. 1999. Santizing effects of fruit waxes at high pH and temperature on orange surfaces inoculated with *Escherichia coli*. *Journal of Food Science*, **64**: 359–362.

Pennock, W. and Maldonaldo, G. 1962. Hot water treatment of mango fruits to reduce anthracnose decay. *Journal of Agriculture, University of Puerto Rico*, **46**: 272–283.

Phatak, H.C. and Summanwar, A.S. 1967. Detection of plant viruses in seeds and seed stocks. *Proceedings International Seed Testing Association*, **32**: 625.

Plaza, P., Usall, J., Torres, R., Lamarca, N., Asensio, A. and Viñas, I. 2003. Control of green and blue mold by curing on oranges during ambient and cold storage. *Postharvest Biology and Technology*, **285**: 195–198.

Porat, R.B., Lers, A., Dori, S., Cohen, L., Weiss, B., Ddaus, A., Wilson, C.L. and Droby, S. 1999. Induction of chitinase and β-1,3-endoglucanase proteins by UV irradiation and wounding in grapefruit peel tissue. *Phytoparasitica*, **27**: 233–238.

Porat, R., Daus, A., Weiss, B., Cohen, L., Fallik, E. and Droby, S. 2000a. Reduction of postharvest decay in organic citrus fruit by a short hot water brushing treatment. *Postharvest Biology and Technology*, **18**: 151–157.

Porat, R., Pavoncello, D., Peretz, J., Weiss, B., Daus, A., Cohen, L., Ben-Yehoshua, S., Fallik, E., Droby, S. and Lurie, S. 2000b. Induction of resistance to *Penicillium digitatum* and chilling injury in 'Star Ruby' grapefruit by a short hot water rinse and brushing treatment. *Journal of Horticultural Science and Biotechnology*, **75**: 428–432.

Prusky, D., Falik, E., Kobiler, I., Fuchs, Y., Zauberman, G., Pesis, E., Roth, I., Weksler, A., Akerman, M., Ykutiely, O., Waisblum, A., Keinan, A. and Ofek, G. 2003. Hot water brush: a new method for the control of postharvest disease caused by *Alternaria* rot in mango fruits. *V International Mango Symposium. Acta Horticulturae*, **455**.

Prusky, D., Ilag, L., Ackerman, M., Kobiler, I. and Fuchs, Y. 2004. New approaches for the control of postharvest diseases of mango fruits. *Acta Horticulturae*, **645**.

Quimio, A.J. and Quimio, T.H. 1974. Postharvest control of Philippine mango anthracnose by hot water treatment. *Philippine Agriculturist*, **58**: 138–146.

Rennie, W.J. 1993. Control of seedborne pathogens in certification schemes. In: *Plant Health and the European Single Market* (ed.) D. Ebbels, pp. 61–68, BCPC Monograph No.54, BCPC Publications, Farnham, UK.

Retamales, J., Defilippi, B.G., Arias, M., Castillo, P. and Manriquez, D. 2003. High-CO_2 controlled atmospheres reduce decay incidence in Thompson Seedless and Red Globe table grapes. *Postharvest Biology and Technology*, **29**: 177–182.

Rodov, V., Agar, T., Peretz, J., Nafussi, B., Kim, J.J. and Ben-Yehoshua, S. 2000. Effect of combined application of heat treatments and plastic packaging on keeping quality of Oroblanco fruit (*Citrus grandis x C. paradise* Macf.). *Postharvest Biology and Technology*, **20**: 287–294.

Rong, R.F. and Feng, S.Q. 2001. Effect of UV-C light irradiation on ripening and disease infection of postharvest tomato. *Journal of China Agricultural University*, **6**: 68–73.

Ryder, E.J. and Johnson, S.A. 1974. A method for indexing lettuce seeds for seedborne *Lettuce mosaic virus* by air-stream separation of light from heavy seeds. *Plant Disease Reporter*, **58**: 1037.

Saaiman, W.C. 2003. Short-wave infra-red radiation as an alternative to the hotwater bath to control postharvest decay in mangoes. *V International Mango Symposium. Acta Horticulturae*, **455**.

Schaad, N.W., Sitterly, W.R. and Humaydan, H. 1980. Relationship of incidence of seedborne *Xanthomonas campestris* to blackrot in crucifers. *Plant Disease*, **64**: 91–92.

Seo, Y., Saeed, A., Oshita, S. and Kawagoe, Y. 1998. Effect of hot water dipping for control of postharvest microbial disease of pear (*Pyrus communis* L.). In: *Mathematical and Control Applications in Agriculture and Horticulture* (eds) A. Munack and H.J. Tantau, pp. 273–276, Pergamon/Elsevier Science Ltd., Oxford, U.K.

Shahbake, M.A. 1999. Effects of heat disinfestations treatment and modified atmosphere packaging on the storage life of Washington Navel and Valencia oranges. *Iranian Journal of Agricultural Sciences*, **30**: 93–102.

Sharma, R.M. and Singh, R.R. 2000. Harvesting, postharvest handling and physiology of fruits and vegetables. In: *Postharvest Technology of Fruits and Vegetables* (eds) L.R. Verma and V.K. Joshi, pp. 94–147, Indus Publishing Co., New Delhi.

Smilanick, J.L., Sorenson, D., Mansour, M., Aieyabei, J. and Plaza, P. 2003. Impact of a brief postharvest hot water drench treatment on decay, fruit appearance and microbe populations of California lemons and oranges. *HortTechnology*, **13**: 333–338.

Smoot, J.J. and Melvin, C.F. 1963. Hot water as a control for decay of oranges. *Proceedings of the Florida State Horticultural Society*, **76**: 322–327.

Smoot, J.J. and Melvin, C.F. 1965. Reduction of citrus decay by hot water treatment. *Plant Disease Reporter*, **49**: 619–623.

Smoot, J.J. and Segall, R.H. 1963. Hot water as a postharvest control of mango anthracnose. *Plant Disease Reporter*, **47**: 739–742.

Spalding, D.H. and Reeder, W.F. 1972. Postharvest disorders of mangoes as affected by fungicides and heat treatments. *Plant Disease Reporter*, **56**: 751–753.

Stapleton, A.E. 1992. Ultraviolet radiation and plants: Burning questions. *Plant Cell*, **4**: 1353–1358.

Stevens, C., Khan, V.A., Tang, A.Y. and Lu, J.Y. 1990. The effect of ultraviolet irradiation on mold rots and nutrients of stored sweet potatoes. *Journal of Food Protection*, **53**: 223–226.

Stevens, C., Lui, J., Khan, V.A., Lu, J.Y., Wilson, C.L., Igwegbe, E.C.K., Kabwe, M.K., Chalutz, E. and Droby, S. 1998. Application of hormetic UV-C for delayed ripening and reduction of *Rhizopus* soft rot in tomatoes: the effect of tomatine on storage rot development. *Journal of Phytopathology*, **146**: 211–221.

Stevenson, W.R. and Hagedorn, D.J. 1970. Effect of seed size and condition on transmission of *Pea seedborne mosaic virus*. *Phytopathology*, **60**: 1148.

Tanaka, M.A. de S., Maeda, J.A. and Plazas, I.H. de A.Z. 2001. Fungi associated to corn seeds under storage conditions. *Scientia Agricola*, **58**: 501–508.

Teitel, D.C., Aharoni, Y. and Barkai-Golan, R. 1989. The use of heat treatments to extend the shelf life of "Galia" melons. *Journal of Horticultural Science*, **64**: 367–372.

Tian, S.P., Fan, Q., Xu, Y., Wang, Y. and Jiang, A.L. 2001. Evaluation of the use of high CO_2 concentrations and cold storage to control *Monilinia fructicola* on sweet cherries. *Postharvest Biology and Technology*, **22**: 53–60.

Tonini, G., Barberini, K., Bassi, F. and Proni, R. 1999. Effects of new curing and controlled atmosphere storage technology on *Botrytis* rots and flesh firmness in kiwifruit. *Acta Horticlturae*, **498**: 285–291.

Vartanian, V.G. and Endo, R.J. 1985. Survival of *Phytophthora infestans* in seeds extracted from tomato fruits. *Phytopathology*, **75**: 375.

Vincente, A.R., Martínez, G.A., Civello, P.M. and Chaves, A.R. 2002. Quality of heat-treated strawberry fruit during refrigerated storage. *Postharvest Biology and Technology*, **25**: 59–71.

Wallen, V.R. 1964. Host–parasite relations and environmental influences in seed-borne diseases. In: *Microbial Behaviour in vivo* and *in vitro* (eds) H. Smith and J. Taylor, pp. 187–212, Cambridge University Press, Cambridge, U.K.

Waller, J.M. 2002. Postharvest diseases. In: *Plant Pathologist's Pocketbook* (eds) J.M. Waller, J.M. Lenné and S.J. Waller, pp. 39–54, CAB International, Wallingford, Oxon, U.K.

Wang, H., Reitmeier, C.A. and Glatz, B.A. 2004. Inactivation of *Escherichia coli* 0157:H7 and other naturally occurring microorganisms in apple cider by electron beam irradiation. *Journal of Food Protection*, **67**: 1574–1577.

Wells, J.M. 1972. Heated wax-emulsions with benomyl and 2,6-dichloro-4-nitroaniline for control of postharvest decay of peaches and nectarines. *Phytopathology*, **62**: 129–133.

Wells, J.M. and Cooley, T.N. 1973. Control of *Pythium* and *Sclerotinia* rots of snap beans with postharvest hot water and chemical dips. *Plant Disease Reporter*, **57**: 234–236.

Wells, J.M. and Stewart, J.K. 1968. Heat pasteurization and chemical fungicides for control of fursarium rot California cantaloupes. *Plant Disease Reporter*, **52**: 262–264.

Wilson, C.L., El Ghaouth, A., Chalutz, E., Droby, S., Stevens, C., Yu, J.Y., Khan, V. and Arul, J. 1994. Potential of induced resistance to control postharvest diseases of fruits and vegetables. *Plant Disease*, **78**: 837–844.

Winter, W., Bänziger, L., Krebs, H., Rüegger, A., Frei, P. and Gindrat, D. 1998. Alternative methods of control of cereal bunts and barley stripe disease. *Agrarforschung*, **5**: 29–32.

Wszelaki, A. and Mitcham, E.J. 2000. Effects of super atmospheric oxygen on strawberry fruit quality and decay. *Postharvest Biology and Technology*, **20**: 125–133.

Yıldız, F., Yildiz, M., Kinay, P., Sen, F. and Karacali, I. 2001. Effect of some pre-and post-harvest treatments on decay of green mold Satsuma mandarins. *Acta Horticlturae*, **553**: 443–445.

8

GENETIC RESISTANCE OF HOST PLANTS FOR DISEASE MANAGEMENT

Successful management of postharvest diseases caused by microbial pathogens may be possible by adopting different strategies that offer varying degrees of protection. Development of cultivars with built-in resistance has been acknowledged as the most economical and ecofriendly strategy for combating not only field crop diseases, but also for managing postharvest diseases. Improvement of host resistance to postharvest diseases may be achieved through approaches involving the transfer of resistance genes from wild relatives of plant species. Identification of sources of resistance by employing reliable screening methods and utilization of selected sources of resistance in the breeding programs are the basic steps followed for the development of cultivars with resistance to targeted pathogens. It is essential to understand the mechanisms of resistance operating in host plants, so that the genes whose products have a distinct role in the development of host resistance to microbial pathogens causing postharvest diseases are identified. Then the possibilities of transferring such genes to susceptible cultivars can be explored.

The nature of the interaction between plants and microbial plant pathogens is the result of the responses of both organisms. These responses are primarily genetically controlled, although the environment may significantly influence the interaction. Generally, there may be a four-way "yes" or "no" interaction. The host plant may exhibit a resistant (R) or susceptible (r) reaction. On the other hand, the pathogen too may be virulent (V) or avirulent (v) on the particular host plant species. The combination "rV" results in infection and symptom expression. In contrast, the combination "Rv" leads to expression of

Postharvest Pathogens and Disease Management, by P. Narayanasamy

resistance to the pathogen in question. In the latter interaction (Rv), the pathogenic threat is perceived by the host plant tissue and it may react by producing a variety of antimicrobial compounds, strengthening the cell walls through lignification and the formation cellular barriers to restrict the pathogen spread. A quiescent infection may represent a transient resistant response of the host tissue for varying duration depending on the host – pathogen combination, stage of tissue maturity, and environmental conditions (Joyce and Johnson, 1999).

Resistance to diseases caused by microbial pathogens is the ability of plants to withstand or arrest the activities of, or to lessen or overcome the effects of, pathogens. Plant disease resistance has enormous practical value as the foundation for the development of disease management systems. Attempts have been made to integrate evolutionary, genetic, epidemiologic, and economic conceptual frame works for crop disease resistance around an emerging understanding of the molecular, genetic, and physiologic mechanisms of resistance. Frequently, use of a disease resistant cultivar has been demonstrated to be the most desirable disease management strategy, since it can reduce or eliminate the undesirable effects of the application of other cultural, chemical, physical, biological, and regulatory methods.

8.1 MOLECULAR BIOLOGY OF HOST PLANT RESISTANCE

8.1.1 Quiescent Infections

Quiescent infections are commonly observed in a wide variety of fruit, vegetable, and flower crops. The commencement of pathogen quiescence and its maintenance on or within the host implies a dynamic equilibrium between the host, the pathogen, and their environment. The equilibrium may be altered by physical and/or physiological changes in the host and its environment, leading to resumption of pathogen aggression. Quiescence may be imposed on the pathogen at different stages of host – pathogen interaction: spore germination, appressorial formation, penetration, symptomless (latent) internal infection, visible nonexpanding lesions (such as ghost spot in tomato), and finally expanding and sporulating lesions that may coalesce to cover larger areas on fruits and vegetables. In compatible (susceptible) interactions, the pathogen development proceeds without any hindrance, although the immature fruits exhibit different levels of resistance until senescence sets in. In contrast, in an incompatible interaction, the preformed antimicrobial compounds present in the superficial or internal tissues may act on the pathogen preventing the penetration of the epidermis. These compounds constitute the passive defense mechanism of the host plant. As soon as the presence of pathogen propagules is recognized by the surveillance system of the plant, the active resistance mechanism is triggered. Various physiological processes are activated, resulting in the production of various antimicrobial compounds that may inhibit the development of pathogen.

The plant cell wall is divided into three regions – the middle lamella, primary cell wall, and secondary cell wall – based on the composition and design. However, no distinct boundary for these region can be recognized and, indeed, the cell wall is somewhat of a continuum. Pectic polysaccharides present in the middle lamella acts as a cementing material holding different cells together. After the initial defense response offered by passive resistance mechanism is breached by the appressorium and infection structures are formed, fungal pathogens produce pectic enzymes capable of degrading the pectic polymers present in the cell walls, leading to loss of structural integrity and separation and disruption of cells. Pathogens causing postharvest diseases of fruits and vegetables are known to produce one or more pectic enzymes, such as pectin esterase, endo- and exopolyglacturonases, endo- or exopectate lyase, and endopectin lyase. The endopectin lyases are produced exclusively by fungal pathogens, whereas bacterial pathogens produce primarily endopectate lyases (Chesson, 1980; Dennis, 1983). It is considered that the pathogen-produced enzymes have a role in the colonization of host tissues. The requirement of pectate lyase for the infection of avocado fruit by *Colletotrichum gloeosporioides* and polygalacturonase 1 (PG1) by *Botrytis cinerea* to infect tomato fruit was demonstrated by an immunoassay using specific antibodies and a partial gene replacement technique (Wattad et al., 1997; ten Have et al., 1998). Many microbial pathogens causing postharvest diseases are capable of degrading the polysaccharides of the cell wall by pectolytic enzymes that are produced inductively (Bateman and Bashman, 1976). The cell walls of the immature fruits and vegetables may not be able to provide the substrate for the action of these enzymes. Factors and conditions which can delay the production or reduce the activity of the pectolytic enzymes may reduce infection and decay. The presence of proteinaceous inhibitors of pectolytic enzymes in pear and pepper has been reported (Abu-goukh and Labavitch, 1983; Brown and Adikaram, 1982).

The cell-wall degrading enzymes of pathogen origin have a different function in the incompatible (resistant) interactions. They are involved in the generation of signals which constitute the components of the recognition process by which the resistant plant perceives the pathogen's presence on its surface. Perception of the pathogen is a prerequisite for activation of the defense responses. The extracellular enzymes produced by microbial pathogens are not directly recognized by plant cell walls which, rather, respond to the products generated by the enzymatic action. Induction of defense responses are initiated following the recognition of the pathogen by the host plant. Plants may protect themselves against pathogen invasion by utilizing a wide range of defense mechanisms.

8.1.1.1 Mechanisms of Quiescence

A. Nutritional Requirements of Pathogens: A pathogen causing a quiescent infection may be activated by the availability of nutrients following changes in sugar concentrations that occur in parallel with physiological processes induced during ripening. The susceptibility to *Botryosphaeria ribis* of apple fruit infused with sucrose or fructose through the petiole was enhanced (Sitterly and Shay, 1960). In contrast, similar treatment of mango fruits did not result in an increase in susceptibility of resistant fruit tissue to *Alternaria alternata*, although higher levels of sucrose sustained maximal growth of the pathogen in vitro (Droby et al., 1987). The involvement of fruit volatiles, such as ethanol and acetaldehyde formed during fruit ripening, was suggested as a possible mechanism of activation of *Monilinia fructicola* infecting apricots from latent to invasive phase (Cruickshank and Wade, 1992). The development of *Colletotrichum gloeosporioides* in flesh of both unharvested unripe and ripe avocado fruit progressed in a similar manner, suggesting that the nonavailability of nutrients to the pathogen may not be a critical factor in the transformation from quiescent to aggressive phase (Prusky et al., 1984; Kobiler et al., 1994). These results do not provide conclusive evidence for the lack of nutrients functioning as a critical factor for the occurrence of quiescent infections.

B. Preformed Antimicrobial Compounds: The concept of preformed toxic compounds present in unripe fruits being responsible for inhibiting the development of postharvest pathogens has been tested in some pathosystems. Chlorogenic acid, present in the peel of unripe apple fruit, was considered to confer resistance to *Gloeosporium perennans* (Hulme and Edney, 1960). However, the concentration of chlorogenic acid present was much lower than that required to inhibit the pathogen, as determined in vitro. Furthermore, the changes in chlorogenic acid levels did not coincide with changes in resistance/susceptibility to the pathogen (Schultz, 1978; Noble, 1981).

The involvement of preformed antifungal compounds in quiescent infections of green tomato by *Botrytis cinerea* seems to have been more convincingly demonstrated. The glycoalkaloid (saponin) tomatine was detected in the peel of green tomatoes at concentrations sufficient to inhibit the pathogen growth (Verhoeff and Liem, 1975). However, the lesions did not develop even after ripening, although mature fruit did not have detectable concentrations of tomatine (Prusky, 1996). A study on genetically related strains of fungal pathogens with varying levels of tolerance to, or ability to degrade, antifungal compounds provided strong support for the view that preformed antifungal compounds have a role in disease resistance. The mutants of *Nectria haematococca* more tolerant to tomatine were able to develop on green tomato fruit, whereas the wild type was pathogenic only on ripe tomato fruit which had little or no tomatine (De Fago et al., 1983).

The resistance of unripe mango fruit to *Alternaria alternata* was considered to be due to the presence in the peel of a mixture of antifungal compounds consisting of 5–12-*cis*-heptadecenyl resorcinol and 5-pentadecenyl resorcinol at fungitoxic concentrations. Such high concentrations of the five-substituted resorcinols were detected in the peel of several mango cultivars. Further, the depletion of these fungitoxic compounds occurred at a faster rate in the susceptible cultivars during ripening compared with resistant cultivars. In addition, the flesh of unripe mango fruit containing subfungitoxic concentrations of the antifungal compounds were susceptible to infection by *A. alternata*, irrespective of the stage of ripening (Droby et al., 1986, 1987).

Two major antifungal compounds, that is 1-acetoxy-2-hydroxy-4-oxo-heneicosa-12,15-diene and 1-acetoxy-2,4-dihydroxy-*n*-heptadeca-16-ene, were detected in unripe avocado fruit whose resistance to *Colletotrichum gloeosporioides*, causing anthracnose disease, was related to on the concentrations of these compounds (Prusky et al., 1982; Prusky et al., 1991). The diene was more fungitoxic than the monoene and it seems to account for most of the antifungal activity. Susceptible cultivars exhibited more rapid loss of these compounds during ripening. The expression of the gene encoding D12 fatty acid desaturase (*avfad 12*), involved in the biosynthesis of (*Z,Z*)-1-acetoxy-2-hydroxy-4-oxo-heneicosa-12,15-diene (AFD) in unripe avocado fruits was studied. The transcripts of *avfad12* were detected by northern analysis of RNA. The transcripts of *avfad12* gene family, the enzymatic activity of D12 fatty acid desaturase, and the level of AFD in unripe avocado fruits showed transient increases when inoculated with *C. gloeosporioides* or exposed to ethylene (40 μl/l), low temperature (4°C), or 1 mm H_2O_2. In contrast, no significant changes could be detected in ripe fruits subjected to these treatments. Reactive oxygen species was produced in unripe–resistant fruits exposed to H_2O_2. This was found to be an important effect of H_2O_2 on the transcripts for the *avfad12* gene family. The treatments which enhanced D12 fatty acid desaturase activity increased the concentration of AFD precursor, linoleic acid, and its incorporation in AFD, resulting in slower decay development. The results revealed temporal relationships among the production of *avfad 12* transcripts, linoleic acid, AFD content, and quiescence of *C. gloeosporioides* (Beno-Moualem et al., 2004). The conditions such as accelerated lipoxygenase activity and treatment with methyl jasmonate, leading to faster degradation of the diene, resulted in increased fruit susceptibility and decay. Epicatechin, present in the peel of avocado fruit, might act as a competitive inhibitor of lipoxygenase activity. During fruit ripening the concentration of epicatechin decreased, thus allowing acceleration of enzymatic activity and consequent faster degradation of antifungal diene (Prusky and Keen, 1993).

Infection of immature persimmon fruit by *Alternaria alternata* induced limited blackspot symptoms as a result of quiescent infection. Treatment of immature fruit with a preparation of purified cellulase induced similar black spot symptoms. Higher peroxidase activity was observed in immature green fruits following both infection and cellulase treatment. Histological analysis of hyphal development using a genetically transformed pathogen expressing green fluorescent protein suggested that the increase in the activity of a new isoform of peroxidase induced by the directly penetrating *A. alternata* may possibly be involved in the quiescent infection in immature persimmon fruit (Eshel et al., 2002).

Proanthocyanidins (PAs) (flavan-3-ol dimers and oligomomers) in strawberry may act as both antifungal compounds and antioxidants to enhance the quality after storage PAs were extracted from six strawberry genotypes and their ability to inhibit germination and growth of *B. cinerea* was assessed. Strawberry PAs significantly inhibited the mycelial growth and exhibited significant correlation with duration of protection to fruits. The results suggested that PA content could be used as an indicator of the level of resistance of strawberry selections to gray mold and for longer shelf life and better quality of fruits (Hébert et al., 2002).

The presence of preformed antifungal compounds inhibitory to *B. cinerea* causing gray mold disease in strawberry was detected by thin layer chromatography (TLC) bioassays. The antifungal activity remained at a higher level in the tissues of green stage fruit I (7 days after anthesis) and it declined progressively with advancing fruit maturity. Two preformed antifungal compounds, with Rf values 0.44 and 0.37, could be detected in the green stage I fruit but not in white or red stage fruit. The results suggested that the antifungal compounds might be phenolics (Terry et al., 2004).

The antifungal activity of the sweetpotato periderm components was assessed. Resin glycosides isolated from Regal periderm inhibited the development of *Fusarium oxysporum*, suggesting that the periderm components might provide protection against fungal pathogens present in the soil (Harrison et al., 2001). *Rhizopus stolonifer* causes soft rot in sweetpotato roots. The presence of antifungal compounds in the external tissues of sweetpotato roots may offer resistance to infection through shallow injuries. Caffeic acid and 3,5-dicaffeoylquinic acid (3,5-DCQA) were isolated from the fresh interior flesh of four sweetpotato cultivars. 3,5-DCQA was more biologically active with an EC_{50} of 2.2 g/l. These compounds present in the external tissues may protect the sweetpotato against fungal invasion through shallow injuries. Furthermore, it was demonstrated that internal flesh tissues could accumulate antifungal compounds when elicited and incubated under curing condition at 30°C and 90 to 95 RH for 24 h (Stange Jr. et al., 2001).

C. Production of Defense-Related Compounds: In response to infection by microbial plant pathogens, several defense-related compounds are produced at different periods after recognition of the pathogen by the host plant. Both qualitative and quantitative variations in the compounds formed, as well as the temporal differences, are observed depending on the nature of interactions between the host and the pathogen. The inducible mechanisms of defense are of four types: (1) generation of signals (active oxygen species, ethylene, salicylic acid, jasmonates, and elicitors); (2) structural barriers (callose, cell wall proteins, cutin, suberin, lignin, phenolics, and waxes); (3) hypersensitive cell death (hypersensitivity response, HR); and (4) inhibitiors of pathogen growth (phytoalexins and pathogenesis-related proteins). In the case of quiescent infections, observed both in compatible and incompatible interactions, the formation of phytoalexins has been reported in certain pathosystems.

Nectria galligena, a wound pathogen of the apple cultivar Bramley's Seedling, invades the wounds and lenticels and colonizes the invaded tissue to a limited extent before harvest. Production of benzoic acid in the necrotic tissue occurred following the initial invasion. A protease produced by *N. galligena* was shown to be the elicitor of benzoic acid synthesis. The secretion of proteases in vivo by *Diaporthe perniciosa* and *Gloeosporium perennans* infecting apple fruits, has been detected. In contrast, *Penicillium expansum*, *Botrytis cinerea*, and *Sclerotinia fructigena* did not secrete protease and hence they could cause decay in immature fruit without inducing accumulation of benzoic acid. The quiescent infection observed in the cultivar Bramley's Seedling might be due to the production of benzoic acid elicited by protease secreted by the pathogen (Swinburne, 1975; Brown and Adikaram, 1983).

The association of phytoalexin synthesis with quiescent infection was established in the pepper – anthracnose interaction. Induction of quiescence in the pepper (*Capsicum annuum*) fruit was shown to be due to formation of the capsicannol phytoalexins, 1-deoxy capsidion and endesmadienol. In the immature fruit inoculated with *Glomerella cingulata*, capsicannol accumulated whereas in the ripening fruit both capsicannol and capsidiol accumulated. As the lesions expanded, both compounds disappeared. Further, the presence of these compounds in high concentrations could be observed only at the sites of arrested lesions but not in the progressive lesions caused by *B. cinerea* (Adikaram et al., 1982, 1988; Swinburne, 1983).

The compatible and incompatible interactions of *Colletotrichum gloeosporioides*, respectively, with unripe and ripe fruit of pepper (*Capsicum annuum*) were investigated by isolating a defensin gene *j1-1* and a thionin like gene, *PepThi* expressed in the incompatible interaction. During fruit ripening, both genes were developmentally and organ-specifically regulated and differentially induced in the compatible and incompatible interactions. The pathogen rapidly induced the expression of the *PepThi* gene in the incompatible, ripe fruit. On the other hand, salicylic acid induced the pathogen-inducible *PepThi* gene in the unripe fruit. Expression of the *j1-1* gene was induced by jasmonic acid in the unripe fruit. In contrast, its expression was suppressed in the ripe fruit. During fruit ripening both small and cysteine-rich protein genes are induced via different signal transduction pathways (Oh et al., 1999). In a further study, the expression of the gene *PepTLP* encoding a protein homologous to other thaumatin-like proteins was studied. Thaumatin-like proteins (TLPs) are a group of pathogenesis-related (PR)-proteins with antimicrobial properties. Their elicitation in several incompatible (resistant) interactions has been demonstrated. The accumulation of *pepTLP* mRNA and PepTLP protein in the incompatible interaction was higher than that in compatible interaction. The PepTLP protein was localized to the intercellular spaces among cortical cells. *PepTLP* gene seems to have the potential for use as a molecular marker in probing for disease resistance, ripening, and sugar accumulation in nonclimacteric pepper fruits (Kim et al., 2002).

Various types of antimicrobial compounds have been detected in fruits and vegetables. The resistance of carrots to rotting by *B. cinerea* was attributed to the accumulation of phytoalexin compound, 3-methyl-6-methyoxy-8-hydroxy-3,4-dihydroisocoumarin (Goodliffe and Heale, 1977). The potential of carrot roots to accumulate the phytoalexin 6-methoxymellein was progressively reduced with increase in storage time and this factor may be responsible for the general decline in resistance to *B. cinerea* (Goodliffe and Heale, 1978). Green bananas inoculated with *Colletotrichum musae* exhibited a growing necrotic reaction within the peel. In the restricted lesions, the presence of five antifungal compounds that were absent in healthy tissue, could be observed, indicating a role for these newly synthesized compounds in the resistance of green bananas. Induction of antifungal compounds is likely to result in the formation of an antimicrobial barrier (Brown and Swinburne, 1980, 1981). The quiescent infection in citrus was considered to be due to the antifungal compound present in the "woundgum" which accumulated in injured citrus exocarp. This compound, 3-[4-hydroxy-3 (methyl-2-butenyl) phenyl]-2-(*E*)-propenal accumulated progressively with increase in time after injury to the fruit (Stange et al., 1993). Fourteen phenylphenalenone-type phytoalexins, including three newly synthesized compounds with antifungal activity, were isolated from the peel of unripe banana fruits injured and inoculated with *C. musae* (Kamo et al., 1998).

Aspartic proteases (APs) have different physiological functions in animals, fungi, and bacteria. An AP was purified from potato tuber and leaves. The changes in the levels of AP in response to infection by *Phytophthora infestans* and wounding were determined in the intercellular washing fluids from tuber disks of two potato cultivars resistant and susceptible to *P. infestans*. In the resistant cultivar, there was higher and faster induction of AP in infected tissues compared with the wounded control. On the other hand, in the susceptible cultivar, a lower and delayed accumulation of AP occurred compared with the resistant cultivar. The AP directly inhibited the germination of cysts of the *P. infestans* and conidia of *Fusarium solani*. Based on the pattern of accumulation and in vitro activity of AP, a role for this enzyme in the defense response of potato has been suggested (Guevara et al., 2002).

The Jaerla potato tubers contained a peptide designated snakin (StSN2) which caused inhibition of fungal and bacterial pathogens. This peptide caused aggregation of cells of both Gram-positive and Gram-negative bacteria. The *StSN2* gene governing the synthesis of the peptide was developmentally expressed in tubers, stem, flowers, shoot apex, and leaves. The expression of this gene was up-regulated by infection of tubers by *Botrytis cinerea*, but down-regulated by the virulent strain of *Ralstonia solanacearum* and *Erwinia chrysanthemi*. The results provide evidence in support of the hypothesis that StSN2 is a component of both constitutive and inducible defense barriers (Berrocal-Lobo et al., 2002). The wild relative of potato *Solanum bulbocastanum* has been shown to have durable resistance to *P. infestans*. With a view to developing potato cultivars resistant to *P. infestans*, a somatic hybrid (the Wisconsin J series) between potato (*S. tuberosum*) and *S. bulbocastanam* was produced. The hybrid varieties (J 138, J 138A 12, J 101K12, J103K12, and J101K9) showing durable resistance were examined to determine the basis of the disease resistance. The total proteins extracted from these hybrid varieties inhibited the spore germination in vitro, the hybrid J138 exhibiting the maximum percent inhibition. A 40-kD protein from J138 tubers was purified and the purified protein inhibited the spore germination of *P. infestans* by 70%. The N-terminal amino acid sequence of this protein (termed patatin-J) had exactly the same amino acid sequences as patatin, the major storage protein of potato tubers. On incubation of J138 plants with the spores of *P. infestans* patatin was expressed in the stem tissue at 72h after inoculation. (Sharma, et al., 2004).

8.1.2 Resistance to Postharvest Diseases

8.1.2.1 Fungal Diseases Pathogen quiescence represents the early phase in disease development, when a dynamic equilibrium between host, pathogen, and environment is maintained, resulting in the absence of visible symptoms. The pathogen may be activated as the postharvest physiological and biochemical responses of the host change, due to the processes of ripening or senescence. In a compatible interaction, these changes may provide favorable conditions for pathogen development. Simultaneously, the pathogen, kept at low metabolic level during the quiescent stage, may activate the pathogenicity factors, leading to colonization of susceptible tissues and expression of visible symptoms. In contrast, in the incompatible interaction, a wide range of defense-related compounds are synthesized following activation of various defense mechanisms, resulting in restriction of development of the pathogen and consequent absence of the symptoms that are characteristic of the disease in question. The progress of pathogen development may be arrested at various stages, such as spore germination, appressorium formation, penetration, and colonization of host tissues.

A. Passive Mechanisms: Passive or pre-existing defense mechanisms involve structural barriers, such as waxy cuticle or reservoirs of antimicrobial compounds, that are strategically positioned to act on the invading pathogens. Pores present in the grape berry surface are vulnerable sites of infection by *Botrytis cinerea*. The number of pores in the berry skin was highly correlated with susceptibility to gray mold disease. Resistance to *B. cinerea* increased with increase in the amount of wax on the surface of the berries (Percival et al., 1993). Water repellency increased with wax content, resulting in a reduction in pathogen adhesion. Thicker cuticle in the berry was associated with increased resistance to *B. cinerea* (Karadimtcheva, 1981). Resistance of berries to *B. cinerea* increased with increase in thickness and number of epidermal and hypodermal cell layers. Cytokinin application led to rapid cell division, which in turn caused an increase in the total thickness of epidermal and hypodermal layers. These changes ultimately reduced incidence of decay due to *B. cinerea* (Sarig et al., 1998; Gabler et al., 2003). Grapes berries acquire ontogenic resistance at 3 to 4 weeks after bloom, but they are highly susceptible to powdery mildew disease at 1 week after bloom. The development of germinating conidia of *Uncinula necator* on the older, resistant berries was arrested before penetration of the cuticle. Although the cuticle thickness of the berries increased progressively with age, the ingress by the pathogen was stopped even before formation of a visible penetration pore. Phenolic compounds with autofluorescence accumulated rapidly beneath appressoria on highly susceptible berries than on highly resistant berries. The pathogenesis-related gene *VvPR-1* in susceptible berries was highly induced following inoculation compared to resistant berries. On the other hand, in resistant berries, a germin-like protein (VvGLP3) was formed within 16h of inoculation. The results suggested that a preformed physical or biochemical barrier near the cuticle surface, or the rapid synthesis of an antifungal compound, in older berries showing resistance during the early stages of pathogenesis may be the key factor in halting the pathogen invasion (Ficke et al., 2004).

The epidermis of peach lines resistant to brown rot, caused by *Monilinia fructicola*, typically possessed thicker cuticles and greater amounts of epicuticular waxes and/or higher levels of pectin, phenolics, chlorophylls, and other biochemical compounds associated with immature tissue. Structural components, particularly cuticular and epicuticular waxes, were significantly affected by environmental conditions in the growing season (Gradziel et al., 2003).

The role of polygalacturonase inhibiting proteins (PGIPs) in plant defense against fungal pathogens by inhibiting the polygalacturonases (PGs) of pathogen origin was studied. The cantaloupe PGIP (CmPGIP) was isolated from 5 to 15 day postanthesis cantaloupe fruit. CmPGIP inhibited crude extracts of PG from fungal pathogens of cantaloupe. The CmPGIP was differentially expressed during fruit development, as shown by northern blot analysis (Fish and Davis, 2004). The PGIP cloned from strawberry (Fragaria × ananassa) showed a high degree of homology to other fruit PGIPs. The expression of strawberry PGIP in healthy leaves, flowers, and fruit at different maturity stages was tracked by analyzing PGIP transcript levels. Following inoculation of strawberry fruits with *Botrytis cinerea*, the transcript levels in various cultivars with different grades of susceptibility were determined. The constitutive PGIP gene expression was maximum in healthy mature fruits compared to leaves, flowers, and immature fruit, indicating that the gene is developmentally regulated. The level of expression of PGIP in less susceptible cultivars was higher compared with other susceptible cultivars. However, all cultivars showed significant induction of PGIP gene expression to a similar level, suggesting a role for PGIP in the development of resistance to *B. cinerea* in strawberry (Mehli et al., 2004).

Carrot cavity spot disease is mainly due to *Pythium violae* and *P. sulcatum*. The typical cavities in carrot roots caused by *P. violae* were due to the activities of pectate lyase and cellulase. Although *P. sulcatum* also produced these enzymes, the effect was much less severe. Cell wall breakdown was observed near the area of hyphal penetration, as the production of cell wall-degrading enzymes was induced (Campion et al., 1997, 1998). Following pathogen invasion a variety of responses, including production of defense-related proteins capable of protecting the plant tissues, was discernible. Synthesis and deposition of defense-related compounds occurred around the site of infection. The cell walls of the more resistant cultivars were protected by the accumulation of fungitoxic compounds synthesized in response to invasion (Guérin et al., 1998). To study the basis of susceptibility/resistance, changes in the internal structure of carrot and activities of defense-related enzymes were determined. By using scanning electron microscopy, the colonization by *P. violae* of tissues of highly susceptible Bertan, moderately susceptible Narbonne, and less susceptible Bolero and Purple Turkey carrots was visualized. The fungal proliferation was observed in the first 2 days of colonization of Narbonne tissue, but the fungus could not be seen in the lesions by day 7 after inoculation. In the case of Purple Turkey, lesions did not develop, although the pathogen had penetrated the root tissue (Cooper et al., 2004).

B. Inducible Mechanisms: Active defense mechanisms are induced when structural barriers are breached by the invading pathogens, resulting in the restriction of further colonization of plant tissues. Active defense mechanisms may be induced by the microbial pathogens in resistant or nonhost plants. Active defense responses of plant tissues may be of three kinds, that is primary, secondary, and systemically acquired responses. Primary responses are restricted to the cells that are infected or in close proximity with the pathogen and the specific signal molecules presented by the pathogens are recognized by the host cells. As a result, the phenomenon known as programmed cell death (PCD) becomes operative in incompatible interactions. Secondary responses are expressed by adjacent cells in response to diffusible signal molecules (elicitors) released by pathogens. Systemically acquired response is the resultant of the action of defense-related compounds that can be translocated throughout the plant, leading to systemic acquired resistance (Hutcheson, 1998). Molecular recognition events leading to the induction of active defense response against postharvest diseases is discussed.

The inducible mechanisms triggered in resistant plants may result in the generation of signals involving the synthesis of active oxygen species, ethylene, salicylic acid, jasmonates, and elicitors as soon as the presence of the pathogen is recognized. Structural barriers such as callose, cell wall proteins, cutin, suberin, lignin, phenolics, and waxes are formed to restrict the rate of pathogen development. Hypersensitive cell death indicates that metabolic activities of the host are accelerated in response to the activities of the pathogen resulting in necrosis of cells or tissues which contain inhibitory compounds resulting in the localization of the pathogens, especially obligate pathogens, in the dead tissues. Production of phytoalexins and pathogenesis-related (PR) proteins capable of inhibiting the pathogen growth may also be a mechanism of development of resistance in some pathosystems. Most of the defense responses have been demonstrated to result in transcriptional activation of specific resistance genes (Kombrink and Somssich, 1995).

Resistance-related structural modifications and programmed cell death (PCD) in postharvest (detached) and in planta (attached) fruits of pepper plants (*Capsicum annuum* cv. Jejujaerae (susceptible) and *Capsicum baccatum* cv. PBC 80 (resistant), inoculated with *Colletotrichum gloeosporioides*, were studied by using light, confocal laser scanning, and electron microscopy. A significant increase was observed in the thickness of the cuticle layer in the tissues of resistant pepper fruit inoculated without wounding, compared with that of susceptible pepper tissues. Cytological features of PCD were present in the resistant pepper fruit following postharvest inoculation. Cleavage of genomic DNA during programmed cell death (PCD) may yield double-stranded, low-molecular-weight DNA fragments (mono- and oligonulcleosomes) and single-stranded breaks in high molecular weight DNA. dUTP nick labeling is performed to detect these fragments of DNA. Terminal deoxynucleotidyl transferase-mediated dUTP nick end labeling (TUNEL)-positive nuclei were observed in the epidermal and subepiermal cells of inoculated resistant chilli pepper fruit beginning 3 days after inoculation. On the other hand, no TUNEL-positive nuclei could be noted in infected susceptible or healthy resistant postharvest fruits. Furthermore, cell enlargement and cell division during later stages of infection, leading to the formation of a periderm-like boundary layer around the inoculation site in the resistant pepper fruit, were also observed. In addition, nuclear modifications and structural changes of hypersensitivity, including separation of plasma membrane from the cell wall, dilation of the endoplasmic reticulum, accumulation of electron-dense inclusions in vacuoles and cytoplasmic vacuolization accompanying fragmentation of the cytoplasm, were discernible (Kim et al., 2004).

After the entry of pathogens, following penetration of the cuticle, several enzymes are secreted depending on the nature of host – pathogen interaction. *B. cinerea* secreted aspartate proteinase, cellulase, polygalacturonase, pectin methylesterases, and β-glucosidase (Staples and Mayer, 1995). Since most chemical responses are initiated in the skin layers, cultivars with a thicker skin and more skin cell layers may be expected to possess greater capacity to synthesize defense-related compounds and may be able to respond more actively to restrict pathogen invasion. A negative correlation between soluble solids content and resistance of berries to infection by *B. cinerea* was observed (Gabler et al., 2003). Tannin-like compounds from the berry skin inhibited hydroxystilbene-degrading laccase and polygalacturonase from *B. cinerea* (Vail and Marois, 1991; Nair and Gill, 1992). The constitutive berry phenols and catechols were inhibitory to *B. cinerea* laccases (Goetz et al., 1999). But no correlation between catechin content or total phenolics in the skin of mature berries and resistance to gray mold could be established by examining resistant cultivars (Gabler et al., 2003).

Abscisic acid (ABA) is a plant hormone involved in the interaction between plant host and microbial pathogens. Mutants of tomato cv. Moneymaker with reduced ABA levels (*sitiens* plants) showed resistance to gray mold pathogen *Botrytis cinerea*. Exogenous application of ABA enhanced the susceptibility of wild type (WT) plants, whereas the *sitiens* plants became susceptible as the wild type plants. The results indicated that restoration of susceptibility of *sitiens* plants was due to an increase in ABA level. ABA seemed to interact with a functional plant defense response against *B. cinerea*. Infection with *B. cinerea* caused a marked increase in phenylalanine ammonia lyase (PAL) activity in the *sitiens* plants, but not in the WT-plants, indicating that ABA levels in healthy WT-plants partly repressed PAL activity. In addition, *sitiens* plants were more sensitive to treatment of roots with benzol (1,2,3) thiadiazole-7, carbothioic acid, resulting in the production of pathogenesis-related (PR) proteins. The threshold values for PR-1a gene expression declined by a factor of 10 to 100 in *sitiens* plants compared with WT plants. ABA seems to negatively modulate the salicylic acid (SA)-dependent defense pathway in tomato (Audenaert et al., 2002).

In carrot, infection by *Pythium violae* causes cavity spot lesions which are brown in color, indicative of the presence of oxidized phenolics in the lesion. A three-fold increase in PAL activity within 2 weeks after inoculation was a characteristic feature and this increase in enzymatic activity was considered to result in deposition of lignin around the lesion and accumulation of phenols, thus offering physical and chemical barriers to restrict pathogen spread (Zamski and Peretz, 1996). In cultured carrot cells following elicitation with heat-stable components of fungal mycelia from *Chaetomium globosum*, significant amounts of phenolic compounds (*p*-hydroxy benzoic acid, caffeic acid, and ferulic acid) were formed along with a corresponding increase in PAL and chitinase activity in elicited cells (Kurosaki et al., 1987). In a later study, four carrot cultivars with different levels of susceptibility were inoculated with *P. violae*. Examination of the activities of enzymes during in vitro colonization showed that PAL and chitinase activities were low during the first 7 days after inoculation of cvs. Bolero (less susceptible) and Bertan (highly susceptible). Peroxidase and β-glucosidase acitivity in Bolero increased briefly at 3 days after inoculation, but the activities generally remained at low level uniformly. In Purple Turkey (least susceptible), the enzymatic activities registered significant increases and were at higher levels generally. It was suggested that the small cell size within the root and higher constitutive levels of enzymes may enhance the levels of resistance in Purple Turkey carrot (Cooper et al., 2004) ((Appendix 8(i)).

A two-component system of development of resistance in grapes to fungal pathogens has been described. Macerating enzymes produced by fungal pathogens infecting berries may be inhibited by proanthocyanidins (PPRA) that are located in berry skin (Hill et al., 1981). The quiescence and development of *Botrytis cinerea* in artificially inoculated berries of grape cultivars Gamay (susceptible) and Gamaret (resistant) were studied. The biochemical parameters, such as constitutive and induced antifungal compounds, polymeric PPRA, and lipid peroxidation products were quantified as markers of senescence. The concentration and mean degree of polymerization of PPRA were always higher in berries of the resistant variety. The inhibitory effect of

Gamaret PPRA on the enzyme activity remained until harvest, whereas Gamay PPRA lost their inhibitory activity at the beginning of véraison. The results indicate that resistance to *B. cinerea* appears to be linked to the maintenance of the pathogen in its latent form in the berries, primarily due to the ability of Gamaret PPRA to inhibit macerating fungal enzyme activities (Pezet et al., 2003).

The involvement of the phytoalexin resveratrol in the resistance of grape berries to *B. cinerea* has been suggested. Large concentrations of the stilbene, resveratrol, were synthesized in immature berries which showed resistance to *B. cinerea*. As the synthesis of reveratrol progressively reduced with maturity, the berries became susceptible to the pathogen (Langcake and McCarthy, 1979). The defense reaction in grape flowers inoculated with *B. cinerea* was studied. In surface-sterilized pea-size grape berries, latent *B. cienerea* was detected predominantly in the receptacle area. Stilbene stress metabolites in the flowers were estimated by high performance liquid chromatography (HPLC) techniques. Resveratrol accumulated mainly after prebloom and full-bloom inoculation, but it did not prevent infection of the susceptible cultivar (Gamay). Another metabolite, ε-viniferin, was present in the necrotic tissues only, but no change in levels of piceid could be noted. On the other hand, pterostilbene and α-viniferin were not detectable at all. Constitutive soluble phenolic compounds, primarily derivatives of quercetin and hydroxyl-cinnamic acid, were present in high concentrations in the calyptra but at low levels in the receptacle area. The results suggest that stilbenes may have a limited role in protecting the grape flowers from infection by *B. cinerea* and that ε-viniferin may be a by product, rather than a deterrent, of infection (Keller et al., 2003). The *trans* and *cis*-resveratrol were induced in grapes inoculated with *B. cinerea* at 2 days prior to sampling. Resveratrol content was not correlated with resistance exhibited by cultivars to varying degrees. No resveratrol was induced in the berry skin of highly resistant grapes following inoculation with *B. cinerea*. The finding that laccase produced by *B. cinerea* would detoxify stilbene phytoalexins diminish the importance of stilbenes in relation to pathogen resistance (Pezet et al., 1991; Staples and Mayer, 1995). Thus the resistance of grapes appears to be associated with the balance between the ability of the host plant to produce phytoalexins and the ability of the pathogen to detoxify the inhibitory compounds (Sarig et al., 1997; Gabler et al., 2003).

Nonspecific lipid transfer proteins (nsLTPs) are small, basic, cystein-rich proteins considered to be involved in plant defense mechanisms. Three cDNAs coding *nsLTPs* genes were studied in 41B-rootstock grape cell suspension in response to defense-related signal molecules. Ergosterol (a fungus-specific sterol) and a proteinaceous elicitor purified from *B. cinerea* induced a strong and rapid accumulation of *nsLTP* mRNAs, whereas other signal molecules, jasmonic acid (JA), cholesterol, and sitosterol, induced accumulation of *nsLTP* mRNA to a lesser degree. Salicylic acid (SA) had no effect on the formation of specific mRNA. Of the four LTP isoforms only the P4 isoform reduced the mycelial growth of *B. cinerea* in vitro in calcium-free medium (Gomés et al., 2003).

Elicitors may be components of the microbial cell surface or they may be released by the enzymatic activity of the host. Ethylene synthesis is induced by pectolytic enzymes and their products in citrus and tomato fruits (Baldwin and Biggs, 1988; Baldwin and Pressey 1988, 1990). Ethylene accelerates fruit ripening/senescence and the consequent increase in the susceptibility of fruit to decay. In contrast, ethylene may also induce defense responses, as seen in the reduced growth of lesions caused by *Penicillium italicum* and *P. digitatum* in citrus exposed to ethylene (Brown, 1973; El-Kazzaz et al., 1983). Treatment of grapefruit with 1-methyl cyclopropene (1-MCP), an irreversible inhibitor of ethylene action, did not influence the disease development, although production of ethylene was enhanced in the absence or presence of *P. digitatum* (Mullins et al., 2000).

The involvement of chitinase and glucanase in resistance of grapefruit to infection by fungal pathogens has been indicated. Young grapefruits with high levels of chitinase and low levels of glucanase activities were found to be resistant to *P. digitatum*. As the chitinase levels decrease with increase in activity of glucanase during ripening, the fruits became increasingly susceptible (McCollum et al., 1997). The increase in the concentrations of chitinase proteins induced by ultraviolet irradiation was reported to be responsible for the development of resistance in UV-treated fruits (Porat et al., 1999), suggesting a role for the chitinases in activation of defense mechanism in fruits (Mc Collum, 2002).

Contributions of ethylene, jasmonate, and salicylate to the development of resistance of tomato to *Botrytis cinerea* were investigated. Treatment of plants with ethylene reduced susceptibility to *B. cienrea*, whereas pretreatment with 1-MCP increased the susceptibility. Expression of several PR-protein genes was induced by pretreatment with ethylene. Protein inhibitor I expression was repressed by ethylene, but induced by 1-MCP. Induction of resistance in the mutant 'Neverripe' following exposure to ethylene was also observed. Similar effects were seen also in the mutant 'Epinastic', constitutively activated in a subset of ethylene responses, and a transgenic line producing negligible ethylene. The mutant 'Defenseless' impaired in jasmonate biosynthesis showed increased susceptibility to *B. cinerea*. A transgenic line with reduced prosystemin expression also responded in a similar manner, whereas a prosystemin overexpressing transgenic line was highly resistant. Salicylate and ethylene acted synergistically in defense gene expression, but antagonistically in the development of resistance to *B. cinerea* (Díaz et al., 2002).

The effect of continuous exposure to exogenous ethylene of table grapes and stone fruits during long-term cold storage on the development of postharvest diseases was assessed. The incidence and severity of brown rot in stone fruits (peach, plum, nectarine, and apricot) due to *Monilinia fructicola* were increased by constant ethylene treatment. However, ethylene exposure did not alter gray mold nesting ability on table grapes artificially inoculated with

Botrytis cinerea. Further, none of the external and internal quality characteristics showed any significant change following exposure to ethylene in all fruits except the enhancement of flesh softening in apricot and flesh mealiness in peaches cv. Elegant Lady (Palou et al., 2003; Palou and Crisosto, 2003).

The defense responses of grapevine following inoculation with *B. cinerea* were assessed by studying: (1) phenylalanine ammonia lyase (PAL) and stilbene synthase (StSy); (2) an acid chitinase (VCH3) and a basic chitinase (VCHIT1b); and (3) a polygalacturonase inhibiting protein (PGIP). In infected leaves the expression of *PAL, StSy*, *PGIP*, and *VCH3* genes was observed 6h post inoculation, whereas there was a delay in the increase in the expression of *VCHITIb* gene. In berries, *VCH3* expression could not be detected, whereas *PAL* and *PGIP* genes were induced to the maximum extent at the stage when loss of berry color occurred and *StSy* and *VCHI* 1b genes were induced still later. The results showed that activation of these genes was not enough to restrict the spread of *B. cinerea* (Bézier, 2002). The expression of pathogenesis-related (PR) proteins and their effect on fungal pathogens of grapevine *Botrytis cincerea, Uncinula necator*, and *Phomopsis viticola* was assessed. Two proteins, identified as PR-proteins by immunological techniques and by N-terminal sequencing as osmotin and thaumatin-like protein, strongly inhibited the spore germination and germ tube growth of *B. cinerea* and *P. viticola.* The proteins showed synergistic effect. The accumulation of these two PR-proteins in infected leaves and berries, as revealed by immunoblots, indicated a role in the development of resistance in grapevine against invasion of fungal pathogens (Monteiro et al., 2003).

The involvement of phenolic compounds in disease resistance has been suggested in many pathosystems. Periderm and cortex tissues of 14 genetically diverse sweetpotato (*Ipomoea batatos*) clone were grown under low stress conditions. The scopoletin (7-hydroxy-6, methoxycoumarin) and scopolin (7-glucosyl scopoletin) content of the stressed tissues were estimated. The effect on the growth of sweetpotato pathogens *Rhizopus stolonifer*, *Lasidiplodia theobromae*, and *Fusarium oxysporum* f.sp. *batatos* was assessed. The glycoside scopolin showed moderate inhibition of growth of *F. O. batatas.* On the other hand, the aglycone scopoletin inhibited the growth of all pathogenic fungi at concentrations present in some sweetpotato clones (Peterson et al., 2003).

Active oxygen is produced and metabolized as part of several processes during normal growth. Production of active oxygen is enhanced by stress, including invasion of plants by various pathogens. Active oxygen species are considered "active", because they react with other molecules without the input of energy. The active oxygen species generated within plants result either from the excitation of an outer electron, forming singlet oxygen ($^1O_2$), or from the successive addition of electrons to molecular oxygen yielding superoxide (O^-_2), hydrogen peroxide (H_2O_2), and the hydroxyl radical ($OH^\cdot$) respectively. During periods of pathogen infection, active oxygen production generally increases and exceeds the scavenging capacity of the plant and pathogen, resulting in increased levels of active oxygen and the deleterious expenditure of energy to nullify the oxidative stress. During early phases of pathogenesis, active oxygen seems to aid in the activation of defense mechanism. In contrast, active oxygen products may be a result of, and contribute to, tissue degeneration during the later stages of pathogenesis.

Production of active oxygen may be one of the first quantifiable responses as the contact between plant cells and pathogen is established. This is referred to as a "respiratory or oxidative burst." As the process of pathogen recognition is accomplished signal transduction events are triggered, and the mechanisms responsible for active oxygen production are activated. Of the various roles suggested for active oxygen, its antimicrobial activity would depend on the scavenging ability of the pathogen. The oxidative reinforcement of the cell wall, involving cross-linking of proteins and lignin production, may be another direct effect of active oxygen. If H_2O_2 is added to plant cells, enhancement of gene transcription of several defense-related processes, in addition to antioxidant mechanisms, has been observed. The involvement of active oxygen in a cascade of events leading to the hypersensitive response and systemic-acquired resistance has also been indicated in some pathosystems. On the other hand, active oxygen may have a role in the tissue degeneration through membrane peroxidation and chlorophyll breakdown (Baker, 2001).

Apples cv. Golden Delicious, picked 7 days before (harvest 1) or after (harvest 2) the commercial harvest time, were wounded or inoculated with *Penicillium expansum*. Fruit from harvest 1 showed less disease incidence and severity compared with fruit from harvest 2. A significant increase in H_2O_2 levels with concomitant increase in activity of superoxide dimutase (SOD) was observed in fruit from harvest 1. No variation was seen in the activities of H_2O_2-scavenging enzyme catalase and unspecific peroxidase. In contrast, in the fruit from harvest 2, with greater susceptibility to *P. expansum*, activities of catalase and peroxidase registered significant increases following wounding and inoculation, whereas H_2O_2 levels and SOD activity were similar to those in control fruit, which were wounded but not inoculated. The results indicated that H_2O_2 and associated metabolism induced by wounding may have a role in the development of resistance in apple fruits against postharvest pathogens (Torres et al., 2003).

The biochemical mechanisms involved in resistance in potato to *P. infestans* were investigated. The oxidative burst and lipid peroxidation, induced by elicitor from the pathogen in suspension-cultured cells of *Solanum* genotypes that differed in the type and level of resistance, were studied. In all genotypes tested, the elicitor-induced reactions were the same, but they showed variation in respect of kinetics and intensity. The reaction of *S. tuberosum* cv. Buzura and Clone H-8105, polygenically resistant and susceptible, respectively, to *P. infestans* were tested by using the culture filtrate of the pathogen. In both genotypes, significant increases in lipoxygenase (LOX) activity were delayed and unaccompanied by changes in the level of lipid peroxidation. In contrast, lipid peroxidation increased in the nonhost *S. nigrum* and this increase coincided with enhanced LOX activity. The results suggested that both production of reactive oxygen species and LOX activity may be involved in the defense strategy in the *Solanum* – *P. infestans* pathosystem (Polkowska-Kowalczyk et al., 2004).

The extracellular, 10 kDa proteins produced by most species of *Phytophthora* are collectively known as elicitins. They show structural similarity and induce a hypersensitive reaction (HR) in specific plant species. *Phytophthora infestans*, causing late blight disease of potato, secretes an elicitor, INF1. Expression of the *inf1* gene was down-regulated in planta during the biotrophic stage and the highest levels of expression were recorded in the late stages of infection, when profuse sporulation and necrosis occurred. The cDNAs of two other genes *inf2A* and *inf2B* were isolated from a cDNA library made from potato tissues infected by *P. infestans*. Based on the multiple sequence alignments and phylogenetic analysis, 19 elicitins were grouped into five distinct classes within the elicitor family (Kamoun et al., 1997b). Resistance to *P. infestans* may not be mediated by a defense response elicited by INF1 in the genus *Solanum* (Kamoun et al., 1997a).

When the tuber disks of Irish Cobbler potato were treated with a fungal elicitor composed of hyphal wall components from *P. infestans*, synthesis of a 51-kDa protein kinase (p51-PK) occurred. Furthermore, SA and arachidonic acid were also able to activate a protein kinase with similar migration as p51-K on sodium dodecyl sulfate – polyacrylamide gel electrophoresis (SDS-PAGE), indicating the possible involvement of p51-PK in the induction of defense responses of potato (Katou et al., 1999).

The activity of lipoxygenase (LOX) has frequently been detected during pathogen-induced defense responses. The potato *LOX* gene, designated *POTLX-3*, showed greatest sequence similarity to the tobacco LOX gene that was earlier reported to contribute to a resistant mechanism against *Phytophthora parasitica* var. *nicotianae. POTLX-3 mRNA* accumulated following treatment with ethylene or methyl jasmonate or inoculation with virulent or avirulent strains of *P. infestans*. During the resistant response, *POTLX*-3 was induced within 6 h, registering steady increase through 24 h, and accumulation of its mRNA continued for a week after inoculation. On the other hand, inconsistent and delayed accumulation of mRNA occurred in a susceptible reaction. Inoculation of potato with the nonpathogen *Pseudomonas syringae* pv. *phaseolicola* also resulted in the induction of *POTLX3* during development of a hypersensitive response, due to the incompatible interaction. The results indicated the specific involvement of POTLX-3 in defense responses against pathogen infection (Kolomiets et al., 2000).

8.1.2.2 Bacterial Diseases Immature, unripe fruits exhibit resistance to microbial pathogens. As the ripening or senescence process commences, there is a progressive increase in their susceptibility to pathogens. Ethylene evolution accelerates ripening of harvested produce. The progression of disease symptoms induced by bacterial pathogens is associated with the production of ethylene. A tomato mutant impaired in ethylene perception – Neverripe –

showed significant reduction in the development of disease symptoms in comparison with wild type after inoculation with bacterial pathogens *Xanthomonas campestris* pv. *vesicatoria* (*Xcv*) and *Psueodmonas syringae* pv. *tomato* (*Pst*). Development of disease symptoms was also hampered in tomato genotypes impaired in ethylene synthesis. The reduction in disease symptoms in the mutant Neverripe plants was a specific effect of ethylene insensitivity, but was not due to reduction in bacterial population or decreased ethylene synthesis. As PR-1B1 mRNA accumulation in response to *Xcv* infection was not influenced, induction of this defense gene (coding for a PR-protein involved in disease resistance) was not dependent on ethylene evolution. It is suggested that by engineering ethylene insensitivity, tolerance to microbial pathogens may be incorporated in tomato (Lund et al., 1998).

In another study, the ripening process of tomato fruit was followed. The Stay-green mutant, green flesh, was deficient in the chloroplast machinery and displayed a different ripening profile. In the wild type, the ripening process is characterized by a substantial induction in glutamate dehydrogenase (GDH, EC-1.4.1.3) and a reduction in glutamine synthetase (GS, EC 6.3.1.2) levels, associated with an increase in the relative glutamine content. The total free amino acid content of green flesh mature fruits decreased during ripening, in contrast to the trend observed in several cultivars of tomato. Although GDH activity and polypeptides were detected in green gf fruits, GDH transcripts could not be detected in these fruits. The results indicated that the expression of the enzymes involved in the primary glutamate metabolism appeared to be differently regulated during ripening process (Bortolotti et al., 2003). Such variations in the physiological process occurring during ripening may possibly account for the resistance status of the fruits to microbial invasion.

Lipopolysaccharides (LPS) form an ubiquitous, indispensable component of the cell surface of Gram-negative plant pathogenic bacteria and they have been shown to have diverse roles in the bacterial pathogenesis of plants. LPS may form a protective coating and contribute to the exclusion of plant-derived antimicrobial compounds, thus promoting the pathogenic potential of the bacterial pathogen. On the other hand, LPS may be recognized by plants and directly trigger some of the plant defense-related responses. LPS may induce defense-related responses which have been found to be weak in many plant – pathogen interactions. The resistance of pepper (*Capsicum annuum*) to incompatible strains of *Xanthomonas axonopodis* (*campestris*) pv. *vesicatoria* (*Xcv*) or *X. campestris* pv. *campestris* (*Xcc*) was due to enhanced synthesis of the hydroxycinnamoyl tyramine conjugates, feruloyl tyramine (FT) and coumaroyl tyramine (CT). In contrast, in response to compatible strains of *Xcv*, only trace amounts of FT and CT were produced. Treatment of incompatible tissues with LPS induced the response more rapidly (within 4 h) and the potentiated state was maintained for at least 38 h. The expression of other defense responses, such as transcription of genes encoding acidic β-1,3 glucanase, was also potentiated by LPS treatment, indicating a wider role for LPS beyond its limited activity as a direct inducer of plant defenses against bacterial infections (Newman et al., 2002). LPS may also sensitize plant tissue to respond more rapidly by: (1) an accelerated synthesis of hydroxycinnamoyl lyramine conjugates; (2) the expression patterns of genes coding for some PR-proteins; and (3) prevention of the hypersensitive reaction caused by avirulent bacteria. The signal transduction mechanisms through which LPS triggers these responses have to be studied thoroughly to assess the contribution of LPS signaling in the development of resistance in plants infected under natural conditions (Erbs and Newman, 2003).

Recognition of bacterial plant pathogens is the key factor in the ability of plants to defend themselves against pathogen invasion. Plants have to recognize a wide array of pathogen-derived molecules, including different oligosaccharides, lipids, peptides, and proteins. Identification of the receptors for the various elicitors in the plant cell surface has been a major challenge to researchers. Biochemical studies have helped in the characterization of some elicitor-binding proteins that may form a component of the recognition complex. In the process of pathogen perception, transmembrane receptor-like protein kinases (RLKs) constitute one of the most likely groups of receptors. Some of the serine/threonine kinases have been identified as resistance or R genes, others as induced by pathogens or elicitors. One of the RLKs belonging to a leucine-rich repeat (LRR) class of putative receptor kinases has been shown to be a receptor for bacterial flagellin and the underlying signal pathway resulting in the activation of defense genes was elucidated (Montesano et al., 2003).

The interaction of extracellular polysaccharides (EPS) of *Clavibacter michiganensis* subsp. *sepedonicus* (*Cms*), causing potato ring rot disease, with protoplasts of isolated host plant cells and microsomal fractions obtained from cell suspension cultures of resistant and susceptible cultivars was studied. The EPS firmly bound to protoplast surfaces and microsomal membranes of a susceptible cultivar. In contrast, no binding of EPS with protoplast surfaces of microsomal membranes of a resistant cultivar occurred. The results suggested that: (1) many receptors to EPS *Cms* were present in the plasma membranes of cells of susceptible potato cultivars; (2) these receptors may contain proteinaceous sites exposed on the external side of the plasma membrane which participate in EPS binding; and (3) the plasma membranes of cells of resistant cultivars may contain a small, but adequate quantity of receptors of EPS able to induce defense responses in potatoes (Romanenko et al., 2003).

A potato gene encoding a putative WRKY protein was isolated from a cDNA library enriched by suppression hybridization for sequences upregulated 1h postinoculation with *Erwinia carotovora* subsp. *atroseptica* (*Eca*). WRKY proteins comprise a family of plant-specific zinc-finger-type factors implicated in the regulation of genes associated with pathogen defense. In vitro, these proteins bind specifically to functionally defined TGAC-containing W box promoter elements of the pathogenesis-related class 10 (PR-10) genes. The cDNA encoded a putative polypeptide of 172 amino acids, containing a single WRKY domain. *St-WRKY1* was strongly induced by *Eca* culture filtrate, indicating this gene to be an elicitor-induced gene. But salicylic acid, methyl jasmonate, ethylene, or wounding did not induce this gene. However, the treatment with culture filtrates from recombinant *Escherichia coli* containing plasmids expressing *Eca* pectate lyase genes, *pelB* and *pelD*, resulted in the upregulation of *St-WRKY1*. The results suggest that either proteins encoded by these genes or oligogalacturonides generated by their activity may elicit a potato defense pathway associated with *St-WRKY1* (Dellagi et al., 2000).

The virulence determinants harpin (HrpN) and polygalacturonase (PehA) from hrp-positive strain of the softrot pathogen *E. carotovora* subsp. *carotovora* (*Ecc*) were employed as tools to elucidate plant responses. In the nonhost *Arabidopsis*, establishment of resistance by HrpN was accompanied by the expression of salicylic acid (SA)-dependent, but also jasmonate/ethylene (JA/ET)-dependent, marker genes PR1 and PDF1.2, respectively. Apparently both SA-dependent and JA/ET-dependent defense pathways were activated. These two elicitors, HrpN and Peh, also cooperated in triggering increased production of superoxide and lesion formation (Kariola et al., 2003).

Host cells in intimate association with, or infected by, a pathogen may respond with a form of programmed cell death (PCD) to different degrees in compatible and incompatible interactions. The HR is a macroscopic manifestation of pathogen-induced PCD. Initiation of the HR process commences with recognition of the pathogen by the host plant, which is primarily mediated by the *avr* genes and the plant resistance (R) genes. This is followed by complex signal transduction pathways involving changes in protein phosphorylation, production of reactive oxygen species, and modification of ion fluxes. It is likely that many of the responses observed in plant tissues inoculated either with fungal or bacterial pathogen or elicitor molecules may be due to signal molecules produced by pathogen-induced PCD. Salicylic acid and/or active oxygen have been suggested as signal molecules (Gilchrist, 1998).

Evidence suggesting that resistance of tomato and pepper to T3 strains of *X. campestris* pv. *vesicatoria* (*Xcv*) was specified by a plant-inducible avirulence (*avr*) gene was presented by Astua-Monge et al. (2000). Hypersensitive response in certain tomato and pepper genotypes was elicited by tomato race 3 (T3) of *Xcv*. The avrXv3 gene encoding a 654-bp open reading frame (ORF) was sequenced and it was found to be plant inducible and controlled by the hypersensitivity and pathogenicity (*hrp*) regulatory system. AvrXv3 had transcription activation activity in yeast. The *Agrobacterium*-mediated transient expression confirmed that AVrXv3 had a direct role in eliciting the HR in tomato (NIL 216), providing support to the view that Avr proteins must be located inside the host plant cell to trigger the HR. Another avirulence gene, designated avrXv4, was isolated from *Xcv* strain ME 90, virulent on *Lycopersicon pennelli*. The predicted AvrXv4 showed high homology with a family of bacterial proteins from plant and mammalian pathogens comprising Avr-Rxv, Avr-BsT, and AvrA. (Astua-Monge et al., 2000). Specific cDNAs showing differential expression were isolated from a pepper cDNA library from HR lesions caused by an avirulent strain of *Xcv*. Ten *Capsicum annuum*-induced (*CAI*) genes encoding putative thionin, lipid transfer protein I and II, osomotin (PR5), class I chitinase, β-1,3-glucanase, SAR8.2, stellacyanin, leucine-rich repeat (LRR) protein, and auxin-repressed protein were identified. Transcripts of the *CAI* genes were strongly or preferentially induced in pepper tissues by infection with *Xcv* or *Phytophthora capsici* and by abiotic elicitor treatment. Etephon and methyl jasmonate strongly induced most of the *CAI* genes in pepper tissues (Jung and Hwang, 2000).

Recognition of the pathogen effector proteins AvrPto or AvrPtoB of *Pseudomonas syringae pv tomato*, to be the pathway of development of resistance conferred by the tomato Pto kinase. Pto-mediated resistance depends on multiple signal transduction pathways and many defense responses, including an oxidative burst, rapid changes in the expression of over 400 genes, and localized death. The role in Pto-mediated resistance in tomato of 21 genes from other plant species involved in defense-related signaling was studied. Virus-induced gene silencing suppressed the expression of the genes tested. Pto-mediated resistance was compromised by silencing of genes encoding two mitogen-activated protein (MAP)-kinases, MEK1 and MEK2, and a key regulator of systemic acquired resistance, NPR1, in addition to two transcription factors. The results indicated that many defense signaling compounds are conserved among diverse plant species (Ekengren et al., 2003).

The resistance (R) gene, *Bs4*, in tomato, specifies recognition of *Xanthomonas campestris* pv. *vesicatoria* (*Xcv*) strains which express cognate Avr Bs4 avirulence protein. A nucleotide-binding leucine-rich repeat (NB-LRR) protein was the predicted protein encoded by *Bs4*. Inoculation of tomato with *Xcv* showed that *Bs4* conferred perception of AvrBs4, but not the AvrBs3 protein, which shares 97% homology. Nevertheless, when both *avrBs4* and *avrBs3* were introduced by *Agrobacterium* T-DNA-mediated transfer, both avirulence genes triggered a *Bs4*-dependent hypersensitive response. The results indicated that the naturally occurring AvrBs3 homologues may be employed to determine recognition specificity (Schornack et al., 2003).

Inoculation of pepper with a nonpathogenic *hrcC* mutant of *X. campestris* pv. *vesicatoria* induced localized deposition of phenolics and callose in papillae and alterations to the plant cell wall leading to increased electron density. Elicitors of callose deposition and other wall changes could be isolated from the *hrcC* mutant and the elicitor activity was attributed to the presence of high molecular weight lipopolysaccharides (LPS). In contrast, wild type *Xcv* suppressed the induction of structural changes induced by LPS. It was suggested that effector proteins produced by phytopathogenic bacteria and delivered by the type III secretion system may have a pivotal role in suppressing the basal defense responses activated by bacterial LPS which lead to restricted multiplication of nonpathogens such as *hrp* mutants (Keshavarzi et al., 2004).

The resistance to diseases conferred by major R genes is not durable. They become ineffective when the pathogens are able to form new strains, races, or biotypes with virulence to nullify the effects of R genes. In order to overcome thc lack of durability of resistance to bacterial diseases conferred by major *R* genes, the efficacy of an inducer of systemic acquired resistance, acibenzolar-S-methyl (ASM), was assessed. Pepper plants carrying one or more *R* genes were sprayed with ASM and challenged with incompatible strains of *Xanthomonas campestris* pv. *vesicatoria*, under greenhouse conditions. ASM-treated plants with *R* genes developed significantly fewer lesions induced by both incompatible and compatible responses than on nonsprayed plants. The disease lesions yielded race-change mutants on isolation. Under field conditions also, a reduction in disease severity was observed and such a decrease was associated with a reduction in the number of race-change mutants. The results seem to indicate that inducers of systemic acquired resistance may be useful for enhancing the durability of genotype-specific resistance conferred by major *R* genes (Romero and Ritchie, 2004).

8.1.2.3 Viral Diseases Viral proteins may provide, during infection, the stimuli that may be recognized as foreign in the host cell, resulting in the elicitation of a defense response in the incompatible (resistant) interactions. Three protein products of *Tobacco mosaic virus* (TMV) have been reported to function as avirulent determinants in the gene for gene interactions between virus strains and host genotypes. The replicase protein, coat protein (CP), and movement protein (MP) of TMV acted as avirulent determinants in various host plant species (Saito et al., 1987; Padget and Beachy, 1993; Webber et al., 1994). The structural property, but not the enzymatic activity, of viral proteins seems to determine their ability to function as elicitors. The three-dimensional structure of the CP of TMV generated avirulence activity (Culver, 1996). The resistance of potato cultivars with the *Rx* gene was elicited by a threonine residue at position 121 in the CP of *Potato virus X* (PVX) (Goulden and Baulcombe, 1993). The nature of the CP determines whether PVX isolates are virulent or avirulent on Rx potato cultivars. The CP gene of PVX has also been shown to have an exclusive effect on the nature of interaction between potato cultivars with the *Nx* resistence gene. The CP gene of the isolate of PVX in question may act as a determinant of avirulence and initiate the sequence of molecular events resulting in HR on cultivars carrying the incompatible resistance gene (Santa Cruz and Baulcombe, 1993).

The constitutive resistance of a clone of the diploid potato species *Solanum phureja* cv. Egg Yolk, Clone 5010 to *Potato leafroll virus* (PLRV) was shown to be due to strong resistance to accumulation of PLRV determined by a quantitative ELISA. The resistance is expressed very strongly in leaf tissue but less strongly in petiole and stem tissue of infected plants. The concentration (titer) of PLRV reached in leaves of the resistant clone was just 2% or less of the virus titer in susceptible plants (cv. Maris Piper). The plants of *S. phureja* transformed with the CP gene of PLRV did not show a greater level of resistance to PLRV than nontransformed plants, when inoculated using viruliferous aphids (Franco-Laro and Brker, 1999). The responses of 27 potato cultivars, including Desirée, to eight isolates of *Cucumber mosaic virus* (CMV) were determined. Primary symptoms, such as necrosis and/or malformation of leaves developed on only 17 cultivars, although all cultivars were infected by one or more CMV isolates. Furthermore, CMV was transmitted to tubers only in five cultivars, including Desirée, and the virus could be subsequently detected in plants of the first and second vegetative progeny. This study indicated the possibility of selecting the potato cultivar(s) that did not allow the transmission of CMV to subsequent generations (Chrzanowska et al., 2004).

The identity of viral determinants of the HR-response and avirulence (*avr*) in *Bean dwarf mosaic virus* (BDMV) was established based on the abilities of this begomovirus BDMV to differentially infect some of the common bean (*Phaseolus vulgaris*) cultivars. The BDMV avr determinant in the bean hypocotyls tissue was mapped to *BDMV-BV1* open reading frame (ORF) and most likely to the BV1 protein. The BV1 also was identified as the HR determinant in the resistant cultivar Othello. In the case of cv. Othello, the resistance to BDMV was not associated with the avrBV1. Hence, a mechanism not dependent on avrB1 appears to function in cv. Othello. The BV1 protein, representing a new class of viral *avr* determinant, has been shown to be a nuclear shuttle protein that mediates viral DNA export from the nucleus (Garrido-Ramirez et al., 2000).

Active defenses, which depend on preformed surveillance systems for detecting and responding to pathogens during primary interaction, have been elucidated in virus infections. Post-transcriptional gene silencing (PTGS) in plants inactivates some aberrant or highly expressed RNAs in a sequence-specific manner in the cytoplasm. Many RNA and DNA viruses stimulate PTGS shortly after infection. Genes silenced by PTGS continue to be transcribed. However, their polyadenylated transcripts become nearly undetectable due to the action of a sequence-specific RNA degradation mechanism. Failure of infection by Caulimoviruses and Nepoviruses following induction of PTGS mechanism has been reported, despite the fact that these viruses exhibit no homology to the host genome. Such an efficient virus-induced silencing of plant genes, leading to resistance, has been demonstrated to be potential, general, antiviral defense strategy in other host–virus interactions (Hammond-Kosack and Jones, 2000).

A natural antiviral system operating in plants and animals has been identified as the RNA interference (RNAi), which is a double-stranded (ds) RNA-inducible, sequence-specific RNA degradation mechanism. Virus infection is established successfully by evasion or suppression of RNAi. The RNA viruses have been reported to encode ds RNA-binding proteins (ds RBPs). The accumulations of ds RNAs results in triggering of RNAi. Heterologous ds RBPs were found to suppress RNAi in plants, indicating the possibility of pathogen encoded ds RBPs inactivating RNAi-mediated host defenses in natural host–virus interactions (Lichner et al., 2003).

Recovery (that is infected plants, after exhibiting severe disease symptoms, recovered from the severe phase of disease and showed mild or no visible symptoms and resistance to reinfection by the same or related viruses) of transgenic peas containing a copy of the *Pea seed-borne mosaic virus* (PSbMV) isolate DPD1 NIb sequence was observed, when challenged with either the homologous (DPD1) or a heterologous (NY) PSbMV isolate. The recovered tissue showed the presence of low molecular RNA molecules (siRNA) derived from degradation of the NIb transgene mRNA. The results, based on experiments using graft inoculations, indicated recovered plant tissues contained extremely low levels of infectious virus with the potential, directly or indirectly, to confer the observed resistance specificity. The viral genome may possibly act, indirectly, as a source of specific siRNA molecules which were detected in the recovered tissues. These could act in tandem with siRNA molecules derived from the transgene mRNA to maintain a level of PTGS activity required to suppress symptom expression. The accumulation of homologous or related viruses may be hampered by PTGS activity (Boogaart et al., 2004). The RNA silencing is a conserved eukaryotic pathway involved in the suppression of gene expression, that is sequence-specific interactions that are mediated by 21 to 24 nt long RNA molecules. In plants, cell-autonomous and cell-mediated steps of RNA silencing form the important mode of activation of an elaborate immune system directed against viral pathogens. In a compatible interaction, as a counter-defensive strategy, viruses induce production of suppressor proteins that can inhibit various steps in the silencing process (Moissiard and Voinnet, 2004). Down-regulation of endogenes via PTGS is a key to the establishment of gene functions in plants. Double stranded RNA (dsRNA) has been demonstrated to be an effective trigger of gene silencing. It has been shown that strong genetic interference can be achieved in a chemically inducible fashion, allowing for temporal and spatial control of gene silencing in transgenic plants (Chen et al., 2003).

The cDNA clone from hot pepper (*Capsicum annuum*) encoding PR-protein 10 (CaPR-10) was isolated from pepper leaves inoculated with *Tobacco mosaic virus* pathotype (TMV-PO). Inoculation with TMV-PO or *X. campestris* pv. *vesicatoria* (*Xcv*) in the incompatible interaction resulted in the induction of CaPR-10 transcripts. The recombinant protein had ribonuclease (RNase) activity against TMV-RNA, as well as against pepper total RNA. Such putative antiviral activity was expressed under varied conditions. The antiviral activity was associated with the recombinant protein that had RNase activity against TMV-RNA. The dramatic increase in CaPR-10 levels from the initial very low concentration, soon after inoculation with TMV-PO, was correlated with increase in the RNAse activity. The results indicated that induction and subsequent phosphorylation of CaPR-10, as shown by immunoblot analysis and pull-down assays, increased its ribonucleolytic activity to cleave invading viral RNAs and that this may be important for its antiviral activity in vivo (Park et al., 2004).

The resistance of capsicum conferred by the L^2 gene was effective against all of the pepper-infecting tobamoviruses except *Pepper mild mottle virus* (PMMoV). On the other hand, resistance due to L^4 gene was effective against all tobamovirus. The coat proteins (CPs) of PMMoV and *Paprika mild mottle virus* (PaMMV) were expressed in *Capsicum frutescens* (L^2L^2) and *C. chacoense* (L^4L^4) plants using the heterologous *Potato virus* X (PVX) basal expression system. In *C. frutescens*, PaMMV-CP was localized in the inoculated leaves which produced necrotic, local lesions, indicating that either PaMMV or PMMoV CP was required to elicit the L^4gene-mediated host response (Gilardi et al., 2004).

Genetic resistance of host plants to viruses may be due to factors operating against the viruses directly or indirectly, affecting the transmission of the viruses by the natural vectors. Virus resistance genes may either restrict virus replication or virus movement to other tissues. Localization and movement of *Plum pox virus* (PPV) in stem tissues of susceptible and resistant apricot (*Prunus armeniaca*) cultivars was studied by employing an immuno-tissue printing technique. PPV was primarily localized in the xylem and less frequently in the cortex and pith. Limited movement of PPV from the inoculation point was observed in the cultivars tested. However, only the susceptible cultivars (Screara, Bebeco, and Colmer) permitted long-distance movement of PPV, suggesting that movement of virus in the xylem is the critical factor determining the susceptibility/resistance of apricot to infection by PPV (Dicenta et al., 2003b). The *Vat* gene in melon has been shown to confer resistance to non-persistent virus transmission by the aphid *Aphis gossypii.* A temporary blockage of aphid stylet tips, preventing release of virus particles, due to the action of the *Vat* gene product has been suggested. The efficiency of inoculation of *Cucumber mosaic virus* (CMV) by *A. gossypii* in a susceptible melon cultivar was not altered, irrespective of the virus source. In contrast, the ability of *Myzus persicae* to transmit CMV was markedly reduced after probing on resistant plants, providing evidence that the *Vat* gene in resistant plant may influence transmission of CMV depending on the vector species (Martin et al., 2003).

8.2 RESISTANCE TO SEED INFECTION

8.2.1 Constitutive Resistance

Constitutive proteins, composed of the proteinases and amylases, are known to be the inhibitors of hydrolytic enzymes and they are found in high concentrations as storage proteins in seeds and other storage organs. Many other types of proteins, in addition to these enzyme inhibitors, are also present in seeds. Barley seeds contain proteins with protective functions such as thionins, endochitinases, ribosome-inhibiting proteins (RIPs), β-glucanase phospholipids transfer protein (LTP), lectin, and thaumatin-like proteins (TLPs) (Shewry, 1995). These proteins have been shown to possess antimicrobial properties.

Resistance of seeds to decay caused by microbial pathogens depends on several constitutive factors. The permeability of cotton seed coat was found to be an important barrier to infection by *Aspergillus flavus*. The seeds of most wild cottons have high levels of impermeability to water through the chalaza, whereas water permeates through the seed coat of cultivars. Immunity to infection by *A. flavus* as well as production of aflatoxin is associated with impermeability of the seed coat to water (Mayna et al., 1969). However, no contribution of cotton seed permeability to resistance against infection by *A. flavus* was revealed by the experiments conducted by Halloin et al. (1991), and protection of embryos against microbial infection was considered to be due to the presence of localized tannin in the nucellus (Halloin, 1983). A protease present in barley embryos was inhibitory to proteases produced by *Aspergillus* spp. (Krishi 1974). A negative relationship between chitinase activity in maize seeds and infection by *A. flavus* was observed. The endogenous chitinase in the resistant genotype arrested the fungal invasion by its action on the fungal cell wall (Chokethaworn et al., 1991). The gene *lipA*, encoding a lipase involved in the breakdown of lipids from aflatoxin-producing *A. flavus* and *A. parasiticus*, was cloned. The expression of the lipase gene (mature mRNA) under substrate-induced conditions correlated well with aflatoxin production in the aflatoxigenic species *A. flavus* and *A. parasiticus* (Yu et al., 2003). It is possible that the resistant maize genotypes may not provide suitable substrate for the expression of the lipase gene of *A. flavus*.

In the attempt to identify markers associated with resistance to *A. flavus* in maize kernel, the potential of maize inbred Tex6 showing resistance to colonization and aflatoxin accumulation by *A. flavus* was studied. A protein inhibitory to growth of *A. flavus* was identified in aqueous extracts of mature Tex6 seeds. This inhibitory protein had a MW of 29,000 and was shown to be an endochitinase with exochitinase activity. The purified Tex6 chitinase inhibited the growth of *A. flavus* by 50% at a concentration of 20 μg/ml. It was considered to make a major contribution to the antifungal activity in Tex6 maize and to be different from previously reported chitinases based on the partial peptide sequence (Moore et al., 2004) (Appendix 8(ii)). The proteomic comparisons of maize kernel embryo proteins of resistant and susceptible genotypes indicated that several proteins spots were unique or up-regulated in resistant embryos. One of these protein spots was identified as glyoxalase I (GLX-I; EC 4.4.1.5). The constitutive activity of GLX-I in resistant maize lines was significantly higher compared with susceptible ones. The GLX-I activity in susceptible genotypes, following infection by *A. flavus*, was at a low level, whereas the methylglyoxal (MG) levels registered significant increases in two of three susceptible genotypes. MG at a concentration of 5.0 μM could induce aflatoxin production in *A. flavus* culture, suggesting a possible stimulation of expression of *aflR*, an aflatoxin biosynthesis regulatory gene. This gene *aflR* was significantly up-regulated in the presence of 5 to 20 μM MG. It is possible that GLX-1 may have an important role in controlling MG levels inside kernels, resulting in lower levels of aflatoxins present in resistant maize genotypes (Chen et al., 2004).

The presence of thionins in plant tissues and their antimicrobial properties have been demonstrated. Thionins from the endosperm of wheat, rye, oat, and barley have been isolated and characterized. They are basic, cysteine-rich polypeptides with a molecular weight of about 5 kDa (Wada, 1982; García-Olmedo et al., 1989). The thionins of wheat (purothionins) and barley (hordothionins) were inhibitory to both fungal and bacterial pathogens, such as *Phytophthora infestans*, *Colletotrichum lagenarium*, *Clavibacter michiganensis*, and *Pseudomonas solanacearum* (García-Olmedo et al., 1992; Bohlmann, 1994; Florack and Stiekema, 1994). Thionins may inhibit protein synthesis in cell-free experimental system. Methyl jasmonate may induce some thionin genes involved in the induction of proteinase inhibitors in tomato (Andersen et al., 1992; Brümmer et al., 1994). Rice seedlings expressing and accumulating a high level of oat thionin in cell walls grew normally and showed resistance to seed-borne bacterial pathogens *Burkholderia plantarii* and *B. glumae* (Iwai et al., 2002).

Several plant species have been reported to contain ribosome-inhibiting proteins (RIPs) which are toxic N-glycosides, capable of modifying ribosomes by cleaving a specific adenine residue on a highly conserved sequence of 28S rRNA. The demonstration of inhibition of infection by *Tobacco mosaic virus* (TMV) by the extracts of pokeweed (*Phytolacca americana*) provided evidence for the involvement of RIPs in plant defense against microbial pathogens (Irvin et al., 1980). The RIP isolated from barley grains showed antifungal activity. The growth of *Botrytis cinerea*, *Rhizoctonia solani*, and *Trichoderma reesi* was inhibited by this RIP in vitro (Roberts and Selitrennikoff, 1986). The activity of barley RIP was markedly increased when it was combined with two other barley seed proteins – β-1,3 glucanase and an endochitinase (Leah et al., 1991). A novel protein named Crip-31, possessing properties similar to RIPs, was isolated from *Clerodendrum inerme*. Crip-31 protected the tomato plants against *Cucumber mosaic virus* (CMV), *Potato Virus Y* (PVY), and *Tomato mosaic virus* (Shelly et al., 2001).

Microbial plant pathogens synthesize a variety of extracellular, hydrolytic enzymes capable of acting on host plant cell wall components, resulting in collapse of cell structure and ultimately causing necrosis of cells or tissues. Polygalacturonase inhibiting proteins (PGIPs) have been isolated from different plant species. The PGIPs isolated from bean (*Phaseolous vulgaris*) increased the period of activity of phytoalexin elicitor-active oligogalacturonides generated by endo-PGs produced by *Colletotrichum lindemuthianum* causing anthracnose disease (Lorenzo et al., 1990). The presence of PGIP in tomato fruits (Stotz et al., 1994), pears (Sharrock and Labavitch, 1994), and apples (Yao et al., 1995) has been reported. The PGIPs appear to inhibit endopolygalacturonases (endoPGs) produced by fungi and they may not be effective against fungal exoPGs, pectate lyases, or pectolytic enzymes from bacteria (Johnson et al., 1993).

Variations in the inhibitory activities of PGIPs from different plant sources against endoPGs produced by fungal pathogens have been observed. Likewise the isoforms (isozymes) of the same endoPG differ in the levels of inhibition. Purified bean PGIP inhibited PGs produced by several fungi, but pear PGIP inhibited only the activity of PG produced by *Botrytis cinerea* (Stotz et al., 2000). The PGIP inhibited only the activity of PG produced by extracted from 'Oroblanco' grapefruit type altered tissue was purified and partially characterized. The grape fruit PGIP with MG1 of 44 kDa inhibited the endoPG (EC 3.2.1.15) produced by *Penicillium italicum* and *B. cinerea*. The radial diffusion and reducing sugar assays showed that endoPGs of *P. italicum* and *B. cinerea* were affected, whereas no effect on *P. digitatum* could be noted, since no endoPG activity was detected in the culture filtrate of this pathogen. The growth of *P. italicum* and *B. cinerea* was inhibited, but no influence on *P. digitatum* was discernible. The inhibitory activity of the PGIP was reduced by heating at 65°C (for 10 min) by 43% and inactivated entirely at 100°C (for 10 min) (D'hallewin et al., 2004). The PGIPs are a family of relatively heat stable glycoproteins occurring in the cell walls of many dicotyledonous plants. They belong to the class of leucine-rich repeat (LRR) proteins which have an important role in the process of recognition of signals derived from protein–protein interactions between host plant and pathogen. It has been hypothesized that binding of fungal PGs by PGIP may contribute to constitutive resistance to soft rot pathogens, whereas activation of PGIP genes may form a component of inducible (active) plant defense (Shewry and Lucas, 1997; Machinandiarena et al., 2001).

Another kind of compounds implicated in plant defense against microbial pathogens are collectively designated defensins. They have been found in a wide range of plants. Immunological or in situ nucleic acid hybridization techniques have been applied for detecting the constitutively expressed plant defensins in different plant tissues. Accumulation of defensins in very high concentrations, preferentially in peripheral cell layers of epidermis, cotyledons, hypocotyls, and endosperm, of radish seeds, has been revealed by the molecular methods. Defensin gene transcription occurred to the maximum extent in the epidermis and leaf primordia of potato tubers. Plant defensins have been shown to have inhibitory effect on hyphal elongation. They were exuded from germinating seeds of radish, accounting for about 30% of the total released proteins, and they may provide protection to the germinating seeds. Defensins induced in potato tubers, flowers, stems, and leaves were effective against *Clavibacter michiganensis* subsp *sepedonicus* causing ring disease, *Ralstonia solanacearum* causing brown rot wilt disease, and *Fusarium solani* causing cortical decay in tubers (Moreno et al., 1994; Terrass et al., 1995). Two cysteine-rich peptides with MWs of 6.8 and 10.0 kDa, present in cowpea seeds, inhibited the growth of *Fusarium solani* and *F. oxysporum*. Sequence analysis of these peptides indicated the presence of a defensin and a lipid transfer protein (LTP) with a high degree of homology to other antifungal peptides derived from plants. These peptides were localized in the cell wall and cytosolic compartments, as revealed by immunofluorescence assays (Carvalho et al., 2001).

A defensin gene (*PpDfn1*) from a cDNA library made from peach (*Prunus persica*) winter bark tissue was cloned and characterized. The open reading frame (ORF) of 237 bp codes for a 79 amino acid peptide related to the defensin family of proteins and showed homology to defensins from other plant species. The gene was seasonally expressed in bark tissues of 1-year-old shoots and in early fruit development stage. A recombinant version, rDFN1, inhibited germination of the fungal pathogens *Penicillium expansum* and *Botrytis cinerea*, but not the bacterial pathogen *Erwinia amylovora* (Wisniewski et al., 2003).

Lectins, proteinaceous compounds, are able to recognize and bind to specific sugar sequences present on the cell surface or precipitate glycoproteins with one or more carbohydrates groups. Chitin-binding lectins are considered to play a role in plant defense against microbial plant pathogens, particularly fungal pathogens that have chitin on the cell surface. Lectin present in the seeds of many graminaceous plants specifically bind the sugar N-acetylgulcosamine (G/C NAC), its oligomers, and chitin, a polymer of G/C NAC residues (Chrispeels and Raikhel, 1991). Wheat germ agglutin and potato lectin have been reported to have antimicrobial properties (Mirelman et al., 1975; Garas and Kuć, 1981; Andreu and Daleo, 1988). Hevein from rubber inhibited the growth of *B. cinerea* at concentrations of 300 to 500 μg/ml (van Parijs et al., 1990).

The presence of proteins that share sequence homology with well-characterized pathogenesis-related (PR)-proteins (PRs) has been detected in many plant species. These proteins, designated PR-like proteins (PRLs), are found in seeds of many plants. PR-like chitinases and thaumatin-like proteins (TLPs) have been isolated from seeds of wheat, barley, oats, sorghum, and maize and these proteins have been shown to have antifungal properties. The PRLs appear to exert a synergistic effect, resulting in the enhancement of resistance to a level significantly broader than any one of them may provide (Roberts and Selitrennikoff, 1986; Hejgaard et al., 1991; Vigers et al., 1991; Radha Jeyalakshmi et al., 2000). Localization and accumulation of rye seed chitinases (RSC-a and RSC-c) with antifungal property were monitored by employing enzyme-linked immunosorbent assay (ELISA) and immunoblot techniques. Both RSC-a and RSC-c chitinases were detected in the endosperm of rye seeds and they accumulated in the seed during a late stage of development (Taira et al., 2001). Many seed protein fractions with potent antifungal activity were generated at high levels in wheat seeds incubated in vitro, indicating that at least part of the antifungal protein generation was not controlled by gene expression. The synthesis of antifungal protein in the early stage of seed germination may be required to offer protection to the seedlings, when the defense system of plants may not have become operational (Wang et al., 2002).

Many seed proteins possess amino acid sequences similar to that of TLPs. Zeamatin, an antifungal protein in maize flour, caused rupture of fungal hyphal tips (Huynh et al., 1992). Tolerance to *Phytophthora infestans* causing potato late blight diseases was attributed to constitutive expression of high levels of osmotin-like protein (Zhu et al., 1996). Over-expression of TLP gene in rice led to increased resistance to *Rhizoctonia solani*, causing sheath blight disease for which no resistance gene has been located (Datta et al., 1999). By employing specific antibodies generated against permatins, another set of TLPs with antifungal activity, expression of permatin gene and protein were detected, initially in the ovary wall and then in the aleurone and ventral furrow of developing barley and oat seeds (Skadsen et al., 2000).

Plants contain several other kinds of antimicrobial compounds. The spring wheat cultivar Frontana showed resistance against initial infection by *Fusarium culmorum*. High amounts of phenolic compounds were present in the glumes, lemmas, and paleas of Frontana prior to, and at all sampling times after, inoculation in comparison to susceptible winter wheat cultivar Agent. Furthermore, structural defense responses, such as cell wall appositions in the infected lemma tissue, were more pronounced. An immunogold labeling technique to detect phenolic compounds revealed increased densities of gold particles on the cell walls of infected lemma tissue of Frontana adjacent to the fungal cells at 3 days after inoculation. In contrast, the susceptible cultivar Agent exhibited little or no response to infection (Siranidou et al., 2002). A hydrophobic amylase inhibitor with MW of 19.7 kDa, purified from maize kernels, inhibited the amylase from *Fusarium verticillioides* which produces the mycotoxin fumonisin. In addition, conidial germination was inhibited by the amylase inhibitor. Since the elimination of fumonisin contamination from cereal kernels is possible by this method, the use of amylase inhibitor has the potential for practical exploitation as the control strategy for the management of grain infection in cereal crops (Figueira et al., 2003). A trypsin inhibitor with associated antifungal activity was isolated from sunflower seeds by affinity chromatography on a trypsin agarose matrix. This protein inhibited spore germination of *Sclerotinia sclerotiorum*. The germination of ascorpores was entirely inhibited at concentration of 14 mg/ml. The results suggest that the trypsin inhibitor protein with a potent antifungal activity may be effectively employed to restrict the fungal invasion in seeds (Mendicta et al., 2004).

A wheat line was transformed with constitutively expressed genes encoding a class IV acidic chitinase and an acidic β-1,3-glucanase. There was a significant delay in the spread of *Fusarium graminearum* causing *Fusarium* head blight (scab) disease. A lesion-mimic phenotype (a phenotype that reacts to the toxin by producing visible lesions similar to that induced by the pathogen itself) in this transgenic line, when homozygous for transgene loci, contained apoplastic fluid consisting of PR-proteins belonging to families of β-1,3-glucanases, chitinases, and thaumatin-like proteins (TLPs). The apoplastic fluid was inhibitory to *F. graminearum* and *Gaeumannomyces graminis* var. *tritici*. Three enzymes and a barley class II chitinase with antifungal activities were purified from apoplastic fluid. Mixtures of proteins exhibited synergistic or additive inhibitory activity against *F. graminearum* and *Pyrenophora tritici-repentis*. The apoplastic fluid of cells showing a hypersensitive response is likely to contain PR-proteins at concentrations required to inhibit the hyphae of the fungal pathogen (Anand et al., 2004).

8.2.2 Active Resistance

A large increase in antifungal activity during seed germination and in seed protein extract has been demonstrated. In maize infected by *Fusarium moniliforme*, induction of defense-related ultrastructural modification, such as the formation of appositions on the outer host cell wall surface, occlusion of intracellular spaces, and formation of papillae, was observed. The pathogenesis-related maize seed (PRms) proteins accumulated at very high levels in those cell types representing the first barrier for fungal penetration, such as the aleurone layer of the germinating seeds, as well as the subcuteller epithelial cells of isolated germinating embryos, as revealed by immunolocalization techniques. A large number of abnormal fungal cells showing PRms-specific labeling could be recognized, indicating a functional role for PRms in the plant defense responses (Murillo et al., 1999). A basic heme-peroxidase (WP1) present in wheat (*Triticum aestivum*) kernels had a MW of 36 kDa and a pI of 8.0. This protein showed a high similarity in N-terminal amino acid sequence with that of wheat flour peroxidase allergen. The WP1 protein exhibited antifungal activity against *Fusarium culmorum* and *Botrytis cinerea*, suggesting a role for peroxidases in the defense against invading pathogens (Caruso et al., 2001).

The expression of defense response genes systemically within wheat spikes inoculated with *Fusarium graminearum* was studied. The accumulation of transcripts of four defense response genes, that is *PO*, *PR-1*, *PR-3* and *PR-5*, in colonized and uncolonized portions of wheat spikes of resistant (R) and susceptible (S) wheat genotypes was monitored. An *F. graminearum* isolate genetically transformed to express the β-glucuronidase (GUS) marker gene was shown to be useful for visualizing and following the spread of the fungus and for quantifying the fungal biomass in spike tissues. At 48 h following inoculation, accumulation of transcripts of all four defense response genes in both colonized and uncolonized regions of spikes of R and S genotypes was evident. Activation of defense response genes occurred in wheat genotypes irrespective of their level of resistance/susceptibility. In addition, direct contact with the pathogen was not required for the induction of defense response genes in both R and S genotypes of wheat (Pritsch et al., 2001).

Chromatographic and electrophoretic analysis showed that during germination of wheat (*T. durum*) seed, more protein fractions with potent antifungal activity were generated and the antifungal activity shifted from small molecules to high molecular proteins. A rapid germination-related increase in antifungal activity was also seen with incubation of seed proteins in vitro. The results suggested that at least part of antifungal protein generation was independent of gene expression. Antifungal proteins active against *F. graminearum* (*Gibberella zeae*) were isolated. During seed germination, a regulated biochemical process may be initiated resulting in the generation of multiple peptides or proteins with antifungal activities. It is likely that this onset of antifungal proteins is transitional in nature, and may have a significant role in the development of resistance in the early stage of seedling growth, when the more sophisticated defense system is yet to develop (Wang et al., 2002).

Resistance of peanut (groundnut) seeds to infection by *Aspergillus flavus* and *A. parasiticus* and subsequent production of the mycotoxin aflatoxin has been reported to be associated with cuticular wax accumulation, seed coat structure, and concentrations of low molecular weight peptide-like compounds and tannin (Sanders and Mixon, 1978; Agarwal and Sinclair, 1996). Maize genotypes showing resistance to aflatoxin production by *A. flavus* were studied in comparison with susceptible cultivars. Five additional constitutive proteins were associated with resistance of eight of 10 genotypes tested. Globulin-1 and globulin-2 with MWs of 58 and 46 kDa, respectively, were among the eight new proteins detected in resistant genotypes. Induction of specific antifungal proteins, such as zeamatin (22 kDa), reaching high levels in resistant genotypes, was conspicuously evident. Embryo-killed maize kernels were unable to synthesize new proteins and consequently they supported high levels of aflatoxin production. The results provide support to the view that synthesis of new proteins by the embryo is required for the development of resistance to *A. flavus*.

Another study examined the possibility that resistance to aflatoxin contamination may be conferred by altering the fatty acid composition of seeds. Seven breeding lines of peanut with relatively low linoleic acid were tested under field conditions. The plots were artificially inoculated with a mixture of *A. flavus* and *A. parasiticus* at about 60 days after planting and subjected to drought and heat stress for the 40 days immediately preceding harvest. The low linoleic acid content of genotypes had no effect on preharvest aflatoxin contamination in peanut (Holbrook et al., 2000).

Aspergillus flavus infects almond (*Prunus dulcis*) kernels both under field and storage conditions and marketability of almond kernels is greatly reduced because of the presence of aflatoxins. The genotypes (40) of almond were evaluated for their relative susceptibility to *A. flavus* by inoculation under controlled conditions at 26°C after 18 days of incubation. Differences in the percentage of kernel surface colonized by *A. flavus* were observed. The Spanish cultivar Ramillette was the least susceptible among the genotypes tested (Dicenta et al., 2003a).

The temporal expression patterns for defense response genes encoding PR-1, PR-2 (β-1, 3-glucanase), PR-3 (chitinase), PR-4, PR-5 (thaumatin-like proteins, TLPs), and peroxidase (PR-9) were determined in wheat spikes of resistant and susceptible cultivars inoculated with *F. graminearum* (*Gibberella zeae*). Accumulation of transcripts for these six defense response genes showed no discernible difference during early stages of pathogenesis (6–12 h after inoculation). However, a more rapid accumulation of two acidic isoforms of β-1, 3-glucanases, Class IV and Class VII chitinases, could be detected in the resistant cultivar (Suman 3) than in the susceptible mutant during the first 24 h after inoculation with *F. graminearum* (Pritsch et al., 2000; Li et al., 2001). Guaiacol peroxidase (POX) and polyphenol oxidase (PPO) activities in wheat cultivars resistant (Suman # 3 and Wangshuibai) and susceptible (Falat and

Golestan) to *F. graminearum* were monitored by spectrophotometric assessment. POX-specific activity in resistant and susceptible wheat cultivars inoculated with *F. graminearum* showed a significant increase during the milk stage compared with control plants. PPO-specific activity in wheat heads attained a maximum level during the milk stage and subsequently declined. Resistant cultivars exhibited three- and two-fold increases in PPO activity over the comparable controls and susceptible cultivars, respectively. Resistance against *Fusarium* head blight (FHB) was induced in susceptible Falat heads by pretreatment with an autoclaved mycelial wall preparation (Mohammadi and Kazemi, 2002).

Barley germplasm showed a useful range of variability for the accumulation of (mycotoxins) deoxynivalenol (DON) and 15-acetyl-deoxynivalenol (15-A DON) produced by *Fusarium graminearum*. The concentrations of DON and 15-A DON present in spikelets inoculated with macroconidia of *F. graminearum* appeared to be an indicator of the level of resistance of germplasm lines (Evans et al., 2000). The presence of low concentrations of DON, consistently, in resistant lines of barley, such as Maris Mink, Symko, and Suvenir, and Canadian cultivars AC Sterling and Marrison, was observed by Mc Callum et al. (2004). Genotypic variation of barley lines in DON accumulation when infected by *F. graminaearum* was studied by conducting a quantitative trait locus (QTL) analysis using a genetic mapping population. A single QTL on chromosome 3 for DON accumulation was identified. Disease symptom development was similar between near-isogenic lines (Smith et al., 2004).

The involvement of PR-proteins in the development of resistance in wheat spikes against *Fusarium culmorum* producing mycotoxins was demonstrated. The subcellular localization of β-1,3-glucanase and chitinase in *F. culmorum*-infected wheat spikes was detected by employing immunogold labeling technique. The distribution of β-1,3-glucanase and chitinase was similar in both healthy and infected wheat spikes. The enzymes were localized primarily in the cell walls of various tissues, including the lemma, ovary, and rachis, while the cytoplasm and organelles of cells in these tissues showed no labeling. But β-1,3-glucanase and chitinase in infected wheat spikes reached higher levels in resistant wheat cultivars compared with the susceptible cultivars. The results indicate that accumulation of β-1,3-glucanase and chitinase may result in the development of resistance to the spread of *F. culmorum* in the spike tissue (Kang and Buchenauer, 2002).

Efforts were made to develop suitable greenhouse inoculation technique to differentiate types of resistance exhibited by wheat genotypes to *F. graminearum*. Central floret infection and atomizing macroconidia onto spikes were compared to determine the percentage of spikelets per spike developing symptoms and area under the disease progress curve. The inoculation techniques employed could not differentiate between different types of partial resistance (Engle et al., 2003). The levels of resistance of barley cultivars to FHB and DON accumulation were assessed in field conditions under artificial inoculation in Eastern Canada and China. None of the 64 cultivars tested was immune. However, three cultivars, that is Island, AC Alberte, and Chevron, showed maximum resistance, with low incidence and DON concentration in grains. Susceptibility to DON accumulation did not lead to a reduction in yield under natural infection conditions in Ontario, Canada (Choo et al., 2004).

The trichothecenes produced by *F. graminearum* are considered to contribute to aggressiveness in wheat infection. Six species (bread and durum wheat, triticale, rye, barley, and oats) were inoculated with two isogenic strains of *F. graminearum*: a wild strain (*Tri 5*$^+$) producing trichothecens and the mutant strain (*Tri 5*$^-$), a nonproducer of trichothecenes. The aggressiveness of *Tri 5*$^+$ and *Tri 5*$^-$ strains varied according to crop species, though *Tri5*$^+$ strain was generally more aggressive. The difference in aggressiveness was less pronounced in rye, which was highly resistant. The level of resistance was high in oats because of the large space in between florets. In barley, the pathogen was able to move externally from one floret to another within the dense spike, without penetrating the rachis, although barley exhibited a moderate Type II resistance to FHB disease. Bread wheat and triticale showed low level of resistance to *Tri 5*$^+$ strain. Durum wheat was highly susceptible to *Tri 5*$^+$ strain and suffered to the maximum extent, even when infected by the *Tri 5*$^-$ strain. The results indicate that the role of trichothecenes may differ depending on the crop species (Langevin et al., 2004). Accumulation of DON, growth of *F. culmorum*, and expression of *Tri* genes involved in trichothecene biosynthesis were monitored. In liquid culture, DON accumulated soon after maximum expression of *Tri6* and coincident with expression of *Tri5*. RT-PCR assays were employed to monitor the expression of *Tri5* and *Tri6* genes (Covarelli et al., 2004).

Fumonisins are mycotoxins produced by *F. moniliforme* (*Gibberella fujikuroi*) and *F. verticillioides*, which cause *Fusarium* ear and kernel rot of corn (maize). A positive correlation between fumonisin production and ear rot was observed. Low fumonisin concentration in grain and low ear rot severity were found to be associated with several F_1 hybrids, indicating the possibility of using the fumonisin level as a measure of resistance to ear and kernel rot disease (Clements et al., 2004).

Inheritance of resistance to *Aspergillus* ear rot and aflatoxin production in corn caused by *Aspergillus flavus* was studied. The progeny derived from crosses between the resistant inbred cv. Tex6 and susceptible inbred cvs B73 and MO 17 were evaluated by inoculating the primary ears with a pinboard. Analysis of generation means indicated that the resistance to both ear rot and aflatoxin production in the B73 × Tex6 cross was the most important. One cycle of selection of resistance within B73 × Tex6 F(3) families was estimated to reduce the percentage of ear rot severity by 8.5% and aflatoxin concentration by 19 ng/g (Hamblin and White, 2000). The cycles of selection of four maize genotypes for ear- and grain-quality characteristics, interactions with *A. flavus* and *Fusarium verticillioides* infection, and ear infestation by insect were compared in two seasons. The maize rows inoculated with *F. verticillioides* had significantly greater infestation by coleopteran beetles and lepidopteran borers than controls and rows inoculated with *A. flavus*. A loss of grain density and lower grain weight were the results of infection by both pathogens. Aflatoxin levels (AFB1 and AFB2) averaged 327 ppb and 589 ppb in the two seasons. Fumonisin concentration in *F. verticillioides* inoculated rows varied from 3.1 to 5.3 ppm. Ear-rot scoring was generally correlated with incidence of *F. verticillioides* in kernels and grain weight loss, but not with *A. flavus* in the grain (Cardwell et al., 2000).

Gibberella zeae, causing wheat *Fusarium* head blight, produces several mycotoxins, inducing mycotoxicoses in humans and animals. The loci associated with pathogenicity and aggressiveness were identified on an amplified fragment length polymorphism (AFLP)-based genetic map of *G. zeae*. Two reproducible, quantitative loci (QTL) for aggressiveness were detected on linkage group I. The QTL linked to the *TR15* locus (trichodiene synthase in the trichothecene pathway gene cluster) accounted for 51% variation, while a second QTL (50 centimorgans away) was tightly linked to the locus governing the type of trichothecene synthesized (Cumagan et al., 2004).

8.3 DEVELOPMENT OF DISEASE RESISTANT CULTIVARS

Breeding programs, with the objective of developing cultivars with sufficient built-in resistance to diseases, aim to ensure that disease levels remain below the threshold for economic loss. The existence of resistance genes against the target pathogens in the gene pool of the crop and availability of effective screening methods are the primary requirements for the development of disease-resistant cultivars. If resistance to a disease is governed by more than a single gene, the breeding and selection process may be complicated, protracted, and expensive. In certain cases, the resistance genes may be tightly linked to agronomically undesirable traits. Screening for disease resistance requires an efficient and reliable method to challenge plants with the pathogen concerned under conditions suitable for disease development, followed by evaluation of plants for infection frequency and disease severity.

Formulation of a disease management system to prolong the useful life of resistance genes depends on the knowledge of the genetics of host resistance, population genetics, evolutionary biology of the pathogen, and the interaction of crop management practices with host-plant resistance. Studies on molecular genetics have been useful for the identification and mapping of resistance genes on plant chromosomes, as well as for the rapid testing of individuals for specific DNA sequence. These developments provided the much-needed solution to overcome the formidable obstacles faced by conventional breeding methods. The procedure known as marker-assisted selection has been applied for prescreening progenies to identify those carrying a known resistance gene or chromosomal sequences closely linked to such gene(s) (Henry, 1997). Disease resistance may be governed by genes at one or a few loci (monogenic, oligogenic, or major gene resistance encoded by R genes). Resistance conferred by the R genes may be overcome by the pathogen by producing a new race, biotype, or strain with a matching virulence gene.

The need for differentiation of host resistance was first suggested by VanderPlank (1963). The potato variety with the *R1* gene shows resistance to race 0 of the late blight pathogen *Phytophthora infestans*, which does not have the matching virulence gene. The unmatched *R1* gene triggers the resistance mechanism in this potato cultivar. None of the isolates of *P. infestans* could infect potato plants with the *R1* gene for about 10 years following emergence of the R1 line (during the late 1920s). However, *P. infestans* eventually produced a physiologic race that could colonize potato plants and produced typical lesions, indicating the capacity of the new race to overcome the *R1* resistance gene. Further research was undertaken, and an additional nine resistance genes in *Solanum demissum* were identified (Black, 1960; Malcolmson and Black, 1966). However, the pathogen was also able to overcome the resistance offered by newly identified R genes by forming new populations armed with virulence genes to match the R genes of the potato plants. The pathogen appeared to produce new races at a faster rate than breeders and plant pathologists could produce potato cultivars with unmatched resistance genes. A shift in the objective in the breeding programs was necessary and potato lines with polygenic or nonspecific resistance (field resistance) were considered to be a better alternative (Holden, 1977).

Horizontal resistance of a potato cultivar may be recognized by exposing cultivars in the field to infection by virulent races to which cultivars are susceptible. The resistance which remains after elimination of vertical resistance, is horizontal. Horizontal resistance is essentially due to polygenic action and quantitative inheritance. The grades of horizontal resistance are not distinctly defined and vary continuously, and it has been variously named as nonspecific, field, general, or polygenic resistance by different researchers. Horizontal resistance may be due to a combination of many mechanisms, leading to a decreased rate of infection and invasion of host tissues and/or sporulation of the pathogen. The level of horizontal resistance is proportional to the concentration (frequency) of polygenes in the test cultivar. A number of parents with low levels of resistance to the desired pathogen may be crossed in all combinations and the genes of resistance to the target pathogen may be gradually accumulated by repeated crossing the selected progenies over a number of generations.

Introduction of single-gene resistance has provided long-lasting resistance against the target pathogens in only a few cases. The rapid loss of protection offered by monogenic resistance necessitated the use of combinations of effective resistance genes which could protect the cultivars for longer periods, as the pathogen would have to make multiple changes simultaneously to be compatible with the different combinations of resistance genes. The concept of durable resistance, as proposed by Johnson (1981, 1983), requires that resistance remains effective in a cultivar when used over a wide area, for a considerable period of time, in an environment favorable for the development of disease.

The resistance genes may have different functions. Some resistance genes have been shown to encode a cellular receptor, capable of recognizing a product or component of a particular pathogen known as an elicitors. The pathogen gene essential for the production of the elicitor is designated avirulence gene or *avr* gene. The plant cell receptor, following interaction with the elicitor, triggers a massive switch in the pattern of plant gene expression that produces a biochemical cascade, resulting in the rapid accumulation of defensive chemicals and structures (Lamb et al., 1992). The hypersensitive response (HR), due to rapid necrosis of the challenged cell and other surrounding cells, is frequently seen as the result of the biochemical cascade. This is followed by reinforcement of structural and biochemical defenses in surrounding cells, resulting in the restriction of multiplication and spread of the pathogen (Goodman and Novacky, 1994). Among the R genes characterized, those encoding proteins with a characteristic region known as a leucine-rich repeat (LRR) have been analyzed in depth. The hypervariable sequences present in LRR regions are considered to be involved in the specificity of elicitor recognition.

Molecular techniques have been applied for the characterization of different stages of interaction between the host plant and potential pathogens. Molecular markers allow the dissection of monogenic and quantitative resistance and they can be used for the preservation and exploitation of germplasm, marker-assisted selection of resistance genes, and generating desired combinations of resistance genes and gene deployment for increasing levels of resistance in desired genotypes. Using molecular makers, all regions of the plant genome can be assayed for linkage to disease resistance. Linkage of disease resistance with undesirable agronomic traits is a major obstacle observed in breeding programs. Mapping of the genome using molecular markers may aid in identifying quantitative trait loci (QTL) for several characteristics including disease resistance. Some of the molecular markers linked to disease resistance genes are presented in Table 8.1.

A random amplified polymorphic DNA (RAPD) marker directly linked with a gene governing resistance to *Beet curly top virus* (BCTV) in common bean was identified. Resistance was conditioned by a single dominant allele tentatively designated *Bct*. The linked RAPD marker was converted into a sequence characterized amplified region (SCAR) marker SAS8(1550). The examination of BCTV resistant and susceptible snap and dry bean genotypes showed that the SCAR was highly useful for marker-assisted selection of *Bct* in snap and dry bean (Larsen and Miklas, 2004). The dominant *Am* gene from *Lycopersicon hirsutum* f.sp. *glabratum* PI 134417 has been reported to confer resistance to most strains of *Alfalfa mosaic virus*, including necrotic strains. *Am* was mapped genetically to the short arm of tomato chromosome 6 in the resistance hot spot. The other R-genes, *Mi* and *Cf-2/Cf-5*, and the quantitative resistance factors *Ty-1*, *OL-1*, and *Bw-5* were located in the same cluster (Parrella et al., 2004).

TABLE 8.1 Molecular Markers Linked to Crop Disease Resistance Genes

Crop / pathogen	Resistance gene	Type of marker	Reference
Bean/*Common bean mosaic virus*	*I*	RAPD	Haley et al., 1994
Bean/*Colletotrichum lindemuthianum*	*Co 4*	RAPD	Cardosode Arruda et al., 2000
Pea/*Pea seed-borne mosaic virus*	*sbm L*	RFLP, RAPD	Timmerman et al., 1993
Pepper/*Pepper veinal mottle virus*	*pvro*	RFLP	Caranta et al., 1996
Potato/*Phytophthora infestans*	*R1 and R3* *Ph3*	RFLP AFLP	El-Kharbotly et al., 1994; Chunwongse et al., 2002
Tomato/*Tomato yellow leafcurl virus*	*Ty-1*	RFLP	Zamir et al., 1994
Tomato/*Phytophthora infestans*	–	–	Bagirova et al., 2003
Banana/*Banana streak virus*	–	AFLP	Lheareux et al., 2003
Grapevine/*Uncinula necator*	–	RFLP	Reisch, 1998
Bean/*Beet curly top virus*	*SAS 8*	RAPD SCAR	Larsen and Milklas, 2004
Yam/*Colletotrichum gloeospoiorides*	*Dcg-1*	RAPD	Mignoua et al., 2003

AFLP, amplified fragment length polymorphism; RAPD, random amplified polymorphic DNA; RFLP, random amplified fragment length polymorphism; SCAR, sequence characterized amplified region.

8.3.1 Sources and Genetics of Resistance to Diseases of Fruits and Vegetables

For any breeding program with the primary aim of developing disease-resistant cultivars, an effective and reliable screening method is required to identify suitable sources of resistance. All available cultivars and germplasm entries have to be screened both under natural and controlled conditions to select the donors of resistance for the effecting transfer of resistance genes to susceptible cultivar(s). By challenging the test entries to the target pathogen at the appropriate inoculum potential under controlled conditions or, when adequate disease pressure is available, under natural conditions, sources of resistance for several pathogens have been identified (Figs 8.1 and 8.2).

Figure 8.1 Serverity index of green mold of mandarin caused by *Penicillium digitatum*. 0-Healthy; 1-resistant; 2-moderately resistant; 3-moderately susceptible; 4-susceptible; 5-highly susceptible. (Courtesy of T. Ashokkumar and A. Palaniswami, Tamil Nadu Agricultural University, Coimbatore, India.)

Figure 8.2 Severity index of blue mold of mandarin caused by *Penicillium italicum*. 0-Healthy; 1-resistant; 2-moderately resistant; 3-moderately susceptible; 4-susceptible; 5-highly susceptible. (Courtesy of T. Ashokkumar and A. Palaniswami, Tamil Nadu Agricultural University, Coimbatore, India.)

Solanum demissum was selected as a source of "R" genes conferring resistance to *Phytophthora infestans*, causing potato late blight disease (Malcolmson and Black, 1966). *Lycopersicon hirsutum* and *L. pimpenellifolium*, wild relatives of tomato, have provided resistance genes to protect tomatoes against early blight disease caused by *Alternaria solani* and *Tomato leafcurl virus* (TLCV) disease (Maiero and Ng, 1989; Green and Kalloo, 1994). A gene, *Ph-3*, in *L. pimpinellifolium* accession L3708 conditioning race-specific resistance to *P. infestans* was mapped. Restriction fragment length polymorphism (RFLP) markers mapped to chromosome 9 and the RFLP markers were employed to select resistant progenies of the cross between CLN 657 (susceptible) and 2L 3708 (resistant) accessions (Chunwongse et al., 2002).

The level of resistance of cultivars is assessed to identify resistant or less susceptible cultivars, to provide basic information for planning further searches for resistance genes. Various parameters are used to determine the level of susceptibility/resistance. Evaluation of resistance to *P. infestans* tuber rot were made at 15 days after inoculation, based on infected area and depth of necrosis. The cultivars and lines Vityaz, PSK5, 153-101, 112-252, and 662-231 showed high resistance to tuber rot (Popkova et al., 1999). Differential interactions in potato tuber blight infection between potato cultivars and *P. infestans* isolates were determined, using whole tuber and tuber slices, by assessing tuber blight incidence and severity, necrosis, and mycelium coverage. The cultivars Kartel and Producent were resistant, while Bintje, and to a lesser extent Astarte, exhibited more susceptibility after inoculation with aggressive strains of *P. infestans.* The quantitative nature of resistant responses exhibited by the cultivars suggested the involvement of quantitative trait loci (QTL) governing resistance to tuber blight (Flier et al., 2001). Tubers of eight potato cultivars grown commercially in the Columbia Basin and 29 advanced clones from Tri State Potato Variety Development Program were evaluated for their resistance to *P. infestans.* Tubers of Umatilla Russet, Russet Legend, Gem Russet, and nine advanced clones, including A-90586-11, showed high levels of resistance under in vitro conditions (Table 8.2). Clones with high resistance expressed quantitative differences compared to susceptible potato cultivars and it is possible that they have a high level of partial resistance (Porter et al., 2004) (Appendix 8(iii)).

A high-resolution genetic map at the R3 locus of potato, conferring full resistance to avirulent isolates of *P. infestans,* was constructed. The R3 locus was composed of two genes, *R3a* and *R3b*, with distinct specificities, which were 0.4 cM apart. They were introgressed from *S. demissum*, the donor species of most characterized race-specific R genes to *P. infestans*. One accession of *S. dimissum* showed the presence of natural recombination between *R3a* and *R3b* genes (Huang et al., 2004). Quantitative resistance to *P. infestans* available in tetraploid cultivars is correlated with late maturity in temperate climates and this characteristic is considered undesirable. By employing polymerase chain reaction (PCR), 30 DNA-based markers from tetraploid potatoes for linkage with QTLs for late blight resistance were tested. Most of the markers were found to originate from within, or be physically closely linked to, candidate genes for quantitative resistance factors. Assessment of interactions between unlinked resistance QTL may provide testable strategies for marker-assisted selection (Bormann et al., 2004). In another study, the asso-

TABLE 8.2 Mean Percentage of External Tuber Rot Observed in Artificially Inoculated Tubers of Nine Potato Clones with Either US-8 or US-11 Clonal Lineages of *Phytophthora Infestans*

Clones	External tuber rot percentage				
	1998	1999 trial 1		1999 trial 2	
	US-8 (1799)	US-8 (1799)	US-11 (110B)	US-8 (1799)	US-11 (110B)
Ranger Russet	41.9 a	80.4 a	95.5 a	73.3 a	24.3 ab
Bannock Russet	29.5 ab	12.1 cd	8.8 c	24.6 b	22.1 ab
Russel Norkotah	17.8 cb	44.6 b	30 b	61.3 a	42.9 a
Russet Burbank	8.4 cb	41.3 bc	O c	17.9 b	23.4 ab
Russet Legend	6.5 c	O d	O c	O b	0.5 b
Gem Russet	0.1 c	O d	16.7 bc	O b	O b
Umatilla Russet	0.1 c	O d	5.8 c	O b	O b
A90586-11	0.1 c	O d	O c	O b	O b
LSD (P = 0.65)	21.8	30.3	17.3	27.7	37.7

Values accompanied by the same letter are not significantly different at P = 0.05 according Fisher's least significant difference.
Source: Porter et al., 2004.

ciation between DNA markers and agronomic characteristics in a gene bank collection of 600 potato cultivars, bred between 1950 and 1990 in different countries, was examined. The cultivars were genotyped with five DNA markers linked to previously mapped QTL for resistance to late blight and plant maturity. By using PCR markers specific for *R1*, a major gene for resistance to late blight, a QTL for disease resistance and plant maturity was detected. The marker alleles associated with greater resistance were traced to *Solanum demissum* as the donor of resistance genes (Gebhardt et al., 2004).

The potato germplasm entries were evaluated for their resistance to pink rot disease caused by *Phytophthora erythroseptica* by inoculating disease-free plantlets with zoospore suspensions of the pathogen. Among the 20 cultivars tested, cultivars Goldrush and Yukon Gold were highly susceptible, whereas Butte and Russet Burbank potatoes were the least susceptible. The cultivars with late-season field maturity were more resistant than those with early or midseason maturity (Peters and Struz, 2001). In a later study, tubers of 34 potato cultivars were evaluated for their resistance to *P. erythroseptica.* The percentage of infected tubers and penetration rot (in mm) were the parameters used to assess the level of resistance/susceptibility. The cv. Atlantic appeared to have some resistance to infection and colonization by *P. erythroseptica.* Snowden was highly susceptible to *P. erythroseptica* but it was most resistant to *Pythium ultimum* (Salas et al., 2003). Optimum storage conditions to identify resistance to potato silver scurf disease, caused by *Helminthosporium solani*, among commercially grown potato cultivars, were determined. Field-grown potato tubers were inoculated with conidial suspension (10^4

conidia/ml) and incubated at 15°C with 95% RH for 1 month followed by 2 months at 85% RH. Under these conditions the cultivars exhibited maximum differences in disease severity, indicting the reliability of the screening procedure. Results with greenhouse-grown tubers were comparable with those from field-grown tubers. This procedure has the potential for use in the identification of resistance sources in wild relatives of potato and to understand the pattern of inheritance of resistance to disease (Hilton et al., 2000).

The resistance of serial tubers of potato cultivars Désirée, Epicure, and King Edward to storage pathogens, *Fusarium sulphureum* (*Gibberella cyanogena*), *F. solani* var. *coeruleum*, *Phoma-exigua* var. *foveata*, and *Polyscytalum pustulans*, was assessed in vitro by using freeze-dried powder derived from subterranean and aerial tubers as a culture medium. Growth rates and numbers of conidiophore and conidia produced by the storage pathogens were significantly reduced when cultured on freeze-dried powders from aerial tubers. The higher concentrations of glycoalkaloids, chlorogenic acid, calcium, and magnesium present in aerial tubers in comparison with subterranean tuber tissue probably contribute to enhanced levels of resistance and effects on size and morphology of spores of storage pathogens. The results indicated the need for exploring the possibility of using aerial tubers as an alternative technique for the production of seed potatoes (Percival et al., 1999). The resistance of 20 potato cultivars to powdery scab disease caused by *Spongospora subterranea* was determined by measuring tuber surface covered by, and symptoms induced by, the pathogen. Among the potato cultivars tested, Elkana, Santé, and Nicola had smaller infected areas, while Désirée, Santé, Karnico, Elles, Producent Florijn, and Elkana rarely exhibited symptoms of infection under field and pot culture conditions (Bus, 1999).

Because of the long longevity of the resting spores (cystosori) of *S. subterranea* it has been difficult to control this pathogen. Screening under field condition provided unreliable results due to unpredictable disease incidence and development. Hence, attempts were made to develop a bioassay with potato tissue-cultured plantlets and cystosori as inoculum. This laboratory assay revealed clear differences between potato cultivars in the severity of zoosporangial root infection which correlated better with ranked tuber infection data, compared to root galling. This bioassay procedure can be advantageously employed at an early breeding stage to screen and select resistant lines, thus avoiding expensive field trials and saving considerable time. Furthermore, this assay allows the use of a standard set of pathogen isolates and facilitates testing their virulence levels in infested soils (Merz et al., 2004).

Potato cultivars and selections were evaluated for their resistance to bacterial soft rot due to *Erwinia carotovora* subsp. *carotovora* (*Ecc*) and *E. carotovora* subsp. *atroseptica* (*Eca*) in a moist chamber. The test entries (34) were inoculated with bacterial suspension (10 tubers/entry), using the puncture inoculation method, and incubated for 4 days in a moist chamber at 21°C and 100% RH in the darkness. Most of the entries were susceptible to tuber soft rot, although some differences in disease incidence and in amount of weight reduction were observed. The line AF 522-5 was the most resistant among the entries tested. The soft rot incidence was positively correlated ($r = 0.65$; $P = 0.05$) with reduction in tuber weight (%), indicating that disease incidence was a fair measure of soft rot potential. Late maturing varieties had slightly less weight loss than early maturing varieties (Reeves et al., 1999). Tetraploid potato clones with germplasm from *Solanum chacoense* and *S. phureja* were examined for their resistance to *Ecc*. Tuber soft rot resistance contributed by *S. chacoense* was successfully transferred to the tetraploid (4x) level. The rotted area in the tuber after inoculation with *Ecc* was estimated. One tetraploid clone proved to be resistant and three clones were partially resistant (Capo et al., 2002).

Screening for resistance to diseases has paved the way for the identification of sources of resistance to some of the economically important diseases of tomato. *Alternaria solani*, causing early blight disease, infects both the foliage and fruits of tomato. Tolerance to the foliar phase of early blight was at high levels in tomato lines USDA 67B833, 68B134, and a selection 71B2. The level of resistance shown in the foliar phase is positively correlated to the level of resistance in the fruit rot phase. Hence, the plants can be tested earlier without waiting until fruits are produced to assess their resistance. This is a distinct advantage and lot of time can be saved. These resistance sources have been used in breeding programs (Barksdale and Stoner, 1977). Both race-specific (vertical) and polygenic (horizontal) resistance to tomato late blight disease caused by *Phytophthora infestans* have been identified in tomato accessions. The small-fruited Red cherry and West Virginia accessions 36 and 106 possess race-specific resistance. The tomato varieties Southland, Wisconsin 55, Australian Earliana, and Danish Extra Early have been found to exhibit moderate level of resistance against most races present in different locations. A Florida variety, Floradade, was reported to show resistance to *Botrytis cinerea*, causing gray mold and ghost spot disease in tomato fruits (Sherf and Macnab, 1986). *Solanum lycopersicoides*, the wild night shade cross-compatible with tomato, has been shown to be a potential source of resistance to *Botrytis cinerea*, causing gray mold disease in hundreds of plant species. The accession *S. lycopersicoides* LA 2951 was the most resistant among the accessions tested. Expression of resistance in the intergeneric hybrid (*L. esculentum* cv. VF36 × *S. lycopersicoides* LA 2951) suggested the resistance to be partially dominant in tomato. By using a green fluorescent protein tagging technique to label the pathogen and confocal microscopy, evidence for hyphal lysis and death of *B. cinerea* at 3 days after inoculation of *S. lycopersicoides* was obtained, suggesting the possible mechanism of resistance operating in this plant species (Guimarães et al., 2004).

The availability of sources of resistance to diseases of pepper (capsicum) has been indicated by different investigations. Resistance to anthracnose disease caused by *Colletotrichum capsici* and *C. piperatum* has been identified in Red Cherry Casabella. Mexican Chili and accessions in *Capsicum sinensis* var *panca*, *C. fasciculatum* and *C. annuum* such as Artakis, Paprika, and Wonder Top (Ullasa et al., 1981; Sherf and Macnab, 1986). An accession of *Capsicum chinense* was highly resistant to *C. capsici* and *C. gloeosporioides*, causing anthracnose/fruit rot disease. By using a QTL mapping approach, the pattern of inheritance of resistance in an F_2 population of cross between *C. chinense* (resistant) and an Indonesian hot pepper variety (*C. annuum*) (susceptible) was studied. Three resistance-related traits, that is the infection frequency, true lesion diameter (averaged over all lesions that were formed), and the overall lesion diameter (averaged over all inoculation points), were considered to determine the levels of resistance. One main QTL was identified with highly significant and large effects on all three traits following inoculation with *C. gloeosporioides* and on true lesion diameter after inoculation with *C. capsici* (Voorrips et al., 2004).

The varietal susceptibility of onion cultivars to *Fusarium oxysporum* f. sp. *cepae* was assessed in the field, at harvest, and after storage. The cultivar Takmark was found to be tolerant to the *Fusarium* basal rot disease of onion (Visser de, 1999). A screening protocol to determine the levels of resistance of onion cultivars to black mold disease caused by *Aspergillus niger* was developed. The protocol includes wound inoculation of onion slices with 40 μl of spore suspension (1×10^5 spores/ml) and incubation at 25°C with 12 h photoperiod, and rating lesion diameter and sporulation at 4 days of after inoculation. Among 42 onion cultivars tested, Red Pinoy, Serrana, Dehydrator No. 3, and Moonlight were tolerant to *A. niger*. These cultivars reacted to inoculation with *A. niger* by producing small lesions which did not sporulate. The susceptible cultivars, such as Explorer, produced larger lesions with profuse sporulation. This screening protocol and the identified tolerant cultivars are being used as donors of tolerance in the Asian Vegetable Research and Development Center (AVRDC), Taipei, in breeding programs for resistance to black mold disease of onion (Ko et al., 2002). The pungency levels, based on the pyruvic acid contents of onion cultivars, was positively correlated with the of incidence of soft rot disease caused by *Erwinia carotovora* subsp. *carotovora*. Arka Pragati and Country onion were found to be resistant to soft rot disease (Sendhil Vel et al., 2001).

Attempts to identify sources of resistance to diseases of cruciferous crops have been successful in certain cases. The resistance of seven Russian and F1 hybrid varieties of cabbage to storage pathogens *Botrytis cinerea* and *Alternaria brassicola* was evaluated. Leaf discs (15 mm) were inoculated with spore suspensions and incubated at 22°C for 7 days. The areas of necrotic tissue in the inoculated leaf discs were measured. The cultivars Galaxy and Monarch were resistant to both pathogens. The results indicated that the resistance to *A. brassicola* can be assessed even during the vegetative growth and one need not wait until harvest. In the case of *B. cinerea*, the dry matter content of the leaves covering the head before harvest may be used an index of resistance to this pathogen (Dzhalilov et al., 1999).

The *Tsw* locus from *Capsicum chinense* governing resistance to *Tomato spotted wilt virus* (TSWV) has been utilized in breeding programs. Accessions of several *Capsicum* spp. were mechanically inoculated with TSWV to select new sources of resistance, since isolates of TSWV that can overcome resistance offered by *Tsw* locus have been detected. Accessions with resistance to virus replication and with a low percentage of infection were identified. A *C. chinense* accession ECU-973 showed resistance to TSWV inoculated both by mechanical and vector inoculation (Cebolla-Cornejo et al., 2004). The resistance-breaking strains of TSWV that could overcome single gene hypersensitive resistance in pepper were found to be stable. As the possibility of a break down of vertical resistance because of the production of virus strains with a matching virulence gene is known, efforts have to be directed towards the development of durable resistance (Thomas-Carroll and Jones, 2003). Among the resistance genes (>40) identified in the genus *Lycopersicon*, the gene *Sw-5* conferring resistance to TSWV was detected in *L. peruvianum* and transferred to the cv. Stevens in South Africa (Stevens et al., 1992). The gene *Sw-5* also confers resistance to two related tomato-infecting tospoviruses, *Groundnut ringspot virus* (GRSV) and *Tomato chlorotic spot virus* (TCSV) (Boiteux and Giordano, 1993). This gene was mapped on the long arm of chromosome 9 of tomato. The use of molecular markers linked to resistance genes of interest has contributed to the improvement of conventional breeding methods. It is possible to select the resistance genes directly through markers tightly linked to the resistance genes. The transfer of *Sw-5* into tomato lines was achieved through molecular assisted backcross breeding schemes performed with the aid of one cleaved amplified polymorphic sequence marker tightly linked to the gene *Sw-5*. The codominant nature of this marker was highly useful in selecting resistant heterozygous/homozygous plants. Successful introgression of resistance gene in elite tomato lines was confirmed by artificial inoculation with TSWV (Langella et al., 2004).

Summer squash (*Cucurbita pepo*), when infected early by papaya ring spot virus – watermelon strain (PRSV-W) suffers severely, resulting in marketable yield losses of up to 100%. Accessions of *C. pepo* with genetic resistance to PRSV were rare. Hence, interspecific hybridization between *C. pepo* and a PRSV-resistant *C. moschata* squash accession was made and the early stages of introgression of this resistance into *C. pepo* were studied. The F2 population were mechanically inoculated with PRSV-W. They exhibited a response intermediate between those of the parental accessions and the estimates of the mean degree of dominance indicated incomplete dominance in the direction of susceptibility. It was suggested that introgression of resistance to PRSV-W of *C. moschata* into *C. pepo* appeared to be technically feasible (de Oliveira et al., 2003).

The levels of resistance of lettuce to *Lettuce mosaic virus* (LMV), which is transmitted through seeds, were assessed by employing an in vitro inoculation and propagation method. In vitro cultivated seedlings or newly generated plantlets were inoculated with various natural isolates of LMV, as well as with green fluorescent protein-tagged recombinant virus isolates. The results of in vitro screening assay were well correlated to known resistance status of cultivars. The efficiency of resistance assay could be significantly improved by the use of green fluorescent protein-tagged virus isolates. The chief advantages of this resistance assay are reduced space requirements and an improved environmental safety while handling recombinant virus and quarantine of virus or transgenic plants (Mumford et al., 2004).

Perennial plants are exposed to the pathogens for a long time, when compared to annuals. Varying levels of pathogen populations will be present in different growth stages of the perennial plants resulting in different disease pressures and consequent variations in disease scores. This situation reflects the wide variations in disease assessments reported by researchers in different geographical locations. This applies even more to postharvest diseases, where only a few investigations have been directed at developing resistant cultivars. Screening of cultivars and germplasm types has revealed the availability of sources of resistance that may be exploited for the development of resistant cultivars of fruits.

The levels of resistance to decay caused by *Botrytis cinerea*, *Penicillium expansum*, *Mucor piriformis*, and *Pezicula malicorticis* based on the force required to break the epidermis and the extent of open sinuses as measures of potential decay resistance, were assessed for apple cultivars. The cultivar Royal Gala was the most resistant to wound pathogens, *B. cinerea*, *P. expansum*, and *M. piriformis*, whereas Fuji and Oregon Spur II were more resistant to *P. malicorticis* than other cultivars (Spotts et al., 1999). The relative susceptibility of 23 apple cultivars to white rot pathogen *Botryosphaeria dothidea* was assessed by inoculating the fruits 2 to 3 weeks prior to harvest and also in vitro. Two parameters, that is disease severity of attached fruit in the field based on lesion growth (mm/degree-day) and disease severity of detached fruit following laboratory inoculations of wounded fruit (mean lesion diameter after 5 days) were used. Based on both assessments made during 2 years, the cultivars were classified as most susceptible (Fortune and Pristine), moderately susceptible (Golden Supreme, Golden Delicious), and least susceptible (Honeycrisp, Yataka, Suncrisp, and PioneerMac). Some new apple cultivars showed greater resistance to *B. dothidea* than current, standard cultivars (Biggs and Miller, 2003) (Appendix 8(iv)).

Grapefruit cultivars and mandarin cultivars were evaluated for their relative susceptibility to *Alternaria alteranata* pv. *citri* causing brown spot disease by inoculation with conidia. They were found to be highly susceptible, whereas lemon and lime cultivars, with the exception of Mexican lime, were not susceptible (Vincent et al., 2004). Improving the resistance of sweet orange and grapefruit to *Citrus tristeza virus* (CTV) disease by conventional breeding methods has been found to be difficult, since they are highly heterozygous hybrids derived by introgression of the true *Citrus* species mandarin (*Citrus reticulata*) and pummelo (*C. grandis*) (Moore, 2001). Expression of CTV-resistance at single cell level is one of the effective methods of characterizing resistance mechanisms operating in the host genotype. Such a resistance mechanism can be easily analyzed in protoplasts from callus prior to regeneration or from plantlets prior to acclimatization. This procedure may help reduce the time and glasshouse space needed for screening adult plants for resistance to CTV. The mechanism of resistance to CTV in citrus relatives *Poncirus trifoliata*, *Swinglea glutinosa*, *Severinia buxifolia*, and selected *Citrus* × *Poncirus* hybrids was studied by isolating leaf and/or cultured cell protoplasts and inoculating them with four biologically distinct CTV isolates. The RNAs of newly produced CTV virions in protoplasts were analyzed by northern blot hybridization and immunoelectron microscopy. Detection of the presence of CTV virions and RNAs in protoplasts from all resistant plants suggested that resistance to CTV observed at plant levels may be due to lack of virus movement and/or some induced resistance response, rather than the failure of virus replication at the cellular level (Albiach-Marti et al., 2004).

Tolerance of peach cultivars to brown rot disease caused by *Monilina laxa* was determined by spray-inoculating 30 tree-ripe fruits for each cultivar with a conidial suspension (10^5 conidia/ml) on the skin and incubating them at 25 ± 2°C and 95 to 100% RH for 7 days. The percentage of fruits infected and percentage of skin rotted were the parameters used to grade the tolerance levels of test cultivars. Glohaven (yellow flesh peach) was the least susceptible followed by Contender, Maria Aurelia (yellow nectarine), and Maria Bianca (white peach). The spray inoculation method was recommended for screening peach cultivars to eliminate highly susceptible ones in breeding programs (Bassi et al., 1998). Mango cultivars were tested for their resistance to postharvest diseases stem-end rot caused by *Lasiodiploida theobromae* and blackrot caused by *Aspergillus niger*. The cultivar Makaram was resistant to both stem-end rot and black rot diseases, whereas Sardar and Vanraj showed resistance to black rot only (Jadeja et al., 2000). Papaya cultivars were evaluated for the tolerance to anthracnose disease caused by *Colletotrichum gloeosporioides* by inoculating the fruits and storing them at 18 ± 3°C. The cultivar Maradol was the most tolerant to anthracnose disease during storage (Ramos, et al., 2000). The major apple scab resistance gene *Vf* has been used extensively in apple breeding programs for development resistant cultivars. The new races 6 and 7 of the pathogen *Venturia inaequalis* have overcome the resistance gene, necessitating the search for more durable resistance genes. A partially resistant apple cv. Discovery and apple hybrid TN 10-8 were inoculated in the greenhouse with eight isolates of *V. inaequalis*, including isolates capable of overcoming *Vf*. One major resistance gene, *Vg*, and seven QTLs resistant to these isolates were identified (Calenge et al., 2004).

Genes conferring disease resistance, when cloned, show remarkable similarities in general structure. Specific conserved domains play a role in protein–protein interaction and signal transduction. Based on the predicted structure of the protein products, the resistance genes have been grouped into several classes. Genes which show similarity to resistance genes have been termed resistance gene analogs (RGAs). It is possible, due to the conservative sequences of RGAs, to design primers for the identification of further RGAs by PCR. Putative RGAs of apricot cv. Stark Early Orange were amplified, cloned, and characterized. Specific primers designed on the basis of hypervariable regions of the sequences identified were used to identify molecular markers for resistance to Sharka disease of apricot caused by *Plum pox potyvirus* (PPV). The RGA marker SEOBT 101, which amplified a band only in the resistant genotypes, was considered to be putatively associated with Sharka resistance (Dondini et al., 2004).

8.3.2 Resistance to Mycotoxigenic Pathogens

Wheat and barley are infected by *Fusarium graminearum* causing *Fusarium* head blight (FHB) disease considered as a continual world-wide problem because of quantitative reduction in yield and qualitative loss due to the contamination of grains with deoxynivalenol (DON) produced by the pathogen. As tetraploid wheat sources of resistance to FHB are scarce, 151 *Triticum dicoccoides* genotypes originating from 16 habitats in Israel and one habitat in Turkey were evaluated for reaction to fungal spread in replicated greenhouse experiments. Significant genetic diversity was found among the tested genotypes. Only a few exhibited moderate resistance. However, none of the *T. dicoccoides* lines possessed the level of resistance found in the common wheat cultivar Sumai 3 (Buerstmayr et al., 2003). Attempts were made to identify the QTL for FHB severity, DON level, and other traits, so that the basis for marker-assisted selection may be provided. Five FHB QTLs were found to be associated with traits such as heading date and plant height. The investigation by Dahleen et al. (2003) offered a starting point for manipulating Zhedar 2-derived resistance by marker-assisted selection in barley to develop cultivars that may exhibit effective resistance under disease pressure.

Many of the QTL for FHB resistance in barley have been found to coincide with the QTLs for plant height, heading date, and spike characteristics. The relationship of morphological and physiological traits to FHB infection and DON accumulation in a barley double-haploid population derived from a Léger × C19831 cross was examined. The plants were inoculated with *F. graminearum* under field conditions in three locations (Ottawa, Charlotte Town, and Hangzhou Zhejiang). FHB incidence and DON contents were positively correlated. FHB resistance was associated with two-row spike, purple lemma, long glume awn, tall stature, and resistance to lodging (Choo et al., 2004). Studies were taken up to identify FHB resistance in wheat (*Triticum aestivum*) – *Lophopyrum* genetic lines that might complement FHB resistance in common wheat and to identify DNA markers that may be useful to tag the resistance genes. The results indicated that the transferability of wheat simple sequence repeat markers to *Lophopyrum* was low. A cleaved amplified polymorphic sequence marker derived from the RFLP probe PSR 129 showed the potential to serve as a dominant marker of resistance (Shen et al., 2004).

A new in vitro assay, involving the inoculation of detached leaves with *Microdochium nivale*, was employed to assess the level of resistance of wheat cultivars to *Fusarium* head blight disease and the results were compared with whole plant resistance ratings of 21 wheat genotypes. An incubation temperature of 10°C and isolates of *M. nivale* var. *majus* of intermediate pathogenicity were found to be useful to assess the differential expression of several components of partial disease resistance. Positive correlations between incubation period and latent period and negative correlations between incubation period and lesion length were established. Commercial cultivars identified with high resistances across all three partial disease resistance components (incubation period, latent period, and lesions length) in the detached leaf assay also exhibited high whole plant FHB resistance ratings except one cultivar Tanker. The most resistant Irish and UK commercial cultivars could be compared to the genotype Frontana and the most resistant CIMMYT germplasm evaluated in the leaf assay (Browne and Cooke, 2004).

A genetically diverse collection of corn inbreds were evaluated for their potential as sources of resistance to fumonisin production and accumulation in grain and *Fusarium* ear and kernel rot caused by *F. verticilioides* and *F. proliferatum*. The F1 hybrids developed with inbred FR1064, and 1589 and 1030 inbreds, were evaluated under controlled and natural conditions. Low fumonisin content in grain and low ear rot severity were observed in several F1 hybrids and their distinct F2 and backcross to FR1064 generations. The results suggest that resistance was governed by several dominant genes and it might be possible to transfer alleles for resistance from these inbreds to FR1064 (Clements et al., 2004). A diverse collection of 158 barley lines were evaluated for their resistance to *F. graminearum* and deoxynivalenol (DON) accumulation. The genotypes resistant to FHB and low DON content following artificial inoculation with *F. graminearum* were identified for use in breeding programs (Mc Callum et al., 2004) (Appendix 8(v)).

It is essential to apply an efficient inoculation technique to assess the levels of resistance of corn genotypes to *Fusarium* ear rot disease. Among the four inoculation techniques tested, injection of inoculum through the husk leaves was more effective in increasing the concentration of fumonisin in grain and severity of ear rot compared with control. Furthermore, this technique also effectively differentiated corn hybrids identified earlier as resistant or susceptible to Fusarium ear rot. Hence this technique has the potential for evaluating corn genotypes for *Fusarium* ear rot resistance and fumonisin concentration (Clements et al., 2003). The inheritance of resistance to aflatoxin production and *Aspergillus* ear rot of corn was studied using corn inbred Oh516 as a source of resistance. Molecular markers associated with resistance were identified from backcross progeny (1) self (1) (BCP(1) S(1)) families developed from the cross of the susceptible inbred B73 and resistant Oh516. The progenies in different generations were evaluated for aflatoxin concentration in grain, percent bright greenish yellow fluorescence (BGYF) (a test for the presence of aflatoxin), and severity of *Aspergillus* ear rot following inoculation. Dominant gene action was related to aflatoxin concentration in grain and percent BGYF. QTLs associated with aflatoxin accumulation in grain were identified on chromosomes 2, 3, and 7 (Busboom and White, 2004).

SUMMARY

Among various strategies adopted for the management of postharvest diseases caused by microbial pathogens, development of resistant cultivars through breeding for resistance has been considered as the most desirable option. The responses of host plants to compatible and incompatible pathogens have been studied at the molecular level to have an insight into the interplay of various physiological processes activated by the pathogen or activation of various defense mechanisms of the host. Distinct differences have been observed in the differential responses reflected by the presence structural barriers that are strategically positioned due to constitutive resistance and production of various defense-related compounds following development of active defense mechanisms. Several pre-existing substances and compounds produced following activation of resistance mechanisms have been characterized and the role of some of them in the development of resistance has been established. Various investigations with the primary objective of developing resistant cultivars using reliable screening methods have been followed. The possibility of containing the postharvest diseases through the use of host resistance is discussed.

APPENDIX 8(I): ASSESSMENT OF ENZYMATIC ACTIVITIES IN CARROT (COOPER ET AL., 2004)

A. β-glucosidase Activity

i. Prepare the enzyme extract (0.2 ml aliquots) and the substrate (0.8 ml) of 0.2 mg/ml *p*-nitrophenyl β-D-glucoside in 0.1 M sodium acetate buffer, pH 5.2 and incubate the reactants at 34°C for 3 h, with a blank (0.2 ml of water).
ii. Stop the reaction by heating in a boiling water bath (100°C) for 7 min and cool.
iii. Add 0.1 M Tris buffer (1.0 ml) and dilute the solution to 3 or 5 ml depending on the intensity of color.
iv. Determine the color intensity of the blank and the samples at 400 nm in a spectrophotometer.
v. Prepare a standard curve using different concentrations of *p*-nitrophenol (between 0 and 50 μg/ml) and convert the values to enzyme units. One unit of glucosidase will liberate 1 μ mol of *p*-nitrophenol per min.

B. Chitinase Activity

i. Incubate samples (0.1 ml) of enzyme solution with 0.4 ml of chitin-azure substrate solution (1.83 mg/ml chitin-azure in 0.1 M sodium acetate buffer, pH 5.2) at 37°C for 4 h.
ii. Add 0.5 ml of sodium acetate buffer and make up the volume to 1 ml.
iii. Incubate the blank with 0.1 ml of water in place of enzyme solution.
iv. Read the color intensity in a spectrophotometer at 575 nm.
v. Calculate the enzyme units based on a standard curve.

APPENDIX 8(II): ASSESSMENT OF CHITINASE ACTIVITY IN MAIZE KERNEL EXTRACT (MOORE ET AL., 2004)

A. Preparation of Maize Kernel Extract

i. Harvest the maize ears (inbred line Tex6, resistant to infection and aflatoxin accumulation by *A. flavus*) at desired intervals after pollination; freeze them in liquid nitrogen; shell the kernels onto dry ice and store at −80°C until required.
ii. Grind the kernels at 4°C in a coffee grinder; suspend the powder (1 kg) in 3 l of citrate buffer (50 mM, pH 5.5); stir well for 1 h and filter through cheese cloth.
iii. Centrifuge at 6000 g for 30 min; filter through Miracloth and use the filtrate to determine the chitinase activity.

B. Chitinase Assay

i. Incubate colloidal chitin and Tex6 kernel extract in citrate buffer (50 mM, pH 5.5) at 28°C in the reaction vessel.
ii. Remove aliquots at selected intervals after incubation and inject directly onto HPLC column.
iii. Integrate peak area associated with the standard curves for the concentrations of mono-, di-, and tri-N-acetyl glucosamines to get a direct reading of chitinase activity at 28°C.

C. Purification of Tex6 Chitinase

i. Perform all steps at 4°C, unless otherwise indicated.
ii. Apply the extracts from mature kernels to Sephadex G-100 GFC Column (1.5 × 1.5 cm) and elute at a flow rate of 9 ml/h.
iii. Pool the fractions exhibiting antifungal activity; concentrate with a Centriprep-10 Unit (Millipore, Billerica, MA) and subject the pooled sample to chitin affinity chromatography, using an affinity column (10 × 0.5 cm) containing purified powdered chitin from crab shells (Sigma Chemical) suspended in 50 mM citrate, pH 5.5.
iv. Wash the column with citrate buffer (50 mM, pH 5.5) and elute with NaCl (2 M) by gravity flow.
v. Add active fractions from the affinity column to a carboxy methyl column (Bio-Rad Laboratories, Hercules, CA); wash with citrate buffer (50 mM, pH 5.5) and elute sequentially with NaCl (75 mM) and NaCl (125 mM); carry out sodium dodecyl sulfate–polyacrylamide gel electrophoresis (SDS-PAGE); stain the gels with Coomassie brilliant blue G-250 and visualize the bands.

APPENDIX 8(III): ASSESSMENT OF RESISTANCE OF POTATO TUBERS TO TUBER ROT CAUSED BY *PHYTOPHTHORA INFESTANS* (PORTER ET AL., 2004)

A. Preparation of Inoculum

i. Multiply the inoculum on detached potato leaflets placed on a fiberglass screen over moistened paper towels in sealed plastic containers and incubate at 15°C with an 18-h photoperiod.
ii. Wash the sporangia from sporulating lesions on the leaflets; adjust sporangial concentration to 1×10^4 sporangia/ml, using a hemocytometer and chill the suspension at 4°C for 2h before inoculation.

B. In Vitro Assessment of Resistance

i. Consider individual tubers as experimental unit and lightly mist the tubers with distilled water.
ii. Inoculate on single eyes located in the middle of the tuber; arrange the inoculated tubers in a suitable statistical design (randomized design) and maintain four to six replicates for each experiment.
iii. Place the tubers in a moist chamber for 18h followed by a 24-h dry period.
iv. Transfer the tubers to store at 9°C after drying.
v. Peel the tubers to an approximate depth of 3mm and assess the severity of symptoms after 36 days by estimating visually the percentage of external surface area with late blight symptom.
vi. Use the following scale for determining resistance category: Highly resistant <2% external rot; Moderately resistant 3 to 6% external rot; Susceptible >7% external rot.

APPENDIX 8(IV): ASSESSMENT OF REACTIONS OF APPLE CULTIVARS TO FRUIT ROT (BIGGS AND MILLER, 2003)

A. Cultures of Fungal Pathogen

i. Select suitable isolates of the pathogen (*Botryosphaeria obtusa*); subculture at required intervals (biweekly) and maintain them on potato dextrose agar (PDA) in Petri plates at 20°C under continuous fluorescent light during the experimental period.

B. Field Evaluation

i. Inoculate the fruit at 2 to 3 weeks preharvest, as determined by average ripening date and quality assessments; make 1-mm deep wound through epidermis of arbitrarily selected fruit using a sterile cork borer (5 mm diameter); remove the circumscribed epidermis; place an agar plug with pathogen (mycelium) over the wound and wrap the wounds in parafilm to maintain moisture.

ii. Remove the parafilm at 4 days after inoculation.
iii. Maintain three replicates of eight fruits each and as many fruits as control.
iv. Reisolate the pathogen from the inoculated fruit periodically to monitor the development of pathogen.

C. Laboratory Evaluation

i. At 2 to 3 weeks prior to normal harvest date, detach the fruit and wash them well with tap water.
ii. Inoculate the fruit as in step B (i) and place in plastic trays with lids and incubate at 21 to 23°C in the laboratory.
iii. Maintain replications as in field evaluation and reisolate the pathogen from inoculated fruits as in field evaluation.
iv. Determine flesh firmness using hand-held penetrometer (Effigi nc., Bologna, Italy) and soluble solids using a hand-held refractometer (Fisher Scientific, Pittsburgh).
v. Determine the disease severity of attached fruit in the field and disease severity of detached fruit in the laboratory inoculations, based on measurement of length and width of each lesion at 5 days postinoculation.
vi. Determine the final cultivar ranks by averaging the mean ranks for field and laboratory assessments.

APPENDIX 8(V): SCREENING BARLEY LINES FOR RESISTANCE TO FUSARIUM HEAD BLIGHT (FHB) DISEASE (MC CALLUM ET AL., 2004)

A. Preparation of Pathogen Inoculum

i. Isolate the pathogen (*Fusarium graminearum*) from FHB-infected barley spikes and cultivate the isolates in Petri plates containing potato dextrose agar.
ii. When the pathogen growth reaches optimal level cut the colonized agar into small squares (2 × 2 cm) and transfer them to a flask containing liquid medium consisting of carboxy methyl cellulose (15.0 g), NH_4CO_3(1.0 g), KH_2PO_4 (1.0 g), $MgSO_4.7H_2O$ (0.5 g), yeast extract (1.0 g), and distilled water (1 liter), incubate at room temperature for 5 days and gently agitate the liquid culture.

iii. Filter through the cheese cloth; dilute the filtrate with distilled water to have a concentration of 5.0×10^4 macroconidia/ml and add a drop of surfactant Tween-20.

B. Inoculation of Barley Lines

i. Plant single 1.5 m rows for each test entry and spray-inoculate the barley spikes at heading, using approximately 100 ml of the conidial suspension (5.0×10^4/l)
ii. Repeat inoculation at 7 days after heading.

C. Assessment of Reactions of Barley Lines

i. Harvest the spikes from a 60-cm section in the middle of each row at 14 days after final inoculation; place them at –20°C to freeze the plant tissues; thaw the frozen spikes and remove the immature spikes that could have escaped inoculation.
ii. Calculate FHB index (disease incidence × average severity) for each line on approximately 30 spikes, using the parameters as detailed below:
iii. a. isease incidence = No. of infected spikes / total no. of spikes examined (infected spikes exhibit characteristic tan brown or dark brown or black color compared with healthy green spikelets.);
   b. Average severity = Mean number of infected spikelets/mean number of total spikelets.

REFERENCES

Abu-goukh, A.A. and Labavitch, J.M. 1983. The *in vivo* role of 'Bartlett' pear fruit polygalacturonase inhibitors. *Physiological Plant Pathology*, **23**: 123–125.

Adikaram, N.K.B., Brown, A.E. and Swinburne, T.R. 1982. Phytoalexin involvement in latent infection of *Capsicum annuum* L. fruit by *Glomerella cingulata* (Stonem). *Physiological Plant Pathology*, **21**: 161–170.

Adikaram, N.K.B., Brown, A.E. and Swinburne, T.R. 1988. Phytoalexin induction as a factor in the protection of *Capsicum annuum* L. fruits against infection by *Botrytis cinerea* Pers. *Phytopathologische Zeitschrift*, **122**: 267–273.

Agarwal, V.K. and Sinclair, J.B. 1996. *Principles of Seed Pathology, 2nd edition*. CRC–Lewis Publishers. Boca Raton, FL, USA.

Albiach-Marti, M.R., Grosser, J.W., Gowda, S., Mawassi, M., Satyanarayana, T., Garnsey, S.M. and Dawson, W.O. 2004. *Citrus tristeza virus* replicates and forms infectious virions in protoplasts of resistant relatives. *Molecular Breeding*, **14**: 117–128.

Anand, A., Lei, Z., Sumner, L.W., Kirankumar, S.M., Arakane, Y., Bockus, W.W. and Muthukrishnan, S. 2004. Apoplastic extracts from a transgenic wheat line exhibiting lesion-mimic phenotype have multiple pathogenesis-related proteins that are antifungal. *Molecular Plant–Microbe Interactions*, **17**: 1306–1317.

Andersen, I., Becker, W., Schlüter, K., Burges, J., Parthier, B. and Apel, K. 1992. The identification of leaf thionin as one of the main jasmonate-induced proteins of barley (*Hordeum vulgare*). *Plant Molecular Biology*, **19**: 193.

Andreu, A. and Daleo, G.R. 1988. Properties of potato lectin fractions isolated from different parts of the tuber and their effect on growth of *Phytophthora infestans. Physiological and Molecular Plant Pathology*, **32**: 323–333.

Astua-Monge, G., Minsavage, G.V., Stall, R.E., Vallejos, C.E., Davis, M.J. and Jones, J.B. 2000. *Xv4-vrxv4*: A new gene-for-gene interaction between *Xanthomonas campestris* pv. *vesicatoria* Race T3 and the wild tomato relative *Lycopersicon pennelli. Molecular Plant–Microbe Interactions*, **13**: 1346–1355.

Audenaert, K., Meyer, G.B. de and Höfte, M.M. 2002. Abscisic acid determines basal susceptibility of tomato to *Botrytis cinerea* and suppress salicylic acid-dependent signal mechanisms. *Plant Physiology*, **128**: 491–501.

Bagirova, S.F., Ignatova, S.I. and Tereshonkova, T.A. 2003. Tomato resistance to *Phytophthora infestans* as revealed by electronmicroscopy. *Acta Horticulturae*, **613.**

Baker, C.J. 2001. Active Oxygen. In: *Encyclopedia of Plant Pathology* (eds) O.C. Maloy and T.D. Murray, Vol.1, pp. 6–9. John Wiley and Sons, New York.

Baldwin, E.A. and Biggs, R.H. 1988. Cell wall-lysing enzymes and products of cell wall digestion elicit ethylene in citrus. *Physiologia Plantarum*, **73**: 58–64.

Baldwin, E.A. and Pressey, R. 1988. Tomato polygalacturonase elicits ethylene production in tomato fruit. *Journal of the American Society of Horticultural Science*, **113**: 92–95.

Baldwin, E.A. and Pressey, R. 1990. Exopolygalacturonase elicits ethylene production in tomato. *HortScience*, **25**: 779–780.

Barksdale, T.H. and Stoner, A.K. 1977. A study of the inheritance to tomato early blight resistance. *Plant Disease Reporter*, **61**: 63–65.

Bassi, D., Rizzo, M. and Cantoni, L. 1988. Assaying brown rot [*Monilinia laxa* Aderh etluhl. (Honey)] susceptibility in peach cultivars and progeny. *Acta Horticulturae*, **465**: 715–721.

Bateman, D.F. and Basham, H.G. 1976. Degradation of plant cell walls and membranes by microbial enzymes. In: *Encyclopedia of Plant Physiology* (eds) R. Heitefuss and P.H. Williams, pp. 316–355. Springer–Verlag, New York.

Wang, X., Beno-Moualem, D., Kobiler, I., Leikin-Frenkel, A., Lichter, A. and Prusky, D. 2004. Expression of 12 fatty acid desaturase during the induced accumulation of antifungal diene in avocado fruits. *Molecular Plant Pathology*, **5**: 575–585.

Berrocal-Lobo, M., Segura, A., Moreno, M., López, G., García-Olmedo, F. and Molina, A. 2002. Snakin–2, an antimicrobial peptide from potato whose gene is locally induced by wounding and responds to pathogen infection. *Plant Physiology*, **128**: 951–961.

Bézier, A., Lambert, B. and Baillieul, F. 2002. Study of defense-related gene expression in grapevine leaves and berries infected with *Botrytis cinerea. European Journal of Plant Pathology*, **108**: 111–120.

Biggs, A.R. and Miller, S.S. 2003. Relative susceptibility of selected apple cultivars to *Botryosphaeria dothidea. HortScience*, **38**: 400–403.

Black, W. 1960. Races of *Phytophthora infestans* and resistance problems in potatoes. *Report of Scottish Society of Research in Plant Breeding*, pp. 29–38.

Bohlmann, H. 1994. The role of thionins in plant protection. *Plant Sciences*, **13**: 1–16.

Boiteux, L.S. and Giordano, L.B. 1993. Genetic basis of resistance against two tospoviruses species in tomato (*Lycopersicon esculentum*). *Euphytica*, **71**: 151–154.

Boogaart, T., van den, Maule, A.J., Davies, J.W. and Lomonossoff, G.P. 2004. Sources of target specificity associated with the recovery against *Pea seedborne mosaic virus* infection mediated by RNA silencing in pea. *Molecular Plant Pathology*, **5**: 37–43.

Bormann, C.A., Rickert, A.M., Ruiz, R.A.C., Paal, J., Lübeck, J., Strahwald, J., Buhr, K. and Gebhardt, C. 2004. Tagging quantitative trait loci for maturity-corrected late blight resistance in tetraploid potato with PCR-based candidate gene markers. *Molecular Plant–Microbe Interactions*, **17**: 1126–1138.

Bortolotti, S., Boggio, S.B., Delgado, L., Orellano, E.G. and Vale, E.M. 2003. Different induction patterns of glutamate metabolizing enzymes in ripening fruits of the tomato mutant green flesh. *Physiologia Plantarum*, **119**: 384–391.

Brown, A.E. 1973. Development of green mold in degreened oranges. *Phytopathology*, **63**: 1104–1107.

Brown, A.E. and Adikaram, N.K.B. 1982. The differential inhibition of pectin enzymes from *Glomerella cingulata* and *Botrytis cinerea* by cell wall protein from *Capsicum annuum* fruit. *Phytopathologische Zeitschrift*, **105**: 27–38.

Brown, A.E. and Adikaram, N.K.B. 1983. A role for pectinase and protease inhibitors in fungal rot development in tomato fruits. *Phytopathologische Zeitschrift*, **106**: 239–251.

Brown, A.E. and Swinburne, T.R. 1980. The resistance of immature banana fruits to anthracnose [*Colletotrichum musae* (Berk and Curt.) Arx]. *Phytopathologische Zeitschrift*, **122**: 267–273.

Brown, A.E. and Swinburne, T.R. 1981. Influence of iron and iron chelators on formation of progressive lesions by *Colletotrichum musae* on banana fruits. *Transactions of British Mycological Society*, **77**: 119–124.

Browne, R.A. and Cooke, B.M. 2004. Development and evaluation of an *in vitro* detached leaf assay for pre-screening resistance to *Fusarium* head blight in wheat. *European Journal of Plant Pathology*, **110**: 91–102.

Brümmer, J., Thole, H. and Kloppstech, K. 1994. Hordothionins inhibit protein synthesis at the level of initiation in the wheat germ system. *European Journal of Biochemistry*, **219**: 425.

Buerstmayr, H., Stierschneider, M., Steiner, B., Lemmens, M., Grisser, M., Nevo, E. and Fahima, T. 2001. Variation for resistance to head blight caused by *Fusarium graminearum* in wild emmer *Triticum dicoccoides* originating from Israel. *Euphytica*, **130**: 17–23.

Bus, C.B. 1999. Powdery scab: symptoms and cultivar differences. *PAV-Bulletin Akkerbouw*, October 16–17, [N1] PAV-Lelystad, Lelystad, Netherlands.

Busboom, K.N. and White, D.G. 2004. Inheritance of resistance to aflatoxin production and *Aspergillus* ear rot of corn from the cross of inbreds B73 and Oh 516. *Phytopathology*, **94**: 1107–1115.

Calenge, F., Faure, A., Goerre, M., Gebhardt, C., Van de Weg, W.E., Parisi, L. and Durel, C.E. 2004. Quantitative trait loci (QTL) analysis reveals both broad-spectrum and isolate-specific QTL for scab resistance in an apple progeny challenged with eight isolates of *Venturia inaequalis*. *Phytopathology*, **94**: 370–379.

Campion, C., Massiot, P. and Rouxel, F. 1997. Aggressiveness and production of cell wall degrading enzymes by *Pythium violae, Pythium sulcatum* and *Pythium ultimum* responsible for cavity spot on carrot. *European Journal of Plant Pathology*, **103**: 725–735.

Campion, C., Vian, B., Nicole, M. and Rouxel, F.A. 1998. A comparative study of carrot root tissue colonization and cell wall degradation by *Pythium violae* and *Pythium ultimum* two pathogens responsible for cavity spot. *Canadian Journal of Microbiology*, **44**: 221–230.

Capo, A., Cammareri, M., Rocca, F.D., Errico, A., Zoina, A. and Conicella, C. 2002. Evaluation for chipping and tuber soft rot (*Erwinia carotovora*) resistance in potato clones from unilateral sexual polyploidization (2x X 4x). *American Journal of Potato Research*, **79**: 139–145.

Caranta, C., Palloix, A., Gebre-Selassie, K., Lefebvre, V., Moiny, B. and Daubeze, A.M. 1996. A complementation of two genes operating from susceptible *Capsicum annuum* lines confers a new and complete resistance to pepper veinal mottle virus. *Phytopathology*, **86**: 739–743.

Cardoso de Arruda, M.C., Alzate-Marim, A.L., Chagas, J.M., Moreira, M.A. and de Barros, E.G. 2000. Identification of random amplified polymorphic DNA markers linked to the *CO4* resistance gene to *Colletotrichum lindemuthianum* in common bean. *Phytopathology*, **90**: 758–761.

Cardwell, K.F. Kling, J.G., Maziya-Dixon, B. and Bosque-Pérez, N.A. 2000. Interactions between *Fusarium verticillioides, Aspergillus flavus* and insect infestation in four maize genotypes in lowland Africa. *Phytopathology*, **90**: 276–284.

Carvalho, A.O., Machado, O.L.T., Cunha, M. da, Santos, I.S. and Gomes, V.M. 2001. Antimicrobial peptides and immunolocalization of a LTP in *Vigna unguiculata* seeds. *Plant Physiology and Biochemistry*, **39**: 137–146.

Caruso, C., Chilosi, G., Leonardi, L., Bertini, L., Magro, P., Buonocore, V. and Caporale, C. 2001. A basic peroxidase from wheat kernel with antifungal activity. *Phytochemistry*, **85**: 743–750.

Cebolla-Cornejo, J., Soler, S., Gomar, B., Soria, M.D. and Nuez, F. 2004. Screening *Capsicum* germplasm for resistance to *Tomato spotted wilt virus* (TSWV). *Annals Applied Biology*, **143**: 143–152.

Chen, S., Hofius, D., Sonnewald, U. and Börnke, F. 2003. Temporal and spatial control of gene silencing in transgenic plants by inducible expression of double-stranded RNA. *Plant Journal*, **36**: 731–740.

Chen, Z.Y., Brown, R.L., Damann, K.E. and Cleveland, T.E. 2004. Identification of a maize kernel stress-related protein and its effect on aflatoxin accumulation. *Phytopathology*, **94**: 938–945.

Chesson, A. 1980. Maceration in relation to the postharvest handling and processing of plant material. *Journal of Applied Bacteriology*, **48**: 1–45.

Chokethaworn, N., Suttajit, M., Vanitanakome, N., Limtraku, P., Iamsupasit, C. and Chutkaew, C. 1991. Chitinase activity in maize seeds and their fungal resistance. In: *Fungi and Mycotoxin in Stored Products* (eds) B.R. Champ, E. Highley, A.D. Hocking and J.I. Pitt, Proceedings of International Conference, ACIAR Proceedings No. 36, Bangkok.

Choo, T.M., Martin, R.A., Ho, K.M., Shen, Q., Fedak, G., Savard, M., Voldeng, H., Falk, D.E., Etienne, M. and Sparry, E. 2004. Fusarium head blight and deoxynivalenol accumulation in Estern Canada: cultivar response and correlation analysis. *Plant Disease*, **88**: 837–844.

Choo, T.M., Vigier, B., Shen, Q.Q., Martin, R.A., Ho, K.M. and Savard, M. 2004. Barley traits associated with resistance to *Fusarium* head blight and deoxynivalenol accumulation. *Phytopathology*, **94**: 1145–1150.

Chrispeels, M.J. and Raikhel, N.V. 1991. Lectins, lectin genes and their role in plant defenses. *Plant Cell*, **3**: 1–9.

Chrzanowska, M., Zielinska, B. and Kaminska, M. 2004. Responses of potato cultivars (*Solanum tuberosum*) to infection with *Cucumber mosaic virus* isolated from different host species. *Annals of Applied Biology*, **144**: 197–203.

Chunwongse, J., Chunwongse, C., Black, L. and Hanson, P. 2002. Molecular mapping of the *Ph-3* gene for late blight resistance in tomato. *Journal of Horticultural Science and Biotechnology*, **77**: 281–286.

Clements, M.J., Kleinschmidt, C.E., Maragos, C.M., Pataky, J.K. and White, D.G. 2003. Evaluation of inoculation techniques for *Fusarium* ear rot and fumonisin contamination of corn. *Plant Disease*, **87**: 147–153.

Clements, M.J., Maragos, C.M., Pataky, J.K. and White, D.G. 2004. Sources of resistance to fumonisin accumulation in grain and *Fusarium* ear and kernel rot of corn. *Phytopathology*, **94**: 251–260.

Cooper, C., Isaac, S., Jones, M.G., Crowther, T., Smith, B.M. and Collin, H.A. 2004. Morphological and biochemical response of carrots to *Pythium violae*, causative agent of cavity spot. *Physiological and Molecular Plant Pathology*, **64**: 27–35.

Covarelli, L., Turner, A.S. and Nicholson, P. 2004. Repression of deoxynivalenol accumulation and expression of *Tri* genes in *Fusarium culmorum* by fungicides *in vitro*. *Plant Pathology*, **53**: 22–28.

Cruickshank, R.H. and Wade, G.C. 1992. The activation of latent infections of *Monilinia fructicola* on apricots by volatiles from the ripening fruit. *Journal of Phytopathology*, **136**: 107–112.

Culver, J.N. 1996. Viral avirulence genes. In: *Plant–Microbe Interactions* (eds) G. Stacey and N.T. Keen, pp. 196–219. Chapman and Hall, N.Y.

Cumagun, C.J.R., Bowden, R.L., Jurgenson, J.E., Leslie, J.F. and Miedaner, T. 2004. Genetic mapping of pathogenicity and aggressiveness of *Gibberella zeae* (*Fusarium graminearum*) toward wheat. *Phytopathology*, **94**: 520–526.

Dahleen, L.S., Agrama, H.A., Horsley, R.D., Steffenson, B.J., Schwarz, P.B., Mesfin, A. and Franckowiak, J.D. 2003. Identification of QTLS associated with *Fusarium* head blight resistance in Zhedar 2 barley. *Theoretical and Applied Genetics*, **108**: 94–104.

Datta, K., Muthukrishnan, S. and Datta, S.K. 1999. Expression and function of PR-protein genes in transenic plants. In: *Pathogenesis-Related Proteins in Plants* (eds) S.K. Datta and S. Muthukrishnan, pp. 261–277, CRC Press, Boca Ratton, FL, USA.

De Fago, G., Kern, H. and Sedlar, L. 1983. Genetic analysis of tomatine insensitivity, sterol content and pathogenicity for green tomato fruits in mutants of *Fusarium solani*. *Physiological Plant Pathology*, **22**: 39–43.

Dellagi, A., Heilbronn, J., Avrova, A.O., Montesano, M., Palva, E.T., Sewart, H.E., Toth, I.K., Cooke, D.E.L., Lyon, G.D. and Birch, P.R. 2000. A potato gene encoding a WRKY-like transcription factor is induced in interactions with *Erwinia carotovora* subsp. *atroseptica* and *Phytophthora infestans* and is coregulated with class I endochitinase expression. *Molecular Plant–Microbe Interactions*, **13**: 1092–1101.

Dennis, C. 1983. Soft fruits. In: *Post-Harvest Pathology of Fruits and Vegetables* (ed.) C. Dennis, pp. 23–48, Academic Press, U.K.

de Oliveira, A.C.B. Maluf, W.R., Pinto, J.E.B.P. and Azevedo, S.M. 2003. Resistance to *Papaya ringspot virus* in summer squash *Cucurbita pepo* introgressed from an interspecific *C. pep X C. moschata* cross. *Euphytica*, **132**: 211–215.

D'hallewin, G., Schirra, M., Powell, A.L.T., Greve, L.C. and Labavitch, J.M. 2004. Properties of a polygalacturonase-inhibiting protein isolated from 'Droblanco' grapefruit. *Physiologica Plantarum*, **120**: 395–404.

Díaz, J., Have, A. ten and Kan, J.A.L. van. 2002. The role of ethylene and wound signaling in resistance of tomato to *Botrytis cinerea. Plant Physiology*, **129**: 1341–1351.

Dicenta, F., Martinez-Gómez, P., Martinez-Pato, E. and Gradziel, T.M. 2003a. Screening for *Apergillus flavus* resistance in almond. *HortScience*, **38**: 266–268.

Dicenta, F., Martinez-Gómez, P. Rubio, M. and Audergon, J.M. 2003b. Localization and movement of *Plum pox virus* in apricot stem tissues. *Annals of Applied Biology*, **142**: 99–105.

Dondini, L., Costa, F., Tataranni, G., Tartarini, S. and Sansavini, S. 2004. Cloning of apricot RGAs (resistance gene analogs) and development of molecular markers associated with Sharka (PPV) resistance. *Journal of Horticultural Science and Biotechnolgy*, **79**: 729–734.

Droby, S., Prusky, D. and Jacoby, D. 1987. Lack of involvement of nutrients in the latency of *Alternaria alteranta* in the unripe mango fruits. *Journal of Phytopathology*, **120**: 85–89.

Droby, S., Prusky, D., Jacoby, B. and Goldman, A. 1986. Presence of an antifungal compound in the peel of mango fruits and their relation to latent infections of *Alternaria alteranta. Phsiological and Molecular Plant Pathology*, **29**: 173–183.

Droby, S., Prusky, D., Jacoby, B. and Goldman, A. 1987. Induction of antifungal resorcinols in flesh of unripe mango fruits and its relation to latent infection by *Alternaria alteranta. Physiological and Molecular Plant Pathology*, **30**: 285–292.

Dzhalibov, F.S., Monakhos, G.F. and Seminov, P.A. 1999. Evaluation of resistance of cabbage hybrids to storage diseases. *Izvestiya Timiryazeveskoĭ Sel'sko-khozyaĭstvennoĭ Akademii*, **2**: 107–113.

Ekengren, S.K., Liu, Y., Schiff, M., Dinesh-Kumar, S.P. and Martin, G.B. 2003. Two MAPK cascades, NPR1 and TGA transcription factors play a role in Pto-mediated disease resistance in tomato. *Plant Journal*, **36**: 905–917.

El-Kharbotly, A., Leonards-Schippers, C., Huigen, D.J., Jacobsen, E., Pereira, A. et al., 1994. Segregation analysis and RFLP mapping of the *R1* and *R3* alleles conferring race-specific resistance to *Phytophthora infestans* in the progeny of diploid potato plants. *Molecular Genetics*, **242**: 749–754.

El-Kazzaz, M.K., Chordas, A. and Kader, A.A. 1983. Physiological and compositional changes in orange fruit in relation to modification of their susceptibility to *Penicil-*

lium italicum by ethylene treatment. *Journal of American Society of Horticultural Science*, **108**: 618–621.

Engle, J.S., Madden, L.V. and Lipps, P.E. 2003. Evaluation of inoculation methods to determine resistance reactions of wheat to *Fusarium graminearum*. *Plant Disease*, **87**: 1530–1535.

Erbs, G. and Newman, M.A. 2003. The role of lipopolysaccharides in induction of plant defence responses. *Molecular Plant Pathology*, **4**: 421–425.

Eshel, D., Beno-Moualem, D., Lorang, J.M., Dinoor, A. and Prusky, D. 2002. Induction of peroxidase during infection of unripe persimmon fruit by *Alternaria alternata*: a possible quiescence mechanism. *Journal of Phytopathology*, **150**: 357–362.

Evans, C.K., Xie, W., Dill-Macky, R. and Mirocha, C.J. 2000. Biosynthesis of deoxynivalenol in spikelets of barley inoculated with macroconidia of *Fusarium graminearum*. *Plant Disease*, **84**: 654–660.

Ficke, A., Gadoury, D.M., Seem, R.C., Godfrey, D. and Dry, I.B. 2004. Host barriers and responses to *Uncinula necator* in developing grape berries. *Phytopathology*, **94**: 438–445.

Figueira, E.L.Z., Hirooka, E.Y., Mendiolaolaya, E. and Balnco-Labra, A. 2003. Characterization of a hydrophobic amylase inhibitor from corn (*Zea mays*) seeds with activity against amylase from *Fusarium verticillioides*. *Phytopathology*, **93**: 917–922.

Fish, W.W. and Davis, A.R. 2004. The purification, physical/chemical characterization and cDNA sequence of cantaloupe fruit polygalacturonase-inhibiting protein. *Phytopathology*, **94**: 334–337.

Flier, W.G., Turkensteen, L.J., Bosch, G.B.M. van den, Vereijken, P.F.G. and Mulder, A. 2001. Differential interaction of *Phytophthora infestans* on tubers of potato cultivars with different levels of blight resistance. *Plant Pathology*, **50**: 292–301.

Florack, D.E.A. and Stiekema, W.J. 1994. Thionins: Properties, possible biological roles and mechanisms of action. *Plant Molecular Biology*, **26**: 25–37.

Franco-Lara, L. and Barker, H. 1999. Characterization of resistance to *Potato leaf roll virus* accumulation in *Solanaum phureja*. *Euphytica*, **108**: 137–144.

Gabler, F.M., Smilanick, J.L., Mansour, M., Ranning, D.W. and Mackey, B.E. 2003. Correlations of morphological anatomical and chemical features of grape berries with resistance to *Botrytis cinerea*. *Phytopathology*, **93**: 1263–1273.

Garas, N.A. and Kuć, J. 1981. Potato lectin lyses zoospores of *Phytophthora infestans* and precipitates elicitors of terpenoid accumulation produced by the fungus. *Physiological Plant Pathology*, **18**: 227–237.

García-Olmedo, F., Carmona, M.J., Lopéz-Fando, J.J., Fernandez, J.A., Castagnaro, A., Molina, A., Hernández-Lucas, C. and Carbonero, P. 1992. Characterization and analysis of thionin genes. In: *Plant Gene Research Series: Genes Involved in Plant Defense* (eds) T. Boller and F. Meins, pp. 283–302, Springer-Verlag, Wien.

García-Olmedo, F., Rodriguez-Palenzuela, P., Hernández-Lucas, C., Ponz, F., Marana, C., Carmona, M.J., López-Fando, J., Fernández, J.A. and Carbonero, P. 1989. The thionins: a protein family that includes purothionins, viscotoxins and crambins. *Oxford Surveys of Plant Molecular and Cell Biology*, **6**: 31–60.

Garrido-Ramirez, E.R., Sudarshana, M.R., Lucas, W.J. and Gilbertson, R.L. 2000. *Bean dwarf mosaic virus* BV1 protein is a determinant of the hypersensitive response and avirulence in *Phaseolus vulgaris*. *Molecular Plant–Microbe Interactions*, **13**: 1184–1194.

Gebhardt, C., Ballvora, A., Walkemeier, B., Oberhagemann, P. and Schüler, K. 2004. Assessing genetic potential in germplasm collections of crop plants by marker-trait association: a case study for potatoes with quantitative variation of resistance to late blight and maturity type. *Molecular Breeding*, **13**: 93–102.

Gilardi, P., García-Luque, I. and Serra, M.T. 2004. The coat protein of tobamo virus acts as elicitor of both L^2 and L^4 gene-mediated resistance in *Capsicum*. *Journal of General Virology*, **85**: 2077–2085.

Gilchrist, D.G. 1998. Programmed cell death in plant disease: The purpose and promise of cellular suicide. *Annual Review of Phytopathology*, **36**: 393–414.

Goetz, G., Fkyerat, A., Metais, N., Kunz, M. Tabacchi, R., Pezet, R. and Pont, V. 1999. Resistance factors to grey mould in grape berries: Identification of some phenolics inhibitors of *Botrytis cinerea* stilbene oxidase. *Phytochemistry*, **5**: 759–767.

Gomés, E., Sagot, E., Gaillard, C., Laquitaine, L., Poinssot, B., Sanejouand, Y., Delrot, S. and Coutos-Thévenot, P. 2003. Nonspecific lipid-transfer protein genes expression in grapes (*Vitis* sp.) cells in response to fungal elicitor treatments. *Molecular Plant–Microbe Interactions*, **16**: 456–464.

Goodliffe, J.P. and Heale, J.B. 1977. Factors affecting the resistance of carrot roots to *Botrytis cinerea*. *Annals of Applied Biology*, **85**: 163 (Abst.).

Goodliffe, J.P. and Heale, J.B. 1978. The role of 6-methoxymellein in the resistance and susceptibility of carrot root tissue to the cold storage pathogen, *Botrytis cinerea*. *Plant Pathology*, **12**: 27–43.

Goodman, R.N. and Novacky, A.J. 1994. *The Hypersensitive Reactions in Plants to Pathogens*. APS Press, St. Paul, MN, USA.

Goulden, M.G. and Baulcombe, D.C. 1993. Functionally homologous host components recognize *Potato virus X* in *Gompherena globosa* and potato. *Plant Cell*, **5**: 921–930.

Gradziel, T.M., Bostock, R.M. and Adaskaveg, J.E. 2003. Resistance to brown rot disease in peach is determined by multiple structural and biochemical components. *Acta Horticulturae*, **622.**

Green, S.K. and Kalloo, G. 1994. *Leafcurl and Yellow Viruses of Pepper and Tomato: an Overview.* Technical Bulletin No.21, Asian Vegetable Research and Development Center, Taipei.

Guérin, L., Benhamou, N. and Rouxel, F. 1998. Ultrastructural and cytochemical investigation of pathogen development and host reactions in susceptible and partially resistant carrots infected by *Pythium violae*, the major causal agent of cavity spot. *European Journal of Plant Pathology*, **104**: 653–665.

Guevara, M.G., Oliva, C.R., Huarte, M. and Daleo, G.R. 2002. An aspartic-protease with antimicrobial activity is induced after infection and wounding in intercellular fluids of potato tubers. *European Journal of Plant Pathology*, **108**: 131–137.

Guimarães, R.L., Chetelat, R.T. and Stotz, H.U. 2004. Resistance to *Botrytis cinerea* in *Solanum lycopersicoides* is dominat in hybrids with tomato and involves induced hyphal death. *European Journal of Plant Pathology*, **110**: 13–23.

Haley, S.D., Afanador, L. and Kelly, J.D. 1994. Identification and application of random amplified polymorphic DNA marker for the *I* gene (potyvirus resistance) in common bean. *Phytopathology*, **84**: 157–160.

Halloin, J.M. 1983. Deterioration of resistance mechanisms in seeds. *Phytopathology*, **73**: 335.

Halloin, J.M., Lee, L.S. and Cotty, P.J. 1991. Pre-ripening damage to cotton seed by *Aspergillus flavus* is not influenced by seed coat permeability. *Journal of American Oil Chemical Society*, **68**: 522.

Hamblin, A.M. and White, D.G. 2000. Inheritance of resistance to Aspergillus ear rot and aflatoxin production of corn from Tex6. *Phytopathology*, **90**: 292–296.

Hammond-Kosac, K. and Jones, J.D.G. 2000. Responses to plant pathogens. In: *Biochemistry and Molecular Biology of Plants* (eds) B.B. Buchanan, W. Gruissem and R.L. Jones, pp. 1102–1156, American Society of Plant Physiologists, Rockville, Maryland, USA.

Harrison, H.F. Jr., Peterson, J.K., Clark, C.A. and Snook, M.E. 2001. Sweetpotato periderm components inhibit *in vitro* growth of root rotting fungi. *HortScience*, **36**: 927–930.

Hébert, C., Charles, M.T., Willemot, C., Gauthier, L., Khanizadeh, S. and Cousinearu, J. 2002. Strawberry proanthocyanidins: biochemical markers for *Botrytis cinerea* resistance and shelf-life predictability. *Acta Horticulturae*, **567**: 659–662.

Hejgaard, J., Jacobsen, S. and Svendsen, J. 1991. Two antifungal thaumatin-like proteins from barley grains. *FEBS Letters*, **291**: 121–131.

Henry, R.J. 1997. *Practical Applictions of Plant Molecular Biology*, Chapman & Hall, London.

Hill, G.K., Stellwaag-Kittler, F., Huth, G. and Schlösser, E. 1981. Resistance of grapes in different development stages to *Botrytis cinerea. Phytopathologische Zeitschrift*, **102**: 328–338.

Hilton, A.J., Stewart, H.E., Linton, S.L., Nicolson, M.J. and Lees, A.K. 2000. Testing the resistance to silver scurf in commercial potato cultivars under controlled environmental conditions. *Potato Research*, **43**: 263–272.

Holbrook, C.C., Wilson, D.M., Matheron, M.E., Hunter, J.E., Knauft, D.A. and Gorbet, D.W. 2000. Aspergillus colonization and aflatoxin contamination in peanut genotypes with reduced linoleic acid compositions. *Plant Disease*, **84**: 148–150.

Holden, J.H.W. 1977. Potato breeding at Pentlandfield. *Report of Scottish Plant Breeding Station*, 1976–1977, pp. 66–87.

Huang, S., Vleeshouwers, V.G.A.A., Werji, J.S., Huttan, R.C.B., van Eck, H.J., Visser, R.G.P. and Jacobsen, E. 2004. The R3 resistance to *Phytophthora infestans* in potato is conferred by two closely linked R genes with distinct specificities. *Molecular Plant–Microbe Interactions*, **17**: 428–435.

Hulme, A.C. and Edney, K.L. 1960. Phenolic substances in the peel of Cox's orange pippin apples with reference to infection by *Gloeosporium perennans*. In: *Phenolics in Plants in Health and Disease* (ed.) J.B. Pridham, pp. 87–94, Pergamon Press, London.

Hutcheson, S.W. 1998. Current concepts of active defense in plants. *Annual Review of Phytopathology*, **36**: 59–90.

Huynh, Q.K, Borgmeyer, J.R. and Zobel, J.F. 1992. Isolation and characterization of a 22 kDa protein with antifungal properties from maize seeds. *Biochemistry and Biophysics Research Communications*, **182**: 1–5.

Irvin, J.D., Kelly, T. and Robertus, J.D. 1980. Purification and properties of a second antiviral protein from *Phytolacca americana* which inactivates eukaryotic ribosomes. *Archives of Biochemistry and Biophysics*, **200**: 418–428.

Iwai, T., Kaku, H., Honkura, R., Nakamura, S., Ochiai, H., Sasaki, T. and Ohashi, Y. 2002. Enhanced resistance to seed-transmitted bacterial diseases in transgenic rice plants over-producing an oat cell wall-bound thionin. *Molecular Plant–Microbe Interactions*, **15**: 515–521.

Jadeja, K.B. and Vaishnav, M.O. 2000. Screening of mango cultivars to postharvest rots and biochemical changes during decay. *Gujarat Agricultural University Research Journal*, **25**: 98–100.

Johnson, R. 1981. Durable resistance: Definition of genetic control and attainment in plant breeding. *Phytopathology*, **71**: 567–568.

Johnson, R. 1983. Genetic background durable resistance. In: *Durable Resistance in Crops* (eds) F. Lamberti, J.M. Waller and N.A. Van der Graaff, pp. 5–26, Plenum Press, NY.

Johnston, D.J., Ramanathan, V. and Williamson, B. 1993. A protein from immature raspberry fruits which inhibits endopolygalacturonases from *Botrytis cinerea* and other microorganisms. *Journal of Experimental Botany*, **44**: 971–976.

Joyce, D.C. and Johnson, G.I. 1999. *Prospects for Exploitation of Natural Disease Resistance in Harvested Horticultural Crops*. Postharvest convention, 1999, Cranfield University, Silsoe, Bedfordshire, UK.

Jung, H.W. and Hwang, B.K. 2000. Isolation, partial sequencing and expression of pathogenesis-related cDNA genes from pepper leaves infected by *Xanthomonas campestris* pv.*vesicatoria*. *Molecular Plant–Microbe Interactions*, **13**: 136–142.

Kamo, T., Kato, N., Hirai, N., Tsuda, M., Fujioka, D. and Ohigashi, H. 1998. Phenylphenalenone-type phytoalexins from unripe Buñgulan banana fruit. *Bioscience, Biotechnology and Biochemistry*, **62**: 95–101.

Kamoun, S., Lindquist, H. and Govers, F. 1997a. A novel class of elicitin-like genes from *Phytophthora infestans*. *Molecular Plant–Microbe Interactions*, **10**: 1028–1030.

Kamoun, S., van West, P., de Jong, A.J., de Groot, K.E., Vleeshouwers, V.G.A.A. and Govers, F. 1997b. A gene encoding a protein elicitor of *Phytophthora infestans* is down-regulated during infection of potato. *Molecular Plant–Microbe Interactions*, **10**: 13–20.

Kang, Z. and Buchenauer, H. 2002. Immunocytochemical localization of β-1,3-glucanase and chitinase in *Fusarium culmorum*-infected wheat spikes. *Physiological and Molecular Plant Pathology*, **60**: 141–153.

Karadimtcheva, B. 1981. Characteristics of the anatomical structure of the grape skin in relation to resistance to grey mold. *Gradinar Lozar Nauka*, **18**: 94–99.

Kariola, T., Palomäki, T.A., Brader, G. and Palva, T. 2003. *Erwinia carotovora* subsp. *carotovora* and *Erwinia*-derived elicitors Hrp and Peh trigger distinct but interacting defense responses and cell death in Arabdopsis. *Molecular Plant–Microbe Interactions*, **16**: 179–187.

Katou, S., Senda, K., Yoshioka, H., Doke, N. and Kawakita, K. 1999. A 51 kDa protein kinase of potato activated with hyphae wall components from *Phytophthora infestans*. *Plant Cell Physiology*, **40**: 825–831.

Keller, M., Viret, O. and Cole, F.M. 2003. *Botrytis cinerea* infection in grape flowers: Defense reaction, latency and disease expression. *Phytopathology*, **93**: 316–322.

Keshavarzi, M., Soylu, S., Brown, I., Bonas, U., Nicole, M., Rossiter, J. and Mansfield, J. 2004. Basal defenses induced in pepper by lipopolysaccharides are suppressed by

Xanthomonas campestris pv. *vesicatoria. Molecular Plant–Microbe Interactions*, **17**: 805–815.

Kim, K.H., Yoon, J.B., Park, H.G., Park, E.W. and Kim, Y.H. 2004. Structural modifications and programmed cell death of chilli pepper fruit related to resistance responses to *Colletotrichum gloeosporioides* infection. *Phytopathology*, **94**: 1295–304.

Kim, Y.S., Park, J.Y., Kim, K.S., Ko, M.K., Cheong, S.J. and Oh, B.J. 2002. A thaumatin-like gene nonclimacteric pepper fruits used as molecular marker in probing disease resistance, ripening and sugar accumulation. *Plant Molecular Biology*, **49**: 125–135.

Kirisi, M. 1974. Proteinase inhibitors in germinating barley embryos. *Physiologia Plantarum*, **32**: 89.

Ko, S.S., Huang, J.W., Wang, J.F., Shanmugasundaram, S. and Chang, W.N. 2002. Evaluation of onion cultivars for resistance to *Aspergillus niger*, the causal agent of black mold. *Journal of the American Society for Horticultural Science*, **127**: 697–702.

Kobiler, I., Prusky, D., Midland, S.L., Sims, J.J. and Keen, N.T. 1994. Compartmentation of an antifungal compounds in oil cells of avocado fruit mesocarp and its effect on susceptibility to *Colletotrichum gloeosporioides. Physiological and Molecular Plant Pathology*, **43**: 319–328.

Kolomiets, M.V., Chen, H., Gladon, R.J., Braun, E.J. and Hannapel, D.J. 2000. A leaf lipoxygenase of potato induced specifically by pathogen infection. *Plant Physiology*, **124**: 1121–1130.

Kombrink, E. and Somssich, I.E. 1995. Defense responses of plants to pathogens *Advances in Botanical Research*, **21**: 1–34.

Kurosaki, F., Tashiro, N. and Nishi, N. 1987. Secretion of chitinase from cultured carrot cells treated with fungal mycelial walls. *Physiological and Molecular Plant Pathology*, **31**: 211–216.

Lamb, C.J., Ryals, J.A., Ward, E.R. and Dixon, R.A. 1992. Emerging strategies for enhancing crop resistance to microbial pathogens. *Biotechnology*, **10**: 1436–1445.

Langcake, P. and McCarthy, W.V. 1979. The relationship of resveratrol production to infection of grapevine leaves by *Botrytis cinerea. Vitis*, **18**: 244–253.

Langella, R., Ercolano, M.R., Monti, L.M., Frusciante, L. and Barone, A. 2004. Molecular marker assisted transfer of resistance to TSWV in tomato elite lines. *Journal of Horticultural Science and Biotechnology*, **79**: 806–810.

Langevin, F., Eudes, F. and Comeau, A. 2004. Effect of trichothecenes produced by Fusarium head blight development in six cereal species. *European Journal of Plant Pathology*, **110**: 735–746.

Larsen, R.C. and Miklas, P.N. 2004. Generation and molecular mapping of a sequence characterized amplified region marker linked with the *Bct* gene for resistance to *Beet curly top virus* in common bean. *Phytopathology*, **94**: 320–325.

Leah, R., Tommerup, H., Svendsen, I. and Mundy, J. 1991. Biochemical and molecular characterization of three barley seed proteins with antifungal properties. *Journal of Biological Chemistry*, **226**: 1564–1573.

Lheareux, F., Carreel, F., Jenny, C., Lockhart, B.E.L. and Iskra-Caruana, M.L. 2003. Identification of genetic markers linked to banana streak disease expression in interspecific *Musa* hybrids. *Theoretical and Applied Genetics*, **106**: 594–598.

Li, W.L., Faris, J., Muthukrishnan, S., Liu, D.J., Chen, P.D. and Gill, B.S. 2001. Isolation and characterization of novel cDNA clones of acidic chitinases and β-1,3-glucanases from wheat spikes infected by *Fusarium graminearum*. *Theoretical and Applied Genetics*, **102**: 353–362.

Lichner, Z., Silhavy, D. and Burgýan, J. 2003. Double-stranded RNA-binding proteins could suppress RNA interference-mediated antiviral defences. *Journal of General Virology*, **84**: 975–980.

Lorenzo, G. de, Ito, Y., D'Ovidio, R., Cervone, F., Albersheim, P. and Darvill, A.G. 1990. Host-pathogen interactions XXXVII. Abilities of the polygalacturonase-inhibiting proteins from four cultivars of *Phaseolus vulgaris* to inhibit the endopolygalacturonases from three races of *Colletotrichum lindemuthianum*. *Physiological and Molecular Plant Pathology*, **36**: 421–435.

Lund, S.T., Stall, R.E. and Klee, H.J. 1998. Ethylene regulates the susceptible response to pathogen infection in tomato. *Plant Cell*, **10**: 371–382.

Machinandiarena, M.F., Olivieri, F.P., Daleo, G.R. and Oliva, C.R. 2001. Isolation and characterization of a polygalacturonase-inhibiting protein from potato leaves. Accumulation in response to salicylic acid, wounding and infection. *Plant Physiology and Biochemistry*, **39**: 129–136.

Maiero, M. and Ng, T.J. 1989. Combining ability estimates for early blight resistance in tomato. *Journal of American Society of Horticultural Science*, **114**: 118–121.

Malcolmson, J.F. and Black, W. 1966. New R genes in *Solanum demissum* Lindl and their complementary races of *Phytophthora infestans* (Mont.) de Bary. *Euphytica*, **15**: 199–203.

Martin, B., Rahbé, Y. and Fereres, A. 2003. Blockage of stylet tips as the mechanism of resistance to virus transmission by *Aphis gossypii* in melon lines bearing the *Vat* gene. *Annals of Applied Biology*, **142**: 245–250.

Mayna, R.Y., Harper, G.A., Franz, A.O., Lee, L.S. and Goldblatt, L.A. 1969. Retardation of the elaboration of aflatoxin in cotton seed by impermeability of the seed coats. *Crop Science*, **9**: 147.

Mazier, M., German-Retana, S., Flamain, F., Dubois, V., Botton, E., Sarnette, S., Le Gall, O., Candresse, T. and Maisonneuve, B. 2004. A simple and efficient method of testing lettuce mosaic resistance in *in vitro* cultivated lettuce. *Journal of Virological Methods*, **116**: 123–131.

McCallum, B.D., Tekauz, A. and Gilbert, J. 2004. Reaction of a diverse collection of barley lines to *Fusarium* head blight. *Plant Disease*, **88**: 167–174.

McCollum, T.G. 2002. Molecular biology of host-pathogen interactions in harvested horticultural crops. *HortScience*, **37**: 12–14.

McCollum, T.G., Doosdar, H., Mayer, R.T. and McDonald, R.E. 1997. Characterization of chitinases and β-1,3, glucanases in grapefruit flavedeo during fruit development. *Physiologia Plantarum*, **99**: 486–494.

Mehli, L., Schaart, J.G., Kjellsen, T.D., Tran, D.H., Slentijn, E.M.J., Shouten, H.J. and Iversen, T.H. 2004. A gene encoding a polygalacturonase inhibiting protein (PGIP) shows developmental regulation and pathogen-induced expression in strawberry. *New Phytologist*, **163**: 99–110.

Mendieta, J.R., Giudici, A.M. and Canal, L. 2004. Occurrence of antimicrobial serine-proteinases in sunflower seeds. *Journal of Phytopathology*, **152**: 43–47.

Merz, U., Martinez, V. and Schwärzel, R. 2004. The potential for the rapid screening of potato cultivars (*Solanum tuberosum*) for resistance to powdery scab (*Spongospora subterranea*) using a laboratory bioassay. *European Journal of Plant Pathology*, **110**: 71–77.

Mignouna, H.D., Abang, M.M., Onasanya, A. and Asiedu, R. 2002. Identification and application of RAPD markers for anthracnose resistance in water yam (*Dioscorea alata*). *Annals of Applied Botany*, **141**: 61–66.

Minsavage, G.V., Stall, R.E., Davis, M.J., Bonas, U. and Jones, J.B. 2000. Resistance of tomato and pepper to T3 strains of *Xanthomonas campestris* pv. *vesicatoria* is specified by a plant-inducible avirulence gene. *Molecular Plant–Microbe Interactions*, **13**: 911–921.

Mirelman, D., Galun, E., Sharon, N. and Lotan, R. 1975. Inhibition of fungal growth by wheat germ agglutin. *Nature*, **256**: 414–416.

Mohammadi, M. and Kazemi, H. 2002. Changes in peroxidase and polyphenol oxidase activities in susceptible and resistant wheat heads inoculated with *Fusarium graminearum* and induced resistance. *Plant Science*, **162**: 491–498.

Moissiard, G. and Voinnet, O. 2004. Viral suppression of RNA silencing in plants. *Molecular Plant Pathology*, **5**: 71–82.

Monteiro, S., Barakat, M., Picarra-Pereira, M.A., Teixeira, A.R. and Ferrieira, R.B. 2003. Osmotin and thaumatin from grape: A putative general defense mechanism against pathogenic fungi. *Phytopathology*, **93**: 1505–1512.

Montesano, M., Brader, G. and Palva, E.T. 2003. Pathogen-derived elicitors: searching for receptors in plants. *Molecular Plant Pathology*, **4**: 73–79.

Moreno, M., Segura, A. and Garcia-Olmedo, F. 1994. Pseudothionin, St 1. A potato peptide active against potato pathogens. *European Journal of Biochemistry*, **223**: 135–139.

Moore, K.G., Price, M.S., Boston, R.S., Weissinger, A.K. and Payne, G.A. 2004. A chitinase from Tex6 maize kernels inhibits growth of *Aspergillus flavus. Phytopathology*, **94**: 82–87.

Mullins, E.D., McCollum, T.G. and McDonald, R.E. 2000. Consequences on ethylene metabolism of inactivating the ethylene receptor sites in diseased non-climacteric fruit. *Postharvest Biology and Technology*, **19**: 155–164.

Murillo, I., Cavallarin, L. and San Segundo, B. 1999. Cytology of infection of maize seedlings by *Fusarium moniliforme* and immunolocalization of the pathogenesis-related PRms protein. *Phytopathology*, **89**: 737–747.

Nair, N.G. and Hill, G.K. 1992. Bunch rot of grapes caused by *Botrytis cinerea*. In: *Plant Diseases of International Importance* (eds) J. Kumar, H.S. Chaube, U.S. Singh and A.N. Mukhopadhyay, pp. 147–169. Prentice Hall, NJ, USA.

Newman, M.A., Roepenack-Lahaye, E. von, Parr, A., Daniels, M.J. and Dow, J.M. 2002. Prior exposure to lipopolysaccharide potentiates expression of plant defenses in response to bacteria. *Plant Journal*, **29**: 487–495.

Noble, J.P. 1981. *The Biochemical Basis of Latency in Bitter-rot of Apple*. Doctoral Thesis, University of Birmingham, UK.

Oh, B.J., Ko, M.K., Kostenyuk, I., Shin, B.C. and Kim, K.S. 1999. Coexpression of a defensin gene and a thionin-like gene via different signal transduction pathways in pepper and *Colletotrichum gloeosporioides* interactions. *Plant Molecular Biology*, **41**: 313–319.

Padget, H.S. and Beachy, R.N. 1993. Analysis of *Tobacco mosaic virus* strain capable of overcoming N gene-mediated resistance. *Plant Cell*, **5**: 577–586.

Palou, L. and Crisosto, C.H. 2003. The influence of exogenous ethylene application during cold storage on stone fruit quality and brown rot development. *Acta Horticulturae*, **628**.

Palou, L., Crisosto, C.H., Garner, D. and Basinal, L.M. 2003. Effect of continuous exposure to exogenous ethylene during cold storage on postharvest decay development and quality attributes of stone fruits and table grapes. *Postharvest Biology and Technology*, **27**: 243–254.

Park, C.J., Kim, K.J., Shin, R., Park, J.M., Shin, Y.C. and Paek, K.H. 2004. Pathogenesis-related protein 10 isolated from hot pepper functions as a ribonuclease in antiviral pathway. *Plant Journal*, **37**: 186–198.

Parrella, G., Moretti, A., Gognalons, P., Lesage, M., Marchoux, G., Gebre-Selassie, K. and Caranta, C. 2004. The *Am* gene controlling resistance to *Alfalfa mosaic virus* in tomato is located in the cluster of dominant resistance genes on chromogome 6. *Phytopathology*, **94**: 345–350.

Percival, D.C., Sullivan, J.A. and Fisher, K.H. 1993. Effect of cluster exposure, berry contact and cultivar on cuticular membrane formation and occurrence of bunch rot (*Botrytis cinerea* Pers: Fr) with 3 *Vitis vinifera* L. cultivars. *Vitis*, **32**: 87–97.

Percival, G.C., Karim, M.S. and Dixon, G.R. 1999. Pathogen resistance in aerial tubers of potato cultivars. *Plant Pathology*, **48**: 768–776.

Peters, R.D. and Struz, A.V. 2001. A rapid technique for the evaluation of potato germplasm for susceptibility to pink rot. *Plant Disease*, **85**: 833–837.

Peterson, J.K., Harrison, H.F., Jackson, D.M. and Snook, M.E. 2003. Biological activities and contents of scopolin and scopoletin in sweetpotato clones. *Hort-Science*, **38**: 1129–1133.

Pezet, R., Pont, V. and Hoang-Van, K. 1991. Evidence for oxidative detoxification of pterostilbene and resveratrol by a laccase-like stilbene oxidase produced by *Botrytis cinerea. Physiological and Molecular Plant Pathology*, **39**: 441–450.

Pezet, R., Viret, O., Perret, C. and Tabacchi, R. 2003. Latency of *Botrytis cinerea* Pers. Fr. and biochemical studies during growth and ripening of two grape berry cultivars, respectively susceptible and resistant to gray mould. *Journal of Phytopathology*, **151**: 208–214.

Polkowska-Kowalczyk, L., Wielgat, B. and Maciejewska, U. 2004. The elicitor-induced oxidative processes in leaves of *Solanum* species with differential polygenic resistance to *Phytophthora infestans. Journal of Plant Physiology*, **161**: 913–920.

Popkova, K., Romero, I.I. and Kukushkina, L. 1999. Evaluation of cultivars for resistance to *Phytophthora. Mezhdunarodnyi Sel'skokhozyaistvennyi Zhurnal*, **5**: 60–62.

Porat, R.A., Lers, A., Dori, S., Cohen, L., Weiss, B., Ddaus, A., Wilson, C.L. and Droby, S. 1999. Induction of chitinase and β-1,3-endoglucanase proteins by UV-irradiation and wounding in grapefruit peel tissue. *Phytopathology*, **27**: 233–238.

Porter, L.D., Inglis, D.A. and Johnson, D.A. 2004. Identification and characterization of resistance to *Phytophthora infestans* in leaves stems, flowers and tubers of potato clones in the Pacific Northwest. *Plant Disease*, **88**: 965–972.

Pritsch, C., Muehlbauer, G.J., Bushnell, W.R., Somers, D.A. and Vance, C.P. 2000. Fungal development and induction of defense response genes during early infection of wheat spikes by *Fusarium graminearum. Molecular Plant–Microbe Interactions*, **13**: 159–169.

Pritisch, C., Vance, C.P., Bushnell, W.R., Somers, D.A., Hohn, T.M. and Muehlbauer, G.J. 2001. Systemic expression of defense response genes in wheat spikes as a response to *Fusarium graminearum* infection. *Physiological and Molecular Plant Pathology*, **58**: 1–12.

Prusky, D. 1996. Pathogen quiescence in postharvest diseases. *Annual Review of Phytopathology*, **34**: 413–434.

Prusky, D., Ascarelli, A. and Jacoby, B. 1984. Lack of involvement of nutrients in the latency of *Colletotrichum gloeosporioides* in unripe avocado fruits. *Phytopathologische Zeitschrift*, **110**: 106–109.

Prusky, D. and Keen, N.T. 1993. Involvement of preformed antifungal compounds in the resistance of subtropical fruits to fungal decay. *Plant Disease*, **77**: 114–119.

Prusky, D., Keen, N.T., Sims, J.J. and Midland, S.L. 1982. Possible involvement of an antifungal diene in the latency of *Colletotrichum gloeosporioides* on unripe avocado fruits. *Phytopathology*, **72**: 1578–1582.

Prusky, D., Kobiler, I., Fishman, Y., Sims, J.J., Midland, S.L. and Keen, N.T. 1991. Identification of an antifungal compounds in unripe avocado fruits and its possible involvement in the quiescent infection of *Colletotrichum gloeosporioides. Journal of Phytopathology*, **132**: 19–27.

Radha Jeyalakshmi, R., Meena, B., Thangavelu, R., Deborah, S.D., Vidhyasekran, P. and Velazhahan, R. 2000. A 45 kDa chitinase purified from pearl millet [*Pennisetum glacum* (L) R. Br] shows antifungal activity. *Journal of Plant Diseases and Protection*, **107**: 605–616.

Ramos, A.M., Nieto-Angel, D., Dominguez-Alvarez, J.L. and Delgadillo-Sanchez, F. 2000. Quality and tolerance in papaya fruits (*Carica papaya* L.) to the inoculation of the fungus *Colletotrichum gloeosporioides* Penz., in postharvest. *Revista Chapingo, Serie Horticulturae*, **7**: 119–130.

Reeves, A.F., Olanya, O.M., Hunter, J.H. and Wells, J.M. 1999. Evaluation of potato varieties and selections for resistance to bacterial soft rot. *American Journal of Potato Research*, **76**: 183–189.

Reisch, B.I. 1998. Molecular markers: The foundation for genetic mapping, DNA fingerprinting and genomics. *Proceedings of 7th International Symposium on Grapevine and Breeding*, Montpellier, France.

Roberts, W.K. and Selitrennikoff, C.P. 1986. Isolation and partial characterization of two antifungal proteins from barley. *Acta Biochimica Biophysica*, **880**: 161–170.

Romanenko, A.S., Lomovatskaya, L.A., Shafikova, T.N., Borovskii, G.B. and Krivolpova, N.V. 2003. Potato cell plasma membrane receptors to ring rot pathogen extracellular polysaccharides. *Journal of Phytopathology*, **151**: 1–6.

Romero, A.M. and Ritchie, D.F. 2004. Systemic acquired resistance delays race shifts to major resistance genes in bell pepper. *Phytopathology*, **94**: 1376–1382.

Saito, T., Meshi, T., Takamatsu, N. and Okada, Y. 1987. Coat protein gene sequences of *Tobacco mosaic virus* encodes host response determinant. *Proceedings of National Academy of Sciences, USA*, **84**: 6074–6077.

Salas, B., Secor, G.A., Taylor, R.J. and Gudmestad, N.C. 2003. Assessment of resistance to tubers of potato cultivars to *Phytophthora erythroseptica* and *Pythium ultimum*. *Plant Disease*, **87**: 91–97.

Sanders, T.H. and Mixon, A.C. 1978. Effect of peanut tanins on percent seed colonization and *in vitro* growth by *Aspergillus parasiticus. Mycopathology*, **66**: 169.

Santa Cruz, S. and Baulcombe, D.C. 1993. Molecular analysis of *Potato virus X* isolates in relation to the potato hypersensitivity gene *Nx. Molecular Plant–Microbe Interactions*, **6**: 707–714.

Sarig, P., Zutkhi, Y., Lisker, N., Shkelerman, Y. and Ben-Arie, R. 1998. Natural and induced resistance of table grapes to bunch rots. *Acta Horticulturae*, **464**: 65–70.

Sarig, P., Zutkhi, Y., Monjauze, A., Lisker, N. and Ben-Arie, R. 1997. Phytoalexin elicitation in grape berries and their susceptibility to *Rhizopus stolonifer. Physiological and Molecular Plant Pathology*, **50**: 337–347.

Schornack, S., Ballvora, A., Gürlebeck, D., Peart, J., Ganal, M., Baker, B., Bonas, U. and Lahaye, T. 2003. The tomato resistance protein Bs4 is a predicted non-nuclear TIR-NB-LRR protein that mediates defense responses to severely truncated derivatives of AvrBs4 and overexpressed AvrBs3. *Plant Journal*, **36**: 46–60.

Schultz, F.A., 1978. Some physiological and biochemical aspects of the action mechanisms of fungal parasites during fruit storage. *Fruits*, **33**: 15–21.

Sendhilvel, A., Palaniswami, A., Thiruvudainambi, S., Rajappan, R. and Raguchander, T. 2001. Varietal reaction of onion bulbs on the incidence of soft rot *Erwinia carotovora* var. *carotovora. Newsletter–National Horticultural Research and Development Foundation*, **21**: 4–8.

Sharma, N., Grusezwski, H.P., Park, S.W., Holm, D.G. and Vivanco, J.M. 2004. Purification of an isoform of patatin with antimicrobial activity against *Phytophthora infestans. Plant Physiology and Biochemistry*, **42**: 647–655.

Sharrock, K.R. and Labavitch, J.M. 1994. Polygalacturonase inhibitors of Bartlett pear fruits: differential effects on *Botrytis cinerea* polygalacturonase isozymes and influence on products of fungal hydrolysis of pear cell walls and on ethylene induction in cell culture. *Physiological and Molecular Plant Pathology*, **45**: 305–319.

Shelly, P., Savarni, T. and Varma, A. 2001. Isolation and characterization of an inducer protein (Crip-31) from *Clerodendrum inerme* leaves responsible for induction of systemic resistance against viruses. *Plant Science*, **161**: 453–459.

Shen, X., Kong, L. and Ohm, H. 2004. Fusarium head blight resistance in hexaploid wheat (*Triticum aestivum*)–*Lophopyrum* genetic lines and tagging of the alien chromatin by PCR markers. *Theoretical and Applied Genetics*, **108**: 808–813.

Sherf, A.F. and Macnab, A.A. 1986. *Vegetable Diseases and Their Control.* Second edition. John Wiley & Sons, Inc., New York.

Shewry, P.R. 1995. Plant storage proteins. *Biological Reviews*, **70**: 375–426.

Shewry, P.R. and Lucas, J.A. 1997. Plant proteins that confer resistance to pests and pathogens. *Advances in Botanical Research*, **26**: 136–192.

Siranidou, E., Kang, Z. and Buchenauer, H. 2002. Studies on symptom development, phenolic compounds and morphological defence responses in wheat cultivars differing in resistance to *Fusarium* head blight. *Journal of Phytopathology*, **150**: 200–208.

Sitterly, W.R. and Shay, J.R. 1960. Physiological factors affecting the onset of susceptibility of apple fruit to rotting by fungus pathogens. *Phytopathology*, **50**: 91–93.

Skadsen, R.W., Sathish, P. and Kaeppler, H.F. 2000. Expression of thaumatin-like permatin PR-5 genes switches from ovary wall to aleurone in developing barley and oat seeds. *Plant Science* (*Limerick*), **156**: 11–22.

Smith, K.P., Evans, C.K., Dill-Macky, R., Gustus, C., Xie, W. and Dong, Y. 2004. Host genetic effect on deoxynivalenol accumulation in *Fusarium* head blight of barley. *Phytopathology*, **94**: 766–771.

Spotts, R.A., Cervantes, L.A. and Mielke, E.A. 1999. Variability in postharvest decay among apple cultivars. *Plant Disease*, **83**: 1051–1054.

Stange, R.R., Midland, S.L., Eckert, J.W. and Sims, J.J. 1993. An antifungal compound produced by grapefruit and Valencia orange after wounding of the peel. *Journal of Natural Products*, **56**: 1627–1629.

Stange, R.R. Jr., Midland, S.L., Holmes, G.J., Sims, J.J. and Mayer, R.T. 2001. Constituents from the periderm and outer cortex of *Ipomoea batatas* with antifungal activity against *Rhizopus stolonifer*. *Postharvest Biology and Technology*, **23**: 85–92.

Staples, R.C. and Mayer, A.M. 1995. Putative virulence factors of *Botrytis cinerea* acting as a wound pathogen. *FEMS Microbiology Letters*, **134**: 1–7.

Stevens, M.R., Scott, S.J. and Gergerich, R.C. 1992. Inheritance of gene for resistance to *Tomato spotted wilt virus* (TSWV) from *Lycopersicon esculentum* Mill. *Euphytica*, **59**: 9–17.

Stotz, H.U., Bishop, J.G., Bergmann, C.W., Koch, M., Albersheim, P., Darvill, A.G. and Labavitch, J.M. 2000. Identification of target amino acids that affect interactions of fungal polygalacturonases and their plant inhibitors. *Physiological and Molecular Plant Pathology*, **56**: 117–130.

Stotz, H.U., Contos, J.J.A., Powell, A.L.T., Bennett, A.B. and Labavitch, L.M. 1994. Structure and expression of an inhibitor of fungal polygalacturonases from tomato. *Plant Molecular Biology*, **25**: 607–617.

Swinburne, T.R. 1971. The infection of apples cv. Bramley's Seedling by *Nectria galligena* Bres. *Annals of Applied Biology*, **68**: 253–262.

Swinburne, T.R. 1975. Microbial proteases as elicitors of benzoic acid accumulation in apples. *Phytopathologische Zeitschrift*. **82**: 152–162.

Swinburne, T.R. 1983. Quiescent infections in postharvest diseases. In: *Postharvest Diseases of Fruits and Vegetables* (ed.) C. Dennis, pp. 1–22, Academic Press, UK.

Taira, T., Yamagami, T., Aso, Y., Ishiguro, M. and Ishihara, M. 2001. Localization, accumulation and antifungal activity of chitinases in rye (*Secale cereale*) seed. *Bioscience, Biotechnology and Biochemistry*, **65**: 2710–2718.

Ten Have, A., Mulder, W., Visser, J. and van Kan, J.A.L. 1998. The endopolygalacturonase gene *Bcpg1* is required for full virulence of *Botrytis cinerea*. *Molecular Plant–Microbe Interactions*, **11**: 1009–1016.

Terras, F.R.G., Eggermont, K., Kovaleva, V., Raikhel, N.V., Osborn, R.W., Kester, A., Rees, S.B., Vanderlayden, J., Cammue, B.P.A. and Broekaert, W.F. 1995. Small cysteine-rich antifungal proteins from radish: their role in host defense. *Plant Cell*, **7**: 573–588.

Terry, L.A., Joyce, D.C., Adikaram, N.K.B. and Khambay, B.P.S., 2004. Preformed antifungal compounds in strawberry fruit and flower tissues. *Postharvest Biology and Technology*, **31**: 201–212.

Thomas-Carroll, M.L. and Jones, R.A.C. 2003. Selection, biological properties and fitness of resistance-breaking strains of *Tomato spotted wilt virus* in pepper. *Annals of Applied Biology*, **142**: 235–243.

Timmerman, G.M., Frew, T.J., Miller, A.L., Weeden, N.F. and Jermyn, W.A. 1993. Linkage mapping of *sbm-1*, a gene conferring resistance to *Pea seedborne mosaic virus* using molecular markers in *Pisum sativum*. *Theoretical and Applied Genetics*, **85**: 609–615.

Torres, R., Valentines, M.C., Usall, J., Viñas and Larrigaudiere, C. 2003. Possible involvement of hydrogen peroxide in the development of resistance mechanisms in Golden Delicious apple fruit. *Postharvest Biology and Technology*, **27**: 235–242.

Ullasa, B.A., Rawal, R.D., Sohi, H.S., Singh, D.P. and Joshi, M.C. 1981. Reactions of sweet pepper genotypes to anthracnose, *Cercospora* leafspot and powdery mildew. *Plant Disease*, **65**: 600–601.

Vail, M.E. and Marois, J.J. 1991. Grape cluster architecture and the susceptibility of berries to *Botrytis cinerea*. *Phytopathology*, **81**: 188–191.

Vanderplank, J.E. 1963. *Plant Diseases: Epidemics and Control*. Academic Press, N.Y.

van Parijs, J., Broekaert, W.F., Goldstein, I.J. and Peumans, W.J. 1990. Hevein: An antifungal protein from rubber tree (*Hevea brasiliensis*) latex. As quoted by Chrispeels and Raikhel, 1991.

Verhoeff, K. and Liem, J.I. 1975. Toxicity of tomatine to *Botrytis cinerea* in relation to latency. *Phytopathologische Zeitschrift*, **82**: 333–338.

Vigers, A.J., Roberts, W.K. and Selitrennikoff, C.P. 1991. A new family of plant antifungal proteins. *Molecular Plant–Microbe Interactions*, **4**: 315.

Vincent, A., Badal, J., Asensi, M.J., Sanz, N., Armengol, J. and García-Jiménez, J. 2004. Laboratory evaluation of citrus cultivars susceptibility and influence of fruit size on Fortune mandarin to infection by *Alteranria alteranta* pv. *citri*. *European Journal of Plant Pathology*, **110**: 245–251.

Visser, C.L.M. de. 1999. *Fusarium* in onions and varietal differences in infection: evaluation of a biotest. *PAV-Bulletin Vollegrondsgroenteteelt*, December, 4–7, Netherlands.

Voorrips, R.E., Finkers, R., Sanjaya, L. and Groenwold, R. 2004. QTL mapping of anthracnose (*Colletotrichum* spp.) resistance in a cross between *Capsicum annuum* and *C. chinense*. *Theoretical and Applied Genetics*, **109**: 1275–1282.

Wada, K. 1982. Localization of purothionins and genome expression in wheat plants. *Plant Cell Physiology*, **23**: 1357.

Wang, X., Thoma, R.S., Carroll, S.A. and Duffin, K.L. 2002. Temporal generation of multiple antifungal proteins in primed seeds. *Biochemical and Biophysical Research Communications*, **292**: 236–242.

Wattad, C.D., Kobiler, A. Dinoor, A. and Prusky, D. 1997. Pectate lyase of *Colletotrichum gloeosporioides* attacking avocado fruits: cDNA cloning and involvement in pathogenicity. *Physiological and Molecular Plant Pathology*, **50**: 197–212.

Weber, H., Schultze, S. and Pfitzner, A.J.P. 1994. Two amino acid substitutions in the *Tomato mosaic virus* 30 kilodalton movement protein counter the ability to overcome the $Tm2^2$ resistance gene in tomato. *Journal of Virology*, **67**: 6432–6438.

Wisniewski, M.E., Bassett, C.L., Artlip, T.M., Webb, R.P., Janisiewicz, W.J., Norelli, J.L., Goldway, M. and Droby, S. 2003. Characterization of a defensin in bark and fruit tissues of peach and antimicrobial activity of a recombinant defense in the year *Pichia pastoris*. *Physiologia Plantarum*, **119**: 563–572.

Yao, C., Conway, W.S. and Sams, C.E. 1995. Purification and characterization of a polygalacturonase-inhibiting protein from apple fruit. *Biochemistry and Cell Biology*, **85**: 1373–1377.

Yu, J., Mohawed, S.M., Bhatnagar, D. and Cleveland, T.E. 2003. Substrate-induced lipase gene expression and aflatoxin production in *Aspergillus parasiticus* and *Aspergillus flavus*. *Journal of Applied Microbiology*, **95**: 1334–1342.

Zamir, D., Ekstein-Mischelson, I., Zakay, Y., Novot, N. and Zeiden, M. et al., 1994. Mapping and introgression of a *Tomato yellow leafcurl virus* tolerance gene, *Ty-1*. *Theoretical and Applied Genetics*, **88**: 141–146.

Zamski, E. and Peretz, I. 1996. Cavity spot of carrots, 2. Cell-wall degrading enzymes secreted by *Pythium* and pathogenesis-related proteins produced by the root cells. *Annals of Applied Biology*, **128**: 195–207.

Zhu, B., Chen, T.H.H. and Li, P.H. 1996. Analysis of late blight disease resistance and freezing tolerance in transgenic potato plants expressing sense and antisense genes for osmotin-like protein. *Planta*, **198**: 70–77.

9

BIOCONTROL AGENTS FOR DISEASE MANAGEMENT

Various disease management strategies have been adopted to reduce incidence and spread of diseases of harvested produce caused by microbial pathogens, with varying degrees of success. Production of disease-free seeds and planting materials, application of preventive and physical methods, modification of storage conditions and handling methods, in addition to adoption of proper cultural practices, have been demonstrated to reduce the pathogen population or sources of inoculum, resulting in a significant reduction in the incidence of postharvest diseases. Traditionally, synthetic fungicides have been used to protect fruits, vegetables, and seeds against fungal and bacterial pathogens causing spoilage of the produce at different stages after harvest. However, the mounting concern and awareness about the health hazard and environmental pollution due to chemical use compelled the United States Government (U.S. Environmental Protection Agency) to ban the use of several chemicals that were frequently applied on fruits and vegetables (Rendall-Dunn, 1991). Furthermore, the possibility of the development of resistance in pathogens to fungicides was another critical factor in the restriction of available chemicals. The need to develop a viable alternative strategy for containing postharvest pathogens was realized, resulting in unprecedented developments in biological control. Several programs were initiated internationally aimed at the development of methods of management of various postharvest diseases of fruits, vegetables, and flowers from temperate, tropical, and subtropical climates (Janisiewicz, 1998). Biological control as defined by Cook and Baker (1983) is the reduction in the amount of inoculum or disease-producing activity of a pathogen accomplished by or through one or more organisms other than man. Some microorganisms present along with the microbial pathogens in the same environment have been found to be our allies in the fight against the pathogens causing postharvest diseases.

The biological control of postharvest diseases may be achieved by: (1) use of microorganisms; (2) application of naturally-derived compounds; and (3) enhancing the natural resistance of products (induction of resistance). The postharvest environments present some unique advantages, and challenges, for the use of biocontrol methods. The postharvest environment provides the following advantages for biological control measures: (1) the partially controlled environment in storage may result in a shift in the balance of interactions between host, pathogen, and antagonistic microbe in favor of antagonist; (2) the efficacy of antagonist may be enhanced because the biocontrol product can be applied directly onto the site where needed in the harvested product; (3) the harvested commodity may be protected relatively free of potential interfering factors; (4) protection is needed for a relatively short period as compared with period of protection required for field crops; and (5) as the harvested fruits and vegetables have high market value, use of a relatively high-cost biocontrol product may be justified. On the other hand, certain obstacles have to be overcome in the attempts to protect the harvested commodity by using microbial agents: (1) the level of control to be achieved is extremely high (95–98%), since the market value of commodity depends on the blemish-free, attractive appearance; (2) very strict food safety considerations demand careful examination of products for the possible presence of toxic or unacceptable substances produced by biocontrol agents; and (3) the potential market for the use of biocontrol organisms against postharvest diseases (biofungicide) is relatively small when compared with that for the control of field crop diseases.

There is a world-wide spurt in the research efforts to identify suitable microorganisms and to improve their effectiveness in controlling the postharvest diseases. It is expected that at least some of the economically important diseases may be tackled effectively by the use of biocontrol agents (BCAs) (Table 9.1).

9.1 BIOLOGICAL CONTROL OF POSTHARVEST DISEASES OF FRUITS

The process of selection of effective BCAs includes several steps as follows.

9.1.1 Isolation of Microorganisms

Extensive and improper application of fungicides resulted in increased incidence of minor fruit diseases, apparently due to a reduction in competition

TABLE 9.1 Microorganisms Exhibiting Biocontrol Potential Against Postharvest Pathogens

Crop	Pathogens	Biocontrol agent	References
Apple	*Botrytis cinerea*	*Trichoderma pseudokoningii*	Tronsmo and Raa, 1977
		Pseudomonas syringae pv. *lachrymans*	Janisiewicz, 1988
	Penicillium expansum	*Acremonium breve*	Janisiewicz, 1988
	B. cinerea *P. expansum* *Rhizopus nigricans*	*Candida sake*	Viñas et al., 1998
	P. expansum	*Candida sake*	Teixidó et al., 1998
		Rhodotorula glutinis (LS-11) and *Cryptococcus laurentii* (LS-28)	Lima et al., 1998
	B. cinerea *P. expansum*	*Pseudomonas viridiflava*	Bryk et al., 1999
	B. cinerea *P. expansum*	*Cryptococcus albidus*	Fan and Tian, 2001; *Zhou* et al., 2001
	B. cinerea *P. expansum* *R. stolonifer*	*Pantoea agglomerans* (CPA-2)	Nunes et al., 2002a
	Botryosphaeria berengeriana f.sp. *piricola*	*Trichoderma atroviride* *T. harzianum*	Kexiang et al., 2002
	B. cinerea	*Candida saitoana*	El-Ghaouth et al., 2003
Grapes	*Botrytis cinerea* *Aspergillus niger* and *Rhizopus stolonifer*	*Candida guilliermondii*, *Acremonium cephalosporium*	Zhavi et al., 2000
Mango	*Aspergillus flavus, A. niger, Colletotrichum gloeosporioides, Lasiodiplodia theobromae, Macrophomina phaseolina, Rhizopus stolonifer*	*Trichoderma viride*	Bhuvaneswari and Rao, 2001.
Banana	*Colletotrichum musae*	*Pseudomonas* sp.	Coste, de and Subasinghe, 1998.

TABLE 9.1 (*Continued*)

Crop	Pathogens	Biocontrol agent	References
Strawberry	*Botrytis cinerea*	*Candida fructus*, *C. glabrata* and *C. oleophila*	El-Neshawy and Shetaia, 2003
		C. reukaufii, *Enterobacteria-strain 5B4*	Guinebretiere et al., 2000
Avocado	*Colletotrichum gloeosporioides*	*Cryptococcus* sp. *Aureobasidium* sp.	Stirling et al., 1998
Kiwifruit	*B. cinerea*	*Bacillis subtilis* *Cryptococcus laurentii*	Duncan, 1991
Pear	*B. cinerea*	*Cryptococcus laurentii*	Teixido et al., 1998; Zhang et al., 2003
	Monilinia laxa	*Trichoderma harzianum* (Trichodex)	Guizzardi et al., 1995
	B. cinerea *P. expansum* *R. stolonifer*	*Pantoea agglomerans*	Nunes et al., 2001
Peach	*Monilinia fructicola* *Rhiopus stolonifer*	*Pseudomonas syringae* *P. fluorescens* *Candida* sp.	Zhou et al., 1999
Citrus	*Geotrichum candidum*	*Pichia guilliermondii* *(Debaryomyces hansenii)*	Mehrotra et al., 1998
	Penicillium digitatum *P. italicum*	Epiphytic yeasts (not identified)	Kinay et al., 1998
	P. digitatum	*Pantoea agglomerans*	Usall et al., 2001
	P. digitatum *P. italicum*	*Bacillus subtilis*	Fan et al., 2000

from saprophytic organisms negatively affected by fungicides. Iatrogenic diseases, resulting from a high levels of chemical use, have been reported on strawberry and citrus (Dennis, 1975; Singh, 1980). Hence, the use of naturally occurring microorganisms to combat postharvest pathogens was considered to be an useful approach. The search for sources of potential biocontrol agents with marked antagonistic activity against target pathogens was undertaken. *Bacillus subtilis*, an antibiotic-producing bacterium isolated from soil was evaluated for its efficacy against *Monilinia fructicola*, causing brown rot disease of peach (Pusey and Wilson, 1984). However, the use of antibiotic-producing BCAs directly on fruits was not favored, because of the possible residue left in the fruit. Further search was directed to identify antagonists that did not produce antibiotics. The yeast *Pichia guilliermondii* (isolate US-7, initially known as *Debaromyces hansenii*) was demonstrated to be effective against several postharvest diseases occurring in various environmental conditions (Chalutz and Wilson, 1990). *P. guilliermondii* was patented as a BCA for use against postharvest diseases (Wilson et al., 1994; Chalutz and Droby, 1998).

Natural microflora were later screened to select the microorganisms with required antagonistic potential. The microorganisms were evaluated for their activity in parallel in vitro and in situ on fruit. Frequently, many bacteria and yeasts with significant biocontrol potential have been isolated by direct screening on fruit, though many of them did not inhibit the development of the pathogen in vitro, indicating the importance of in situ tests on fruits and vegetables. Isolations of microorganisms have been made from fresh fruits or leaves, or fruits stored under cold storage for several months, by washing them with a phosphate buffer at pH 6.5 plus mild sonication. Imprinting on media of fruit peels or leaves or plating ground tissue may also give fairly good recovery of microorganisms. The washings may be plated on various media, especially on nutrient yeast dextrose agar, potato dextrose agar + 10% fruit (apple) juice, and malt yeast agar. Plates are then incubated at 24°C. One set of plates with colonies of microorganisms isolated is seeded with conidial suspension of test pathogens (*Botrytis cinerea* or *Penicillium expansum*) and incubated for 48 h. The colonies producing inhibition zones are marked, isolated, and identified (Janisiewicz, 1987, 1991).

The microorganisms with potential biocontrol activity have to be precisely and rapidly identified. Both traditional and modern molecular methods have been employed. In principle, the methods that are applicable for the detection, identification, and differentiation of microbial pathogens and their strains/varieties/pathotypes can also be used for the identification of antagonistic fungi (Chapter 2). Potato dextrose agar medium has been found to be useful for the isolation and maintenance of fungal cultures. Yeasts have been grown successfully on nutrient yeast dextrose broth or agar media (Janisiewicz et al., 2000, 2001; El Ghaouth et al., 2003).

Molecular methods have been shown to be more sensitive, rapid, and precise. The genetic diversity of 205 isolates of *Aureobasidium pullulans*, effective against many postharvest pathogens, was investigated by employing the random amplified polymorphic DNA (RAPD) technique. Colonization of wounds in fruits and dispersal of the BCAs in the environment could be studied. A specific fragment of DNA (L4) of *A. pullulans* was cloned, sequenced, and used to design two sequence-characterized amplification region (SCAR) primers and a 242-bp riboprobe. Both SCAR primers and the 242-bp digoxigenin (DIG)-labeled riboprobe were highly specific for the L-47 strain of *A. pullulans*. The limit of detection of was 10^5 cells of *A. pullulans* per ml, which is 10 times lower than the limit of the method based on colony forming units (CFU) (Schena et al., 2002). The plating method did not differentiate strain K of the yeast antagonist *Pichia anomala* from other isolates/strains. RAPD amplification with primer OPN 13 produced a fragment of about 2000 bp, which was specific to strain K. A SCAR marker of 262 bp, based on this fragment, was amplified with K1 and K2 primers for strain K as confirmed by Southern blot assay. By combining a plating technique on a semiselective medium, followed by a direct strain K-SCAR amplification without DNA extraction, the colonies of *P. anomala* on apples were quantified at 24 h after treatment. The population densities decreased at 1 week after application in cold storage conditions (De Clercq et al., 2003).

9.1.2 Screening

Various criteria may be used as the basis for selecting the microorganisms with required biocontrol potential. Reduction in the number of wounds infected to less than 50% and inhibition of rot expansion by more than 75% were the criteria for the selection of prospective BCAs in several investigations. After shortlisting the BCAs by primary screening, further fine tuning is done by determining the minimum effective concentration of each potential antagonist. About 20% of the most effective antagonists are advanced for further testing using larger numbers of fruits. In the next phase of testing, the most efficient isolates are tested for their survival on fruit at the wound site under different storage conditions of relative humidity, temperature, and composition of gases and for compatibility with other antagonists (since BCA mixtures may be more effective than a single agent), postharvest treatments, and additives (Janisiewicz, 1991; 1998).

Pseudomonas syringae pv. *lachrymans* isolate L-22-64 controlled *Penicillium expansum* effectively, but it had little effect against *Botrytis cinerea* on Golden Delicious apples. However, both pathogens could be controlled by combining *P. syringae* pv. *lachrymans* with *Acremonium breve* (Janisiewicz, 1988). Epiphytic microorganisms isolated from apples, pears, and apple leaf surfaces were screened for biocontrol potential against *P. expansum* (blue mold), *B. cinerea* (gray mold), and *Rhizopus nigricans* (*R. stolonifer*) (Rhizopus rot) on apple. After primary and secondary screening of 933 bacteria and yeasts, the yeast isolate CPA-1 of *Candida sake* was selected as the most effective against all the three diseases. This strain also offered excellent protection against rot development under cold storage conditions. Populations of *C. sake* increased by more than 50-fold during the first 24 h at 20°C reaching the same maximum level both at 20°C and 1°C (Viñas et al., 1996; 1998). The isolate LS-11 of *Rhodotorula glutinis* and LS-28 of *Cryptococcus laurentii* were the most effective against *P. expansum*, among more than 200 yeasts tested. Furthermore, these BCAs were active against *B. cinerea*, *P. expansum*, *R. stolonifer*, and *Aspergillus niger*, when applied on inoculated apples, pears, strawberries, kiwifruits, and table grapes. The isolate LS-28 had greater biocontrol potential than LS-11 and both were effective against *P. expansum* in cold storage conditions (Lima et al., 1998).

By screening epiphytic populations of bacteria on apple leaves and fruits to identify potential antagonists, the three most promising isolates were tentatively identified as *Pseudomonas viridiflava*, exhibiting antagonistic activity against *B. cinerea* and *P. expansum* (Bryk et al., 1999). Likewise, the effectiveness of several further antagonistic fungi and bacteria against postharvest pathogens has been demonstrated: *Paenibacillus polymyxa* against *B. cinerea* in strawberries (Helbig, 2001); *Metschnikowia pulcherrima* strains against *P. expansum* in apples (Janisiewicz et al., 2001); *Cryptococcus laurentii* (strain 317) and *Candida ciferrii* (strain 283) against *P. expansum* in apples (Vero et al., 2002; He et al., 2003) and in pears (Zhang et al., 2003); *Aureobasidium pullulans* against *B. cinerea* and *Monilinia laxa* in sweet cherries and table grapes (Schena et al., 2003); *Trichoderma viride* against *Sclerotium cepivorum* in onion (Clarkson et al., 2002); and several bacterial and a fungal antagonist against *Helminthosporium solani* causing silver scurf disease in potato (Michaud et al., 2002). In order to select BCAs with a broad spectrum of activity, more suitable for commercial development, it is desirable to carryout evaluation of the biocontrol potential of the BCAs against several pathogens infecting different fruit crops (Lima et al., 1999).

9.1.3 Characterization of Selected Biocontrol Agents (BCAs)

The success of biocontrol of postharvest diseases depends largely on the potential of the BCAs for acting on the microbial pathogens under different environmental conditions and in combination with other factors, such as the cultivar, compatibility with regular postharvest practices, treatment, and additives and storage conditions. The strain 75-A2 of *Metschnikowia pulcherrima* was the most resistant to diphenylamine (DPA), common postharvest antioxidant used to treat apples (Janisiewicz et al., 2001). The time and frequency of application at optimal concentrations of BCAs will also have a significant influence on the effectiveness of treatment with BCAs. The selected BCAs should not be a health hazard due to the production of antibiotics or any toxic metabolite.

Postharvest pathogens may infect the host plant parts at any time after flowering and proceed through various stages of fruit development in the field and after harvest (Dennis, 1983). Early field infection results in quiescence until after harvest. When fruits ripen, resumption of pathogen activity leads to symptom expression (Chapter 4). Thus, biological control of fruit diseases resulting from floral infections has to commence at bloom with the colonization by the BCA of senescent flower parts before, or at the time of, infection. In the case of wound pathogens, the fruits have to be protected at harvesting or at preharvest stage. Field application of BCA may result in early colonization of fruit surfaces offering protection to the fruits against infection by postharvest pathogens entering through wounds caused by improper handling. For the BCAs to be effective must be able to tolerate the low nutrient availability, UV radiation, high temperature, and dry conditions that occur frequently during growing seasons. Attempts have been made to evaluate the biocontrol potential of various BCAs against the postharvest pathogens infecting apples, pear, grapes, and strawberry based on these parameters.

The biocontrol potential of *Aureobasidium pullulans* (isolate L47) and yeast *Candida oleophila* against grey mold of table grapes was evaluated. The BCAs were sprayed on bunches in the field before harvest and the grapes were either cold-stored at 0°C or left on the vines under plastic covering. *A. pullulans* applied 5 days prior to harvest provided sustained protection against *B. cinerea*, under both field and storage conditions, while *C. oleophila* showed only marginal activity in the first year of study. The greater activity of *A. pullulans* was considered to be due to its higher survival rates on berries under different conditions of the trials (Lima et al., 1997). Unmodified cells (that is the cells were not subject to treatment with glycerol, glucose or trehalose to modify the endogenous sugar alcohol and sugar content) and low water activity (a_w)-tolerant cells of *Candida sake* (CPA-1), were applied to apples before harvest and the efficacy for control blue mold of apples (cv. Golden Delicious) caused by *P. expansum* under commercial storage conditions was examined. The population of the unmodified cells did not show variation under cold storage conditions, whereas the population of the low-a_w-modified CPA-1 cells increased. However, under cold storage conditions both types of cells showed similar patterns of population changes – an initial increase followed by a decline later. Both unmodified and low a_w-tolerant cells were found to provide protection to the treated apples to the same level (greater than 50% reduction in size of infected wounds) (Teixidó et al., 1998).

The biocontrol efficacy of *C. sake* applied at pre- and postharvest stages on Golden Delicious apples, wounded before and after harvest and inoculated with *Penicillium expansum* prior to cold storage, was assessed. Postharvest treatment with *C. sake* resulted in effective control of the blue mold disease irrespective of wounding. Following application of the antagonistic yeast, reduction in disease incidence was up to 50%, whereas reduction in lesion diameter was by a maximum of 80%. There was no added advantage from the preharvest application of *C. sake*. High population levels of *C. sake* were detected in apples receiving the postharvest application, even after 90 days in cold storage, whereas the yeast population in apples receiving the preharvest application declined rapidly (Teixidó et al., 1999) (Fig. 9.1).

The yeasts *Candida oleophila*, *Cryptococcus infirmo-miniatus*, *C. laurentii*, and *Rhodotorula glutinis* were applied to Bosc and d'Anjou pear fruit in the field 3 weeks before harvest and their rate of survival was determined at intervals. *C. infirmo-miniatus* was the most effective and consistent in decay control (Benbow and Sugar, 1999). Another yeast, *C. albidus*, originally isolated from mature strawberry fruit, was tested for its ability to protect strawberry fruits against grey mold disease. The blooms were treated with a suspension of *C. albidus*. The grey mold incidence was reduced by 33, 28, and 21% during the 3 years of field evaluations. Addition of alginate, xanthan, or cellulose to the

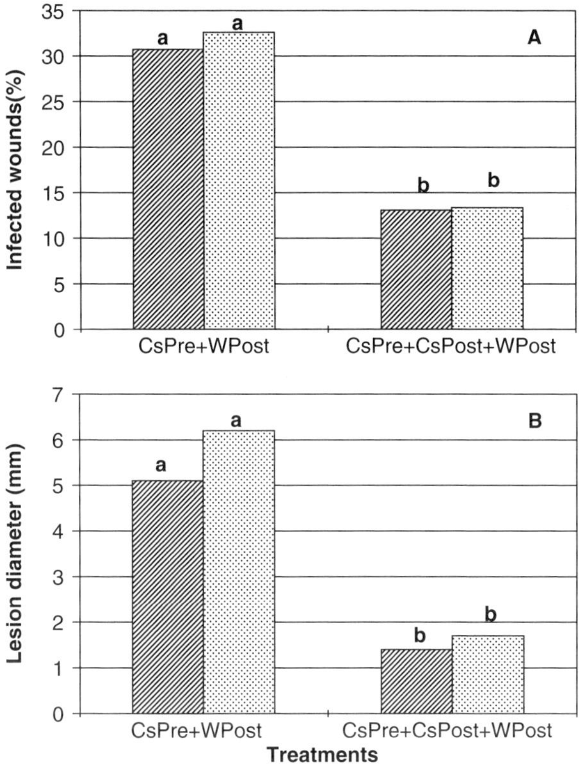

Figure 9.1 Effect of pre- and postharvest application of *Candida sake* on postharvest wounds and both pre- and postharvest development of blue mold (*Penicillium expansum*) in Golden Delicious apples during 1994–1995 (■) and 1995–1996 (□) growing seasons. Different letters in the bars indicate significant differences between means as per Duncans's Multiple Range Test ($P < 0.01$). CsPre, preharvest application of *C. sake*; CsPost, postharvest application of *C. sake*; WPost, postharvest wounds. (Courtesy of Teixidó *et al.*, 1999; Elsevier, Oxford, United Kingdom.)

yeast cell suspension enhanced the effectiveness of disease control (Helbig, 2002). In a more elaborate study, the biocontrol fungus *Ulocladium atrum* was applied as conidial suspension (2×10^6 conidia/ml) on flowers and young strawberry fruits at six stages of development. *U. atrum* suppressed sporulation of *B. cinerea* on flower petals by 15 to 54% but did not prevent the incidence of *B. cinerea* on stamens. The efficacy of the BCA was rated as low to moderate and disease incidence was reduced by 21% on ripe fruits. The low efficacy of *U. atrum* was attributed to possible poor coverage with, or colonization of, floral parts by the BCA. Application of *U. atrum* at late flowering or early fruit stages was considered to be most suitable (Boff et al., 2002a, b). These studies indicate that some BCAs have the capacity to tolerate the environmental conditions existing in the orchards and to provide significant control of diseases that cause serious losses after harvest. Such an attribute is one of the most important requirements for a biocontrol agent to be successful.

The attributes of BCAs that are desirable may vary depending on the biocontrol system. Some attributes are, however, essential for the BCAs to be considered as an effective alternative for the synthetic fungicides and chemicals that have been used for the management of postharvest diseases:

1. Capacity to tolerate and survive variable environmental conditions;
2. Ability to multiply rapidly in the wound sites and to reach the optimum concentration;
3. Possibility for use along with current practices followed during commercial storage;
4. Effectiveness against multiple infection by several pathogen;
5. Potential for replacing, and insensitivity to, the chemicals that are currently being used.

The yeast *Candida sake* was effective for the control of blue mold caused by *Penicillium expansum* in apples (cv. Golden Delicious) when applied as preharvest spray as well as application on fruits under commercial storage conditions (Teixidó et al., 1998). Three strains of *Trichoderma harzianum*, that is T-39 (TRICHODEX a commercial product), T-161, and T-166 were evaluated under field conditions for the control of strawberry anthracnose disease caused by *Colletorichum acutatun* and gray mold disease caused by *Botrytis cinerea*, using different intervals of application and dosages. At higher concentration (0.8%), all strains were more effective in reducing the severity of anthracnose, whether they were applied alone or in combination with other strain (at 0.4% each). Strains T-39 (0.4%) or T-161 + T-39 (0.4% each) were as effective as the fungicide fenhexamid against gray mold disease (Freeman et al., 2004). The yeast antagonist *Metschnikowia fructicola* was evaluated for the control of *B. cinerea* infecting strawberry pre- and postharvest stages and *Rhizopus stolonifer* causing storage rot. Application of *M. fructicola* against preharvest rots was equally effective as the fungicide fenhexamid in two growing season in greenhouse trials. In the field experiment the disease incidence was also reduced to commercially acceptable levels. Field application reduced the fruit rot during postharvest storage by 64 to 72%. The population density of *M. fructicola* on fruits treated at weekly intervals, was about 1×10^5 CFU/fruit (Karabulut et al., 2004) (Table 9.2).

Candida sake provided excellent control of rot development in apples under cold storage conditions. The populations of *C. sake* increased by more than 50-fold at 20°C and the population reached the same levels at 20° and 1°C after 3 and 20 days respectively (Viñas et al., 1998). The ability of *Candida sake* (Strain CPA-1) to colonize the surface of apples under various storage conditions and its greater capacity to colonize apple wounds were demonstrated by Usall et al. (2001a) (Fig 9.2). Application of *C. sake* (CPA-1), at

TABLE 9.2 Effectiveness of *Metchnikowia Fructicola* in Controlling Postharvest Rots of Strawberry in Storage

Location of tests	Treatments	Decayed fruit (%) Harvest number						Mean/ harvest
		1	2	3	4	5	6	
Greenhouse	Yeast (0.5 g/l)	49.5 ab	49.5 a	41.0 b	57.7 ab	49.0 ab	53.9 a	50.1c
	Yeast (1.5 g/l)	24.7 b	16.1 b	16.5 b	24.1 bc	16.5 c	12.5 b	18.4 b
	Fenhexamid	66.0 a	33.0 ab	33.0 b	8.2 c	0.0 c	8.9 b	24.8 b
	Control	90.7 a	82.5 a	74.2 a	82.5 a	82.5 a	76.3 a	81.5 a
Field	Yeast (0.5 g/l)	33.5 b	23.6 b	12.6 b	23.4 b	25.6 b	–	23.7 c
	Fenhexamid	57.3 a	26.0 b	45.9 ab	42.8 ab	19.4 b	–	38.3 b
	Control	63.9 a	65.2 a	70.4 a	57.3 a	72.1 a	–	65.8 a

Values followed by the same letter are not significantly different at P 0.05 as per Duncan's multiple range test.
Source: Karabulut et al., 2004.

10^7 CFU/ml as a drench in commercial trials for three seasons, resulted in a reduction in the incidence of decay to a level equal to that of imazalil (375 ppm) and higher than that of thiabendazole (425 ppm) + folpet (1000 ppm). An increase in the population of *C. sake* on the surface of wounded fruits by five fold was observed, while the population declined by more than 10 fold on the surface of unwounded fruits during the first 60 days of storage at 1°C. The BCA was not affected by immersion in solution of benomyl, sulfur, flusilazol, ziram, TBZ, or diphenylamine. However, it was not compatible with captan, imazalil, and ethyoxyquin at recommended rates. The ability of *C. sake* to grow in culture at a wide range of temperature (1 to 34°C) was another desirable attribute, as reported by Usall et al. (2001a).

Preharvest application of the yeasts *Trichosporon pullulans*, *Cryptococcus laurentii*, and *Rhodotourla glutinis* resulted in colonization of the surface of sweet cherry. However, the period of their survival varied significantly. *C. laurentii* showed strong survival ability on fruit surfaces under field conditions as well as adaptability to postharvest storage conditions of low temperature, low O_2, and high CO_2 concentrations (Tian et al., 2004). A talc-based formulation of *Pseudomonas fluorescens* (FP7) amended with chitin was sprayed on mango trees preharvest at fortnight intervals. Induction of flowering and reduction in latent infection by the anthracnose pathogen *Colletotrichum gloeosporioides* were observed to a greater extent compared to treatment with *Bacillus subtilis* and *Saccharomyces cerevisiae* (Vivekananthan et al., 2004). The yeasts *Rhodotorula glutinis* and *Cryptococcus laurentii* were able to grow at wide range of temperatures (0–35°C) and they were effective against *P. expansum* under cold storage conditions. *R. glutinis* showed higher and greater stability in biocontrol activity (Lima et al., 1998). *Aureobasidium pullulans* exhibited a high and durable activity against *B. cinerea* infecting grapes, under both field and cold storage conditions, when compared to *Cryptococcus oleophila* (Lima et al., 1997).

The yeasts *Trichosporon* sp. and *Cryptococcus albidus* were evaluated for their biocontrol potential against *B. cinerea* and *P. expansum* infecting apple (cv. Golden Delicious) and pear (cv. Jingbai) fruits stored at 1°C in air and under controlled atmospheres (CA) with 3% O_2 + 3% CO_2 or 3% O_2 + 8% CO_2. *Trichosporon* sp. was more effective against both pathogens and decay control was better in apples than in pear, indicating that biocontrol efficiency may vary depending on the nature of host–pathogen combination (Tian et al., 2002). *Crytococcus laurentii* was evaluated for its efficacy in reducing blue mold decay of pear caused by *P. expansum*. The disease development was entirely inhibited by treatment of pears with *C. laurentii* cell suspension at 10^9 cells/ml at 25°C after an incubation for 7 days after inoculation. The treatment was most effective when the yeast cells were applied simultaneously or prior to inoculation with *P. expansum*. The efficacy was reduced if applied after inoculation with *P. expansum*. The wounds were rapidly colonized by the yeast during the first 3 days after application at 20°C and then the population stabilized for the rest of the storage period (Zhang et al., 2003).

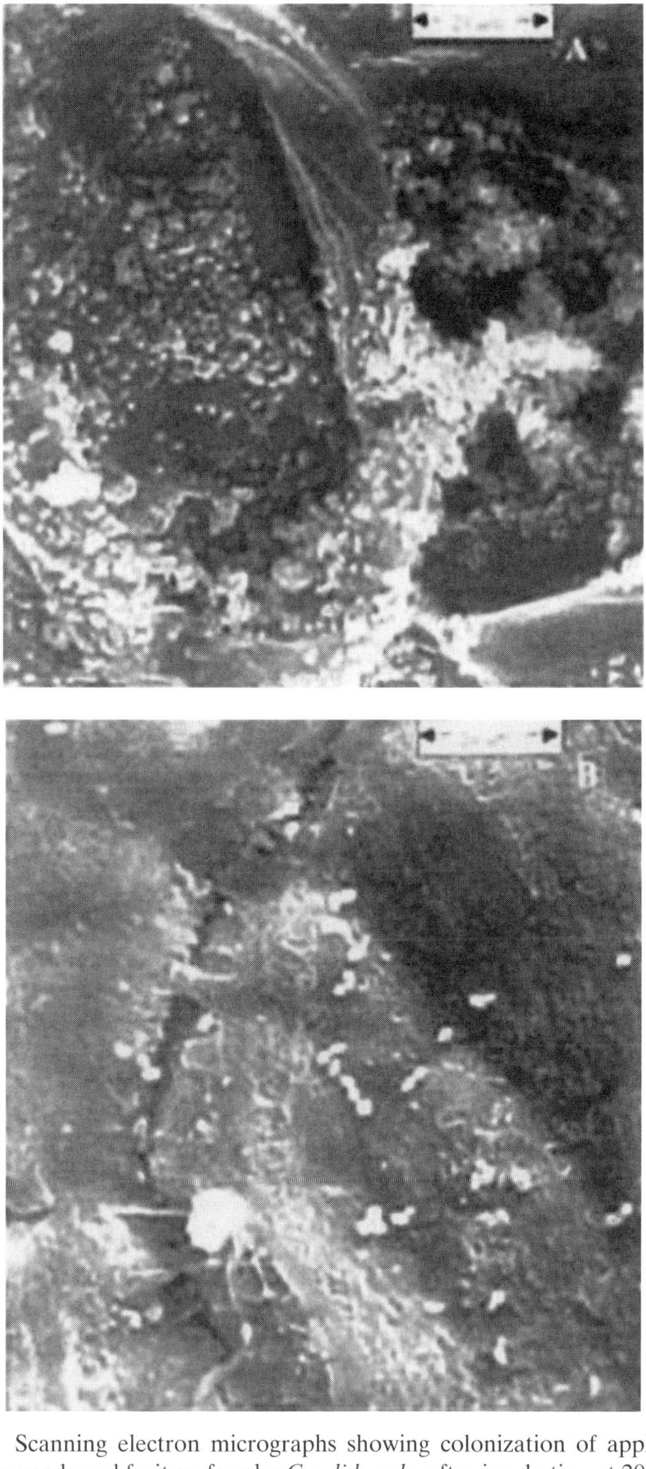

(a)

(b)

Figure 9.2 Scanning electron micrographs showing colonization of apple (Golden Delicious) wounds and fruit surface by *Candida sake* after incubation at 20°C for 24 h. (a) Apple wound; (b) apple fruit surface. (Courtesy of Usall *et al.*, 2001a; Elsevier, Oxford, United Kingdom.)

The efficiency and rapidity of colonization of wound sites and fruit surface by BCAs is another desirable attribute. The ability of BCAs to protect strawberry fruit wounds after harvest against *B. cinerea* was assessed. *Candida reukaufi* and *C. pulcherrima*, when applied at 10^3 CFU/wound, protected fruit wounds. These BCAs effectively colonized fruit wounds and strongly inhibited spore germination of *B. cinerea* in vitro (Guinebretiere et al., 2000). It is advantageous to use the BCA that is capable of protecting fruits against two or more postharvest pathogens. *C. sake* isolate CPA-1 was found to be very effective against three important pathogens, *B. cinerea*, *P. expansum*, and *R. nigricans*, infecting apples (Viñas et al., 1998). The wide spectrum of antagonistic activity of the BCA against several pathogens may indicate the operation of multiple mechanisms in the ideal BCA.

The persistence and ability to survive in the various environments may be a desirable attribute for the microbial biocontrol agents. An effective monitoring system to detect and quantify the BCA applied on the fruit is very useful to determine the fate of the BCA. A quantitative competitive polymerase chain reaction (QC-PCR) method and enzyme-linked oligosorbent assay (ELOSA) were applied to monitor *Pichia anomala* strain K effective against postharvest diseases on apples. The amounts of strain K cell suspension could be quantified in the phosphate buffer used for recovering cells from the apple surface. Strain K populations of 10^3 to 10^6 yeast per apple could be accurately estimated. The threshold of this detection technique was 10^3 CFU/apple, which was about 100 times more sensitive than a conventional isolation method for monitoring strain K (Pujol et al., 2004). The bacterial antagonist *Pantoea agglomerans* reduced the decay due to gray mold of apples at three atmospheric conditions tested. The populations of the BCA followed similar patterns under all three atmospheric conditions. The possibility of using *P. agglomerans* on apples at a rate of 8×10^7 CFU/ml, by growers in packing houses, has been demonstrated (Nunes et al., 2002b).

There is the possibility of utilizing new technologies, such as a new harvester which which removes apples from the tree by a quick pull . The stem loss (stempulls) during harvest could create potential entry points for the postharvest pathogen *Penicillium expansum*, thus predisposing the stem cavity to decay. However, the apple fruits could be protected by employing the bacterial antagonist *Pseudomonas syringae* (used in BioSve 110), without resorting to the use of synthetic fungicides (Janisiewicz and Peterson, 2004).

In order to quantify loss of viability of the yeast *Aureobasidium pullulans*, a real-time in situ procedure was developed. *A. pullulans* (PRAFS 8) was transformed with green fluorescent protein (GFP). The transformed yeast cells were exposed to chemicals or other adverse conditions. A strong positive relationship between fluorescence and number of viable yeast cells was established and fluorescence was used as a direct indicator of yeast cell viability. The loss of fluorescence as a measure of yeast viability was quantified by image analysis. The results revealed the potential use of GFP fluorescence for determining yeast cell viability applied for the control of microbial plant pathogens (Sabev et al., 2004).

The use of chemicals for the management of postharvest diseases has to be reduced, if possible avoided, because of obvious undesirable adverse effects on consumers and environment. Nevertheless, no alternative method with equal effectiveness, if not more, has yet been developed for the management of most of the postharvest diseases. However, the possibility of developing biocontrol systems that can effectively replace chemicals has been revealed by some investigations. The development of resistance in microbial pathogens to chemicals is known, posing serious problems in tackling diseases caused by fungicide-resistant strains of the pathogen. The bacterial antagonist *Pantoea agglomerans* (CPA-2 strain) was shown to provide protection against *B. cinerea* in Golden Delicious apples as effectively as the fungicide imazalil (Nunes et al., 2002). A combination of yeasts *Cryptococcus laurentii* and *C. infirmo-miniatus* provided effective control of blue mold caused by *P. expansum* in pear to the extent of 91% as against 88% in control obtained by applying thiabendazole (TBZ) at a high dose of 528µg/ml. Application of *C. infirmo-miniatus* + TBZ at half dose (264µg/ml) was the most effective treatment in a commercial testing on apple (Goyal and Spotts, 1997). In addition, *C. infirmo-miniatus* could also suppress brown rot disease caused by *Monilinia fructicola* applied as dip in a conidial suspension of the BCA (Spotts et al., 1998).

Thiabendazole (TBZ) has been commonly applied against postharvest pathogens resulting in the development of TBZ-resistant strains of *P. expansum*. Control of wound infection by TBZ-resistant strains could be achieved by employing yeasts *Rhodotorula glutinis*, *Cryptococcus infirmo-miniatus*, *C. laurentii* and the registered biocontrol products Aspire (containing *Candida oleophila)* and Bio-Save 11 (containing *Pseudomonas syringae*) (Sugar and Spotts, 1999). BCAs with low sensitivity towards fungicides frequently used on fruits and vegetables may be applied along with low doses of fungicides. *R. glutinis* and *C. laurentii* strains with low sensitivity to fungicides have been identified (Lima et al., 1998). *Trichoderma harzianum* (T88) and *T. atroviride* (T 95) were found to be equally effective in providing control against *Botryospharia berengeriana* f.sp. *spiricola* causing apple ring rot disease as the routine chemical (Kexiang et al., 2002).

9.1.4 Biocontrol Activity of Selected Biocontrol Agents under Controlled Storage Conditions

The biocontrol potential of *Erwinia* sp. for the control of gray mold disease of apple under cold storage at 5°C and modified-atmosphere packaging was assessed. The apples were inoculated and stored in polyethylene bags with an initial gas composition of five levels of O_2 (1–15%) and CO_2 (0–15%). The antagonist and O_2 acted synergistically to reduce development of the pathogen. High CO_2 atmosphere reduced the growth of *B. cinerea*. *Erwinia* sp. was found to be effective in controlling the disease under ambient conditions (Floros et al., 1998). Two isolates of *Pantoea agglomerans* (B66 and B90) and two isolates of *Pseudomonas* sp. (B 194 and B224) were the most effective in providing protection to apples against *B. cinerea* (gray mold) and *P. expansum* (blue mold).

The protective activity of the BCAs on apples stored in controlled atmosphere (CA) was not reduced, and occasionally was more effective, than in normal storage conditions. These BCAs may act in several ways: by inhibiting spore germination, reducing germ tube elongation, and through parasitism and competition for nutrients (Sobiczewski and Bryk, 1999). The bacterial antagonist *Pantoea agglomerans* CPA-2 was able to inhibit the growth of the pathogens *B. cinerea*, *P. expansum*, and *Rhizopus stolonifer* on pear cv. Blanquilla during cold storage in normal air or low-oxygen atmosphere, by colonizing the wounded fruits rapidly at both 20°C or 1°C (Nunes et al., 2001a). Among the eight strains of the yeast *Metschnikowia pulcherrima* evaluated for their biocontrol potential against apple blue mold caused by *P. expansum*, the strain T5-A2 was the most effective on harvested mature apples treated with the lowest concentration of the antagonist and stored for 3 months at 0.5°C (Janisiewicz et al., 2001).

Application of BCA by itself may not always provide commercially acceptable level of control of postharvest diseases. However, it is possible to increase the efficacy of BCA by combining with other postharvest treatments. Controlled atmospheres (CA) have been shown to reduce fungal growth (Ahmadi et al., 1999). A combination of CA with a BCA provided beneficial effects against gray and blue mold diseases of apple (Tian et al., 2002). Four antagonistic yeasts (*Trichosporon pullulans*, *Cryptococcus laurentii*, *Rhodotorula glutinis*, and *Pichia menmbranefaciens*) effectively controlled the postharvest pathogens *Alternaria alternata*, *Penicillium expansum*, *B. cinerea*, and *Rhizopus stolonifer* on sweet cherries at 25°C, *T. pullulans* being the most effective. The antagonistic activities of *C. laurentii* and *R. glutinis* against *A. alternata* and *P. expansum* were significantly increased by combination with controlled atmospheres with 10% O_2 + 10% CO_2. These BCAs protected the sweet cherries under CA conditions more effectively for 60 days than the fruits stored at 0°C for 30 days (Quin et al., 2004).

Continuous search for the BCAs with greater biocontrol potential becomes essential in order to replace the synthetic fungicides. An endophytic fungus isolated from a cinnamon tree, *Muscodor albus* has been shown to act in a different manner. *M. albus* produces about 28 volatile compounds which could inhibit or kill several species of fungi, Oomycetes, and bacteria (Strobel et al., 2000; Worapong et al., 2001). *Rhizoctonia solani*, *Sclerotinia sclerotiorum*, and *Verticillium dahliae* are important fungal pathogens that were killed by the volatile compounds from *M. albus*. In another study, *Botrytis*, *Colletotrichum*, *Geotrichum*, *Monilinia*, *Penicillium*, and *Rhizopus*. *P. expansum* and *B. cinerea*, causing blue mold and gray mold diseases of apple, were entirely controlled by fumigation for 7 days with a culture of *M. albus*. Fumigation for 24 to 72 h immediately or 24 h after inoculation also controlled the blue and gray mold diseases of apples. Similar exposure to volatiles of *M. albus* was also effective against brown rot disease of peaches caused by *Monilinia fructicola*. Two major volatile compounds, 2-methyl-1-butanol and isobutyric acid, were detected in the head space. As *M. albus* is a sterile mycelium and does not need direct contact with the commodities to be protected, this biofumigant fungus offers a potential protective mechanism for the control of postharvest diseases. This investigation has brought to light a biocontrol agent with a different biocontrol mechanism not reported earlier (Mercier and Jiménez, 2004).

9.2 BIOLOGICAL CONTROL OF POSTHARVEST DISEASES OF VEGETABLES

Biological control for containing postharvest diseases and spoilage of vegetables has advanced from its primitive origin in food preservation (preparation of pickles) to application of selected and well-characterized microorganisms (as commercial products) during the past two decades. Some of the microbes with biocontrol potential have also been found to provide additional benefit because of their ability to inhibit different bacterial species of human health concern such as *Clostridium botulinum*, *Escherichia coli*, *Listeria monocytogenes*, *Salmonella* spp. *Staphylococus aureus*, and *Yersinia enterocolitica* (Hanlin and Evancho, 1992). The acceleration of research efforts to find alternative disease management strategies to replace synthetic fungicides have resulted in the development of commercial products registered by U.S. Environmental Protection Agency (EPA) for postharvest application against microbial plant pathogens. As in the case of fruits, the environment in which the vegetables are to be stored and the target pathogens (based on their economic importance) have been clearly determined. The parameters for selection of microorganisms, present in the natural environment of plants and soil, for use against vegetable pathogens are also similar to the biocontrol agents selected for use against fruit pathogens. These parameters include the adaptation of the agent to the expected storage environment and competitiveness, in terms of cost and efficacy, compared to the chemical control. Furthermore, activity against a range of postharvest pathogens infecting the same fruit or vegetable is important, since a BCA that is effective against one pathogen may provide conditions for a minor disease to become a major one. Hence it is preferable to use mixtures of mutually compatible BCAs that exhibit activity against a wide spectrum of pathogens.

Vegetables are grown under different agroecosystems and both vegetative and propagative plant parts, such as roots, leaves, and fruits, are consumed as vegetables. The pathogens present in the soil and other plant parts may reach the vegetables, necessitating the provision of a protective umbrella for longer periods. Various kinds of BCAs have been employed for the control of postharvest diseases of vegetables. Bacteria and yeast species have been frequently shown to be effective. The success of biocontrol agents on vegetables depends on the methods of marketing, handling, and packaging. Preference for packaged vegetables apparently increases due to consumer appeal, convenient handling, and possible longer maintenance of high quality. Microbial populations are influenced by packaging because of alterations of environmental conditions.

9.2.1 Hypogeal Vegetables

Among the vegetables developing below soil level, potato is the most important and the potato tubers are infected by a large number of pathogens. *Erwinia carotovora* subsp. *carotovora* (*Ecc*) causes postharvest soft rot of tubers, in addition to seed piece decay and aerial stem rot disease in potato. Application of two different strains of fluorescent *Pseudomonas* spp. reduced the disease in vitro (Burr and Schroth, 1977). Application of *Pseudomonas putida* strain M17, in conjunction with proper harvesting and handling practices, reduced the postharvest soft rot due to *Ecc*, when applied as a seed piece treatment or as postharvest treatment by 50% and 75%, respectively (Coyler and Mount, 1984). Postharvest application of *P. putida* was found to be more effective because of greater colonization of infection points on intact tubers. This BCA may have multiple biocontrol mechanisms of antibiosis and induction of resistance. Two plant growth-promoting rhizobacteria (PGPR) *P. fluorescens* and *P. putida* controlled the soft rot disease effectively. Production of siderophore and antibiotics by the BCAs was demonstrated by the procedure developed by Xu and Gross (1986). The effectiveness of PGPR against potato blackleg and soft rot diseases was reported by Kloepper (1983). Suppression of deleterious root-colonizing microorganisms may be a result of growth promoting effects and shifts in populations of root zone microorganisms induced by PGPR (Suslow and Schroth, 1982; Burr and Caeser, 1984). Ringrot disese of potato caused by *Clavibacter michiganens* subsp. *sepedonicus* could be controlled by dipping the roots of potato seedlings in suspensions of *P. aureofaciens* and *P. fluorescens* in glasshouse trials. Combinations of these two PGPR did not provide greater protection compared to treatment with a single strain (De la Cruz et al., 1992).

A novel approach to the control of bacterial pathogens causing postharvest diseases that has been attempted is the use of bacteriophages to cause lysis of susceptible bacterial cells. Bacteriophages isolated from freshwater lakes were able to cause lysis of 16 of the 23 serogroups of *E. caorotovora* subsp. *carotovora*, indicating the possibility of employing them for the control of soft rot diseases (Eayre et al., 1995). The lytic response following the interaction between the bacterial species and specific phages has been used as a basis for detection and identification of the bacteria. A highly virulent and polyvalent *Streptomyces* phage isolated from a potato field in Western Australia was evaluated for its ability to disinfest seed potato tubers artificially inoculated with *Streptomyces scabies* causing potato powdery scab disease. Infected tubers were immersed in a phage suspension (1×10^9 plaque forming units (PFU)/ml for 24 h. The phage-treated potatoes had reduced levels of scab surface lesions

(1.2%) compared with untreated control tubers (23%). There was also a significant reduction in the number of lesions following phage treatment of mother tubers (McKenna et al., 2001).

Nonpathogenic isolates of *Streptomyces* spp. showed biocontrol activity against *S. scabies* (Liu et al., 1995). However, no consistent level of control of the powdery scab could be obtained during different seasons or locations. The existence of multiple mechanisms of pathogen inhibition, such as antibiosis and resource competition, might have contributed to the inconsistent results. The efficacy of an antibiotic-producing *Streptomyces* spp. was evaluated for control of potato powdery scab disease. The effectiveness of disease control was not correlated with in vitro sensitivity to antibiotic inhibition. Further, variation in the susceptibility/resistance of cultivars may also be a factor influencing the effectiveness of control of the powdery scab disease (Ryan et al., 2004).

The possibility of degrading the toxins produced by microbial plant pathogens by employing other microorganisms leading to suppression of, or development of resistance to, diseases has been reported by Toyoda et al. (1988) and Pedras et al. (2002). The phytotoxins produced by *S. scabies* have been identified as 4-nitroindol-3-yl containing dioxopiperazines. Thaxtomin A is the most important member of this group of compounds. The fungus *Aspergillus niger* var. *niger*, isolated from the bark of a decomposing deciduous tree, exhibited the ability to transform thaxtomin A to much less toxic metabolites. Under optimum conditions, *A. niger* var. *niger* converted thaxtomin A into two major and five minor metabolites which were characterized by mass spectral and nuclear magnetic resonance techniques. The major metabolites were much less toxic than thaxtomin A when tested on aseptically produced mini-tubers (Lazarovits et al., 2004).

Avirulent strains of the microbial pathogens have been shown to induce resistance in treated plants against the virulent strains of the same pathogen. Hence they can be employed as biocontrol agents for providing protection against the microbial pathogens. This principle is advantageously exploited in the case of potato brown rot disease caused by *Ralstonia solanacearum*. Dipping potato tuber seed pieces in a suspension of different bacterial species reduced the severity of the disease (Kempe and Sequeira, 1983; Martin and French, 1985). A combined application of avirulent strains of *R. solanacearum* and pectolytic strains of *Pseudomonas fluorescens* to seed pieces reduced the severity of brown rot disease significantly. The efficiency of bacterial antagonist could be increased by coating the antagonist with calcium carbonate (Ciampi-Panno et al., 1989). The increased resistance following application of calcium salts demonstrated in other pathosystems may also contribute to the effectiveness of brown rot disease control. The avirulent strains of *R. solanacearum* exhibit dual functions. They can enhance the level of resistance of potato tubers to virulent strains of *R. solanacearum*. Furthermore, due to the suppression of the activity of root knot nematode, damage to tubers is considerably reduced, resulting in a significant reduction in incidence of brown rot disease. Thus higher level of disease control can be achieved by using the avirulent strains of *R. solanacearum* compared with untreated control plots. The commercially available avirulent strain of *R. solanacearum* (PSSOL) has been suggested for use against diseases caused by *R. solanacearum* in vegetables (Natural Plant Protection Route d'-Artix, Nogueres, France) (Janisiewicz and Korsten, 2002b).

Among the fungal pathogens, *Rhizoctonia solani* (sclerotial stage), *Macrophomina phaseolina* (conidial stage), and *Thanatephorus cucumeris* (teleomorph) may infect potato plants in the field, causing charcoal rot of tubers. The percentage of infection by *M. phaseolina* at harvest was reduced by treating seed tubers with the bacterial antagonist *Bacillus subtilis* (Thirumalachar and O'Brien, 1977). Potato black scurf disease (*Rhizoctonia solani*) is initiated by both seed and soil-borne inocula, posing difficulties for protecting the plants and tubers during the entire growing season. Pathogen sclerotia were parasitized by *Verticillium biguttatum* (Velvis and Jager, 1983). Formation of sclerotia on newly produced potato tubers was inhibited by *V. biguttatum*, although symptoms of infection were not affected. The extent of reduction in black scurf symptoms on tubers by the antagonist was comparable to, and in some cases greater than, that obtained by using chemical disinfectants of soil (Velvis and Jager, 1983). *Hormiactis fimicola*, another fungal antagonist, could grow at temperatures below the minimum for growth of *V. biguttatum*. When these antagonists were applied simultaneously, *H. fimicola* reduced the antagonistic activity of *V. biguttatum* and hence there was no increase in the effectiveness of control of *R. solani* (Jager et al., 1979; Velvis and Jager, 1983). In contrast, the bacteria *Azotobacter chroococcum* in combination with *V. biguttatum* provided effective protection to potato plants, sprouts, and stolons. Production of sclerotia on new tubers could be significantly reduced by inoculating seed tubers with three isolates of *V. biguttatum* separately or mixed (Meshram, 1984; Jager and Velvis, 1986). Preparations of *Trichoderma viride* and *Gliocladium virens* produced in a fermentor were applied as dusts to seed potatoes inoculated with *R. solani*. Disease severity and incidence was reduced appreciably, since both soil- and tuber-borne inocula were significantly reduced. The toxic metabolites of *T. viride* might contribute to its antagonistic potential, in addition to its ability to parasitize the pathogen (Beagle-Ristaino and Papavizas, 1985). The presence of gliotoxin and viridin was detected in potato tubers treated with *G. virens* (Aluko and Hering, 1970). The biocontrol products formulated from *Trichoderma* spp. *T. harzianum* strain KRL-AG2 and *Streptomyces griseoviridis* strain K61 have also been found to be effective against *R. solani* infecting vegetables (Janisiewicz and Korsten, 2002b).

Several species of *Fusarium*, such as *F. solani* var. *coeruleum*, *F. sulphureum* (syn *F. sambucinum*), *F. thrichothecioides*, *F. equiseti*, *F. oxysporum*, *F. sporotrichoides*, and *F. avenaceum*, have been reported to be associated with dry rot disease of potato, a serious world-wide storage disease (Snowden, 1992). All commonly cultivated potato cultivars are susceptible to the dry rot disease. A high percentage of virulent strains of *Gibberella pulicaris* (teleomorph of *F. sambucinum*) are able to detoxify sesquiterpene phytoalexins produced by potatoes in response to interaction with the pathogen (Desjardins et al., 1992). Application of *Burkholderia* (*Pseudomonas*) *cepacia* reduced the dry rot disease, but the effect seemed to be inconsistent (Burkhead et al., 1994). However, *B. cepacia* strain B 37W (NRRLB 14858) was found to inhibit the growth of *G. pulicaris* to the maximum extent. Pyrrolnitrin was the major antifungal compounds produced by *B. cepacia*. Some of the bacterial antagonists when used in pairs showed additive effects, resulting in enhancement of efficiency of biocontrol, possibly because of complementary antibiotic production. In comparison, yeasts were less effective and inconsistent (Schisler et al., 1995, 1997, 1998). In another study, species of *Bacillus* were tested for their potential for controlling *Fusarium* dry rot disease. *B. cereus* (X16) was the most effective for the control of the dry rot on seed tubers and increased the yield parameters as well. During the traditional and cold storage for 6 and 8 months respectively, dry rot incidence was reduced significantly in potato boxes treated with each antagonist, when compared with treatment with the fungicide carbendazim (Safdi et al., 2002).

The synthetic chemical sprout inhibitor, 1-methyl ethyl-3-chlorophenyl carbamate, is the only chemical registered for postharvest sprout control of stored potatoes. In the attempt to find alternative sprout control methods, bacterial strains with superior dry rot disease suppression were evaluated. Of the six strains tested, *Pseudomonas fluorescens* bv. V S11:P:12 and two strains of *Enterobacter* sp. S11:T:07 and S11:P:08 exhibited maximum effectiveness in controlling sprouting of tubers, indicating that these bacterial strains provide this advantage in addition to control of dry rot disease (Burkhead et al., 2003).

Production of a mycotoxin trichothecene by *G. pulicaris*, implicated in mycotoxicoses of humans and animals, emphasized the need for developing more effective biocontrol systems (Senter et al., 1991). In addition, many thiabendazole (TBZ)-resistant strains were detected, indicating that TBZ, the only registered postharvest chemical for control of *G. pulicaris*, may become ineffective, as the TBZ-resistant strains are likely to persist (Desjardins et al., 1993). *Pseudomonas fluorescens*, *Pantoea agglomerans*, and *Enterobacter cloacae* were evaluated and the bacterial BCAs reduced the disease incidence by 17 to 25%, in addition to being more effective than TBZ (Schisler et al., 2000).

Silver scurf caused by *Helminthosporium solani* is another important postharvest disease of potato. The presence of thiabendazole-resistant strains of the pathogen is widespread, underscoring the need for alternative strategy for checking the incidence and spread of silver scurf disease. Application of *Pseudomonas corrugata*, as a postharvest treatment, reduced the disease severity and secondary transfer of the pathogen to daughter tubers under glasshouse conditions (Chun and Shetty, 1994).

The storage rot caused by *Botryodiplodia theobromae* in sweetpotato and yams is responsible for appreciable loss. Three isolates of the yeasts *Saccha-*

romyces cerevisiae, two of *Pichia anomala*, and one of *Debaryomyces hansenii* were effective against storage rot, restricting the spread of the disease (Ray and Das, 1998). In carrots, licorice caused by *Mycocentrospora acerina* and crater rots caused by *Rhizoctonia carotae* could be controlled by application of a cold-tolerant isolate of *Trichoderma harzianum*. Freshly harvested carrots were immersed in the suspension of conidia (10^7 conidia/ml) of *T. harzianum* for 5 min and this treatment resulted in the decrease in the percentages of disease incidence by 47% or 75% after storage at 0 to 0.5°C for 6 months or 8.5 months despite the inability of the antagonist to develop at this storage temperature (Tronsmo, 1989, 1993). Of the 10 fungal species tested, only *Trichoderma viride* inhibited the growth of *Sclerotinia sclerotiorum* causing soft rot and *Alternaria radicina* causing black rot in carrots in dual culture procedure for assessing the biocontrol potential of the test microorganisms (Sesan, 1993).

Onions suffer due to the attack by basal and neck rots caused by *Fusarium oxysporum* and *Aspergillus* sp. or *Botrytis allii* respectively. The neck rot is generally, seen after the plants have been topped. Hence, application of biocontrol agents soon after topping can be expected to be effective, since the BCAs can exclude the pathogen from wounds (Köhl et al., 1991). *Trichoderma harzianum* was the most effective in protecting onions from infection by the pathogens. The incidence of basal rot disease was reduced from 16% (in control) to 4% in treated onions. *Trichoderma harzianum* also effectively controlled neck rot disease. It was suggested that the BCA may be applied at the time of transplanting in the field or at harvesting (Lee et al., 2001).

9.2.2 Flower, Leafy, and Stem Vegetables

Over 100 microorganisms isolated from cabbage leaves were evaluated for their biocontrol potential against *B. cinerea*, causing gray mold in stored Dutch White cabbage. *Serratia plymuthica* strain CL43, *S. liquefaciens* strain CL80, and *Pseudomonas fluorescens* strains CL42, CL66, and CL82 were promoted for evaluation in biocontrol tests with cabbage heads in commercial cold storage. The heads were dipped into either bacterial suspensions (10^7 to 10^8 CFU/ml) or fungicide metalaxyl solution followed by spray inoculation with *B. cinerea* and storage at 4 to 6°C or 1 to 3°C. The level of protection provided by the strains CL80 and CL82 was consistent, while the strain CL82 proved to be more effective than the fungicide treatment. Furthermore, strain CL82 was able to survive very well and dominated totally the phyllosphere during the entire period of testing (32 weeks) (Leifert et al., 1993a, b; Stanley et al., 1994).

The bacterial soft rot caused by *Erwinia carotovora* subsp. *carotovora* is another major postharvest disease prevalent all over the world. No chemical giving effective protection against the disease has been found. Further, the high risk of residual chemicals is an undesirable effect associated with chemical use, in addition to the development of resistance in pathogen to the chemicals. Two antagonistic bacterial species, *Pseudomonas aeruginosa* and *Acinetobacter genospecies* 15, inhibited the pathogen development in vitro. In addition both BCAs reduced the disease severity by up to 25% following treatment with 10^6 to 10^7 CFU/ml of the BCAs (Adeline and Sijam, 1999). The possibility of using *Bacillus* spp. for the control of cabbage black rot caused by *Xanthomonas campestris* pv. *campestris* was indicated by Massomo et al. (2003).

Sclerotinia minor and *S. sclerotiorum* cause watery soft rot and lettuce drop diseases of lettuce. As the lettuce leaves are consumed as fresh raw product, the most desirable disease management strategy seems to be the use of effective microbes. Under glasshouse conditions, application of *Coniothyrium minitans* to the soil resulted in a reduction in disease incidence and consequent increase in yield. *C. minitans* could not only survive in the field for a year, but also even spread to adjacent plots, indicating its potential for saprophytic survival and competitive ability with other soil-borne microbes. Application of *Coniothyrium minitans* was shown to be effective against pink rot disease of celery caused by *S. minor* and *S. sclerotiorum*, under both glasshouse and field conditions. However, repeated applications of the BCA may be required to reduce the sclerotial population in the soil below the inoculum potential (Budge and Whipps, 1991). The ability of the mycoparasite *Sporidesmium sclerotivorum* to protect plants and destroy sclerotia of the pathogen produced on diseased plants showing drop disease symptoms was demonstrated by Adams and Ayers (1982).

9.2.3 Fruity Vegetables

Some preharvest diseases may seriously affect the quality of fruity vegetables, necessitating control measures in the field. Furthermore, vegetables such as tomatoes are likely to come in contact with the soil and have a longer time in the field. The decay may spread during transit and storage, because of direct contact between diseased and healthy fruits. Gray mold and *Alternaria* rot of tomatoes were controlled effectively by adding the yeast *Pichia guilliermondii* to wounds of tomatoes after harvest (Chalutz et al., 1988). Fruit rot (soil rot) caused by *Rhizoctonia solani* (telemorph: *Thanatephorus cucumeris*) may infect fruits through seed and soil-borne inocula. This disease is common in tomatoes intended for processing and the problem becomes worse due to the secondary invaders causing bacterial soft rot or *Rhizopus* rot. *Trichoderma harzianum*, when added to soil or applied as a coating on tomato fruits, at a concentration of 10^9 conidia/ml, provided significant control. *T. harzianum* applied in the soil showed significant saprophytic survival, maintaining high population levels during the growing season, although the population of *T. harzianum* on fruit surfaces declined considerably during the storage period of 2 weeks (Strashnov et al., 1985). Formulations of *T. harzianum* KRL-AG2, *Trichoderma* spp., and *Pseudomonas cepacia* are commercially available for use against *R. solani* (Janisiewicz and Korsten, 2002b). Application of *Trichoderma* spp. resulted in partial control of *R. solani* causing soil rot disease in cucumber (Lewis and Papavizas, 1980).

The possibility of using bacteriophages for the control of bacterial diseases has been proposed. Various combinations of bacteriophages, harpin protein, and acibenzolar-S-methyl (ASM) were evaluated for the control tomato bacterial spot disease under field conditions. In three consecutive seasons, bacteriophage in combination with ASM or harpin reduced the bacterial spot. Comparison of disease severity assessments or harvested yield showed that treatment with bacteriophage provided significantly better disease control and increase in yield of marketable fruit compared with nonbacteriophage treatments (Obradovic et al., 2004).

9.2.4 Minimally Processed Fruits and Vegetables

The preparation and distribution of fresh-cut produce is a rapidly developing industry, providing the consumer with a preferred and nutritious food. The food-borne disease caused by nontyphoidal salmonellae is widespread and accounts for an estimated 1.4 million cases of infection annually in the United States (Mead et al., 1999). Fruits and vegetables have been reported to be a possible reservoir for *Salmonella* (Anon, 2000). Fruits and vegetables contain the nutrients required for the rapid development of food-borne pathogens. However, the external barriers, such as the peel and rind, prevent access by pathogens to the internal tissues. Wounded fruit or fresh-cut fruit slices without the natural barrier may be colonized by the pathogens. For pathogens to grow on minimally processed products they must be able to overcome several obstacles (Breidt and Fleming, 1997). Since the minimally processed produce is usually preserved by refrigeration, the pathogens capable of growing at low temperatures, such as *Listeria monocytogenes* and *Yersinia enterocolita*, may be able to reach high concentrations. Beneficial lactic acid bacteria (LAB) that can survive under refrigerated conditions may counteract the activities of microbial pathogens offering the required product safety (Thomas and Ó Beirne, 2000).

Lactic acid bacteria (LAB) have been found in many foods and they have been employed mostly in food fermentation, where the growth of pathogens may be precluded because of the activities of LAB. The populations of members of *Enterobacteriaceae* present on vegetables are markedly reduced by LAB. LAB are safe to consume and are present as a dominant group in the natural flora of many foods (Stiles, 1996). LAB may act on food-borne pathogens through several mechanisms, including competition for space and nutrients, production of bacteriocins that are active against Gram-positive bacteria, and acidification of the environment making the condition unfavorable for the development of pseudomonads and *Enterobacteriaceae* (Leverentz et al., 2003).

Lytic bacteriophages have been demonstrated to be an effective alternative for decontamination of fresh cut fruits and vegetables. They are environmentally safe, as they are natural and ubiquitously present in the environment. Further, they are highly specific and selective against bacterial pathogens with no effect on other desirable microflora present on the surface. Fresh-cut fruits and vegetables are prone to colonization by pathogenic bacteria such as *Salmonella enteritidis*. They can survive on fresh-cut melons and apples stored at 5°C. *Salmonella*-specific phages cause lysis of the pathogen cells. Phages mixture reduced *Salmonella* populations present on honeydew melon slices stored at 5 and 10°C, the extent of reduction by phage mixtures was greater than the maximal reduction achieved by using chemical sanitizers. However, the phages were not effective in reducing *Salmonella* populations on apple slices at the different temperatures of storage tested. This differential effect may be due to the phage titer remaining relatively stable on melon slices, while phage titer on apple slices sharply declined to reach undetectable levels at 48 h after application (Leverentz et al., 2001).

Yeasts occur in high populations naturally on fresh produce and often they may be present along with LAB in natural food habitats (Deak and Beuchat, 1996). Many yeast species have been reported to be effective against fungal pathogens, such as *Botrytis cinerea* and *Penicillium* spp causing postharvest diseases (Janisiewicz, 1987; Filonow et al., 1996; Ippolito et al., 1998; Ippolito et al., 2000). The yeasts rapidly colonize the fruit wound and prevent decay development. Bacterial antagonists, such as *Pseudomonas sringae*, have been shown to be effective against fungal pathogens causing postharvest diseases. Fresh cells of *P. syringae* applied at 2.4×10^8 CFU/ml into the wounds of apple cv. Golden Delicious prevented the increase of a dreadful human pathogen *Escherichia coli* 0157:H7. The BCA, thus, in addition to controlling postharvest pathogens may also be effective in preventing the development of food-borne human pathogens in the freshly wounded tissue of intact or freshcut fruits (Janisiewicz et al., 1999).

9.2.5 Food Products

Contamination of food products with mycotoxins has been a constant threat to the health of humans and animals (Chapter 6). The efficacy of removing ochratoxin A (OTA) from yeast peptone glucose (YPG) (artificial medium), synthetic grape juice medium (SGM), and natural grape juice by viable and dead (heat- or acid-treated) oenological *Saccharomyces cerevisiae* (five strains) and *S. bayanus* (one strain) was compared with a commercial yeast wall additive. The yeast strains removed OTA up to a maximum of 45% from YPG and SGM media. Heat- and acid-treated yeast cells were more efficient than viable cells in removing OTA, indicating that reduction in OTA concentration was through adsorption and not by catabolism. The test with OTA-spiked grape juice showed that addition of heat-treated cells was the most effective, since the entire quantity of OTA in the grape juice was rapidly removed by adsorption within 5 min. The results clearly demonstrated that heat-treated yeast cells may be advantageously used for decontamination of grape juice and must with no negative impact on human health (Bejaoui et al., 2004).

9.3 BIOLOGICAL CONTROL OF SEED SPOILAGE

Seeds, when infected by microbial pathogens, form the primary sources of infection, carrying infection to the next generation and to new areas. Seed treatment employing physical, chemical, and biological methods has been attempted and each one has its own merits and drawbacks. Biocontrol agents have been demonstrated to be more desirable and advantageous in certain pathosystems. The grains and feed materials that are contaminated by mycotoxins produced by microbial pathogens are particularly suitable for treatment with BCAs. For seed treatment with BCAs to be effective, genetically superior strains should be used and a conducive environment for the BCAs has to be provided by minimizing the competition from the seed microflora. Furthermore, the BCAs must retain viability and efficacy for long periods (about 1 year) at room temperature until the seeds are sown or used for consumption.

Wheat loose smut caused by *Ustilago segetum* var. *tritici* is an important disease causing considerable losses. The BCAs *Trichoderma viride*, *T. harzianum*, *Pseudomonas fluorescens*, and *Gliocladium virens* when applied as a seed treatment controlled the loose smut disease. These BCAs were also compatible with the fungicide carboxin whose concentration could be reduced to half of normal dose to achieve total control of the disease (Dharam Singh and Maheshwari, 2001).

The efficacy of the yeast *Pichia anomala* for the postharvest control of *Penicillium roquefortii* during storage of feed wheat grain under airtight conditions was evaluated in outdoor silos. Winter wheat (cv. Kosack) in 160-kg portions were inoculated with *P. anomala* and stored at a water activity of 0.93 for 1 week. During the first 2 months, the population of *P. anomala* increased to about 10^7 CFU/g from 10^4 CFU in the control. The spoilage-causing fungus *P. roquefortii* did not develop during the storage period, probably because of high numbers of *P. anomala* in combination with high CO_2 and low O_2 concentrations in the silos. The in vitro assessments of survival of *P. anomala* showed that this BCA could survive long-term storage in airtight sealed tubes better at 15°C than at –20°C. In addition, the aerobic stability of moist wheat after 10 and 12 months of silo storage was significantly enhanced by an initial inoculation with *P. anomala*. The effectiveness of *P. anomala* to protect rye, barley, and oats, also stored under airtight conditions, has been observed. (Petersson and Schnürer, 1998; Petersson et al., 1999). *P. roquefortii* has been reclassified into three species *P. roquefortii*, *P. carneum*, and *P. paneum* based on differences in ribosomal DNA sequences and nature of secondary metabolites, including mycotoxins. *P. anomala* inhibited the growth of all three species to similar levels, as reflected by dilution plating and identification of individual colonies by random amplified polymorphic-DNA (RAPD) fingerprinting. A pronounced inhibition of growth of pathogens by, and fungicidal effects of, *P. anomala* were observed at 10^4 CFU/g and 10^5 CFU/g of grain respectively (Boysen et al., 2000).

Application of BCAs to seeds may reduce seed-borne fungi and produce more vigorous seedlings. Corn (maize) seeds infected by *Fusarium moniliforme* and *F. graminearum* treated with *Bacillus subtilis* or *Chaetomium globosum* produced healthy seedlings as the seedling blight disease could be effectively controlled to a level equal to that obtained by treatment with captan or thiram (Mew and Kommedahl, 1968, 1972).

The biocontrol potential of 36 bacterial isolates from maize rhizoplane was evaluated in vitro. *Serratia plymuthica* inhibited the mycelial growth of *Fusarium moniliforme* (*Gibberella fujikuroi*) (Lucon and Melo, 2000). Isolates of *Trichoderma* sp. and *Bacillus subtilis* were able to inhibit the growth of *G. fujikuroi* as well as the mycotoxin accumulation (Bacon et al., 1999). Fusarium head blight (FHB) caused by *Gibberella zeae* inflicts extensive yield and quality losses in wheat and barley. Bacterial and yeast strains reduced FHB severity on cultivar Renville. *Bacillus subtilis* strain AS 43-4 reduced the severity of FHB by as much as 90% in the greenhouse assays. However, the field assessment of biocontrol potential, showed that *Cryptococcus* sp. OH 71-4 and *C. nodaensis* OH 182-9 could reduce the disease severity to a greater extent than other antagonists tested. The mycotoxin deoxynivalenol (DON) content of the grains, however, showed no significant variation due to treatment with BCAs (Schisler et al., 2002).

The seeds of tomato (cv. Krasnaya Strela and Majyskiv) and cucumber (cv. TCXA-575, Belaya, Dacha, Gladiator, and Estafata) were found, in Russia, to be colonized by several fungi capable of producing mycotoxins. *Aspergillus* spp. and *Penicillium* spp. were more frequently recorded in both tomato and cucumber seeds. *Trichoderma harzianum* G-43 exhibited high biocontrol potential against several pathogens such as *Fusarium moniliforme*, *Phoma* spp., *Aspergillus niger*, *Ascochyta pisi*, and *Acremonium fuscum* (Alimova et al., 2002). The efficacy of different strains of *Pseudomonas fluorescens* and *Bacillus subtilis* in inhibiting the development of *Erwinia carotovora* subsp. *carotovora* (*Ecc*) in onion seeds was evaluated. *P. fluorescens* strain FP7 and *B. subtilis* exhibited greater biocontrol activity against *Ecc* in addition to improving the percentage of seed germination (Sendhilvel et al., 2001). Application of *Pseudomonas fluorescens* strain 96.578 to sugarbeet seeds resulted in the production of a new cyclic peptide tensin which inhibited the growth of *Rhizoctonia solani* and reduced seed infection by *R. solani* (Nielsen et al., 2000). Seed priming inoculation of sugarbeet seeds with rifampicin resistant (Rif^+) strains of both Gram + and – bacterial isolates (mostly *Pseudomonas* spp.) was found to be more effective in retaining the bacterial isolates than other application methods such as seed soaking, encapsulation in alginate, and pelleting using an inoculated peat carrier. After pelleting with fungicides and drying at 40°C, *P. marginalis/ putida* P1W1 maintained populations of more than 6.6 log 10 CFU/g of seed during 4-month storage at 15°C (Walker et al., 2004).

Aspergillus flavus and *A. parasiticus* are known to infect peanut (groundnut) plants and invade pods. The infected kernels contain the mycotoxin afla-

TABLE 9.3 Effect of Applications of Nontoxigenic Strains of *Aspergillus Flavus* and *A. Parasiticus* on Aflatoxin Content of Peanut Kernels

Treatments	Concentration of aflatoxin (ppb) (AF)	
	1998	1999 (contaminated with AF)
Field treatment	1.4	374.2
Field treatment + postharvest application	0.8	54.1
Postharvest application	48.8	9145.1
No treatment (control)	78.0	516.8

Source: Dorner and Cole, 2002

toxin at high levels, resulting in adverse reactions in humans and animals when contaminated kernels or other products are consumed. The nontoxigenic strains of *A. flavus* and *A. parasiticus* were evaluated for their biocontrol activities against toxigenic strains. The peanuts cv. Florunner plants were treated with the conidial suspension of nontoxigenic strains in field plots and also after harvest one half portions of pods were treated. Peanuts were stored for 3 to 5 months under high temperature and relative humidity conditions designed to promote aflatoxin contamination. The concentrations of aflatoxins in different treatments were determined (Table 9.3). Treatment of pods with nontoxigenic strains prior to storage did not offer any additional protection against aflatoxin contamination. Field application of the nontoxigenic strain reduced aflatoxin contamination because of apparent carryover effect (Dorner and Cole, 2002).

Biocontrol agents have been reported to offer effective protection against seedborne pathogens that cause seedling mortality. Biopriming of carrot seeds infected by *Alternaria dauci* and *A. radicinia* with the fungal antagonist *Clonostachys rosea* IK726 resulted in higher seedling stand compared with nonprimed (control) and hydroprimed (with water) seeds. Microscopic examination of bioprimed seeds with *C. rosea* IK726 transformed with green fluorescent protein (GFP) reporter gene revealed that carrot seeds were covered with a fine web of sporulating mycelium of *C. rosea*. The study demonstrated that biopriming techniques employing suitable BCAs have the potential to counter the adverse effects of pathogens on seedling establishment (Jensen et al., 2004).

Botrytis cinerea together with *Sclerotinia sclerotiorum* was found to be a serious threat to alfalfa seed production in western Canada. Young blossoms, pods, and seeds are infected by *B. cinerea*. The BCAs *Clonostachys rosea*, *Gliocladium catenulatum*, *Trichoderma atroviride*, and *Trichothecium roseum* inhibited the sporulation of *B. cinerea* on detached alfalfa florets. Infection of alfalfa pods by *B. cinerea* was more effectively suppressed by *C. rosea* and *G. catenulatum* than the other BCAs, when applied on fresh petals at anthesis stage. Furthermore, these BCAs suppressed the development of *B. cinerea* on pods and seeds of alfalfa when they were applied on senescent petals at the pod development stage. In the three field trials conducted during 2000 to 2002, application of *C. rosea* to the upper parts of alfalfa plants significantly reduced pod rot and seed rot caused by *B. cinerea*, resulting in enhanced production of healthy seeds. The results clearly show the potential of *C. rosea* for use as an effective alternative to the synthetic fungicides (Li et al., 2004).

9.4 MECHANISMS OF BIOCONTROL

The mechanisms of action of BCAs involved in the control of postharvest diseases are not clearly understood, probably because of the difficulty in studying the complex interactions occurring between the pathogen, BCA, host, and possibly other microorganisms present on the plant surface. Furthermore, precise methods to study the microbial interactions in wounds in fruits and vegetables are yet to be developed. However, various mechanisms of action of BCAs, such as antibiosis, production of lytic enzymes, parasitism, competition for limiting nutrients and space, and induction of resistance, have been suggested. It is very likely that the BCAs may act on the postharvest pathogen through more than one mechanism.

9.4.1 Antibiosis

The biocontrol potential of some of the BCAs may depend on their ability to produce antibiotics. *Bacillus subtilis* B-30 isolated from soil inhibited the growth of *Monilinia fructicola*, causing peach brown rot disease, on culture medium by producing the antibiotic iturin (Gueldner et al., 1988). Likewise, pyrrolnitrin produced by *Pseudomonas cepacia* LT-4-12-W also reduced the in vitro growth and conidial germination of *M. fructicola* (Janisiewicz and Roitman, 1988). Both BCAs were effective against the fruit pathogens *Penicillium expansum* and *Botrytis cinerea* (Pusey and Wilson, 1984; Janisiewicz and Roitman, 1988). However, the production of antibiotics as a critical factor in biocontrol potential of these BCAs is not established unequivocally, since *P. cepacia* could be employed for the control of green mold disease of lemon caused by a pyrrolnitrin-resistant strain of *P. digitatum* (Smilanick and Denis-Arrue, 1992). Likewise, use of the strain LT-4-12W of *P. cepaciae* resulted in significant control of blue mold decay on oranges inoculated with pyrrolnitrin-resistant mutants of *P. italicum* (Janisiewicz and Korsten, 2002a).

Pseudomonas syringae strains ESC-10 and ESC-11 contained in BioSave products produced syringomycin E and effectively controlled green and blue molds of citrus caused by *P. digitatum* and *P. italicum*, respectively. Although the purified syringomycin could inhibit the growth of a variety of fungi, the presence of syringomycin was not detected in the fruit wounds treated with *P. syringae*, raising a doubt as to the role of the antibiotic in disease control (Bull et al., 1997), and suggesting the operation of a different mechanism not depent on the production of syringomycin. The *syrB* mutants of strains ESC-10 and ESC-11 were generated by disrupting the *syrB* biosynthesis gene by a *lacZ* reporter gene coding for β-galactosidase. In cultures inoculated with the *syrB* mutant of strain ESC-10, β-galactosidase activity was higher in media containing albedo tissue after 4 days of incubation. The reporter-gene system has been used to determine the production of syringomycin E by these strains in wounds on lemons or oranges (Bull et al., 1998). However, it may not be feasible for the antibiotic-producing BCAs to be registered for postharvest use on food products because of the concern relating to the introduction of antibiotics into our food, which may have an adverse effect on the resistance of humans to antibiotics. Furthermore, inhibition of microbial pathogens by a single chemical compound may prove to be ineffective when a strain of the pathogen resistant to the chemical develops.

The ability of microorganism to produce antibiotics has to be determined by performing antibiosis tests on agar plates (Appendix 9(i)) (Janisiewicz et al., 2000). Application of cell suspensions of the yeast *Metschnikowia pulcherrima* (10^8 cells/ml) to wounds on apples reduced the growth of *B. cinerea* and *P. expansum* during storage at 23°C. The development of *Monilinia* sp. in apple wounds stored at 23°C and *B. cinerea* and *P. expansum* in apple wounds stored at 4°C was entirely suppressed by *M. pulcherrima*. The culture filtrates and autoclaved cells of the BCA had no effect on the pathogens, indicating the requirement of viable cells for the antagonistic activity. The BCA reduced the mycelial growth of the pathogens in solid medium, indicating the production of some diffusible toxic metabolites by the BCA (Spadaro et al., 2002).

9.4.2 Competition for Nutrients and Space

Generally, antagonists are very good colonists and competition for substrate or site has been considered to be a general mode of action of biocontrol agents and other mechanisms (antagonism, direct action, and induction of resistance, described below) may apparently facilitate a competitive edge for the antagonist to establish early in the wounds or on the plant surface. The BCAs effective against postharvest pathogens may also act predominantly through this mechanism. Many investigations on different biocontrol systems have concluded that the successful competition of BCAs with the pathogens infecting fruits for nutrients and space may be the possible mechanism of biocontrol (Droby et al., 1989; Wilson and Wisniewski, 1989; Wisniewski et al., 1989; Roberts, 1990, 1991; Mercier and Wilson, 1994; Arras et al., 1998). Many successful antagonists that do not produce antibiotics are able to grow rapidly at the wound sites and are better adapted to extreme nutritional and environmental conditions compared with postharvest pathogens.

Pichia guilliermondii utilized nutrients more effectively when cocultured with the pathogen *Penicillium digitatum* infecting grapefruit (Droby et al., 1989). Rapid multiplication and efficient colonization of wound sites by *P. guilliermondii* in grapefruit (Droby et al., 1989), by *Pseudomonas cepacia* in apple and pear (Janisiewicz and Roitman, 1988), and *Pseudomonas* spp. in lemons

(Smilanick Denis-Arrue, 1992) and in pineapple (Reyes et al., 2004) have been reported. The differential effects of low temperatures favoring the BCAs, resulting in their enhanced efficacy, demonstrated in some biocontrol systems (Droby and Chalutz 1994; Droby et al., 1993) also lends indirect support to the view that competition for nutrients and space may be the possible mechanism of biocontrol. The utilization ^{14}C glucose by the cells of the antagonistic pink yeast *Sporobolomyces roseus* was higher to such an extent as to prevent germination of conidia of *Botrytis cinerea* by nutrient deprivation (Filonow et al., 1996). A nondestructive method, using tissue culture plates with cylinder inserts containing a defusing membrane at one end, to study competition for nutrients without competition for space, was developed by Janisiewicz et al. (2000) (Fig. 9.3) (Appendix 9(ii)). In a further study, the role of competition for sugars by the yeasts *Cryptococcus laurentii* BSR-Y22 or *Sporobolomyces roseus* FS 43-238 that effectively reduced gray mold caused by *B. cinerea* in apples was studied at 22°C. The increase in populations of *C. laurentii* and *S. roseus* in wounds in apples was six to nine times from 1 to 7 days following inoculation, compared with *Sacharomyces cerevisiae* which had less antagonist activity against *B. cinerea*. The BCAs utilized greater amounts of ^{14}C-labeled fructose, glucose, or sucrose than the conidia of *B. cinerea* during 48 h. The BCAs mixed with conidia in sterile, dilute solutions of fructose,

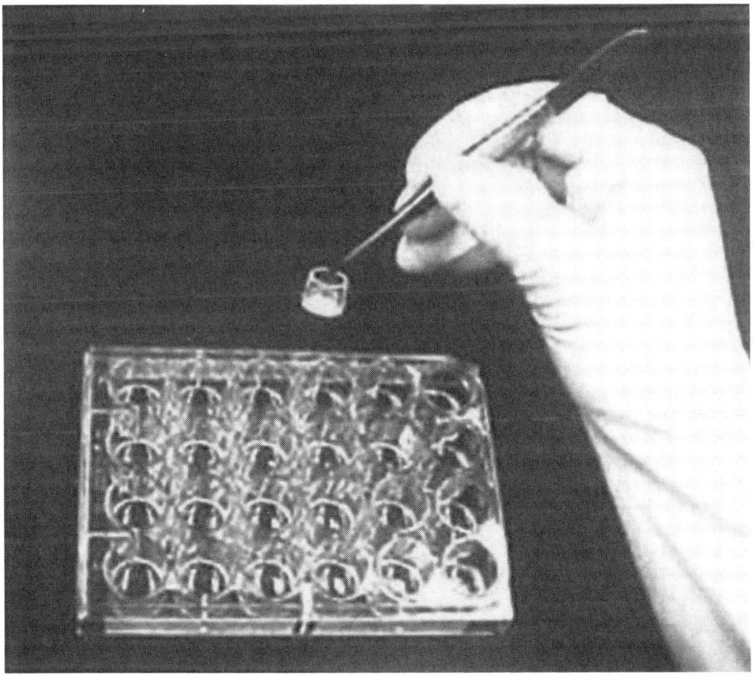

Figure 9.3 Experimental tissue culture plate for demonstration of competition for nutrients between biocontrol agent and fungal pathogen. (Courtesy of Janisiewicz *et al.*, 2000; The American Phytopathological Society, St. Paul, MN, USA.)

glucose, or sucrose or in dilute apple juice inhibited conidial germination compared with control (without BCA). The BCAs rapidly depleted sugars from juice or sugar solutions. The results suggested that these BCAs might act on the pathogens primarily by competing for nutrients (Filonow, 1998). Among the yeasts tested, *Pichia guilliermondii* was the most effective in inhibiting the growth of *Ceratocystis paradoxa*, causing black rot of pineapple fruit. The mode of action of *P. guilliermondii* seemed to be competition for space and nutrients (Reyes et al., 2004).

An important factor involved in competition for nutrients appears to be the attachment of antagonists to pathogen hyphae, as in the case of *Enterobacter cloacae* to *Rhizopus stolonifer* (Wisniewski et al., 1989) and *Pichia guilliermondii* to *Penicillium italicum* (Arras et al., 1998). In contrast, no direct physical interaction seems to be required for the antagonistic activity of *Aureobasidium pullalans* against *B. cinerea*, *P. expansum*, *Rhizopus stolonifer*, and *Aspergillus niger* infecting table grapes and *B. cinerea* and *P. expansum* on apple fruits (cv. Royal Gala) (Castoria et al., 2001). The need for direct contact between the antogonist and the pathogen for the antagonistic activity may be assessed by the method developed by Janisiewicz et al. (2000).

The involvement of metabolic products of BCAs in their biocontrol activities against postharvest pathogens has been indicated by some investigations. *P. guilliermondii* produced β-glucanase which was considered to cause lysis of *B. cinerea* (Wisniewski et al., 1991). The activity of exo-β-1,3-glucanase of *Pichia anomala* strain K increased by three fold in the presence of cell wall preparations of *B. cinerea* in apple wounds, resulting in the reduction of lesion size by more than half compared with control without the cell wall preparation of the pathogen. The results indicated that the exo-β-1,3 glucanase might be involved in the antagonistic activity of *P. anomala* strain K, resulting in effective disease control (Jijakli and Lepoivre, 1998; Grevesse et al., 1998 a,b). *Aureobasidium pullulans* was also able to increase the activities of β-1,3-glucanase and chitinase that could act on postharvest pathogens causing various decays on apple, table grapes, and other fruits (Ippolito et al., 2000). Production of extracellular exochitinanse (N-acetyl-β-D-glucosaminidase and β-1,3 glucanase) by *Aureobasidicum pullulans* was detected both in vitro and in apple wounds, which are known to be the primary sites of penetration by postharvest pathogens, suggesting that these enzymes may have a role in the biocontrol activity of *A. pullulans*. Competition for nutrients appeared to be the possible mode of action of this BCA and no evidence for antibiosis or direct physical interaction with the pathogen could be observed (Castoria et al., 2001). It is possible that enhanced activities of these extracellular enzymes may be of antagonist origin, produced in response to the elicitors generated by the pathogen activity, or they may be induced in the fruits as defense responses to the presence of pathogen. The importance of these enzymes in the biocontrol of potential of microorganisms may be revealed by the investigations aimed to evaluate disease suppression by mutants with a disrupted β-1,3 glucanase gene.

Biocontrol agents may produce a wide range of compounds in addition to extracellular enzymes and the contribution of these compounds to the biocontrol potential of the BCA has to be assessed. *Aureobasidium pullulans* was reported to produce a new antibiotic designated aureobasidins (Takesako et al., 1991). The involvement of siderophores in the biocontrol activity of *Rhodotorula glutinis* against blue rot disease of apple caused by *Penicillium expansum* was investigated by in vitro and in vivo assays. Rhodotorulic acid produced by *R. glutinis* enhanced the biocontrol potential of the strains of *R. glutinis*. Rhodotorulic acid reduced the growth of *P. expansum*, whereas in the presence of iron no adverse effect on growth of *P. expansum* was noted, indicating that antagonism due to the siderophore was related to competition for iron. On apple wounds, blue mold development was more effectively curtailed by the combination of BCA and siderophore than by the BCA alone. The disease incidence (DI) percentage was significantly reduced in fruits protected by the application of *R. glutinis* plus rhodotorulic acid than by *R. glutinis* alone (Calvente et al., 1999).

Erwinia herbicola strains B66 and B90, an effective biocontrol agent against *B. cinerea* (gray mold) and *P. expansum* (blue mold) infecting apples, exhibited chemotaxis to the conidia and germ tubes of the pathogens resulting in inhibition of germination of conidia and lysis of germ tubes in diluted apple juice. That there was no detectable interaction in undiluted apple juice, suggested that the biocontrol activity of *E. herbicola* strains may depend on the successful competition for nutrients in this system (Sobiczewski and Bryk, 1996; Bryk et al., 1998). *Candida oleophila* present in the registered biocontrol product Aspire has to efficiently compete with the pathogen *Penicillium digitatum*, causing green mold disease in citrus, for the nutrients released by injuries. The BCA should rapidly colonize wounds on the fruit surface, including minor injuries involving only oil vesicles. *C. oleophila* colonized puncture-related injuries to oil glands within 1 to 2 days after application at 21°C and 30°C, but at 13°C no colonization by the BCA could be seen. Ruptured oil glands were colonized more effectively if *C. oleophila* was applied at 7h after injury rather than immediately, probably due to the toxicity of peel oil to the BCA, but not to *P. digitatum* (Brown et al., 2000).

Pantoea agglomerans CPA-2 provides effective control against *P. digitatum* and *P. italicum* infecting citrus fruits. The study to understand its mode of action showed that CPA-2 did not produce antibiotics and also did not induce the key enzymes involved in the development of resistance in treated fruits. The BCA was effective only when it was in close contact with the fungal pathogens. By using tissue culture plates with cylinder inserts that allowed competition for nutrients to be studied without competition for space, it was observed that the BCA acted on the pathogens by reducing germination of conidia present in the cylinder. Conidial germination was entirely inhibited when the BCA and pathogen conidia were in physical contact. The results indicated that the competition for nutrients was one mechanisms of action of the strain CPA-2. However, physical contact between the BCA and pathogen was an important factor for effective control (Poppe et al., 2003).

9.4.3 Direct Interaction between the Biocontrol Agent and Pathogen (Parasitism)

Parasitism or direct interaction as a mechanism of biocontrol has been demonstrated in the case of soil-borne diseases and grapevine diseases infecting both foliage and fruits. *Coniothyrium minitans* may act as a mycoparasite on *Sclerotinia sclerotiorum* by penetrating the sclerotia and also by producing cell wall-degrading enzymes such as chitinase and β-1,3-glucanase (Jones and Watson, 1969; Jones et al., 1974). *Fusarium proliferatum* parasitized the grapevine downy mildew pathogen *Plasmopara viticola* which infects both leaves and berries. The hyphae of *F. proliferatum* could be seen coiling around and also inside the sporangiophores of *P. viticola* as revealed by electron micrographs (Falk et al., 1996). Parasitism of grapevine powdery mildew pathogen *Uncinula necator* by *Ampelomyces quisqualis* (Daoust and Hofstein, 1996) and strawberry powdery mildew disease caused by *Sphaerotheca macularis* f.sp. *fragariae* by *Lecanicillium lecanii* (Miller et al., 2004) are also known. On the other hand, no conclusive evidence is available indicating that the biocontrol activity of microbes depends on the direct parasitism of any postharvest pathogen. A few cases, however, suggest such a possible direct interaction between the BCA and the postharvest pathogen.

Attachment of the cells of the yeast *Pichia guilliermondii* to the mycelium of the pathogen *B. cinerea* was observed. Exposure to compounds that may affect protein integrity and respiration blocked this lectin-type binding between the BCA and pathogen (Wisniewski et al., 1991). The enhanced activity of β-1,3-glucanase shown by *P. guilliermondii* might have resulted in the BCA–pathogen attachment and consequent degradation of pathogen cell walls (Wisniewski et al., 1991; Droby et al., 1993). The antifungal activity of extracellular polysaccharide present on this BCA cells was demonstrated by Droby et al. (1993). By using transmission electron microscopy (TEM) and gold cytochemistry, the intracellular interaction between the antagonist *Verticillium lecanii* and *Pencillium digitatum* causing green mold has been studied. The growth of *P. digitatum* was inhibited by *V. lecanii* and this effect was correlated with striking changes in the cells of *P. digitatum*, including retraction of the plasma membrane and cytoplasm disorganization. Deposition on the inner host cell surface of a chitin and cellulose-enriched material, considered as a host structural defense reaction, then occurred. The accumulation of a new chitin correlated with a decrease in the amount of wall-bound chitin in the pathogen (Benhamou and Brodeur, 2000; Benhamou, 2004).

9.4.4 Induction of Host Plant Resistance

Biocontrol agents (BCAs) grow rapidly and colonize wound sites where infections by postharvest pathogens frequently occur. The BCAs out-compete the pathogens for space and nutrients and some of them have been demonstrated to induce (enhance) resistance levels of host tissues, resulting in a significant reduction in decay development. Antagonistic yeasts are capable of inducing resistance responses in various fruits, as in the case of *Pichia guilliermondii* strain US-7 in apples (Wisniewski and Wilson, 1992), *Aureobasidium pullulans* in apples (Ippopolito et al., 2000), and in strawberry (Adikaram et al., 2002). Induction of resistance to *B. cinerea* causing gray mold in apples following treatment of wounds of fresh apples with *Candida saitoana* was studied in detail. Significant increases in the activities of chitinase and β-1,3-glucanase involved in the development of resistance were observed in the treatment site (wounds) (Fig. 9.4). In the case of stored apples, enhancement of activities of defense-related enzymes was also noted (El-Ghaouth et al., 2003). Induction of resistance to several postharvest pathogens by biocontrol agents is discussed in more detail in Section 10.2.3.

Phytoalexins produced by the host plant tissues, following the interaction between host and pathogen, have been considered to have a role in the development of resistance to diseases. Reduction of green mold decay (*Penicillium digitatum*) in oranges due to the application of the yeast species *Candida famatum* was attributed to the enormous increase (12-fold) in the phytoalexins scoparone and scopoletin in the wounds at 4 days after inoculation. As the production of the phytoalexins was relatively slow, their effect on the pathogen development may not be significant. Nevertheless, the rapid colonization and partial lysis of the hyphae of *P. digitatum* by the BCA, *C. famatum*, suggests possible multiple mechanisms of biocontrol in this pathosystem (Arras, 1996). Ultrastructural investigations of citrus fruits treated with *Verticillium lecanii* and challenged with *Penicillium digitatum* causing green mold, revealed the restriction of pathogen growth and alterations in the hyphae that penetrated the mesocarp tissue. Accumulation of callose and lignin-like compounds at sites of pathogen colonization in the exocarp tissue was indicated by cytochemical studies. *V. lecanii* and chitosan appeared to function in a similar manner in inducing resistance in citrus fruits against blue mold pathogen (Benhmou, 2004).

9.5 ENHANCEMENT OF EFFICIENCY OF BIOCONTROL AGENTS

The efficiency of the selected biocontrol agent can be increased by altering postharvest environments to favor the development of the BCA, leading to rapid colonization of target fruit vegetable surface (wounds). Additives to the preparations of BCAs, provision of suitable physical and chemical environment, and mechanical/ physical barriers to prevent access by the pathogen have been shown to provide favorable conditions for the BCAs. Manipulation of the genetic constitution by the transfer of genes from other organisms and use of mixtures of BCAs with varying biocontrol potential and survival abilities under different environmental conditions have also been demonstrated to enhance the efficiency of the BCAs.

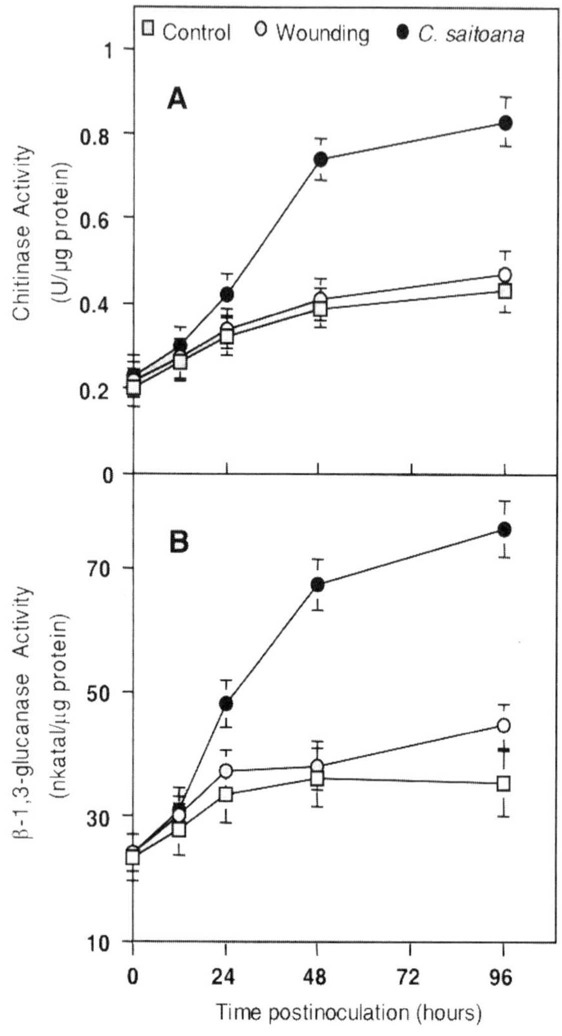

Figure 9.4 Time course of changes in activities of defense-related enzymes in extracts from wounds of fresh apples treated with *Candida saitoana* (●) or sterile water (○) and of nonwounded control fruit (□) at 0, 12, 24, 48 and 96 h after treatment. (a) Chitinase activity; (b) β-1,3-glucanase activity (Courtesy of El Ghaouth *et al.*, 2003; The American Phytopathological Society, St. Paul, MN, USA.)

9.5.1 Provision of a Favorable Environment

Fruits and vegetables may be stored at predetermined temperature, humidity, and in gas compositions for varying periods with the primary objective of maintaining the quality to meet the market demands. The BCAs are screened for their ability to develop rapidly under the required storage conditions and only the BCAs that satisfy this requirement are selected. The biocontrol microbes may have different mechanisms of action for the control of postharvest pathogens. Most of them appear to compete for the available nutrition resulting in the restriction of disease development.

The availability of nitrogen may be a limiting factor in the carbon-rich environment of apple and pear wounds. The bacterial BCA *Pseudomonas syringae* strain L-59-66) developed at a faster rate in wounds of mature apple fruits following the addition of L-asparagine and L-proline, attaining a 10-fold increase during the crucial period of first 24 h at room temperature and the first month of storage at 1°C. This treatment prevented the blue mold decay completely as against 50% decay in the control (Janisiewicz et al., 1992). The efficacy of 22 nitrogenous compounds were evaluated for their ability to enhance the efficiency of *Candida sake*. L-serine and L-aspartic acid improved the biocontrol potential of *C. sake* against *P. expansum* on apples. The minimum required concentration of *C. sake* could be lowered by the addition of ammonium molybdate (1 mM) without reducing the level of disease control. In cold storage, addition of ammonium molybdate to *C. sake* entirely eliminated the incidence of blue mold on pears and reduced the severity and incidence of the disease by more than 80% on apples (Nunes et al., 2001b). In a further study to assess the effect of ammonium molybdate on the biocontrol potential of *C. sake*, it was observed that application of *C. sake* (2×10^6 CFU/ml) plus ammonium molybdate (5 mM/l) markedly reduced *P. expansum*, *B. cinerea*, or *R. stolonifer* in apples stored at 20°C for 7 days and reduced blue and gray molds by more than 90% in apples stored at 1°C for 60 days, indicating the effectiveness of the treatment at both storage conditions, in addition to semicommercial trials at 1°C in air and in a low oxygen atmosphere for 120 days. The addition of ammonium molybdate reduced substantially the amount of *C. sake* biomass required to achieve control of postharvest diseases with a consequent reduction in the cost of production of BCA in large scale (Nunes et al., 2002c). In a similar study with pears cv. Banquilla under semicommercial trial, the efficacy of *C. sake* at 2×10^6 CFU/ml with addition of ammonium molybdate (5 mM) was increased by more than 88% at 1°C for 5 months. However, the population of the BCA in pear wounds was reduced significantly in the presence of ammonium molybdate at 1 or 5 mM concentrations at 20° and 1°C (Nunes et al., 2002a).

Preferential stimulation of growth of the biocontrol agent by nutrient analog 2-deoxy-D-glucose (2DOG) has been demonstrated. The development of the antagonists *P. syringae*, *Sporobolomyces roseus*, and *Candida saitoana* was favored by the addition of 2-deoxy-D-glucose which could not be metabolized, though it was absorbed by the pathogens causing blue mold in apples and pears. Addition of DOG at 4 mg/ml led to a large reduction (10 fold) in the concentration of antagonists required to have the desired level of disease control (Janisiewicz, 1994 a, b) (Table 9.4). The efficacy of application of *Candida saitoana* in combination with 2DOG was evaluated. The combined application to fruit wounds before inoculation was more effective in controlling decay of apple, orange, and lemon caused by *B. cinerea*, *P. expansum*, and

TABLE 9.4 Effect of 2-Deoxy-D-Glucose (2DOG) on The Population Dynamics of *Pseudomonas Syringae* in Wounds in Apples (cv. Golden Delicious) and Pears (cv. Anjou)

2DOG concentration (mg/ml)	Population of antagonist (log CFU/wound)			
	Apples		Pears	
	0 h	48 h	0 h	48 h
0	5.38 (0.21)	6.8 (0.90)	4.88 (0.37)	7.45 (0.29)
1	5.27 (0.05)	6.72 (0.11)	5.30 (0.62)	7.36 (0.23)
2	5.47 (0.13)	7.01 (0.08)	5.47[a]	7.38 (0.18)
3	5.57 (0.06)	6.65 (0.42)	4.99 (0.21)	7.27 (0.43)
4	5.61 (0.09)	6.95 (0.32)	5.09 (0.32)	7.61 (0.04)

[a] Only one observation. Numbers in parenthesis represent standard deviation.
Source: Janisiewicz, 1994a.

P. digitatum respectively. When applied within 24 h after inoculation, the combination of *C. saitoana* and 2DOG was as effective as the fungicide imazalil in controlling blue mold of apple and green mold or orange and lemon (El Ghaouth et al., 2000). The suspension of *C. saitoana* (10^8 CFU/ml) plus 2DOG (0.2%) was sprayed at 15 ml/kg of wounded apple (cvs. Rome and Empire), orange (cvs. Washington, Valencia and Hamlin), and lemon (cv. Eureka) and stored under semicommercial conditions. The decay by *B. cinerea* and *P. expansum* on apples was controlled more effectively than with thiabendazole (TBZ). On other hand, the level of control of *P. digitatum* on oranges was to a similar extent or lower than imazalil treatment (El Ghaouth et al., 2001).

The antagonism of *Rhodotorula glutinis* against *Penicilium expansum* was indicated to be due to a siderophore which prevented iron utilization by the pathogen. The addition of siderophores may reduce disease incidence in apples by sequestering iron needed for germination of spores of *P. expansum*. It is possible that production of siderophores by the antagonist may be stimulated by the addition of siderophores and consequently an iron-deficient environment may be created at the wound site. Such an environment may prove to be unfavorable for disease development (Calvente et al., 1999). These studies show that by screening various nutrients it may be possible to select nutrients or compounds that may stimulate antagonist populations, while limiting the development of postharvest pathogens.

9.5.2 Addition of Chemical Supplements

Chemicals which do not function as nutrients to the BCAs have been evaluated for their ability to enhance the biocontrol potential of the BCAs. In a test of 52 yeast strains, *Rhodotorula glutinis* (strain 21A) was found to colonize the surface of orange fruits, wounds and also the fungal pathogen *Penicillium italicum*, probably by competing for space and nutrients in wound sites. Strain

21A arrested the rot development by 98.3 and 83.2% at a concentration of 10^8 and 10^7 cells/ml, indicating the requirement for high concentrations of BCA. The possibility of enhancing the antagonistic activity of strain 21A by addition of calcium chloride and lower cell concentration of BCA was explored. Calcium at 2% markedly enhanced the biocontrol potential of strain 21A inoculated at 10^7 and 10^6 cells/ml, compared with strain 21A alone. The enhancement in the effectiveness of biocontrol may be due to the toxicity of calcium to *P. italicum* by affecting the osmotic balance in the fungal cells and by inhibition of pectinolytic enzymes in the wound site (Arras et al., 1998). The differential effect of calcium chloride on *Rhizopus stolonifer* and yeasts *Pichia guilliermondii* and *P. membranefaciens* was studied. $CaCl_2$ (2%) inhibited the growth of *R. stolonifer*, but not the colony forming units of the yeast species in potato dextrose broth. The addition of calcium reduced the spore germination and growth of germ tubes in vitro. Furthermore, the percent disease incidence and lesion diameter on the fruits treated with BCA + $CaCl_2$ were significantly reduced compared with the BCA alone. At a concentration of 5×10^8 CFU/ml of BCA, complete inhibition of the growth of *R. stolonifer* and total absence of infection by the pathogen on peach and nectarine fruits could be achieved, irrespective of the presence or absence of calcium (Tian et al., 2002).

The effect of combined application of silicon (Si) as sodium metasilicate and *Cryptococcus laurentii* on the development of decay caused by *Penicllium expansum* and *Monilinia fructicola* in sweet cherry at 20°C was assessed. Results of scanning electron microscopic observations revealed that silicon strongly inhibited the growth of both pathogens in the wounds of sweet cherry fruit. In addition, Si treatment enhanced the activities of defense-related enzymes, PAL, polyphenol oxidase, and peroxidase in treated fruits, compared with untreated controls. No increase in the levels of lignin was noticed in the tissues of treated fruits. However, a cytochemical reaction and tissue browning near the site of wounding could be seen. Furthermore, marked increase in the populations of the biocontrol yeast was evident following the combined application of the BCA and Si, resulting in a synergistic effect on the restriction of decay caused by *P. expansum* and *M. fructicola*. The results indicate an improvement in the biocontrol efficacy of *C. laurentii* against these postharvest diseases of sweet cherry (Qin and Tian, 2005).

Different coating formulations such as shellac, sucrose ester, and cellulose that support populations of bacterial and yeast biocontrol agents effective against postharvest decay of grapefruit were evaluated. The surface populations of *Pseudomonas syringae*, *P. fluorescens*, and *P. putida* were stable between 10^3 and 10^4 CFU/cm^2 on shellacked fruit over 4 months at 13°C. On the other hand, the population of *Candida oleophila* registered a steady increase from 2.35×10^3 CFU/cm^2 to nearly 10^5. Incorporation of *P. syringae* and *C. oleophila* in ester and cellulose coatings initially increased the population in wounds reaching a concentration of about 10^6 CFU/cm^2 of yeast suspension alone than the preliminary dip application could. Significant differences in the populations were evident for several weeks. There was no significant influence due to method of application on the efficacy of *P. syringae* under cold storage conditions. In contrast, the efficacy of *C. oleophila* was enhanced after 3 and 4 months, when applied in the shellac than when applied by a preliminary immersion and subsequent drying of fruit prior to shellacking (McGuire, 2000).

9.5.3 Use of Mixtures of Biocontrol Agents

A knowledge of microbial ecology of fruit surfaces is essential to study microbial community structure. Most of the work relates to grape and apples, with limited studies on other fruits, including stone fruits and soft fruits. However, very little is known about the interaction of the microorganisms on fruit surfaces (Dennis, 1983). In most studies, the efficacy of a single BCA has been assessed, with the expectation of a high level of control against a spectrum of pathogens under different environmental conditions. Almost all BCAs exhibite antagonistic activity under a given set of conditions against a particular pathogen. The limitations of the BCAs were understood, such as of variations in performance on various cultivars or at different stages of maturity of fruits. To obtain the desired level of disease control, a higher concentration of the BCA may be required. The use of antagonist mixtures to widen the spectrum of activity was suggested primarily for the control of soil-borne pathogens (Cook, 1993; Raupach and Kloepper, 1998).

Application of mixtures of BCAs may provide certain advantages:

1. Widening of spectrum of activity resulting in the control of two or more postharvest diseases;
2. Increase in the effectiveness under different situations such as cultivars, maturity stages, and locations;
3. Enhancement, in efficiency and reliability, of biocontrol as the components of the mixture act through different mechanisms, that is antagonism, parasitism, and induction of resistance in host;
4. Reduction in application rate and cost of treatment;
5. Combination of different biocontrol traits without the transfer of alien genes through genetic transformation.

The enhancement of efficiency of biocontrol agents may be due to: better utilization of substrate, resulting in acceleration of the growth rate; removal of substances inhibitory to one organism by the other BCA; production by one microbe of nutrients that may be used by another; and formation of more stable microbial community that may exclude other microbes, including pathogens (Janisiewicz, 1998).

While selecting the components of antagonist mixtures, certain attributes have to be considered, including: (1) absence of antagonism between one BCA against another component; and (2) selection of components with positive interactions (mutualism) that allow more effective utilization of resources. The practical approach to select the components of mixtures is to evaluate the biocontrol agents with a mixture of many antagonists and to remove the ineffective or incompatible ones (Fukui et al., 1999). The components of mixtures may be selected by pairing antagonists at random or after screening for minimum mutual niche overlap in the nutrient utilization assessed on BIOLOG (Biolog Inc., Hayward, CA) standard plates or on customized BIOLOG plates containing nutrients present at the wound site (Janisiewicz, 1996; Janisiewicz and Korsten, 2002).

Random combinations of antagonists isolated from apple and pear fruit and leaves were evaluated in comparison with individual antagonist for the control of blue mold disease of apple caused by *Penicillium expansum*. A mixture of *Pseudomonas syringae* and *Sporobolomyces roseus* was more effective than either of the antagonist individually, resulting in a reduction of both incidence of wound infection and lesion size (Janisiewicz and Bors, 1995).

Antagonists suitable for application as mixtures may be obtained from the microbial succession at wound site. The sequence of antagonists colonizing freshly made wounds in apple fruit in the orchard just before harvest is determined, followed by the selection of microbes for the mixtures from the succeeding microbes. By following this procedure, the antagonists, selected for mixtures to be used in the control of blue mold disease, were found to be ecologically suited to the chemical environment of apple fruit. The antagonist mixtures could also adapt to the stable physical environment in storage (Janisiewicz, 1996). The nutrient composition at the wound site must be precisely determined to predict the competitive interactions between antagonists in a mixture and between the pathogen and the antagonists.

The habitat compatibility of fruit surface compared to the inner parts of the fruit for the growth of yeasts and bacterial antagonists has to be assessed. The natural coexistence of microorganisms in fruit wounds was determined by exposing the flesh tissue to colonization by natural microflora in a nonmanaged orchard of Golden Delicious apple before harvest. Flesh tissues were exposed to natural colonization for 7 days after the removal of skin in a small area and the procedure was repeated for 5 consecutive weeks until regular harvest time. Yeasts with high biocontrol potential were isolated at each sampling and those isolated at normal harvest showed the highest biocontrol potential. The nutrient utilization pattern, apple tissue colonization by microbes, and biocontrol potential of the microbes isolated were the parameters considered when determining antagonist mixtures (Janisiewicz, 1994b, 1996). The effectiveness of antagonist mixtures growing simultaneously or in succession has to be assessed at temperatures at which fruits are stored, since the population dynamics may change with temperature (Janisiewicz, 1998). Although the use of antagonist mixtures offers more effective control, the economic viability of this approach appears to be a major obstacle for its adoption.

The black rot of pineapple caused by *Ceratocystis paradoxa* could be controlled by the yeast *Pichia guilliermondii*. But the combined application of *P. guilliermondii* and a mixture of five yeast isolates was still more effective and the level of control was comparable to current industry practice of holding fruit at a low temperature (8–10°C) and fungicide (Reyes et al., 2004). By applying mixtures of *Aureobasidium pullulans*, *Rhodotorula glutinis*, and *Bacillus subtilis* three times before harvest, the populations of the BCAs were increased on the surface of apple. With a mixture of *A. pillulans* (10^6 ml) and *B. subtilis* (10^8 ml) control of *P. expansum* and *B. cinerea* to the level provided by a fungicide, could be obtained (Leibinger et al., 1997).

The antagonist mixture consisting of *Candida sake* CPA-1 (at 2×10^6 and 2×10^7 CFU/ml) and *Pantoea agglomerans* CPA-2 (at 2×10^7 and 8×10^7 CFU/ml) was evaluated for their efficacy in controlling *P. expansum* and *B. cinerea* on pear (cv. Blanquilla) and apple (cv. Golden Delicious). At room temperature, a mixture (50:50 of both BCAs) arrested the rot development completely in pears and the maximum control of blue mold in apples was seen at this proportion. Under cold storage, the combination of *C. sake* at 2×10^7 CFU/ml + *P. agglomerans* at both concentrations, at a 50:50 proportion, was the effective against pear decay. Blue mold and gray mold of apple were also reduced to the maximum extent. Blue mold disease was absent completely, while gray mold lesion size was reduced by more than 95%, indicating the effectiveness of the antagonist mixtures against postharvest diseases (Nunes et al., 2002a).

Potato dry rot is caused by several species of *Fusarium* or strains of individual species. A microbial antagonist may exhibit different effect on these pathogens. The bacterial antagonists *Pseudomonas* sp., *Enterobacter* sp., and *Pantoea* sp. showed antagonistic activity against the fusaria to different levels. Of all possible pairing within two sets of 10 of the most active strains against dry rot tested, 16 of 90 pairs controlled the dry rot disease better than predicted level from averaging the performance of individual strains making up the pair (Schisler et al., 1998).

9.5.4 Genetic Manipulation

Improvement in the biocontrol potential of microorganisms may be achieved by adopting different approaches as discussed above. The principal obstacle facing widespread use of BCAs is that the level of control provided under commercial conditions frequently falls short of chemical control. A superior antagonist has not yet been identified through conventional screening of the naturally occurring microflora. Enhancing the biocontrol potential of microbes could be the most important factor in their acceptance for management of postharvest diseases on commercial scale. The genetic potential of a BCA will have the pivotal role in its antagonistic activity, which may be either positively or negatively affected by other factors. The genetic diversity of a BCA has to be studied by collecting all available strains and isolates and subjecting them to analysis by different molecular techniques, such as random amplification of polymorphic DNA (RAPD), DNA finger printing, and polymerase chain reaction (PCR). If genes controlling the production of antimicrobial compounds or enzymes can be identified, such information can facilitate the identification of methods to enhance the biocontrol potential of the microorganism selected for genetic manipulation.

9.5.4.1 Assessment of Genetic Diversity of Biocontrol Agents The isolates of *Aureobasidum pullulans* from the surfaces of several fruits and vegetables were subjected to molecular analysis by using arbitrarily primed PCR (AP-PCR). This study indicated the presence of a high genetic variability within *A. pullulans* and the biocontrol potential of 41 isolates varied widely (Schena et al., 1999). Similar searches for available isolates of the BCAs should be made to select the most potent candidate whose biocontrol potential can be further enhanced by other molecular techniques.

The activity of an exo-β-1,3-glucanase of the yeast *Pichia anomala* strain K effective against the gray mold pathogen *B. cinerea* was studied in a synthetic medium supplemented with either laminarin, *B. cinerea* cell wall preparation (CWP), or glucose. Production of the enzyme was accelerated by the CWP of *B. cinerea* as a sole carbon source. The purified exo-β-1,3-glucanase from culture filtrates of strain k(paexg2) inhibited the germ tube growth of *B. cinerea* strongly, in addition to inhibiting conidial germination and induction of morphological changes. The exo-β-1,3-glucanase activity detected on apples treated with strain K was similar to paexg2 in several properties. Two genes *PAEXG1* and *PAEXG2* coding for exo-β-1,3-glucanase were identified in the genome of strain K by using PCR with degenerant primers designed on the basis of conserved amino acid regions and on the N-terminal sequence of paexg2 (Grevesse et al., 1998a,b). The segregation of *PAEXG1* and *PAEXG2* alleles in haploid segregants indicated that there was no relationship between exo-β-1,3-glucanase activity in vitro and their biocontrol potential against *B. cinerea* in apples (Grevesse et al., 1998b). The *PAEXG2* gene encoding the exo-β-1,3-glucanase was isolated from *P. anomala* strain K and the gene product was characterized. *PAEXG2* codes for an acidic protein consisting of 427 amino acids with MW of 45.7 kDa. Disruption of *PAEXG2* by insertion of the *URA3* marker gene encoding orotidine monophosphate decarboxylase resulted in a reduction in biocontrol potential, as well as in reduced colonization of wounds in apples. Disruption of *PAEXG2* led to loss of all detectable exo-β-1,3-glucanase activity in vitro and in situ. However, the biocontrol potential in wounded apples against *B. cinerea* was not affected due to disruption this gene, indicating that the biocontrol activity of strain K did not depend on the production of exo-β-1,3-glucanase (Grevesse et al., 2003).

9.5.4.2 Inducing Variability by Mutation The possibility of inducing variants exhibiting lack of pathogenicity that may be employed as biocontrol agents against the same pathogen was suggested. Mutants of *Colletotrichum gloeosporioides* isolate Cg-14 infecting avocado were generated by insertional mutagenesis by restriction enzyme-mediated integration transformation. Among the 14 isolates showing reduced virulence compared with wild type strain, the strain Cg-M-142, when preinoculated, delayed symptom development by the wild type strain. Furthermore, this mutant induced resistance to the wild type strain and this phenomenon was accompanied by an increase in the levels of preformed antifungal compound, diene, from 760 to 1200 μg/g fresh weight at 9 days after inoculation. Another mutant, Cg-M-1150, failed to produce appressoria, but it exhibited reduced macerating ability on mesocarp and no symptoms on the pericarp. This mutant neither altered the concentration of diene produced nor delayed the appearance of decay symptoms. This study indicated that mutants with reduced pathogenicity may be used for biological control of anthracnose disease of avocado (Yakoby et al., 2001).

9.5.4.3 Transformation of Biocontrol Agents From the wild-type strain 1-182 of *Candida oleophila*, histidine auxotrophs were produced by using methanesulfonate and they were transformed with plasmids containing the *HIS3*, *HIS4*, and *HIS5* genes of *Saccharomyces cerevisiae*. *HIS5* gene was transformed to complement histidine auxotrophy of *C. oleophila*. Integration of transgenes was verified by DNA-gel-blot analysis. No detectable physiological differences between the transformants and wild type were observed. Further, the biocontrol potential of the transformants was not altered. A genetically marked transformant with a β-glucuronidase gene, could colonize wounds on oranges followed by increase in population under field conditions. PCR amplification of a portion of the β-glucuronidase gene established the identity of this transformant with good survival under field conditions (Chand-Goyal et al., 1999).

Over-expression of indigenous genes encoding lytic enzymes in the biocontrol agents may enhance the biocontrol potential of the transformed BCAs. *CoEXG1* gene encoding β-1,3-glucanase (glucan 1,3-β-glucosidase) of *Candida oleophila* was cloned by colony hybridization of *Escherichia coli* harboring a partial genomic library of *C. oleophila*. A fragment of *CoEXG1* obtained using degenerate oligomers, constructed from *Saccharomyces cerevisiae* and *C. albicans* exoglucanase gene sequences, was used as a probe. *C. oleophila* secreted β-glucanase continuously in vitro and the putative protein belonged to a family of highly conserved proteins in yeasts (Segal et al., 2001).

The possibility of transforming biocontrol agents by incorporating genes that code for antimicrobial compounds from diverse sources such as insects (animal kingdom) has been suggested. *Colletotrichum coccodes* causes fruit rot of tomato. Cecropin is an antimicrobial protein isolated from the giant silk moth *Hyalophora cecropia*. A cecropinA-based peptide inhibited the germination of conidia of *C. coccodes* at 50 μM. The DNA sequence encoding the peptide was cloned in the plasmid pRS413, using a *Saccharomyces cerevisiae* invertase (β-fructo-furanosidase) leader sequence for secretion of the peptide, and expressed in yeast. The transformants inhibited the growth of germinated conidia of *C. coccodes*. The decay development in tomatoes cv. Roma inoculated with *C. coccodes* was also inhibited. This study opens up a new approach for effective control of postharvest diseases by employing BCAs whose biocontrol potential can be enhanced by expression of antifungal peptides from other organisms (Jones and Prusky, 2002).

Trichoderma harzianum, a soil fungus has been demonstrated to be an efficient biocontrol agent against several phytopathogens. The antagonism of *T. harzianum* correlates well with the production of antifungal activities, including secretion of fungal cell wall degrading enzymes such as chitinases. The chitinases Chit 42 and Chit 33 from *T. harzianum* CECT 2413 do not have a chitin-binding domain. By incorporating a cellulose-binding domain (CBD) from cellobiohydrolase II of *T. reesei* to these enzymes, hybrid chitinases Chit 33-CBD and Chit 42-CBD with stronger chitin-binding capacity were engineered. The transformants of *T. harzianum* over-expressing the native chitinases exhibited higher levels of chitinase-specific activity, resulting in greater inhibition of growth of pathogens such as *Botrytis cinerea*, *Rhizoctonia solani*, and *Phytophthora citrophtora*. The results reveal the importance of endochitinase in the antagonistic activity of *T. harzianum* strains and the effectiveness of incorporation of a CBD to enhance the biocontrol potential of the microorganisms against fungal pathogens (Limón et al., 2004).

9.6 FORMULATIONS OF BIOCONTROL AGENTS

The development of a shelf-stable formulated product that can retain biocontrol activity at a level which is similar to that of freshly isolated microorganisms is the basic requirement for commercialization of biocontrol products. The microbial biocontrol agents may be either formulated as dry powder or suspension in liquid (oil), so that they can be handled in commercial channels of distribution and storage (Rhodes, 1993). Few studies have been published on the formulation of microorganisms for biocontrol of postharvest pathogens and hence only proprietory formulations of commercial products for postharvest applications are available. Various aspects, such as the development of suitable medium for mass multiplication, shelf life, ability to grow and survive after application under different conditions, effectiveness of disease control, and ease and cost of application, need to be studied and documented.

The microorganisms with high biocontrol potential need to be selected and their nutritional requirements, optimum temperature, pH, and other conditions for their rapid multiplication have to be determined. The compounds that differentially stimulated growth of the antagonist with no detectable stimulating effect on the test pathogen were identified for the antagonist *Pseudomonas syringae* (strain L-59-66) that was found to be effective against *Penicillium expansum* causing apple blue mold disease. This beneficial bacterium occurred naturally on fruit trees and was capable of controlling postharvest diseases of pear and citrus fruits. A formulation of *P. syringae* compatible with practices in packing houses was developed (Janisiewicz et al., 1992; Janisiewicz, 1994a,b).

The carbon and nitrogen sources that provided maximum biomass production of *Pantoea agglomerans* strain CPA-2 were determined. A combination of nitrogen sources such as yeast extract (5 g/l) and dry beer yeast (10 g/l) with inexpensive carbohydrates such as sucrose (10 g/l) and molasses (20 g/l) was found to be suitable for the mass multiplication of this bacterial BCA which was effective against *P. digitatum* and *P. italicum* infecting oranges. This study provided a reliable basis for a scale-up of the fermentation process to an industrial project (Costa et al., 2001). The procedure for the formulation of *Pichia anomala* (strain K) and *Candida oleophila* (strain O) which protected apples against gray mold and blue mold diseases caused by *B. cinerea* and *P. exapansum* respectively was standardized by Jijakli (2000).

A mass production system for *Pichia onychis* which provided effective control of *Botrytis allii* and *Rhizopus* sp. was developed and the total cost of production/litre was also calculated (Diaz et al., 2001). The optimal growth conditions required for the growth of *Candida sake* strain CPA-1 in a labscale fermenter were determined. Maximum growth (about 8×10^8 CFU/ml) was attained after 30 h at 400 rpm, 150 l of air/h, and an initial concentration of 10^6 CFU/ml. These findings will be useful for further scaling up of *C. sake* production (Abadias et al., 2003a,b).

The fungi and bacteria employed as biocontrol agents have to be protected against freeze-drying injury in order to preserve the viability and to enhance the shelf stability of the biocontrol agents. The viability of the biocontrol agents may be enhanced by the use of freeze-drying protective agents and rehydration media. Maximum protection from freeze-drying injury to the bacterial antagonist *Pantoea agglomerans* strain CPA-2 was provided by 5% trehalose with a survival of over 60% of the bacterial cells. Skimmed milk (10%) was found to be the most effective rehydration medium, from which 100% of the freeze-dried bacterium could be recovered. *P. agglomerans* CPA-2 reduced the blue mold and gray mold diseases of pome fruits significantly (Costa et al., 2000). Further study showed that freeze-dried cells of *P. agglomerans* could be stored in glass vials or in high barrier plastic bags at 4°C for 3 months with no reduction in the biocontrol of potential against *P. digitatum* (Costa et al., 2002). The viability, storage stability, and biocontrol potential of freeze-dried *Candida sake* strain CPA-1 against *Penicillium expansum*, causing blue mold in Golden Delicious apple, were assessed. Survival of *C. sake* was the highest (85%), when the yeast cells were protected with lactose (10%) and rehydrated with skimmed milk. However, the yeast cells freeze-dried with lactose (10%) + skimmed milk (10%) as protectant and peptone (1%) as a rehydration medium reduced the incidence of blue mold disease to the maximum extent. In general, freeze-drying reduced the biocontrol potential significantly compared to fresh yeast cells (Abadias et al., 2001). The biocontrol potential of *Aureobasidium pullulans* was enhanced by the addition of xanthan gum. When applied on strawberry in the field from bloom to fruit at green stage the incidence of storage rot caused by *B. cinerea* was significantly reduced, due to enhanced survival rate of the BCA (Ippolito et al., 1998). A shelf-stable biological product containing *Epicoccum nigrum* which was effective against brown rot disease was prepared by freeze-drying the conidia with skimmed milk or fluidized-bed drying alone. These treatments maintained an initial viability for 30 and 90 days, respectively, for storage at room temperature (Larena et al., 2003).

Suspensions of eight yeast isolates were mixed with cellulose and dried. The product was then milled into a fine powder. The efficacy of this yeast–cellulose formulation, applied as a dry powder to sporulating colonies of *B. cinerea* infecting kiwifruit, was assessed. The formulation significantly suppressed liberation of conidia by binding directly to the pathogen spores. By using α-cellulose prepared with *Candida pulcherrima*, a reduction of about 50% in the number of conidia released from the lesions induced by *B. cinerea* could be achieved. Such suppression of liberation by a yeast formulation may provide an effective management tool for biological control of sporulating postharvest fungal pathogens (Cook, 2002).

The effect of modifying water activity (a_w) and the addition of protective substances to the preservation medium of liquid formulations of *Candida sake* stored at 4 and 20°C was evaluated. The a_w of the preservation medium was altered to have a range from 0.72 to 0.95 by adding glycerol or polyethylene glycol (PEG). The viability of yeast cells was improved by a_w levels of 0.93 to 0.95 but not to the desired level. When sugars, such as trehalose, and polyols, such as glycerol or PEG, were used as protectants cells of *C. sake* maintained viabilities of greater than 60%. Storage at 4°C was preferable to 20°C because of maintenance of greater viability of yeast cells at the lower temperature. The biocontrol efficiency of liquid formulation of *C. sake* in controlling *Penicillium expansum* in wounds on apples was similar to the fresh yeast cells. Use of 10% sucrose as protectant in the liquid formulation of *C. sake* would be an acceptable proposition, as the cost would not be a limiting factor for industrial production (Torres et al., 2003). In another study, preservation of *C. sake* CPA1 strain cells in trehalose solution (0.96 M) which was isotonic with yeast cells yielded best results. After storage at 4°C for 7 months, yeast cells that were grown in the sorbitol modified medium and preserved with isotonic solution of trehalose maintained the viability at high levels. The formulated BCA effectively controlled the blue mold disease of apples (Abadias et al., 2003b).

Systematic assessments of the efficacy of all available isolates and strains should be made to select the most potent microbial agents that can be commercially developed as a biofungicide. Some BCAs may require specific set of conditions for their antagonistic activity. In the case of spore-forming *Bacillus* sp, the termination of endogenous dormancy is an important factor for rapid action. The formulations include wetting agents, to facilitate reabsorption of moisture from the air, and oil carriers, enabling microbial antagonists to reach depressions, stoma, and lenticels through which the pathogens may penetrate into the host. Longer shelf life of biological products may be desirable, since the products with a short shelf life (less than 6 months) may require a cold chain and extensive field services to ensure retention of biocontrol activity to the required level (Jones and Burges, 1998; Janisiewicz and Korsten, 2002a).

A biocontrol product was commercialized under the name Bio-Save11, from a formulation containing *Pseudomonas syringae*, by EcoScience Corporation. *P. syringae* (strain L-59-66 renamed as strain ESC-11) controlled blue mold caused by *Penicillium expansum*, gray mold caused by *B. cinerea*, and mucor rot caused by *Mucor* spp. on apple and pear (Janisiewicz and Jeffers, 1997; Janisiewicz and Marchi, 1992). *P. syringae* ESC 11 is sold under the name Biosave ™110 and it is recommended for postharvest decays of pear and apple. *P. syringae* strain ESC-10, effective against citrus rots, is marketed under the name Biosave ™100 (Janisiewicz and Jeffers, 1997). The strain ESC-11 was able to reduce the crown rot of banana caused by a complex of fungi including *Fusarium semitectum* and *F. moniliforme*. In addition, potato dry rot caused by *F. sambucinum* could also be controlled by strain ESC-11. Furthermore, it is possible to prevent growth of the food-borne pathogen, *Escherichia coli* 0157 : H7 in apple wounds by applying the strain ESC-11 (Kenwick and Jacobsen, 1998; Janisiewicz et al., 1999; Wiliamson et al., 1999). The powder formulation Bio-Save11 was as effective as the unformulated laboratory wet preparation in the control of blue mold and gray mold in Golden Delicious and Red Delicious apples. Formulating *P. syringae* strain ESC-11 into a powder did reduce its survival and growth in fruit wounds, but it was effective in protecting the fruits against postharvest pathogens. Hence, the commercial product has the potential to replace synthetic fungicides to control fruit rots (Janisiewicz and Jeffers, 1997).

A white yeast, *Candida oleophila* was shown to be effective against blue mold and green mold on citrus fruit and blue mold and gray mold on apple. A formulation containing *C. oleophila* has been commercialized and marketed under the name Aspire by Ecogen, Inc. (Droby et al., 1998). Commercially available biocontrol products that have been recommended for use against microbial plant pathogens infecting fruit and vegetable crops are presented in Table 9.5.

9.7 NATURAL COMPOUNDS

Natural plant- and animal-derived compounds possessing antimicrobial properties may offer safe alternatives to synthetic fungicides that have been applied for the control of postharvest diseases. In addition, some of these compounds have been shown to induce resistance to postharvest diseases in fruits and vegetables. This aspect is discussed in detail in Chapter 10.

9.7.1 Compounds of Plant Origin

A wide range of secondary metabolites produced by plants, such as essential oils, are endowed with antimicrobial, allelopathic, antioxidant, and bioregula-

TABLE 9.5 Biocontrol Products Commercially Available for Application Against Microbial Pathogens

Microbial agent	Product	Crops	Target diseases	Formulation/ application method	Manufacturer/ distributor
Ampelomyces quisqualis	AQ 10 Bio-fungicide	Grapes, apples, strawberries, tomatoes and cucurbits	Powdery mildews	Water-dispersible granule/ spray	Ecogen, Inc, USA
Candida oleophila strain I-182	Aspire	Citrus, pome fruits	Blue mold, gray mold, green mold	Wettable powder/ drench, dip or spray for postharvest application	Ecogen, Inc, USA
Fusarium oxysporum (nonpathogenic)	Biofox C	Tomato	Seed rot	Dust or alginate granule/seed treatment	S.I.A. PA. Italy
Pseudomonas syningae (strain 10 LP, 110)	Biosave 10LP, 110	Pome fruits, citrus, cherries, potatoes	Molds, mucor rot and sour rot	Lyophilized product, frozen cell concentrated pellets/ postharvest application	Eco Science Corporation, USA
Pseudomonas fluorescens A 506	Blight Ban A506	Pome fruits, strawberry, potato	Fire blight, soft rots	Wettable powder/ spray	Nu Farm, Inc., USA
Bacillis subtilis GB 03	Companion	Greenhouse crops	Root rots, damping off	Liquid/drench or seedling treatment	Growth products USA
Coniothyrium minitans	Contans WG, Intercept WG	Onion	Basal and neck rots	Water dispersible granules/spray	Prophyta Biologischer Pflanzenschutz GmbH, Germany

TABLE 9.5 (*Continued*)

Microbial agent	Product	Crops	Target diseases	Formulation/ application method	Manufacturer/ distributor
Burkholderia cepacia	Deny	Bean, peas, other vegetable crops	Root rots	Dried biomass/ aqueous suspension/ seed treatment	Stine Microbial Products, USA
Bacillus subtilis	HiStick N/T	Peanuts, soybean	Root rots	Slurry/seed treatment	Becker Underwood, USA
Erwinia amylovora Hrp N harpin protein	Messenger	Vegetable and ornamental crops	Fireblight	Powder/spray or drench	EDEN Bioscience Corporation, USA
Streptomyces griseoviridis strain K61	Mycostop	Vegetable and ornamental crops	Seed rot, stem rot	Powder/spray or drench	Kemira Agro Oy Finland
Bacillus subtilis FZB 24	Rhizo-plus	Potatoes, vegetables crops	Powdery scab/ root rot	Suspension/seed treatment or soil drench	KFZB Biotechnick GmbH, Germany
Bacillus subtilis	Serenade	Grapes, vegetables, pome fruits, peanuts	Powdery mildew, late blight brown rot, fireblight	Wettable powder/spray	Agra Quess Inc., USA.

Source: Fravel, 2002.

tory properties (French, 1985; Elakovich, 1988). The presence of essential oils in plants belonging to the genera *Ocimum*, *Thymus*, *Origanum*, *Anethum*, *Eucalyptus*, *Foeniculum*, and *Citrus* is well known. The fungicidal property of the essential oils was attributed to carvacrol present in thyme and origanum oil and to *p*-anisaldehyde formed due to oxidation of anethole present in anise oil (Caccioni and Guizzardi, 1994). During ripening of fruits, volatile aromatic compounds with antifungal property are produced. Acetaldehyde formed during fruit ripening provided protection to apples, stone fruits, and cherries against postharvest pathogens (Stadelbacher and Prasad, 1974; Mattheis and Roberts, 1993; Caccioni et al., 1994). The fungistatic/fungicidal activity of hexanal and benzaldehyde, produced during etheric metabolism in stone fruits, against *Monilinia laxa* and *Rhizopus stolonifer* has also been demonstrated (Caccioni et al., 1995).

The fungitoxic activity of 12 essential oils (EOs) distilled from medicinal plants was assessed. The oil from *Thymus capitatus* exhibited strong fungicidal activity against *Penicillium digitatum*, *P. italicum*, *B. cinerea*, and *Alternaria citri* in vitro at 250 ppm. The fungitoxicity of *T. capitatus* EOs sprayed on healthy orange inoculated with *P. digitatum* was weak at atmospheric pressure, but in vacuum conditions conidial mortality on fruit exocarp was high (90–97%). The efficacy of this EO was comparable to that of thiabendazole (TBZ) at 2000 ppm concentration. The morphology of *P. digitatum* hyphae and conidia was markedly affected by exposure to *T. capitatus* EO vapors as revealed by scanning electron microscopic observations. Carvacrol was found to be the predominant compound, accounting for 81 to 83% of EO vapors present in *T. capitatus* and exhibited fungitoxicity (Arras and Usai, 2001). Application of Mentha oil (*Mentha arvensis*) reduced blue mold rot caused by *P. italicum* in orange and lime and enhanced the shelf life by 6 to 8 days (Tripathi et al., 2004).

Botrytis cinerea infects leaves and berries of grapevine. The effects of EOs from thyme (*Thymus vulgaris*) and clove (*Syzygium aromaticum*) and massoilactone extracted from bark of *Cryptocarya massoia*, were determined on the growth of *B. cinerea*. The sporulation of *B. cinerea* on artificially induced necrotic lesion was markedly reduced by thyme (Thyme R) and massoialactone oils at 0.33% which was not phytotoxic. The bunch rot and leaf colonization by *B. cinerea* was significantly controlled by a single application of either EO at veraison. Spray applications of Thyme R oil at 0.33% from flowering to harvest effectively controlled bunch rot disease. Senescence of floral tissues following treatment with Thyme R was also observed (Walter et al., 2001). The effectiveness of tea tree oil vapor in inhibiting the growth of *B. cinerea* was demonstrated. The growth was entirely inhibited by tea tree oil at 500 ppm, the inhibitory effect decreasing with increase in dilution of the oil. It may be useful to apply this essential oil in carefully designed packaging. Essential oils which have been registered as food additives may be much easier to register for postharvest uses in place of synthetic fungicides (Jobling, 2000). Treatment of Embul banana with *Ocimum basilicum* oil (0.16% v/v) controlled crown rot and anthracnose caused by *Colletotrichum musae* effectively, enabling bananas to be stored for up to 21 days at 13 ± 1°C with no adverse effect on their organoleptic properties. This treatment compared well with application of benomyl. Hence spraying emulsions of essential oil of *O. basilicum* prior to cool storage was recommended as a safe, cost-effective method with commercial potential for controlling postharvest diseases and extending storage life (Anthony et al., 2003).

Fumigation of apricots (*Prunus armeniaca*) with 2 mg/l of thymol vapor reduced the conidial germination of *Monilinia fructicola* to 2% as against 98% germination on untreated fruit. Shrinkage and protoplast collapse of conidia due to the treatment were observed by microscopy. The disease incidence following exposure of Manch apricots to fumigation with thymol (5 mg/l) was reduced to 3% as against 64% in untreated control fruits. Similar beneficial effect of thymol on plums also was demonstrated. Fumigation with thymol resulted in increased fruit firmness in both apricot and plum. Thymol fumigation was phytotoxic to apricots but not to plum (Liu et al., 2002).

Sweet cherries are seriously affected by gray mold disease caused by *B. cinerea*. Fumigation with thymol (30 mg/l) for 25 min before sealing in modified atmosphere packages prior to cold storage reduced disease incidence to 0.5% as against 36% in untreated control. The treated fruits had lower total soluble solids, higher titrable acidity, and greater stem browning than the controls (Chui et al., 1999). Sweet cherries inoculated with the conidia of *Monilinia fructicola*, causing brown rot disease, and *Penicillium expansum*, causing blue mold disease, and fumigated with thymol for 10 min were stored at 10°C. After storage for 13 days sweet cherries were fumigated with thymol (10 mg/l). Brown rot disease incidence was significantly reduced but there was no perceptible effect on blue mold disease (Chu et al., 2001). The possibility of using essential oils for the control of mold diseases and sour rot of citrus fruits (El-Mohamedy et al., 2002) and crown rot and anthracnose diseases of banana (Ranasinghe et al., 2002) has been indicated.

The effectiveness of volatiles from cv. Isabella (*Vitis labrusca*) on the development of *Botrytis cinerea* was assessed in vitro as well as in situ. The biological activities of the volatiles were quantified by using the closed Mariotte system. The Isabella volatiles inhibited the sporulation and sclerotia formation of the pathogen. In contrast, the volatiles from Roditis grapes (*V. vinifera*) stimulated the sporulation. The in situ determination showed that the Isabella volatiles had antifungal properties against *B. cinerea* as reflected by the reduction in the amount of inoculum produced and pathogenicity. The antibiotic activity of the volatiles was more pronounced at 21°C indicating the possibility of using *Isabella* volatiles as biocontrol agents of *B. cinerea* (Kulakiotu et al., 2004). The effects of volatiles from 'Isabella' grapes on the development of *B. cinerea* and gray mold disease incidence on kiwifruit c. Hayward (*Actinidia deliciosa*) were determined. Reduction in both the inoculum density and the activity of the pathogen following exposure to 'Isabella' volatiles resulted in the lower level of disease incidence. The inhibitory action of volatiles reached the maximum at 21°C, irrespective of the interval between wounding and inoculation of the kiwifruits with the pathogens (Kulakiotu et al., 2004). The results of the investigations on grapes and kiwifruit reveal the potential of the 'Isabella' volatiles for effective control of gray mold diseases.

The plant species belonging to the family Cruciferae produce glucosinolates, a large class of approximately 100 compounds, which have antimicrobial properties effective against many microbial plant pathogens (Fenwick et al., 1983). The glucosinolates are brought into contact with the enzyme myrosinase when the plants are injured. Myrosinase catalyzes hydrolysis, producing D-glucose, sulfate ion, and a series of compounds such as isothiocyanate, thiocyanate, and nitrile, depending on the substrate and other reaction conditions. Among the six glucosinolates tested, glucoraphenine isothiocyanate inhibited the development of *Monilinia laxa* in inoculated pear cvs. Conference and Kasier. Allyl-isothiocyanate as a volatile substance provided effective control of green mold disease in Conference pear inoculated with a TBZ-resistant strain (Mari and Guizzardi, 1998). Kiwifruit cv. Hayward were inoculated with *B. cinerea* causing gray mold disease and treated with δ-decalactone, δ-dodecalactone, or β-ionone at 5 μl/fruit (Ward et al., 1998). The percentage decay due to *B. cinerea* was less in most of the treated fruits compared with untreated controls after storage for 18 weeks. Treatment with δ-dodecalactone provided a higher level of control and was comparable with that of the fungicide vinclozolin.

Numerous plant species have been screened to identify those containing compounds with antimicrobial property. Most of these studies provide results of in vitro experiments, aimed at assessing the effect on conidial germination and mycelial growth of pathogens causing postharvest diseases including: storage rots of apples (Sharma and Bhardwaj, 2000), mango (Singh et al., 2000), mango and papaya (Ribeiro and Bedendo, 1999; Bautista-Baños et al., 2002), and carrots (Hörberg, 1998; Prakasam et al., 2001). Only a few reports indicate the effectiveness of natural products in controlling the disease in vivo. Aqueous extracts of *Vitex negundo* var. *purpurescens* (10%) offered protection to carrots against postharvest diseases *Cladosporium* rot (*Caldosporium oxysporum*), *Fusarium* rot (*Fusarium solani* f.sp. *radicicola*), and *Geotrichum* rot (*Geotrichum candidum*) (Prakasam et al., 2001).

Seed extracts and powders of *Pithecellobium dulce* (huamuchil) exhibited fungicidal and fungistatic activity against *Botrytis cinerea*, *Penicillium digitatum*, and *Rhizopus stolonifer*. Partial purification indicated that triglycerol and triterpene saponins could be the compounds that were effective against these postharvest pathogens (Bautista-Baños et al., 2003; Barrera-Necha et al., 2003). Ethanol extracts of garlic cloves (0.1%), alone or in combination with sunflower cooking oil or fruit wax, were evaluated for their efficacy in controlling green and blue molds in grape fruit (cvs. Valencia and Shamouti) caused by *P. digitatum* and *P. italicum*, respectively. Clove extracts when combined with oil provided 100% control of both green and blue mold diseases and treatment (1% extract + oil) was equally effective as the fungicide application (imazalil 500 ppm + quazatine 1000 ppm) on Valencia oranges (Obagwu and Korsten et al., 2003). Extracts of *Solanum torvum* was reported to be more effective in controlling the banana anthracnose disease caused by *Colletotrichum musae*, compared with a standard fungicide, benomyl (0.1%). Furthermore, an increase in shelf life by 16 to 20 days over control was the additional benefit of treatment with extract of *S. torvum* (Thangavelu et al., 2004).

Plant-derived compounds have been evaluated for their efficacy in controlling some of the seed-borne and storage fungi. Although many plant species have been screened to identify the plants containing effective antimicrobial compounds, efforts to characterize the antimicrobial compounds are limited. Of the five plant extracts tested, pure garlic clove extract completely suppressed the growth of *Aspergillus candidus*, *A versicolor*, *Penicillium aurantiogriseum*, *P citrinum*, *P. brevicompactum*, and *P. griseofulvum* associated with wheat grains. However, the concentrations of extract required to suppress the fungal growth was toxic to the germ. Nevertheless, no significant reduction in germination of seeds sown in the soil was evident, indicating the feasibility of using the natural antifungal product for treatment of wheat seeds (Scholz et al., 1999). Common bunt caused by *Tilletia tritici* in wheat and stem smut of rye due to *Urocystis occulta* may pose problems for organic production. Seed treatment with mustard flour controlled seed-borne infection by *T. tritici* in wheat with no adverse effect on seed germination or vigor. The rye stem smut could be controlled effectively by the application of both mustard flour and milk powder. Mustard flour, effective against both diseases, has potential as a seed treatment in organic agriculture (Borgen and Kristensen, 2001). The incidence of common bunt (*T. caries*) in winter wheat was markedly reduced by seed treatment with skimmed milk powder, to 3 wheat ears/m^2 from 52 infected wheat ears/m^2 when untreated seeds were used. The combination of seed treatment with warm water (45°C for 2 h) and skimmed milk powder controlled the seed-borne infection by common bunt, snowmold (*Gerlachia nivalis*), head blight (*Fusarium graminearum*), and damping-off (*Septoria nodorum*) in winter wheat. Yellow mustard meal (Tillecur) at 60 ml/kg of seed was found to be an effective alternative to chemical treatment for the control of common bunt disease (Winter et al., 2001).

Application of essential oils, such as oregano, cinnamon, lemongrass, clove and palmarose, reduced the growth rate and production of mycotoxins zearalenone (ZEA) and deoxynivalenol (DON) by *Fusarium graminearum* in maize. The antifungal and antimycotoxigenic activity of the essential oils suggest that they may be considered as a potential alternative to the fungicides for the control of this pathogen infecting maize grains (Velluti et al., 2004). The effectiveness of neem (*Azadirachta indica*) and pungam (*Ponganiaglabra*) oil-based emulsifiable concentrate formulations for the control of grain discoloration of rice, caused by *Helminthosporium oryzae* and *Magnaporthe grisea* (*Pyricularia oryzae*), was reported by Rajappan et al. (2001).

Infection of groundnut (peanut) kernels by *Aspergillus* spp. results in production of the mycotoxin aflatoxin (AFB). Powders from the leaves of *Ocimum gratissimum* and cloves of *Syzygium aromaticum* were evaluated for their efficacy for the control of kernel infection by *Aspergillus*. Packaging groundnut kernels with jute bags and interlaced polypropylene bag, with or without plant powders, provided 100% protection against fungi. Application of *S. aromaticum* powder at 3% and in combination with jute bag packaging effectively suppressed cross-infection of healthy kernels (with 12% moisture level) by fungi from diseased kernels when both infected and healthy kernels were present in the same lot (Awuah and Ellis, 2002).

9.7.2 Compounds of Animal Origin

Chitosan is a biodegradable fiber (polymer) derived from crab-shell chitin and it offers great potential as an antifungal preservative for fresh fruits and vegetables. Chitosan forms films that can be used to coat the surface of fruits and vegetables. It has fungicidal properties and when applied for coating, fruit ripening may be delayed, consequently increasing the shelf life of fruits and vegetables. In addition, chitosan has been demonstrated to be an inducer of resistance to postharvest diseases in treated fruits and vegetables (Wilson et al., 1994) (Chapter 10).

Strawberry plants cv. Seascape were sprayed twice at 10 days interval with chitosan (2, 4, or 6 g/l) and harvested fruits were challenged with *B. cinerea* and stored at 3° or 13°C. Chitosan reduced postharvest fungal rot significantly and maintained the keeping quality of fruits compared with water sprayed controls. The incidence of decay decreased with increase in concentration of chitosan applied and increased with storage period and temperature. Chitosan sprays (6 g/l) applied twice protected the fruits from decay and maintained fruit quality at an acceptable level throughout the storage period of 4 weeks at 3°C (Reddy et al., 2000b). Preharvest sprays of chitosan, three times, reduced rots due to *B.cinerea* and *R. stolonifer* with greatest reductions observed at 1.0% concentration. Likewise, dipping strawberries in chitosan (1.0%) for 3 to 4 seconds resulted in maximum protection against *B. cinerea*, where the incidence of *R. stolonifer* was negligible (Romanazzi et al., 2000). Chitosan coating of strawberries and raspberries stored at 13°C was as effective as thiabendazole in controlling the decay of berries caused by *B. cinerea* and *R. stolonifer*. In addition, chitosan coating exerted beneficial effects on firmness, titrable acidity, vitamin C content and anthocyanin content of strawberries and raspberries stored at 4°C (Zhang and Quantick, 1998).

Chitosan treatment (5 and 10 mg/ml) reduced the incidence of brown rot disease of peaches caused by *Monilinia fructicola* significantly and delayed the disease development compared with control. Treated peaches were firmer and had higher titrable acidity and vitamin C content than the fungicide prochloraz-treated or control peaches. Chitosan has the potential to control brown rot, preserve the valuable attributes, and prolong the shelf life of postharvest peaches (Liand Yu, 2001). Stem scar application of chitosan (400 µl of a 10 g/l solution) to tomatoes stored at 20°C provided protection against blackmold rot caused by *Alternaria alternata* (Reddy et al., 2000a).

SUMMARY

Application of synthetic fungicides has been the traditional strategy for the management of diseases caused by microbial pathogens both at pre- and postharvest stages. The increasing concerns for health hazards and environmental pollution due to chemical use has necessitated the development of alternative strategies for the control of postharvest diseases. Management of postharvest diseases by employing biocontrol agents has been demonstrated to be most suitable to replace the chemicals which are either being banned or recommended for limited use.

Various factors have to be considered in the screening and selection of microorganisms with biocontrol potential. The identity of the microorganism with a wide spectrum of activity, capacity for survival in various conditions and ability to compete with other microorganism and pathogens for available nutrients and space, induction of resistance to postharvest pathogens, and with multiple mechanisms of biocontrol has to be precisely and rapidly established by using conventional and molecular methods. The efficiency of the selected microbial agents may be enhanced by the addition of chemical supplements, using mixtures of biocontrol agents, and physiological and genetic manipulation. The efficiency of formulated products has to be assessed under varied conditions and against different pathogens infecting the same fruit or vegetable. So far, only a few products with high biocontrol potential have been made available on a commercial scale. With intensive research being carried out in various laboratories, the possibility of identifying potent microbes and developing suitable biocontrol products for commercial marketing appears to be bright.

APPENDIX 9 (I): ANTIBIOSIS TESTS FOR PRODUCTION OF ANTIFUNGAL COMPOUNDS (JANISIEWICZ ET AL., 2000)

A. Multiplication of Pathogen

i. Cultivate the pathogen (*Penicillium expansum*) in potato dextrose agar medium consisting of boiled potatoes (200 g), dextrose (20 g), agar (15 g), and distilled water (1 l).
ii. Collect conidia from 7 to 10-day-old culture by transferring small amount of sterile water and adjust the concentration of conidia to 2×10^5 or 1×10^6 per ml using a hemocytometer.

B. Multiplication of the Biocontrol Agent (BCA)

i. Grow the BCA (*Auerobasidium pullulans*) in Erlenmeyer flasks (250 ml) containing of 50 ml of nutrient yeast dextrose broth (NYDB) (Difco) medium consisting of nutrient broth (8 g), yeast extract (5 g), and dextrose (10 g) per liter and place the culture on a shaker at 50 rpm for 20 h.

ii. Harvest the yeast cells by centrifugation at 6500 g; wash in sterile distilled water (SDW) twice and resuspend the cells in SDW.

iii. Adjust the concentration of yeast cells to an optical density of 0.13 (75% transmittance or 8.7×10^6 CFU (colony forming units/ml) at 420 nm.

C. Dual Culture Tests on Agar Plates

i. Dispense water agar (20%) + apple juice (fruit juice) at 10,1, 0.1, or 0% and allow the medium to set.

ii. In one set of plates, streak the conidial suspensions of BCA and pathogen across each other on the dish, taking care not to place inoculum at the point of crossing to prevent carry over of inoculum of one test organism into another. Both the BCA and the pathogen may be inoculated simultaneously or the pathogen may be inoculated at 2 days after placing the BCA to allow the diffusion of antifungal compounds, if any, from the BCA into the medium.

iii. In another set of plates, streak the BCA and the pathogen simultaneously in parallel lines with a space of about 3 cm between and incubate the plates at 22°C for 2 weeks.

iv. Determine the extent of growth inhibition.

APPENDIX 9 (II): BIOCONTROL OF POSTHARVEST PATHOGEN THROUGH COMPETITION FOR NUTRIENTS (JANISIEWICZ ET AL., 2000)

A. Cultivation of Pathogen and Biocontrol Agent (BCA)

i. Follow the procedure described in A and B of Appendix 9(i).

B. Preparation of Culture Plates

i. Use tissue culture plates with 24 wells per plate Costar, Corning Inc., Corning, N.Y) and culture plate inserts Multicell-CM (Millipore Corp., Bedford, MA) consisting of polystyrene cylinder and a hydrophilic polytetrafluoroethylene (PTFE) membrane, with a pore size of 0.45 μm, attached to the bottom of the cylinder; the spacers present at bottom allow free liquid movement below the membrane.

ii. Check the movement of liquid through membrane using crystal violet solution placed inside the cylinder inserted into the water in the well outside the cylinder; the movement in reverse direction can be checked by changing the location of dye solution and water.

C. Preparation of Test Reactants

i. Prepare the suspension of BCA in apple juice by mixing apple juice at concentrations of 10 or 1% and conidial suspension at 75% turbidity (8.7×10^6 CFU/ml) in equal volumes, the BCA concentration being similar to that of commercial products.
ii. Prepare conidial suspension of the pathogen (*P. expansum*) in water (2×10^5 conidia/ml).

D. Competition Tests

i. Dispense water and either 5 or 0.5% apple juice alone or with BCA (4.35×10^6 CFU/ml) at a rate of 0.6 ml/well.
ii. Transfer conidial suspension of the pathogen inside the cylinder inserts at 0.4 ml/cylinder.
iii. Place the cylinders in the wells of the plates and incubate at 26°C for 24 h.
iv. Remove the cylinders from the wells; blot the membrane from the bottom with tissue paper until all liquid from inside the cylinder is absorbed.
v. Cut the membrane with the sharp scalpel; transfer it to glass slide and stain with lactophenol-cotton blue.
vi. Observe under the microscope.
vii. Determine the BCA populations in the wells turbidimetrically using the spectrophotometer.
viii. Filter the suspension of BCA through a 0.22 μm Millipore filter.
ix. Determine the depletion of amino acids using the high-pressure liquid chromatography and fluorescence detection (Agilent Technologies, Wilmington, DE).

E. Assessment of Pathogen Viability

i. Prepare a parallel set of culture plates with inserts as described above.
ii. Remove the inserts containing pathogen conidial suspension at the end of 24 h incubation period and blot the membrane as done earlier.
iii. Insert them into new wells containing only water or apple juice at similar concentrations used earlier and incubate for 24 h at 26°C.
iv. Remove the inserts and follow the procedure as in steps (iv) and (v) under D above.

F. Evaluation of Antifungal Activity of BCA

i. Prepare Petri dishes containing nutrient yeast dextrose agar medium and punch out agar discs to form wells using sterile cork borer (0.5 cm diameter).
ii. Dispense the filtrate (step viii of D above) into the wells at 100 μl/well and incubate the plates at 4°C for 24 h.
iii. Seed the plates after incubation with an aqueous suspension of conidia of pathogen (10^6 conidia/ml) and incubate at 24°C for 48 h.
iv. Measure the inhibition zone around the wells.

REFERENCES

Abadias, M., Teixidó, N., Usall, J., Benabarre, A. and Viñas, I. 2001. Viability, efficacy and storage stability of freeze-dried biocontrol agent *Candida sake* using different protective and rehydration media. *Journal of Food Protection*, **64**: 856–861.

Abadias, M., Teixidó, N., Usall, J. and Viñas, I. 2003a. Optimization of growth conditions of the postharvest biocontrol agent *Candida sake* CPA-1 in a labscale fermenter. *Journal of Applied Microbiology*, **95**: 301–309.

Abadias, M., Teixidó, N., Usall, J. and Viñas, I. 2003b. Liquid formulation of the postharvest biocontrol agent *Candida sake* CPA-1 in isotonic solutions. *Phytopathology*, **93**: 436–442.

Adams, P.B. and Ayers, W.A. 1982. Biological control of *Sclerotinia* lettuce drop in the field *by Sporodesmium sclerotivorum*. *Phytopathology*, **72**: 485–488.

Adeline, T.S.Y. and Sijam, K. 1999. Biological control of bacterial soft rot of cabbage. *Proceedings of the Symposium on Biological Control in the Tropics*, Serdang, Malaysia (eds) L.W. Hong, S.S. Sastroutomo, I.G. Caunter, J. Ali, L.K. Yeang, S. Vijaysegaran and Y.H. Sin, CABI Publishing, Wallingford, UK.

Adikaram, N.K.B., Joyce, D.C. and Terry, L.A. 2002. Biocontrol activity and induced resistance as a possible mode of action for *Auerobasidium pullulans* against gray mold of strawberry fruit. *Australian Plant Pathology*, **31**: 223–229.

Ahmadi, H., Biasi, W.V. and Mitcham, E.J. 1999. Control of brown rot decay of nectarines with 15% carbon dioxide atmospheres. *Journal of American Horticultural Science*, **124**: 708–712.

Alimova, F.K., Zakharova, N.G., Asghar, M.G., Fattahkova, A.N. and Garusov, A.V. 2002. The quarantine control and the methods of disinfection of tomato and cucumber seeds in different regions of Russian Federation. *Pakistan Journal of Scientific and Industrial Research*, **45**: 191–196.

Aluko, M.O. and Hering, T.F. 1970. The mechanisms associated with the antagonistic relationship between *Corticium solani* and *Gliocladium virens*. *Transactions of British Mycological Society*, **153**: 173–179.

Anonymous, 2000. Safe handling practices for melons (6/8/00). In: *Retail Food Safety Program Information Manual*. Food and Drug Administration, Washington, D.C.

Anthony, S., Abeywickrama, K. and Wijeratnam, W.S. 2003. The effect of spraying essential oils of *Cymbopogon nardus*, *Cymbopogon flexuosus* and *Ocimum basilicum* on postharvest diseases and storage life of Embul banana. *Journal of Horticultural Science and Biotechnology*, **78**: 780–785.

Arras, G. 1996. Mode of action of an isolate of *Candida famata* in biological control of *Penicillium digitatum* in orange fruits. *Postharvest Biology and Technology*, **8**: 191–198.

Arras, G., DeCicco, V., Arru, S. and Lima, G. 1998. Biocontrol by yeasts of blue mold of citrus and the mode of action of an isolate of *Pichia guilliermondii*. *Journal of Horticultural Science and Technology*, **73**: 413–418.

Arras, G., Sanna, P., Astone, V. and Arru, S. 1998. Effect of $CaCl_2$ on the inhibitory activity of *Rhodotorula glutinis* against *Penicillium italicum* on orange fruits. *Italus Hortus*, **5**: 67–70.

Arras, G. and Usai, M. 2001. Fungitoxic activity of 12 essential oils against four postharvest citrus pathogens: chemical analysis of *Thymus capitatus* oil and its effect in subatmospheric pressure conditions. *Journal of Food Protection*, **64**: 1025–1029.

Awuah, R.T. and Ellis, W.O. 2002. Effects of some groundnut packaging methods and protection with *Ocimum* and *Syzygium* powders on kernel infection by fungi. *Mycopathologia*, **154**: 29–36.

Bacon, C.W., Yates, I.E., Hinton, D.M. and Meredith, F. 1999. Biological control of *Fusarium moniliforme* in maize. *Environmental Health Perspectives*, **109** (Suppl. 2), 325–332.

Barrera-Necha, L., Bautista-Baños, L., Bravo-Luna, L., Bermudez-Torres, K., Jimenez-Estrada, M., Reyes-Chilpa, R. and García, Suárez, F.J. 2003. Antifungal activity against postharvest fungi by extracts and compounds of *Pithecellobium dulce* seeds (Huamuchil). *Acta Horticulturae*, **628.**

Bautista-Baños, S., Barrera-Necha L.L., Bravo-Luna, L. and Bermúdez-Torres, K. 2002. Antifungal activity of leaf and stem extracts from various plant species on the incidence of *Colletotrichum gloeosporioides* of papaya and mango fruit after storage. *Revista Mexicana de Fitopatologia*, **20**: 8–12.

Bautista-Baños, S., Garcia-Dominguez, E., Barrera-Necha, L.L., Reyes-Chilpa, R. and Wilson, C.L. 2003. Seasonal evaluation of the postharvest fungicidal activity of powders and extracts of huamuchil (*Pithecellobium dulce*): action against *Botrytis cinerea*, *Penicillium digitatum* and *Rhizopus stolonifer* of strawberry fruit. *Postharvest Biology and Technology*, **29**: 81–92.

Beagle-Ristaino, J.E. and Papavizas, G.C. 1985. Biological control of *Rhizoctonia* stem canker and black scurf of potato. *Phytopathology*, **75**: 560–564.

Bejaoui, H., Mathieu, F., Taillandier, P. and Lebrihi, A. 2004. Ochratoxin A removal in synthetic and natural grape juices by selected oenological *Saccharomyces* strains. *Journal of Applied Microbiology*, **97**: 1038–1044.

Benbow, J.M. and Sugar, D. 1999. Fruit surface colonization and biological control of postharvest yeast applications. *Plant Disease*, **83**: 839–844.

Benhamou, N. 2004. Potential of the mycoparasite, *Verticillium lecanii*, to protect citrus fruit against *Penicillium digitatum*, the causal agent of green mold: A comparison with the effect of chitosan. *Phytopathology*, **94**: 693–705.

Benhamou, N. and Brodeur, J. 2000. Evidence for antibiosis and induced host defense reactions in the interaction between *Verticillium lecanii* and *Penicillium digitatum*, the causal agent of green mold. *Phytopathology*, **90**: 932–943.

Bhuvaneswari, V. and Rao, M.S. 2001. Evaluation of *Trichoderma viride* antagonistic to postharvest pathogens on mango. *Indian Phytopathology*, **54**: 493–494.

Boff, P., Köhl, J., Gerlagh, M. and Kraker, J. de. 2002a. Biocontrol of green mold by *Ulocladium atrum* applied at different flower and fruit sprays of strawberry. *Biocontrol*, **47**: 193–206.

Boff, P., Köhl, J., Jansen, M., Horsten, P.J.F.M., van der Plas, C.L. and Gerlagh, M. 2002b. Biological control of gray mold with *Ulocladium atrum* in annual strawberry crops. *Plant Disease*, **86**: 220–224.

Borgen, A. and Kristensen, L. 2001. Use of mustard flour and milk powder to control common bunt (*Tilletia tritici*) in wheat and stem smut (*Urocystis occulta*) in rye in organic agriculture. In: *Seed Treatment: Challenges and Opportunities* (ed.) A. Biddle, British Crop Protection Council, Farkham, U.K.

Boysen, M.E., Bjornehoi, M.S. and Schnürer, J. 2000. Effect of the biocontrol yeast *Pichia anomala* on interactions between *Penicillium roquefortii*, *Penicillium carneum* and *Penicillium paneum* in moist grain under restricted air supply. *Postharvest Biology and Technology*, **19**: 173–179.

Breidt, F. and Fleming, H.P. 1997. Using lactic acid bacteria to improve the safety of minimally processed fruits and vegetables. *Food Technology*, **51**: 44–51.

Brown, G.E., Davis, C. and Chambers, M. 2000. Control of citrus green mold with Aspire is impacted by the type of injury. *Postharvest Biology and Technology*, **18**: 57–65.

Bryk, H., Dyki, B. and Sobiczewski, P. 1998. Antagonistic effect of *Erwinia herbicola* on *in vitro* spore germination and germ tube elongation of *Botrytis cinerea* and *Penicillium expansum*. *Biocontrol*, **43**: 97–106.

Bryk, H., Sobiczewski, P. and Berczyński, S. 1999. Evaluation of protective activity of epiphytic bacteria against gray mold (*Botrytis cinerea*) and blue mold (*Penicillium expansum*). *Phytopathologia Polonica*, **18**: 69–79.

Budge, S.P. and Whipps, J.M. 1991. Glasshouse trials of *Coniothyrium minitans* and *Trichoderma* species for the biological control of *Sclerotinia sclerotiorum* in celery and lettuce. *Plant Pathology*, **40**: 59–66.

Bull, C.T., Stack, J.P. and Smilanick, J.L. 1997. *Pseudomonas syringae* strains ESC-10 and ESC-11 survive in wounds on citrus and control of green and blue molds of citrus. *Biological Control*, **8**: 81–88.

Bull, C.T., Wadsworth, M.L., Pogge, T.D., Le, T.T., Wallace, S.K. and Smilanick, J.L. 1998. Molecular investigations into mechanisms in the biological control of postharvest diseases of citrus. *Bulletin OILB/SROP*, **21**: 1–6.

Burkhead, K.D., Bothast, R.J., Slininger, P.J. and Schister, D.A. 2003. Postharvest biological control of potato sprouting by *Fusarium* dry rot suppressive bacteria. *Biocontrol Science and Technology*, **13**: 477–494.

Burkhead, K.D., Schisler, D.A. and Slininger, P.J. 1994. Pyrrolnitrin production by biological control agent *Pseudomonas cepacia* B37w in culture and in colonized wounds of potatoes. *Applied and Environmental Microbiology*, **60**: 2031–2039.

Burr, T.J. and Caesar, A. 1984. Beneficial plant bacteria. *CRC Critical Reviews of Plant Science*, **2**: 1–20.

Burr, T.J. and Schroth, M.N. 1977. Inhibition of soft-rot of potato by antagonistic *Pseudomonas* spp. *Proceedings of American Phytopathological Society*, **4**: 158 (Abst.).

Caccinoni, D.R.L. and Guizzardi, M. 1994. Inhibition of germination and growth of fruit and vegetable postharvest pathogenic fungi by essential oil components. *Journal of Essential Oil Research*, **6**: 173–179.

Caccioni, D.R.L., Tonini, G. and Guizzardi, M. 1994. Postharvest treatments with acetaldehyde vapors for *Monilinia laxa* (Aderh. And Ruhl) Honey control in stone fruits. *Proceedings of Environmental Biotic Factors in Integrated Plant Disease Control. Poznan, Poland*, pp. 185–187.

Caccioni, D.R.L., Tonini, G. and Guizzardi, M. 1995. Antifungal activity of stone fruit aroma compounds against *Monilinia laxa* (Aderh et Ruhl) Honey and *Rhizopus stolonifer* (Ehrenb.) *in vivo* trials. *Journal of Plant Disease and Protection*, **102**: 518–525.

Calvente, V., Benuzzi, D. and Tosetti, M.I.S., de. 1999. Antagonistic action of siderophores from *Rhodotorula glutinis* upon the postharvest pathogen *Penicilllium expansum*. *International Biodeterioration and Biodegradation*, **43**: 167–172.

Castoria, R., Curtis, F. de, Lima, G., Caputo, L., Pacifico, S. and Cicco, V. de. 2001. *Aureobasidium pullulans* (LS-30), an antagonist of postharvest pathogens of fruits: study on its modes of action. *Postharvest Biology and Technology*, **32**: 717.

Chalutz, E. and Droby, S. 1998. Biological control of postharvest disease. In: *Plant–Microbe Interactions and Biological Control* (eds) G.J. Boland and L.D. Kuykendall, pp. 157–170, Marcel Dekker, Inc., New York.

Chalutz, E. and Wilson, C.L. 1990. Biocontrol of green and blue mold and sour rot of citrus by *Debaromyces hansenii*. *Plant Disease*, **74**: 134.

Chand-Goyal, T., Eckert, J.W., Droby, S., Glickmann, E. and Atkinson, K. 1999. Transformation of *Candida oleophila* and survival of a transformant on orange fruit under field conditions. *Current Genetics*, **35**: 51–57.

Chu, C.L., Liu, W.T., Zhou, T. and Tsao, R. 1999. Control of postharvest gray mold rot of modified atmosphere packaged sweet cherries by fumigation with thymol and acetic acid. *Canadian Journal of Plant Science*, **79**: 685–689.

Chu, C.L., Liu, W.T. and Zhou, T. 2001. Fumigation of sweet cherries with low levels of thymol and acetic acid to reduce postharvest brown rot and blue mold rot. *Fruits (Paris)*, **56**: 123–130.

Chun, W.W.C. and Shetty, K.K. 1994. Control of silver-scurf disease of potatoes caused by *Helminthosporium solani* Dur. and Mont. with *Pseudomonas corrugata*. *Phytopathology*, **84**: 1090 (Abst.).

Ciampi-Panno, L., Fernandez, C., Bustamante, P., Andrade, N., Ojeda, S. and Contreras, A. 1989. Biological control of bacterial wilt of potatoes caused by *Pseudomonas solanacearum*. *American Potato Journal*, **66**: 315–332.

Clarkson, J.P., Payne, T., Mead, A. and Whipps, J.M. 2002. Selection of fungal biological control agents of *Sclerotium cepivorum* for control of white rot by sclerotial degradation in a UK soil. *Plant Pathology*, **51**: 735–745.

Cook, D.W.M. 2002. Effect of formulated yeast in suppressing the liberation of *Botrytis cinerea* conidia. *Plant Disease*, **86**: 1265–1270.

Cook, R.J. 1993. Making greater use of introduced microorganisms for biological control of plant pathogens. *Annual Review of Phytopathology*, **31**: 53–80.

Cook, R.J. and Baker, K.F. 1983. *The Nature and Practice of Biological Control of Plant Pathogens*. The American Phytopathological Society, St. Paul, MN, USA.

Costa, E., Teixidó, N., Usall, J., Atarés, E. and Viñas, I. 2001. Production of the biocontrol agent *Pantoea agglomerans* strains CPA-2 using commercial products and by products. *Applied Microbiology and Biotechnology*, **56**: 367–371.

Costa, E., Usall, J., Teixidó, N., García, N. and Viñas, I. 2000. Effect of protective agents, rehydration media and initial cell concentration on viability of *Pantoea agglomerans* strain CPA-2 subjected to freeze drying. *Journal of Applied Microbiology*, **89**: 793–800.

Costa, E., Usall, J., Teixidó, N., Torres, R. and Viñas, I. 2002. Effect of package and storage conditions on viability and efficacy of freeze-dried biocontrol agent *Pantoea agglomerans* strain CPA-2. *Journal of Applied Microbiology*, **92**: 873–878.

Coste, D.M. de and Subasinghe, S.S.N.S. 1999. Antagonistic bacteria associated with the fruit skin of banana in controlling its postharvest diseases. *Tropical Science*, **38**: 206–212.

Coyler, P.D. and Mount, M.S. 1984. Bacterization of potatoes with *Pseudomonas putida* and its influence on postharvest soft rot diseases. *Plant Disease*, **68**: 703–706.

Daoust, R.A. and Hofstein, R. 1996. *Ampleomyces quisqualis* a new biofungicide to control powdery mildew in grapes. *Brighton Crop Protection Conference: Pests and Diseases*, **1**: 33–40.

Deak, T. and Beuchat, L. 1996. *Handbook of Food Spoilage Yeasts.* CRC Press, New York.

De Clercq, D., Cognet, S., Pujol, M., Lepoivre, L. and Jijakli, M.H. 2003. Development of a SCAR marker and a semi-selective medium for specific quantification of *Pichia anomala* strain K on apple fruit surfaces. *Postharvest Biology and Technology*, **29**: 237–247.

De la Cruz, A.R., Poplawsky, A.R. and Wiese, M.V. 1992. Biological suppression of potato ringrot by fluorescent pseudomonads. *Applied Environmental Microbiology*, **58**: 1986–1991.

Dennis, C. 1975. Effect of preharvest fungicides on the spoilage of soft fruit after harvest. *Annals of Applied Biology*, **81**: 227–234.

Dennis, C. (ed.) 1983. *Post-Harvest Pathology of Fruits and Vegetables.* Academic Press, London.

Desjardins, A.E., Christ-Harned, E.A., McCormick, S.P. and Secor, G.A. 1993. Population structure and genetic analysis of field resistance to thiabendazole in *Gibberella pulicaris* from potato tubers. *Phytopathology*, **83**: 164–170.

Desjardins, A.E., Gardner, H.W. and Wettring, K.M. 1992. Detoxification of sesquiterpene phytoalexins by *Gibberella pulicaris* (*Fusarium sambucinum*) and its importance for virulence on potato tubers. *Journal of Industrial Microbiology*, **9**: 201–211.

Dharam Singh and Maheswari, V.K. 2001. Biological seed treatment for the control of loose smut of wheat. *Indian Phytopathology*, **54**: 457–460.

Diaz, A. and Marina Cotes, A. 2001. Design of a culture method for mass production of yeats with potential for biological control of postharvest pathogens. *ASCOLFI Informa*, **27**: 35–37.

Dorner, J.W. and Cole, R.J. 2002. Effect of application of nontoxigenic strains of *Aspergillus flavus* and *Aspergillus parasiticus* on subsequent aflatoxin contamination of peanuts in storage. *Journal of Stored Products Research*, **38**: 329–339.

Droby, S. and Chalutz, E. 1994. Mode of action of biocontrol agents of postharvest diseases. In: *Biological Control of Postharvest Diseases–Theory and Practice* (eds) C.L. Wilson and M. Wisniewski, pp. 63, CRC Press, Boca Raton, FL, USA.

Droby, S., Chalutz, E., Wilson, C. and Wisniewski, M. 1989. Characterization of the biocontrol activity of *Debaryomyces hansenii* in the control of *Penicillium digitatum* on grapefruit. *Canadian Journal of Microbiology*, **35**: 794–800.

Droby, S., Cohen, L., Daus, A., Weiss, B., Horev, B., Chalutz, E., Kalz, H., Keren-Tzun, M. and Shachnai, A. 1998. Commercial testing of Aspire: a yeast preparation for the biological control of postharvest decays of citrus. *Biological Control*, **12**: 97–101.

Droby, S., Hofstein, R., Wilson, C.L., Wisniewski, M.E., Fridlender, B., Cohen, L., Weiss, B., Timar, D. and Chalutz, E. 1993. Pilot testing of *Pichia guilliermondii*: A biocontrol agent of postharvest diseases of citrus fruit. *Biological Control*, **3**: 47.

Droby, S., Robin, D., Chalutz, E. and Chet, I. 1993. Possible role of glucanase and extracellular polymers in the mode of action of yeast antagonist of postharvest diseases. *Phytoparasitica*, **21**: 167 (Abst.).

Duncan, R. 1991. *Biological Control of Botrytis cinerea on Kiwifruit Through Field Antagonistic Microorganisms*. MS Thesis California State University, Fresno.

Eayre, C.G., Bartz, J.A. and Concelmo, D.E. 1995. Bacteriophages of *Erwinia ananas* from fresh water lakes. *Plant Disease*, **79**: 199–206.

Elkovich, S.D. 1988. Terpenoids as models for new agrochemicals. *ACS Symposium Series*, **380**: 2580–2610.

El Ghaouth, A., Smilanick, J.L., Brown, G.E., Ippolito, A. and Wilson, C.L. 2001. Control of decay of apple and citrus fruits in semicommercial tests with *Candida saitoana* and 2-deoxy-D-glucose. *Biological Control*, **20**: 96–101.

El Ghaouth, A., Smilanick, J.L. Wisniewski, M. and Wilson, C.L. 2000. Improved control of apple and citrus fruit with a combination of *Candida saitoana* and 2-deoxy-D-glucose. *Plant Disease*, **84**: 249–253.

El Ghaouth, A., Wilson, C.L. and Wisniewski, M. 2003. Control of postharvest decay of apple fruit with *Candida saitoana* and induction of defense responses. *Phytopathology*, **93**: 344–348.

El-Mohamedy, R.S.R., Abd-El-Kareem, F. and Abd-Alla, M.A. 2002. Effect of some constitutents of citrus essential oil against postharvest pathogenic fungi of citrus fruits. *Arab Universities Journal of Agricultural Sciences*, **10**: 335–350.

Errampalli, D. and Crnko, N. 2004. Control of blue mold caused by *Penicillium expansum* on apples 'Empire' with fludioxonil and cyprodinil. *Canadian Journal of Plant Pathology*, **26**: 70–75.

El-Neshawy, S. and Shetaia, Y.M.H. 2003. Biocontrol capability of *Candida* spp. against *Botrytis* rot of strawberries with respect to fruit quality. *Acta Horticulturae*, **604**.

Falk, S.P., Pearson, R.C., Gadoury, D.M., Seem, R.C. and Sztenberg, A. 1996. *Fusarium proliferatum* as a biocontrol agent aginst grape downy mildew. *Phytopathology*, **86**: 1010–1017.

Fan, Q. and Tian, S.P. 2001. Postharvest biological control of gray mold and blue mold on apple by *Cryptococcus albidus* (Saito) Skinner. *Postharvest Biology and Technology*, **21**: 341–350.

Fan, Q., Tian, S.P., Li, Y.X., Wang, Y., Xu, Y. and Li, J.D. 2000. Postharvest biological control of green mold and blue mold of citrus fruits by *Bacillus subtilis*. *Acta Phytopathologica Sinica*, **30**: 343–348.

Fenwick, G.R., Heaney, R.K. and Mullin, W.J. 1983. Glucosinolates and their breakdown products in food and food plants. *Critical Reviews of Food Science and Nutrition*, **18**: 123–201.

Filonow, A.B. 1998. Role of competition for sugars by yeasts in the biocontrol of gray mold of apple. *Biocontrol Science and Technology*, **8**: 243–256.

Filonow, A.B., Vishniac, H.S., Anderson, J.A. and Janisiewicz, W.J. 1996. Biological control of *Botrytis cinerea* in apple by yeasts from various habitats and their putative mechanisms of antagonism. *Biological Control*, **7**: 212.

Floros, J.D., Dock, L.L. and Nielsen, P.V. 1998. Biological control of *Botrytis cinerea* growth on apples stored under modified atmospheres. *Journal of Food Protection*, **61**: 1661–1665.

Fravel, D. 2002. Commercial biocontrol products available for use against plant pathogens. *APS Biological Control Committee*, St. Paul, MN, USA, p 9.

Freeman, S., Minz, O., Kolesnik, I., Barbul, O., Zveibil, A., Maymon, M., Nitzani, Y., Kirshner, B., Rav-David, D., Bilu, A., Dag, A., Shafir, S. and Elad, Y. 2004. *Trichoderma* biocontrol of *Colletotrichum acutatum* and *Botrytis cinerea* and survival in strawberry. *European Journal of Plant Pathology*, **110**: 361–370.

French, R.C. 1985. The bioregulatory action of flavour compounds on fungal spores and other propagules. *Annual Review of Phytopathology*, **23**: 173–199.

Fukui, R., Fukui, H. and Alvarez, A.M. 1999. Comparison of single versus multiple bacteria on biological control of anthurium blight. *Phytopathology*, **89**: 366–373.

Goyal, T.C. and Spotts, R.A. 1997. Biological control of postharvest diseases of apple and pear under semi-commercial conditions using three saprophytic yeasts. *Biological Control*, **10**: 199–206.

Grevesse, C. Jijakli, M.H., Duterme, O., Colinet, D. and Lepoivre, P. 1998a. Preliminary study of exo-β-1,3-glucanase encoding genes in relation to the protective activity of *Pichia anomala* (strain K) against *Botrytis cinerea* in postharvest apples. *Bulletin OILB/SROP*, **21**: 81–89.

Grevesse, C., Jijakli, M.H. and Lepoivre, P. 1998b. Study of exo-β-1,3-glucanase activity production by the yeast *Pichia anomala* in relation to its antagonistic properties against *Botrytis cinerea. Mededelingen–Faculteit Landbouwkundige en Toegepaste Biologische Wetenschappen, Universitiet Gent*, **63**: 1685–1692.

Grevesse, C., Jijakli, M.H. and Lepoivre, P. 1998b. Study of the exo-β-1,3-glucanase activity production by yeast *Pichia anomala* in relation to its antagonistic properties against *Botrytis cinerea* on postharvest apples. *Meded. Fac. Landbouwwet–Rijksuni V. Gent*, **63**: 1682–1685.

Grevesse, C., Lepoivre, P. and Jijakli, M.H. 2003. Characterization of the exoglucanase-encoding gene *PaEXG2* and study of its role in the biocontrol activity of *Pichia anomala* strain K. *Phytopathology*, **93**: 1145–1152.

Gueldner, R.C., Reilly, C.C., Pusey, P.L., Costello, C.E., Arrendale, R.F. et al. 1988. Isolation and identification of iturins as antifungal peptides in biological control of peach brown rot with *Bacillus subtilis. Journal of Agricultural and Food Chemistry*, **36**: 366–370.

Guinebretiere, M.H., Nguyen-The., C., Morrison, N., Reich, M. and Nicot, P. 2000. Isolation and characterization of antagonists for the biocontrol of the postahrvest wound pathogen *Botrytis cinerea* on strawberry fruits. *Journal of Food Protection*, **63**: 386–394.

Guizzardi, M., Elad, Y. and Mari, M. 1995. Treatments against postharvest fruit diseases using Trichodex (*Trichoderma harzianum*). *Proceedings of 5th International Trichoderma / Gliocladium Workshop*, Beltsville, MD, USA, p. 39.

Hanlin, J.H. and Evancho, G.M. 1992. The beneficial role of microorganism in the safety and stability of refrigerated foods. In: *Chilled Food: A Comprehensive Guide* (eds) C. Dennis and M. Stringer, pp. 229–259, Ellis Horwood Ltd., Chichester, UK.

He, D., Zheng, X.D., Yin, Y-M., Sun, P. and Zhang, H.Y. 2003. Yeast application for controlling apple postharvest diseases associated with *Penicillium expansum*. *Bulletin of Botanica Academia Sinica*, **43**: 211–216.

Helbig, J. 2001. Biological control of *Botrytis cinerea* Pers. Ek Fr. in strawberry by *Paenibacillus polymyxa* (isolate 18191). *Journal of Phytopathology*, **149**: 265.

Helbig, J. 2002. Ability of the antagonistic yeast *Cryptococcus albidus* to control *Botrytis cinerea* in strawberry. *Biocontrol*, **47**: 85–89.

Hörberg, H. 1998. Influence on volatile plant extracts on storage pathogens of carrots *in vitro*. *Växtskyddsnotiser*, **62**: 87–89.

Ippolito, A., El Ghaouth, A., Wisniewski, M. and Wilson, C. 2000. Control of postharvest decay of applefruit by *Aureobasidium pullulans* and induction of defense responses. *Postharvest Biology and Technology*, **19**: 265–272.

Ippolito, A., Nigro, F., Romanazzi, G. and Campanella, V. 1998. Field application of *Aureobasidium pullulans* against *Botrytis cinerea* storage rot of strawberry. In: *Non-Conventional Methods for the Control of Postharvest Disease and Microbial Spoilage* (eds) B. Bertolini, P.C. Sijmons, M.E. Guerzoni and F. Serra, pp. 127–133, European Community, Luxembourg.

Jager, G., Hoopen, A. and Velvis, H. 1979. Hyperparasities of *Rhizoctonia solani* in Dutch potato fields. *Netherlands Journal of Plant Pathology*, **85**: 253–268.

Jager, G. and Velvis, H. 1986. Biological control of *Rhizoctonia solani* on potatoes by antagonists: The effectiveness of three isolates of *Verticillium biguttatum* as inoculum for seed tubers and of a soil treatment with a low dosage of pencycuron. *Netherlands Journal of Plant Pathology*, **92**: 231–238.

Janisiewicz, W.J. 1987. Postharvest biological control of blue mold on apples. *Phytopathology*, **77**: 481.

Janisiewicz, W.J. 1988. Biocontrol of postharvest diseases of apples with antagonist mixtures. *Phytopathology*, **78**: 194–198.

Janisiewicz, W.J. 1991. Control of postharvest diseases of fruits with biocontrol agents. In: *The Biological Control of Plant Diseases* (ed.) J. Bay-Petersen, FFTC Book Series No. 42, pp. 56–68, Food and Fertilizer Technology Center for the Asian and Pacific Region, Taipei, Taiwan.

Janisiewicz, W.J. 1994a. Enhancement of biocontrol of blue mold with the nutrient analog 2-deoxy-D-glucose on apples and pears. *Applied Environmental Microbiology*, **68**: 2671–2676.

Janisiewicz, W.J. 1994b. Nutritional characterization of natural microflora of apple and its possible role in selection of biocontrol agents. *Phytopathology*, **84**: 1152.

Janisiewicz, W.J. 1995. Development of a microbial community of bacterial and yeast antagonists to control wound-invading postharvest pathogens of fruits. *Applied Environmental Microbiology*, **61**: 3261–3267.

Janisiewicz, W.J. 1996. Ecological diversity, niche overlap and coexistence of antagonists used in developing mixtures for biocontrol of postharvest diseases of apples. *Phytopathology*, **86**: 473–479.

Janisiewicz, W.J. 1998. Biocontrol of postharvest diseases of temperate fruits-challenges and opportunities. In: *Plant–Microbe Interactions and Biological Control* (eds.) G.J. Boland and L.D. Kuykendall, pp. 171–197, Marcel Dekker, Inc., New York.

Janisiewicz, W.J. and Bors, B. 1995. Development of a microbial community of bacterial and yeast antagonists to control wound-invading postharvest pathogens of fruits. *Applied and Environmental Microbiology*, **61**: 3261–3267.

Janisiewicz, W.J., Conway, W.S. and Leverentz, B. 1999. Biological control of postharvest decays of apple can prevent growth of *Escherichia coli* 0157:H7 in apple wounds. *Journal of Food Protection*, **62**: 1372–1375.

Janisiewicz, W.J. and Jeffers, S.N. 1997. Efficacy of commercial formulation of two biofungicides for control of blue mold and gray mold of apples in cold storage. *Crop Protection*, **16**: 629–633.

Janisiewicz, W.J. and Korsten, L. 2002a. Biological control of postharvest diseases of fruits. *Annual Review of Phytopathology*, **40**: 411–441.

Janisiewicz, W.J. and Korsten, L. 2002b. Microbial control of postharvest disease and spoilage. In: *Postharvest Physiology and Pathology of Vegetables* (eds) J.A. Bartz and J.K. Brecht, pp. 543–562, Marcel Dekker, Inc., New York.

Janisiewicz, W.J. and Marchi, A. 1992. Control of storage rots on various pear cultivars with saprophytic strain of *Pseudomonas syringae*. *Plant Disease*, **76**: 555–560.

Janisiewicz, W.J. and Peterson, D.L. 2004. Susceptibility of the stempull area of mechanically harvested apples to blue mold decay and its control with a biocontrol agent. *Plant Disease*, **88**: 662–664.

Janisiewicz, W.J. and Roitman, J. 1988. Biological control of blue mold and gray mold on apple and pear with *Pseudomonas cepacia*. *Phytopathology*, **78**: 1697–1700.

Janisiewicz, W.J., Tworkoski, T.J. and Kurtzman, C.P. 2001. Biocontrol potential of *Metchnikowia pulcherrina* strains against blue mold of apple. *Phytopathology*, **91**: 1098–1108.

Janisiewicz, W.J., Tworkoski, T.J. and Sharer, C. 2000. Characterizing the mechanism of biological control of postharvest diseases on fruits with a simple method to study competition for nutrients. *Phytopathology*, **90**: 1196–1200.

Janisiewicz, W.J., Usall, J. and Bors, B. 1992. Nutritional enhancement of biocontrol of blue mold on apples. *Phytopathology*, **82**: 1364–1370.

Jensen, B., Knudsen, I.M.B., Madsen, M. and Jensen, D.F. 2004. Biopriming of infected carrot seed with an antagonist, *Clonostachys rosea* selected for control of seedborne *Alternaria* spp. *Phytopathology*, **94**: 551–560.

Jijakli, M.H. 2000. Apples: Storage diseases – biological control based on two yeast strains. *Arboriculture Fruitiére*, **539**: 19–23.

Jijakli, M.H. and Lepoivre, P. 1998. Characterization of an exo-β-1,-3-glucanase produced by *Pichia anomala* strain K an antagonist of *Botrytis cinerea* on apples. *Phytopathology*, **88**: 335–343.

Jobling, J. 2000. Essential oils: a new idea for postharvest disease control. *Good Fruit and Vegetables*, **11**: 50.

Jones, D., Gordon, A.H. and Bacon, J.S.D. 1974. Cooperative action by endo- and exo-b (1,3)-glucanases from parasitic fungi in the degradation of cell wall glucans of *Sclerotinia sclerotiorum* (Lib.) de Bary. *Biochemical Journal*, **140**: 47–55.

Jones, D. and Watson, D. 1969. Parasitism and lysis by soil fungi of *Sclerotinia sclerotiorum* (Lib.) de Bary, a phytopathogenic fungus. *Nature*, **224**: 287–288.

Jones, K.A. and Burges, H.D. 1998. Technology of formulation and application. In: *Formulations of Microbial Biopesticides* (eds) H.D. Burges, pp. 7–30, Kluwer, Boston.

Jones, R.W. and Prusky, D. 2002. Expression of an antifungal peptide in *Saccharomyces:* a new approach for biological control of the postharvest disease caused by *Colletotrichum coccodes. Phytopathology*, **92**: 33–37.

Karabulut, O.A., Tezcan, H., Daus, A., Cohen, L., Wiess, B. and Droby, S. 2004. Control of preharvest and postharvest fruit rot in strawberry by *Metschnikowia fructicola. Biocontrol Science and Technology*, **14**: 513–521.

Kempe, J. and Sequeira, L. 1983. Biological control of bacterial wilt of potatoes: Attempts to induce resistance by treating tubers with bacteria. *Plant Disease*, **67**: 499–503.

Kenwick, S. and Jacobsen, B.J. 1998. Biological control of *Fusarium* dry rot on potato with antagonistic bacteria in commercial formulation. *Phytopathology*, **88**: S47.

Kexiang, G., Xiaoguang, L., Yonghong, L., Tianbo, Z. and Shuliang, W. 2002. Potential of *Trichoderma harzianum* and *T. atroviride* to control *Botryosphaeria berengeriana* f.sp. *piricola*, the cause of apple ring rot. *Journal of Phytopathology*, **150**: 271–276.

Kinay, P., Yildiz, M., Droby, S., Yildiz, F., Cohen, L. and Weiss, B. 1998. Evaluation of antagonistic activity of epiphytic yeasts against rot pathogens of mandarin orange and grapefruit. In: *Molecular Approaches in Biological Control* (eds) B. Duffy, U. Rosenberger and G. Defago, Delmont, Switzerland. IOBC/WPRS Bulletin OILB/SROP, **21**: 291–296.

Kloepper, J.W. 1983. Effect of seed piece inoculation with plant growth-promoting rhizobacteria on populations of *Erwinia carotovora* potato roots and daughter tubers. *Phytopathology*, **73**: 217–219.

Köhl, J.W., Molhoek, M.L. and Fokkema, N.J. 1991. Biological control of onion neck rot (*Botrytis aclada*): Protection of wounds made by leaf topping. *Biocontrol Science and Technology*, **7**: 261–269.

Kulakiotu, E.K., Thanassoulopoulos, C.C. and Sfakiotakis, E.M. 2004a. Biological control of *Botrytis cinerea* by volatiles of 'Isabella' grapes. *Phytopathology*, **94**: 924–931.

Kulakiotu, E.K., Thanassoulopoulos, C.C. and Sfakiotakis, E.M. 2004b. Postharvest biological control of *Botrytis cinerea* on kiwifruit by volatiles of 'Isabella' Grapes. *Phytopathology*, **94**: 1280–1285.

Larena, I., De Cal, A., Liñán, M. and Melgarejo, P. 2003. Drying of *Epicoccum nigrum* conidia for obtaining a shelf-stable biological product against brown rot disease. *Journal of Applied Microbiology*, **94**: 508–514.

Lazarovits, G., Hill, J., King, R.R. and Calhoun, L.A. 2004. Biotransformation of the *Streptomyces scabies* phytotoxin thaxtomin A by the fungus *Aspergillus niger. Canadian Journal of Microbiology*, **50**: 121–126.

Lee, J.T., Bae, D.W., Park, S.H., Shim, C.K., Kawk, Y.S. and Kim, H.K. 2001. Occurrence and biological control of postharvest decay in onion caused by fungi. *Plant Pathology Journal*, **17**: 141–148.

Leibinger, W., Breuker, B., Hahn, M. and Mengden, K. 1997. Control of postharvest pathogens and colonization of the apple surface by antagonistic microorganisms in the field. *Phytopathology*, **87**: 1103–1110.

Leifert, C., Epton, H.A.S., Knight, C., Poole, N., Rogerson, M., Sigee, D.C. and Stanley, R. 1993a. Reduction of losses due to fungal rotting during cold storage of Dutch White cabbages by postharvest treatment with *Pseudomonas serratia* antagonists. In: *Biological Control of Foliar and Postharvest Diseases* (eds) N.J. Fokkema, J. Köhl and Y. Elad, pp. 127–130, IOBC / WPRS Bulletin No. 16(11), IOBC / WPRS, France.

Leifert, C., Sigee, D.C., Stanley, R., Knight, C. and Epton, H.A.S. 1993b. Biocontrol of *Botrytis cinerea* and *Alternaria brassicola* on Dutch White cabbage by bacterial antagonists at cold store temperatures. *Plant Pathology*, **42**: 270–279.

Leverentz, B., Janisiewicz, W. and Conway, W.S. 2003. Biological control of minimally processed fruits and vegetables. In: *Microbial Safety of Minimally Processed Foods*, pp. 319–332, CRC Press LLC, FL, USA.

Leverentz, B., Conway, W.S., Alavidze, Z., Janisiewicz, W.J., Fuchs, Camp, M.J., Chighladze, E. and Sulakvelidze, A. 2001. Examination of bacteriophage as a biocontrol method for *Salmonella* on fresh-cut fruit: A model study. *Journal of Food Protection*, **64**: 1116–1121.

Lewis, J.A. and Papavizas, G.C. 1980. Integrated control of *Rhizoctonia* fruit rot of cucumber. *Phytopathology*, **70**: 85–89.

Li, G.Q., Huang, H.C., Acharya, S.N. and Erickson, R.S. 2004. Biological control of blossom blight of alfalfa caused by *Botrytis cinerea* under environmentally controlled and field conditions. *Plant Disease*, **88**: 1246–1251.

Li, H.Y. and Yu, T. 2001. Effect of chitosan on incidence of brown rot, quality and physiological attributes of postharvest peach fruit. *Journal of the Science Food and Agriculture*, **81**: 269–274.

Lima, G., Curtis, F. de, Castoria, R. and Cicco, V. de. 1998. Activity of the yeasts *Cryptococcus lauerentii* and *Rhodotorula glutinis* against postharvest rots on different fruits. *Biocontrol Science and Technology*, **8**: 257–267.

Lima, G., Arru, S., Curtis, F. de and Arras, G. 1999. Influence of antagonist, host fruit and pathogen on the biological control of postharvest fungal diseases by yeasts. *Journal of Industrial and Biotechnology*, **23**: 223–229.

Lima, G., Ippolito, A. Nigro, F. and Salerno, M. 1997. Biological control of gray mold of stored table grapes by preharvest applications of *Aureobasidium pullulans* and *Candida oleophila*. *Difesa delle Piante*, **20**: 21–28.

Limón, M.C., Chacón, M.R., Mejías, R., Delgado-Jarana, J., Rincón, A.M., Cadón, A.C. and Benítz, T. 2004. Increased antifungal and chitinase specific activities of *Trichoderma harzianum* CECT 2413 by addition of a cellulose binding domain. *Applied Microbiology and Biotechnology*, **64**: 675–685.

Liu, D. Anderson, N.A. and Kinkel, L.L. 1995. Biological control of potato scab in the field with antagonistic *Streptomyces scabies*. *Phytopathology*, **85**: 827–831.

Liu, W.T., Chu, C.L. and Zhou, T. 2002. Thymol and acetic acid vapors reduce postharvest brown rot of apricots and plums. *Hort Science*, **37**: 151–156.

Lucon, C.M.M. and Melo, I.S. 2000. Effect of seed bacterization on the development of maize plants and *Fusarium moniliforme* control. *Fitopatologia Brasileira*, **25**: 529–537.

Mari, M. and Guizzardi, M. 1998. The postharvest phase: Emerging technologies for the control of fungal diseases. *Phytoparasitica*, **26**: 59–66.

Martin, C. and French, E.R. 1985. *Bacterial Wilt of Potato, Pseudomonas solanacearum. Technical Information Bulletin*, No. 13, International Potato Center, Lima, Peru, 16 pp.

Massomo, B.M.S., Mortensen, C.N., Mabagala, R.B., Newman, M.A. and Hockenhull, J. 2003. Biological control of blackrot (*Xanthomonas campestris* pv. *campestris*) of cabbage in Tanzania with *Bacillus* strains. *Journal of Phytopathology*, **151**: 98–105.

Mattheis, J.P. and Roberts, R.G. 1993. Fumigation of sweet cherry (*Prunus avium* 'Bing') fruit with low molecular weight aldehydes for postharvest decay control. *Plant Disease*, **77**: 810–814.

McGuire, R.G. 2000. Population dynamics of postharvest decay antagonists growing epiphytically and within wounds on grapefruit. *Phytopathology*, **90**: 1217–1223.

McKenna, F., El-Tarabily, K.A., Hardy, G.E.ST.J. and Dell, B. 2001. Novel *in vivo* use of a polyvalent *Streptomyces* phage to disinfest *Streptomyces scabies* infected seed potatoes. *Plant Pathology*, **50**: 666.

McLaughlin, R.J., Sequeira, L. and Weingartner, D.P. 1990. Biocontrol of bacterial wilt of potato with an avirulent strain of *Pseudomonas solanacearum*: Interaction with root-knot nematodes. *American Potato Journal*, **67**: 93–107.

Mead, P.S., Slutsker, L., Dietz, V., McCaig, L.F., Bresee, J.S., Shapiro, C., Griffin, P.M. and Tauxe, R.V. 1999. Food-related illness and death in the United States. *Emerging Infectious Disease*, **5**: 607–625.

Mehrotra, N.K., Neeta Sharma, Nigam, M. and Ghosh, R. 1998. Biological control of sour rot of citrus fruits by yeast. *Proceedings of the National Academy of Sciences India, Section B, Biological Sciences*, **68**: 135–139.

Mercier, J. and Jiménez, J.I. 2004. Control of fungal decay of apples and peaches by the biofumigant fungus *Muscodor albus. Postharvest Biology and Technology*, **31**: 1–8.

Mercier, J. and Wilson, C.L. 1994. Colonization of apple wounds of naturally occurring microflora and introduced *Candida oleophila* and their effect on infection by *Botrytis cinerea* during storage. *Biological control*, **4**: 138–144.

Meshram, S.U. 1984. Suppressive effect of *Azotobacter chroococcum* on *Rhizoctonia solani* infestation of potatoes. *Netherlands Journal of Plant Pathology*, **90**: 127–132.

Mew, I.C. and Kommedahl, T. 1968. Biological control of seedling blight of corn by coating kernels with antagonistic microorganisms. *Phytopathology*, **58**: 1395.

Mew, I.C. and Kommedahl, T. 1972. Interaction among microorganisms occurring naturally and applied to pericarps of corn kernels. *Plant Disease Reporter*, **56**: 861.

Michaud, M., Martinez, C., Bélanger, R.R. and Tweddell, R.J. 2002. Selection of antagonist against *Helminthosporium solani*, causal agent of potato silver scurf. *Plant Disease*, **86**: 717–720.

Miller, T.C., Gubler, W.D., Laemmlen, F.F., Geng, S. and Rizzo, D.M. 2004. Potential for using *Lecanicllium lecanii* for suppression of strawberry powdery mildew. *Biocontrol Science and Technology*, **14**: 215–220.

Nielsen, T.H., Thrane, C., Christophersen, C. Anthoni, U. and Sørensen, J. 2000. Structure, production characteristics and fungal antagonism of tensin – a new fungal cyclic lipopetide from *Pseudomonas fluroescens* strain 96–578. *Journal of Applied Microbiology*, **89**: 992–1001.

Northover, J. and Zhou, T. 2003. Control of *Rhizopus* rot of peaches with postharvest treatments of tebuconazole, fludioxonil and *Pseudomonas syringae*. *Canadian Journal of Plant Pathology*, **25**: 144–153.

Nunes, C., Teixidó, N., Usall, J. and Viñas, I. 2001a. Biological control of major postharvest diseases on pear fruits with antagonistic bacterium *Pantoea agglomerans* (CPA-2). *Acta Horticulturae*, **553**: 403–404.

Nunes, C., Usall, J., Teixidó, N. Abadias, M. and Viñas, I. 2002a. Improved control of postharvest decay of pears by the combination of *Candida sake* (CPA-1) and ammonium molybdate. *Phytopathology*, **92**: 281–287.

Nunes, C., Usall, J., Teixidó, N., Fons, E. and Viñas, I. 2002b. Postharvest biological control by *Pantoea agglomerans* (CPA-2) on Golden Delicious apples. *Journal of Applied Microbiology*, **92**: 247–255.

Nunes, C., Usall, J., Teixidó N., Miró, M. and Viñas, I. 2001b. Nutritional enhancement of biocontrol activity of *Candida* sake (CPA-1) against *Penicillium expansum* on apples and pears. *European Journal of Plant Pathology*, **107**: 543–551.

Nunes, C., Usall, J., Teixidó N., Torres, R. and Viñas, I. 2002a. Control of *Penicllium expansum* and *Botrtis cinerea* on apples and pears with a combination of *Candida sake* and *Pantoea agglomerans*. *Journal of Food Protection*, **65**: 178–184.

Nunes, C., Usall, J., Teixidó, N. and Viñas, I. 2001c. Biological control of postharvest pear diseases using a bacterium *Pantoea agglomerans* CPA-2. *International Journal of Food Microbiology*, **70**: 53–61.

Nunes, C., Usall, J., Teixidó, N. and Viñas, I. 2002 c. Improvement of *Candida sake* biocontrol activity against postharvest decay by the addition of ammonium molybdate. *Journal of Applied Microbiology*, **92**: 927–935.

Obagwu, J. and Korsten, L. 2003. Control of citrus green and blue molds with garlic extracts. *European Journal of Plant Pathology*, **109**: 221–225.

Obradovic, A., Jones, J.B., Momol, M.T., Balogh, B. and Olson, S.M. 2004. Management of tomato bacterial spot in the field by foliar applications of bacteriophages and SAR inducers. *Plant Disease*, **88**: 736–740.

Pedras, M.S.C., Zaharia, I.L. and Ward, D.E. 2002. The destruxins: synthesis, biosynthesis, biotransformation and biological activity. *Phytochemistry*, **59**: 579–596.

Petersson, S., Jonsson, N. and Schnürer, J. 1999. *Pichia anomala* as a biocontrol agent during storage of high-moisture feed grain under airtight conditions. *Postharvest Biology and Technology*, **15**: 175–184.

Peterson, S. and Schnürer, J. 1998. *Pichia anomala* as a biocontrol agent of *Penicillium roquefortii* in high-moisture wheat, rye barley and oats stored under airtight conditions. *Canadian Journal of Microbiology*, **44**: 471–476.

Poppe, L., Vanhoulte, S. and Höfte, M. 2003. Mode of action of *Pantoea agglomerans* CPA-2, an antagonist of postharvest pathogens of fruits. *European Journal of Plant Pathology*, **109**: 963–973.

Prakasam, V., Abraham, S. and Kannan, C. 2001. Management of postharvest fungal diseases of carrot using botanicals. *South Indian Horticulture*, **49**: 271–274.

Pujol, M., De Clercq, D., Cognet, S., Lepoivre, P. and Jijakli, M.H. 2004. Monitoring system for the biocontrol agent *Pichia anomala* strain K using quantitative competitive PCR-ELISA. *Plant Pathology*, **53**: 103–109.

Pusey, P.L. and Wilson, C.L. 1984. Postharvest biological control of stone fruit brown rot by *Bacillus subtilis*. *Plant Disease*, **68**: 753.

Quin, G., Tian, S. and Xu, Y. 2004. Biocontrol of posthavest diseases on sweet cherries by four antagonistic yeasts in different storage conditions. *Postharvest Biology and Technology*, **31**: 51–58.

Qin, G.Z. and Tian, S.P. 2005. Enhancement of biocontrol activity of *Cryptococcus laurentii* by silicon and the possible mechanism involved. *Phytopathology*, **95**: 69–75.

Rajappan, K., Ushamalini, C., Subramanian, N., Narsimhan, V. and Abdul Kareem, A. 2001. Management of grain discoloration of rice with solvent-free EC formulations of neem and pungam oils. *Phytoparasitica*, **29**: 171–174.

Ranasinghe, L., Jayawardena, B. and Abeywickrama, K. 2002. Fungicidal acitivty of essential oils of *Cinnamomum zeylanicum* (L.) and *Syzygium aromaticum* (L.) Merr et L.M. Perry against crown rot and anthracnose pathogens isolated from banana. *Letters in Applied Microbiology*, **35**: 208–211.

Raupach, G.S. and Kloepper, J.W. 1998. Mixtures of plant growth-promoting rhizobacteria enhance biological control of multiple cucumber pathogens. *Phytopathology*, **88**: 1158–1164.

Ray, R.C. and Das, B.B. 1998. Yeasts as possible microbial antagonists of posthravest spoilage of tuber crops. *Journal of Root Crops*, **24**: 99–104.

Reddy, M.V.B., Angers, P., Castaigne, F. and Arul, J. 2000a. Chitosan effects on black mold rot and pathogenic factors produced by *Alternaria alternata* in postharvest tomatoes. *Journal of the American Society for Horticultural Science*, **125**: 742–747.

Reddy, M.V.B., Belkacemi, K., Corcuff, R., Castaigne, F. and Arul, J. 2000b. Effect of preharvest chitosan sprays on posthravest infections of *Botrytis cinerea* and quality of strawberry fruit. *Postharvest Biology and Technology*, **20**: 39–51.

Rendall-Dunn, A.J. 1991. General news. *Postharvest News and Information*, **2**: 3.

Reyes, M.E.Q., Rohrbach, K.G. and Paull, R.E. 2004. Microbial antagonists control postharvest blackrot of pineapple. *Postharvest Biology and Technology*, **33**: 193–203.

Rhodes, D.J. 1993. Formulations of biocontrol agents. In: *Exploitation of Microorganisms* (ed.) D.G. Jones, pp. 411–439, Chapman and Hall, London.

Ribeiro, L.F. and Bedendo, I.P. 1999. Inhibitory effect of plant extracts on *Colletotrichum gloeosporioides* – the causal agent of postharvest rot in papaya fruits. *Scientia Agricola*, **56** (4 supl.): 1267–1271.

Roberts, R.G. 1990. Postharvest biological control of gray mold of apple by *Cryptococcus lauerntii. Phytopathology*, **80**: 526–530.

Roberts, R.G. 1991. Characterization of postharvest biocontrol of deciduous fruit diseases by *Cryptococcas* spp. *Biological Control of Postharvest Diseases of Fruits and Vegetables*, USDA Publication, ARS-92, 32 pp.

Romanazzi, G., Nigro, R. and Ippolito, A. 2000. Effectiveness of pre-and post-harvest chitosan treatments on storage decay of strawberries. *Rivista di Frutticoltura edi Ortofloricoltura*, **62**: 71–75.

Ryan, D.A., Kinkel, L.L. and Schottel, J.L. 2004. Effect of pathogen isolate, potato cultivar and antagonist strain on potato scab severity and biological control. *Biocontrol Science and Technology*, **14**: 301–311.

Sabev, H.A., Handley, P.S. and Robson, G.D. 2004. *In situ* quantification of biocide efficacy using GFP transformed *Aureobasidium pullulans. Journal of Applied Microbiology*, **97**: 1132–1139.

Sadfi, N., Chérif, M., Hajlaoui, M.R. and Boudabbous, A. 2002. Biological control of the potato tubers dry rot caused by *Fusarium roseum* var. *sambucinum* under greenhouse, field and storage conditions using *Bacillus* spp. isolates. *Journal of Phytopathology*, **150**: 640–648.

Schena, L., Ippolito, A., Zahavi, T. Cohen, L., Nigro, F. and Droby, S. 1999. Genetic diversity and biocontrol activity of *Aureobasidium pullulans* isolates against postharvest rots. *Postharvest Biology and Technology*, **17**: 189–199.

Schena, L., Nigro, F., Pentimone, I., Ligorio, A. and Ippolito, A. 2003. Control of postharvest rots of sweet cherries and table grapes with endophytic isolates of *Aureobasidium pullulans*. *Postharvest Biology and Technology*, **30**: 209–220.

Schena, L., Sialer, M.F. and Gallitelli, D. 2002. Molecular detection of strain L47 of *Aureobasidium pullulans*, a biocontrol agent of postharvest diseases. *Plant Disease*, **86**: 54–60.

Schisler, D.A., Burkhead, K.D., Slininger, P.J. and Bothast, R.J. 1998. Selection, characterization and use of microbial antagonists for the control of *Fusarium* dry rot of potatoes. In: *Plant–Microbe Interaction and Biological Control* (eds) G.J. Boland and L.D. Kuykendall, pp. 199–222, Marcel-Dekker, New York.

Schisler, D.A., Khan, N.I., Boem, M.J. and Slininger, P.J. 2002. Greenhouse and field evaluation of biological control of *Fusarium* head blight on Durum wheat. *Plant Disease*, **86**: 1350–1356.

Schisler, D.A., Kurtzman, C.P., Bothast, R.J. and Slininger, P.J. 1995. Evaluation of yeasts for biological control of *Fusarium* dry rot of potatoes. *American Potato Journal*, **72**: 339–353.

Schisler, D.A., Slininger, P.J. and Bothast, R.J. 1997. Effects of antagonist cell concentration and two-strain mixtures on biological control of *Fusarium* dry rot of potatoes. *Phytopathology*, **87**: 177–183.

Schisler, D.A., Slininger, P.J., Kleinkopf, G., Bothast, R.J. and Ostrowski, R.C. 2000. Biological control of *Fusarium* dry rot of potato tubers under commercial storage conditions. *American Journal of Potato Research*, **77**: 29–40.

Scholz, K., Vogt, M. and Kunz, B. 1999. Application of plant extracts for controlling fungal infestation of grains and seeds during storage. In: *Modern Fungicides and Antifungal Compounds* (eds) H. Lyr, P.E. Russell, H.W. Dehne and H.D. Sisler, pp. 429–435, Intercept Ltd., Andover, U.K.

Segal, E., Yehuda, H., Arad, T., Droby, S., Wisniewski, M. and Goldway, M. 2001. Cloning of 1,3-beta-glucanase of the yeast biocontrol agent *Candida oleophila*. *Acta Horticulturae*, **553**: 397–398.

Sendhilvel, V., Palaniswami, A., Angappan, K. and Raguchander, T. 2001. Effect of seed treatment for the management of onion soft rot (*Erwinia carotovora* subsp. *carotovora*). *South Indian Horticulture*, **49**: 278–280.

Senter, L.H., Sanson, D.R., Corley, D.G., Tempesta, M.S., Rottinghaus, A.A. and Rottinghaus, G.E. 1991. Cytotoxicity of trichothecene mycotoxins isolated from *Fusrium sporotrichioides* (MC-72083) and *Fusarium sambucinum* in bay hamster kidney (BHK-21) cells. *Mycopathologia*, **113**: 127–131.

Sesan, T.S. 1993. Relationships between the main pathogenic fungi occurring on stored carrots and some saprophytic fungi. In: *Biological Control of Foliar and Postharvest Diseases* (eds) N.J. Fokkema, J. Köhl and Y. Elad, pp. 103–104, IOBC/WPRS Bulletin No. 16, IOBC/WPRS, France.

Sharma, I.M. and Bhardwaj, S.S. 2000. Use of plant extracts and yeast antagonists in the management of storage scab and rots of apple fruits. *Journal of Biological Control*, **14**: 17–23.

Singh, J.N., Pinaki Acharya and Singh, B.B. 2000. Effect of GA_3 and plant extracts on storage behavior of mango (*Mangifera indica* L.) cv. Langra. *Haryana Journal of Horticultural Sciences*, **29**: 199–200.

Singh, V. 1980. Control of *Alternaria* rot in citrus. *Australian Plant Pathology*, **9**: 12–13.

Smilanick, J.L. and Denis-Arrue, R. 1992. Control of green mold of lemons with *Pseudomonas* species. *Plant Disease*, **76**: 481–485.

Snowdon, A.L. 1992. A color atlas of postharvest diseases and disorders of fruits and vegetables. *Vegetables*, Vol. 2, Wolfe Publishing, London.

Sobiczewski, P. and Bryk, H. 1996. Biocontrol of *Botrytis cinerea* and *Penicillium expansum* on postharvest apples by an antagonistic bacteria. In: *International Conference on Integrated Fruit Production* (eds) F. Polesny and W. Muller, Olszak, pp. 344–345.

Sobiczewski, P. and Bryk, H. 1999. The possibilities and limitations of biological control of apples against gray mold and blue mold with bacteria *Pantoea agglomerans* and *Pseudomonas* sp. *Progress in Plant Protection*, **39**: 139–147.

Spadaro, D., Vola, R., Piano, S. and Gullino, M.L. 2002. Mechanisms of action and efficacy of four isolates of the yeast *Metchnikowia pulcherrima* active against postharvest pathogens on apples. *Postharvest Biology and Technology*, **24**: 123–134.

Spotts, R.A., Cervantes, L.A., Facteau, T.J. and Chand Goyal, T. 1998. Control of brown rot and blue mold of sweet cherry with preharvest iprodione postharvest *Cryptococcus infirmo-miniatus* and modified atmospheric packaging. *Plant Disease*, **82**: 1158–1160.

Stadelbacher, G.J. and Prasad, K. 1974. Postharvest decay control of apple by acetaldehyde vapor. *Journal of American Society of Horticultural Science*, **99**: 364–368.

Stanley, R., Brown, M., Poole, N., Rogerson, M., Sigee, D.C., Knight, C., Irvin, C.C. Epton, H.A.S. and Leifert, C. 1994. Biocontrol of postharvest fungal diseases on Dutch White cabbage by *Pseudomonas* and *Serratia* antagonists in storage trials. *Plant Pathology*, **43**: 605–611.

Stiles, M.E. 1996. Biopreservation by lactic acid bacteria. *Antonie van Leeuwenhoek*, **70**: 331–345.

Stirling, A.M., Pegg, K.G. and Hayward, A.C. 1998. Interaction of *Colletotrichum gloeosporioides*, epiphytic microorganism and nutrients on avocado leaves and fruit. *Australasian Plant Pathology*, **27**: 169–179.

Strashnov, Y., Elad, Y., Sivan, A., Rudich, Y. and Chet, I. 1985. Control of *Rhizoctonia solani* fruit rot of tomatoes by *Trichoderma harizianum* Rifai. *Crop Protection*, **4**: 359–364.

Strobel, G.A., Dirkse, E., Sears, J. and Markworth, C. 2001. Volatile antimicrobials from *Mucor albus*, a novel endophytic fungus. *Microbiology*, **147**: 2943–2950.

Sugar, D. and Spotts, R.A. 1999. Control of postharvest decay in pear by four laboratory-grown yeasts and two registered biocontrol products. *Plant Disease*, **83**: 155–158.

Suslow, T.V. and Schroth, M.N. 1982. Rhizobacteria of sugar beets: Effects of seed application and root colonization on yield. *Phytopathology*, **72**: 199–206.

Takesako, K., Ikai, K., Haruna, F., Endo, M., Shimanaka, K. et al. 1991. Aureobasidins, a new antifungal antibiotics. Taxonomy, fermentation, isolation and properties. *Journal of Antibiotics*, **44**: 919–924.

Teixidó, N., Usall, J. and Viñas, I. 1999. Efficacy of preharvest and postharvest *Candida sake* biocontrol treatments to prevent blue mold on apples during cold storage. *International Journal of Food Microbiology*, **50**: 203–210.

Teixidó, N., Viñas, I. Usall, J. and Magan, N. 1998. Control of blue mold of apples by preharvest application of *Candida sake* grown in media with different water activity. *Phytopathology*, **88**: 960–964.

Thangavelu, P., Sundaraju, P. and Sathiamoorthi, S. 2004. Management of anthracnose disease of banana caused by *Colletorichum musae* using plant extracts. *Journal of Horticultural Science and Biotechnology*, **79**: 664–668.

Thirumalachar, M.J. and ÓBrien, M.J. 1977. Suppression of charcoal rot in potato with a bacterial antagonist. *Plant Disease Reporter*, **61**: 543–546.

Thomas and ÓBeirne, D. 2000. Evaluation of the impact of short-term temperature abuse on the microbiology and shelf-life of a model ready-to-use vegetable combination product. *International Journal of Food Microbiology*, **59**: 47–57.

Tian, S., Qin, G. and Xu, Y. 2004. Survival of antagonistic yeasts under field conditions and their biocontrol ability against postharvest diseases of sweet cherry. *Postharvest Biology and Technology*, **33**: 327–331.

Tian, S.P., Fan, Q., Xu, Y. and Jiang, A.L. 2002. Effect of calcium on biocontrol activity of yeast antagonists against the postharvest fungal pathogen *Rhizopus stolonifer. Plant Pathology*, **51**: 352–358.

Tian, S.P., Fan, Q., Xu, Y. and Liu, H.B. 2002. Biocontrol efficacy of antagonist yeasts to gray mold and blue mold on apples and pears in controlled atmospheres. *Plant Disease*, **86**: 848–853.

Torres, R. Usall, J., Teixidó, M., Abadias, M. and Viñas, I. 2003. Liquid formulation of the biocontrol agent *Candia sake* by modifying water activity or adding protectants. *Journal of Applied Microbiology*, **94**: 330–339.

Toyoda, H., Hashimoto, H., Utsumi, R., Kobayashi, H. and Ouchi, S. 1988. Detoxification of fusaric acid by a fusaric acid-resistant mutant of *Pseudomonas solanacearum* and its application to biological control of *Fusarium* wilt of tomato. *Phytopathology*, **78**: 1307–1311.

Tripathi, P., Dubey, N.K., Banerji, R. and Chansouria, J.P.N. 2004. Evaluation of some essential oils as botanical fungitoxicants in management of postharvest rotting of citrus fruits. *World Journal of Microbiology and Biotechnology*, **20**: 317–321.

Tronsmo, A. 1989. *Trichoderma harzianum* used for biological control of storage rot on carrots. *Norwegian Journal of Agricultural Science*, **3**: 157–161.

Tronsmo, A. 1993. Biological control of postharvest diseases on carrots. In: *Biological Control of Foliar and Postharvest Diseases* (eds) N.J. Fokkema, J. Köhl and Y. Elad pp. 99–102, IOBC/WPRS Bulletin No. 16(11) IOBC / WPRS, France.

Tronsmo, A. and Raa, J. 1977. Antagonistic action of *Trichoderma pseudokoningii* against the apple pathogen *Botrytis cinerea. Phytopathologische Zeitschrift*, **89**: 216–220.

Usall, J., Teixidó, M., Fons, E. and Viñas, I. 2000. Biological control of blue mold on apple by a strain of *Candida sake* under several controlled atmosphere conditions. *International Journal of Food Microbiology*, **58**: 83–92.

Usall, J., Teixidó, M., Torres, R., de Eribe, X.O. and Viñas, I. 2001a. Pilot tests of *Candida sake* (CPA-1) applications to control postharvest blue mold on apple fruit. *Postharvest Biology and Technology*, **21**: 147–156.

Usall, J., Teixidó, N. and Viñas, I., 2001b. Biological control of *Penicillium digitatum* on citrus fruits with the antagonistic bacterium *Pantoea agglomerans. Acta Horticulturae*, **553**: 377–381.

Velluti, A., Sanchis, V., Ramos, A.J., Turon, C. and Marín, S. 2004. Impact of essential oils on growth rate, zearalenone and deoxynivalenol production by *Fusarium graminearum* under different temperature and water activity conditions in maize grain. *Journal of Applied Microbiology*, **96**: 716–724.

Velvis, H. and Jager, G. 1983. Biological control of *Rhizoctonia solani* on potatoes by antagonists: Preliminary experiments with *Verticillium biguttatum*, a sclerotium-inhabiting fungus. *Netherlands Journal of Plant Pathology*, **89**: 113–123.

Vero, S., Mondino, P., Burgueno, J., Soubes, M. and Wisniewski, M. 2002. Characterization of biocontrol activity of two yeast strains from Uruguay against blue mold of apple. *Postharvest Biology and Technology*, **26**: 91–98.

Viñas, I., Usall, J., Teixidó, N., Fons, E. and de Eribe, O. 1996. Successful biological control of the major postharvest diseases of apple and pear with a new strain of *Candida sake. Brighton Crop Protection Conference: Pests and Diseases*, **2**.

Viñas, I., Usall, J., Teixidó, N. and Sanchis, V. 1998. Biological control of major postharvest pathogens on apple with *Candida sake. International Journal of Food Microbiology*, **40**: 9–16.

Vivekananthan, R., Ravi, M., Ramanathan, A. and Samiyappan, R. 2004. Lytic enzymes induced by *Pseudomonas fluorescens* and other biocontrol organisms mediate defence against the anthracnose pathogen in mango. *World Journal of Microbiology and Biotechnology*, **20**: 235–244.

Walker, R., Rossall, S. and Asher, M.J.C. 2004 Comparison of application of methods to prolong the survival of potential biocontrol bacteria on stored sugarbeet seed. *Journal of Applied Microbiology*, **97**: 293–305.

Walter, M., Jaspers, M.V., Eade, K., Frampton, C.M. and Stewart, A. 2001. Control of *Botrytis cinerea* grape using thyme oil. *Australian Plant Pathology*, **30**: 21–25.

Ward, B.G., Poole, P.R., Hill, R.A. and Cutler, H.G. 1998. Postharvest application of natural products to control *Botrytis* storage rot in kiwifruit. *Acta Horticulturae*, **464**: 225–228.

Williamson, S.M., Guzman, M., Anas, O., Marin, D.H., Jin., X. and Sutton, T.B. 1999. Evaluation of potential biocontrol agents for crown rot of banana. *Phytopathology*, **89**: S85.

Wilson, C.L., El Ghaouth, A., Chalutz, E., Droby, S., Stevens, C.S., Lu, J.Y., Khan, V. and Arul, J. 1994. Potential of induced resistance to control postharvest diseases of fruits and vegetables. *Plant Disease*, **78**: 837–844.

Wilson, C.L. and Wisniewski, M.E. 1989. Biological control of postharvest diseases of fruits and vegetables: an amerging technology. *Annual Review of Phytopathology*, **27**: 425–441.

Winter, W., Bänziger, I., Rüegger, A., Schachermayr, G., Krebs, H., Frei, P. and Gindrat, D. 2001. Skim milk powder and yellow mustard meal treatment: alternatives to the chemical seed-dressing for the control of common bunt in wheat. *Agararforschung*, **8**: 118–123.

Wisniewski, M., Biles, C., Droby, S., Mc Laughlin, R., Wilson, C. and Chalutz, E. 1991. Mode of action of the postharvest biocontrol yeast, *Pichia guilliermondii*, 1. Characterization of attachment to *Botrytis cinerea*. *Physiological and Molecular Plant Pathology*, **39**: 245–258.

Wisniewski, M.E. and Wilson, C.L. 1992. Biological control of postharvest diseases of fruits and vegetables: Recent advances. *HortScience*, **27**: 94–98.

Wisniewski, M.E., Wilson, C. and Hershberger, W. 1989. Characterization of inhibition of *Rhizopus stolonifer* germination and growth by *Enterobacter cloacae*. *Canadian Journal of Botany*, **67**: 2317–2323.

Worapong, J., Stroel, G., Ford, E.J., Li, J.Y., Bairol, G. and Hess, W.M. 2001. *Muscodor albus* anamorph sp. Nov. an endophyte from *Cinnamomum zeylanicum*. *Mycotaxon*, **79**: 67–79.

Xu, G.W. and Gross, D.C. 1986. Selection of fluorescent pseudomonads antagonistic to *Erwinia carotovora* and suppressive of potato seed piece decay. *Phytopathology*, **76**: 414–422.

Yakoby, N., Zhou, R., Kobiler, I., Dinoor, A. and Prusky, D. 2001. Development of *Colleotrichum gloeosporioides* restriction enzyme-mediated integration mutants as biocontrol agents against anthracnose disease in avocado fruits. *Phytopathology*, **91**: 143–148.

Zahavi, T., Cohen, L., Weiss, B., Schena, L., Daus, A., Kaplunov, T., Zutkhi, J., Ben-Arie, R. and Droby, S. 2000. Biological control of *Botrytis*, *Aspergillus* and *Rhizopus* rots on table and wine grapes in Israel. *Postharvest Biology and Technology*, **20**: 115–124.

Zhang, D.L. and Quantick, P.C. 1998. Antifungal effects of chitosan coating on fresh strawberries and raspberries during storage. *Journal of Horticultural Science* and *Biotechnology*, **73**: 763–767.

Zhang, H.Y., Zheng, X.D. and Xi, Y.F. 2003. Biocontrol of postharvest blue mold rot of pear by *Cryptococcus laurentii*. *Journal of Horticultural Science and Biotechnology*, **78**: 888–893.

Zhou, T., Chu, C.L., Liu, W.T. and Schaneider, K.E., 2001. Postharvest control of blue mold and gray mold on apples using isolates of *Pseudomonas syringae*. *Canadian Journal of Plant Pathology*, **23**: 246–252.

Zhou, T., Northover, J. and Schneider K.E. 1999. Biological control of postharvest diseases of peach with phyllosphere isolates of *Pseudomonas syringae*. *Canadian Journal of Plant Pathology*, **21**: 375–381.

10

BIOTECHNOLOGY FOR THE IMPROVEMENT OF RESISTANCE TO POSTHARVEST DISEASES

Development of cultivars genetically resistant to postharvest diseases is known to be the most desirable disease management strategy (Chapter 8). However, the major obstacle to achieving this goal is the lack of durability of resistance. This is due to the selection pressure exerted on the microbial pathogens, which are likely to adapt themselves to newly introduced resistance (R) genes by producing new strains. Furthermore, the available germplasm may not contain dependable sources of resistance. In addition, the loss or dilution of desirable characteristics when, with great difficulty, resistance genes from wild relatives are transferred, has limited the efforts to develop genetically resistant cultivars. Rapid developments in molecular biology and genetic engineering during recent decades have opened up amazing possibilities for the development of crop cultivars resistant to diseases that have, so far, eluded the crop protection umbrella.

Plants are endowed with natural protective systems provided by a wide range of barriers already in place for defense against pathogen invasion. This kind of resistance, known as constitutive resistance, is due to the combined effect of various natural barriers. On the other hand, plants, following recognition of a pathogen or its by-products, frequently mount a highly coordinated cascade of defense responses to restrict development of the pathogen. This type of resistance, known as induced or acquired resistance, consists of: reinforcement of the cell wall by deposition of lignin, callose, and hydroxyproline-rich glycoproteins; accumulation of phytoalexins; and synthesis of proteinase inhibitors and lytic enzymes capable of attacking fungal cell walls (Lamb

Postharvest Pathogens and Disease Management, by P. Narayanasamy

et al., 1987; Dixon and Lamb, 1990). The defense responses can be activated by application of abiotic and biotic elicitors (inducers). The transcriptional activation that is required for the activation of defense responses may be restricted to the tissue close to the site of stimulus (localized acquired resistance) or may be expressed systemically, far away from the site of elicitor application (systemic acquired resistance, SAR). Studies on the molecular biology of disease resistance provide the possibility of a clear insight into the plant–microbe interactions that result in the expression of resistant or susceptible reactions of the host plant to the target pathogen. The tools of genetic engineering have been used to isolate disease resistance genes from diverse sources and express them in the desired crops. These options, which were not available earlier, can be advantageously exploited for the production of transgenic plants expressing disease resistance genes and for the enhancement of levels of host resistance through the use of elicitors.

Biotechnology may be defined as any technique that uses living organisms (or parts of organisms) to make or modify products, to improve plants or animals, or to develop microorganisms for specific use. Recombinant DNA techniques, cell fusion, monoclonal antibodies, and novel bioprocessing techniques are some of the important procedures followed widely (Vidaver and Tolin, 2001). The biotechnological approaches for improving resistance of fruit and vegetable crops is of two kinds: (1) genetic manipulation; and (2) activation of natural host defense mechanisms.

10.1 GENETIC MANIPULATION

Appreciable progress has been made in our understanding of the genetic control of disease resistance genes by creating variability through mutation or transformation of plants through the transfer of desired genes.

10.1.1 Mutation Breeding

With a view to enhancing level of resistance of garlic (*Allium sativum*) to white rot disease caused by *Sclerotium cepivorum* and to improving its storability under natural conditions, a mutation breeding program was carried out. Cloves of two garlic cvs, Kisswany and Yabroudy, were irradiated with gamma ray. The M3 and M4 generations were tested by artificial inoculation with *S. cepivorum* followed by planting in infested soil. Twelve lines of cv. Kisswany showed only 3% infection as against 29% infection in the controls. In 12 lines of cv. Yabroudy, the incidence of disease was reduced to less than 5% compared with 20% incidence in the control. Furthermore, improvement in storability under natural conditions and appreciable decreases in weight loss during storage were also noted in the mutant lines of both garlic cultivars (Al-Safadi et al., 2000).

The resistance of cauliflower (*Brassica oleracea* var. *botrytis*) could be enhanced by inducing somaclonal variation. Calli of cauliflower were treated with ethylene methane sulfonate (EMS) or gamma irradiation and then they were screened for their resistance to black rot disease caused by *Xanthomonas campestris* pv. *campestris* (*Xcc*). The calli were exposed to the culture filtrate of *Xcc*. The calli surviving the exposure to culture filtrate at 30% concentration were selected. Esterase isozyme pattern analysis revealed significant differences between the plants derived from selected and unselected calli and also between those derived from EMS or gamma ray-treated and selected calli. There was strong correlation between resistance of calli to the culture filtrate and the resistance of regenerated plants to *Xcc*, indicating the usefulness of the in vitro screening technique for selecting resistant lines (Mangal and Sharma, 2002).

Somatic hybrid plants can be obtained by fusing two protoplasts with different characteristics, such as resistance to diseases. Protoplast fusion was effected between several clones of *Solanum tuberosum* (potato) (Sueprior, Dejima, and dihaploid of Superior, susceptible to blackleg and tuber soft rot caused by *Erwinia carotovora* subsp. *carotovora* and *E. carotovora* subsp. *atroseptica*) and *S. brevidens* (resistant to the bacterial pathogens). Following inoculation with the bacteria, tubers and stems of potato clones exhibited soft macerated areas around inoculation sites. In contrast, localized wound response was observed in the stems of tubers of somatic hybrids, indicating a high level of protection against the bacterial pathogens (Ahn et al., 2001). This investigation demonstrates the wider application of biotechnological techniques for the early development of cultivars resistant to postharvest pathogens.

Concomitant evolution of ethylene is associated with the progress of symptom expression in the response by plants to infection by many microbial pathogens. A tomato mutant (cv. 'Never ripe'), impaired in ethylene perception, showed reduction in disease intensity, compared with the parent line, following inoculation with *Xanthomonas campestris* pv. *vesicatoria, Pseudomonas syringae* pv. *tomato*, and *Fusarium oxysporum* f.sp. *lycopersici*. Reduction in disease symptoms was associated with ethylene insensitivity in this mutant. The results suggested that broad tolerance to diverse vegetable diseases might be achieved via engineering of ethylene insensitivity in tomato (Lund et al., 1998). Involvement of abscisic acid (ABA) in the interaction between plants and microbial pathogens has been indicated. Tomato (cv. Moneymaker) mutants with reduced levels of ABA (*sitiens* plants) were shown to be much more resistant to gray mold pathogen *Botrytis cinerea* than the wild (parent) type (WT) plants. Exogenous application of ABA restored susceptibility to *B. cinerea* in *sitiens* plants, while susceptibility of WT plants was further increased, indicating that ABA has a major role in the level of susceptibility/resistance of tomato genotypes (Audenaert et al., 2002a).

10.1.2 Genetic Transformation of Plants

Transgenic plants constitutively expressing defense-related genes from various sources have been generated. The exogenous DNA, introduced artificially via genetic engineering techniques, should be able to be passed through the germ line of the plant, following incorporation of the foreign DNA into the plant's nuclear material. A plant may also be manipulated to produce, or overproduce, one of its own proteins by specifically mutating the gene's endogenous control mechanisms. Such bioengineered plant may not be technically placed under the category of transgenic. However, these plants have an inheritably altered germline. New crop cultivars with resistance to diseases may be developed more rapidly by employing the various biotechnological methods as compared to classical breeding procedures. Furthermore, multiple, diverse genes conferring resistance to diseases may be incorporated simultaneously by transgenic plant technology, which allows a wider genetic diversity to be exploited. Transformation of plants by transferring beneficial genes from bacteria, fungi, viruses, or varoius species of plants has been achieved most frequently by using the phytopathogenic bacterium *Agrobacterium tumefaciens*. The Ti (tumor inducing) plasmid of *A. tumefaciens* containing the T-DNA has been employed for gene transfer to plants. The T-DNA region can be disarmed, rendering the plasmid nononcogenic, but still retaining its ability to integrate with plant host genome and to synthesize opines. Furthermore, it is possible to trim down to only absolutely essential functions, providing more room for cloning exogenous DNA.

It is essential to accumulate knowledge on the cause of the postharvest disease and the temporal regulation of defense response genes in relation to the physiological changes that occur in harvested fruits and vegetables. Activation of different kinds of host genes during the transition of reproductive organs from the developmental stage to harvesting, and their subsequent storage, may be expected. Identification of the function and regulation of expression of these genes may provide the basis for the development of practical methods of controlling postharvest diseases. As the susceptibility of harvested produce to infection by microbial pathogens is known to increase when the process of ripening is initiated, it may be appropriate to employ developmentally regulated promoters to drive the expression of defense genes. This approach may result in enhancement of resistance in the harvested tissues at specific stages, when they become most vulnerable to pathogen invasion.

10.1.2.1 Fungal Diseases The fungal pathogen *Botrytis cinerea* causes gray mold disease in several fruits and vegetables. Tomato plants were transformed with pear fruit polygalacturonase inhibitor protein (pPGIP). Transgenic expression of pPGIP induced abundant accumulation of the heterologous protein in all tissues and it did not influence the expression of the endogenous tomato fruit PGIP (tPGIP). All transgenic tissues showed the presence of pPGIP in the cell wall protein fraction. The expressed pPGIP was active in both leaf and fruit tissues as an inhibitor of endo-PGs from *Botrytis cinerea*. The expression of pPGIP resulted in reduction of growth of *B. cinerea* on ripe tomato fruit and tissue break down was reduced by as much as 15% compared to control (nontransformed) fruits. Likewise, expression of pPGIP in leaves, decreased the lesions of macerated tissue by about 25%. The results show that the expansion of lesions and tissue maceration by *B. cinerea* could be arrested by inhibiting the fungal PGs by the expression of heterologous PGIP (Powell et al., 2000).

Transformed grapevine plants were produced by introducing the rice chitinase gene (*RCC* 2), classified as a Class I chitinase, into the somatic embryos of grapevine cv. Neo Muscut by agroinfection. Two transformants showed enhanced resistance to powdery mildew disease caused by *Uncinula necator.* Observations under a scanning electron microscope revealed that conidial germination, mycelial growth, and conidial formation were suppressed in the transformants showing resistance. In addition, the transgenic lines showed resistance to anthracnose disease caused by *Elsinoe ampelina*, resulting in a reduction in disease lesions (Yamamoto et al., 2000).

By applying an *Agrobacterium tumefaciens*-mediated transformation technique, the class I chitinase cDNA (*RCC2*) of rice driven by the cauliflower mosaic virus (CaMV) 35S promoter was also introduced into cucumber. Lines resistant (CR32), intermediate resistant (CR3), and susceptible (CR 20) to gray mold disease were selected. The rice chitinase levels in the lines with resistance (CR 32 and CR3) were higher than in a susceptible line (CR 20), as revealed by an enzyme-linked immunosorbent assay (ELISA). Studies with in situ indirect fluorescent antibody technique and three-dimensional fluorescence microscopy showed a homogeneous distribution of cell expression of *RCC2* to a high level in the epidermal and mesophyll cells of resistant (CR 32) plants, whereas the degree of expression of *RCC2* in susceptible lines was lower. The growth of *B. cinerea* within the leaf tissue of resistant lines was suppressed. It was suggested that the high level of expression and intracellular localization of the rice chitinase may result in enhancement of resistance to *B. cinerea* in transgenic cucumber plants (Kishimoto et al., 2002).

Transgenic apple lines expressing the endochitinase gene (*ech42*) of the biocontrol agent *Trichoderma harzianum* were more resistant to apple scab disease, caused by *Venturia inaequalis*, than nontransformed Marshall McIntosh apple. The transfer of the foreign genes was verified by ELISA (for the detection of NPTII protein), polymerase chain reaction (PCR), and Southern blot analysis. Disease severity was reduced by 0 to 99.7% (number of lesions), 0 to 90% (percentage of leaf area affected), and 1 to 56% (conidia recovered) in the transgenic lines. However, the plant growth was adversely affected in proportion to the level of expression of endochitinase (Bolar et al., 2000). The human lysozyme has been demonstrated to have lytic activity against fungal and bacterial plant pathogens. A human lysozyme gene, under the control of a constitutive CaMV 35S promoter, expressed in a plasmid, was introduced into two carrot cultivars by *Agrobacterium tumefaciens*-mediated transformation procedure. One transgenic line exhibited enhanced resistance to *Alternaria dauci*. Accumulation of synthesized human lysozyme protein correlated well with the higher level of resistance observed (Takaichi and Oeda, 2000).

Physiological functions may be altered by manipulation of appropriate genes of the host. The effect of such gene manipulation on disease resistance has been studied in tomatoes. Expansins are proteins that are involved in cell wall loosening, through cell wall modification, during fruit development. The expansin gene (*LeExp1*) governs ripening-related accumulation of mRNA and protein. Transgenic silencing of the expression of this gene leads to production of firmer fruits, a characteristic required for longer shelf-life. The shelf-life of tomato fruits from these transgenic plants was extended by 5 to 10 days, depending on the type of packaging. However, the enhancement of fruit firmness did not influence the level of resistance of tomato fruits to *B. cinerea* and *A. alternata* (Brummel et al., 2002). Exposure of tomato to ethylene induced expression of several PR-protein genes before *B. cinerea* infection. Ethylene induced resistance in the mutant *Neverripe*. A transgenic line with reduced prosystemin expression exhibited susceptibility, whereas a prosystemin-over expressing transgene was highly resistant. The mutant Defenceless, impaired in jasmonate biosynthesis, had a higher level of susceptibility (Díaz et al., 2002). The peptide snakin (StSN2), isolated from potato tubers, exhibited antifungal and antibacterial properties. Rapid aggregation of both Gram positive and Gram negative bacterial cells occurred in the presence of snakin. The *StSN2* gene was developmentally expressed in tubers, stems, flowers, shoot apex, and leaves. Expression of this gene was up-regulated after infection of tubers by *B. cinerea* and down-regulated by the virulent *Ralstonia solanacearum* and *Erwinia chrysanthemi*. StSN2 may be a component of both constitutive and inducible defense barriers (Berrocal-Lobo et al., 2002).

Messenger RNA (mRNA) with a sequence that is antisense to the mRNA normally formed following transcription of a gene, may bind with the sense mRNA, resulting in the failure of synthesis of the protein. Thus an endogenous plant mRNA can be turned off by transforming the plant with the complementary antisense sequence. It may be possible to use antisense mRNA to block the synthesis of compounds that are formed in plants following the recognition of elicitors from microbial pathogens. By using antisense RNA the fruit ripening process can be regulated and consequently postharvest losses may be reduced. Repression of tomato polygalacturonase (PG) could help in increasing fruit shelf-life. When the antisense RNA to mRNA of tomato PG was introduced into the plant, the resulting tomatoes had a longer shelf-life with no adverse effect on fruit texture or flavor. As the tissues of these tomatoes softened at a slower rate compared with normal tissues, the enhanced fruit firmness may provide a greater level of resistance to postharvest pathogens (Kramer et al., 1990). In another investigation, tomatoes transformed with an antisense gene to ethylene, a hormone that hastens ripening, have also been developed. These transformants also had a reduced rate of fruit ripening and a greater level of resistance (Hamilton et al., 1990).

Salicylic acid (SA) is known to function as an endogenous transduction signal (Malamy et al., 1990; Métraux et al., 1990). Synthesis and accumulation of SA are the basic requirements for the activation of many plant defense responses, including expression of a diverse group of defense-related genes (Klessig and Malamy, 1994; Durner et al., 1997). The genes encoding pathogenesis-related (PR) proteins are induced by exogenous application of SA. Most of the PR-proteins are considered to enhance the level of host-plant resistance to diseases (Ryals et al., 1996; Wobbe and Klessig, 1996). The promoter of PR-2d, a β-1,3-glucanase gene of *Nicotiana tabacum*, has been well characterized (Hennig et al., 1993). The sequences from 364 to 288 bp of the same promoter conferred a high level of activation of SA in transgenic tobacco plants. In order to study the action of SA and salicylic acid glucoside (SAG), transgenic cucumber plants bearing a reporter system consisting of RR-2d (-1706) promoter fused to the *uidA* gene (GUS) were produced. Treatment of transgenic cucumber plants with SA, showed increased GUS activity two to 11 fold over that of the control. Endogenous SA and its conjugate salicylic acid glucoside (SAG) rose in parallel after inoculation with *Pseudoperonospora cubensis*. Inoculation of transgenic cucumber with *Erysiphe polyphage* increased GUS activity four to 44 fold over that of the control. The gene construct (PR-2d promoter/*uid A* (GUS)) was expressed according to the signaling pattern of the native species and was stably transmitted to progeny over four generations (Yin et al., 2004).

10.1.2.2 Bacterial Diseases The soft rot pathogen *Erwinia carotovora* (*Ec*) produces pectate lyase (PL) enzymes, functioning as major virulence factors. In the process of degrading plant cell wall pectin by PLs, unsaturated oligogalacturonates (OG) are formed and the OGs in turn may elicit plant defense responses. The potato cv. Désirée was transformed with a gene encoding the isoenzyme PL3 of *E. carotovora atroseptica* (*Eca*) via an *A. tumefaciens*-transformation protocol. Field evaluation showed that four PL transgenic potato lines exhibited enhanced resistance to *Erwinia* soft rot in tubers. The threshold density of *Eca* causing progressive soft rot was up to 19-fold higher on tuber tissue containing PL-enzyme. The enhancement of the level of resistance of transgenic tubers was attributed to higher resistance of tuber cell walls to *Erwinia*-derived enzymes, an increased activities of polyphenol oxidase (PPO), and phenylalanine ammonia lyase (PAL) in tuber tissues, and reinforcement of necrotic tissue in the wound surface of tubers following infection with *Eca* (Wegener, 2002).

Citrus canker disease caused by *Xanthomonas axonopodis* pv. *citri* is an endemic disease occurring in all countries. Introduction of genes conferring resistance to canker into susceptible cultivars is the best strategy to control the disease. The 'Early Gold' sweet orange and 'Murcott' taugor were transformed. The DNA encoding for a non-destructive selectable marker – enhanced green fluorescent protein gene (p524 EGFP1) – and a potential canker resistance gene (*pC822*) from the *Xa21* gene family of rice were introduced into protoplasts. The incorporation of *Xa21* gene in the regenerated shoots of Early Gold sweet orange was confirmed by PCR analysis. It is expected that the *Xa21* gene, providing broad spectrum *Xanthomonas* resistance in rice, may also offer protection to sweet orange against canker disease (Guo and Grosser, 2004).

Seed rot in rice is caused by seed borne bacterial pathogens *Burkholderia glumae* and *B. plantarii*. The endogenous rice seed thionins are insufficient to protect the seeds against the invasion by bacterial pathogens. Thionins purified from seed and etiolated seedlings of barley exhibit antimicrobial activity against necrotrophic pathogens. Transgenic rice plants expressing oat thionin accumulated high concentrations of oat thionin in cell walls and grew almost normally with bacterial staining only on the surface of stomata. In contrast, the bacterial staining revealed the presence of bacterial pathogens around stomata and also in intercellular spaces of nontransformed plants, which showed severe blight symptoms. The results indicate that oat thionin has the potential to provide protection to rice against bacterial infection (Iwai et al., 2002).

10.1.2.3 Viral Diseases Viruses infecting crop plants generally cause systemic infection, leading to retarded growth and considerable reduction in both quantity and quality of harvested produce. The appearance of virus-infected apples, papaya, tomatoes, potatoes, and cucumber is adversely affected, resulting in appreciable economic losses. Pathogen-derived resistance (PDR) is an approach involving the generation of transgenic plants expressing the viral gene coding for coat protein (CP), movement protein (MP), or replicase, with the aim of producing transformed plants with resistance to the virus.

Papaya ringspot virus (PRSV) causes serious losses in most countries. Nearly all of papaya hectarage in the Puna district of Hawaii was destroyed by this disease. The virus exists in the form of different strains, in many geographical locations. Transgenic papaya plants expressing the CP gene of PRSV have been demonstrated to be highly resistant under field conditions. The transformed plants remained completely free of infection, while 91 to 100% of the nontransformed plants were infected, indicating the reliability of the field evaluation of resistance of transgenic cultivars SunUp and Rainbow, homozygous and hemizygous for the coat protein transgene, respectively. Cultivation of such transgenic papaya cultivars offers a practical solution to the problem posed by PRSV (Ferriera et al., 2002). By employing the same technology, the resistance of local papaya cultivars grown in Venezuela was improved. Papaya plants were transformed with CP gene from two different geographical PRSV isolates (VE and LA). Local papaya cultivars were effectively protected against homologous and heterologous isolates of PRSV. The presence of CP gene in the transformed papaya plants was demonstrated by amplification by PCR (Fig. 10.1) (Fermin et al., 2004).

In another study, four transgenic papaya lines expressing the CP gene of PRSV were evaluated under field conditions. The transgenic lines did not exhibit severe symptoms of PRSV. In contrast, all nontransformed control plants were infected and showed severe symptoms at 3 to 5 months after planting. Although some transformed plants showed mild symptoms, no apparent

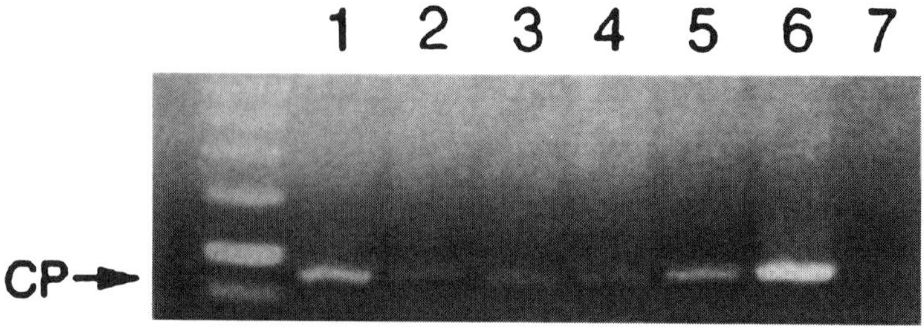

Figure 10.1 Detection of the coat protein (CP) gene of *Papaya ringspot virus* amplified by polymerase chain reaction (PCR) in transformed papaya plants. 1. Cl; 2. C2; 3. C3; 4. T2; 5. T3; 6. CP-transgenic Rainbow papaya; 7. nontransgenic Sunrise papaya. (C1, C2, C3, T2, and T3 plants were derived from five PRSV-resistant transgenic R0 plants, from three independent transformation experiments.) (Courtesy of Fermin et al., 2004; The American Phytopathological Society, St. Paul, MN, USA.)

TABLE 10.1 Fruit Yields of Transgenic Papaya Lines Expressing Coat Protein (CP) Gene of *Papaya Ringspot Virus* (PRSV)

Transgenic lines	Yield (kg/plot)	
	Trial 1	Trial 2
16-0-1	500.0 ± 26.4[a]	375.6 ± 5.8[a]
17-0-1	518.4 ± 18.1[a]	–
17-0-5	482.0 ± 4.1[a]	377.9 ± 43.3[a]
18-2-4	–	362.0 ± 95.5[a]
Control	187.0 ± 45.3[b]	118.0 ± 15.0[a]

Means with same letter did not differ significantly at P = 0.05.
Source: Bau et al., 2004.

adverse effect on fruit yield and quality in transgenic plants with mild symptoms were noted. The yields of transgenic plants were enhanced by 10.8 to 11.6 and 54.3 to 56.7 fold, respectively, in the first and second experiments. The fruits were of marketable quality with no ringspots or distortion (Table 10.1) (Bau et al., 2004). Transgenic papaya plants expressing CP genes in the sense orientation (S-CP), antisense orientation (AS-CP), sense orientation with frame-shift mutation (FS-CP), or sense orientation mutated with three in-frame stop codons (SC-CP) were generated and mechanically inoculated with PRSV HIK (Florida isolate). Highly resistant transgenic lines were identified and crossed with six papaya genotypes. Plants derived from 54 crosses and representing 17 transgenic lines were evaluated under field conditions for their resistance to PRSV. Transgenic lines showed 23.3% infection, whereas nontransgenic control plants had 96.7% infection by PRSV (Table 10.2) (Davis and Ying, 2004).

TABLE 10.2 Incidence of Papaya Ringspot Disease Under Field Conditions on R1 Generation of Transgenic Papaya Lines Expressing Coat Protein (CP) Gene of *Papaya Ringspot Virus* (PRSV)

Transgenic lines	Number of infected plants/ total number of plants	Resistant (%)
SC-CP-3	2/16	87.5
SC-CP-4	2/16	87.5
SC-CP-6	45/202	87.8
SC-CP-75	29/278	87.3
SC-CP-95	17/162	89.5
FS-CP-17-1	6/32	81.2
FS-CP-17-2	35/178	80.3
FS-CP-26	11/106	89.6
FS-CP-33	6/32	81.2

SC-CP = sense orientation mutated with three in-frame stop codons; FS-CP = sense orientation with a frame-shift mutation.
Source: Davis and Ying, 2004.

Trifoliate orange plants (*Poncirus trifoliata*) were transformed with a binary vector containing the capsid polyprotein (pCP) gene of *Citrus mosaic virus* (CiMV). The level of tolerance of the transformants was evaluated by artificial inoculation with CiMV. Reverse transcription polymerase chain reaction (RT-PCR) assay was employed to detect the presence of CiMV in the transformants. The transgenic line 24 showed 7.1% infection, whereas 65.17% of nontransformed plants were infected at 60 days after inoculation, indicating the effectiveness of incorporation of the CP gene of CiMV for the control of the disease (Iwanami et al., 2004).

The performance of tomato plants expressing CP gene from *Tobacco mosaic virus* (TMV) under field conditions was evaluated. Transgenic plants exhibited visible symptoms in only about 5%, as against 99% infection in nontransformed control plants. The yield of transgenic plants was not affected, whereas the control plants suffered a reduction of 26 to 35% (Nelson et al., 1988). Tomato lines expressing CP gene of *Cucumber mosaic virus* (CMV) strain WL showed high levels of resistance. The productivity of transgenic tomato plants was increased by 17 fold and the fruit weight showed an increase of 44% compared to control (Fuchs et al., 1996). Tomato plants expressing a benign variant of CMV satellite RNA (CMV Tfn-satRNA) did not exhibit symptoms following inoculation with a satRNA-free strain of CMV (CMV-FL). Two different mechanisms may operate in the transgenic tomato plants resulting in resistance to CMV: (1) down-regulation of CMV by Tfn-satRNA amplified from the transgenic transcripts; and (2) RNA silencing, as suggested by the events leading to the degradation of satRNA sequences (Cillo et al., 2004).

The utility of the strategy of using transgenic plants for disease management was demonstrated in the case of potato leafroll disease. Extensive evaluation of the performance of transgenic Russet Burbauk potato lines expressing the CP or replicase genes of *Potato leafroll virus* (PLRV) demonstrated the desirability of growing transgenic potato lines which were not infected under either field or glasshouse conditions (Thomas et al., 1998).

The potato cv. Igor was severely infected by NTN isolates of *Potato virus Y* (PVY^{NTN}) causing potato tuber necrotic ringspot disease (PTNRD). To save the potato cultivar, the cultivation of which had ceased, the CP gene sequence of PVY^{NTN} was transformed into this cultivar. Several transgenic lines showed resistance to PTNRD to different levels. The highly resistant lines developed no foliage or tuber symptoms when they were graft-inoculated. The virus could not be detected by either enzyme-linked immunosorbent assay (ELISA) or infectivity test. Among the 34 transgenic lines, two showed resistance to PVY^{NTN} and PVY^{O} and one line was resistant to PVY^{O} (Racman et al., 2001).

Posttranscriptional gene silencing (PTGS) or RNA silencing is a sequence-specific mRNA degradation mechanism in plants that does not affect transcription, at least not initially. This approach of RNA silencing was applied to engineer potato plants resistant to PVY. Double-stranded (ds) RNA derived from the 3′ terminal part of the CP gene of PVY was expressed in transgenic potatoes of the commercial cultivar 'Spunta'. Transgenic plants were analyzed for generation of transgene-derived short interfering RNAs (siRNAs) prior to PVY inoculation. Among the 15 transgenic lines generated, 12 lines produced siRNAs and they were highly resistant to three strains of PVY, each belonging to three different subtypes of PVY, that is PVY^{N}, PVY^{O}, and PVY^{NTN} (Missiou et al., 2004). The transgenic lines of squash (*Cucurbita pepo* var. *melopepo*) expressing the CP genes of CMV, *Zucchini yellow mosaic virus* (ZYMV) and *Watermelon mosaic virus* (WMV) exhibited higher levels of resistance, compared to nontransgenic lines, to all three viruses under field conditions. The transgenic plants gave a 50-fold increase in the marketable yield and high monetary gains (Fuchs et al., 1998).

10.2 ACTIVATION OF NATURAL HOST DEFENSE MECHANISMS

Progressive decline in the use of synthetic fungicides or other chemicals is the result of growing awareness of the possible hazards to humans and animals. To find suitable alternatives to the fungicide, intensive efforts were taken up to identify effective biocontrol agents or products, leading to the development of two commercially available biofungicides, Biosave and Aspire (Chapter 9). However, these products showed wide variations in their efficacy, depending on the crops, cultivars, and physical and chemical environments. Enhancing the plant's resistance by genetic manipulation, though most desirable, has been shown to be time consuming and limited by the unavailability of dependable sources of resistance genes. In this context, the strategy of exploitation of natural disease resistance in harvested produce holds promise for the management postharvest diseases of fruits and vegetables.

Interactions between plants and microbial pathogens have been demonstrated to be genetically controlled. However, only limited information is available regarding the nature and regulation of defense mechanisms of harvested fruits and vegetables. The natural defense responses of plants can be turned on by the application of abiotic or biotic elicitors, rendering them resistant. Several studies, with a variety of fruits and vegetables, have revealed that the strategy of inducing resistance to microbial diseases is an important, manageable form of protection to harvested produce. It has to be recognized that defense mechanisms operating in harvested fruits and vegetables may be different from those in the growing plant, since the harvested commodities are senescing. The process of senescence is known to reduce the defense responses of tissues, making them vulnerable to invasion by microbial pathogens. The activation of defense responses may occur only in a localized manner, in tissues close to the site of stimulus, or it may be expressed systemically in tissues far away from the site of stimulus. Resistance to postharvest diseases may be induced by using both physical, chemical, and biological agents. The extent of protection against microbial pathogens afforded by, and the feasibility of using, elicitors of resistance for large-scale application are discussed in this chapter.

10.2.1 Physical Agents

The usefulness of various physical agents, such as ultraviolet light, gamma irradiation, and heat treatment, in protecting commodities against postharvest pathogens has been demonstrated by several studies (Chapter 7).

10.2.1.1 Low-Dose Ultraviolet (UV) Light The UV-light present in sunlight has been shown to occur in three ranges of wave lengths that are designated UV-C (<280 nm), UV-B (280–320 nm), and UV-A (320–390 nm). All types of UV radiation can damage plant genomic DNA and physiological processes at different doses and periods of exposure (Luckey, 1980; Stapleton, 1992). Exposure to low doses of UV-C light and gamma radiation treatments reduced the storage rots of onions (Lu et al., 1987) and sweetpotatoes (Stevens et al., 1990). A wide array of fruits and vegetables have been shown to respond to treatment with UV-C light, resulting in enhancement of resistance to diseases (Fig. 10.2) (Table 10.3).

In treated citrus fruits, the onset of UV-induced resistance coincided with induction of phenylalanine ammonia lyase activity (PAL), a key enzyme in the phenylpropanoid pathway, as well as peroxidase (PO) activity (Chalutz et al., 1992; Droby et al., 1993). PAL activity in the peel of grapefruit increased within 24 h after treatment and remained at higher levels for 72 h, whereas PO activity reached its peak at 72 h after treatment. It is well known that both enzymes have a role in inducible resistance in plants against microbial pathogens (Wilson et al., 1994).

Enhancement of resistance of grapefruit (*Citrus paradisi*) cv. Marsh Seedless following UV irradiation against development of green mold decay caused

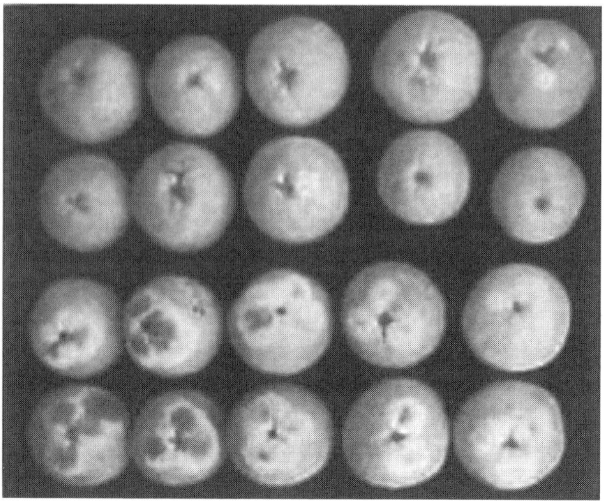

Figure 10.2 Effect of UV-C treatment ($3.2\,kJ/m^2$) on development of *Penicillium digitatum* on citrus fruits inoculated at three points on the stem-end at 24h after UV-C treatment. Disease incidence was recorded at 7 days after inoculation. (Courtesy of Wilson et al., 1994; The American Phytopathological Society, St. Paul, MN, USA.)

by *Penicillium digitatum* was observed by Porat et al. (1999). Immunoblotting analysis using citrus-specific chitinase and β-1,3 endoglucanase antibodies revealed that UV irradiation, wounding of fruits, or a combination of these two treatments induced accumulation of a 24-kDa chitinase protein in the fruit peel tissue. In contrast, UV irradiation or wounding alone did not induce accumulation of 39 and 43 kDa β-1,3 endoglucanase proteins, but the combination of the two treatments increased these protein levels. Both chitinase and β-1-3-endoglucanase may play a role in the development of resistance induced by UV treatment in grapefruits against *P. digitatum* (Fig. 10.3).

Orange cultivars (Washington Navel, Biondo Commune, Tarocco, and Valencia Late) and grapefruit cv Star Ruby treated with UV-C light at $0.5\,kJ/m^2$ exhibited resistance to decay development compared with untreated control fruits. Following UV-C irradiation, accumulation of the phytoalexins scoparone (6,7-dimethoxycoumarin) and scopoletin occurred in the flavedo tissue, their concentrations depending on the dosage of UV-C light. Doses above $0.5\,kJ/m^2$ caused damage on the fruit. Both phytoalexins showed similar accumulation patterns, although scopoletin was present in higher concentrations in grapefruit and in lower concentration in oranges. No detectable levels of scoparone and scopoletin were present in control fruits. No changes in the contents of soluble solids and titrable acidity level were evident due to UV-irradiation (D'hallewin et al., 1999, 2000). Induction of scoparone (6,7-dimethyxycoumarin) in flavedo tissue of lemon cv. Eureka was detected by histological analysis. This fluorescent phytoalexin accumulated only on the four to five layers of cells adjacent to the inoculation site. Development of

TABLE 10.3 Induction of Resistance to Postharvest Diseases by UV-C Light

Crop	Pathogen	References
Apple	*Penicillium expansum*	Capdeville et al., 2002
Beans	*Colletotrichum lindemuthianum*	Andebrhan and Wood, 1980
Cabbage	*Xanthomonas campestris* pv. *campestris*	Brown et al., 2001
Carrot	*Botrytis cinerea* *Sclerotinia sclerotiorum*	Mercier et al., 1993, 2000
Citrus (grapefruit, lemon, and orange)	*Penicillium digitatum*	Ben-Yehoshua et al., 1992; Wilson et al., 1994; Droby et al., 1993; Porat et al., 1999; D'hallewin et al., 1999, 2000; Yildiz et al., 2001; Kim and Ben-Yehoshua, 2001.
Grapes	*Botrytis cinerea*	Cia et al., 2000
Onion	*Sclerotium cepivorum*	Al-Safadi et al., 2000
Peaches	*Monilinia fructicola* *Botrytis cinerea*	Lu et al., 1993 Crisosto et al., 1998; El Ghaouth et al., 2003a
Strawberry	*Botrytis cinerea* *Monilinia fructicola*	Nigro et al., 2000; Marquenie et al., 2002a, 2003
Tomato	*Rhizopus stolonifer* *Alternaria alteranta*	Liu et al., 1993; Stevens et al., 1998; Rong and Feng, 2001

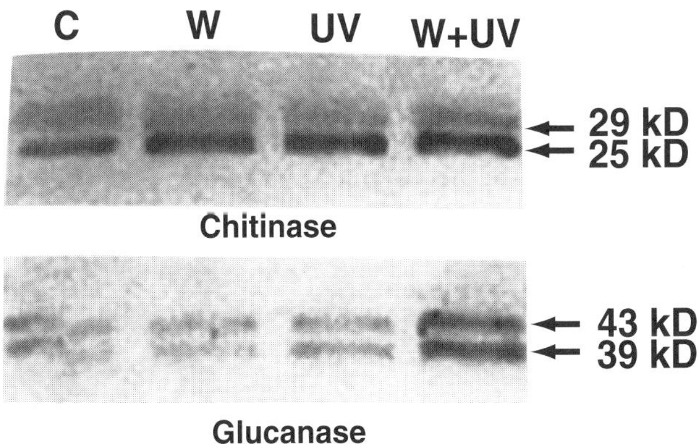

Figure 10.3 Influence of wounding, UV, and the combination of wounding and UV on accumulation of chitinase and β-1.3-endoglucanase protein in grapefruit flavedo tissue at 72 h after treatment. C-Control; W-wounding; UV-UV-irradiation. (Courtesy of Porat et al., 1999; Ariel Publishers, Rehovot, Israel.)

resistance to *P. digitatum* appears to be parallel with production of scoparone (Kim and Ben-Yehoshua, 2001).

Induction of resistance in apples was reported to be associated with stimulation of antifungal hydrolases (Wilson and El Ghaouth, 1993). Resistance to *Penicillium expansum* causing blue mold could be induced by UV-C light and also by chitosan, harpin, and yeasts. Among these inducers of resistance, UV-C light treatment was the most effective in reducing intensity of blue mold disease as reflected by area under the disease progress curve (AUDPC) assessment. Treatment with UV-C was more effective on freshly harvested fruits than on fruits stored under controlled atmosphere (CA). There was a clear time-dependent response of the fruit to the treatments, in which treatments applied 96h before inoculation provided maximum protection against the pathogen. This study showed conclusively that these agents acted by inducing resistance in apple, rather than by inhibiting the development of the pathogen, since there was no direct contact of the pathogen with any of the agents tested (Capdeville et al., 2002).

The storage rot of strawberry caused by *Botrytis cinerea* was significantly reduced by treating the fruits with UV-C light (0.01 J/cm^2), whereas temperatures ranging from 40° to 48°C (for 3 to 15 min.) did not affect the pathogen development. Moreover, the highest temperature damaged the berry surface and reduced fruit firmness. In contrast, no damage was noticeable on the fruit surface even when high doses, up to 6 J/cm^2 were applied (Marquenie et al., 2002b). The effects of UV-C light and heat treatments on two major postharvest pathogens, *B. cinerea* and *Monilinia fructigena* infecting strawberries and cherries, were studied. *M. fructigena* was more sensitive than *B. cinerea* to both treatments (Marquenie et al., 2002a). Exposure of strawberry fruits to UV-C light at low doses reduced storage decay by *Botrytis*. An increase in PAL activity at 12h after irradiation indicated the activation of a metabolic pathway related to the biosyntheses of phenolic compounds, which have antifungal activity. In addition, UV-C light increased ethylene production proportional to the doses applied, reaching the peak at 6h after treatment. The results indicate that reduction in disease incidence in strawberry was due to induction of resistance by UV-C light, though a germicidal effect of reducing external contamination of pathogens cannot be ruled out (Nigro et al., 2000).

Several factors may influence the effectiveness of UV-C treatment on different fruits and vegetables. Grapefruits harvested at various times during the growing season responded differently to UV-C treatments. Likewise, the temperature at which fruits are stored following UV-C treatment influenced the levels of resistance (Droby et al., 1993). Freshly harvested apple fruits were more responsive to UV-C treatment than CA-stored fruits (Capdeville et al., 2002). The mechanism of induction of resistance was studied by determining the effects of UV-C light treatment in peach. UV-C light treatment caused a rapid induction of chitinase, β-1,3 glucanase, and PAL activities starting 6h after treatment reaching the peak (two fold) at 96h after treatment, compared with control. The enhancement of enzymatic activities was preceded by a gradual activation of the corresponding genes (El Ghaouth et al., 2003a) (Fig. 10.4).

Use of UV-C light for disease management may also lead to some negative effects if high doses are used, as was found with strawberries. Doses greater than 1.0 J/cm^2 may affect the color of the leaves (Marqueine et al., 2002b). Higher doses caused greater damages in grapefruit harvested in November than the damage in fruits harvested in February and May (D'hallewin et al., 2000). Treatment of strawberry with UV-C alone induced resistance to *B. cinerea* while pulsed white light and heat treatment did not alter the resistance level (Marquenie et al., 2003).

The possibility of inducing resistance in vegetables has been explored. Stimulation of activity of antifungal hydrolases in bell peppers and tomato fruits after UV-C treatment has been observed (El Ghaouth et al., 1992a). UV-C treatment was found to be more effective in reducing soft rot of tomatoes caused by *Rhizopus stolonifer* in mature green and breaker stage tomato than in ripe red tomatoes (Liu et al., 1993). Production of the phytoalexin 6-methoxymellein was induced in carrot slices exposed to UV-C light and the resistance of treated carrot tissues to infection by *B. cinerea* and *Sclerotinia sclerotiorum* was increased significantly (Mercier et al.,1993). Concentration

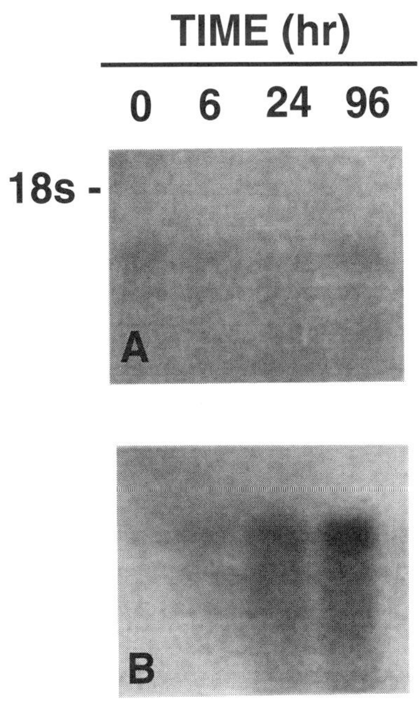

(a)

Figure 10.4 (a) RNA blot analysis of peach chitinase on peach fruit untreated (**A**) and treated (**B**) with UV-C at different times after treatment. The position of lower ribosomal fragment (18S) on the stained gel is indicated.

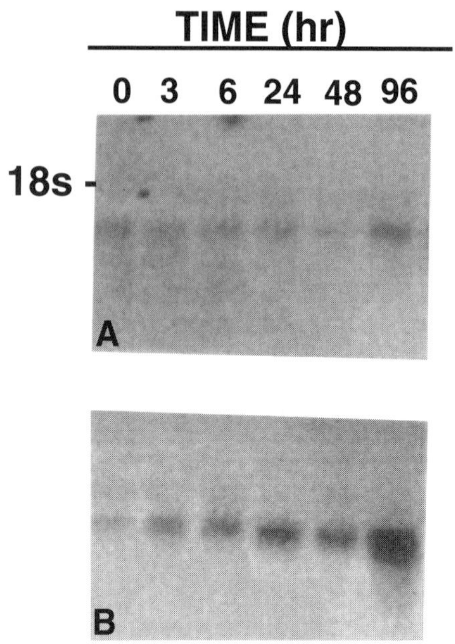

(b)

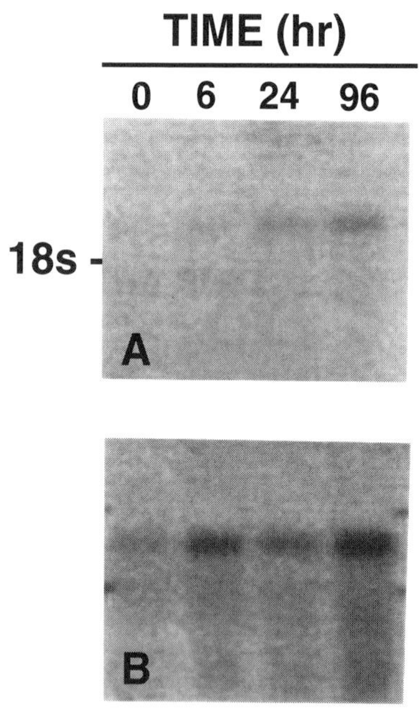

(c)

Figure 10.4 (*continued*) (b) RNA blot analysis of peach β-1,3-glucanase on peach fruit untreated (**A**) and treated (**B**) with UV-C at different times after treatment. The position of lower ribosomal fragment (18S) on the stained gel is indicated. (c) RNA blot analysis of peach phenylalanine ammonia lyase (PAL) on peach fruit untreated (**A**) and treated (**B**) with UV-C at different times after treatment. The position of lower ribosomal fragment (18S) on the stained gel is indicated. (Courtesy of El Ghaouth et al., 2003a; The American Phytopathological Society, St. Paul, MN, USA.)

of 6-methoxymellein in UV-treated carrot exceeded the ED_{50} necessary for inhibition of *B. cinerea* (El Ghaouth and Wilson, 1995). UV radiation had no systemic effect and disease resistance was induced only in tissues directly exposed to radiation. Although UV-treatment and pathogen-induced resistance in carrot may involve the same defense mechanism, the responses are probably mediated differently, because UV radiation has an essentially local effect (Mercier et al., 2000). Delayed ripening and color development in mature green tomato fruits was observed when the tomatoes were treated with UV-C light and inoculated with *Alternaria alternata*. Infection by the pathogen was reduced, indicating the induction of resistance in treated tomatoes (Rong and Feng, 2001).

The effect of UV-C light treatment on the development of black rot disease of cabbage caused by the bacterial pathogen *Xanthomonas campestris* pv. *campestris* was studied. Seeds treated with the optimum dose of UV-C ($3.6\,kJ/m^2$) produced plants with the most desirable color, maximum weight, largest head diameter, and delayed maturity. UV-C light treatment resulted in black rot reduction of 90%, 40%, 60%, and 60% in plants from treated seeds stored for 2 days, 1, 5, and 8 months respectively at 8 weeks after transplanting cabbage plants (Brown et al., 2001).

10.2.1.2 Heat Treatments Heat treatment of fresh fruits and vegetables for the control of decay caused by microbial pathogens differs from uses of heat for wound healing and suppression of viruses, insects and nematodes. Postharvest heat treatments are employed for only short periods (3 to 5 min) because of the presence of pathogens on the surface or within a few outer layers of cells of the commodities. Generally, fruits and vegetables can tolerate temperatures of 50 to 60°C for 5 to 10 min and even shorter exposures at this temperature range have been effective against several microbial pathogens (Barkai-Golan and Phillips, 1991). Two factors, that is location of the target pathogen in or on the produce and thermal sensitivity of the target pathogen are important in the development of effective heat treatments. Postharvest heat treatment is a potential nonchemical disease management strategy acting by: (1) directly inhibiting pathogen growth; (2) activating the natural resistance of the host; and (3) slowing down the ripening process.

Management of postharvest diseases by heat treatment has been reported to be effective for several fruits and vegetables, but heat treatment has been demonstrated to increase levels of resistance of treated fruits and vegetables in only some instances (Ben Yehoshua et al., 1988; Klein and Lurie, 1991). A new hot water brushing (HWB) technique, comprising of rinsing hot water on the fruits as they move along a belt of brush rollers, was developed by Porat et al. (2000a). A 20-s HWB treatment at 59° or 62°C reduced the decay of grapefruit cv. Star Ruby after artificial inoculation of wounded fruits with *Penicillium digitatum* causing green mold disease, by 52 or 70%, respectively, compared with untreated control fruits. The HWB treatments applied 1 to 3 days prior to inoculation were the most effective in enhancing the resistance of fruits to green mold disease. In addition, HWB treatments at 59° or 62°C for 20s significantly reduced the chilling injury index and the percentage of fruits displaying chilling-injury symptoms was reduced by 42 and 58%, respectively, after 6 weeks storage at 2°C and an additional week at 20°C. Moreover, the additional of benefits of HWB treatments were cleaning of fruits and improvement of general appearance without causing surface injury. Scanning electron microscopy revealed that HWB treatment at 56°C for 20s had smoothened the fruit epicuticular waxes and thus covered and sealed stomata and cracks on the fruit surface which would have served as potential sites of pathogen entry (Porat et al., 2000b).

The hot water dip (HWD) treatment for 2min at 52° to 53°C prevented decay for at least 1 week in lemon fruit inoculated with *P. digitatum.* The growth of the pathogen was arrested for 24 to 48h due to a transient inhibitory effect of the HWD treatment. During this lag period the combined effects of the pathogen and HWD treatment induced the build up of resistance in the peel tissue. Production of lignin in the sites of inoculation commenced within 24h after inoculation or wounding. Lignin accumulation continued for a week, when inoculation was followed by HWD. In the case of untreated lemon fruits, no increase in lignin content was seen and the fruits rotted entirely within 3 days after inoculation. The concentration of the phytoalexin scoparone at inoculation sites of HWD fruits started to increase at 24h after treatment and reached a level sufficient to inhibit the pathogen within 2 days after treatment. Scopoletin, another phytoalexin, was detected in inoculated and heat-treated lemons. Heat treatment, as such, was not able to induce the resistance response unless it is combined with wounding or challenge by pathogen (Nafussi et al., 2001).

10.2.2 Chemicals

Enhancing natural mechanisms of resistance to postharvest diseases by the application of both inorganic and organic chemicals has been demonstrated to be a potential strategy for disease management.

10.2.2.1 Calcium Calcium (Ca) has been shown to be associated with disease resistance and increasing the amount of Ca in plant storage organs by different methods is considered as an effective method of increasing natural resistance. Ca may be applied at both pre- and postharvest stages. Preharvest sprays of Ca reduced storage losses caused by *Gloesoporium spp.* in apples (Sharples and Johnson, 1977), whereas postharvest treatments of apples with Ca resulted in a reduction of decay caused by *Penicillium expansum* (Conway and Sams, 1983). However, applying Ca directly to fruits and vegetables may be more effective in enhancing resistance to postharvest diseases. Vacuum infiltration of Ca (250 to 1500μg Ca/g dry weight applied as $CaCl_2$) was the most successful in increasing apple tissue Ca level. There was a negative correlation between the Ca concentration in apple tissue and extent of decay caused by *P. expansum*, the reduction in decay area being more than 50% in apple with maximum Ca content (Conway and Sams, 1983). Infiltration of $CaCl_2$ reduced the decay caused by three isolates of *B. cinerea* in apple and grapes by three to five fold by inhibiting the polygalacturonase activity of the pathogen (Chardonnet et al., 2000). $CaCl_2$ (4.0%) solution at 46.1°C was used for dipping mango fruits for 90 min and the treated fruits were stored at 10°C. This treatment reduced anthracnose disease caused by *Glomerella cingulata* (Freire and Chitarra, 1999). Calcium application reduced the enzymatic activity of *Monilinia fructicola* infecting peach cv. Bucti and increased the concentrations of neutral sugars, in addition to reducing in the degree of cell wall pectin esterification, resulting in enhanced resistance to brown rot disease (Souza et al., 1999). The Ca contents of potato tuber tissue were increased by fertilization with $Ca(NO_3)_2$ and $CaSO_4$. The percent surface area of tuber decay caused by the bacterial pathogen *Erwinia carotovora* pv. *atroseptica* decreased as the tuber Ca content increased, the maximum reduction being 50% of the decay in tubers with lowest Ca content. In tubers vacuum infiltrated with $Ca(NO_3)_2$ solution and inoculated with *E. carotovora* pv. *atroseptica*, the percentage of decayed surface area was reduced from 93% to 15% over the range of infiltrated Ca concentrations. No visible injury on tuber surface was seen following Ca treatments (McGuire and Kelman, 1984).

The Ca-induced resistance of storage organs to postharvest diseases is considered to be due to an interaction between certain cell wall components and Ca ions. The mechanism of resistance associated with tissues high in Ca possibly involves a reduced rate of cell wall maceration as a result of the enhanced structural integrity. Furthermore, Ca is known to inhibit polygalacturonase activity at low concentrations. Calcium treatments increased the Ca content both in the whole grape berries and in the peel, resulting in a reduction of rot caused by *B. cinerea* in the bunches. An increase in the cellulose content of cell wall and increase in protopectins in the cell wall, but a decrease of the same in cytoplasmic fractions following Ca-application indicated the involvement of calcium ions in the stabilization of the cell wall structure. These changes result in enhancement of resistance to *B. cinerea* (Buescher et al., 1979; Miceli et al., 1999)

10.2.2.2 Phosphates Phosphates of potassium and sodium have been shown to be potential inducers of resistance to *Colletotrichum lagenarium* causing anthracnose disease of cucumber (Gottstein and Kuć, 1989). Cucumber was protected against powdery mildew disease caused by *Sphaerotheca fuliginea* by boric acid (0.005 M), copper sulfate (0.0025 M), manganese chloride (0.0025 M), or dipotassium phosphate (0.1 M) 2 h before inoculation. None of the chemicals tested induced phytotoxic symptoms (Reuveni et al., 1996, 1997). Phosphates were considered to generate an endogenous systemic acquired resistance (SAR) signal due to calcium sequestration at the points of phosphate application. The activities of peroxidase and β-1,3 glucanase were enhanced in leaves protected by phosphate application (Gottstein and Kuć,

1989; Reuveni et al., 1997). A later investigation on induction of systemic protection by K_2HPO_4 in cucumber against anthracnose showed that high level of SAR induction was associated with localized cell death in cucumber leaves. The phosphate-mediated cell death was preceded by a rapid generation of superoxide and hydrogen peroxide. In addition, a local and systemic increase in free and conjugated salicylic acid (SA) levels was detected. Induction of SAR by phosphate showed similarity with the biological resistance inducer, *Tobacco necrosis virus* (TNV), in the early steps in signal transduction leading to SAR, but differed from the steps involved in the resistance induced by the commercial plant activator benzo (1,2,3) thiadiazole-7-carbothioic acid-*S*-methyl ester (BTH) (Orober et al., 2002).

10.2.2.3 Chitosan Chitosan, a naturally derived biodegradable fiber (polymer), offers great potential as an antifungal preservative for fresh fruits and vegetables. Chitosan forms films that can be used to coat the surface of fruits and vegetables and it regulates gas and moisture exchange. It has fungicidal properties and an ability to induce resistance to postharvest diseases. Furthermore, when applied as a coating, chitosan may delay ripening, thus increasing the shelf-life, resulting in a significant reduction of the disease incidence of fruits and vegetables (Fig. 10.5) (Wilson et al., 1994). Chitosan has been demonstrated to stimulate defense enzymes and the formation of physical barriers in harvested fruits and vegetables.

Bell pepper treatment with chitosan resulted in a retardation of *B. cinerea* development and reduction in the degradation of cell wall components.

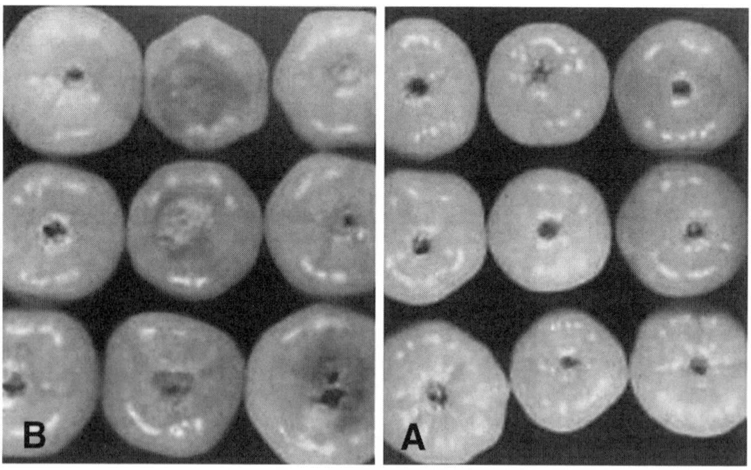

Figure 10.5 Effect of treatment of mature green tomato with chitosan on the development of gray mold disease caused by *Botrytis cinerea*. Disease incidence was recorded at 10 days after inoculation of tomatoes held in plastic containers at 20°C, 95% RH. (a)-treated; (b)-control (untreated). (Courtesy of Wilson et al., 1994; The American Phytopathological Society, St. Paul, MN, USA.)

Various levels of cellular disorganization, ranging from wall loosening to cytoplasm disintegration, due to chitosan treatment was observed in the pathogen. Chitosan also induced cellular alterations in *Rhizopus stolonifer*-infected strawberries (El Ghaouth et al., 1992a; Wilson and El Ghaouth, 1993). The ability of chitosan to elicit synthesis of antifungal hydrolases, such as chitinases, β-1,3-glucanases, and chitosanases, was demonstrated in strawberry fruit, bell pepper, and tomato fruit. The activities of these enzymes in these commodities remained elevated for up to 14 days after treatment, giving a head start in restricting fungal colonization (Wilson and El Ghaouth, 1993). Prestorage treatment of bell pepper with chitosan resulted in the induction of chitinase and β-1,3-glucanases, followed by a substantial reduction of chitin labeling of the walls of invading hyphae. There was also thickening of the host cell wall, formation of papillae, and plugging of some intercellular spaces with fibrillar material partially impregnated with amorphous electron-opaque substances which are likely to be antifungal phenolic compounds (Wilson et al., 1994).

Strawberry plants cv. Seascape were sprayed twice, at a 10-day interval, with chitosan (2, 4 or 6 g/l) and harvested fruits were challenged with *B. cinerea* and stored at 3° or 13°C. Chitosan significantly reduced postharvest fungal rot and maintained the keeping quality of fruits compared with water-sprayed controls. The incidence of decay decreased with increased chitosan concentration and increased with storage period and temperature. Chitosan sprays (6 g/l) applied twice protected the fruits from decay and maintained fruit quality at an acceptable level throughout the storage period of 4 weeks at 3°C (Reddy et al., 2000b). Three preharvest sprays of chitosan reduced rots due to *B. cinerea* and *R. stolonifer*, with greatest reductions observed at 1.0% concentration. Likewise, dipping strawberries in chitosan solution (1.0%) for 3 to 4 s resulted in maximum protection against *B. cinerea*, where the incidence of *R. stolonifer* was negligible (Romanazzi et al., 2000). Chitosan coating of strawberries and raspberries stored at 13°C was as effective as thiobendazole in controlling the decay of berries caused by *B. cinerea* and *R. stolonifer*. In addition, chitosan coating exerted beneficial effects on firmness, titrable activity, vitamin C content, and anthocyanin content of strawberries and raspberries stored at 4°C (Zhang and Quantick, 1998).

Chitosan treatment (5 and 10 mg/ml) reduced the incidence of brown rot of peaches caused by *Monilinia fructicola* significantly and delayed the disease development compared with control. Treated peaches were firmer and had higher titrable acidity and vitamin C content than the fungicide prochloraz-treated or control peaches. Chitosan has the potential to control brown rot, preserve valuable attributes, and prolong the shelf-life of postharvest peaches (Li and Yu, 2001). Stem scar application of chitosan (400 μl of a 10 g/l) solution to tomatoes stored at 20°C provided protection against black mold caused by *Alternaria alternata* (Reddy et al., 2000a). In carrot roots treated postharvest with a 0.2% chitosan hydrolysate prepared by using *Streptomyces* N-174 chitosanase, the frequency and size of rot caused by *Sclerotinia sclerotiorum* were reduced, compared to untreated controls. When carrots were treated with either chitosan hydrolysate or high molecular weight chitosan and inoculated at intervals over the following 5 days, the decline in infection was greater with the hydrolysate treatment. The results indicated that chitosan treatment induced host resistance (Molly et al., 2004).

In chitosan treated tomato fruits, the activity of macerating enzymes, such as polygalacturonase (PG), pectate lyase, and cellulase, in tissue in the vicinity of lesions caused by *Alternaria alternata* was about 50% less than in control fruits. Chitosan treatment also inhibited production of oxalic acid, fumaric acid (chelating agents), and host-specific toxins, such as alternariol and alternariol monomethylether, by the fungus. The pH of the infected tissue decreased from 4.7 to 4.0 in the control fruit, the optimum for PG activity, while pH of chitosan-treated fruit remained at 4.6. Furthermore, chitosan also induced production of rishitin (a phytoalexin) in tomato tissue (Reddy et al., 2000a). Chitosan-treated peaches showed lower respiration rate, less ethylene and malondialdehyde (MDA) production, higher superoxide dimutase activity, and better membrane integrity (Li and Yu, 2001). The multiple modes of action of chitosan may be particularly useful in providing effective protection to fruits and vegetables against postharvest diseases.

The systemic nature and persistence of defense enzymes in plant tissue, on elicitation by chitosan, could be important in retarding the resumption of quiescent infection, which typically becomes active when tissue resistance declines. Since many postharvest diseases arise from latent infections that become active upon the decline of the biosynthetic potential of the tissue to produce antimicrobial compounds, use of chitosan may prove to be very effective.

10.2.2.4 Jasmonates Postharvest application of jasmonic acid (JA) and methyl jasmonate (MJ), at 10 μmol/l, reduced decay caused by *Penicillium digitatum* (green mold) after either natural infection or artificial inoculation of grapefruit cv. Marsh Seedless. This treatment also protected the treated fruit against chilling injury. As JA and MJ did not inhibit spore germination or germ tube elongation in *P. digitatum*, they reduce green mold decay in grape fruits indirectly by enhancing natural resistance of the fruits to the pathogen both at high (24°C) and low temperatures (2°C) (Droby et al., 1999). Application of MJ vapors to mango cv. Tomny Atkins for 24 h at 25°C reduced chilling injury during subsequent storage at 7°C for 21 days and after 5 days of shelf-life at 20°C. Exposure to MJ vapors did not influence the ripening process (Gonzalez-Aguilar et al., 2000).

10.2.2.5 Methyl Salicylate (MSA) The suppressive effect of MSA fumigation on gray mold rot caused by *B. cinerea* in harvested strawberry fruits was assessed. In MSA-treated plots, fruit decay was reduced by one-third compared with untreated controls. The inhibitory effect was more conspicuous when MSA was applied to fruits. MSA was converted into salicylic acid (SA) and increased the activity of chitinase, which is one of the pathogenesis-related

(PR)-proteins. MSA is one of the natural volatile compounds in strawberry fruits and could be applied as a nontoxic alternative to fungicide application (Kim and Choi, 2002).

10.2.2.6 Laminarin β-1,3-glucan laminarin derived from the blue green algae, *Laminaria digitata*, efficiently elicited defense responses in grapevine cells and plants against *Botrytis cinerea* and *Plasmopara viticola*. The development of these pathogens in infected grapevine plants was effectively curtailed by laminarin. Calcium influx, alkalinization of the extracellular medium, an oxidative burst, activation of two mitogen-activated protein kinases, expression of 10 defense-related genes with different kinetics and intensities, increases in chitinase and β-1,3-glucanase activities, and production of two phytoalexins, resveratrol and epsilon-viniferin, were the defense responses elicited by laminarin. Application of laminarin reduced infection by *B. cinerea* and *P. viticola* which could infect berries later, by about 55 and 75% respectively. Laminarin was applied to grapevine plants and infection by *Botrytis cinerea* and *Plasmopara viticola* was reduced, by about 55 and 75% respectively. By reducing the pathogen populations on the grapevine plants, the incidence of these diseases on berries may be reduced. The results indicate that activation of natural defense responses using elicitors could be advantageously exploited as an effective strategy to protect grapevine plants against pathogens occurring at pre- and postharvest stages (Aziz et al., 2003).

10.2.2.7 Acibenzolar The plant activator acibenzolar *S*-methyl benzo (1,2,3) thiadiazole-7-carbothioate (BTH) applied to strawberry plants at 0.25 to 2.0 mg active ingredient (a.i.)/ml delayed the development of gray mold disease (caused by *Botrytis cinerea*) on harvested fruits by about 2 days when stored at 5°C. The delay was equivalent to a 15 to 20% increase in storage life of the fruit (Terry and Joyce, 2000).

Application of BTH (as Bion WG 50) and acetyl salicylic acid (ASA) at 100 mg a.i./l and 400 mg a.i./l, respectively, reduced the severity of dry rot disease caused by *Fusarium semitectum* in wound-inoculated potato tubers at postharvest stage. BTH treatment increased the activity of β-1,3-glucanase in leaves, stem, tubers, and stolons of potato plants (Bokshi et al., 2003). Pepper plants sprayed with acibenzolar-*S*-methyl (ASM, a derivative of BTH) and copper hydroxide were protected against infection of fruits by *Xanthomonas campestris* pv. *vesicatoria* (*Xcv*). As the protection offered by ASM was observed when the inducer was entirely degraded, it appears that the protection was due to activation of systemic acquired resistance (SAR) (Buonaurio et al., 2002). Pepper plants carrying one or more major genes for resistance (R) to *Xcv* were sprayed with ASM followed by challenge with incompatible strains of *Xcv*. At 3 weeks after inoculation, the ASM-treated plants showed significantly fewer lesions induced by the both incompatible (hypersensitive) and compatible (susceptible) responses than the number of lesions formed on untreated plants. Disease lesions contained race-change mutants. The results indicate the potential usefulness of SAR inducers to increase the durability of genotype-specific resistance conferred by R genes (Romero and Ritchie, 2004).

10.2.2.8 Aminobutyric Acid Application of aminobutyric acid at 20 mM to specific wound sites on the grapefruit peel tissue induced resistance to *Penicillium digitatum* causing green mold disease. Induction of resistance by aminobutyric acid was local and limited to the vicinity (1 to 2 cm) of treated site. Direct antifungal activity at higher concentrations of aminobutyric acid was observed, as reflected by inhibition of conidial germination and elongation of germ tube. Activation of defense responses in grape peel tissue following treatment, including activation of chitinase gene expression and protein accumulation after 48 h was observed. An increase in PAL activity after 72 h was also evident (Porat et al., 2003).

10.2.2.9 Miscellaneous Compounds Several compounds of diverse origin have been evaluated for their ability to induce resistance in plants against postharvest pathogens. Red Delicious apple fruits after harvest were sprayed with harpin (40, 80 and 160 mg/l), as a commercial formulation, and wound-inoculated with *P. expansum* causing blue mold disease. Treatment with harpin resulted in reduced disease incidence in addition to a slower rate of disease development. Similar beneficial effect of harpin applied preharvest on disease incidence in storage was also observed. The results showed that spraying apple trees with harpin a few days prior to harvest is a promising strategy for reducing blue mold decay in storage (de Capdeville et al., 2002).

10.2.3 Biocontrol Agents

Several antagonistic microorganisms have the potential to protect fruit and vegetables against pathogens and they have been suggested as alternatives to synthetic fungicides. These organisms may grow rapidly and colonize wound sites where infections occur and out-compete postharvest pathogens for space and nutrients. Some of them have been shown to induce resistance in host tissues resulting in significant reduction in decay development. Antagonistic yeasts are capable of inducing resistance responses as in the case of *Pichia guilliermondii* (US-7) and *Candida saitoana* (Wisniewski and Wilson, 1992). The yeast species stimulated the production of defense-related enzymes and antimicrobial substances. The yeast *Aureobasidium pullulans* multiplied rapidly and controlled decay caused by either *B. cinerea* or *P. expansum*. In addition, *A. pullulans* caused a transient increase in β-1,3 glucanase, chitinase, and peroxidase activities commencing from 24 h after treatment and reaching maximum levels at 48 and 96 h after treatment (Ippolito et al., 2000). Likewise, *A. pullulans* was reported to increase natural resistance to *B. cinerea* causing grey mold in strawberry (Adikaram et al., 2002).

The yeast *Candida oleophila*, forming the basic component of the commercial product "Aspire," when applied to surface wounds or to intact Marsh

Seedless grapefruit, elicited systemic resistance to *Penicillium digitatum*. The induction of resistance was pronounced 24h after elicitation, which required viable yeast cells at concentrations of 10^8 to 10^9 cells/ml. Application of *C. oleophila* cell suspension to grapefruit peel tissue increased ethylene biosynthesis, PAL activity, and phytoalexin accumulation. Enhancement of chitinase and β-1,3 endoglucanase protein levels was indicated by western immunoblot analysis. Scanning electron microscopy (SEM) showed that spore germination and germ tube growth of *P. digitatum* were markedly inhibited in wounds made near the yeast-treated sites. This evidence indicated that induced resistance against *P. digitatum* could be an important component of the multiple modes of action of *C. oleophila* (Droby et al., 2002). The efficacy of *Candida saitoana* in inducing systemic resistance in apple fruit against *B. cinerea* was assessed (Appendix 10(i)). When *C. saitoana* was applied 48 to 72h preinoculation with *B. cinerea*, the lesion diameter was reduced by more than 50 and 70% respectively, compared with wounding. In addition to inducing systemic resistance, *C. saitoana* increased chitinase and β-1,3 glucanase activities, with a higher accumulation in fresh than in stored apples. In fresh apples, the onset of systemic resistance to *B. cinerea* coincided with increase in chitinase and β-1,3 glucanase activity in systemically protected tissue. It appears that antifungal hydrolases may be involved in the systemic protection induced by *C. saitoana* (El Ghaouth et al., 2003a, b) (Fig. 10.6).

The possibility of microbial agents other than yeasts capable of inducing resistance to postharvest pathogens has been indicated. The mycoparasite *Verticillium lecanii* has been demonstrated to have potential to protect citrus fruit against *P. digitatum* causing green mold disease. This BCA could protect the treated citrus fruits by a combination of mechanisms including direct antimicrobial activity (Chapter 9) and induced resistance. *Verticillium lecanii* inhibited the growth of *Penicillium digitatum* causing green mould disease. The intercellular interaction between *V. lecanii* and *P. digitatum*, was studied by gold cytochemistry and electron microscopy. Retardation of the plasma membrane and disorganization of cytoplasm of the fungal pathogen could be visualized. These alterations were associated with the deposition on the inner host cell surface of a chitin and cellulose-enriched material which appeared to be laid down as a structural defense reaction. The accumulation of chitin correlated with a decrease in the amount of wall-bound chitin (Benhamou and Bordeur, 2000).

Evidence obtained by using cytochemical, scanning and transmission electron microscopic techniques indicated that *V. lecanii* could trigger defense-related reactions in citrus fruits. The mechanism of induction of resistance by the antagonist was compared with that of chitosan, which has been reported to be an effective inducer of resistance against several postharvest pathogens. Restriction of pathogen colonization at the host-cell surface correlated well with marked disorganization of pathogen hyphae. Both *V. lecanii* and chitosan exhibited a similar capacity for inducing a striking response in *P. digitatum*-infected citrus fruits. Labeling with gold-complexed laccase indicated that

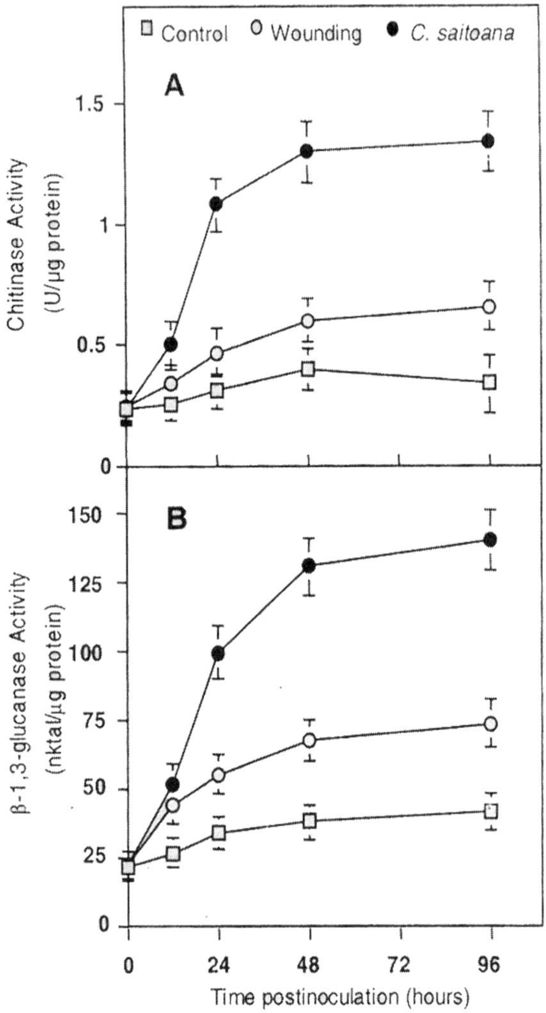

Figure 10.6 Time course of changes in chitinase (**A**) and β-1,3-glucanase (**B**) activities in nontreated tissues close to wounds of fresh apples that have been treated with *Candida saitoana* (●) or sterile water (○) and nonwounded control fruit (□). (Courtesy of El Ghaouth et al., 2003b; The American Phytopathological Society, St. Paul, MN, USA.)

treatment with either *V. lecanii* or chitosan stimulated the de novo synthesis of phenolic compounds such as lignin and/or the polymerization of pre-existing free, soluble phenols. Furthermore, accumulation of callose in citrus exocarp tissues following treatment with *V. lecanii* or chitosan was also observed. The results suggest that *V. lecanii* and chitosan may induce transcriptional activation of defense gene, leading to the accumulation of structural and biochemical compounds at strategic sites (Benhamou, 2004).

Phytophthora capsici causes root rot, crown rot, leaf and stem blight, and fruit rot of cucumber, inflicting serious losses at all stages. *Trichoderma hamatum* 382 (TC382) in compost-amended potting mix significantly reduced the severity of the disease and it was equally effective as BTH or mefenoxam (Subdue MAXX). As TC382 remained spatially separated from the pathogen in the infected plant tissue, the protection offered by the BCA was attributed to induction of systemic resistance (Khan et al., 2004).

The growth-promoting rhizobacteria (PGPR) *Pseudomonas aeruginosa* 7NSK2 produces pyochelin (Pch), its precursor salicylic acid (SA), and pyocyanin as secondary metabolites. The wild-type (7NSK2) and its mutant KMPCH (Pch negative, SA-positive) could induce resistance to *B. cinerea* in tomato (parent) but not in transgenic NahG tomato expressing the *NahG* gene of *Pseudomonas putida*. The wild-type PGPR possibly acted by efficiently converting SA to Pch. A pyocyanin mutant of 7NSK2 was generated by a mutation in the *phzM* gene encoding an *O*-methyltransferase. The mutant PHZ 1 could not induce resistance to *B. cinerea* indicating a role for pyocyanin in induction of resistance. Complementation for pyocyanin or coinoculation with another mutant 7NSK2-562 (Pch negative, SA-negative, pyocyanin positive) resulted in enhancement of resistance to *B. cinerea* (Audenaert et al., 2002). Preharvest application of PGPR, *Pseudomonas fluroescens* (strain FP7) amended with chitin not only induced resistance to mango anthracnose disease caused by *Colletotrichum gloeosporioides*, but also enhanced the yielding capacity of mango trees. Lytic enzymes such as chitinase and β-1,3 glucanase involved in defense responses of plants showed marked increase, following treatment with the combination of FP7 and chitin (Vivekananthan et al., 2004) (Appendix 10(ii)).

The plant tissues, upon infection by microbial pathogens or treatment with elicitors, often react by activating a highly coordinated biochemical and structural defense system that restricts the spread of the pathogens. Though harvested fruits and vegetables also possess inducible defense responses, this potential has not been exploited adequately for the management of postharvest diseases. Activating biochemical defense responses in harvested fruits and vegetables through prestorage treatment with UV-C light, heat, chemicals, and antagonistic yeasts has been demonstrated to be a potential management strategy for reducing losses caused by postharvest pathogens. As more information becomes available about the function and regulation of disease resistance genes in harvested tissue, it would be possible to engineer the expression of complex resistance in postharvest commodities. The results obtained with different combination of biological products demonstrated the potential of this multifaceted approaches as a viable alternative to synthetic fungicides. The complexity of this mode of action, displayed by combined alternatives, may make the development of pathogen resistance more difficult, and present a more complex disease deterrence barrier than the approach relaying on a single biological agent (Chapter 12).

10.3 MYCOTOXIN MANAGEMENT THROUGH TRANSFORMATION

Fungal pathogens infecting cereals such as *Fusarium* spp. cause direct quantitative and qualitative losses of grains. In addition, contamination of grains with the mycotoxins produced by these pathogens is a potential risk for humans and animals due to the carcinogenic properties of mycotoxins. Zearalenone (ZEN) produced by different species of *Fusarium* (Chapter 6) has frequently been detected at high concentrations in maize, wheat, and barley. ZEN can induce severe morphological and functional disorders of reproductive organs in animals, especially female swine. Only limited success has been achieved by using chemicals to reduce the ZEN levels in foods. Food grade *Lactobacillus* strains could remove ZEN to some extent by binding to the cell surface. A more efficient method of detoxifying ZEN, by employing genetically modified organisms, has been developed. ZEN can be converted into a nontoxic compound by the enzymatic action of lactonohydrolase encoded by the gene *zhd 101* isolated from *Clonostachys rosea* (Takashi-Ando et al., 2002). The recombinant *Escherichia coli* strains expressing the detoxifying gene *Zhd 101* were able to remove ZEN and derivatives more efficiently than the recombinant yeast *Saccharomyces cerevisiae*. The transformed *E. coli* entirely degraded ZEN in liquid media, whereas 75% of ZEN was detoxified by the yeast transformant, which was significantly more effective than the food grade *Lactobacillus* strains (Takahashi-Ando et al., 2004). This research effort has provided an exciting and effective line of approach for the management of dreadful mycotoxins contamination of foods and feeds.

The possibility of generating transgenic plants capable of adapting to deoxynivalenol (DON) has been indicated. Using yeast as a model system, several mutations have been caused in the gene encoding ribosomal protein L3 (Rpl 3) which could confer semidominant resistance to trichothecenes. Tobacco plants transformed with an engineered tomato (Le RPL3) cDNA with one of the amino acid changes identified in yeast, exhibited higher level of resistance to DON. In the presence of wild type Rpl 3 protein, the engineered Rpl 3 protein was not utilized, unless the transgenic plants were challenged with sublethal amounts of DON (Mitterbauer et al., 2004). A wheat line transformed with constitutively expressing genes encoding a class IV acidic chitinase and an acidic β-1,3-glucanase exhibited significant delay in the spread of *Fusarium graminearum* causing *Fusarium* head blight disease under greenhouse conditions. The apoplastic fluid from the transgenic plants contained PR-proteins belonging to the families of β-1,3-glucananses, chitinases, and thaumatin-like proteins. The apoplastic fluid showed growth inhibiting activity against *F. graminearum*, in addition to two other fungal pathogens infecting wheat. The antimicrobial potential of apoplastic fluid may depend on concentrations and combination that reach sufficient levels to inhibit the pathogens (Anand et al., 2004). These reports open up a new approach for the management of mycotoxins in agricultural commodities.

SUMMARY

Developments in molecular biology and genetic engineering have provided new options and opportunities which are not available for producing crop cultivars with built-in resistance through conventional breeding techniques. It is possible, by genetic engineering, to identify, isolate, and incorporate disease resistance genes from wild relatives, microbes, and even animals in the desired cultivars which are to be protected. Furthermore, the cultivars with the desired traits, but susceptibility to diseases, may be protected by enhancing the natural disease resistance mechanisms operating in those cultivars by employing physical, chemical, and biological inducers/elicitors of resistance. The comparative usefulness of various techniques applied on a wide range of fruits and vegetables to improve the levels of resistance to postharvest diseases is discussed.

APPENDIX 10(I): ASSESSMENT OF INDUCED RESISTANCE BY MICROBIAL AGENT (*CANDIDA SAITOANA*) (EL GHAOUTH ET AL., 2003B)

A. Culturing the Biocontrol Agent (BCA)

i. Streak the suspension of the BCA (*C. saitoana*) on nutrient yeast dextrose agar plates and incubate at 24°C for 48 h.
ii. Transfer the BCA to nutrient yeast dextrose broth (10^8 CFU of the BCA) in Erlenmeyer flask (125 ml); incubate for 48 h at 24°C on a rotary shaker (200 rpm); centrifuge at 3000 × g for 20 min; resuspend the pellet in sterile distilled water and centrifuge again.
iii. Disperse the pellet in sterile distilled water and adjust the yeast cell concentration to 10^8 CFU/ml.

B. Culturing the Fungal Pathogen (*Botrytis cinerea*)

i. Isolate the pathogen from infected fruit (apple) and cultivate the pathogen on potato dextrose agar (PDA) by following standard isolation techniques.
ii. Collect the conidia from 2-week old PDA cultures incubated at 27°C in the dark; prepare conidial suspension by flooding the culture plates with 10 ml aliquots of of sterile distilled water containing 0.1% Tween (v/v) and adjust the conidial concentration to 10^5 spores/ml using a hemocytometer.

C. Preparing Test Fruits (Apple)

i. Hand pick the test fruits at maturity from the trees; sort them to remove fruits with apparent injuries or infection and store at 4°C under high humidity (95% RH) for the required period (1 week to 5 months), depending on the experimental design.

D. Testing the Resistance-inducing Capacity

i. Make wounds (3 mm wide by 5 mm deep) individually on fresh and stored fruits at two different locations (2 cm apart), to prevent contact between pathogen and BCA; treat the initial wound on each fruit with (yeast) BCA suspension (35 μl of 10^8 CFU/ml) or sterile water.
ii. Make superficial wounds (5 mm wide by 1 mm deep) with a razor blade at different time interval(0, 24, 48, and 72 h) after treatment with BCA; challenge inoculate the wounds with the spore suspension (10 μl) of the pathogen at 10^4 spores/ml and store the fruits at 20°C under 95% RH in enclosed plastic trays.
iii. Maintain four replicates of 10 fruits each; arrange the treatment in a randomized block design and repeat the experiment twice.
iv. Evaluate the fruits under different treatments including control (water treated) daily for the rot development and superficial wounds challenged with the pathogen.
v. After 7 days of storge, determine the lesion diameters at the wound site; analyze the tissue samples containing superficial wounds for the contamination with BCA.
vi. Remove tissue samples with a cork borer (6 mm diameter) from four fruits randomly selected from each treatment; homogenize the tissue in sterile water (5 ml) and vortex; plate dilutions of the tissue suspension in triplicates on yeast maltose agar medium and incubate the plates at 24°C. Pathogen populations as CFU/ml are recorded.
vii. Analyze the data statistically using analysis of variance (ANOVA).

APPENDIX 10(II): DETECTION OF CHITINASE AND B-1,3-GLUCANASE ACTIVITIES (PAN ET AL., 1991)

A. Enzyme Preparations

i. Freeze the test plant tissue in liquid nitrogen soon after collecting the samples and extract by grinding in 0.05 M sodium acetate buffer (pH 5.0) added at 1 ml/g of tissue with a mortar and sea sand at 4°C.
ii. Dialyze the extract against two changes of water, followed by two changes of 0.01 M sodium acetate buffer (pH 5.0) over night at 4°C.

B. Polyacrylamide Isoelectrofocusing (IEF)

i. Run 2.0 mm thick 7.5% polyacrylamide gels (pH 3.5–9.5) using an LKB 2117 Multiphor system for 3 h at 6°C (constant power 10 W); apply the samples to the gel surface on sample application pieces (LKB).

C. Native Polyacrylamide Gel Electrophoresis (PAGE)

i. Prepare polyacrylamide resolving gels (15%, 1.5 mm thick) by mixing 15 ml of acrylamide/Bisacrylamide; 30:0.8), 3.75 ml of 3.0 M Tris-HCL (pH 8.8) for anodic PAGE or 3.75 ml of 3.0 M acetic acid KOH (pH 4.3) for cathodic PAGE, 1.5 ml of 1.5% ammonium persulfate and 9.75 ml of distilled water, degas for 10 min, then add 15 µl of *N*, *N*, *N′*, *N′* tetramethylethylene diamine (TEMED) and pour.

ii. Prepare the stacking gels by using 2.5 ml of 30% acrylamide, 1.0 ml of 1.5% ammonium persulfate, 11.5 ml of distilled water and 5.0 ml of 0.5 M acetic acid-KOH (pH 6.7) for cathodic PAGE; degas the gel for 10 min; add 15 µl of TEMED, and pour the gel.

iii. Run the gels for 4 h at 8°C (constant current: 30 mA) with an LKB 2001 Vertical Electrophoresis Unit.

D. Detection of Enzymes

i. For staining to visualize chitinase, incubate after electrophoresis or isoelectrofocusing, the PAGE or IEF gels attached to supporting glass plates, in 1.0 M sodium acetate (pH 5.0) for 5 min; cover the gels with a 7.5% (0.75 mm thick) polyacrylamide overlay gel (attached to another supporting glass plate) containing glycol chitin in 0.1 M sodium acetate (pH 5.0) and eliminate the bubbles between PAGE or IEF gels and overlay gels by gently sliding overlay gels on the top of the gels.

ii. Incubate the gels at 40°C for 1.5 h under moist conditions; incubate overlay gels in freshly prepared 0.01% fluorescent brightener 28 (w/v) in 500 mM Tris-HCl (pH 8.9) at room temperature for 5 min; discard the brightness solution and incubate the overlay gels in water at room temperature in darkness for 2 h or in the fridge overnight.

iii. Place overlay gels on a UV-transilluminator and observe the presence of chitinase isozymes as clear zones and store in the refrigerator for photographing.

iv. For staining to visualize β-1,3 glucanase, wash the PAGE or IEF gels with water; incubate with 0.05 M sodium acetate (pH 5.0) for 5 min and again at 40°C for 30 min in a mixture containing 75 ml of 0.05 M sodium acetate (pH 5.0) and 1 g of laminarin dissolved in 75 ml of water by heating in a boiling water bath.

v. Incubate the gels in a mixture of methanol, water, and acetic acid (5:5:2) for 5 min and wash with water.

vi. Stain the PAGE gels with the staining solution containing 0.3 g of 2,3,5-triphenyl-tetrazolium chloride in 200 ml of 1.0 M NaOH in a boiling water bath until the appearance of red bands (about 10 min may be required).

vii. Stain the IEF gels in the same manner and heat in a microwave oven for 3 min (shaking once in every 30 seconds) instead of using water bath.

viii. The gels are stained for proteins with Coomassie Brilliant Blue R 250 then dried.
ix. Run blank gels as controls without substrates glycol chitin and laminarin to verify the activities of chitinase and β-1,3 glucanase enzymes.

REFERENCES

Adikaram, N.K.B., Joyce, D.C. and Terry, L.A. 2002. Biocontrol activity and induced resistance as a possible mode of action for *Aureobasidium pullulans* against grey mould of strawberry fruit. *Australian Journal of Plant Pathology*, **31**: 223–229.

Ahn, Y.K., Kang, J.C., Kim, H.Y., Lee, S.D. and Park, H.G. 2001. Resistance to blackleg and tuber soft rot in interspecific somatic hybrids between *S. brevidens* and *S. tuberosum* (Superior, Dejima and Dihaploid of Superior). *Journal of the Korean Society for Horticultural Science*, **42**: 430–434.

Al-Safadi, B., Mir Ali, N. and Arabi, M.I.E. 2000. Improvement of garlic (*Allium sativum* L.) resistance to white rot and storability using gamma irradiation induced mutations. *Journal of Genetics and Breeding*, **54**: 175–181.

Andebrhan, T. and Wood, R.K.S. 1980. The effect of ultraviolet radiation on the reaction of *Phaseolus vulgaris* to species of *Colletotrichum. Physiological Plant Pathology*, **17**: 105–110.

Anand, A., Lei, Z., Summer, L.W., Kirankumar, S.M., Arakane, Y., Bockus, W.W. and Muthukrishnan, S. 2004. Apoplastic extracts from a transgenic wheat line exhibiting lesion mimic phenotype have multiple pathogenesis-related proteins that are antifungal. *Molecular Plant–Microbe Interactions*, **17**: 1306–1317.

Audenaert, K. Meyer, G.B. de and Höfte, M.M. 2002a. Abscisic acid determined basal susceptibility of tomato to *Botrytis cinerea* and suppresses salicylic acid-dependent signal mechanisms. *Plant Physiology*, **128**: 491–501.

Audenaert, K., Pattery, T., Cornelis, P. and Höfte, M. 2002b. Induction of systemic resistance to *Botrytis cinerea* in tomato by *Pseudomonas aeruginosa* 7NSK2: role of salicylic acid, pyochelin and pyocyanin. *Molecular Plant–Microbe Interaction*, **15**: 1147–1156.

Aziz, A., Poinssot, B., Daire, X., Adrian, M., Bézier, A., Lambert, B., Joubert, J. and Pugin, A. 2003. Laminarin elicits defense responses in grapevine and induces protection against *Botrytis cinerea* and *Plasmopara viticola. Molecular Plant–Microbe Interactions*, **16**: 1118–1128.

Barkai-Golan, R. and Phillips, D.J. 1991. Postharvest heat treatment of fresh fruits and vegetables for decay control. *Plant Disease*, **75**: 1085–1089.

Bau, H.J., Cheng, Y.H., Yu, T.A., Yang, J.S., Liou, P.C., Hsiao, C.H., Lin, C.Y. and Yeh, S.D. 2004. Field evaluation of transgenic papaya lines carrying the coat protein gene of *Papaya ringspot virus* in Taiwan. *Plant Disease*, **88**: 594–599.

Benhamou, N. 2004. Potential of the mycoparasite *Verticillium lecanii*, to protect citrus fruit against *Penicillium digitatum*, the causal agent of green mold, a comparison with the effect of chitosan. *Phytopathology*, **94**: 693–705.

Benhamou, N. and Brodeur, J. 2000. Evidence for antibiosis and induced host defence reaction in the interaction between *Verticillium lecanii and Penicillium digitatum*, the causal agent of green mold. *Phytopathology*, **90**: 932–943.

Ben-Yehoshua, S., Rodov, V., Kim, J.J. and Carmeli, S. 1992. Preformed and induced antifungal materials of citrus fruits in relation to the enhancement of decay resistance by heat and ultraviolet treatments. *Journal of Agricultural and Food Chemistry*, **40**: 1217–1221.

Ben-Yehoshua, S., Shapiro, B., Kim, J.J., Sharoni, J., Carmeli, S. and Kashman, Y. 1988. Resistance of citrus fruit to pathogens and its enhancement by curing. In: *Proceedings of the Sixth International Citrus Congress* (eds) R. Goren and K. Mendel, pp. 1371–1379, TelAviv, Israel.

Berrocal-Lobo, M., Segura, A., Moreno, M., López, G., Garcia-Olmedo, F. and Molina, A. 2002. Snakin–2, an antimicrobial peptide from potato whose gene is locally induced by wounding and responds to pathogen infection. *Plant Physiology*, **128**: 951–961.

Bokshi, A.I., Morris, S.C. and Deerall, B.J. 2003. Effects of benzothiadiazole and acetyl salicylic acid on β-1,3-glucanase activity and disease resistance in potato. *Plant Pathology*, **52**: 22–27.

Bolar, J.P., Norelli, J.L., Wong, K.W., Hayes, C.K., Harman, G.E. and Aldwinckle, H.S. 2000. Expression of endochitinase from *Trichoderma harzianum* in transgenic apple increases resistance to apple scab and reduces vigor. *Phytopathology*, **90**: 72–77.

Brown, J.E., Lu, T.Y., Stevens, C., Khan, V.A., Lu, J.Y., Wilson, C.L., Collins, D.J., Wilson, M.A., Igwegbe, E.C.K., Chalutz, E. and Droby, S. 2001. The effect of low dose ultraviolet-C seed treatment induced resistance in cabbage to black rot (*Xanthomonas campestris* pv. *campetris*). *Crop Protection*, **20**: 873–883.

Brummel, D.A., Howie, W.J., Ma, C. and Dunsmuir, P. 2002. Postharvest fruit quality of transgenic tomatoes suppressed in expression of a ripening-related expansin. *Postharvest Biology and Technology*, **25**: 209–220.

Buescher, R.W., Hudson, J.M. and Adams, J.R. 1979. Inhibition of polygalacturonase softening of cucumber pickles by calcium chloride. *Journal of Food Science*, **44**: 1786–1787.

Buonaurio, R., Scarponi, L., Ferrara, M., Sidoti, P. and Bertona, A. 2002. Induction of systemic acquired resistance in pepper plants by acibenzolar-S-methyl against bacterial spot disease. *European Journal of Plant Pathology*, **108**: 41–49.

Capdeville, G. de, Wilson, C.L., Beer, S.V. and Aist, J.R. 2002. Alternative disease control agents induce resistance to blue mold in harvested 'Red Delicious' apple fruit. *Phytopathology*, **92**: 900–908.

Chalutz, E., Droby, S., Wilson, C.L. and Wisniewski, M.E. 1992. UV-induced resistance to postharvest diseases of citrus fruit. *Journal of Phytochemistry and Photobiology*, **15**: 367–374.

Chardonnet, C.O., Sams, C.E., Trigiano, R.N. and Conway, W.S. 2000. Variability of three isolates of *Botrytis cinerea* affects the inhibitory effects of calcium on this fungus. *Phytopathology*, **90**: 769–774.

Cia, P., Benato, E.A., Anjos, V.D.A. and Vieites, R.L. 2000. Effects of irradiation on storage of grapes cv. Italia. *Revista Brasileira de Fruticultura*, **22**: 62–67.

Cillo, F., Finethi-Sialer, M.M., Papanice, M.A. and Gallitelli, D. 2004. Analysis of mechanisms involved in the *Cucumber mosaic virus* satellite RNA-mediated transgenic resistance in tomato plants. *Molecular Plant – Microbe Interactions*, **17**: 98–108.

Conway, W.S. and Sams, C.E. 1983. Calcium infiltration of Golden Delicious apples and its effect on decay. *Phytopathology*, **73**: 1068–1071.

Crisosto, C., Seguel, X. and Michailides, T.J. 1998. Comparing pulsated and ultraviolet light postharvest fungicide for peach fruit decay control. *Acta Horticulturae*, **465**: 471–479.

Davis, M.J. and Ying, Z. 2004. Development of papaya breeding lines with transgenic resistance to *Papaya ringspot virus*. *Plant Disease*, **88**: 352–358.

D'hallewin, G., Schirra, M., Manueddu, E., Piga, A. and Ben-Yehoshua, S. 1999. Scoparone and Scopoletin accumulation and ultraviolet-C induced resistance to postharvest decay in oranges as influenced by harvest date. *Journal of the American Society for Horticultural Science*, **124**: 702–709.

D'hallewin, G., Schirra, M., Pala, M. and Ben-Yehoshua, S. 2000. Ultraviolet irradiation at $0.5\,kJ\,m^{-2}$ reduces decay without causing damage or affecting postharvest quality of Star Ruby grapefruit (*C. paradisici* Macf.). *Journal of Agricultural and Food Chemistry*, **48**: 4571–4575.

Díaz, J., Have, A. ten and Kan, J.A.L. van. 2002. The role of ethylene and wound signaling in resistance of tomato to *Botrytis cinerea*. *Plant Physiology*, **129**: 1341–1351.

Dixon, R.A. and Lamb, C. 1990. Molecular communication in interaction between plants and microbial pathogens. *Annual Review of Plant Physiology and Plant Molecular Biology*, **41**: 349–367.

Droby, S., Chalutz, E., Horev, B., Cohen, L. Gaba, V., Wilson, C.L. and Wisniewski, M.E. 1993. Factors affecting UV-induced resistance in grapefruit against the green mold decay caused by *Penicillium digitatum*. *Plant Pathology*, **42**: 418–424.

Droby, S., Porat, R., Cohen, L., Weiss, B., Shapiro, B., Philosop-Hadas, S. and Meir, S. 1999. Suppressing green mold decay in grapefruit with postharvest jasmonate application. *Journal of the American Society for Horticultural Science*, **124**: 184–188.

Droby, S., Vinokur, V., Weiss, B., Cohen, L., Daus, A., Goldschmidt, E.E. and Porat, R. 2002. Induction of resistance to *Penicillium digitatum* in grapefruit by the yeast biocontrol agent *Candida oleophila*. *Phytopathology*, **92**: 393–399.

Durner, J., Shah, J. and Klessig, D.F. 1997. Salicylic acid and disease resistance in plants. *Trends in Plant Science*, **2**: 266–274.

El Ghaouth, A., Arul, J. and Asselin, A. 1992a. Potential use of chitosan in postharvest preservation of fresh fruits and vegetables. In: *Advances in Chitin and Chitosan* (eds) J.B. Brines, P.A. Sandford and J.P. Zikakis, pp. 440–445, Elsevier Applied Science, Amsterdam, the Netherlands.

El Ghaouth, A., Arul, J., Asselin, A. and Benhamou, N. 1992b. Antifungal activity of chitosan on postharvest pathogens: Induction of morphological and cytological alterations in *Rhizopus stolonifer*. *Mycological Research*, **96**: 769–779.

El Ghaouth, A. and Wilson, C. 1995. Biologically-based technologies for the control of postharvest diseases. *Postharvest News and Information*, **6**: 5–11.

El Ghaouth, A., Wilson, C.L. and Callahan, A.M. 2003a. Induction of chitinase, β-1,3-glucanase and phenyl alanine ammonia lyase in peach fruit by UV-C treatment. *Phytopathology*, **93**: 349–355.

El Ghaouth, A., Wilson, C.L. and Wisniewski, M. 2003b. Control of postharvest decay of apple fruit with *Candida saitoana* and induction of defense responses. *Phytopathology*, **93**: 344–348.

Fermin, G., Inglessis, V., Garboza, C., Rangel, S. and Dagert, M. 2004. Engineered resistance against *Papaya ringspot virus* in Venezuelan transgenic papayas. *Plant Disease*, **88**: 516–522.

Ferreira, S.A., Pitz, K.Y., Manshardt, R., Zee, F., Fitch and Gonsalves, D. 2002. Virus coat protein transgenic papaya provides practical control of *Papaya ringspot virus* in Hawaii. *Plant Disease*, **86**: 101–105.

Freire, M. Jr. and Chitarra, A.B. 1999. Effect of calcium chloride application on hydrothermally treated mango cv. Tommy Atkins fruits. *Pesquisa Agropecuaria Brasileira*, **34**: 761–769.

Fuchs, M., Prouvidenti, R., Slighton, J.L. and Gonsalves, D. 1996. Evaluation of transgenic tomato plants expressing the coat protein gene of *Cucumber mosaic virus* strain WL under field condition. *Plant Disease*, **80**: 270–275.

Fuchs, M., Tricoli, D.M., Carney, K.J., Schesser, M., McFerson, J.R. and Gonsalves, D. 1998. Comparative virus resistance and fruit yield of transgenic squash with single and multiple coat protein genes. *Plant Disease*, **82**: 1350–1356.

González-Aguilar, G.A., Fortiz, J., Cruz, R., Baez, R. and Wang, C.Y. 2000. Methyl jasmonate reduces chilling injury and maintains postharvest quality of mango fruit. *Journal of Agricultural and Food Chemistry*, **48**: 515–519.

Gottstein, H.D. and Kuć, J. 1989. Induction of systemic resistance to anthracnose in cucumber by phosphates. *Phytopathology*, **79**: 176–179.

Guo, W.W. and Grosser, J.W. 2004. Transfer of a potential canker resistance gene into *Citrus* protoplasts using GFP as the selectable marker. *Acta Horticulturae*, **632.**

Hamilton, A.J., Lycett, G.W. and Grierson, D. 1990. Antisense gene that inhibits synthesis of the hormone ethylene in transgenic plants. *Nature*, **346**: 284–287.

Henning, J., Dewey, R.E., Cutt, J.R. and Klessig, D.F. 1993. Pathogen, salicylic acid and developmental dependent expression of β-1,3-glucanase/GUS gene fusion in transgenic tobacco plants. *Plant Journal*, **4**: 481–493.

Ippolito, A., El Ghaouth, A., Wisniewski, M. and Wilson, C. 2000. Control of postharvest decay of apple fruit by *Aureobasidium pullulans* and induction of defense responses. *Postharvest Biology and Technology*, **19**: 265–272.

Iwai, T., Kaku, H., Honkura, R., Nakamura, S., Ochiai, H., Sasaki, T. and Ohashi, Y. 2002. Enhanced resistance to seed transmitted bacterial diseases in transgenic rice plants over-producing an oat cell-wall bound thionin. *Molecular Plant-Microbe Interactions*, **15**: 515–521.

Iwanami, T., Shimizu, T., Ito, T. and Hirabayashi, T. 2004. Tolerance to *Citrus mosaic virus* in transgenic trifoliate orange lines harboring capsid polyprotein gene. *Plant Disease*, **88**: 865–868.

Khan, J., Ooka, J.J., Miller, S.A., Madden, L.V. and Hoitink, H.A.J. 2004. Systemic resistance induced by *Trichoderma hamatum* 382 in cucumber against *Phytophthora* crown rot and leaf blight. *Plant Disease*, **88**: 280–286.

Kim, J.J. and Ben-Yehoshua, S. 2001. Histological detection of phytoalexin scoparone from heat-treated and UV-irradiated lemon fruit after inoculation with *Penicillium digitatum*. *Plant Pathology Journal*, **17**: 271–275.

Kim, M.A. and Choi, S.J. 2002. Induction of gray mold rot resistance by methyl salicylate application in strawberry fruits. *Journal of the Korean Society for Horticultural Science*, **43**: 29–33.

Kishimoto, K., Nishizawa, Y., Tabei, Y., Hibi, T., Nakajima, M. and Akutsu, K. 2002. Detailed analysis of rice chitinase gene expression in transgenic cucumber plants showing different levels of disease resistance to gray mold (*Botrytis cinerea*). *Plant Science*, **162**: 655–662.

Klein, J.D. and Lurie, S. 1991. Postharvest heat treatment and fruit quality. *Postharvest News and Information*, **2**: 15–19.

Klessig, D.F. and Malamy, J. 1994. The salicylic acid signal in plants. *Plant Molecular Biology*, **26**: 1439–1458.

Kramer, M., Sanders, R.A., Sheehy, R.E., Melis, M., Kuehn, M. and Hiatt, W.R. 1990. Field evaluation of tomatoes with reduced polygalacturonase binding antisense RNA. In: *Horticultural Biotechnology* (eds) A.B. Bennelt and S.D. O'Neill, Wiley-Liss, New York.

Lamb, C.J. Lawton, M.A., Dron, M. and Dixon, R.A. 1989. Signals and transduction mechanisms for activation of plant defence against microbial attack. *Cell*, **56**: 215–224.

Li, H.Y. and Yu, T. Effect of chitosan on incidence of brown rot, quality and physiological attributes of postharvest peach fruit. *Journal of the Science of Food and Agriculture*, **81**: 269–274.

Liu, J., Stevens, C., Khan, V.A., Lu, J.Y., Wilson, C.L., Adeyeye, O., Kabwe, M.K., Pusey, P.L., Chalutz, E., Sultana, T. and Droby, S. 1993. Application of ultraviolet-C light on storage rots and ripening of tomatoes. *Journal of Food Protection*, **56**: 868–872.

Lu, J.Y., Lukombo, S.M., Stevens, C., Khan, V.A., Wilson, C.L., Pusey, P.L. and Chalutz, E. 1993. Low dose of UV and gamma radiation on storage rot and physicochemical changes in peaches. *Journal of Food Quality*, **16**: 301–309.

Lu, J.Y., Stevens, C., Yakubu, P. and Loretan, P.A. 1987. Gamma electron beam and UV radiation on control of storage rots and quality of Walla Walla onions. *Journal of Food Processing and Preservation*, **12**: 53–62.

Luckey, T.D. 1980. *Hormesis with Ionizing Radiation*. CRC Press, Boca Raton, FL, USA.

Lund, S.T., Stall, R.E. and Klee, J.H. 1998. Ethylene regulates the susceptible response to pathogen infection in tomato. *Plant Cell*, **10**: 371–382.

Malamy, J., Carr, J.P., Klessig, D.F. and Raskin, I. 1990. Salicylic acid: a likely endogenous signal in the resistance response to tobacco to viral infection. *Science*, **250**: 1002–1004.

Mangal, M. and Sharma, D.R. 2002. *In vitro* mutagenesis and cell selection for the induction of black rot resistance in cauliflowers. *Journal of Horticultural Science and Biotechnology*, **77**: 268–272.

Marquenie, D., Michiels, C.W., Impe, J.F. van, Schrevens, E. and Nicolai, B.N. 2003. Pulsed white light in combination with UV-C and heat to reduce storage rot of strawberry. *Postharvest Biology and Technology*, **28**: 187–194.

Marquenie, D., Lammertyn, J., Geeraerd, A.H., Soontjens, C., Impe, T.F. van, Nicolaï, B.M. and Michiels, C.W. 2002a. Inactivation of conidia of *Botrytis cinerea* and *Monilinia fructigena* using UV-C and heat treatment. *International Journal of Food Microbiology*, **74**: 27–35.

Marquenie, D., Michiels, C., Schenk, A., Impe, J. van and Nicolai, B. 2002b. Control of storage decay in cherries using physical techniques. *Fruitteelt–Nieuws*, **15**: 12–13.

Marquenie, D., Michiels, C.W., van Impe, J.F., Schrevens, E. and Nicolai, B.N. 2003. Pulsed white light in combination with UV-C and heat to reduce storage rot of strawberry. *Postharvest Biology and Technology*, **28**: 455–461.

Marquenie, D., Schenk, A., Nicolaï, B., Michiels, C., Soontjens, C. and Impe J. van. 2002b. Use of UV-C and heat treatment to reduce storage rot of strawberry. *Acta Horticulturae*, **567**: 779–782.

McGuire, R.G. and Kelman, A. 1984. Reduced severity of *Erwinia* soft rot in potato tubers with increased calcium content. *Phytopathology*, **74**: 1250–1256.

Mercier, J., Arul, J., Ponnampalam, R. and Boulet, M. 1993. Induction of 6-methoxymellein and resistance of storage pathogens in carrot slices by UV-C. *Journal of Phytopathology*, **137**: 44–55.

Mercier, J., Roussel, D., Charles, M.T. and Arul, J. 2000. Systemic and local responses associated with UV-and pathogen-induced resistance to *Botrytis cinerea* in stored carrot. *Phytopathology*, **90**: 981–986.

Métraux, J.P., Signer, H., Ryals, J., Ward, E., Wyss-Benz, M., Gaudin, J., Raschdorf, K., Schmid, E., Blum, W. and Inverardi, B. 1990. Increase in salicylic acid at the onset of systemic acquired resistance in cucumber. *Science*, **250**: 1004–1008.

Miceli, A., Ippolito, A., Linsalata, V. and Nigro, F. 1999. Effect of preharvest calcium treatments on decay and biochemical changes in table grape during storage. *Phytopathologia Mediterranea*, **38**: 47–53.

Missiou, A., Kalantidis, K., Boutla, A., Tzortzakaki, S., Tabler, M. and Tsagris, M. 2004. Generation of transgenic potato plants highly resistant to *Potato virus Y* (PVY) through RNA silencing. *Molecular Breeding*, **14**: 185–197.

Mitterbauer, R., Poppenberger, B., Raditschnig, A., Lucyshyn, D., Lemmens, M., Glösse, J. and Adam, G. 2004. Toxin-dependent utilization of engineered ribosomal protein L3 limits trichothecene resistance in transgenic plants. *Plant Biotechnology*, **2**: 329–340.

Molloy, C., Cheah, L.H. and Koolaard, J.P. 2004. Induced resistance against *Sclerotinia sclerotiorum* in carrots treated with enzymatically hydrolysed chitosan. *Postharvest Biology and Technology*, **33**: 61–65.

Nafussi, B., Ben-Yehoshua, S., Rodov, V., Peretz, J., Ozer, B.K. and D'hallewin, G. 2001. Mode of action of hot-water dip in reducing decay of lemon fruit. *Journal of Agricultural and Food Chemistry*, **49**: 107–113.

Nelson, R.S., McCormick, C.M., Delannay, X., Dube, P., Layton, J., Anderson, E.J., Kaniewski, M., Proksch, R.K., Horsch, R.B., Rogers, S.G., Fraley, R.T. and Beachy, R.N. 1988. Virus tolerance, plant growth and field performance of transgenic tomato plants expressing coat protein gene of *Tobacco mosaic virus*. *Bio/Technology*, **6**: 403–409.

Nigro, F., Ippolito, A., Lattanzio, V., Venere, D.D. and Salerno, M. 2000. Effect of ultraviolet-C light on postharvest decay of strawberry. *Journal of Plant Pathology*, **82**: 29–37.

Orober, M., Siegrist, J. and Buchenauer, H. 2002. Mechanisms of phosphate-induced disease resistance in cucumber. *European Journal of Plant Pathology*, **108**: 345–353.

Pan, S.W., Ye, X.S. and Kuć, J. 1991. A technique for detection of chitinase, β-1,3-glucanase and protein patterns after a single separation using polyacrylamide gel electrophoresis or isoelectrofocusing. *Phytopathology*, **81**: 970–974.

Porat, R., Daus, A., Weiss, B., Cohen, L., Fallik, E. and Droby, S. 2000a. Reduction of postaharvest decay in organic citrus fruit by a short hot water brushing treatment. *Postharvest Biology and Technology*, **18**: 151–157.

Poart, R., Lers, A., Dori, S., Cohen, L., Weiss, B., Daus, A., Wilson, C.L. and Droby, S. 1999. Induction of chitinase and β-1,3-endoglucanase proteins by UV-irradiation and wounding in grapefruit peel tissue. *Phytoparasitica*, **27**: 233–238.

Porat, R., Pavoncello, D., Peretz, J., Weiss, B., Daus, A., Cohen, L., Ben-Yehoshua, S., Fallik, E., Droby, S. and Lurie, S. 2000b. Induction of resistance to *Penicillium digitatum* and chilling injury in 'Star Ruby' grapefruit by a short hot water rinse and brushing treatment. *Journal of Horticultural Science and Biotechnology*, **75**: 428–432.

Porat, R., Vinokur, V., Weiss, B., Cohen, L., Daus, A., Goldschmidt, E.E. and Droby, S. 2003. Induction of resistance to *Penicillium digitatum* in grapefruit by aminobutyric acid. *European Journal of Plant Pathology*, **109**: 901–907.

Powell, A.L.T., van Kan, J., ten Have, A., Visser, J., Greve, L.C., Bennett, A.B. and Labavitch, J.M. 2000. Transgenic expression of pear PGIP in tomato limits fungal colonization. *Molecular Plant–Microbe Interactions*, **13**: 942–950.

Qin, G.Z. and Tian, S.P. 2005. Enhancement of biocontrol activity of *Cryptococcus laurentii* by silicon and the possible mechanism involved. *Phytopathology*, **95**: 69–75.

Racman, D.S., McGeachy, K., Reavy, B., Štrukelj, B., Zel, J. and Barker, H. 2001. Strong resistance to potato necrotic ringspot disease in potato induced by transformation with coat protein gene sequences from an NTN isolate of *Potato Virus Y*. *Annals of Applied Biology*, **139**: 269–275.

Reddy, M.V.B., Angers, P., Castaigne, F. and Arul, J. 2000a. Chitosan effects on blackmold rot and pathogenic factors produced by *Alternaria alteranta* in postharvest tomatoes. *Journal of American Society for Horticultural Science*, **125**: 742–747.

Reddy, M.V.B., Belkacemi, K., Corcuff, R. Castaigne, F., Arul, J., Castaigne, F., Arul, T. 2000b. Effect of preharvest chitosan sprays on postharvest infection by *Botrytis cinerea* and quality of strawberry fruit. *Postharvest Biology and Technology*, **20**: 39–51.

Reuveni, M., Agapov, V. and Reuveni, R. 1996. Controlling powdery mildew caused by *Sphaerotheca fuliginea* in cucumber by foliar sprays of phosphate and potassium salts. *Crop Protection*, **15**: 49–53.

Reuveni, M., Agapov, V. and Reuveni, R. 1997. A foliar spray of micronutrient solution induces local and systemic protection against powdery mildew (*Sphacrotheca fuligincea*) in cucumber plants. *European Journal of Plant Pathology*, **103**: 581–588.

Romanazzi, G., Nigro, F. and Ippolito, A. 2000. Effectiveness of pre-and post-harvest chitosan treatments on storage decay of strawberries. *Rivista di Frutticoltura e di Ortofloricoltura*, **62**: 71–75.

Romero, A.M. and Ritchie, D.F. 2004. Systemic acquired resistance delays race shifts to major resistance genes in bell pepper. *Phytopathology*, **94**: 1376–1382.

Rong, R.F. and Feng, S.Q. 2001. Effect of UV-C light irradiation on ripening and disease infection of postahrvest tomato. *Journal of China Agricultural University*, **6**: 68–73.

Ryals, J.A., Nuenschwander, U.H., Willits, M.G., Molina, A., Steiner, H.Y. and Hunt, M.D. 1996. Systemic acquired resistance. *Plant Cell*, **8**: 1809–1819.

Sharples, R.O. and Johnson, D.S. 1977. The influence of calcium on senescence changes in apples. *Annals of Applied Biology*, **85**: 450–453.

Souza, A.L.B. de, Chitarra, M.I.F., Chitarra, A.B. and Machado, J.da C. 1999. Postharvest resistance of peach (*Prunus persica* cv. Biuti) to *Monilinia fructicola*: induction of biochemical responses to the application of $CaCl_2$ at the site of the injury. *Ciência e Agrotechnologia*, **23**: 865–875.

Stapleton, A.E. 1992. Ultraviolet radiation and plants: Burning questions. *Plant Cell*, **4**: 1353–1358.

Stevens, C., Khan, V.A., Tang, A.Y. and Lu, J.Y. 1990. The effect of ultraviolet irradiation on mold rots and nutrients of stored sweetpotatoes. *Journal of Food Protection*, **53**: 223–226.

Stevens, C., Liu, J., Khan, V.A., Lu, J.Y., Wilson, C.L., Igwegbe, E.C.K., Kabwe, M.K., Chalutz, E. and Droby, S. 1998. Application of hormetic CV-C for delayed ripening and reduction of *Rhizopus* soft rot in tomatoes: the effect of tomatine on storage rot development. *Journal of Phytopathology*, **146**: 211–221.

Takahashi-Ando, N., Kimura, M., Kakeya, H., Osada, H. and Yamaguchi, I. 2002. A novel lactonohydrolase responsible for the detoxification of zearalenone: enzyme purification and gene cloning. *Biochemical Journal*, **365**: 1–6 .

Takahashi-Ando, N., Ohsato, S., Shibata, T., Hamamoto, H., Yamaguchi, I. and Kimura, M. 2004. Metabolism of zearalenone by genetically modified organisms expressing the detoxification gene from *Clonostachys rosea. Applied and Environmental Microbiology*, **70**: 3239–3245.

Takaichi, M. and Oeda, K. 2000. Transgenic carrots with enhanced resistance against two major pathogens *Erysiphe heraclei* and *Alternaria dauci. Plant Science* (*Limerick*), **153**: 135–144.

Terry, L.A. and Joyce, D.C. 2000. Suppression of gray mold on strawberry fruit with the chemical plant activator acibenzolar. *Pest Management Science*, **56**: 989–992.

Thomas, P.E., Hassan, S., Kaniewski, W.K. Lawson, E.C. and Zalewski, J.C. 1998. A search for evidence of virus/transgene interactions in potatoes transformed with the *Potato leafroll virus* replicase and coat protein genes. *Molecular Breeding*, **4**: 407–417.

Vidaver, A.K. and Tolin, S.A. 2001. Biotechnology: oversight and regulation. In: *Encylopedia of Plant Pathology* (eds) O.C. Maloy and T.D. Murray, John Wiley and Sons, Inc., New York.

Vivekananthan, R., Ravi, M., Ramanathan, A. and Samiyappan, R. 2004. Lytic enzymes induced by *Pseudomonas fluorescens* and other biocontrol organisms mediate defence against the anthracnose pathogen in mango. *World Journal of Microbiology and Biotechnology*, **20**: 235–244.

Wegener, C.B. 2002. Induction of defense responses against *Erwinia* soft rot by an endogenous pectate lyase in potatoes. *Physiological and Molecular Plant Pathology*, **60**: 91–100.

Wilson, C.L., El Ghaouth, A., Chalutz, E., Droby, S., Stevens, C., Lu, J.Y., Khan, V. and Arul, J. 1994. Potential of induced resistance to control of postharvest diseases of fruits and vegetables. *Plant Disease*, **78**: 837–844.

Wilson, C.L. and El Ghaouth, A. 1993. Multifaceted biological control of postharvest diseases of fruits and vegetables. In: *Pest Management: Biologically. Based Technologies* (eds) E.D. Lumsden and R.L. Vauzhn, pp. 181–185, American Chemical Society, Press, Washington DC.

Wilson, C.L. and Wisniewski, M.E. (eds) 1994. *Biological Control of Postharvest Diseases of Fruits and Vegetables–Theory and Practice.* CRC Press, Boca Raton, FL, USA.

Wisniewski, M.E. and Wilson, C.L. 1992. Biological control of postharvest diseases of fruits and vegetables: Recent advances. *HortScience*, **27**: 94–98.

Wobbe, K.K. and Klessig, D.F. 1996. Salicylic acid-an important signal in plants. In: *Plant Gene Research* (eds) E.S. Dennis, pp. 167–196, Springer, Berlin, Germany.

Yamamoto, T., Iketani, H., Ieki, H. Nishizawa, Y., Notsuka, K., Hibi, T., Hayashi, T. and Matsuta, N. 2000. Transgenic grapevine plants expressing a rice chitinase gene with enhanced resistance to fungal pathogens. *Plant Cell Reports*, **19**: 639–646.

Yildiz, F., Yildiz, M., Kinay, P., Sen, F. and Karacali, I. 2001. Effect of some pre-and post-harvest treatments on decay of green mold Satsuma mandarins. *Acta Horticulturae*, **553**: 443–445.

Yin, Z., Hennig, J., Szwacka, M. and Malepszy, S. 2004. Tobacco PR-2d promoter is induced in transgenic cucumber in response to biotic and abiotic stimuli. *Journal of Plant Physiology*, **161**: 621–629.

Zhang, D.L. and Quantick, P.C. 1998. Antifungal effects of chitosan on fresh strawberries and raspberries during storage. *Journal of Horticultural Science and Biotechnology*, **73**: 763–767.

11

POSTHARVEST DISEASE MANAGEMENT THROUGH CHEMICALS

The development of microbial pathogens, and the consequent incidence of diseases caused by them, can be limited indirectly by host management methods, such as use of disease-free seeds and planting materials, proper cultivation practices, and host resistance, which restrict or prevent the introduction of pathogens and/or build-up of pathogen populations. The effects of these methods may not be readily perceptible. In contrast, the pathogen management methods involving the use of chemicals and biocontrol agents exert direct, negative effects on the pathogen development and increase in population. The effects of chemical application are highly visible and, in some postharvest diseases, the efficiency of control is high. Even though the application of chemicals has been shown to be effective, many chemicals have been withdrawn or their use has been restricted because of the growing concern regarding health hazards and development of resistance to the chemicals by the postharvest pathogens. However, epidemiological studies seem to indicate that the major preventable risk factors for poor health are not pesticide residues in food, but other factors such as smoking, dietary imbalances, endogenous hormones, and inflammation. It has been estimated that about 99.9% of the chemicals that humans ingest are naturally occurring. The amounts of synthetic pesticide in plant foods should be viewed in comparison with the amounts of natural pesticides produced by plant themselves (Gold et al., 2001) The need to develop alternative methods has led to the emergence of biocontrol as a principal nonchemical strategy for postharvest diseases.

Postharvest Pathogens and Disease Management, by P. Narayanasamy

However, the use of chemicals can not be entirely dispensed with, since alternative strategies have not yielded consistent results under varied conditions.

A wide range of chemicals have been tested and their effectiveness has been found to be variable, depending on several factors such as host plant species, cultivar, sensitivity of the microbial pathogens, and storage conditions. The chemicals may be classified as fumigants, treated wraps and waxes, and dips, sprays, or dusts. Both inorganic and organic chemicals have been used on various crops at preharvest and postharvest stages.

Depending on the sources of infection and other conditions favoring disease incidence, the chemicals may be applied prior to and/or after harvest. To prevent field infection, which may remain quiescent, fungicides may be applied at the preharvest stage. Application of fungicides in the field has significant postharvest implications, since residue levels permitted for most pesticides on harvested produce are extremely small. Many countries have imposed zero tolerance limits. In addition, treatment of fruits and vegetables under field conditions is expensive and it may be difficult to entirely eliminate latent infections by fungal pathogens causing postharvest diseases, such as anthracnose disease in bananas and mangoes. In such cases, postharvest chemical control becomes essential. Wounds form the predominant avenue for many postharvest pathogens to gain entry into the harvested commodities. Treatment with chemicals has been demonstrated to protect the product effectively against pathogens, in addition to eradication of pathogens that have already invaded the produce through wounds.

The nature of the tissues to be protected and consumer expectation of residue-free produce (or within permitted levels) have placed more stringent regulations on use of chemicals. Though the same chemical may be used for the field and postharvest treatments, the formulations may differ. The postharvest chemicals have to be applied in strict accordance with the manufacturer's recommendations and food additive regulations of the countries concerned. Good surface distribution, adherence, lack of mammalian toxicity, efficient penetration into the epidermal tissues of fruits and vegetables, ability to eradicate latent infections, and absence of off-flavors and other adverse effects on quality of the product are some of the characteristics required for postharvest chemicals.

11.1 FUMIGANTS

Treatment with fumigants has been found to be preferable for delicate produce and also for produce transported or stored in closed containers. Sulfur dioxide (SO_2) is the most frequently employed fumigant for the control of gray mold disease caused by *Botrytis cinerea* in table grapes. SO_2 may be applied directly from cylinders, or by burning sulfur or by using sodium metabisulfite ($Na_2S_2O_5$) from which SO_2 is released. The conidia present on the fruit surface may be killed by SO_2, while established infection is not affected. Off-flavors and bleached skin spots may be noted, if the treatment is for longer periods or excessive. A novel monolithic type device which releases SO_2 from $Na_2S_2O_5$ at a constant rate over an extended period was developed. The $Na_2S_2O_5$ particles are evenly dispersed throughout the matrix. This new device reduced SO_2 bleaching due to the presence of lower initial levels of SO_2 and over an extended period. Decay caused by *B. cinerea* in table grapes was drastically reduced by increasing the concentration of $Na_2S_2O_5$ incorporated in the new polymer SO_2 devices (Opperman et al., 1999). The efficacy of using SO_2 from generators at low temperature for the control of decay due to *B. cinerea* in table grapes was assessed. A temperature of 4°C in combination with SO_2 treatment was the most effective in preserving fruit quality of two cultivars Italia and Red Globe. High temperatures increased the weight loss, stem color, decay, and SO_2 residue in berries (Muñoz et al., 2003). Use of sulfur pills, introducing SO_2 into the container through a tube and maintenance of internal SO_2 concentrations greater than 0.5% were tested for reducing decay by *B. cinerea* and for better grape flavor (Crisosto and Smilanick, 2002).

Chlorine in the gaseous form is a very potent disinfectant and it is rarely applied directly. It is much safer and easier to use when dissolved in water. Chlorine for disinfection may be employed as pressurized chlorine gas, calcium hypochlorite (a soluble solid), or a solution of sodium hypochlorite. The most common source of chlorine used in postharvest chlorination is calcium hypochlorite and it is available as granulated powder or large tablets. Proper chlorination may be achieved by frequent monitoring of the solution and maintaining proper pH of the solution. Chlorine gas released from a salt mixture, combined with storage for 25 days at 0°C, significantly reduced gray mold decay in table grapes inoculated with *B. cinerea*. Naturally infected Thompson Seedless grapes, from a commercial harvest, developed significantly less decay than in control on arrival in the United States after transportation lasting 25 days. No deleterious effect due to chlorine could be observed (Zoffoli et al., 1999).

Chlorine is used at 100 ppm in the pear industry to reduce or minimize the build-up of fungal conidial concentration in water systems for washing the fruit. Chlorine may be either generated from sodium hypochlorite (bleach) or gaseous chlorine may be used directly. Accumulation of sodium component in the water system may result in damage to the fruit. Hence, changing the water in the dump tank or flume and careful monitoring of the levels are required to minimize the adverse effects. Rinsing the treated fruit with fresh water is necessary. Chlorine has no residual activity and it may be employed to kill the spores in the water rather than on the pear fruit or within wounds, since it is bound to the exposed tissues. At pH close to 7.0, chlorine is most effective. It should not be used along with calcium or sodium ligninsulfonate, applied to water to raise the specific gravity to 1.05 so that pear fruit will float in the solution (Kupferman, 1997).

The market quality of Jonathan and Golden Delicious apples was improved by dipping the fruits in buffered sodium hypochlorite solution. A 7-min dip in

800 ppm chlorine resulted in mean increase from 25% and 55% to 100% in extra fancy grade for Jonathan and Golden Delicious apples, respectively, and increased market value by 31% and 14% respectively. The treatment eliminated all blemishes caused by sooty blotch and flyspeck diseases from the fruits (Batzer et al., 2002).

The effect of chlorine on the development of *Geotrichum candidum* and *Rhizopus stolonifer*, causing sour rot and *Rhizopus* rot of tomato, was assessed. Wounded tomatoes placed in a flume with free chlorine at 30 mg/l and then exposed to spores for 1 min showed reduced decay (50% less) during storage at 24°C for 6 days compared to fruits exposed to spores and water alone (Bartz et al., 2001).

Chlorine dioxide (ClO_2) was tested for its antifungal activity against *Monilinia laxa* causing brown rot disease of nectarines and plums. The germination of conidia was entirely inhibited by ClO_2 at 50 µg/ml after 1 min of contact with conidia. In a semicommercial trial, ClO_2 at 10 µg/ml prevented infection by *M. laxa* in wounded, inoculated nectarines and plums (Mari et al., 1999). Treatment of potato tubers with sodium hypochlorite alone or in combination with thiophanate-methyl reduced the incidence of black scurf disease caused by *Rhizoctonia solani* but there was phytotoxic symptom in treated tubers. Marketable tuber quantity was, however, increased by the combined treatment (Errampalli et al., 2001).

The germicidal effects of ozone (O_3), iodine, and chlorine in water were determined using the fungal pathogens *Colletotrichum gloeosporioides, Pestalotiopsis mangiferae*, and *Lasiodiplodia theobromae*. Ozone (2.2 mg/l, 15 min contact) effectively inhibited the spore germination of all these fungal pathogens. On the other hand, iodine and chlorine were inhibitory to *Fusarium oxysporum*, killing all the spores. At higher concentrations (0.005%), chlorine and iodine were more effective than ozone against *L. theobromae* (Barbosa et al., 2000). The quality of mango fruits was not adversely affected by treatment with ozone and iodine solution, as indicated by physiological weight loss, total soluble solids, firmness, and color (Barbosa et al., 2002). Following application of chlorine, the presence of trihalomethane residues and other carcinogenic by-products has been detected. Hence, ozone could be used as an effective alternative as it leaves no detectable level of residue. It may be used in wash water and during the hydrofreezing process at packing house facilities (Barbosa et al., 2000, 2002).

The effectiveness of ozone in reducing growth of *Penicillium digitatum* and *P. italicum* in vitro and development of postharvest green and blue molds on artificially inoculated Valencia orange and Eureka lemon was determined. Both oranges and lemons were continuously exposed to 1.0 ± 0.05 ppm ozone at 10°C in an export container for 2 weeks. The incidence of both green and blue molds was delayed by about 1 week and the development of disease was slower compared with untreated control fruits. However, the final disease level was not significantly reduced by ozone treatment. The sporulation in fungicide resistant strains of *P. digitatum* and *P. italicum* may be reduced, with a possible increase in the useful life of postharvest fungicides (Palou et al., 2001).

Acetic acid fumigation has been shown to be a viable alternative to sodium hypochlorite dips for sterilizing fruit surfaces. Acetic acid is the principal organic component of vinegar, known for its preservative and flavoring properties. Though the antimicrobial properties of acetic acid were reported earlier (Levine and Fellers, 1940), its use as fumigant for control of microorganisms causing food spoilage has been demonstrated in the recent years. Acetic acid molecules, following vaporization of vinegar, become a gas as undissociated acid that exists as a mixture of monomers, hydrogen-bonded dimers, and trimers (Seaton, 1993). The undissociated acetic acid molecules may easily pass through the membrane of conidia on the fruit surface. The conidia may be either killed or inactivated by lowering the pH of the cell protoplasm (Sholberg et al., 1998).

Fumigation of apples with acetic acid at 2 or 4 mg/l, before wounding, prevented decay due to *Botrytis cinerea* or *Penicillium expansum* causing gray and blue mold diseases respectively (Sholberg and Gaunce, 1995). Germination of conidia of *Monilinia fructicola* causing brown rot disease in stone fruits and strawberries and *P. expansum* was effectively arrested by vapors of several vinegars containing 4.2 to 6.0% acetic acid (2.5 to 3.6 M/l). Vapor from red wine vinegar (1.0 ml) reduced decay by *Botrytis cinerea* in strawberry. Heat vaporized white vinegar (5% acetic acid) reduced decay by *P. expansum* on Jonagold apples. The results suggest that vinegar vapor could be employed as an alternative to liquid biocides such as sodium hypochlorite for sterilizing fruit surfaces. The effectiveness of vinegar vapor may be enhanced if the vinegar is evaporated completely, by exposure for a period of 17 h, and storing the fruits at 20°C (Fig. 11.1) (Sholberg et al., 2000). The feasibility of using acetic acid for fumigation of large quantities of apples to control blue mold disease was determined. Apples were fumigated in standard wooden or plastic apple boxes or small wooden bins. They were either wounded and evaluated for decay after a week at 20°C or stored at 1°C for 3 or more months and then they were evaluated for decay. Acetic acid fumigation was as effective as fungicide thiabendazole and the fruit quality was not affected by acetic acid fumigation (Sholberg et al., 2001). Peracetic acid (PAA) treatment was evaluated for its efficacy against brown rot caused by *Monilinia laxa* and soft rot caused by *Rhizopus stolonifer* affecting peach, sweet cherry, apricot, and nectarine. Dipping the fruits in PAA solution (125 mg/l) for 1 min provided significant control of *Monilinia* rots, whereas a higher concentration of PAA (250 mg/l) was required to protect the fruits against *R. stolonifer*. The efficiency of disease control achieved through hydrorefrigeration was improved by addition of PAA (125 mg/l) in cold water. Application of PAA has been shown to be very useful to control stone fruit diseases that can spread during shipping and marketing (Mari et al., 2004).

Nitrous oxide (N_2O) has been shown to have the potential for use against postharvest fungal pathogens, *B. cinerea, Colletotrichum acutatum, Monilinia*

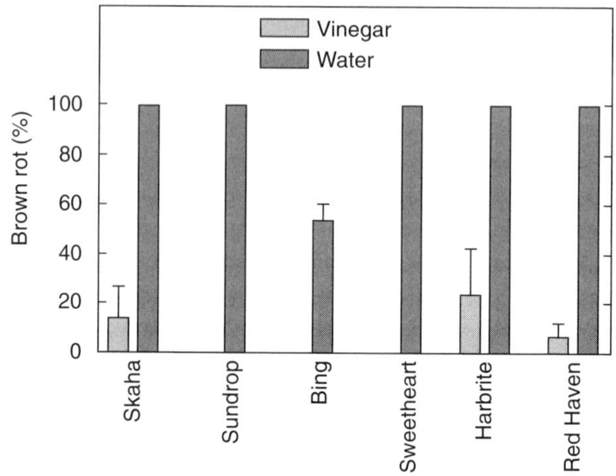

Figure 11.1 Effect of vinegar vapor on the development of brown rot disease (*Monilinia fructicola*) in apricots (cvs. Skaha and Sundrop), cherries (cvs. Bing and Sweetheart) and peaches (cvs. Harbrite and Redhaven) inoculated with conidial suspension (20 μl drops; 1 × 10^5 CFU/ml) and fumigated for 17 h with 1 ml of water or vinegar, wounded with a glass rod and incubated for 5 days at 20°C. (Courtesy of Sholberg et al., 2000; The American Society for Horticultural Science, Alexandria, USA.)

fructicola, and *Penicillium expansum*. *P. italicum* and *Rhizopus stolonifer* were highly sensitive to N_2O. The inhibitory effect was attributed to the biophysical properties of N_2O (Qadir and Hashinaga, 2001a). Apple inoculated with *A. alternata* and *P. expansum*, strawberries inoculated with *B. cinerea* and *R. stolonifer*, and Satsuma mandarin inoculated with *Geotrichum candidum* were exposed to a mixture of N_2O (80%) and O_2 (20%) and stored at 20°C. Nitrous oxide delayed the appearance of disease and reduced lesion expansion rate. It is considered that N_2O may either directly inhibit the pathogen development and/or indirectly enhance the resistance of host tissue to pathogen invasion (Qadir and Hashinaga, 2001b).

Treatment of potato tubers with hydrogen peroxide plus (HPP) was tested for its effectiveness against *Helminthosporium solani* causing silver scurf disease. Tubers were exposed to 10% HPP for 10 h (per treatment) after curing them for 12 days at 13 ± 1°C and 95% RH. The temperature was reduced to 9 ± 1°C in storage rooms after HPP treatment. The 3-year study showed that the treated tubers (10% HPP, five times), after 6 months of storage, had 2% infection as against 38% infection in untreated control tubers. The growth rate of the fungal pathogen was significantly reduced following exposure to HPP (Afek et al., 2001). A combination of lactic acid (1.5%) and H_2O_2 (1.5%) was used to treat apples, oranges, and tomatoes for 15 min at 40°C. These commodities were spot-inoculated near the stem end of the fruit with five strain mixtures of *E. coli* 0157:H7, *Salmonella enteritidis*, and *Listeria monocytogenes*, pathogenic to humans, and were submerged in sterile deionized water containing lactic acid and H_2O_2. Chemical treatment reduced the pathogenic bacterial populations to undetectable levels. Furthermore, the sensory and qualitative characteristics of apples treated with the chemical wash solution were not adversely affected. This chemical treatment has the potential to eliminate *E. coli* 0157:H7, *S. enteritidis*, and *L. monocytogenes* present on apples, oranges, and tomatoes at the processing or packaging level (Venkitanarayanan et al., 2002, McWatters et al., 2002). The lethal dose (LD_{90}) of H_2O_2 to be used for disinfestations of *B. cinerea* causing gray mold diseases may vary depending on the nature of the fruits. The LD_{90} values differed from 4.99 to 40.3 g active ingredient (a.i.)/l (Copes, 2004).

The effectiveness of four volatile compounds, (E)-2-hexenal, (E)-2-hexenal diethyl acetal, benzaldehyde, and methyl benzoate, in controlling gray mold disease of strawberry caused by *Botrytis cinerea* was determined. (E)-2-hexenal (100 µl) was the only treatment that allowed any significant increase in mold populations during storage at 2°C for 7 days followed by removal of the volatile and transfer to 22°C for 3 days (Ntirampemba et al., 1998). Hyphal growth of *Penicillium expansum* and *B. cinerea* was reduced by 50% following exposure to hexenal (100 µl/l) for 48 h. Decay lesion development on Golden Delicious and Jonathan apple slices exposed to the same concentration of hexenal was reduced. Stimulation of aromatic volatile production in treated apple slices was also observed (Song et al., 1998). Fumigation of seedless table grapes cv. Crimson Seedless with (E)-2-hexenal for 2 weeks during storage in containers (1.1 l) at 2°C resulted in significant reduction in mold incidence and the effect persisted throughout 12 days of storage at 20°C. Treatment with hexenal did not affect O_2 or CO_2 concentrations within containers as well as fruit firmness and contents of soluble solids (Archbold et al., 1999).

The fungitoxicity of another volatile compound, allyl-isothiocyanate (AITC), against *P. expansum* causing blue mold of pear cvs. Conference and Kaiser was assessed. Exposure of fruits for 24 h in an AITC (5 mg/l) enriched atmosphere provided the most effective control of the disease. Lesion diameter was reduced with increase in AITC concentration. Treatments with AITC were effective up to 24 h after inoculation for Conference and up to 48 h for Kaiser pear. AITC treatment may be particularly useful in locations where TBZ-resistant strain occur, since the disease incidence of blue mold could be reduced by 90% in both cultivars. The AITC produced from pure sinigrin or from defatted meal of *Brassica juncea* may be an economically viable alternative to synthetic fungicides against *P. expansum* (Mari et al., 2002). The efficacy of volatile substances emanating from Isabella grapes (*Vitis labrusca*) for the control of gray mold disease of kiwifruit (*B. cinerea*) is described in Section 9.7.1.

11.2 CHEMICALS USED AS DIPS

Electrolyzed oxidizing (EO) water containing free chlorine is the product of a new concept developed by Japanese scientists. The fungitoxic activity of EO-water on *Monilinia fructicola* infecting peach was determined under conditions similar to retail conditions (at 2°C and 95% RH for more than 10 days). Wounded and inoculated fruits were immersed in tap water at 26°C for 5 to 10 min (control) or treated with EO-water varying in oxidation reduction potential (ORP), pH, and free available chlorine (FAC). Brown rot infection was not reduced by EO-water in wounded fruit, whereas disease incidence and severity were reduced in nonwounded, inoculated fruit, the reduction being maximum in fruit immersed in EO-water for up to 5 min (Al-Haq et al., 2001). The efficacy of EO-water in controlling *Botryosphaeria* rot on European pear cv. La-France was assessed. Wounded fruits were inoculated with *B. berengeriana*, incubated for 4 h, immersed in EO-water and held at 20°C, and above 90% RH. Disease incidence and severity were reduced and no chlorine toxicity was seen, indicating that EO-water is an effective surface sanitizer (Al-Haq et al., 2002).

Decay of grape berries and/or desiccation of stems and pedicels may lead to marked postharvest deterioration of table grapes. The usefulness of a postharvest dip in ethanol to prevent decay of table grapes was assessed. Immersion of detached berries in 70% ethanol eliminated most of the fungal and bacterial populations on the berry surface. The conidial germination of *Botrytis cinerea* was arrested in vitro. Dipping of grape bunches in 33, 40, or 50% ethanol, prior to packaging, resulted in inhibition of berry decay that was equivalent to, or better than, that which could be achieved with SO_2, released from generator pads. The bunch appearance, berry bloom, or berry firmness was not impaired and ethanol-treated berries received higher organoleptic scores than SO_2-treated berries (Lichter et al., 2002). Combination of ethanol and heat (46.3°C for 2 min) applied onto ungerminated spores of *B. cinerea* retarded mycelial development. Ethanol or heat treatment of mature berries did not influence the expression of the gene that encodes the heat shock protein, HSP 101, or grape berry alcohol dehydrogenase. The results suggested that tolerance of conidia to ethanol or heat may be correlated with their ability to repair damage through the action of chaperones such as HSP 104 (Lichter et al., 2003).

Ethanol (50%) applied at 1 h before harvest significantly reduced decay incidence due to *B. cinerea* in strawberry after storage for 3 days at 1°C followed by 2 days at 24°C (Karabulut et al., 2004a). The germination of conidia of *Penicillium expansum* and *B. cinerea* were entirely inhibited by a 10-s exposure to 40% and 30% ethanol respectively. Immersion of naturally infected sweet cherry fruit in 20, 30, 40, or 50% ethanol reduced decay per cent after storage for 10 days to the same extent (about 60 to 85%). The efficacy of hot water dips was significantly increased by addition of ethanol (0 to 10%) the most effective treatment being ethanol 10% at 60°C. No visible surface injury to the fruit following these treatments could be observed (Karabulut et al., 2004b). The toxicity of brief exposures of *Rhizopus stolonifer*, *Aspergillus niger*, *B. cinerea*, and *Alternaria alternata* to heated aqueous ethanol solutions was determined. A concentration of ethanol that was subinhibitory at ambient temperatures was inhibitory when heated to 40°C, a much lower temperatures than that causing inactivation of spores by water alone. Thus, ethanol and heat were synergistic. Hence the control of these postharvest pathogens may be achieved at much lower temperatures and ethanol concentrations when compared with either used alone (Gabler et al., 2004).

Sodium orthophenylphenate (SOPP) is a chemical registered in the United States for use exclusively on pears and it is labeled for use in a drench, line spray, or dip followed by a rinse with potable water. About 80% of the winter pears in Washington and Oregon are treated with SOPP (3300 to 3800 ppm) as a dip treatment in the initial stage of the packing process. SOPP is applied in the initial water dump of the packing line, at the point in which pears are floated out of the picking bin. SOPP has been found to be effective in controlling/suppressing fungal pathogens causing decay in pears by about 90% and it is considered as an essential treatment for decay control, allowing a longer marketing period (Kupferman, 1997).

The potential of sodium carbonate (SC), potassium carbonate (PC), sodium bicarbonate (SBC), potassium bicarbonate (PBC), and ammonium bicarbonate (ABC) was assessed. The concentrations of these salts that inhibited the germination of conidia of *Botrytis cinerea* by 95% (EC_{95}) were 16, 17, 36, 58, and 163 mM respectively. ABC was more effective in providing protection to grapes against gray mold disease than SBC, PBC, PC, and chlorine. Addition of chlorine (200 µg/ml) to the bicarbonate salts improved their efficacy in controlling gray mold disease. Treatment of grapes with ABC, SBC, or ethanol (70%) did not adversely affect the fruit quality (Gabler and Smilanick, 2001).

11.3 CHEMICALS APPLIED AS SPRAYS OR DUSTS

Fungal pathogens causing postharvest diseases are more numerous than bacterial pathogens. Hence many fungicides have been developed for the control of the diseases caused by them. Fungicides may be grouped based on their activity, target site of activity, and nature of active ingredients (a.i.). Chemicals are more commonly classified based on their chemical constitution. The commercially available fungicides generally contain a mixture of active ingredients, inert diluents, wetting agents, stickers, and emulsifiers. They may be formulated as wettable powder (WP), emulsifiable concentrates (EC), or dusts. Wettable powder is applied as a suspension in water and contains finely ground inert material, such as kaolin, and a wetting agent. Emulsifiable concentrates are formulated in oil that can be diluted to desired levels and they mix with water more easily than wettable powders. Dust formulations have low contents of active ingredients (1 to 10%), while WPs and ECs contain high concentrations (30 to 80%) of active ingredients. Protectant fungicides, such as sulfur or lime sulfur, have to be applied prior to infection. They do not have systemic action, since they are not absorbed by, and translocated into, the plant tissues. The fungicidal activity of nonsystemics is lost after short periods and after exposure to high temperatures and rainfall. In contrast, organic chemicals with systemic action are retained for longer periods and they have eradicative action, resulting in the elimination of fungal pathogens that have established themselves in wounds on the surfaces of fruits and vegetables. However, the major problem with the systemic fungicides has been the development of resistance in fungal pathogens, sooner or later, leading to the removal or restricted use of the systemics. The fungicides may be applied either at preharvest, to prevent infection under field conditions, or as postharvest treatment prior to storage to protect the harvested produce (Table 11.1).

11.4 ASSESSMENT OF FUNGICIDAL ACTIVITY

The toxicity of the fungicides may be assessed by determining the percentage of inhibition of germination of spores and retardation of mycelial growth in media amended with different concentration of test fungicides. Poisoned food technique or radial growth (RG) test is traditionally performed in suitable solid medium, usually potato dextrose agar (PDA), incorporated with test chemicals (Appendix 11(i)). The efficacy of fungicides can be determined by using various methods. Different methods of inoculation may be employed, depending on the nature of the pathogens, crop, and environments to simulate natural conditions required for infection. Optimal concentration of conidia should be provided for successful infection. For certain host–pathogen combinations, specific method(s) may be needed (Appendix 11(ii) and Appendix 11 (iii)).

11.4.1 Nonsystemic Fungicides

The effectiveness of nonsystemic fungicides in controlling postharvest diseases of fruits has been demonstrated. From in vitro tests, green mold disease of lemons, caused by *Penicillium digitatum*, was reduced by more than 80% by immersion of the fruit for 1 min in a lime–sulfur solution (3%) in water. Heat improved the effectiveness of lime–sulfur for the control of both green mold and sour rot (caused by *Geotrichum citri-aurantii*). The incidence of green mold was significantly reduced by immersing inoculated lemons in lime–sulfur solution (2.6%) at 40.4°C for 90 s. No injury due to treatment was observed on the fruit. Lime–sulfur was equally or more effective in reducing the incidence of green mold on oranges and lemons than those exposed to a soda ash tank treatment or soda ash followed by the fungicide imazalil in wax. Significant advantages of using lime–sulfur are the absence of residues in treated fruits and effective disposal of spent solutions as a soil conditioner (Smilanick and Sorenson, 1999). Immersion of lemons or oranges in warm, liquid lime–sulfur solution (40.6 to 43.3°C) that contained 0.75% calcium polysulfide, for 1 to 4 min, reduced the incidence of green mold by 80% or more. This treatment also reduced sour rot incidence by 35 to 70% (Smilanick and Sorenson, 2001).

TABLE 11.1 Management of Postharvest Diseases Using Fungicides/Antibiotics

Crop	Disease	Fungicides	References
Apple	Storage rot (*Gloeosporium*)	Captan, thiram Benomyl	Moore and Edney, 1959 Burchill and Edney, 1977
	Gray mold	Benzimidazoles	Cargo and Dewey, 1970
	Alternaria rot	Benzimidazoles, iprodione, mancozeb	Selmaoui et al., 1997
	Brown rot	Benomyl, carbendazim, thiabendazole, dicloran	Sharma, 2000
Avocado	Fruit rots, stem-end rot, anthracnose	Benomyl, prochloraz	Everett and Korsten, 1996; Everett et al., 1999
Banana	Anthracnose	Prochloraz, carbendazim	Ullasa and Rawal, 1984
Cabbage (white)	Gray mold	Iprodione, thiophanate-methyl	Adamicki and Robak, 2000
Cabbage	Black rot	Chlortetracyline	Bhat et al., 2000
Citrus	Brown rot	Copper, alliette	Ritenour et al., 2003
	Diplodia stem end rot (SER)	Benlate, thiabendazole (TBZ)	
	Anthracnose	Benlate, TBZ	
	Alternaria SER	Imazalil	
Citrus	Green mold	Imazalil WP and EC	Rao, 1984
		Lime sulfur	Smilanick and Sorenson, 1999
		Thiabendazole, sodium bicarbonate	Smilanick, 2003
Grape-vine	Downy mildew	Mancozeb, metalaxyl, azoxystrobin	Wong and Wilcox, 2001
Kiwifruit	Gray mold	Vinclozolin	Michailides and Elmer, 2000
Mango	Anthracnose	Benomyl, prochloraz	Arauz, 2000; Oosthuyse, 2000
	Anthracnose, stem-end rot	Captan, ziram, chlorothalonil, carbendazim, thiophanate-methyl	Banick et al., 1998

TABLE 11.1 (*Continued*)

Crop	Disease	Fungicides	References
Peach	Brown rot	Tebuconazole, cyproconzole	Tonini et al., 2000a, b
	Scab	Micronized wettable sulfur, chlorothalonil, captan, azoxystrobin	Schnabel and Layne, 2004
Pear	Blue mold, gray mold, *Mucor* rot	TBZ, captan	Kupferman, 1997
Papaya	Anthracnose, stem-end rot	Benomyl, iprodione, benzimidazole	Liberato et al., 2001
Pepper (Chili)	*Phytophthora* blight	Mefenoxam (Ridomil Gold)	Parra and Ristaino, 2001
	Fruit rot	Carbendazim, iprodione, propiconazole, hexaconazole	Hegde and Anahosur, 2001
Plum	Brown rot	Tebuconazole, flusilazole, myclobutanil	Northover and Cerkauskas, 1998
Potato	Silver scurf	Fludioxonil, iprodione, imazalil, azoxystrobin	Hervieux et al., 2001
	Tuber rot	Mefenoxam	Taylor et al., 2002
	Pink rot and leak	Mefenoxam	Taylor et al., 2004
	Soft rot	Kasugamycin	Bartz, 1999
Strawberry	Gray mold, anthracnose	Thiram, iprodione, dichlofluanid	Washington et al., 1999
Sweetpotato	Black rot	Thiophanate-methyl, carbendazim	Yang et al., 2000

The control of *Botrytis* fruit rot of strawberry commences with preharvest application of fungicides in the field. Application of full label rate of fungicides throughout the season provided the most effective control of both preharvest and postharvest fruit rot caused by *B. cinerea*. The fruit decay incidence was kept below 5% for up to 10 days, by the application of captan or thiram at full label rate and use of proper cold storage. A 10 day minimum storage life affords a longer shipping interval and access to more distant markets for Florida strawberries (Blacharski et al., 2000). The effect of weekly applications of captan and thiram was assessed. The incidence of postharvest *Botrytis* fruit rot was markedly affected by harvest date, length of time in storage, and fungicide treatment. The highest rate of captan and thiram treatments provided maximum protection and longest storage life. Addition of iprodione to either captan or thiram did not offer additional benefit consistently or increase in yield (Blacharski et al., 2001). On the other hand, half label rate mixtures of thiram + iprodione, applied at 7-day intervals, effectively controlled rots caused by *Botrytis cinerea* and anthracnose caused by *Colletotrichum acutatum*, reducing all rots from 39.1 to 4.9% (Washington et al., 1999).

The cost of the chemicals and application may be a major factor influencing the acceptability of recommendations for disease management. The reduced fungicide program, consisting of applications of chlorothalonil and captan or chlorothalonil, captan, and azoxystrobin, provided excellent control of scab in two peach cultivars, Contender and Cresthaven, when disease pressure was low to moderate. The reduced fungicide programs were considered to be preferable to weekly applications of sulfur (11 applications) because of lower cost with equal effectiveness (Schnabel and Layne, 2004).

The efficacy of captan (Captaf), ziram (CumanL), and chlorothalonil (Kavach) for the control of anthracnose caused by *Colletotrichum gloeosporioides* and stem-end rot caused by *Botryodiplodia theobromae* in mango was assessed. Ziram and captan were more effective than chlorothalonil, but less effective compared to systemic fungicides carbendazim and thiophanate-methyl (Banik et al., 1998). Mancozeb, when applied at 1790 μg/ml, at 1 to 5 days before inoculation of grapevine with the downy mildew pathogen *Plasmopara viticola*, provided complete control of the disease and it showed moderate to high antisporulant activity. Mancozeb did not show curative action, that is it was not effective, when applied after infection, although it may reduce sporulation of the pathogen (Wong and Wilcox, 2001).

11.4.2 Systemic Fungicides

The systemic fungicides, generally, have a selective and narrow spectrum of activity against fungi belonging to specific taxonomic groups. The mobility of these fungicides within plant tissues essentially differentiates them from the nonsystemic fungicides. The specific and selective activity of systemic fungicides has been responsible for the development of fungicide-resistant strains of fungal pathogens causing postharvest diseases.

11.4.2.1 Acylalanine Group These compounds containing (±) methyl-*N*-(2-methoxyacetyl)-*N*-(2,6-xylyl) alaninate as active ingredient all are specifically active against Oomycetes, especially Peronosporales. Inhibition of ribosomal RNA and polymerase activity of the fungal pathogen seems to be the primary mechanism of metalaxyl compounds on sensitive fungi belonging to the genera *Phytophthora, Pythium*, and *Plasmopara*, causing diseases of fruits and vegetable crops. Production and germination of oospores of sensitive strains of *Phytophthora infestans*, infecting potato plants and tubers causing late blight disease, were markedly affected by the application of metalaxyl (Hanson and Shattock, 1998). Metalaxyl (260 μg/ml) entirely prevented development of downy mildew disease of grapevine caused by *Plasmopara viticola*, when applied prior to inoculation as well as 1 day after inoculation. Application at 3 to 5 days after inoculation also reduced disease severity and sporulation substantially (mean reduction of 46 and 94% respectively) (Wong and Wilcox, 2001).

Phytophthora erythroseptica, causing pink rot disease, can directly infect potato tubers, while *Pythium ultimum* needs a wound injury to infect the tubers, which develop leak symptoms. Mefenoxam was more effective in controlling pink rot relative to leak, following challenge inoculation with the respective fungal pathogen. The use of mefenoxam to control pink rot in the field and storage can be effective, since cell isolates tested were found to be sensitive to this compound (Taylor et al., 2004).

11.4.2.2 Benzimidazole and Thiophanate Group Benomyl compounds with methyl 1 (bulyl-carbamoyl) benzimidazol 2-yl carbamate (MBC) as the active ingredient may inhibit mitosis and cause disorganization of the fine structure of cells of *Botrytis cinerea*, an important postharvest pathogen causing gray mold diseases in several fruits and vegetables. Inhibition of DNA synthesis may be observed later, as a result of the blockage of nuclear division in the fungi exposed to benomyl fungicides (Richmond and Philips, 1975). MBC and thiabendazoles (TBZ) are bound to fungal tubulin at the same site of binding, indicating that MBC and TBZ may have similar mode of action on sensitive fungal pathogens (Howard and Aist, 1980). The benzimidazoles primarily act as powerful inhibitors of the microtubule assembly, in addition to adversely affecting respiration of fungi. Benzimidazoles act selectively on certain fungi belonging to Ascomycetes (Davidse, 1986).

Benzimidazole compounds have been reported to be effective against anthracnose and fruit rot diseases of fruits and vegetables caused by *Colletotrichum* spp. and *Botryodiplodia* spp. Spraying benomyl during flowering significantly reduced the stem-end rot of avocado caused by *Phomopsis* sp., *Botryosphaeria* sp., and *Colletotrichum* sp. (Everett et al., 1999). Among the five fungicides tested carbendazim and thiophanate-methyl provided most effective control of anthracnose disease caused by *Colletotrichum gloeosporioides* and stem-end rot disease caused by *Botryodiplodia theobromae* of mango (Banik et al., 1998). Treatment of mangoes with hot benomyl (2.0 g/l) at 50°C for 5 min effectively controlled anthracnose disease (Oosthuyse, 2000). Preharvest application of benomyl (2.2 kg/ha) or thiophanate-methyl (2.2 kg/ha) reduced the decay of citrus fruit after storage in nine out of 10 experiments, frequently reducing the decay by about 50%. The incidence of stem-end rot, caused primarily by *Diplodia natalensis* or *Phomopsis citri*, was effectively reduced only by benomyl and thiophanate-methyl. The anthracnose caused by *Colletotrichum gloeosporioides* was significantly controlled by benomyl and phosphoric acid (Ritenour et al., 2004).

Postharvest treatment of apple cv. Golden Delicious with carbendazim (500 μg/ml) provided maximum protection against brown rot caused by

Monilinia laxa and *M. fructigena*. When used as a protective treatment, carbendazim completely prevented infection by these pathogens (Sharma, 2000). The effectiveness of carbendazim and thiophanate-methyl was assessed for the control of black rot of sweet potato caused by *Ceratocystis fimbriata*. Both fungicides were highly inhibitory to the mycelial growth of *C. fimbriata* in potato dextrose agar medium. Tubers treated with thiophanate-methyl or carbendazim were protected to an extent of more than 90% in storage (Yang et al., 2000). The efficacy of preplanting treatment of potato seed tuber with a combination of thiophanate-methyl (as Easoat 50 g a.i./100 kg seed tubers) and sodium hypochlorite (NaOCl, 500 ppm) for 8 min for the control of black scurf disease caused by *Rhizoctonia solani* was determined. This combined treatment reduced *Rhizoctonia* on progeny tubers at harvest and after storage. Thiophanate-methyl treatment following NaOCl disinfection suppressed the growth of *R. solani*. The marketable tuber yield was also increased in addition to the absence of any phytotoxic symptoms due to the treatment with these compounds (Errampalli and Johnston, 2001). The toxicity of carbendazim against *Colletotrichum capsici* causing pepper (chili) fruit rot was studied along with other systemic and nonsystemic fungicides. Field application of carbendazim (0.1%) resulted in a least percent disease index of 30.96%, as compared to 71.5% in iprodione-treated plots. The significant advantage of using carbendazim was the maximum net return with a cost – benefit ratio of 1:14.39 (Hegde and Anahosur, 2001).

Thiabendazole (TBZ) is one of the few fungicides approved for postharvest use on apples, pears, and citrus as a drench, dip, or line spray. It is chemically 2(4-thiazol-4-yl) benzimidazole. TBZ remains stable in an aqueous suspension throughout the storage life of treated fruit. It is relatively effective against blue mold, gray mold, stem-end rot, and neck-rot diseases. Development of resistance to TBZ in postharvest pathogens, such as *Penicillium expansum*, is known (Kupferman, 1997). The residue tolerances (ppm) for TBZ in the United States, Canada, Japan, and the *Codex Alimentarius* is 10 ppm, while the European Union permits only 5 ppm of TBZ residue in citrus (Ritenour et al., 2003).

The green mold caused by *Penicillium expansum* is one of the important postharvest diseases of citrus, and TBZ has been used on fruit for more than three decades, in combination with commercial wax and spraying. Application of TBZ was effective against stem-end rot caused by *Botryodiplodia theobromae* in Amber, Ambersweet, and Hamlin oranges and mandarins, when applied at 700 and 1000 ppm before and after degreening, respectively (Brown and Chambers, 1999). Immersion of citrus fruit in aqueous TBZ was found to be superior to the general practice of spraying TBZ with wax. Green mold caused by *P. digitatum* was reduced by 50% by a brief dip in TBZ solution (10 μg/ml) which was a much lower concentration compared to 200 μg/ml required for application with wax. Combined application of TBZ and sodium bicarbonate ($NaHCO_3$) at 3% controlled TBZ-resistant isolates of *P. digitatum*. This combination was effective against sour rot caused by *Geotrichum*

TABLE 11.2 Effect of Thiabendazole-Drenching on the Incidence of Stem, Calyx, and Puncture Gray Mold in Controlled Atmosphere (CA) Stored d'Anjou Pear from Orchards in Mid-Columbia Region or Oregon and Washington

Orchard	Stem-end gray mold		Calyx-end gray mold		Puncture gray mold	
	Nondrenched	Drenched	Nondrenched	Drenched	Nondrenched	Drenched
1	2.45	0.63	0.91	0.55	2.21	0.64
3	–	0.51	–	0.23	–	0.31
5	1.04	0.14	0.26	0.06	1.18	0.35
6	0.94	0.39	0.32	0.14	0.97	0.55
7	0.34	0.23	0.14	0.11	0.75	0.73
8	1.37	0.19	1.03	0.21	2.92	0.24
9	0.49	0.05	0.15	0.08	0.48	0.17
11	0.91	0.86	0.14	0.13	0.57	0.60
12	0.19	0.08	0.03	0.08	1.82	0.29
13	0.82	0.62	0.31	0.12	0.41	0.38
14	0.19	0.08	0.32	0.11	0.52	0.09
15	0.46	–	0.32	–	0.64	–
Mean	0.84 b	0.34 c	0.36 c	0.17 c	1.13 a	0.40 c
LSD (P = 0.05)	0.283	–	–	–	–	–

Values are percentage of decayed pear fruit after CA storage in a commercial packing house. Values for mean gray mold incidence followed by the same letter are not significant different at p = 0.1. LSD = least significant difference.
Source: Lennox et al., 2004.

citri-aurantii, which was controlled by TBZ alone and thus in the case of sour rot disease the presence of SBC did not provide an added advantage. Treatment of lemons with a mixture of TBZ (200 μg/ml) + $NaHCO_3$ (3%) and sodium hypochlorite (200 μg/ml) at 40.6°C followed by rinsing with water reduced infection by TBZ sensitive isolate of *P. digitatum* to 1.3% from 83.6% in untreated (control) fruit. The treatment reduced decay by TBZ-resistant isolates to 16% from 99.1% in untreated fruit (Smilanick, 2003). The effectiveness of a prestorage drench of d'Anjou pear with TBZ (525 μg a.i./ml) for the control of gray mold (*Botrytis cinerea*), Mucor rot (*Mucor piriformis*), blue mold (*Penicillium expansum*), bull's-eye rot (*Neofabraea* spp. syn. *Pezizicula* spp.), and side rot (*Phialophora malorum*) in controlled atmosphere (CA) storage was assessed. Incidence of gray mold in CA-storage was reduced by a prestorage TBZ drench. The stem-end and puncture gray mold also were reduced, while the fungicide drench had no effect on the incidence of other diseases (Table 11.2) (Lennox et al., 2004).

11.4.2.3 Imidazole Compounds Prochloraz is an important fungicide employed in packlines for the control of mango anthracnose disease. Treated mangoes were placed in cool storage at 12.5°C for 28 days. All packline fungicidal treatments incorporating prochloraz provided superior control directly after cool storage. On ripening, prochloraz added to the hot water bath (180 ml/100 l of water) was used as dip at 50°C for 5 min, and this treatment reduced the anthracnose significantly in mango cultivars Zill and Kent. This treatment decreased localized skin darkening and internal breakdown in Zill, but not in Kent (Oosthuyse, 2000). In another trial, prochloraz was reported to inhibit mycelial growth of *Colletotrichum gloeosporioides* completely and to provide good control of the anthracnose disease in vivo (Nascimento et al., 2000). Use of prochloraz has been approved by the European Union, based on the guidelines of the FAO/WHO *Codex Alimentarius* and the U.S. Environmental Protection Agency. The acceptable maximum residue limit for prochloraz was 2 ppm as per Codex. Prochloraz may be used as a protectant or as an eradicant spray (Arauz, 2000). Application of prochloraz, either in wax or as an ultra-low-volume spray, reduced the postharvest anthracnose disease of avocado caused by *Colletotrichum acutatum* (Everett and Korsten, 1996). Wax by itself appeared to reduce the incidence of stem-end rot caused by *Botryodiplodia theobromae*. However, waxing was found to increase the incidence of all postharvest diseases on 'Fuerte' avocados (Darvas et al., 1990).

Imazalil [1-(allyloxy-2,4-dichlorophenoxy-ethyl)-imidazole-1-carboxamide] is particularly effective against green mold, including benzimidazole-resistant strains, and reduces sporulation of the fungal pathogens. It exhibited some antifungal activity against *Alternaria* stem-end rot (black rot) (Ritenour et al., 2003). Imazalil was shown to be compatible with biocontrol agent yeasts. Treatment of citrus with yeast isolates and lower concentrations of imazalil or thiabendazole reduced the infection by *Penicillium digitatum* and *P. italicum* causing green and blue mold diseases by 90 to 100% (Kinay et al., 2003). Prochloraz at 400 mg a.i./l gave complete control (100%) of both papaya anthracnose disease caused by *Colletotrichum gloeosporioides* and stem-end rot caused by *C. gloeosporioides* and *Phoma caricae papayae*. The control efficacy was reduced to 95% when treated with prochloraz at 250 mg a.i./l. However, phytotoxicity symptoms were noticed after treatment at these concentrations of prochloraz. The concentration of prochloraz had to be reduced to 100 μg a.i./l to avoid phytotoxicity problem but with a consequent reduction in disease control (Liberato and Tatagiba 2001). Potato seed tubers infected with *Helminthosporium solani* causing silver scurf disease were treated with imazalil (15 g a.i/t of potatoes). Fungicidal treatment had no significant effect on disease severity or on yield of total and marketable tubers (Hervieux et al., 2001). The effectiveness of imazalil, prochloraz, and thiabendazole (TBZ) applied alone or as a mixture for the control of potato silver scurf (*H. solani*) and dry rot (*Fusarium solani* var. *coeruleum*) diseases was assessed. The mixture of TBZ and imazalil was more effective in controlling dry rot disease than imazalil alone. Likewise, the treatment of tubers with this mixture reduced the severity of silver scurf disease to the maximum extent. The fungicides exhibited similar effectiveness when used for tuber treatment before planting (Carnegie et al., 1998).

11.4.2.4 Dicarboximide Compounds These compounds are structurally related. They have been reported to cause several toxic effects, such as mitotic instability (Georgopoulos et al., 1975), inhibition of protein synthesis (Pappas and Fisher, 1979), swelling and lysis of mitochondria, and inhibition of nucleic acid synthesis in *Botrytis cinerea*, *Ustilago maydis*, and *Sclerotinia sclerotiorum* (Hisada et al., 1978). The main metabolic route is hydrolysis of the imide ring. Since hydrolysis of the imide ring also proceeded in sterilized distilled water, the metabolites, by ring-opening followed by decarboxylation, may originate from the hydrolysis in the plant tissue (root). The 4 position of the phenyl ring was substituted with a fluorine atom. Hence, hydrolysis appeared to be the major pathway of fluoroimide on leaves and fruits of apple (Katagi and Mikami, 2000).

Gray mold caused by *Botrytis cinerea* and anthracnose caused by *Colletotrichum acutatum* are the important and frequently occurring diseases in strawberry in Australia. Iprodione, either alone or as a mixture with thiram or dichlofluanid was evaluated for the control of these diseases. Iprodione, when applied alone did not reduce gray mold significantly, suggesting the presence of dicarboximide resistant strains of *B. cinerea*. In contrast, by applying half label rate mixtures of either iprodione + dichlofluanid or iprodione + thiram, most consistent control of all rots could be achieved. Residue analysis indicated that residues of iprodione and dichlofluanid, but not thiram, were within the current approved maximum residue limits (Washington et al., 1999). Iprodione alone or in combination with captan or thiram was applied during two peak bloom periods. Postharvest incidence of gray mold caused by *B. cinerea* was recorded after storage at 4°C for 14 days. Fungicide application reduced preharvest infection by *B. cinerea* and increased marketable yield. Addition of iprodione, either to captan or thiram, did not consistently increase the disease control efficiency significantly (Blacharski et al., 2001).

Incidence of gray mold in kiwifruit could be significantly reduced by preharvest spray of dicarboximide fungicides (Pennycook, 1984). But the widespread emergence of strains of *B. cinerea* resistant to dicarboximide led to banning of the use of dicarboximide fungicides as preharvest treatment in New Zealand. Iprodione and benomyl are the only two fungicides permitted for the control of *Sclerotinia* flower and fruit rot (Michcailides and Elmer, 2000). Application of iprodione alone (382 g a.i./ha) and in combination at reduced rate of 306 g a.i./ha, with thiophanate-methyl (700 g a.i./ha) reduced infection by *Botrytis cinerea* in cabbage to the maximum extent. The outer leaves of cabbage retained their green color in iprodione-treated plots, while those in other treatments exhibited signs of yellowing. Highest percentage of marketable heads (70 to 84%) could be obtained following treatment with iprodione (Adamicki and Robak, 2000).

Vinclozolin, a dicarboximide fungicide, was evaluated as a preharvest application for its efficacy for the control of *Botrytis* gray mold in kiwifruit. Application of vinclozolin twice at bloom and two more preharvest sprays provided the best control of gray mold in storage (Suadi-Hasbun, 1987; Sommer et al., 1994). Use of vinclozolin in vineyards for the control of *B. cinerea* could increase the risk of production of resistant strains of *B. cinerea*. Resistance to dicarboximides in gray mold pathogen was reported in Europe and New Zealand (Lorenz and Pommer, 1985; Manning and Pak, 1995).

11.4.2.5 Triazoles Triazole compounds with wide-spectrum activity have been developed. Tebuconazole marketed as Folicur has been reported to be effective against diseases of apple, pear, peach, and grapes (Angelini, 1996). Among the 12 fungicides tested, symptomless latent infection of European plums by *Monilinia fructicola*, causing brown rot, was effectively suppressed by tebuconazole and flusilazole. The sterol-inhibiting fungicides were applied twice at mid-season to Stanley bearing trees with high incidence of latent infections. Tebuconazole alone could provide temporary suppression of *M. fructicola* in excised immature fruits. Fruits were sprayed with tebuconazole, fusilazole, and fenbuconazole to reduce brown rot infection during postharvest incubation. The treated mature fruit exhibited 45 to 65% infection as against 89% in control fruit sprayed with water. Poor adhesion of fungicide droplets to the waxy fruit surface at the time of application could be a reason for the lower level of disease control obtained compared to fruits with no or less waxy coating (Northover and Cerkauskas, 1998).

Spraying tebuconazole in the last 3 weeks before harvesting gave best control of brown rot disease of peaches and nectarines caused by *Monilinia laxa*. Application of tebuconazole was followed 10 days later by application of cyproconazole or fenbuconazole. The fruits could be stored for a period of 15 to 20 days with a shelf-life of 5 days. Since postharvest application of fungicides was not permitted in Italy, control of diseases at preharvest stage was the only option available (Tonini et al., 2000a, b). The effectiveness of tebuconazole, propiconazole, and triadimenol in controlling anthracnose (*Colletotrichum gloeosporioides*) and stem-end rot (*C. gloeosporioides* and *Phoma caricae-papayae*) was assessed. The fungicide treatment provided complete control (100%) of both diseases. However, phytotoxocity symptoms due to fungicide treatment were observed (Liberato and Tatagiba, 2001).

11.4.2.6 Strobilurins Strobilurins are natural products with fungicidal properties prepared from the Basidiomycete fungi *Strobilurus tenacellus* and *Oudemansiella mucida.* Azoxystrobin derived from *O. mucida* exhibited antifungal activity against a wide range of fungal pathogens belonging to Oomycetes, Ascomycotina, Basidiomycotina, and Deuteromycotina. It inhibits spore germination and also has remarkable curative properties. Azoxystrobin has been reported to provide excellent control of grapevine downy mildew and powdery mildew diseases (Godet et al., 1977; Dacol et al., 1988). Azoxystrobin, along with prochloraz and sodium bicarbonate were evaluated for their efficacy in controlling mango anthracnose disease. Azoxystrobin inhibited the mycelial growth entirely in vitro and effectively prevented the development of anthracnose disease in mango fruits (Nascimento et al., 2000). Likewise, strobilurin fungicides were found to be effective in controlling postharvest diseases of 'Hass' avocado, when applied as preharvest treatment. Amistar and Flint fungicides were superior to Stroby formulation. The incidence of anthracnose caused by *Colletotrichum gloeosporioides* was significantly reduced by 66% and 74% respectively following treatment with Amistar and Flint (Willingham et al., 2000). The impact of a reduced-fungicide program consisting of chlorothalonil, azoxystrobine, and captan (five application each) on peach scab disease control and fruit quality was evaluated. When disease pressure was low to moderate, reduced fungicide programs provided excellent disease control. But their effect on fruit quality was inconsistent (Schnabel and Layne, 2004). Application of strobilurins azoxystrobin, trifloxystrobin, pyraclostrobin, and methoxycrylate either alone or incorporated with copper and mancozeb, provided effective control of black spot of citrus caused by *Guignardia citricarpa*, comparable to the industry standard of copper and mancozeb. Furthermore, rind damage due to strobilurins was much less than the industry standard fungicides currently recommended in Queensland (Miles et al., 2004).

11.4.2.7 Phenylpyrroles and Miscellaneous Compounds Two phenylpyrroles, fenpiclonil and fludioxonil, were synthesized using the secondary metabolite of *Pseudomonas pyrrocinia*, pyrrolnitrin as a lead compound. The photostability of fludioxonil made it suitable for the control of *Botrytis cinerea, Monilinia* spp., and *Sclerotinia* spp. when it was sprayed on plant surface (Nyfeler and Ackermann, 1992; Stensvand, 1998). Fenpiclonil was considered to inhibit glucose metabolism of the pathogen, the target site being a transmembrane sugar carrier associated with glucose (Jespers and deWaard, 1995).

The effectiveness of fenpiclonil, applied either alone or in a mixture, for the control of postharvest diseases of potato was assessed by spraying the fungicides on potato seed tubers using an electrostatic sprayer or a hydraulic sprayer. Fenpiclonil, imazalil, and thiabendazole formed bigger deposits and control of gangrene caused by *Phoma foveata* (*P. exigua var. foveata*) was associated with biggest deposits. Fenpiclonil was effective in controlling the dry rot disease caused by (*Fusarium solani* var. *coeruleum*. Furthermore, fenpiclonil and a mixture of thiabendazole and imazalil were able to reduce the severity of silver scurf disease caused by *Helminthosporium solani* to the maximum extent (Carnegie et al., 1998). Potato seed tubers were treated with fludioxonil (25 g a.i./t of potatoes). The incidence of silver scurf on daughter tubers at harvest and after different storage periods was not significantly affected by treatment with fludioxonil (Hervieux et al., 2001).

For the control of *Rhizopus* rot of peach caused by *Rhizopus stolonifer*, the efficiency of new fungicides was determined. Fludioxonil, tebuconazole, or *Pseudomonas syringae* when applied at inoculation with *R. stolonifer* gave 75 to 100% reduction, comparable with that of dicloran and iprodione. However, applied as postinoculation treatments, the biocontrol agent was ineffective, whereas fludioxonil and tebuconazole provided 90% control at 472 and 718 μg/ml respectively (Northover and Zhou, 2003). Application of fludioxonil (45 μg/ml), cyprodinil (50 μg/ml), and fludioxonil + cyprodinil (50 + 75 μg/ml) provided more than 97.0% control of *P. expansum* causing blue mold disease in 'Empire' apples for up to 1 month. However, higher doses of fludioxonil (100 μg/ml) and fludioxonil + cyprodinil (150 + 225 μg/ml) was necessary to control blue mold disease for 62 days at 4°C. These fungicides are particularly useful where the existence of *P. expansum* strains resistant to thiabendazole was observed, since these fungicides have modes of action different from that of thiabendazole (Errampalli and Crnko, 2004).

The fungicidal effect of fludioxonil on *Colletotrichum lagenarium*, causing cucumber anthracnose disease, has been shown to be through hyperactivation of a fungal signal transduction pathway. The *OSC1* gene encoding a MAP kinase related to yeast *Hog1* was isolated from *C. lagenarium*. The *osc1* knock-out mutants, showing increased resistance to fludioxonil, were sensitive to high osmotic stress, indicating that Osc 1 was involved in responses to hyperosmotic stress and sensitivity to fludioxonil. The wild type strain of *C. lagenarium* was not able to infect cucumber in the presence of fludioxonil, because of a failure of appresssorium-mediated penetration. In contrast, *osc1* mutants infected successfully. Fludioxonil treatment activated phosphorylation of *Osc1*, suggesting improper activation of *Osc1* by fludioxonil caused negative effects on fungal growth. *Osc1* was rapidly translocated to the nucleus in appressorial cells after addition of fludioxonil, suggesting that fludioxonil could impair the function of infection structures. Further, activation of Hog1-type MAPKs in *Botrytis cinerea* and *Cochliobolus heterostrophus* by fludioxonil suggested that it acted, in part, through activation of the MAPK cascade in fungal pathogens (Kojima et al., 2004).

Fenhexamid, belonging to the chemical class of hydroxyanilides, has been reported to be effective against diseases caused by *Botrytis cinerea, Monilinia* spp., and *Sclerotinia sclerotiorum* following foliage application (Rosslenbroick et al., 1998). Fenhexamid was applied at 750 g a.i./ha on strawberries, raspberries, and black currants from early flowering. Gray mold caused by *B. cinerea* was as effectively controlled as the commercial standard treatment. Marked reduction in postharvest and storage rots in all three crops was observed following treatment with fenhexamid, in addition to retention of natural taste of fruit produce from treated crops (Adam and Birch, 1998). Strawberry, black and red-currant, and cherry crops were efficiently protected by the application of fenhexamid (Teldor) against *Botrytis cinerea* and *Monilinia laxa* (blossom wilt), resulting in a significant increase in yield. Application of Teldor during flowering can postpone postharvest *Botrytis* infections in strawberries and black currant (Heltbech et al., 2000).

Another fungicide belonging to a new chemical class of amino acids, amide carbomates (SZ X 72), exhibited excellent antifungal activity against *Plasmopara viticola* causing downy mildew disease, infecting leaves and bunches of grapevine. In addition, this fungicide was effective against *Phytophthora infestans* infecting potato tubers and *Pseudoperonospora cubensis* infecting cucumber (Stenzel et al., 1998). Aqueous application of polyhexamethylene biguanide (PHMB) to citrus fruits (Sunburst mandarins, Orlando tangelo, and Ambersweet and Hamlin orange) effectively controlled the stem-end rot caused by *Botryodiplodia theobromae*, sour rot caused by *Geotrichum citri aurantii*, and green mould caused by *Penicillium digitatum* (Brown and Chambers 1999).

A new mandelamide compound SX 623509 [N-2 (ethoxy-3-methoxy) phenylethyl-3,4-dichloromandelamide] was evaluated for its potential for the control of *Phytophthora infestans*. The pathogen cultured in pea-broth minimal media were exposed to different concentrations of SX 623509. The effect of the chemical on the lipid metabolism was followed with radiolabeled acetate, choline, and ethanolamine. The growth of *P. infestans* and lipid labeling from C^{14} acetate were significantly reduced by the chemical. It was concluded that the mandelamide SX 623509 adversely affected lipid synthesis at concentrations similar to those inhibiting growth. The alterations in lipid-labeling patterns may be due to inhibition of cholinephosphotransferase, which may form a future target site for fungicide development (Griffiths et al., 2003).

11.5 CHEMICALS ALTERNATIVE TO CONVENTIONAL FUNGICIDES

Several postharvest pathogens have developed resistance to commonly used fungicides. Hence, a wide range of chemicals has been evaluated for their potential for use as alternative to the current postharvest fungicides (Table 11.3). The effectiveness of these chemicals varies widely, depending on many factors, such as the sensitivity of the pathogen, crop, cultivar, time of application, and storage conditions. The fungitoxicity of the chemicals have been tested in vitro by determining the inhibition of conidial germination and mycelial development in culture media amended with different concentration of the test chemicals. Among 22 naturally occurring monoterpenoids screened for activity against *Botrytis cinerea* and *Monilinia fructicola* causing gray mold and brown rot diseases, the phenolic monoterpenoids carvacrol and thymol were the most effective in inhibiting spore germination and mycelial growth of the pathogens (Rong and Ting, 2000). Though some chemicals were found to have inhibitory effect on the postharvest pathogens in vitro, their efficacy to control the diseases caused by them in vivo has not been assessed.

Nine antioxidants (ascorbic acid, benzoic acid, butylated hydroxyanisole (BHA), butylated hydroxytoluene, dimethyl sulfoxide, propionic acid, propyl gallate (PG), propyl paraben (PP), and thiourea) were evaluated over a wide range of concentrations for their effectiveness against *Colletotrichum musae* causing crown rot diseases of banana. BHA was the most effective at low concentrations in combination with benzoic acid and PP, causing strong suppression of anthracnose lesions and crown rot development (Khan et al., 1999). It may be possible to use BHA, a food-grade chemical, to enhance the activity of fungicides such as imazalil or thiabendazole (TBZ) currently applied for the control of *C. musae* on bananas, allowing lower concentrations of fungicide to be used (Khan et al., 2001).

The usefulness of ammonium molybdate as potential alternative to synthetic fungicides for the control of *Botrytis cinerea*, *Penicillium expansum*, and *Rhizopus stolonifer* infecting apple has been demonstrated. Ammonium

TABLE 11.3 Chemicals with Potential for Use as Novel Fungicides

Crop	Disease/pathogen	Chemical	Reference
Apple	Blue mold/*Penicillium expansum*	Sodium hypochlorite 2-aminobutane	Baker and Heald, 1934 Maclean and Dewey, 1964
	Gray mold, blue mold, and *Rhizopus* rot/ *Botrytis cinerea, Penicillium expansum, Rhizopus stolonifer*	Ammonium molybdate	Nunes et al., 2001
	Brown rot/*Monilinia fructiola*	Diphenylamine, sodium orthophenylphenate	Sharma, 2000
Banana	Crown rot/anthracnose *Colletotrichum musae*	Butylated hydroxy anisole	Khan et al., 1999, 2001
Citrus oranges, lemons	Green mold *Penicillium digitatum*	Sodium carbonate	Smilanick et al., 1999
	Green and blue molds/ *Penicillium digitatum, P. italicum*	Sodium orthophenylphenate Sodium molybdate Ammonium molybdate	Simmone, 1999 Palou et al., 2002
Orange	*P. digitatum, P. italicum* and *B. cinerea*	Synthetic hexapeptides PAF 26 and PAF 32	López García et al., 2000, 2002
Melon (*Cucumis melo)*	Fruit rot/*Myrothecium roridum*	Calcium chloride, calcium sulfate, calcium carbonate, calcium nitrate	Lima et al., 1998
Papaya	Anthracnose/ *Colletotrichum gloeosporioides*	Calcium chloride, calcium carbonate	Saborío et al., 2000
Mango	Anthracnose/ *Colletotrichum gloeosporioides*	Potassium phosphonate	Zainuri et al., 2001
Pear	Gray mold and fruit rot/ *Botrytis cinerea, Alternaria alternata, Mucor piriformis*	Calcium chloride Sodium hypochlorite, sodium ligninsulfonate	El-Sheikh Aly et al., 1998 Kupferman, 1997
Potato	Silver scurf *Helminthosporium solani*	Aluminum chloride	Hervieux et al., 2002

molybdate (15 mM) reduced lesion diameters of *P. expansum*, *B. cinerea*, and *R. stolonifer* by 84%, 88%, and 100% respectively at 20°C. Among the three pathogens, *R. stolonifer* was the most sensitive to ammonium molybdate (Table 11.4). A significant reduction in severity and incidence of *P. expansum* and *B. cinerea* was noted in the case of apples treated with ammonium molybdate and stored for 3 months at 1°C. As a preharvest application, ammonium molybdate was effective in reducing blue mold decay after 3 months in cold storage. Ammonium molybdate was as effective as imazalil applied at the commercial dose, indicating its potential for use against blue and gray molds in place of the synthetic fungicide (Nunes et al., 2001). Dipping oranges for 150 s in solutions of sodium molybdate (24.2 mM) or ammonium molybdate (1.0 mM) at 48 or 53°C provided satisfactory control of green and blue molds. Molybdenum salts at higher concentrations were phytotoxic and stained the fruit. *Penicillium* decay of citrus fruit could be effectively controlled by potassium sorbate, sodium benzoate, and ammonium molybdate (Palou et al., 2002). The effectiveness of phosphoric acid, applied as a preharvest application, for the control of tuber rots caused by *Phytophthora infestans*, *P. erythroseptica*, and *Pythium ultimum* was assessed. Two foliage sprays of phosphoric acid (9.37 kg a.i./ha) at 2-week intervals reduced late blight tuber rot in Umatilla Russet potato significantly. There was no significant reduction in disease incidence at a lower rate (7.49 kg a.i./ha). Three applications of phosphoric acid were required to reduce the pink rot severity. There was no appreciable decrease in rot due to *P. ultimum* following preharvest application of phosphoric acid. However, total tuber yield was not increased due to phosphoric acid treatment (Johnson et. al. 2004).

Low-toxicity chemicals, particularly common food additives, were evaluated for their efficacy for the control of green and blue mold diseases of citrus. The chemicals were tested as heated solutions. Oranges, artificially inoculated with *P. digitatum* and *P. italicum*, were immersed for 2 min in solutions at 40.6°C and natural pH of potassium sorbate (0.2 M), sodium benzoate (0.2 M), or mixtures (0.1 + 0.1 M) of potassium sorbate with sodium benzoate, sodium propionate, or sodium acetate. The incidence of green mold was reduced by 70 to 80% after 7 days of storage at 20°C. The mixtures did not provide more effective control than individual salts (Palou et al., 2002). The effect of sorbitol on the development of apple fruit rot caused by *Aspergillus niger* was assessed in relation to watercore severity. The enzymes produced by *A. niger* induce softening of flesh tissue of affected apple. Application of sorbitol retarded the softening of flesh tissue by inhibiting the activities of galactosidase and pectin esterase produced by *A. niger*. Enhanced levels of sorbitol in tissues with watercore may inhibit infection by *A. niger* (Kumpoun et al., 2003).

Helminthosporium solani causes silver scurf disease of potato of economic importance. Many strains of *H. solani* have become resistant to thiabendazole which was commonly applied for the control of silver scurf disease. Among the 23 salts tested, aluminum chloride (0.2 M) applied either 2, 4, or 7 days after inoculation with the fungal pathogen, was the most effective. Conidial

TABLE 11.4 Effect of Treatment of Apples cv. Golden Delicious with Ammonium Molybdate Followed by Inoculation with Suspension of Conidia of Three Fungal Pathogens on Lesion Development and Percentage Infection of Wounds

Concentration of	*P. expansum*			*B. cinerea*			*R. stolonifer*		
ammonium molybadate (mM)	Lesion diameter (mm)	Percentage reduction	Wounds infected (%)	Lesion diameter (mm)	Percentage reduction	Wounds infected (%)	Lesion diameter (mm)	Percentage reduction	Wounds infected (%)
0	23.2 a	–	100 a	37.2 a	–	100 a	120 a	–	100 a
1	13.2 b	43.11	89 ab	28.4 b	23.6	100 a	0 b	100	0 b
5	11.3 b	52.30	94 ab	24.2 c	35.0	100 a	0 b	100	0 b
10	5.1 c	78.10	55 ab	6.3 d	83.10	66 b	0 b	100	0 b
15	3.8 c	83.63	44 c	4.3 d	88.4	50 b	0 b	100	0 b

Lesion diameter and per cent wounds infected is the mean of 18 wounds (three replicates, three fruit samples per treatment, and two lesions per fruit). Wounds were inoculated with 20µl of conidial suspension (10^4/ml) and determined after 7 days of incubation at 20°C. Figures followed by the same letter are not statistically significant ($P < 0.05$).

Source: Nunes et al., 2001.

germination and mycelial growth of *H. solani* were strongly arrested using in vitro tests (Hervieux et al., 2002). A mixture of maltodextrin, carboxymethyl cellulose, propylene glycol, and sorbitan esters was coated on mango cv. Manila fruits and stored at 15 and 25°C and 85 ± 5% RH. The coating was removed by washing with water to check for the development of anthracnose caused by *Colletotrichum gloeosporioides* and fruit fly *Anastrepha obliqua*. Development of both anthracnose and proliferation of fruit fly occurred within 6 days in control fruits at both storage temperatures. By retaining the coating for 9 days, the incidence of anthracnose was reduced by 70% and proliferation of fruit fly larvae could be also avoided. Coatings may be applied as a component of surface treatments to guarantee the phytosanitory certification of tropical fruits, such as mangoes (Diaz Sobac et al., 2000).

Aluminum and bisulfite salts have been found to be toxic to several fungi and bacteria. Even at low concentrations, aluminum chloride and sodium metabisulfate were toxic to *Erwinia carotovora* susp. a*troseptica* (*Eca*) and *E. carotovora* subsp. *carotovora* (*Ecc*) causing postharvest soft rot of potato tuber (Yaganza et al., 2001). Various mechanisms of action of aluminum and sulfite on microorganisms have been suggested. Alterations in the ultrastructure of *Eca* following exposure to aluminum chloride and sodium metabisulfite were studied by transmission electron microscopy. The bacterial cells were treated with 0.05, 0.1, and 0.2 M concentrations of both salts for 0 to 20 min. Plasma membrane integrity was determined by using the SYTOX Green fluorochrome that can only penetrate bacterial cells with altered membranes. All concentrations of aluminum chloride (especially 0.2 M) caused loosening of cell walls, cell wall rupture, cytoplasmic aggregation, and absence of extracellular vesicles. On the other hand, sodium metabisulfite induced primarily retraction of plasma membrane and cellular voids, which increased at higher concentrations. Mortality of *Eca* cells was proportional to SYTOX stain absorption, registering increases with increasing salt concentrations (0.2 M) and duration of exposure (20 min). Membrane damage and subsequent cytoplasmic aggregation resulted in the mortality of *Eca* following treatment with aluminum chloride. In contrast, sulfite was transported intracellularly across the membrane by free diffusion of molecular SO_2 with little damage to the cellular membrane, resulting in the death of bacterial cells (Fig. 11.2). Sodium metabisulfite killed the bacterial cells at all concentration tested within 5 min of exposure and it was more effective and faster in causing mortality of *Eca* than aluminum chloride. This rapid killing could be due to free diffusion of molecular SO_2 into cells without seriously damaging cell wall integrity (Yaganza et al., 2004) (Appendix 11(iv)).

The effect of application of silicon (Si), as sodium metasilicate, on the development of decay in sweet cherry caused by *Penicillium expansum* and *Monilinia fructicola* was determined at 20°C. There was positive correlation between the concentrations of Si and extent of inhibition of fruit decay. The spore germination and germ tube were strongly inhibited by Si. Inhibition of growth of both pathogens was revealed by scanning electron microscopic observation. Further, enhancement of defense-related enzyme activities in the Si-treated fruits and biocontrol efficiency of the yeast *Cryptococcus leurenti* when combined with Si has also been observed (Qin and Tian, 2005).

A set of novel synthetic hexapeptides was synthesized and their bioactivity was determined. The peptides PAF 26, PAF 32, and PAF 34 showed stronger activity against *Penicillium digitatum*, *P. italicum*, and *B. cinerea*. PAF 26 exhibited greater activity against orange fruit decay in vivo (López García et al., 2000, 2002). The antibacterial activity of *Xanthomonas campestris* pv. *glycines* (*Xcg*) 8ra against several bacterial plant pathogens, such as *X. axonopodis*, *X. campestris* pv. *campestris*, *X. campestris* pv. *citri*, *X. campestris* pv. *pruni*, and *X. campestris* pv. *vesicatoria*, was reported by Woo et al. (1998). The bacteriocin, glycinecin A produced by *Xcg* is a heterodimeric protein consisting of 39- and14-kDa subunits, which are encoded by the *glyA*, *glyB* genes. Coexpression of the two subunits in the same host is necessary for the antibacterial activity of glycinecin A (Heu et al., 2001). Application of glycinecin A provided effective control of the bacterial fruit rot of pepper caused by *X. campestris* pv. *vesicatoriae*. The level of control was equal to that of copper hydroxide (Jeon et al., 2001). A study to determine the mode of action of

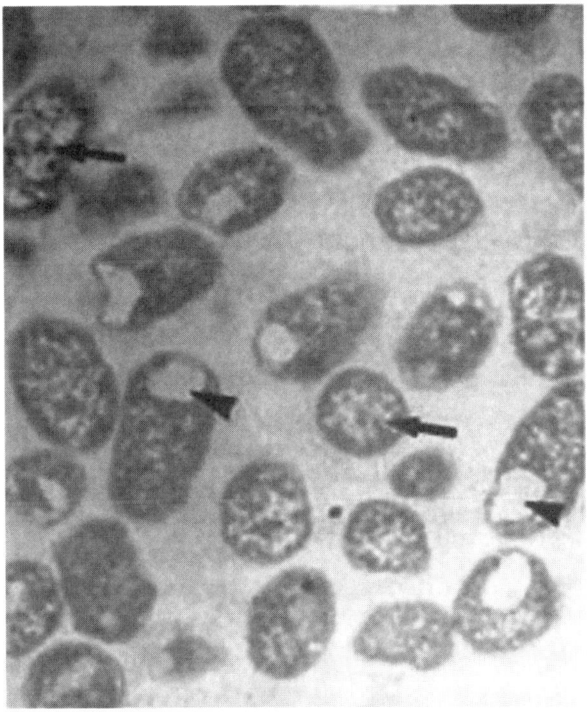

(a)

Figure 11.2 (a) Effect of aluminium chloride ($AlCl_3$) on the ultrastructure of *Erwinia carotovora* subsp. *atroseptica*. Aggregation of cytoplasmic materials (arrows) and empty areas (arrowheads) are present.

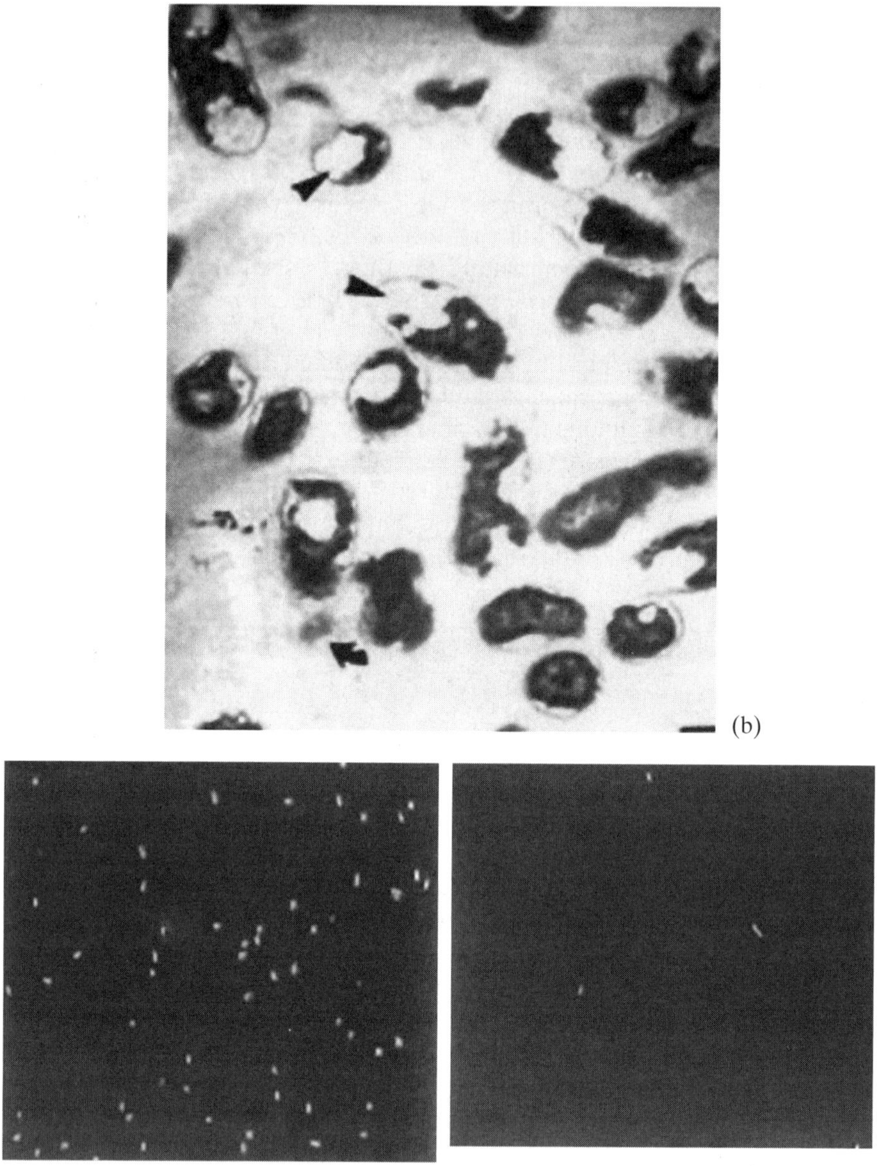

(b)

(c)

Figure 11.2 (*continued*) (b) Effect of sodium metabisulfite ($Na_2S_2O_5$) on the ultrastructure of *E. carotovora* subsp. *atroseptica*. Empty spaces (arrowheads) and extracellular vesicles (arrows) are present. (c) Fluorescence of bacterial cells treated with SYTOX stain. Left-intense yellow-green fluorescence in most bacterial cells treated with $AlCl_3$; right-untreated bacterial cells displaying fluorescence only in a few isolated cells. (Courtesy of Yaganza et al., 2004; The American Society for Microbiology, Washington DC, USA.)

glycinecin A on *X. campestris* pv. *vesicatoria* showed that the bacteriocin caused slow dissipation of membrane potential and rapid depletion of the pH gradient in treated bacterial cells, leading to leakage of potassium ions and ultimately death of sensitive bacterial cells in a dose-dependent manner. The sensitive bacterial cells were killed within 2 h of incubation (Pham et al., 2004).

11.6 SEED TREATMENT WITH CHEMICALS

Treatment of seeds with chemicals has been demonstrated to be the most effective and least expensive method of managing seed-borne diseases of many crops. The fungicides may form a protective zone around the treated seeds, resulting in the elimination of fungal pathogens present in the seeds or protection of emerging seedlings from infection by pathogens present in the soil. The external seed-borne pathogens may be effectively controlled by non-systemic fungicides, such as sulfur, captan, or thiram, whereas the systemic fungicides, which can be absorbed by plant tissues, may be more effective in controlling both externally and internally seed-borne diseases. Carboxin, carbendazim, and metalaxyl have been used for successful control of systemic diseases such as smuts, blights, and downy mildews of cereals and vegetable crops.

Seeds may be treated with dry powder formulation and stored for various periods depending on the crop, nature of disease, and fungicide. Larger seeds, such as legumes, may be subjected to dry seed treatment. Seed soak or dip treatment may be adopted in the case of crops with smaller seeds. Immersion of crucifer seeds in streptomycin (3000 μg/ml for 30 min) was reported to be effective against *Xanthomonas campestris* pv. *campestris* (Klisiewicz and Pound, 1960). The seed-borne inoculum of *X. campestris* pv. *vitians* in lettuce seeds was substantially reduced by treatment with sodium hypochlorite (1%) for 15 min or copper hydroxide + mancozeb, resulting in reduction of seed-borne infection to about 2% or below (Pernezny et al., 2002). *Alternaria dauci* infection of carrot seed was controlled by a seed-soaking treatment with thiram (0.2%) and iprodione (0.25%) for 24 h (Strandberg, 1984). Seed-borne infection of tomato seeds by *Phytophthora infestans* could be reduced substantially by soaking the seeds in 0.05% *o*-hydroxy diphenyl for 15 min. In the of case seeds with irregular surface and small size, pelleting may be more efficient for protecting the seeds. It may be possible to provide larger loads of fungicides around the seeds and to vary the spatial orientation of active materials within the pellet. Gum Arabic, starch, carboxymethyl cellulose (CMC), gelatin, methyl cellulose, polyvinyl alcohol, and polyoxyethylene glycol-based waxes may be used as binders for pelleting (Agarwal and Sinclair, 1996). The limitation of this method of application of fungicide is probably the slow absorption of fungicide by seeds. The uniformity of fungicide application may be improved by slurry treatment of seeds which may help avoid certain problems associated with dry fungicide treatment. Paraffin or vegetable oils may be employed as adhesives and water-dispersible fungicides are mixed in water to form a slurry which is coated on the seeds (Kitamura et al., 1991; Taylor and Harman, 1990).

The efficacy of the fungicides captan, chlorothalonil, thiram, and a commercial mixture of captan + carboxin for control of storage rots affecting wheat seeds was evaluated, under storage temperature of 25°C + 85% RH and dry conditions. Chlorothalonil, captan, and captan + carboxin mixture prevented the activity of storage fungi and protected the germination of wheat seed stored at 85% RH (Moreno Martínez et al., 1998). The effect of iprodione application on the allowable storage time (time to lose 0.5% of original dry matter) for shelled maize at 18 and 22% moisture (wet basis) and 20°C was studied. With 3.29 ml of water per kg of wet maize as carrier, iprodione application increased the allowable storage time by about 45% for 18% moisture and about 16% for 22% moisture (Wilcke et al., 1998). Strobilurin fungicides and one azole were evaluated for their effect on the development of *Fusarium culmorum* causing head blight of wheat. Visual assessment, enzyme linked immunosorbent assay (ELISA) for detection of mycotoxin, 1000-grain weight, and yield were the parameters used for evaluation. The relationship between the parameters was significant and they can be employed for assessing the efficacy of the fungicides (Chala et al., 2003).

The effect of application of fungicides, both in the field and after harvest, was determined for the development of storage fungicides in sorghum CSH 14 seeds. The higher the intensity of grain molds in the field, the lower was the storability of seeds. Field spraying followed by postharvest seed treatment with thiram (0.2%) + carbendazim (0.2%) in a ratio of 1:1 was effective in controlling seed mycoflora and the treated seeds had higher percent germination and vigor index (Padule et al., 1999). The cereal fungicide Charisma, containing flusilazole and famoxate (famoxadone), reduced the infection of wheat grains by *Fusarium culmorum* by 67.0 to 80.3%, whereas tebuconazole, azoxystrobin, and carbendazim could reduce infection by 34 to 67%, 9.6 to 35.8%, and 37 to 44%, respectively. Charisma treatment also improved the percentage of seed germination to a level greater than that could be achieved by treatment with other fungicides. Furthermore, application of Charisma at early and mid-anthesis stage of crop development resulted in the lowest levels of deoxynivalenol (DON, vomitoxin) in wheat grains (Greenfield and Rosall, 2001).

The fungicides metaconazole, azoxystrobin, and tebuconazole were evaluated for the control of *Fusarium* head blight (FHB) pathogens. The effectiveness of the fungicides was assessed by competitive PCR assay based on primers derived from the biosynthetic gene, trichodiene synthase (*Tri5*). No correlation between FHB severity and deoxynivalenol (DON) concentration in harvested grain could be observed. However, the population of trichothecene-producing *Fusarium* was positively correlated with DON in grains. Metaconazole and tebuconazole reduced the *Fusarium* population in grains significantly, while azoxystrobin was ineffective. It was suggested that the fungicides, by altering the proportion of trichothecene-producing *Fusarium* spp., may affect the FHB severity and DON content of the grains (Edwards et al., 2001).

The efficacy of single application of prochloraz, tebuconazole, epoxiconazole, or bromuconazole, at manufacturer's recommended dose at the beginning of anthesis, was assessed. Tebuconazole, prochloraz, and bromuconazole were effective in reducing infection by *F. graminearum* and *F. culmorum* causing wheat head blight disease. These fungicides reduced the DON concentration in grains of wheat artificially inoculated with the pathogens by 43%, while epoxiconazole could neither reduce infection nor DON concentration in grains. Further, under natural conditions of infection, the DON concentration was reduced from 73 to 96% by prochloraz, bromuconazole, and tebuconazole application (Menniti et al., 2003). Though control of *Fusarium* head blight disease can be achieved by chemicals, the need for adopting cultural and biological strategies along with exploitation of host plant resistance has been emphasized by Pirgozliev et al. (2003).

Different kinds of effects of application of fungicides on the production of DON by *F. culmorum* and other *Fusarium* spp., depending on the nature and concentration of fungicides, have been reported. Enhancement of levels of DON in grains following application of azoxystrobin (strobilurin) was observed under glasshouse and field conditions (Ellner and Schröer, 2000; Simpson et al., 2001). However, only limited activity of azoxystrobin against *Fusarium* spp. was noted by Simpson et al. (2001), although it was highly effective against *Microdochium nivale* colonizing wheat heads (Siranidou and Buchenauer, 2001). In ears treated with azoxystrobin, increase in DON concentration was correlated with increase in the population of *F. culmorum* (Edwards et al., 2001). DON accumulation in a defined medium amended with fungicides was monitored, in addition to assessing the effects on growth of *F. culmorum* and expression of genes involved in trichothecene biosynthesis (*Tri* genes) (Fig. 11.3). The strobilurin fungicides trifloxystrobin and azoxystrobin, at a range of concentrations, significantly reduced the accumulation of DON in culture medium. When trifloxystrobin was added to the cultures before initiation of synthesis of trichothecenes, accumulation of DON was markedly reduced. RT-PCR assays, to determine the expression of *Tri5* and *Tri6* genes, indicated that trifloxystrobin inhibited the initiation of trichothecene biosynthesis, resulting in the lower levels of DON (Fig. 11.4) (Covarelli et al., 2004).

Fungal contamination of peanut (groundnut) seeds was determined by using a blotter test at the beginning and at 6-month intervals during storage. The fungi commonly associated with peanut seeds were *Aspergillus niger*, *A. flavus*, *Penicillium* spp., and *Rhizopus* spp., the population of which declined with increase in storage period. Treatment of seeds with thiram (2.5 g/kg) did not affect the seed quality and fungal incidence (Usberti and Amaral, 1999).

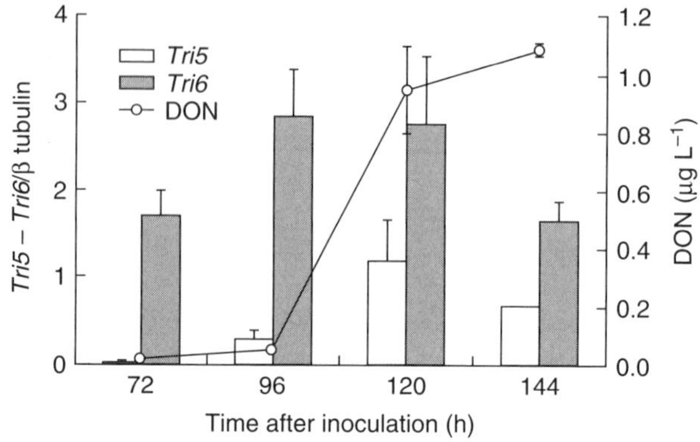

Figure 11.3 Time-course of deoxynivalenol (DON) production and expression of *Tri 5-Tri 6* in *Fusarium culmorum* between 70 and 144 h after inoculation. Bars, standard error of the mean. (Courtesy of Covarelli et al., 2004; Blackwell Publishing Ltd., Oxford, United Kingdom.)

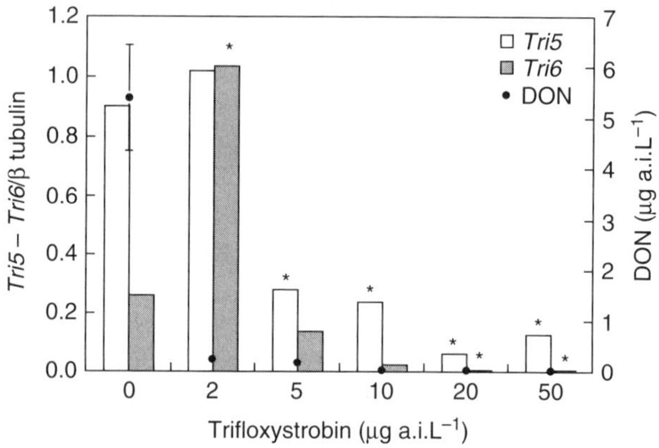

Figure 11.4 Effect of trifloxystrobin treatment at different concentrations on the accumulation of deoxynivalenol (DON) in the culture medium and expression of *Tri 5* and *Tri 6* genes involved in trichothecene biosynthesis in cultures of *Fusarium culmorum*. * Significantly lower than control (P = 0.05); + significantly higher than control (P = 0.05). (Courtesy of Covarelli et al., 2004; Blackwell Publishing Ltd., Oxford, United Kingdom.)

11.7 DEVELOPMENT OF RESISTANCE TO FUNGICIDES

Application of chemicals, among the various disease management strategies, has provided maximum protection against microbial pathogens causing pre- and postharvest diseases of various crops. But the main drawback is the development of resistance to the chemicals in microbial pathogen, even after a few seasons during which they were found to be effective. The presence of fungicide-resistant strains of several pathogens has been reported from many countries. Ogawa et al. (1977) listed different pathogens that showed resistance to fungicides. Benzimidazole compounds are highly prone to induce resistance in fungal pathogens.

Crop protection in general, and production of fruits and vegetables in particular, has, for over three decades, been faced with several problems arising from the development of resistance in fungal pathogens against the fungicides used to control them. As the primary aim of modern agriculture is sustainability – economic, technical, and environmental – there is an imperative need to develop effective systems to overcome the problem of resistance to fungicides. It has been estimated that about 135 different formulated compounds, valued at approximately 4.7 billion US dollars, have been applied on various crops (Brent, 1995). However, with appearance of resistance to various fungicides that were applied for the control of important pathogens such as *Botrytis cinerea*, monitoring is essential to determine whether resistance is the actual cause of the lack of disease control and to check whether the application of resistance management strategies is effective.

Resistance to fungicide builds up through the survival and spread of rare mutants. Such development of resistance can be either discrete, resulting from a single gene mutation, or gradual, due to the action of polygenes. Mechanism of resistance to fungicides may be through: (1) alteration of the biochemical target site so that it may no longer be sensitive; (2) development of an alternative metabolic pathway that avoids the target site; (3) metabolic breakdown of the target fungicide; and (4) exclusion or expulsion of the fungicide. However, the most common mechanism operating in fungal pathogen seems to be the change in the biochemical target site of the fungicide.

The contact fungicides developed earlier, with multisite action, act as general enzyme inhibitors affecting many target sites after penetrating the fungal cell. The chances of the development of resistance to these fungicides are negligible, since many sites in the fungus would have to change simultaneously. In contrast, most of the modern fungicides have selective or single-site action. Hence, a single gene mutation would suffice to cause the target site to alter, making the fungus less sensitive or insensitive to the fungicide concerned. Several major resistance genes have been identified and characterized. Identification of benzimidazole resistance in *Botrytis cinerea*, by employing specific oligonucleotide probes, has been successful. These methods are rapid, sensitive, and economical (Martin et al., 1992).

The results of risk evaluation study have to be carefully interpreted. Strains of *B. cinerea* resistant to the phenylpyrrole fungicide, fludioxonil, produced in the laboratory, showed cross-resistance to dicarboximides. However, dicarboximide-resistant field isolates of *B. cinerea* were found to be sensitive to fludioxonil (Hiber et al., 1994). The phenylamides were specifically applied against Oomycete pathogens which could not be controlled by other fungicides. But resistance to these fungicides occurred suddenly and seriously within a few years after their introduction in 1977. Resistance to metalaxyl in

Pseudopernospora cubensis infecting cucumber, *Phytophthora infestans* infecting potatoes, and *Plasmopara viticola* infecting grapes made the manufacturer withdraw the single product immediately. Recommendation for the use of mixtures with multisite fungicides was offered (Staub, 1984). An effective approach to overcoming benzimidazole resistance was the application of a mixture of the benzimidazole fungicide, carbendazim, with diethofencarb for the control of *B. cinerea* in grapes and other crops. Diethofencarb exhibits negative cross-resistance with respect to benzimidazoles, inhibiting only the benzimidazole resistant strains. The components of this mixture seem to have a synergistic effect, resulting in the control of sensitive and resistant strains of *B. cinerea* (Elad et al., 1992; Leroux and Moncomble, 1994). Dicarboximides were used mainly for the control of fungi belonging to the genera *Botrytis*, *Sclerotinia*, and *Monilinia* to replace benzimidazole fungicides. However, resistant strains were detected, both in the field and laboratory. The field isolates showed varying degrees of resistance and pathogenicity, and other fitness factors decreased as the resistance levels increased. In European grapevines, the problem of resistance increased after intensive use of dicarboximides where *B. cienrea* was most prevalent (Brent, 1995).

Build-up of resistance to benzimidazole in *Botrytis cinerea* in vineyards in Switzerland against is a classical example, reported by Schüepp and Küng (1981). The pathogen could not be controlled within a period of 2 years of benzimidazole application. Resistance to the fungicide was stable, resulting in the withdrawal of benzimidazoles from the market (Leroux and Clerjeau, 1985). The percentage of isolates of *B. cinerea* resistant to benzimidazoles increased with increase in number of applications of the fungicide. Hence, the number of applications against *Botrytis* or *Venturia* should be restricted and the timing of application should considered carefully. Based on the extent of mycelial growth, the sensitivity of *Monilinia fructicola* isolates to thiophanate-methyl was determined. Among the isolates collected during 1992 to 1998, 75% were found to be resistant, whereas 22% of the isolates collected in 2002 showed resistance to the fungicide. Three groups of isolates of *M. fructicola* with distinct ranges of values for 50% effective concentration (EC_{50}) to thiophanate-methyl were distinguished. None of the tested isolates of *M. fructicola* was resistant to either iprodione or tebuconazole (Yoshimura et al., 2004).

Application of vinclozolin as preharvest sprays provided effective control of gray mold caused by *Botrytis cinerea* in storage. However, widespread occurrence of isolates of *B. cinerea* resistant to dicarboximide fungicides was observed in New Zealand. Hence, control of postharvest mold was directed towards methods of reducing the size of preharvest populations of *B. cinerea* in grapevines, such as summer pruning (Michailides and Elmer, 2000). In Italy more than 70% of *Phialophora* isolates were found to be resistant to benzimidazole fungicides, making their application on the kiwifruit ineffective (Piano et al., 2001).

The occurrence of benomyl-tolerant strains of *Penicillium expansum* causing blue mold and *Botrytis cinerea* causing gray mold in apple was detected in mid-Columbia region of Oregon and Washington (Bertrand and Saulie Carter, 1978), in New York (Rosenberger and Meyer, 1979, 1981), and in California (Holmes and Eckert, 1999). Benomyl was found to be effective against postbloom fruit drop disease caused by *Colletotrichum acutatum* and postharvest anthracnose disease caused by *Colletotrichum gloeosporioides* (Peres et al., 2002; Timmer, 2003). But in pure culture, *C. acutatum* was inherently much less sensitive to benomyl, whereas most isolates of *C. gloeosporioides* were entirely inhibited by benomyl at 1 μg/ml. In a further study, isolates of *C. acutatum* from 17 groves and isolates of *C. gloeosporioides* from seven groves, collected from different locations, were examined for their sensitivity to benomyl. Benomyl at 1.0 μg/ml entirely inhibited growth of 133 isolates of *C. gloeosporioides* but seven isolates were highly resistant to the fungicide. On the other hand, all isolates of *C. acutatum* were only partially inhibited at 0.1 and 1.0 μg/ml. The sensitivity of the isolates of *C. acutatum* was not affected by the application of benomyl or location of groves. Partial sequencing of β tubulin gene did not reveal nucleotide substitutions in codons 198 or 200 in *C. acutatum* that are usually associated with benomyl resistance in other fungal pathogens (Peres et al., 2004).

The fungicide tolerance of *P. expansum* causing blue mold disease in stored pears was investigated. Of the 55 isolates of *P. expansum* examined, 76.4% showed tolerance to thiabendazole (TBZ), showing EC_{50} values of more than 100 ppm of TBZ. Cross-tolerance between TBZ, benomyl, and carbendazim, belonging to the benzimidazole group, was observed (Pinilla et al., 1998). Fifty of isolates of *P. expansum* recovered from stored pears were scored for TBZ resistance and for their pathogenic fitness. Nine isolates were sensitive (S), while 41 isolates exhibited resistance (R or RR). RR isolates induced greater severity of symptoms in inoculated fruits, but they did not show variation in growth fitness in vitro compared with sensitive isolates. Mutants generated from S strain by UV-irradiation also showed similar infection potential. The β-tubulin gene of RR and S isolates was amplified and sequenced. Mutations correlating with TBZ resistance could be identified at amino acids Phe 167 and Glu 198. Similar changes were present in the laboratory-induced resistant isolates (Baraldi et al., 2003). In order to overcome the resistance of *P. expansum* to TBZ, application of fludioxonil or cyprodinil, which have modes of action different from each other as well as from that of TBZ, has been suggested to protect apples (Errampalli and Crnko, 2004).

Qualitative assay of 326 isolates of *Penicillium digitatum* collected from California citrus groves and packinghouses was performed to assess their sensitivity to imazalil, TBZ, and O-phenylphenol. The proportion of isolates exhibiting resistance to all three fungicides increased from 43% in 1988 to 77% in 1990 and 74% in 1994. Imazalil-resistant biotypes of *P. digitatum* were frequently isolated in California packinghouses, while resistant *P. italicum* was rare (Holmes and Eckert, 1999). The spores of benzimidazole-sensitive strains were mixed with six different tolerant strains of *P. italicum* in equal proportions to produce the disease in healthy lemon fruits. The competitive behavior of tolerant strains was investigated by plating the spores from these artificially inoculated fruits for four generations. The tolerant strains were capable of surviving for extended periods along with sensitive strains, even in the absence of selection pressure (Ullasa et al., 1998). The variations in utilization of carbon and nitrogen sources by the isolates of *P. digitatum* sensitive and resistant to thiophanate-methyl were determined. Rate of growth of resistant isolate was higher in all nitrogen and carbon sources tested compared to sensitive isolates, indicating the more efficient utilization of various nutrients by resistant isolates (Khilare and Gangawane, 2000).

The blossom blight and brown rot disease of stone fruits caused by *Monilinia fructicola* was effectively controlled by demethylation inhibitor fungicides. These fungicides gradually replaced the benzimidazoles, against which resistance in *M. fructicola* was reported (Zehr et al., 1991). However, reduced sensitivity in *M. fructicola* to demethylation inhibitor fungicides after prolonged, repeated exposure in the field was detected (Zehr et al., 1999). Sensitivity to propiconazole was assessed, based on the concentration of the fungicides in agar media required to suppress radial growth of mycelium by 50% (EC_{50}). AP isolates have high EC_{50} values, while DL isolates have low EC_{50} values. The isolates of *M. fructicola* with reduced sensitivity (high EC_{50} values) to propiconazole could not be controlled in the field (Table 11.5). Disease incidence on peaches inoculated with AP5 and AP6 (with higher EC_{50} values) was greater after curative treatment with propiconazole at 0.15 l/ha. Further protective application of propiconazole at half the label rate resulted in higher disease incidence due to AP isolates compared with DL isolates. Similar trend in disease incidence for AP and DL isolates was seen following protective or curative application of propiconazole at full label rate (0.30 l/ha) (Schnabel et al., 2004).

Resistance of potato pathogens to fungicides was studied by conducting a survey during 1997 to 1999. The level of tolerance to mefenoxam (metalaxyl)

TABLE 11.5 Effect of Propiconazole Treatment on Brown Rot Disease Incidence on Peach Fruit

Isolate	EC_{50} value	Disease incidence (%)					
		Protective treatment (l/ha)			Curative treatment (l/ha)		
		0	0.15	0.30	0	0.15	0.30
DL 71	0.02	100.0	54.5 a	42.2 a	100.0	21.0 a	14.3 a
DL 72	0.02	100.0	58.7 a	42.4 a	100.0	25.2 a	15.0 a
AP 5	0.42	100.0	85.4 a	72.9 ab	100.0	60.4 b	32.7 ab
AP 6	0.43	100.0	89.3 a	86.6 b	100.0	83.7 b	42.2 b

Values within a column followed by the same letter are not significantly different based on Fisher's protected LSD ($P < 0.05$). Values are means of three independent experiments represented as percentages of control.
Source: Schnabel et al., 2004.

in *Phytophthora erythroseptica* causing pink rot disease, one of the most important storage diseases of potato, was determined. The presence of metalaxyl-tolerant isolates of *P. erythroseptica* was detected in Idaho, United States (Gudmestad et al., 2000). In a further extended study (from 1997 to 2000), the sensitivities to mefenoxam of *Pythium ultimum* (causing leak disease), in addition to *P. erythroseptica*, were assessed. The isolates of *P. erythroseptica* (805) and *P. ultimum* (190) were tested for their ability to develop in V8 medium amended with 0.01 to 100 μg/ml of mefenoxam. Isolates of *P. erythroseptica* exhibiting resistance to mefenoxam in Idaho and Minnesota States were identified. The results suggested that pink rot and leak diseases could become serious problems in the future and monitoring of the development of resistant isolates in all potato production areas should be taken up (Taylor et al., 2002). Isolates of *P. infestans* showing resistance to dimethomorph were generated using medium amended with this fungicide. However, these isolates had reduced fitness for infection of leaf disks or whole tubers. The results suggested that development of field resistance to dimethomorph in *P. infestans* was unlikely with currently employed resistant management strategies in Michigan State, United States (Table 11.6) (Stein and Kirk, 2004).

Resistance to mefenoxam in *Phytophthora capsici* causing *Phytophthora* blight disease of bell pepper was assessed by testing 150 isolates, of which 30% were classified as sensitive, 10% as intermediate, and 59% as resistant. The proportion of resistant isolates ranged from 28 to 100% in different fields. The

TABLE 11.6 Percentages of Infected Potato Tubers Inoculated with Wild-Type (WT) Isolates of *Phytophthora Infestans* and the Same Isolates Repeatedly Subcultured on Nonamended (CT) and Dimethomorph Amended Medium (DM)

Pathogen isolates	Percent infection			
	WT	CT	DM	
Pi	88	100 a	33 b	22 b
Pi	955	33 a	22 a	0 b
Pi	458	100 a	89 a	44 b
Pi	670	100 a	67 a	22 c
Pi	671	100 a	100 a	100 a
Pi	213	100 a	100 a	100 a
Pi	94-4	56 a	56 a	44 a
Pi	95-7	100 a	78 b	44 c
Pi	97-2	67 a	67 a	22 b
Pi	98-1	100 a	56 a	33 c
Pi	98-2	100 a	100 a	89 a

Figures followed by the same letter in a row are not significantly different as per Fisher's LSD (P = 0.05).
Source: Stein and Kirk, 2004.

highest proportion of resistant isolates were present in fields which received mefenoxam alone rather than in combination with other fungicides, indicating the need for regulating the schedule of fungicide use (Parra and Ristaino, 2001). Insensitivity to mefenoxam was common among isolates of *P. capsici* occurring in squash fields in Florida, United States. Three resistant isolates had EC_{50} values ranging from 5 μg/ml to more than 60 μg/ml of mefenoxam (Ploetz et al., 2002).

Isolates of *Fusarium sambucinum* resistant to TBZ were recovered from eastern Canada. Later this pathogen, causing dry rot disease in potato tubers in storage, was isolated from potatoes treated with TBZ in three storage bays. All isolates were resistant to TBZ with EC_{50} values ranging from 7 to 82 mg/l, while the sensitive isolates had EC_{50} values of less than 1 mg/l (Peters et al., 2001). Development of resistance to TBZ in *Helminthosporium solani*, causing potato silver scurf disease, was also observed. Higher level of disease incidence was considered to be due to the appearance of isolates of *H. solani* resistant to TBZ (Errampalli et al., 2001).

Nonsystemic fungicides have multisite and broad spectrum action, unlike the systemic fungicides having specific sites of action leading to frequent development of resistance in fungal pathogens and their consequent ineffectiveness against such pathogens. However, instances of fungal pathogens overcoming the effects of nonsystemic fungicides resulting in unsatisfactory disease control and reduced marketable produce have been reported. Resistance to sodium orthophenylphenol (SOPP), diphenyl, and 2-aminobutane in postharvest fungal pathogens infecting citrus fruits during storage has been observed (Ogawa et al., 1977). Hence, it is essential that optimization of use of fungicides, both systemic and nonsystemic, has to be required, along with monitoring of development of strains of pathogens with resistance to fungicides that are applied for several seasons.

Molecular genetics of development of resistance to fungicides has been studied in certain cases. The high level of resistance to benzimidazole may be acquired in a single step and seems to be controlled by major genes. Generally, resistance alleles may exist at very low frequencies in populations not exposed to the fungicides. Fungicides permitting the one-step major change in sensitivity may be divided into two groups: (1) high-risk fungicides (fitness not seriously affected) and (2) moderate risk fungicides (serious effect on fitness). Detection of resistant isolates or strains may have to be done rapidly and efficiently. Polymerase chain reaction (PCR)-based detection methods for identification of isolates showing resistance to fungicide have been demonstrated to be rapid and more specific than traditional cultural methods. Detection of point-mutations (one-step change) has been possible by employing PCR with allele-specific oligonucleotides (ASOs). Allele-specific amplification of gene segments determining resistance or sensitivity may be done using ASOs directly as primers (Williams et al., 1990).

By employing allele-specific amplification of gene segments, most of the strains of *Botrytis cinerea* could be precisely diagnosed (Hollomon and Butters, 1994). Resistance to benzimidazoles in *B. cinerea* is governed by a single major gene. Luck and Gillings (1995) reported that all benzimidazole sensitive isolates (Ben^s) had the sequence GAG (Glu) at codon 198, whereas resistant isolates (ben^{HR}) had a single base substitution to GCG (Ala) at this position. Resistant isolates were identified using the PCR assay. DNA probe techniques are sensitive and rapid, providing results within 24 to 48 h and these techniques may be employed for monitoring resistance in slow-growing organisms that may be difficult to isolate in pure culture.

A rapid method for detecting fungicide tolerant strains formed due to continued application of a fungicide is desirable. McKay and Cooke (1997) developed a PCR-based technique to identify the strain of *Helminthosporium solani* tolerant to TBZ. Conventional methods need several weeks to distinguish the TBZ tolerant strains. Primers designed from the conserved region of the fungal β-tubulin gene were employed for amplifying by PCR and for sequencing a portion of this gene. In the TBZ-tolerant strains, due to a point mutation at codon 198, a change in the amino acid sequence from glutamic acid to alanine or glutamine was observed. By using species-specific PCR primers designed to amplify the desired regions in conjunction with a restriction endonuclease to cause cleavage in sensitive isolates, fungicide-tolerant strains can be rapidly detected, enabling initiation of further actions for fungicide resistance management. The usefulness of PCR-based methods for the rapid and precise detection of TBZ-resistant isolates of *H. solani* was reported by Errampalli et al. (2001).

Development of resistance to dicarboximides in postharvest fungal pathogens has also been observed. Isolates of *Botrytis cinerea* exhibiting resistance were first detected in northern Europe within a few years of application of carboximide. In the Champagne region of France, the disease pressure was high, necessitating the use of enormous amounts of fungicides. Consequently, resistant strains reached high levels in vineyards treated heavily with carboximides (Leroux et al., 1977; Leroux and Besselat, 1984). Compared to benzimidazoles, however, the loss of activity of carboximide was less complete. The resistant isolates were less fit than the wild strains. This fact allowed either the use of carboximide as a mixture with thiram or its restricted application (Moncomble, 1990).

Generally, when a single fungicide is frequently used, selection of a resistant strain is faster and the selection pressure is less intense, if a mixture of fungicides is used. If the use of a fungicide concerned is restricted or withdrawn, the resistant strains recede slowly or rapidly depending on the stability level of resistance. Use of a mixture of systemic and protectant fungicides may help to contain the increase in population of resistant strains. It may be possible to adopt the use of mixtures because they can be packaged together with the acceptance of industry. Close monitoring of the response of pathogens to the current and new fungicides will be useful to plan effective strategies when resistance is detected in the future.

Isolates of *Botrytis cinerea*, showing resistance to fenhexamid, exhibited enhanced sensitivity to the phenylpyrrole, fludioxonil, dicarboximide, and iprodione. In greenhouse assessments, reduced virulence of all highly fenhexamid-resistant isolates of *B. cinerea* on cucumber was observed compared with wild-type. High concentrations of fenhexamid were ineffective, when applied as a preventive measure on cucumber (Ziogas et al., 2003). Two distinct populations of *B. cinerea*, designated Hyd R1 and non-Hyd R1, the former showing natural resistance to the fungicide hydroxanilide, fenhexamid, and the latter being sensitive to the fungicide, were distinguished. In addition, vegetative incompatibility between them was also seen. The 3-keto reductase gene *ERG27* was isolated by using PCR. The gene product displayed striking homology with mammalian 17-HSD 7, indicating a common function between Erg27p-like protein and 17-HSD 7 in sterol biosynthesis. The analysis of genetic polymorphism of gene product revealed 12 amino acid differences between strains of Hyd R1 and non-Hyd R1 types that may account for natural resistance of Hyd R1to fenhexamid. It was hypothesized that Hyd R1 and non-Hyd R1 strains may constitute two different species. Furthermore, sequence analysis of Erg27p-like protein showed that Hyd R3 isolated from treated populations of non-Hyd R1 strains exhibiting high resistance to fenhexamid had two mutations that could be used as population markers (Albertini and Leroux, 2004).

Dicarboximide fungicides have been applied widely for the control of several fungal pathogens, including *B. cinerea* (*Botryotinia fuckeliana*), resulting in the appearance of pathogen strains with resistance to these fungicides in many countries (Faretra and Pollastro, 1993; Leroux et al., 1999; Yourman and Jeffers, 1999). Yet the exact mechanism of resistance development has not been clearly understood, although evidence indicating a possible link with an osmoregulatory pathway was obtained (Cui et al., 2002; Leroux et al., 2002). The resistance to dicarboximide in isolates of *B. cinerea* was conferred by a single locus, *Daf1* (Faretra and Pollastro, 1991). *Daf1* encoded a putative osmosensing histidine kinase orthologous to *os-1* in *Neurospora crassa* (Leroux et al., 2002).

In a further study, the DNA sequence polymorphisms in the putative two-component histidine protein kinase encoded by *Daf1* were identified within a sample of five sensitive and 27 dicarboximide-resistant field strains of *B. cinerea*. All resistant strains exhibited changes in the second amino acid repeat region, which has been hypothesized to be the binding site of dicarboximide fungicide. On the basis of the differences in this repeat region, four classes of *B. cinerea* isolates with dicarboximide resistance were recognized: class 1 (isoleucine to serine mutation); class 2 (isoleucine to asparagine mutation); class 3 (isoleucine to arginine mutation) with mutation occurring at position 365; and class 4 (glutamine to proline mutation) with mutation occurring at position 369 (Cui et al., 2004).

Resistance of *B. cinerea* isolates to various other fungicides has been assessed under greenhouse and field conditions. The sensitivity of isolates of *B. cinerea* (obtained from greenhouses) to pyrimethanil (anilinopyrimidine) was determined based on the growth of the pathogen in defined minimal medium. The sensitive isolates had ED_{50} values ranging from 0.04 to

0.5 mg a.i./l, whereas the resistant isolates could tolerate concentration of the fungicide ranging from 1 to 10 mg a.i./ml. Of the 307 isolates of *B. cinerea* tested, resistance to benzimidazoles, dicarboximides, *N*-phenylcarbamates, and anilinopyrimidines in 90, 77, 23, and 12% of the isolates respectively was observed. Cross-resistance to dicarboximide and benzimidazole accounted for 65.8% of the isolates present in the greenhouses, whereas 14% of the isolates exhibited resistance to *N*-phenylcarbamates in addition to dicarboximide and benzimidazole. Some of the isolates (3%) were found to be resistant to all the four groups of fungicides, indicating the need for development of an effective fungicide resistance management strategy (Moyano et al., 2004).

The single spore isolates (200) of *B. cinerea* were classified into six representative groups of fungicide-resistant phenotypes based on the mode of spore germination and mycelial growth on fungicide-amended media as follows: (1) phenylcarbamate highly resistant (PcmHR, wild type) 68 isolates; (2) dicarboximide moderately and phenylcarbamate highly resistant (DicMR PCMHR) eight isolates; (3) benzimidazole and phenyl carbamate highly resistant (Ben HR Pcm HR) four isolates; (4) dicarboximide and benzimidazole moderately resistant and phenylcarbamate highly resistant (DicMR BenMR-Pcm HR) five isolates; (5) dicarboximide moderately and benzimidazole highly resistant (Dic MRBen MR) 20 isolates; and (6) benzimidazole highly resistant (BEN HR) 95 isolates. To study the relationships among the resistant isolates, two representative isolates of each phenotype were analyzed by DNA RAPD fingerprint technique. Based on the determination by unweighted pair group method arithmetic average (UPGMA) and dendrograms, the six phenotypic resistant groups were classified into three clusters. A positive correlation between conventional and molecular techniques in the differentiation of fungicide-resistant phenotypes was observed (Paplomatas et al., 2004).

Tolerance/resistance of bacterial pathogens to antibiotics is well known, leading to restriction of their use. Application of streptomycin alone resulted in the development of resistant strains of *Erwinia amylovora* causing fireblight of pear (Schroth et al., 1969). Even when mixtures of two antibiotics, streptomycin and oxytetracycline, were used, strains of *Pseudomonas syringae* pv. *syringae* infecting pear resistant to both antibiotics were detected in the orchards (Spotts and Cervantes, 1995). Strains of *Xanthomonas campestris* pv. *vesicatoria* tolerant to copper have earlier been reported to cause severe losses in several countries, including United States, Taiwan, Korea, Barbados, and Mexico. In Australia, development of *X. campestris* pv. *vesicatoria* strains (causing pepper black spot disease) resistant to copper bactericides was observed. About 25% of 75 strains collected were considered copper tolerant (more than 1.0 mM $CuSO_4$). The proportion of copper-tolerant strains increased after 12 weekly sprays of copper and a high proportion of copper-tolerant strains were prevalent after 21 sprays (Fig. 11.5). This report seems to be the first on the occurrence of copper tolerant strains of *X. campestris* pv. *vesicatoria* in Australia (Martin et al., 2004).

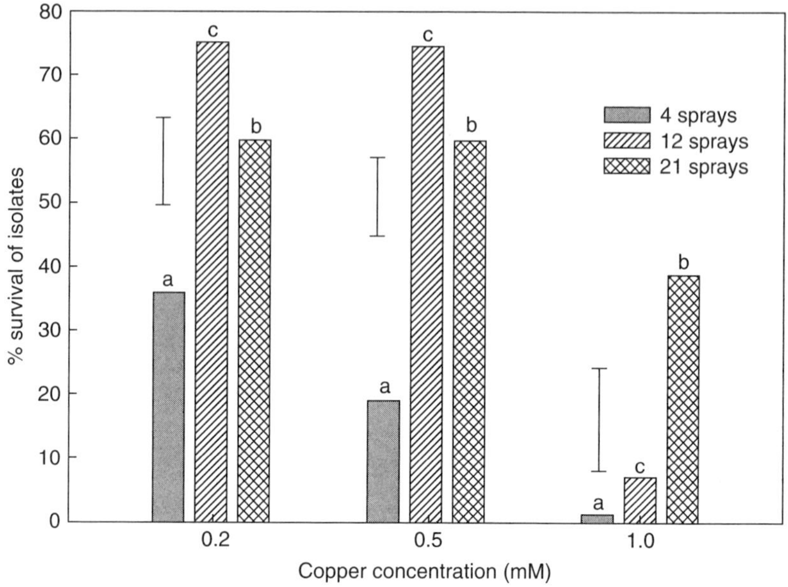

Figure 11.5 Detection of copper-tolerant field isolates of *Xanthomonas campestris* pv. *vesicatoria* from fields that received up to 21 copper sprays. (Courtesy of Martin et al., 2004; The American Phytopathological Society, St. Paul, MN, USA.)

11.8 ASSESSMENT OF FUNGICIDE RESIDUES

Synthetic fungicides are the chief means of controlling postharvest diseases. However, concerns about the presence and persistence of chemical residues in fruits, vegetables, and food products and the development of fungicide-resistant strains of postharvest pathogens resulted in the deregistration of some fungicides. The European Parliament voted in favor of a total ban on the postharvest pesticide treatment of fruits and vegetables with chemicals as soon as this practice becomes feasible (Rendall Dunn, 1991). It may be desirable to treat crops before harvest, since infection occurring in the field may remain quiescent and develop later in storage. In addition, less fruit manipulation may be necessary and there would be less potential for damage through injuries that can occur during any postharvest treatment. When the use of fungicides becomes unavoidable, the possibility of fungicide residues persisting in fruits and vegetables may be greater.

Enzyme-linked immunosorbent assay (ELISA) was employed to detect the residues of thiabendazole (TBZ) in the peels of apple, oranges, grapefruit, bananas, and potato. ELISA can detect 0.1 ppm in peel samples, corresponding to 10 to 40 ppb in the whole fruit or tuber. This technique was shown to be specific and more rapid than other methods (Brandon et al., 1995). The usefulness of ELISA for quantitation of the fungicide tetraconzole in apple fruits and fruit juices was reported by Cairoli et al. (1996) and Chen et al. (1996).

Higher percentages of residues of captan, iprodione, and metalaxyl could be detected in tomato and carrot pulp. Mancozeb residue was within limits in tomatoes, but not in carrot. Residue percentages varied from 56.4 to 75.2 for carrots and 49.7 to 95.4 for tomatoes. Washing removed more residues from carrots than from tomatoes (Burchat et al., 1998).

The persistence of folpet and its metabolic product phthalimide in grapes and wines was studied using gas chromatography (GC) and high performance liquid chromatography (HPLC). A high concentration of folpet, but limited amount of phthalimide, were detected in freshly pressed must from grapes treated with folpet. After 24 h no folpet was detected but appreciable level of phthalimide was present in unprocessed wine, whereas the filtered wine had only traces of phthalimide. Fermentation of wine was delayed by 1 to 5 days by folpet treatment, due to an effect on the yeast (Viviani Nauer et al., 1997). Folpet was entirely degraded during the wine making process. However, phthalimide was present in the wine at the end of fermentation, since it seemed to have no effect on fermentation activities of the yeast. In contrast, folpet totally inhibited alcoholic fermentation by *Saccharomyces cerevisiae* (Cabras et al., 1997).

Postharvest dips/drenches with chemicals are used to protect fruits and vegetables against postharvest diseases. Persistence of residues may depend on the amount of fungicides used and storage temperature. The presence of residues of TBZ was strictly dependent on the amount of fungicide used and treatment temperature. The residue level in lemon fruits was almost the same for fruits dipped in 1200 ppm TBZ at 20°C and those dipped in 150 ppm TBZ at 50°C (Schirra et al., 1998). Orange fruits were dipped for 1 min in aqueous suspension of imazalil (100 mg/l) or TBZ (1000 mg/l) and surfactant (0.002%) and the fruits were stored for 0, 7, 14, 21, and 28 days at 4 ± 2°C, 90% ± 5 RH. The orange peel had the highest concentrations of fungicides, as detected by GLC with flame ionization detection (FID) and flame photometric detection (FPD). Apparently, no migration of the fungicides from peel to the pulp occurred during storage (Oliveira et al., 2000). These studies show that the persistence of fungicides may depend on the nature of fruit, concentration of fungicide applied, and storage conditions. It is necessary to minimize the use of fungicides by adopting integrated disease management systems (Chapter 12).

11.9 CONTROL OF PATHOGENS ON FRESH-CUT PRODUCE

Because of consumer demand for freshness and convenience, the use of fresh-cut produce offering a ready-to eat product with fresh-like quality is increasing rapidly. However, fresh produce may have the potential to become hazardous and a health risk, since some food-borne illness outbreaks might have originated from contaminated fresh produce. It has been demonstrated that fresh-cut fruits and vegetables possess resident microflora consisting of many human pathogens, such as *Listeria monocytogenes*, *Yersinia enterocolit-*

ica, *Staphylococcus aureus*, *Salmonella* sp., *Escherichia coli* 0157:H7, and *Clostridium botulinum* (Hentges, 2003).

Fresh-cut produce may be contaminated at every step of processing, from the field to table. Adoption of recommended good agricultural practices (GAPs) and good manufacturing practices (GMPs) may help to minimize the risk of contamination. A combination of subinhibitory preservation methods, including surface disinfection, modified atmosphere packaging, refrigerated storage, increased acidity, and irradiation, may be useful in reducing the pathogen population substantially.

As a first step, fresh produce is washed in potable water to remove adhering soil and to reduce the number and development of surface microorganisms. The washing procedures may result in a reduction of 1 to $2\log_{10}$ CFU/g in total microflora population (Adams et al., 1989). In order to reduce the food-borne pathogens further, surface sanitization (disinfection) has to be carried out and this can be achieved by various treatments.

11.9.1 Treatment with Water-Dispersible Chemicals

Aqueous solutions of sodium hypochlorite or hypochlorous acid may be sprayed on the produce or they may be immersed in the disinfectant solution. Cantaloupe dipped in hypochlorite solution (200 ppm) had significantly smaller psychrotrophic and aerobic microbial populations compared with water-washed cantaloupe (Ayhan et al., 1998). Organic acids, such as acetic acid, lactic acid, citric acid, propionic acid, sorbic acid, and ascorbic acid, have also been used with varying degrees of success (Adams et al., 1989). Lactic acid was found to be more effective than acetic acid in reducing *L. monocytogenes* on cut lettuce and cabbage (Zhang and Farber, 1996). Calcium lactate, used by the beverage industry as a source of calcium to fortify fruit juice, was evaluated for its effect on spoilage and pathogenic microorganisms. The pH of fortified orange juice was adjusted to 3.6 or 4.1 and stored at 4 and 10°C for 6 weeks. Three strain mixture of *Salmonellae* were killed in all juice stored at 4°C and in low pH juice stored at 10°C. But the pathogens persisted in juice with pH 4.1 stored at 10°C except in samples that contained 20 to 30% dietary reference intake of calcium lactate (Yeh et al., 2004).

11.9.2 Treatment with Gaseous Chemicals

Chlorine has been routinely used as a sanitizer and it has antifungal, antibacterial, and antiviral activities. As it dissociates in water entirely, no toxic residue is left on the food surface. Furthermore, chlorine is economical and suitable for large-scale application (Zhang and Farber, 1996). Chlorine dioxide (ClO_2) gas generated by a dry chemical sachet was evaluated for its effectiveness against food-borne pathogens in lettuce leaves. Lettuce leaves were inoculated with a cocktail of three strains of *Escherichia coli* 0157:H7, *Listeria monocy-*

togenes, and *Salmonella typhimurium* and were exposed to ClO_2 gas for 30 min. The treatment reduced the population of *E. coli* by 3.4 log, *S. typhimurium* by 4.3 log, and *L. monocytogenes* by 5.0 log CFU. The results showed that the ClO_2 gas sachet effectively eliminated the pathogens on lettuce with no adverse effect on the visual quality. Hence, this product may be used during storage and transport of lettuce to improve its microbial safety (Lee et al., 2004). The effectiveness of ozone (3 ppm), chlorine dioxide (3 and 5 ppm), chlorinated trisodium phosphate (100 and 200 ppm), and peroxyacetic acid (80 ppm) against *E. coli* 0157:H7, and *L. monocytogenes* was determined. The produce (apples, lettuce, strawberries, and cantaloupe) was dip-inoculated with suspensions containing the pathogens (10^6 CFU/g) and held overnight followed by immersion in each sanitizer solution for up to 5 min. Ozone and ClO_2 (5 ppm) were most effective in reducing pathogen populations, by 5.6 log (Rodgers et al., 2004).

Acidic electrolyzed water (AcEW, 30 ppm free, available chlorine), ozonated water (5 ppm ozone), and sodium hypochlorite (NaOCl, 150 ppm free, available chlorine) were evaluated for use as potential sanitizers of cucumbers and strawberries. Washing cucumbers in alkaline electrolyzed water for 5 min followed by treatment with AcEW for 5 min resulted in a reduction that was at least 2 log CFU per cucumber greater than that of other treatments. In the case of strawberry all treatments reduced the coliform bacterial and fungal populations by 1.0 to 1.5 log CFU per strawberry. None of the treatments could completely inactivate or remove the microorganisms from the surface of cucumber or strawberry (Koseki et al., 2004).

SUMMARY

Management of diseases of harvested produce by using different kinds of chemicals is considered to be more effective than other disease management strategies. Several factors, such as host plant species, cultivar, sensitivity of microbial pathogens, and storage conditions and practices influence the effectiveness and feasibility of fungicide application. The chemicals have to be applied either prior to and/or after harvest, since field infection of commodities may be more important in certain cases. The fungicides may be applied as fumigants, treated wraps and waxes, dips, sprays, or dusts. The fungicides may either have protective and/or eradicative action and they may be grouped as nonsystemic and systemic fungicides, depending on their ability to penetrate into the internal plant tissues. They have been used to protect seeds, fruits, and vegetables to prevent spoilage in storage. The development of resistance to key fungicides in the fungal pathogens and presence of residues of the chemicals in the harvested produce or products above permissible limits have become major problems, necessitating the formulation of alternative control strategies to keep the postharvest pathogens at bay.

APPENDIX 11(I): POISONED FOOD TECHNIQUE OR RADIAL GROWTH (RG) TEST (CARPENTER, 1942; GOLEMBIEWSKI ET AL., 1995)

i. Cultivate the test pathogen(s) in potato dextrose agar medium (PDA) or any medium required for the rapid development of the pathogen and incubate for 7 to 10 days as required.
ii. Prepare different concentrations of the test fungicide in water or organic solvent such as acetone; mix the fungicide solution with about 20 ml of molten PDA at about 60°C to give a final concentration of 0.1% (v/v); pour the mixture of the medium and the fungicide into each sterile Petri plate and allow the medium to set.
iii. Prepare the agar disks using a sterile cork borer (7 or 8 mm) from the actively growing peripheral zones in the culture plates in which the pathogen is cultivated (step (i)); transfer the agar disks of fungal culture to the centre of the plates containing the medium amended with the fungicides and incubate the plates at a constant temperature (about 20 to 22°C) for optimal growth of the pathogen, for 4 to 7 days.
iv. Measure the colony diameter at two position at a right angle to each other and calculate the mean diameter of the colony in each plate.
v. Calculate the relative growth in different treatments using the formula:

$$\frac{\text{Mean colony diameter in fungicide amended medium}}{\text{Mean colony diameter in unamended medium (control)}} \times 100$$

vi. Calculate the effective concentration (EC_{50}) at which the colony diameter is reduced to 50% of the colony diameter in the unamended medium.

APPENDIX 11(II): FRUIT PEEL DISK ASSAY (KRAUSS, 1996)

i. Multiply the pathogen in potato dextrose agar (PDA) medium; flood the plates containing sufficiently grown culture with sterile water and gently scrap the culture surface with a sterile transfer loop.
ii. Filter the conidial suspension through Whatman No.2 filter paper to remove hyphal fragments and adjust the concentration of conidia to the desired level (10^4 to 10^6).
iii. Surface sterilize the green banana fruits with 70% alcohol and prepare the peel disks using a sterile cork borer (12 mm diameter).
iv. Dip the peel disks in the fungicide solutions at different concentrations for 10 s and dry them in the laminar flow cabinet.
v. Inoculate the peel disks by dipping them in conidial suspension for 10 s, or stab inoculate at the centre of the peel disk on the outer surface using a sterile needle.

vi. Place the inoculated peel disks, outer surface upward, in the Petri plate lined with moist Whatman No.3 filter paper and cover with lid also lined with moist filter paper to maintain high humidity.
vii. Assess disease severity using a 0 to 5 scale; determine the percent area covered (turning brown) for the dip-inoculated disks; determine area of rot radius in mm for stab-inoculated disks and measure the severity daily for 5 consecutive days after inoculation.

APPENDIX 11(III): EVALUATION OF FUNGICIDE LIME–SULFUR FOR THE CONTROL OF POSTHARVEST DISEASES OF CITRUS (SMILANICK AND SORENSON, 1999)

i. Select the fruits (lemons and oranges) before any postharvest treatment is applied and wash the fruits thoroughly avoiding any possible injury.
ii. Cultivate the fungal pathogen in potato dextrose agar (PDA) for 1 to 2 weeks at 20°C; collect the spores from the agar surface using a small glass rod, after adding a small amount of sterile 0.05% Triton X-100 to the culture plate and filter the spore suspension through two layers of cheese cloth.
iii. Dilute the spore suspension with sterile water to have a spore concentration of 1×10^6 spores/ml.
iv. Make puncture injury on the fruit surface (1 mm wide × 2 cm deep) to penetrate the albedo tissue, but not the juice sacs below; inoculate fruit using a stainless steel rod dipped in the spore suspension and incubate the inoculated fruits for 24 h at 16 to 20°C.
v. Dip the fruits into different concentrations of lime–sulfur solution kept in a stainless steel, computer controlled, heated tank system.
vi. Maintain sufficient replicates and determine the percentage of disease incidence after incubation for sufficient duration.

APPENDIX 11(IV): ASSESSMENT OF EFFECTS OF ANTIMICROBIAL CHEMICALS ON THE ULTRASTRUCTURE OF BACTERIAL PATHOGEN (*ERWINIA CAROTOVORA* SUBSP. *ATROSEPTICA*) (YAGANZA ET AL., 2004)

A. Electron Microscopy – Sample Preparation

i. Gently scrap the surface of a 16 h old bacterial culture multiplied on nutrient agar (NA) at 24°C and prepare the suspension of bacterial cells (10^8 CFU/ml).
ii. Suspend in 1 ml of aluminum chloride solutions (at 0.05, 0.1, or 0.2 M concentrations), sodium metabisulfite (0.05, 0.1, or 0.2 M concentrations) or 0.5% NaCl adjusted to pH 2.5, 3.0, 3.5, or 4.0 with HCl kept

in microfuge tubes; maintain the control using 0.5% NaCl solution at pH 7.0 and incubate the bacterial suspensions at 24°C.

iii. Add to each treatment (bacterial suspensions) 200 µl of 15% glutaraldehyde (pH 6.0) in 0.1 M sodium cacodylate buffer (SCB), to minimize the impact of salts on bacterial cells during centrifugation, and centrifuge at 2360 g for 5 min at 4°C with a Biofuge 17R centrifuge (Heraeus Sepatech GmbH, Osterode, Germany).

iv. Fix the pellet with a mixture of 3% glutaraldehyde 0.2% ruthenium red, and 0.05 M $CaCl_2$, followed by fixation in 0.1 M SCB (pH 7.3) for 2.5 h at room temperature.

v. Prepare gels of 1 to 2 mm^3 by adding 10% bovine serum albumin and 5% glutaraldehyde in SCB to the pellet; rinse the gels thoroughly with 0.1 M SCB and postfix with 2% osmium tetroxide in 0.01 M SCB for 2 h at room temperature.

vi. Dehydrate the gels by using a series of increasing concentrations of ethanol and embed in JEMBED 812 resin and cut thin (1 µm) and ultrathin (90 nm) sections.

vii. Stain the sections with toluidine blue O and safranin O; contrast them with uranyl acetate and lead citrate and examine the sections under light and electron microscope.

B. Assessment of Bacterial Cell Viability

i. Cultivate the test bacteria at 24°C in Erlenmeyer flasks (250 ml) containing 100 ml of tryptic soy broth (Difco) under agitation (150 rpm) for 16 h; centrifuge the suspension at 2360 g for 5 min at 4°C and recover the pellet.

ii. Resuspend the pellet to have 10^8 CFU/ml in 1 ml of aluminum chloride (0.05, 0.1, or 0.2 M), sodium metabisulfate (0.05, 0.1, or 0.2 M) or 0.5% NaCl (adjusted to pH 2.5, 3.0, 3.5, or 4.0 with HCl) solutions placed in microcentrifuge tubes; maintain controls in 0.5% NaCl solution at pH 7.0 and centrifuge after 5, 10, and 20 min (including centrifugation time) at 2360 g for 5 min at 4°C.

iii. Recover the pellet containing bacteria; wash with 0.5% NaCl (pH 7.0); concentrate again by centrifugation and resuspend the bacteria in 1 ml of 0.5% NaCl (pH 7.0).

iv. Transfer aliquot of 100 µl of the suspensions to an equal volume of SYTOX Green to attain a final concentration of 5 µM and allow the mixture to remain as such for 5 min.

v. Examine the bacterial cells in different treatments under Leitz Orthoplan microscope with blue light excitation using a BP455-490 exciter filter combined with an RKP 510 separator mirror and an LP 515 barrier filter.

vi. Determine the percentage of fluorescent bacteria (emitting blue light) based on at least 250 bacterial cells counted using a hemocytometer under normal light and carry out plate counts on nutrient agar (NA) in parallel to the SYTOX test to determine bacterial cell viability.
vii. Repeat the experiments at least twice maintaining duplicate for each replicate.

REFERENCES

Adam, N.M. and Birch, P.A. 1998. The development of fenhexamid 50WG for the control of *Botrytis cinerea* (gray mold) on soft, cane and bush fruit crops in Great Britain. *Brighton Crop Protection Conference: Pests and Diseases*, Vol. 3, pp. 849–856.

Adams, M.R., Hartley, A.D. and Cox, L.J. 1989. Factors affecting the efficacy of washing procedures used in the production of prepared salads. *Food Microbiology*, **6**: 69–77.

Adamicki, F. and Robak, J. 2000. Effects of fungicide treatments before harvest on the incidence of gray mould caused by *Botrytis cinerea* Pers. quality and storage ability of white cabbage. *Vegetable Crops Research Bulletin*, **53**: 45–53.

Afek, U., Orenstein, J. and Kim, J.J. 2001. Control of silver scurf disease in stored potato by using hydrogen peroxide plus (HPP). *Crop Protection*, **20**: 69–71.

Agarwal, V.K. and Sinclair, J.B. 1996. *Principles of Seed Pathology*, 2nd edn, Lewis Publishers, Boca Raton, FL, USA.

Albertini, C. and Leroux, P. 2004. A *Botrytis cinerea* putative 3-keto reductase gene (*ERG27*) that is homologous to the mammalian 17-hydroxysteroid dehydrogenase type 7 gene (*17-HSD7*). *European Journal of Plant Pathology*, **110**: 723–733.

Al-Haq, M.I., Seo, Y., Oshita, S. and Kawagoe, Y. 2001. Fungicidal effectiveness of electrolyzed oxidizing water on postharvest brown rot of peach. *HortScience*, **36**: 1310–1314.

Al-Haq, M.I., Seo, Y., Oshita, S. and Kawagoe, Y. 2002. Disinfection effects of electrolyzed oxidizing water on suppressing fruit rot of pear caused by *Botryosphaeria berengeriana. Food Research International*, **35**: 657–664.

Angelini, R. 1996. Folicur (tebuconazole): a new triazole fungicide with a wide spectrum of activity. *Informatore Agarario Supplemento*, **52**: 46–50.

Arauz, L.F. 2000. Mango anthracnose: Economic impact and current options for integrated management. *Plant Disease*, **84**: 600–611.

Archbold, D.D., Hamilton-Kemp, T.R., Clements, A.M. and Collins, R.W. 1999. Fumigating 'Crimson Seedless' table grapes with (E)–2-hexenal reduces mold during long-term postharvest storage. *HortScience*, **34**: 705–707.

Ayhan, Z., Chism, G.W. and Richter, E.R. 1998. The shelf-life of minimally processed fresh-cut melons. *Journal of Food Quality*, **21**: 29–40.

Baker, K.F. and Heald, F.D. 1934. An investigation of factors affecting the incidence of lenticel infection of apples by *Penicillium expansum. Bulletin of Washington Agricultural Experiment Station*, **298**.

Banik, A.K., Kaiser, S.A.K.M. and Dhua, R.S. 1998. Evaluations of some systemic and nonsystemic fungicides against *Colletotrichum gloeosporioides* and *Diplodia natalensis* causing postharvest fruit rot of mango. *Journal of Soils and Crops*, **8**: 119–122.

Baraldi, E., Mari, M., Chierici, M. Pondrelli, M., Bertolini, P. and Pratella, G.C. 2003. Studies on thiabendazole resistance of *Penicillium expansum* of pears: Pathogenic fitness and genetic characterization. *Plant Pathology*, **52**: 362–370.

Barbosa, M.C., Leon-Garcia, L.P. de, Sepilveda-Sánchez, J. and Nieto-Angel, D. 2002. Effects of ozone, iodine and chlorine on spore germination of fungi isolated from mango fruits. *Revista Mexican de Fitopatologia*, **20**: 60–65.

Barbosa, M.C., Ponce de Leon, G.L., Sepulveda, S.J., Guillen, G.E. and Nieto, A.D. 2000. Effect of O_3, I_2 and Cl_2 to *Colletotrichum gloeosporioides* Penz, *Fusarium oxysporum* Schlecht, *Lasidiplodia theobromae* Pat. and *Pestalotiopsis mangiferae* P. Heen. Control. *Acta Horticulturae*, **509**: 737–744.

Bartz, J.A. 1999. Suppression of bacterial soft rot in potato tubers by application of kasugamycin. *American Journal of Potato Research*, **76**: 127–136.

Bartz, J.A., Eayre, C.G., Mahovic, M.J., Concelmo, D.E., Brecht, J.K. and Sargent, S.A. 2001. Chlorine concentration and the inoculation of tomato fruit in packing house dump tanks. *Plant Disease*, **85**: 885–889.

Batzer, J.C., Gleason, M.L., Weldon, B., Dixon, P.M. and Nutter, F.W. Jr. 2002. Evaluation of postharvest removal of sooty blotch and flyspeck on apples using sodium hypochlorite, hydrogen peroxide with peroxyacetic acid and soap. *Plant Disease*, **86**: 1325–1332.

Bertrand, P.F. and Saulie-Carter, J.L. 1978. The occurrence of benomyl-tolerant strains of *Penicillium expansum* and *Botrytis cinerea* in mid-Columbia region of Oregon and Washington. *Plant Disease Reporter*, **62**: 305–320.

Bhat, N.A., Masoodi, S.D. and Sidique, S.H. 2000. Chemical control of blackrot of cabbage under field conditions in Kashmir Valley. *Applied Biological Research*, **2**: 87–89.

Blacharski, R.W., Bartz, J.A., Xiao, C.L. and Legard, D.E. 2001. Control of postharvest *Botrytis* fruit rot with preharvest fungicide applications in annual strawberry. *Plant Disease*, **85**: 597–602.

Blacharski, R.W, Legard, D. and Bartz, J.A. 2000. *The Effect of Preharvest Fungicide Applications on Control of Postharvest Disease*. Research Report No. 1. University of Florida, Department of Plant Pathology and Gulf Coast Research and Education Center, Dover, USA.

Brandon, D.L., Binder, R.G., Bates, A.H. and Montague, W.C. Jr. 1995. Competitive ELISA of thiabendazole residues in produce using indirectly immobilized monoclonal antibodies. *Food and Agricultural Immunology*, **7**: 99–108.

Brent, K.J. 1995. *Fungicide Resistance in Crop Pathogens:* How can it be managed? FRAC Monograph No.1, Global Crop Protection Federation (GCPF) Brussels, Belgium, 52 pp.

Brown, G.E. and Chambers, M. 1999. Evaluation of polyhexamethylene biguanide for control of postharvest diseases of Florida citrus. *Proceedings of the 112th Annual Meeting of the Florida State Horticultural Society*, **112**: 118–121.

Burchat, C.S., Ripley, B.D., Leishman, P.D., Ritcey, G.M., Kakuda, Y. and Stephenson, G.R. 1998. The distribution of nine pesticides between the juice and pulp of carrots and tomatoes after home processing. *Food Additives and Contaminants*, **15**: 61–71.

Burchill, R.T. and Edney, K.L. 1977. The control of *Gloeosporium* storage rot of apple by reduced spray programmes. *Annals of Applied Biology*, **87**: 51–56.

Cabras, P., Angioni, A., Garan, V.L., Melis, M., Pirisi, F.M., Farris, G.A., Sotgiu, C. and Minelli, E.C. 1997. Persistence and metabolism of folpet in grapes and wine. *Journal of Agricultural and Food Chemistry*, **45**: 476–479.

Cairoli, S., Arnoldi, A. and Pagani, S. 1996. Enzyme-linked immunosorbent assay for quantitation of the tetraconazole in fruits and fruit juices. *Journal of Agricultural and Food Chemistry*, **44**: 3849–3854.

Cargo, C.A. and Dewey, D.H. 1970. Thiabendazole and benomyl for the control of postharvest decay of apples. *HortScience*, **5**: 259–260.

Carnegie, S.F., Cameron, A.M., Lindsay, D.A., Sharp, E. and Nevison, I.M. 1998. The effect of treating seed potato tubers with benzimidazole, imidazole and phenylpyrrole fungicides on the control of rot and skin blemish diseases. *Annals of Applied Biology*, **133**: 343–363.

Carpenter, J.B. 1942. A toximetrix study of some eradicant fungicides. *Phytopathology*, **32**: 845–856.

Chala, A., Weinert, J. and Wolf, G.A. 2003. An integrated approach to the evaluation of the efficacy of fungicides against *Fusarium culmorum*, the cause of head blight of wheat. *Journal of Phytopathology*, **151**: 673–678.

Chen, T., Dwyre-Gygax, C., Hadfield, S.T. Willetts, C. and Breuil, C. 1996. Development of an enzyme-linked immunosorbent assay for a broad spectrum triazole fungicide: hexaconazole. *Journal of Agricultural and Food Chemistry*, **44**: 1352–1356.

Copes, W.E. 2004. Dose curves of disinfectants applied to plant production surfaces to control *Botrytis cinerea. Plant Disease*, **88**: 509–515.

Covarelli, L., Turner, A.S. and Nicholson, P. 2004. Repression of deoxynivalenol accumulation and expression of *Tri* genes in *Fusarium culmorum* by fungicides *in vitro*. *Plant Pathology*, **53**: 22–28.

Crisosto, C.H. and Smilanick, J. 2002. New technology to reduce damage by *Botrytis cinerea* in the storage of table grapes. *Rivista di Fruitticoltura e di Ortofloricoltura*, **64**: 30–38.

Cui, W., Beever, R.E., Parkes, S.L. and Templeton, M.D. 2004. Evolution of an osmosensing histidine kinase in field strains of *Botryotinia fuckeliana* (*Botrytis cinerea*) in response to dicarboximide fungicide use. *Phytopathology*, **94**: 1129–1135.

Cui, W., Beever, R.E., Parkes, S.L. Weeds, P.L. and Templeton, M.D. 2002. An osmosensing histidine kinase mediates dicarboximide fungicide resistance in *Botryotinia fuckeliana* (*Botrytis cinerea*). *Fungal Genetics and Biology*, **36**: 187–198.

Dacol, L., Gibbard, M., Hadson, M.O. and Knight, S. 1998. Azoxystrobine: development on horticultural crops in Europe. *Brighton Crop Protection Conference: Pests and Diseases*, **3**: 843–848.

Darvas, J.M., Kotze, J.M. and Wehner, F.C. 1990. Effect of treatment after picking on the incidence of postharvest fruit disease of avocado. *Phytophylactica*, **19**: 489–493.

Davidse, L.C. 1986. Benzimidazole fungicides: mechanism of action and biological impact. *Annual Review of Phytopathology*, **24**: 43–65.

Diaz-Sabac, R., Pérez-Flores, L. and Vernon-Carter, E.J. 2000. Emulsion coatings control fruit fly and anthracnose in mango (*Mangifera indica* cv. Manila). *Journal of Horticultural Science and Biotechnology*, **75**: 126–128.

Edwards, S.G., Pirgozliev, S.R., Hare, M.C. and Jenkinson, P. 2001. Quantification of trichothecene-producing *Fusarium* species in harvested grain by competitive PCR to determine efficacies of fungicides against *Fusarium* head blight of winter wheat. *Applied and Environmental Microbiology*, **67**: 1575–1580.

Elad, Y., Yunis, H. and Katan, T. 1992. Multiple fungicide resistance to benzimidazoles, dicarboximides and diethofencarb in field isolates of *Botrytis cinerea* in Israel. *Plant Pathology*, **41**: 41–46.

El-Sheikh Aly, M.M., Felaifel, M.S.A., Fouad, N.A. and Badawy, H.M.A. 1998. Comparative effectiveness of some fungicides and salts applied preharvest or postharvest for controlling pear fruit rots. *Annals of Agricultural Science*, **1**: 135–149.

Errampalli, D. and Crnko, N. 2004. Control of blue mold caused by *Penicillium expansum* on apples 'Empire' with fludioxonil and cyprodinil. *Canadian Journal of Plant Pathology*, **26**: 70–75.

Errampalli, D. and Johnston, H.W. 2001. Control of tuber-borne black scurf (*Rhizoctonia solani*) and common scab (*Streptomyces scabies*) of potatoes with a combination sodium hypochlorite and thiophanate-methyl preplanting seed tuber treatment. *Canadian Journal of Plant Pathology*, **23**: 68–77.

Errampalli, D., Saunders, J.M. and Holley, J.D. 2001. Emergence of silver scurf (*Helminthosporium solani*) as an economically important disease of potato. *Plant Pathology*, **50**: 141–153.

Everett, K.R. and Korsten, L. 1996. Postharvest rots of avocados: improved chemical control by using different application methods. *Proceedings of the 49th New Zealand Plant Protection Conference*, pp. 37–40.

Everett, K.R., Stevens, P.S. and Cutting, J.G.M. 1999. Postharvest fruit rots of avocado are reduced by benomyl applications during flowering. *Proceedings of 52nd New Zealand Plant Protection Conference* (ed.) M. ÓCallaghan, pp. 153–156, New Zealand Plant Protection Society, Auckland. New Zealand.

Faretra, F. and Pollastro, S. 1991. Genetic basis of resistance to benzimidazole and dicarboximide fungicides in *Botryotinia fuckeliana* (*Botrytis cinerea*). *Mycological Research*, **95**: 943–951.

Faretra, F. and Pollastro, S. 1993. Genetics of sexual compatibility and resistance to benzimidazole and dicarboximide fungicides in isolates of *Botryotinia fuckeliana* (*Botrytis cinerea*) from nine countries. *Plant Pathology*, **42**: 48–57.

Gabler, F.M., Mansour, M.F., Smilanick, J.L. and Mackey, B.E. 2004. Survival of spores of *Rhizopus stolonifer*, *Aspergillus niger*, *Botrytis cinerea* and *Alternaria alternata* after exposure to ethanol solutions at various temperatures. *Journal of Applied Microbiology*, **96**: 1354–1360.

Gabler, F.M. and Smilanick, J.L. 2001. Postharvest control of table grape gray mold on detached berries with carbonate and bicarbonate salts and disinfectants. *American Journal of Enology and Viticulture*, **52**: 12–20.

Georgopoulos, S.G., Chrysayl, M. and White, G. 1975. Carboxin resistance in the haploid, the heterozygous diploid and the plant parasitic dicaryotic phase of *Ustilago maydis. Pesticide Biochemistry and Physiology*, **5**: 543–551.

Griffiths, R.G., Dancer, J., O'Neil, E. and Harwood, J.L. 2003. A mandelamide pesticide alters lipid metabolism in *Phytophthora infestans. New Phytologist*, **158**: 345–353.

Godet, F., Roques, J.F. and Compagnon, J.M. 1997. Azoxystrobine: polyvalent cereal and grapevine fungicide. *Phytoma*, **50**: 52–54.

Gold, L.S., Slone, T.H., Ames, B.N. and Manley, N.B. 2001. Pesticide residues in food and cancer risk: A critical analysis. In: *Handbook of Pesticide Toxicology*, 2nd edn (ed.) R. Krieger, pp. 799–843, Academic Press, San Diego, CA, USA.

Golembiewski, R.C., Varges, J.M. Jr., Jones, A.L. and Detweiler, A.R. 1995. Detection of demethylation inhibitor (DMI) resistance in *Sclerotinia homeocarpa* populations. *Plant Disease*, **79**: 491–493.

Greenfield, J.E. and Rossall, S. 2001. Investigation of a new cereal fungicide for the control of *Fusarium* and *Fusarium* toxins. *Aspects of Applied Biology*, **64**: 237–238.

Gudmestad, N.C., Salas, B., Taylor, R.J. Rivera, V.V. and Secor, G.A. 2000. Assessment of mefenoxam resistance in *Phytophthora erythroseptica* in the USA. *BCPC Conference: Pests and Diseases*, **1**: 435–440.

Hanson, K. and Shattock, R.C. 1998. Effect of metalaxyl on formation and germination of oospores of *Phytophthora infestans. Plant Pathology*, **47**: 116–122.

Hegde, G.M. and Anahosur, K.H. 2001. Evaluation of fungitoxicants against fruit rot of chilli and their effect on biochemical constituents. *Karnataka Journal of Agricultural Sciences*, **14**: 836–838.

Heltbech, K., Jensen, J., Husby, J. and Højer, P. 2000. Teldor® WG-new high-efficacy fungicide for fruit-growing. *DJF Rapport, Havebrug*, **12**: 135–146.

Hentges, D.L. 2003. Safe handling of fresh-cut produce and salads. In: *Food Safety Handbook* (eds) R.H. Schmidt and G.E. Rodrick, pp. 425–442, Wiley-Interscience, USA.

Hervieux, V., Chabot, R., Arul, J. and Tweddell, R.J. 2001. Evaluation of different fungicides applied as seed tuber treatments for the control of potato silver scurf. *Phytoprotection*, **82**: 41–48.

Hervieux, V., Yaganza, E.S., Arul, J. and Tweddell, R.J. 2002. Effect of organic and inorganic salts on the development of *Helminthosporium solani*, the causal agent of potato silver scurf. *Plant Disease*, **86**: 1014–1018.

Heu, S.G., Oh, J.H., Kang, Y.S., Ryu, S.R., Cho, S.K., Cho, Y.S. and Cho, M.J. 2001. *gly* gene cloning and expression and purification of glycinecin A, a bacteriocin produced by *Xanthomonas campestris* pv. *glycines* 8ra. *Applied and Environmental Microbiology*, **67**: 4105–4110.

Hilber, U.W., Scuepp, H. and Schwinn, F.J. 1994. Resistance risk evaluation of fludioxonil, a new phenylpyrrole fungicide. In: *Fungicide Resistance* (eds) S. Heaney, D. Slawson, D.W., Hollomon, M. Smith, P.E. Russell and D.W. Parry, pp. 399–402, British Crop Protection Council, Farnham, Survey, UK.

Hisada, Y., Kato, T. and Kawase, Y. 1978. Mechanism of antifungal action of procymidone in *Botrytis cinerea. Annals of Phytopathological Society, Japan*, **44**: 509.

Hollomon, D.W. and Butters, J.A. 1994. Molecular determinants for resistance to crop protection chemicals. In: *Molecular Biology in Crop Protection* (eds) G. Marshall and D. Walters, pp. 98–117, Chapman and Hall, London.

Holmes, G.J. and Eckert, J.W. 1999. Sensitivity of *Penicillium digitatum* and *P. italicum* to postharvest citrus fungicides in California. *Phytopathology*, **89**: 716–721.

Howard, R.J. and Aist, J.R. 1980. Cytoplasmic microtubules and fungal morphogenesis: ultrastructural effects of methyl benzimidazole 2-yl-carbamate determined by freeze substitution of hyphal cells. *Journal of Cell Biology*, **87**: 55–64.

Jeon, Y.H., Cho, M.J., Cho, Y.S. and Hwang, I.G. 2001. Effect of glycinecin A on the control of bacterial leaf spot pepper and bacterial leaf blight of rice. *Korean Journal of Plant Pathology*, **17**: 249–256.

Jespers, A.B.K. and de Waard, M.A. 1995. Effect of fenpiclonil on phosphorylation of glucose in *Fusarium sulphureum. Pesticide Science*, **44**: 167–175.

Johnson, D.A., Inglis, D.A. and Miller, J.S. 2004. Control of potato tuber rots caused by Oomycetes with foliar applications of phosphoric acid. *Plant Disease*, **88**: 1153–1159.

Karabulut, O.A., Arslan, U. and Kuruoglu, G. 2004a. Control of postharvest diseases of organically grown strawberry with preharvest applications of some food additives and postharvest hot water dips. *Journal of Phytopathology*, **152**: 224–228.

Karabulut, O.A., Arslan, U., Kuruoglu, G. and Ozgenc, T. 2004b. Control of postharvest diseases of sweet cherry with ethanol and hot water. *Journal of Phytopathology*, **152**: 298–303.

Katagi, T. and Mikami, N. 2000. Primary metabolism of agrochemicals in plants. (ed.) T. Roberts, pp. 44–101, John Wiley and Sons, Ltd., Chichester, UK.

Khan, S.H., Aked, J. and Magan, N. 1999. The potential for antioxidant chemicals to control *Colletotrichum musae* on banana fruit. *Medelingen-Faculteit Landbouwkundige en Toegepaste Biologische Wetenschappen Universiteit Gent*, **64**: 531–537.

Khan, S.H., Aked, J. and Magan, N. 2001. Control of the anthracnose pathogen of banana (*Colletotrichum musae*) using antioxidants alone and in combination with thiabendazole or imazalil. *Plant Pathology*, **50**: 489.

Khilare, V.C. and Gangawane, L.V. 2000. Effect of carbon and nitrogen nutrition on the growth of thiophanate-methyl resistant *Penicillium digitatum* causing green mold of sweet orange. *Journal of Phytopathological Research*, **13**: 61–64.

Kinay, P., Yildiz, M., Yildiz, Y. Delen, N. and Tosun, N. 2003. Control of postharvest *Penicillium* decay of citrus fruits with antagonistic yeasts and chemical fungicides. *Acta Horticulturae*, **553**.

Kitamura, S., Watanabe, M. and Nakazama, M. 1991. Process of producing coated seed. *U.S. Patent*, **4**: 250.

Klisiewicz, J.M. and Pound, G.S. 1960. Studies on control of black rot of crucifers with antibiotics. *Phytopathology*, **50**: 642.

Kojima, K., Takano, Y., Yoshini, A., Tanaka, C., Kikuchi, T. and Okuno, T. 2004. Fungicide activity through activation of a fungal signalling pathway. *Molecular Microbiology*, **53**: 1785–1796.

Koseki, S., Yoshida, K., Isobe, S. and Itoh, K. 2004. Efficacy of acidic electrolyzed water for microbial decontamination of cucumbers and strawberries. *Journal of Food Protection*, **67**: 1247–1251.

Krauss, U. 1996. Establishment of a bioassay for testing control measures against crown rot of banana. *Crop Protection*, **15**: 269–274.

Kumpoun, W., Motomura, Y. and Harada, Y. 2003. Inhibition of Aspergillus rot by sorbitol in apple fruit with watercore symptoms. *Postharvest Biology and Technology*, **29**: 121–127.

Kupferman, E. 1997. Postharvest applied chemicals to pears: a survey of pear packers in Washington, Oregon and California. *Report of Winter Pear Control Committee*, 7. Pesticide Environmental Stewardship Program, USA. File: A:/Pear disease control.htm.

Lee, S.Y., Costello, M. and Kang, D.H. 2004. Efficacy of chlorine dioxide gas as a sanitizer of lettuce leaves. *Journal of Food Protection*, **67**: 1371–1376.

Lennox, C.L., Spotts, R.A. and Booyse, M. 2004. Incidence of postharvest decay of d'Anjou pear and control with a thiabendazole drench. *Plant Disease*, **88**: 474–478.

Leroux, P. and Besselat, B. 1984. Pourriture grise: la résistance aux fungicides de *Botrytis cinerea. Phytoma Defense des Cultures*, Juin, pp. 25–33.

Leroux, P., Chapeland, F., Debrosses, D. and Gredt, M. 1999. Patterns of cross-resistance to fungicides in *Botrytinia fuckeliana* (*Botrytis cinerea*) isolates from French vineyards. *Crop Protection*, **18**: 687–697.

Leroux, P. and Clerjeau, M. 1985. Resistance of *Botrytis cinerea* Pers. and *Plasmopara viticola* (Berk. and Cur.) Berl. and de Toni to fungicides in French vineyards. *Crop Protection*, **4**: 137–160.

Leroux, P., Fritz, R., Debieu, A., Albertini, C., Lanen, C., Bach, J., Gredt, M. and Chapeland, F. 2002. Mechanisms of resistance to fungicides in field strains of *Botrytis cienrea. Pest Management Science*, **58**: 876–888.

Leroux, P., Fritz, R. and Gredt, M. 1977. Etudes en laboratoire de souches de *Botrytis cinerea* Pers. resistantes á la Dichlozoline an Dicloran au Quintozene, á la Vinclozoline et au 26019 RPC (Glycophene). *Phytopathologische Zeitschrift*, **89**: 347–358.

Leroux, P. and Moncomble, D. 1994. Resistance of *Botrytis cinerea* to dicarboximides, benzimidazoles and phenylcarbamates in the champagne vineyards. In: *Fungicide Resistance* (eds) S. Heaney, D. Slawson, D.W. Hollomon, M. Smith, P.E. Russell and D.W. Parry, pp. 267–270, British Crop Protection Council, Farnham, Surrey, UK.

Levine, A.S. and Fellers, C.R. 1940. Action of acetic acid on food spoilage microorganisms. *Journal of Bacteriology*, **39**: 499–514.

Liberato, J.R. and Tatagiba, J.S. 2001. *In vitro* and postharvest assessment of fungicides in controlling anthracnose and stem-end rot of papaya fruits. *Summa Phytopathologica*, **27**: 409–414.

Lichter, A., Yochanan, Z., Sonego, L., Orit, D., Kaplunov, T., Sarig, P. and Ben-Arie, R. 2002. Ethanol controls postharvest decay of table grapes. *Postharvest Biology and Technology*, **24**: 301–308.

Lichter, A., Zhou, H.W., Vaknin, M., Dvir, O., Zutchi, Y., Kaplunov, T. and Lurie, S. 2003. Survival and responses of *Botrytis cinerea* after exposure to ethanol and heat. *Journal of Phytopathology*, **151**: 601–605.

Lima, G.S. de A., Assuncão, I.P. and Oliveira, S.M.A. de. 1998. Effect of treatment of melon fruits (*Cucumis melo*) with different calcium sources on rot caused by *Myrothecium roridum*. *Summa Phytopathologica*, **24**: 276–279.

López-García, B., Gonzalez-Candelas, L., Pérez-Payá, E. and Marcos, J.F. 2000. Identification and characterization of a hexapeptide with activity against phytopathogenic fungi that cause postharvest decay in fruits. *Molecular Plant–Microbe Interactions*, **13**: 837–846.

López-García, B., Pérez-Payá,E. and Marcos, J.F. 2002. Identification of novel hexapeptides bioactive against phytopathogenic fungi through screening of a synthetic peptide combinatorial library. *Applied and Environmental Microbiology*, **68**: 2453–2460.

Lorenz, G. and Pommer, E.H. 1985. Morphological and physiological characteristics of dicarboximide sensitive and resistant isolates of *Botrytis cinerea. Bulletin of OEPP/EPPO*, **15**: 353–360.

Luck, J.E. and Gillings, M.R. 1995. Rapid identification of benomyl resistant strains of *Botrytis cinerea* using polymerase chain reaction. *Mycological Research*, **99**: 1483–1488.

Maclean, D.C. and Dewey, D.H. 1964. Reduction of decay of pre-packaged apples with 2-aminobutane. *Quarterly Bulletin of Michigan State University, Agricultural Experiment Station*, **47**: 225–230.

Manning, M.A. and Pak, H.A. 1995. *Botrytis cinerea* storage rot of kiwifruit: efficacy of preharvest sprays in orchards with dicarboximide-resistant *Botrytis* populations. *Proceedings of 48th Plant Protection Conference*, pp. 22–26, Hastings, New Zealand.

Mari, M., Cembali, T., Baraldi, E. and Casalini, L. 1999. Peracetic acid and chlorine dioxide for postharvest control of *Monilinia laxa* in stone fruits. *Plant Disease*, **83**: 773–776.

Mari, M., Gregori, R. and Donati, I. 2004. Postharvest control of *Monilinia laxa* and *Rhizopus stolonifer* in stone fruit by peracetic acid. *Postharvest Biology and Technology*, **33**: 319–325.

Mari, M., Leoni, O., Iori, R. and Cembali, T. 2002. Antifungal vapour-phase of allylisothiocyanate against *Penicillium expansum* on pears. *Plant Pathology*, **51**: 97.

Martin, H.L., Hamilton, V.A. and Kopittke, R.A. 2004. Copper tolerance in Australian populations of *Xanthomonas campestris* pv. *vesicatoria. Plant Disease*, **88**: 921–924.

Martin, L.A., Fox, R.T.V., Baldwin, B.C. and Connerton, I.F. 1992. Use of polymerase chain reaction for the diagnosis of MBC resistance in *Botrytis cinerea. Proceedings 1992 Brighton Crop Protection Conference – Pests and Diseases*, pp. 207–214.

McKay, G.J. and Cooke, L.R. 1997. A PCR-based method to characterize and identify benzimidazole resistance in *Helminthosporium solani. FEMS Microbiology Letters*, **152**: 371–378.

McWatters, K.H., Doyle, M.P., Walker, S.L., Rimal, A.P. and Venkitanarayanan, K. 2002. Consumer acceptance of raw apples treated with an antibacterial solution designed for home use. *Journal of Food Protection*, **65**: 106–110.

Menniti, A.M., Pancaldi, D., Maccaferri, M. and Casalini, L. 2003. Effect of fungicides on *Fusarium* head blight and deoxynivalenol content in durum wheat grain. *European Journal of Plant Pathology*, **109**: 109–115.

Michailides, T.J. and Elmer, P.A.G. 2000. Botrytis gray mold of kiwifruit caused by *Botrytis cinerea* in the United States and New Zealand. *Plant Disease*, **84**: 208–223.

Miles, A.K., Willingham, S.L. and Cooke, A.W. 2004. Field evaluation of strobilurins and a plant activator for the control of citrus black spot. *Australian Journal of Plant Pathology*, **33**: 371–378.

Moncomble, D. 1990. Rapport general sur les activités techniques en. 1989. *Vign. Champen, Fér*, **2**: 9–30.

Moore, M.H. and Edney, K.L. 1959. The timing of spray treatments for the control of storage rot of apple caused by *Gloeosporium* spp. *Report of East Malling Research Station*, 1958, 106–109.

Moreno-Martínez, E., Rivera, A. and Badillo, M.V. 1998. Effect of fungi and fungicides on the preservation of wheat seed stored with high and low moisture content. *Journal of Stored Products Research*, **34**: 231–236.

Moyano, C., Gómez, V. and Melgarejo, P. 2004. Resistance to pyremethanil and other fungicides in *Botrytis cinerea* populations collected on vegetable crops in Spain. *Journal of Phytopathology*, **152**: 484–490.

Muñoz, V., Benato, E.A., Sigrist, J.M.M., Oliveira, J.J. do, V. and Corrêa, A.C.C. 2000. Effect of SO_2 for controlling *Botrytis cinerea* in Italia and Red Globe grapes stored at different temperatures. *Revista Brasileira de Fruticultura*, **22**: 100–105.

Nascimento, S.R.C., Neto, S.E.A. and Hafle, O.M. 2000. Use of prochloraz, azoxystrobin and sodium bicarbonate for the postharvest control of *Colletotrichum gloeosporioides* in mango 'Tommy Atkins'. *Summa Phytopathologica*, **26**: 379–382.

Northover, J. and Cerkauskas, R.F. 1998. Fungicide suppression of symptomless latent infections of *Monilinia fructicola* in European plums. *Canadian Journal of Plant Pathology*, **20**: 234–242.

Ntirampemba, G., Langlois, B.E., Archbold, D.D., Hamilton-Kemp, T.R. and Barth, M.M. 1998. Microbial populations of *Botrytis cinerea*-inoculated strawberry fruit exposed to four volatile compounds. *Journal of Food Protection*, **61**: 1352–1357.

Nunes, C., Usall, J., Teixidó, N., Ochoa de Eribe, X. and Viñas, I. 2001. Control of postharvest decay of apples by pre-and post-harvest application of ammonium molybdate. *Pest Management Science*, **57**: 1093–1099.

Nyfeler, R. and Ackermann, P. 1992. Phenylpyrroles, a new class of agricultural fungicides related to natural antibiotic pyrrolnitrin. In: *Synthesis and Chemistry of Agrochemicals* (eds) D.R. Baker, J.G. Penyes and J.J. Seffens, pp. 395–404, ACS, Washington DC.

Ogawa, J.M., Gilpatrick, J.D. and Chiarappa, L. 1977. Review of plant pathogens resistant to fungicides and bactericides. *FAO Plant Protection Bulletin*, **25**: 97–111.

Oliveira, J.J. do V. and Toledo, M.C. de, F. 2000. Residue levels of thiabendazole and imazalil in orange fruit [*Citrus sinensis* (L) Osbeck cv. Pêra] after treatment by dipping in postharvest. *Revista Brasileira de Fruiticultura*, **22**: 339–344.

Oosthuyse, S.A. 2000. Differences in the effect of various packline fungicidal treatments on the manifestation of disease in mango. *Acta Horticulturae*, **509:** 725–731.

Opperman, W.J., Fourie, J.F. Sanderson, R.D. and Britz, T. 1999. A monolithic device for in-package SO_2 generation for the control of postharvest *Botrytis* decay of table grapes. *Journal of Food Quality*, **22**: 427–438.

Padule, D.N., Mahajan, P.D., Perane, R.R. and Patil, R.B. 1999. Efficacy of fungicides for increasing storability of grain molds-infected sorghum hybrid CSH 14. *Seed Research*, **27**: 95–99.

Palou, L., Smilanick, J.L., Crisosto, C.H. and Mausour, M. 2001. Effect of gaseous ozone exposure on the development of green and blue molds on cold stored citrus fruit. *Plant Disease*, **85**: 631–638.

Palou, L., Usall, J., Smilanick, J.L., Aguilar, M.J. and Viñas, I. 2002. Evaluation of food additives and low toxicity compounds as alternative chemicals for the control of *Penicillium digitatum* and *Penicillium italicum*. *Pest Management Science*, **58**: 459–466.

Paplomatas, E.J., Pappas, A.C. and Antoniadis, D. 2004. A relationship among fungicide resistant phenotypes of *Botrytis cinerea* based on RAPD analysis. *Journal of Phytopathology*, **152**: 503–508.

Pappas, A.C. and Fisher, D.J. 1979. A comparison of the modes of action of vinclozolin, procymidone, iprodione and prochloraz against *Botrytis cinerea*. *Pesticide Science*, **10**: 239.

Parra, G. and Ristaino, J.B. 2001. Resistance to mefenoxam and metalaxyl among field isolates of *Phytophthora capsici* causing *Phytophthora* blight of bell pepper. *Plant Disease*, **85**: 1069–1075.

Pennycook, S.R. 1984. Spraying kiwifruit for control of *Botrytis* storage rot. *Orchardist, New Zealand*, **57**: 251–258.

Peres, N.A.R., Souza, N.L., Peever, T.L. and Timmer, L.W. 2004. Benomyl sensitivity of isolates of *Colletotrichum acutatum* and *C. gloeosporioides* from citrus. *Plant Disease*, **88**: 125–130.

Peres, N.A.R., Souza, N.L., Zitko, S.E. and Timmer, L.W. 2002. Activity of benomyl for control of bloom fruit drop of citrus caused by *Colletotrichum acutatum*. *Plant Disease*, **86**: 620–624.

Pernezny, K., Nagata, R., Raid, R.N., Collins, J. and Carroll, A. 2002. Investigation of seed treatments for management of bacterial leaf spot of lettuce. *Plant Disease*, **86**: 151–155.

Peters, R.D., Macdonald, I.K., MacIssac, K.A. and Woodworth, S. 2001. First report of thiabendazole-resistant isolates of *Fusarium sambucinum* infecting stored potatoes in Nova Scotia, Canada. *Plant Disease*, **85**: 1030.

Pham, H.T., Riu, K.Z., Jang, K.M., Cho, S.K. and Cho, M. 2004. Bactericidal activity of glycinecin A, a bacteriocin derived from *Xanthomonas campestris* pv. *glycines* on phytopathogenic *Xanthomonas campestris* pv. *vesicatoria* cells. *Applied and Environmental Microbiology*, **70**: 4486–4490.

Piano, S., Gilardi, G., Galliano, A. and Gullino, M.L. 2001. Serious attacks of *Phialophora* sp. on kiwifruit. *Informatore Fitopatologica*, **51**: 76.

Pinilla, B., Alvarez, M.M. and Reyes, M.S. 1998. Occurrence of benzimidazole-resistant strains of *Penicillium expansum* in pear packinghouses in Chile. *Acta Horticulturae*, **475**: 471–477.

Pirgozliev, S.R., Edwards, S.G., Hare, M.C. and Jenkinson, P. 2003. Strategies for the control of Fusarirum head blight. *European Journal of Plant Pathology*, **109**: 731–742.

Ploetz, R., Heine, G., Haynes, J. and Watson, M. 2002. An investigation of biological attributes that may contribute to the importance of *Phytophthora capsici* as a vegetable pathogen in Florida. *Annals of Applied Biology*, **140**: 61–67.

Qadir, A. and Hashinaga, F. 2001a. Nitrous oxide inhibits *in vitro* growth of multiple postharvest fungi. *HortScience*, **36**: 1302–1304.

Qadir, A. and Hashinaga, F. 2001b. Inhibition of postharvest decay of fruits by nitrous oxide. *Postharvest Biology and Technology*, **22**: 279–283.

Rao, N.N.R. 1984. Effect of postharvest fungicidal treatment of citrus fruits against the control of green mold rot. *Journal of Food Science and Technology*, **21**: 25–28.

Rendall-Dunn, A.J. 1991. General News. *Postharvest News and Information*, **2**: 3.

Richmond, D.V. and Philips, A. 1975. The effect of benomyl and carbendazim on mitosis in hyphae of *Botrytis cinerea* Pers. Ex. Fr. and roots of *Allium cepa* L. *Pesticide Biochemistry and Physiology*, **5**: 367.

Ritenour, M.A., Pelosi, R.R., Burton, M.S., Stover, E.W., Dou, H. and McCollum, T.G. 2004. Assessing the efficacy of preharvest fungicide applications to control postharvest diseases of Florida citrus. *HortTechnology*, **14**: 58–62.

Ritenour, M.A., Zhang, J., Wardowski, W.F. and Brown, G.E. 2003. Postharvest decay control recommendations for Florida citrus fruit. *Circular 359-A, Horticultural Sciences Department, Florida Cooperative Extension Service, Institute of Food and Agricultural Sciences*, University of Florida, FL, USA.

Rodgers, S.L., Cash, J.N., Siddiq, M. and Ryser, E.T. 2004. A comparison of different chemical sanitizers for inactivating *Escherichia coli* 0157: H7 and *Listeria monocytogenes* in solution and in apples, lettuce, strawberries and cantaloupe. *Journal of Food Protection*, **67**: 721–731.

Rong, T. and Ting, Z. 2000. Antifungal activity of monoterpenoids against postharvest pathogens *Botrytis cinerea* and *Monilinia fructicola. Journal of Essential Oil Research*, **12**: 113–121.

Rosenberger, D.A. and Meyer, F.W. 1979. Benomyl-tolerant *Penicillium expansum* in apple packinghouses in eastern New York. *Plant Disease Reporter*, **63**: 37–40.

Rosenberger, D.A. and Meyer, F.W. 1981. Postharvest fungicides for apples: development of resistance to benomyl, vinclozolin and iprodione. *Plant Disease*, **65**: 1010–1013.

Rosslenbroick, H.J., Brandes, W., Krueger, B.W., Kuck, K.H., Pontzen, R., Stenzel, K. and Suty, A. 1998. Fenhexamid (KBR 2738) – a novel fungicide for control of *Botrytis cinerea* and related pathogens. *Brighton Crop Protection Conference, Pests and Diseases*, **2**: 327–334.

Saborío, D., Sáenz, V., Arauz, L.F. and Bertsch, F. 2000. The effect of pre-and postharvest calcium applications on anthracnoses. (*Colletotrichum gloeopsorioides*) severity, and quality of papaya (*Carica papaya*) fruit. *Agronomía Costarricense*, **24**: 77–88.

Schirra, M., Angioni, A., Ruggiu, R., Minelli, E.V. and Cabras, P. 1998. Thiabendazole uptake and persistence in lemons following postharvest dips at 50°C. *Italian Journal of Food Science*, **10**: 105–107.

Schnabel, G., Bryson, P.K., Bridges, W.C. and Brannen, P.M. 2004. Reduced sensitivity in *Monilinia fructicola* to propiconazole in Georgia and implications for disease management. *Plant Disease*, **88**: 1000–1004.

Schnabel, G. and Layne, D.R. 2004. Comparison of reduced-application and sulfur-based fungicide programs on scab intensity, fruit quality and cost of disease control on peach. *Plant Disease*, **88**: 162–166.

Schroth, M.N., Beutte, J.A., Moller, W.J. and Reil, W.O. 1969. Fire blight of pears in California: Current status and research progress. *California Plant Pathology*, **7**: 1–5.

Schüepp, H. and Küng, M. 1981. Stability of tolerance to MBC in populations of *Botrytis cinerea* in vineyards of northern and eastern Switzerland. *Canadian Journal of Plant Pathology*, **3**: 180–181.

Seaton, W.H. 1993. Acetic acid. In: *Acetic Acid and its Derivatives* (eds) V.H. Agreda and J.R. Zoeller, pp. 73–100, Marcel Dekker, Inc., New York.

Selmaoui, K., Boubaker, A. and Douira, A. 1997. *In vivo* and *in vitro* activity of some fungicides on development of postharvest disease of apple caused by *Alternaria tenuis* Nees. *Al Awamia*, **97**: 441–450.

Sharma, R.L. 2000. Efficacy of postharvest fungicidal treatments against brown rot of apple. *Journal of Mycology and Plant Pathology*, **30**: 94–95.

Sholberg, P.L., Cliff, M. and Moyls, A.L. 2001. Fumigation with acetic acid vapor to control decay of stored apples. *Fruits (Paris)*, **56**: 355–365.

Sholberg, P.L., Delaquis, P.J. and Moyls, A.L. 1998. Use of acetic acid fumigation to reduce the potential for decay in harvested crops. *Recent Research Developments in Plant Pathology*, **2**: 31–42.

Sholberg, P.L. and Gaunce, A.P. 1995. Fumigation of fruit with acetic acid to prevent postharvest decay. *HortScience*, **30**: 1271–1275.

Sholberg, P., Haag, P., Hocking, R. and Bedford, K. 2000. The use of vinegar vapor to reduce postharvest decay of harvested fruit. *HortScience*, **35**: 898–903.

Simone, G.W. 1999. Postharvest disease control in citrus (*Citrus* spp.). *Plant Disease Management Guide, Volume 3*, PDMG-V3-13 University of Florida, Gainesville, USA.

Smilanick, J.L. 2003. Improvement in control of postharvest decay of citrus by thiabendazole with sodium bicarbonate. *Phytopathology*, **93**: S105 (Abst.).

Smilanick, J.L., Margosan, D.A., Mlikota, F., Usall, J. and Michael, I.F. 1999. Control of citrus green mold by carbonate and bicarbonate salts and the influence of commercial postharvest practices on their efficacy. *Plant Disease*, **83**: 139–145.

Smilanick, J.L. and Sorenson, D. 1999. Use of lime-sulfur solution for the control of postharvest mold of citrus fruit. *Pest Management Grants Final Report, USDA–ARS.*

Smilanick, J.L. and Sorenson, D. 2001. Control of postharvest decay of citrus fruit with calcium polysulfide. *Postharvest Biology and Technology*, **21**: 157–168.

Sommer, N.F., Suadi, J.E. and Fortlage, R.J. 1994. Postharvest storage diseases. In: *Kiwifruit Growing and Handling* (eds) J. Hasey, R.S. Johnson, J.A. Grant and W.O. Reid, pp. 116–122. University of California, Oakland, USA.

Song, J., Leepipattwit, R., Deng, W. and Beaudry, R. 1998. Hexanal vapor acts as residueless antifungal agent that enhances aroma biosynthesis in apple fruit. *Acta Horticulturae*, **464**: 219–224.

Spotts, R.A. and Cervantes, L.A. 1995. Copper, oxytetracyline and streptomycin resistance of *Pseudomonas syringae* pv. *syringae* strains from pear orchards in Oregon and Washington. *Plant Disease*, **79**: 1132–1135.

Staub, T. 1984. Early experiences with phenylamide resistance and lessons for continued successful use. In: *Fungicide Resistance* (eds) S. Heaney, D. Slawson, D.W. Hollomon, M. Smith, P.E. Russell and D.W. Parry, pp. 131–138, British Crop Protection Council, Farnham, Surrey, UK.

Stein, J.M. and Kirk, W.W. 2004. The generation and quantification of resistance to dimethomorph in *Phytophthora infestans*. *Plant Disease*, **88**: 930–934.

Stensvand, A. 1998. Evaluation of new fungicides and a biocontrol agent against gray mold on strawberries. *Tests Agrochemicals and Cultivars*, **19**: 70–71.

Stenzel, K., Pontzen, R., Seitz, T., Tiemann, R. and Witzenberger, A. 1998. SZX 722: a novel systemic oomycete fungicide. *Brighton Crop Protection Conference: Pests and Diseases*, **2**: 367–374.

Strandberg, J.O. 1984. Efficacy of fungicides against persistence of *Alternaria dauci* on carrot seed. *Plant Disease*, **68**: 39.

Suadi-Hasbun, J.E. 1987. *Pathological and physiological interactions between Actinidia deliciosa* (kiwifruit) and *Botrytis cinerea* in low temperature environments, Ph.D. Thesis, University of California, Davis, USA.

Taylor, A.G. and Harman, G.E. 1990. Concepts and technologies of selected seed treatments. *Annual Review of Phytopathology*, **68**: 1087.

Taylor, R.J., Salas, B. and Gudmestad, N.C. 2004. Differences in etiology affect mefenoxam efficacy and the control of pink rot and leak tuber diseases of potato. *Plant Disease*, **88**: 301–307.

Taylor, R.J., Salas, B., Secor, G.A., Rivera, V. and Gudmestad, N.C. 2002. Sensitivity of North American isolates of *Phytophthora erythroseptica* and *Pythium ultimum* to mefenoxam (metalaxyl). *Plant Disease*, **86**: 797–802.

Timmer, L.W. (ed.) 2003. *Florida Citrus Pest Management Guide*, University of Florida, Cooperative Extension Service Publication, No. SP-43, Gainesville, Florida.

Tonini, G., Menniti, A.M., Zanoli, R. and Ceroni, M.R. 2000a. *Monilinia laxa* postharvest rot control of peach and nectarine. *Atti, Giornate fitopatologiche, Perugia*, **2**: 63–68.

Tonini, G., Zanoli, R. and Gregori, R. 2000b. Peaches and nectarines: pre- and postharvest strategy to prevent rot. *XXIV Convegno Peschicolo, Perna nuov peschicoltura: produzione, organizzazione, mercato, Cesena, Italia.*

Ullasa, B.A., Om Prakash, Eckert, J.W. and Bretschnelder, B. 1998. Competitive behaviour of benzimidazole resistant strains of *Penicillium italicum* in citrus. *Indian Phytopathology*, **51**: 72–74.

Ullasa, B.A. and Rawal, R.D. 1984. *Annual Report*. Indian Institute of Horticultural Research, Bangalore, India, pp. 168.

Usberti, R. and Amaral, H.M. 1999. Fungicide dressing timing, seed size, seed origin and fungal incidence effects on groundnut (*Arachis hypogaea* L.) storability. *Seed Science and Technology*, **27**: 699–706.

Viñas, I., Usall, J. and Sanchis, V. 1991. Tolerance of *Penicillium expansum* to postharvest fungicide treatment in apple packinghouses in Lleida (Spain). *Mycopathologia*, **113**: 15–18.

Viviani-Nauer, A., Hoffmann-Boller, P. and Gafner, J. 1997. *In vivo* detection of folpet and its metabolite phthalimide in grape must and wine. *American Journal of Enology and Viticulture*, **48**: 67–70.

Venkitanarayanan, K.S., Lin, C.M., Bailey, H. and Doyle, M.P. 2002. Inactivation of *Escherichia coli* 0157: H7, *Salmonella enteritidis* and *Listeria monocytogenes* on apples, oranges and tomatoes by lactic acid with hydrogen peroxide. *Journal of Food Protection*, **65**: 100–105.

Washington, W.S., Engleitner, S. and Shanmugananthan, N. 1999. The effect of longer spray intervals, reduced rates or mixtures of three fungicides on fruit rots of strawberry. *Plant Protection Quarterly*, **14**: 88–91.

Wilcke, W.F., Ng, H.F., Morey, R.V., Meronuck, R.A. and Lang, J.P. 1998. Effect of iprodione fungicide application on deterioration rate of stored shelled corn. *Transactions of the ASAE*, **41**: 1761–1765.

Williams, J.G.K., Kubeilik, A.R., Livak, K.J., Rafalski, J.A. and Tingey, S.V. 1990. DNA polymorphism amplified by arbitrary primers are useful as genetic markers. *Nucleic Acid Research*, **12**: 6531–6535.

Willingham, S.L., Pegg, K.G., Coates, L.M., Cooke, A.W., Dean, J.R., Langdon, P.W.B. and Beasley, D.R. 2001. Field management of avocado postharvest diseases. *Acta Horticulturae*, **553:** 435–438.

Wong, F.P. and Wilcox, W.F. 2001. Comparative physical modes of action of azoxystorbin, mancozeb and metalaxyl against *Plasmopara viticola* (grapevine downy mildew). *Plant Disease*, **85**: 649–656.

Woo, J.S., Heu, S.G. and Cho, Y.S. 1998. Influence of growth conditions on the production of a bacteriocin by *Xanthomonas campestris* pv. *glycines* 8ra. *Korean Journal of Plant Pathology*, **14**: 376–381.

Yaganza, E.S., Arul, J. and Tweddell, R.J. 2001. Postharvest application of organic and inorganic salts for the control of potato tuber soft rot. *Phytopathology*, **91**: S198 (Abst.).

Yaganza, E.S., Rioux, D., Simard, M., Arul, J. and Tweddell, R.J. 2004. Ultrastructural alterations of *Erwinia carotovora* subsp. *atroseptica* caused by treatment with aluminium chloride and sodium metabisulfite. *Applied and Environmental Microbiology*, **70**: 6800–6808.

Yang, X.J., Chen, F.R. and Zhang, L.X. 2000. Screening of fungicides for control of *Ceratocystis fimbriata* Ellis et Halsted. *Plant Protection*, **26**: 38–39.

Yeh, J.Y., Hoogetoorn, E. and Chen, J. 2004. Influence of calcium lactate on the fate of spoilage and pathogenic microorganisms in orange juice. *Journal of Food Protection*, **67**: 1429–1432.

Yoshimura, M.A., Luo, Y., Ma, Z. and Michailides, T.J. 2004. Sensitivity of *Monilinia fructicola* from stone fruit to thiophanate-methyl, iprodione and tebuconazole. *Plant Disease*, **88**: 373–378.

Yourman, L.F. and Jeffers, S.N. 1999. Resistance to benzimidazole and dicarboximide fungicides in greenhouse isolates of *Botrytis cinerea. Plant Disease*, **83**: 569–575.

Zainuri, Joyce, D.C., Wearing, A.H., Coates, L. and Terry, L. 2001. Effects of phosphonate and saliclylic acid treatments on anthracnose disease development and ripening of Kenningston Pride mango. *Australian Journal of Experimental Agriculture*, **41**: 805–813.

Zehr, E.I., Luszez, L.A., Olien, W.C., Newall, W.C. and Toler, J.E. 1999. Reduced sensitivity in *Monilinia fructicola* to propiconazole following prolonged exposure in peach orchards. *Plant Disease*, **83**: 913–916.

Zehr, E.I., Toler, J.E. and Luszcz, L.A. 1991. Spread and persistence of benomyl-resistant *Monilinia fructicola* in South Carolina peach orchards. *Plant Disease*, **75**: 590–593.

Zhang, S. and Farber, J.M. 1996. The effects of various disinfectants against *Listeria monocytogenes* on fresh-cut vegetables. *Food Microbiology*, **13**: 311–321.

Ziogas, B.N., Markoglou, A.N. and Malandrakis, A.A. 2003. Studies on inherent resistance risk to fenhexamid in *Botrytis cinerea. European Journal of Plant Pathology*, **109**: 311–317.

Zoffoli, J.P., Latorre, B.A., Rodríguez, E.J. and Aldunce, P. 1999. Modified atmosphere packaging using chlorine gas generators to prevent *Botrytis cinerea* on table grapes. *Postharvest Biology and Technology*, **15**: 135–142.

12

INTEGRATED SYSTEMS FOR THE MANAGEMENT OF POSTHARVEST DISEASES

Microbial pathogens, depending on their pathogenic potential (virulence), the level of susceptibility/resistance of crop cultivars, and the existence of favorable environmental conditions, may cause losses of marketable produce to a varying extent. In order to counter the adverse effects of microbial pathogens, one or more strategies for disease management that are compatible with each other have been evaluated. Different disease management strategies that have to be applied, both at pre- and postharvest stages, have been integrated to provide more effective disease control than that possible with a single approach. Many economically important crop diseases have been managed by integrating various strategies, such as crop sanitation, certification, crop rotation (crop sequence), adoption of suitable planting/sowing date, and use of resistant cultivars, with nominal use of fungicides or other chemicals. The term integrated disease management (IDM) is increasingly seen in the plant pathology literature, with the term "pest" largely confined to insects (Narayanasamy, 2002). A knowledge of pathogen biology and ecology, sources of inoculum, and epidemiological factors and storage conditions favoring disease incidence and spread may be useful in selecting and integrating suitable strategies for the effective management of postharvest diseases of both durable and perishable commodities.

Postharvest Pathogens and Disease Management, by P. Narayanasamy

12.1 SEED-BORNE DISEASES

12.1.1 Integrated Management of Seed-borne Diseases

Management of seed-borne diseases has been attempted primarily through the application of chemicals. However, the effectiveness of cultural practices and sanitation procedures for reducing seed infection by microbial pathogens has been indicated by several investigations (Maude, 1996; Agarwal and Sinclair, 1996). The desirability of employing biocontrol agents (BCAs) to reduce the incidence of diseases caused by seed-borne pathogens has been emphasized (Narayanasamy, 2002). However, there are only a few reports indicating the advantages of integrating two or more approaches to disease management.

Seed germination and seedling health could be improved by soaking carrot seeds in a suspension of *Bacillus subtilis* strain T99, which was effective against the seed-borne pathogen *Alternaria radicina.* Metiram (as Polyram-combi) and iprodione (as Rovral), when added in small quantities to the BCA suspension enhanced the effectiveness of the antagonistic bacteria (Hentschel, 1991). By integrating the use of the biocontrol agent with an application of the chemical, more effective control of the seed-borne disease may be obtained. The efficacy of polyethylene glycol (PG), metalaxyl, and *Trichoderma harzianum* was evaluated for the control of *Pythium aphanidermatum*, causing damping-off disease of table beets. By integrating metalaxyl, osmoconditioning PG, and BCA, disease incidence could be more effectively suppressed, in addition to reducing in the amount of fungicide required by 50% of the recommended dose when applied separately (Abdalla and El-Gizawy, 2000). In the case of wheat loose smut disease caused by *Ustilago segetum* var. *tritici*, a combination of any one of biological agents *T. viride*, *T. harzianum*, *Pseudomonas fluorescens*, or *Gliocladium virens* with carboxin (Vitavax) at 0.125% provided almost total protection against the disease in three field trials. In addition to more effective control, the combined application resulted in a reduction in the amount of fungicide required by one-half of the recommended dose. The other advantages realized through the integrated approach were higher seed germination, seedling emergence, and seed yield of wheat cultivar cv. HD2189 (Dharam Singh and Maheshwari, 2001).

12.1.2 Integrated Management of Mycotoxin Contamination

Seed microflora may comprise both beneficial and harmful microorganisms. However, the adverse effects caused by seed-borne fungal and bacterial pathogens outweigh the usefulness of beneficial microbes. In addition to a failure of seed germination, reduction in seedling vigor, and seed deterioration caused by the pathogens, production of mycotoxin by some fungal pathogens and contamination of foods and feeds have been the causes of concern, due to the possible health hazards. As mycotoxin contamination is unavoidable and unpredictable, it poses a unique challenge to food safety

(Park and Stoloff, 1981; FAO, 1997). The quality and safety of foods and feeds is of vital importance to ensure that markets are not compromised by the distribution and sale of poor quality or unsafe food. Development of integrated systems of management is considered to be the most effective approach to meet the risks associated with mycotoxin contamination. The Hazard Analysis and Critical Control Point (HACCP) approach identifies, evaluates, and controls hazards that are significant for food safety. It is a structured, systematic approach for the control of food safety throughout the commodity system, from the "seed to spoon" (Wareing, 1999). Mycotoxin production and contamination may be contained by taking measures during pre- and postharvest stages. The control parameters include various factors, such as time of harvesting, temperature and moisture during storage, selection of agricultural products prior to processing, decontamination conditions, addition of chemicals, and final product storage (Lopez-Garcia et al., 2004).

12.1.2.1 Preharvest Control The first step in ensuring a safe final product is prevention through preharvest control. Some seeds are contaminated with mycotoxins in the field. If the infection occurs in the field, as in the case of wheat, barley, and corn, the fungal pathogens (*Fusarium* spp.) will continue to develop during postharvest stages and storage. Mycotoxins, such as fumonisin B_1, are invariably produced preharvest. Aflatoxins may be produced both preharvest and postharvest. Drying the seeds to a safe water activity level is one of the most effective measure that can be applied. By reducing moisture levels to 14% for maize and 9.5% for groundnuts at 20°C it is possible to reduce the growth of *Aspergillus flavus* (Wareing, 1999).

Insect infestation of seeds results in greater level of damaged kernels, favoring higher incidence of *A. flavus* and *A. parasiticus*. Hence, control of insect infestation may help prevent proliferation of *Aspergillus* spp. and aflatoxin production. Proper disposal of infected crop residues that may form the sources of inoculum for infection of the next crop and adoption of a proper crop sequence (rotation) (Chapter 5) have been suggested to reduce infection by fungi producing mycotoxins. A maize–soybean rotation may result in a reduction in the incidence of *Fusarium* spp. compared to monoculture of maize. Soil fertility and drought stress appear to have some influence on the level of preharvest aflatoxin contamination of maize. Drought followed by high moisture conditions have been found to be favorable for the development of *F. moniliforme* and fumonisin production. Development of cultivars resistant to toxigenic fungal pathogens may be the ideal approach for the management of mycotoxin contamination. Some investigations have indicated the possibility of producing wheat and corn cultivars with resistance to the pathogens producing mycotoxins (Chapter 8).

12.1.2.2 Postharvest Control Factors such as timeliness, clean-up, and drying to maintain safe moisture levels are important during harvesting. As crops left on the field for longer periods show higher levels of mycotoxin contamination, it is essential to harvest the crops at the right time, followed by adequate drying. For mycotoxin decontamination, biological methods have been explored. The possibility of degrading aflatoxins using certain fungi which produce peroxidases was reported by Lopez-Garcia and Park (1998). Among the several chemicals evaluated for their ability to inactivate and reduce the hazard of selected mycotoxins, ammoniation has been shown to be the most effective process. Aflatoxin contamination in maize, peanuts, and cotton could be significantly reduced by an ammoniation process (Lopez-Garcia and Park, 1998).

Because of the unpredictable and heterogeneous nature of mycotoxins production and contamination, it may not be possible to achieve 100% destruction of all mycotoxins in all food systems. However, it is considered that the use of a HACCP-based hurdle system, in which contamination is monitored and controlled throughout production and postproduction operations, may be effective. The development of suitable integrated mycotoxin management systems may control at various points from the field to the consumer (Lopez-Garcia et al., 2004).

12.2 POSTHARVEST DISEASES OF PERISHABLES

It is essential to ensure that close coordination, constant surveillance, and efficient technical support for rapid detection and precise identification of microbial pathogens, and feed back on the effectiveness of corrective measures taken to restrict the incidence and subsequent spread of diseases, are available. All methods that may lead to a reduction of pathogen inoculum, such as use of disease-free seeds and planting materials, modification of cultural practices and application of organic amendments and mulches, soil solarization, eradication of alternate and alternative hosts, and proper disposal of infected plant debris, have to be implemented. These methods are ecofriendly, since they will not pollute the environment. Furthermore, avoidance of wounds to the harvested produce is the basic precaution, to be strictly enforced in all cases. Though these methods are nonspecific, they will effectively reduce the chances of infection by microbial pathogens. The strategies that can directly act on the microbial pathogens may be integrated broadly into four combinations: (1) physical + chemical; (2) physical + biocontrol; (3) biocontrol + chemical; and (4) biocontrol + physical + chemical methods.

12.2.1 Integration of Physical and Chemical Methods

Combined application of heat treatments, plastic packaging, and fungicide was evaluated for the control of mold diseases and keeping quality of 'Oroblanco' fruit (*Citrus grandis* × *C. paradisi*). Hot water dip at 52°C for 2 min or 'hot drench brushing' at 52, 56, or 60°C for 10 s and the standard packing house treatment, that is waxing with the addition of thiabendazole (TBZ) and 2,4-

D isopropyl ester, were tested. The treated fruits were stored for 2 weeks at 1°C (simulated quarantine treatment) followed by 12 to 13 weeks at 11°C (simulated sea transportation condition) and one additional week at 20°C (simulated retail shelf-life period). Polyethylene linear packaging increased the risk of incidence of diseases, if decay control measure was not employed. Application of TBZ, hot water dip, or curing (at 36°C for 72 h) controlled the mold diseases caused *Penicillium digitatum* and *P. italicum*. Hot drench brushing at 60°C and hot water dip slowed fruit softening and reduced button abscission (Rodov et al., 2000). When heat treatment was combined with fungicide, the concentration of the fungicide required for the control of decay caused by *Penicillium digitatum* could be substantially reduced. Imazalil solution (400 ppm) heated to 55°C was as effective as nonheated imazalil at 1000 ppm. This finding indicated that the possibility of development of fungicide resistance in postharvest pathogens could be curtailed (Ben-Yehoshua et al., 2000).

The black spot decay of mango caused by *Alternaria alternata* was effectively controlled by a combined hot water spray and fruit brushing (HWB) treatment for 15 to 20 s, in addition to enhancement of keeping quality of fruit. However, after storage for 3 weeks at 12°C and another week at 20°C, HWB treatment alone was less effective than the treatment with the combination of HWB + prochloraz (900 μg/ml), indicating the increased effectiveness of the combination of heat and fungicide treatment (Prusky et al., 1999). By combining heat and fungicide, the period of exposure to higher temperature could be considerably reduced. A combined hot water spray fruit brushing treatment for 15 to 20 s with prochloraz (225 μg/ml) was shown to be highly effective against *Alternaria* rot in fruits with a high relative quiescent infected surface rating of 36 at harvest. In contrast, when prochloroz was used without HWB, a four-fold (900 μg/ml) increase in the concentration of prochloroz was required to have similar level of disease control. Thus the combination of heat and chemical reduced the requirement of synthetic fungicides to achieve satisfactory level of disease control (Prusky et al., 1999). Treatment with hot water (52°C for 15 min) and carbendazim (0.1%) protected mango fruits completely from anthracnose disease caused by *Colletotrichum gloeosporioides*. The treated fruits could be stored for 26 days at 12°C. Thiophanate-methyl (0.1%) treated fruits could be stored for only 12 days at 12°C. The fungicides alone were not effective against anthracnose disease at ambient temperature, indicating that the efficacy of the fungicides against postharvest pathogen may be increased by increasing the temperature (Om Prakash and Pandey, 2000).

The effect of treatment of table grapes with gamma-irradiation and sulfur dioxide (SO_2) on the control of gray mold disease caused by *Botrytis cinerea* was investigated. Irradiation with 1.0 kGY of gamma rays, SO_2 (3 g of Na_2SO_3/5 kg cluster) or a combination of gamma rays and SO_2 were tested. The treated and untreated control fruit clusters were stored in cold storage. Gamma-irradiation in combination with SO_2 was more effective in protecting two cultivars Baladi and Helwani table grapes against rotting caused by gray mold disease than gamma-irradiation or SO_2 applied alone (Al-Bachir, 1998).

Integration of curing and controlled atmosphere (CA) storage and fungicide application was very effective in reducing disease incidence in kiwifruit. Refrigeration curing and CA conditions (CO_2 level at 15%) in combination with application of fenhexamid or fludioxonil + cyprodinil reduced the incidence of gray mold disease to nearly negligible levels in kiwifruit (Tonini et al., 2000).

12.2.2 Integration of Physical and Biocontrol Methods

The effect of heat treatment combined with application of a biocontrol agent (BCA) on the development of blue mold disease caused by *Penicillium expansum* in 'Gala' apples was investigated. Apple fruits were exposed to heat at 38°C for 4 days and wound-inoculated with *P. expansum*, followed by application of the BCA *Pseudomonas syringae*. Disease development was assessed after storage for 7 days at 20°C or 3 months at 1°C. The heat treatment reduced the pathogen population on the apple surface, but provided little residual protection (that is the protection was transient). Application of the BCA further enhanced the level of disease control because of the residual protection, indicating the synergistic action of heat and BCA, and resulting in a substantial reduction in decay of the apples (Leverentz et al., 2000). In another study, the eradicative activity of heat was integrated with the antagonistic activity of heat-tolerant yeasts. Prestorage hot air treatment was applied to apples at 12 h after inoculation with *P. expansum* followed by an application of yeasts. The biocontrol potential was significantly improved by the earlier heat treatment (Leverentz et al., 2001).

The antagonist yeast *Metschnikowia pulcherrima* T5-A2 was employed in combination with heat (38°C for 4 days) and 1-methylcyclopropene (1-MCP) (an ethylene receptor inhibitor) to maximize control of apple fruit decay caused by *P. expansum* (blue mold) and *Colletotrichum acutatum* (bitter rot) on 'Golden Delicious' apples under controlled atmosphere (CA) conditions. Treatment with 1-MCP alone increased both bitter rot and blue mold decays. In contrast, both diseases could be effectively checked on 1-MCP-treated apples by a combination of the heat and BCA application. 1-MCP by slowing down apple maturation, may possibly extend the action of natural mechanisms. The BCA controlled bitter rot more effectively than blue mold, whereas heat treatment effectively controlled blue mold. These treatments, with different mechanisms of actions, were able to provide the desired level of protection to apple against both diseases under CA conditions (Tables 12.1 and 12.2; Fig. 12.1) (Janisiewicz et al., 2003).

The combination of curing (at 37°C for 72 h, 95% RH) of grapefruits inoculated with *Penicillium digitatum* and after incubation for 1, 36, or 72 h at 25°C and application of the yeast *Candida famata* (*Torulopsis candida*) was evaluated for the control of decay. The combined treatments, carried out within 36 h after inoculation with the pathogen, significantly reduced the decay percentage, both during storage and after a simulated marketing period of 1 week (D'hallewin et al., 1999). The interaction between kiwifruit curing and BCA

TABLE 12.1 Lesion Diameter (mm) in Apples Inoculated with *C. acutatum* Followed by Storage for 4 Months at 0.5°C + 2 Weeks at 24°C

Inoculation	Interaction means					
	Incubation		1-MCP		Heat	
	0 h	12 h	Not treated	Treated	Not treated	Treated
C. acutatum	10.70 ay	14.35 ax	10.37 y	14.67 ax	12.16 ax	12–88 ax
C. acutatum + antagonist	3.61 bx	5.50 bx	–	4.56 b	5.86 bx	3.26 bx

Means within columns with different letters (a and b) are significantly different (P = 0.05). Means within rows with different letters (x and y) are significantly different (P = 0.05).
Source: Janisiewicz et al., 2003.

TABLE 12.2 Lesion Diameter (mm) in Apples Inoculated with *P. expansum* with or without Heat Treatment and Stored at 0.5°C

Heat treatment	Period of storage	Inoculation	Interaction means			
			Incubation		1-MCP	
			0 h	12 h	Not applied	Applied
Nonheated	2 months	*P. expansum*	39.80 ay	45.99 ax	37.69 ay	48.10 ax
		P. expansum + antagonist	12.84 bx	11.89 bx	9.90 by	14.83 bx
Heated	2 months + 2 weeks at 24°C	*P. expansum*	1.64 ax	2.78 ax	0.86 ay	43.7 ax
		P. expansum + antagonist	0.67 ax	0.27 bx	0.56 ax	0.36 bx

Means within columns for each storage period with different letters (a and b) are significantly different (P = 0.05). Means within rows for incubation or 1-MCP treatment with different letters (x and y) are significantly different.
Source: Janisiewicz et al., 2003.

application for inhibition of infection by *Botrytis cinerea* was investigated. Yeast antagonists were applied to the wounds resulting from removal of pedicels at the natural abscission layer. The yeasts conferred significant level of protection against the gray mold pathogen. The biocontrol potential of yeasts was improved by fruit curing (incubation at 10°C) and topical yeast application made after 96 h of fruit curing (Cook et al., 1999). The effect of application of *Pantoea agglomerans* (CPA-2 strain) alone or in combination with a curing treatment at 33°C for 65 h on development of green mold (*Penicillium digitatum*) on lemons stored at ambient and cold storage condition was studied. Curing *P. agglomerans*-treated lemons at 33°C for 65 h entirely eliminated 24-h-old infections on artificially inoculated lemons stored at 20°C for 14 days and on naturally infected lemons stored at 10°C for 3 weeks plus 7 additional days at 20°C (Plaza et al., 2004).

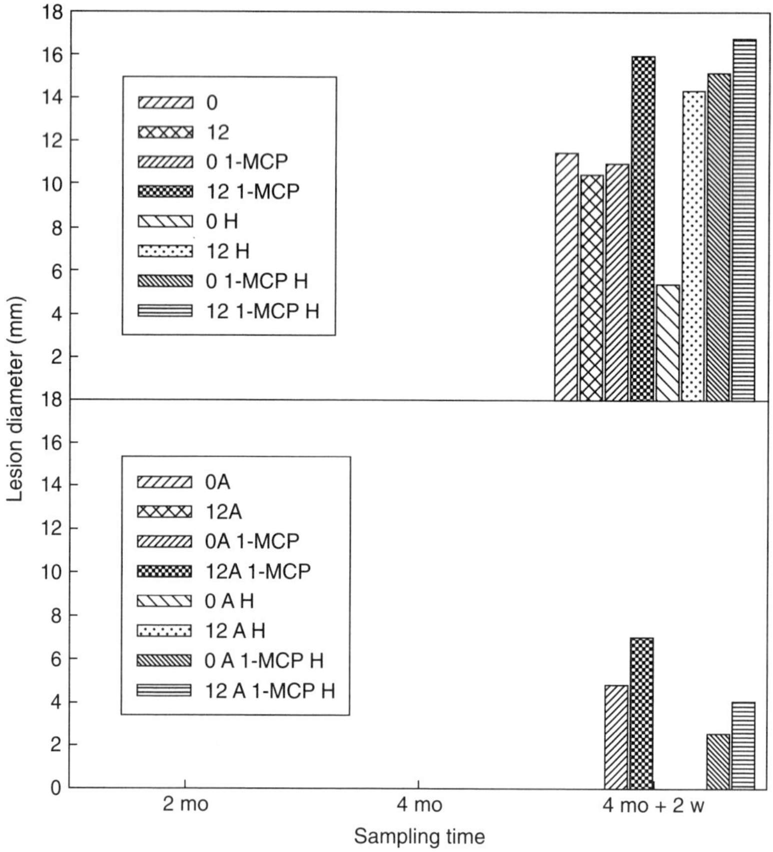

Figure 12.1 Effect of different treatments on the severity of decay of Golden Delicious apples inoculated with *Colletotrichum acutatum*, incubated at room temperature, and stored at 0.5°C under CA conditions for different periods of time. Treatments: A *Metschnikowia pulcherrima* T5-A2; H-heat: 1-MCP- 1-methylcyclopropene. (Courtesy of Janisiewicz *et al.*, 2003; Elsevier, Oxford, United Kingdom.)

Four antagonist yeasts, *Trichosporon pullulans*, *Cryptococcus laurentii*, *Rhodotorula glutinis*, and *Pichia membranefaciens*, were tested for their ability to control four main postharvest pathogens, *Alternaria alternata*, *Penicillium expansum*, *Botrytis cinerea*, and *Rhizopus stolonifer*, infecting sweet cherries. *T. pullulans* was the most effective in controlling all four pathogens at 25°C. However, the activities of *C. laurentii* and *R. glutinis* were substantially increased by combination with CA conditions (10% O_2 + 10% CO_2), resulting in more effective control of *A. alternata* and *P. expansum*. CA conditions suppressed the growth of *T. pullulans* and consequently its biocontrol potential was markedly affected. The results indicated the differential effects of the combination of BCAs and CA conditions and the need to select the suitable combination for achieving the best control of postharvest diseases (Qin et al., 2004).

12.2.3 Integration of Biocontrol and Chemical Methods

Chemicals with two different mechanisms of action on the postharvest pathogens have been combined with biocontrol agents. Some chemicals do not have antimicrobial activity, but they are known to induce resistance in host tissues against microbial pathogens (Chapter 10). In contrast, others act directly on the microbial pathogens, inhibiting spore germination, mycelial growth, or sporulation, resulting in the prevention or reduction in disease incidence or severity (Chapter 11).

The effectiveness of the antagonistic yeast *Candida* spp. could be enhanced by combining with calcium chloride ($CaCl_2$) solution (2%) for the control of gray and blue mold diseases of apples (McLaughlin et al., 1990; Wisniewski et al., 1995). Combined application of $CaCl_2$ (68 mM) and *Pichia guilliermondii* provided more effective protection than when $CaCl_2$ was used alone against green mold disease in grapefruit caused by *P. digitatum* (Droby et al., 1997). Likewise, pressure infiltration of $CaCl_2$ (0.27 M) followed by application of *Pseudomonas syringae* (isolate ESC, component of BioSave 110) reduced blue mold decay on Golden Delicious apples inoculated with *P. expansum* to a greater extent after cold storage for 6 months compared to individual treatment with either $CaCl_2$ or BCA (Janisiewicz et al., 1998). Integration of $CaCl_2$ and BCA provides additional benefits, including reduction in the concentration of both components and alleviation of certain physiological disorders such as bitter pit, without loss of effectiveness of disease control (Janisiewicz and Korsten, 2002).

Natural products, such as chitosan and its derivatives including glycolchitosan, have been demonstrated to possess an ability to induce resistance and antifungal activity (Allan and Hadwiger, 1979; Wilson et al., 1994) (Chapter 10). Spore germination of *B. cinerea* and *P. expansum* was inhibited by glycolchitosan (0.5%) which did not inhibit the growth of the BCA *Candida saitoana*, either in vitro or in apple wounds. The gray and blue molds of apple caused by *B. cinerea* and *P. expansum*, respectively, were more effectively controlled by the combined application of *C. saitoana* and glycolchitosan (0.2%). The effectiveness was similar to that of the fungicide imazalil on oranges and lemons. Sodium carbonate, when applied as a pretreatment, enhanced the level of protection offered by all treatments tested against green mold. Pretreatment with sodium carbonate followed by the combination of *C. saitoana* with 0.2% glycolchitosan gave the best control of green mold in both green and yellow lemons (El Ghaouth et al., 2000a). A bioactive coating, consisting of *C. saitoana* and glycolchitosan (0.2%), was evaluated for the control of postharvest diseases of apple and citrus with natural inoculation that simulated commercial packing house conditions. The decay in several apple cultivars was controlled more effectively by the bioactive coating than either *C. saitoana* or glycolchitosan alone. In addition, the level of control was comparable, or even superior, to thiabendazole in reducing decay, depending on the apple cultivar tested. Likewise, the efficacy of bioactive coating was equal to that of imazalil in controlling rotting in many cultivars of oranges (El Ghaouth et al., 2000b).

Fruit coatings, when applied with BCAs, may help reduce decay caused by postharvest pathogens. Application of a fruit coating containing sodium salts of carboxy-methylcellulose, sucrose esters of fatty acids, and mixed with sucroglycerides and soap (marketed as TALProlong) reduced the spread of postharvest diseases of pome fruits (Bancroft, 1995). TALProlong appears to reduce ripening of fruits, thereby extending the period of natural resistance against pathogens during storage (Janisiewicz and Korsten, 2002). Fruit coating formulations of bleached and unbleached shellac, capable of supporting the survival of *Candida oleophila*, were evaluated for their ability to promote surface colonization of citrus by *C. oleophila.* Coating formulations containing sucrose esters favored the development of yeast populations to a greater extent than those based upon shellac. Development of decays in grapefruit protected with fruit coating was slower (McGuire and Dimitroglou, 1999).

Generally regarded as safe (GRAS) substances, such as sodium carbonate, sodium bicarbonate, and ethanol, in combination with BCAs, were evaluated for their efficacy in controlling postharvest diseases. Citrus green mold caused by *P. digitatum* was more effectively controlled by combining sodium carbonate (3%) and the bacterial antagonist *Pseudomonas syringae* strain ESC 10 than by employing either treatment alone. The combined treatment offered certain advantages over individual treatments, such as the eradicative action of sodium carbonate and residual persistent protection by *P. syringae*, resulting in elimination of existing infection and protection against reinfection by *P. digitatum* (Smilanick et al., 1997, 1999).

Such a combination of desirable effects of sodium bicarbonate and the BCA *Pantoea agglomerans* (strains CPA-2) was demonstrated for the control of the green and blue mold diseases due to *P. digitatum* and *P. italicum*, respectively. The efficacy of *P. agglomerans* for the control of green mold was enhanced by combining with sodium bicarbonate (2%) resulting in a complete (100%) and 97.6% reduction of decay respectively at 3° and 20°C, compared with untreated control fruits. The blue mold disease was also reduced to a satisfactory level following the combined treatment, which could be an effective alternative to synthetic fungicides (Teixido et al., 2001). The effectiveness of the combination of ethanol (10%) with ethanol-resistant *Saccharomyces cerevisiae* strains 1440 and 1749 for the control of gray mold of apple and green mold of lemons was reported. Gray mold decay in apples was almost entirely eliminated (from 90% to nearly 0%), whereas green mold in lemons was reduced to less than 5% (Smilanick et al., 1995; Mari and Carati, 1998).

Ammonium molybdate was evaluated as a fungicide for the control of blue and gray mold diseases infecting pome fruits. In vitro tests revealed its inhibitory effect on spore germination of *P. expansum* and *B. cinerea* and reduced lesion diameters in apples treated with ammonium molybdate. Preharvest treatment resulted in a significant reduction in blue mold decay in cold storage for 3 months (Nunes et al., 2001). Combination of ammonium molybdate with the yeast *Candida sake* enhanced the efficacy of control of blue and gray mold diseases on Blanquilla pears stored at 20°C. In semicommerical trials at 1°C for 5 months, the disease percentage was decreased by more than 88% following application of *C. sake* at 2×10^6 CFU/ml and ammonium molybdate at 5 mM. The concentration of the BCA required could be markedly reduced by the combined treatment (Nunes et al., 2002).

Combined application of silicon (Si) as sodium metasilicate and the yeast *Cryptococcus laurentii* enhanced the biocontrol efficiency of *C. laurentii* against *Penicillium expansum* and *Monilinia fructicola* infecting sweet cherry. In addition, Si exhibited direct antifungal activity in addition to stimulation of activities of defense-related enzymes, phenylalanine ammonia lyase (PAL), polyphenol oxidase (PPO), and peroxidase (PO), resulting in more effective protection. The results indicate the potential of combined application of silicon and microbial biocontrol agent for the control of postharvest diseases of sweet cherry (Qin and Tian, 2005).

Various fungicides that were found to be effective against postharvest pathogens have been tested for their compatibility with the BCAs with the aim of reducing the quantity of fungicides and/or BCA required to achieve satisfactory control of the postharvest disease concerned. The viability of *Candida sake* (strain CPA-1) was not reduced after immersion in benomyl, flusilazole, thiabendazole (TBZ), sulfur, ziram, or diphenylamine, indicating the compatibility of fungicide and the BCA, whereas captan, imazalil, and ethoxyquin decreased the viability of *C. sake.* (Usall et al., 2001). It is possible to enhance the effectiveness of BCA treatment by combining with a compatible fungicide for the control of apple blue mold disease. The possibility of controlling blue and gray mold diseases of apples was explored by integrating the bacterial BCA *Pseudomonas syringae* MA-4 and cyprodinil. The combined application of cyprodinil (20 μg/ml) and strain MA-4 (3×10^7 CFU/ml) provided greater than 90% control against blue mold caused by *Penicillium expansum*. In contrast, gray mold caused by *Botrytis cinerea* could be controlled at a reduced concentration of cyprodinil (2.5 μg/ml) and a higher concentration of strain MA-4 (1×10^8 CFU/ml). As cyprodinil does not have bactericidal activity, it could be safely combined with bacterial antagonist (Ting et al., 2002).

The extracellular polymers (EP) isolated from the yeast *C. oleophila* was inhibitory to the germination of conidia and germ tube elongation of *Penicillium expansum* infecting apricot. Treatment of apricot cv. Amaar with EP + TBZ (200 ppm) provided maximum protection against *P. expansum* (El-Neshawy, 1999). The efficacy of BCAs, *Pichia guilliermondii* (strain 5A), *Candida oleophila* (strain 13L), and *Rhodotorula glutinis* (strain 21A) the commercial product Aspire™, and thiabendazole (TBZ), alone or in combination, was assessed for the control of green mold of citrus caused by *P. digitatum* under commercial conditions. Addition of TBZ (100 ppm) enhanced the effectiveness of disease control by Aspire™ and *C. oleophila* (Arras and Arru, 1999a). The differential response of combinations of BCAs with fungicide was reported by Arras et al. (2001). Application of TBZ (0.1 or 1.2 g/l) following treatment with Aspire™ (containing *C. oleophila*), but not with *P. guilliermondii*, was effective for the control of *P. digitatum* and *P. italicum* infecting oranges cv. Wasington Navel. The yeast *Debaryomyces hansenii* (strain 4E) was able to act directly against the postharvest pathogens *P. digitatum*, *P. italicum*, and *B. cinerea* and also indirectly by inducing resistance in host tissues, as reflected by production of the phytoalexins, scopoletin and scoparone. This BCA was resistant to TBZ (5 g/l) and imazalil (0.2 g/l), allowing the feasibility of integrating the BCA and fungicide for more effective control of the diseases (Arras and Arru, 1999b).

In another study, the combination of yeast isolates with TBZ and imazalil for the control of green and blue molds affecting grapefruit, mandarins, lemons, and oranges was evaluated, under cold storage and laboratory conditions (21 ± 1°C). The disease severity due to mold diseases was reduced by 90 to 100% when the yeasts were combined with lower doses of TBZ and imazalil. However, imazalil reduced the yeast populations to some extent (Kinay et al., 2001). The effect of treatment with *C. oleophila* and TBZ, either individually or in combination, for the control of storage rots of tangelo cv. Minneola (MT) and orange cv. Washintgon Navel was assessed. The fruits were dipped in TBZ (400 ppm), *C. oleophila*, and TBZ + *C. oleophila* prior to storage for 1 to 4 months. The fruits receiving combined treatment showed least rotting (Dundar and Gocer, 2001).

The combination of a preharvest fungicide application under field conditions and biocontrol agent treatment prior to storage was evaluated for the control of brown rot caused by *Monilinia fructicola* infecting sweet cherries. A single preharvest spray of iprodione at 1.13 kg a.i./ha reduced brown rot in stored cherry. The effectiveness of disease control was significantly enhanced when the fruits were also treated with a postharvest dip in a suspension of *Candida infirmo-miniatus* containing (1.5×10^8 CFU/ml). The incidence of brown rot was reduced by the combined treatment from 41.5% (in the control) to 0.4% with modified atmosphere packaging (Spotts et al., 1998).

The aggressiveness of *Botrytis cinerea* was determined in vitro by using a geranium leaf disk method, the diameters of lesions formed on leaf disks reflecting the degree of aggressiveness. The efficiency of the yeast *Rhodotorula glutinis* differed greatly depending on the aggressiveness of *B. cinerea* isolates. *R. glutinis* could inhibit the lesion development following inoculation with 16 isolates of *B. cinerea*, whereas the BCA did not have any effect on three isolates. When the yeast was combined with the fungicide vinclozolin at a reduced rate (50 μg a.i./ml), the lesion diameter formed by five isolates resistant to vinclozolin was significantly reduced, compared with the lesion diameters produced in the presence of BCA alone. Thus, the fungicide resistance of the pathogen isolates is reduced in the presence of the BCA, indicating the advantage of combining the chemical and BCA for fungicide resistance management. The efficacy of *R. glutinis* was enhanced by different concentrations of vinclozolin, disease control being better with higher concentration of the fungicide. The results suggest that variability in the biocontrol potential of *R. glutinis* against isolates of *B. cinerea* resistant to vinclozolin could be reduced by the addition of the fungicide (Buck and Jeffers, 2004).

12.2.4 Integration of Biocontrol, Physical, and Chemical Methods

In order to overcome the adverse effects of currenty employed synthetic fungicides for the control of postharvest pathogens, the integrated application of various safe postharvest treatments was evaluated. The effect of integration of heat treatment, calcium infiltration, and BCA application for the control of blue mold in Gala apples was assessed. Apple fruits were heated at 38°C for 4 days after harvest, pressure-infiltrated with $CaCl_2$ solution (2%), or treated with bacterial antagonist *Pseudomonas syringae*, alone or in combinations, to control decay caused by *P. expansum*. Calcium + BCA and $CaCl_2$ + BCA + heat reduced the blue mold decay by 89% and 91%, respectively. The integration of heating followed by Ca-infiltration and treatment with BCA could be an effective alternative for the control of postharvest decay of apples in place of synthetic fungicides (Conway et al., 1999).

Application of hot water rinsing and brushing (HWB) (62°C for 20s) sodium bicarbonate (baking soda) (2%), and the yeast antagonist *C. oleophila* (10^8 cells/ml) at 24h after inoculation of orange cv. Shamouti and grapefruit cv. Star Ruby reduced decay caused by *P. digitatum* by 68, 61, and 23% respectively. The combination of any two of these treatments or all three together reduced the decay by 87 to 90% over the decay in control (untreated) fruit. HWB and SBC treatment showed sanitizing effect, disinfecting the inoculated wounds, whereas the yeast and sodium bicarbonate could provide residual protection against latent infection by *P. digitatum*. The results indicate that by selecting suitable postharvest treatments, the use of fungicides may be avoided or minimized to a great extent (Porat et al., 2002).

For the effective control of brown rot disease of sweet cherries caused by *Monilinia fructicola*, an integrated system of disease management consisting of a preharvest fungicidal application, postharvest treatment with a yeast antagonist, followed by modified atmosphere packaging (MAP) and cold storage was tested. Propiconazole was applied as a preharvest spray and the fruits were treated with a wettable dispersible granular formulation of the yeast *Cryptococcus infirmo-miniatus* (CIM) followed by storage in modified atmosphere and storage at 2.8°C for 20 days or at –0.5°C for 42 days. There was a significant synergistic action between the fungicidal application and CIM treatment. Further, modified atmosphere reduced brown rot incidence significantly as compared to air-stored fruits. This integrated approach assumes greater importance, as postharvest fungicide options for cherry are limited (Spotts et al., 2002). The effect of a combined application of the yeast antagonist *Metschnikowia pulcherrima* and chemicals, such as acibenzolar-*S*-methyl, ethanol, or sodium bicarbonate, and heat treatment on the incidence of blue mold disease caused by *Penicillium expansum* and gray mold caused by *Botr-*

tytis cinerea was assessed. The yeast applied at 10^8 cells/ml was the most effective, providing a reduction of 56.6 and 97.2% in the lesion diameter of blue and gray mold diseases after a storage period of 5 days at 23°C and 20 days at 4°C. Heat treatment and sodium bicarbonate significantly improved the efficacy of the BCA against blue mold at 23°C. The blue mold disease could be more efficiently controlled by applying ethanol (20%) or sodium bicarbonate (5%) prior to *M. pulcherrima* and also by the application of *M. pulcherrima* in 0.1% sodium bicarbonate. The fungistatic effect of ethanol or sodium bicarbonate might prolong the persistence of the BCA for longer periods (Spadaro et al., 2004).

12.2.5 Integrated Management of Food-borne Pathogens

Outbreaks of salmonellosis have been linked to contamination of cantaloupe melon. Contamination might be introduced into flesh from the rind by cutting or by contact of cut pieces with contaminated rinds. The efficacy of hot water (70 or 90°C) or hot hydrogen peroxide (5%) treatments in reducing the population of native microflora, and inoculated *Salmonella* on cantaloupe rind and transfer to fresh-cut tissue during cutting, was assessed. All treatments significantly reduced aerobic mesophilic bacteria, yeast, and mold on treated whole melon and fresh-cut pieces. Dipping fresh-cut pieces prepared from inoculated whole melon in hot water (97°C) or H_2O_2 (70°C) for 60s resulted in elimination of *Salmonella*, as shown by dilution plating onto agar medium. However, the presence of the pathogen could be detected after enrichment at days 3 and 5 of storage at 5°C. The results indicated that H_2O_2 could be used to reduce the population of *Salmonella* on contaminated whole cantaloupes and it may be useful for increasing microbial safety of the fresh-cut product (Ukuku et al., 2004).

SUMMARY

Various strategies for the management of postharvest diseases have been found to be effective to varying levels under wide range of conditions which interact with each other. The possibility of integrating the different effective strategies to achieve higher level of control of postharvest pathogens and to minimize or replace the use of synthetic fungicides has been explored in certain host–pathogen. The usefulness of integrating different strategies to provide better control of diseases and to obtain safe, disease- and residue-free food products is discussed.

REFERENCES

Abdalla, M.Y. and El-Gizawy, A.M. 2000. Effect of some biological and chemical seed treatments of table beets on incidence of *Pythium aphanidermatum* damping-off in greenhouse. *Alexandria Journal of Agricultural Research*, **45**: 167–177.

Agarwal, V.K. and Sinclair, J.B. 1996. *Principles of Seed Pathology*. CRC Press, Boca Raton, FL, USA.

Al-Bachir, M. 1998. Use of gamma-irradiation and sulphur dioxide to improve storability of two Syrian grape cultivars (*Vitis vinifera*). *International Journal of Food Science and Technology*, **33**: 521–526.

Allan, C.R. and Hadwiger, L.A. 1979. The fungicidal effect of chitosan on fungi of varying cell wall composition. *Experimental Mycology*, **3**: 285–287.

Arras, G., Agabbio, M., Fiori, G.M. and Maltoni, S.L. 2001. Integrated and biological control of pathogens in citrus packinghouses. *Rivista di Frutticoltura e di Ortofloricoltura*, **63**: 17–22.

Arras, G. and Arru, S. 1999a. Inhibitory activity of antagonist yeasts against citrus pathogens in packinghouse treats. *Agricultura Mediterranea*, **129**: 249–265.

Arras, G. and Arru, S. 1999b. Integrated control of postharvest citrus decay and induction of phytoalexins by *Debaryomyces hansenii*. *Advances in Horticultural Science*, **13**: 76–81.

Bancroft, R.D. 1995. The use of surface coating to ameliorate the rate of spread of postharvest fungal diseases of top fruit. *International Journal of Biodeterioration and Biodegradation*, **36**: 385–405.

Ben-Yehoshua, S., Peretz, J., Rodov, V., Nafussi, B., Yekutieli, O., Wiseblum, A. and Reger, R. 2000. Postharvest application of hot water treatment in citrus fruits: the road from laboratory to the packing house. *Acta Horticulturae*, **518**: 19–28.

Buck, J.W. and Jeffers, S.N. 2004. Effect of pathogen aggressiveness and vinclozolin on efficacy of *Rhodotorula glutinis* PM4 against *Botrytis cinerea* on geranium leaf disks and seedlings. *Plant Disease*, **88**: 1262–1268.

Conway, W.S., Janisiewicz, W.J., Klein, J.D. and Sams, C.E. 1999. Strategy for combining heat treatment calcium infiltration and biological control to reduce postharvest decay of ‘Gala’ apples. *HortScience*, **34**: 700–704.

Cook, D.W.M., Long, P.G. and Ganesh, S. 1999. The combined effect of delayed application of yeast biocontrol agents and fruit curing for the inhibition of the postharvest pathogen *Botrytis cinerea* in kiwifruit. *Postharvest Biology and Technology*, **16**: 233–243.

de Capdeville, G., Beer, S.V., Watkins, C.B., Wilson, C.L., Tedeschi, L.O. and Aist, J.R. 2003. Pre-and post-harvest harpin treatments of apples induce resistance to blue mold. *Plant Disease*, **87**: 39–44.

D’hallewin, G., Arras, G., Dessi, R., Dettori, A. and Schirra, M. 1999. Citrus green mould control in stored “Star Ruby” grapefruit by the use of biocontrol yeast under curing conditions. *Acta Horticlturae*, **485**: 111–115.

Dharam Singh and Maheswari, V.K. 2001. Biological seed treatment to control of loose smut of wheat. *Indian Phytopathology*, **54**: 457–460.

Droby, S., Wisniewski, M.E., Cohen, L., Weiss, B., Touito, D. et al. 1997. Influence of $CaCl_2$ on *Penicillium digitatum*, grapefruit peel tissue and biocontrol activity of *Pichia guilliermondii*. *Phytopathology*, **87**: 310–315.

Dundar, O. and Gocer, S. 2001. Control of storage rots of Washington Navel oranges and Minneola by a combination of yeast antagonist and thiabendazole. *Acta Horticulturae*, **553**: 399–401.

El Ghaouth, A., Smilanick, J.L. and Wilson, C.L. 2000a. Enhancement of the performance of *Candida saitoana* by the addition of glycolchitosan for the control of postharvest decay of apple and citrus fruit. *Postharvest Biology and Technology*, **19**: 103–110.

El Ghaouth, A., Smilanick, J.L., Brown, G.E., Ippolito, A., Wisniewski, M. and Wilson, C.L. 2000b. Application of *Candida saitoana* and glycolchitosan for the control of postharvest diseases of apple and citrus fruit under semi-commercial conditions. *Plant Disease*, **84**: 243–248.

El-Neshawy, S.M. 1999. Efficacy of *Candida oleophila* strain 128 in preventing *Penicillium expansum* infection on apricot fruit. *Acta Horticulturae*, **485**: 141–147.

Food and Agriculture Organization. 1997. Worldwide regulations for mycotoxins for 1995. *A Compendium of Food and Nutrition, Paper No. 64*, Rome.

Hentschel, K.D. 1991. Biological control of seedborne *Alternaria radicina* on carrots by antagonistic *Bacillus subtilis*. *Bulletin SROP*, **14**: 73.

Janisiewicz, W.J., Conway, W.S., Glenn, D.M. and Sams, C.E. 1998. Integrating biological control and calcium treatment for controlling postharvest decay of apples. *HortScience*, **33**: 105–109.

Janisiewicz, W.J. and Korsten, L. 2002. Biological control of postharvest diseases of fruits. *Annual Review of Phytopathology*, **40**: 411–441.

Janisiewicz, W.J., Leverentz, B., Conway, W.S., Saftner, R.A., Reed, A.N. and Camp, M.J. 2003. Control of bitter rot and blue mold of apples by integrating heat and antagonist treatments on 1-MCP treated fruit stored under controlled atmosphere conditions. *Postharvest Biology and Technology*, **29**: 129–143.

Kinay, P., Yildiz, M., Yildiz, F., Delen, N. and Tosun, N. 2001. Control of postharvest *Penicillium* decay of citrus fruits with antagonistic yeasts and chemical fungicides. *Acta Horticuolturae*, **553**: 383–387.

Leverentz, B., Janisiewicz, W.J., Conway, W.S. and Saftner, R.A. 2001. Effect of combining biocontrol heat treatment and MCP-treatment on the reduction of postharvest decay of 'Delicious' apples. *Phytopathology*, **91**: S55.

Leverentz, B., Janisiewicz, W.J., Conway, W.S., Saftener, R.A., Fuchs, Y., Sams, C.E. and Camp, M.J. 2000. Combining yeasts or a bacterial biocontrol agent and heat treatment to reduce postharvest decay of "Gala" apples. *Postharvest Biology and Technology*, **21**: 87–94.

Lopez-Garcia, R. and Park, D.L. 1998. Effectiveness of postharvest procedures in management of mycotoxin hazards. In: *Mycotoxins in Agriculture and Food Safety* (eds) D. Bhatnagar and S. Sinha, pp. 407–433, Marcel Dekker, Inc., New York.

Lopez-Garcia, R., Park, D.L. and Phillips, T.D. 2004. *Integrated Mycotoxin Management Systems*. Internet Resources: File: 11A: Mycotoxin management systems food, nutrition and agriculture. Available at www.fao.org/documents/show_cdr.asp.

Mari, M. and Carati, A. 1998. Use of *Saccharomyces cerevisiae* with ethanol in the biological control of grey mould on pome fruits. In: *COST 914–COST 915 Joint Workshop – Non-Conventional Methods for the Control of Postharvest Disease and Microbial Spoilage* (eds) P. Bertolini, P.C. Sijmons, M.E. Guerzoni and F. Serra, pp. 85–91, Luxembourg, Germany.

Maude, R.B. 1996. *Seedborne Diseases and Their Control: Principles and Practice.* CAB International, Wallingford, Oxon, U.K.

McGuire, R.G. and Dimitroglou, D.A. 1999. Evaluation of shellac and sucrose ester fruit coating formulations that support biological control of postharvest grapefruit decay. *Biocontrol Science and Technology*, **9**: 53–65.

Mc Laughlin, R.J., Wisniewski, M.E., Wilson, C.L. and Chalutz, E. 1990. Effect of inoculum concentration and salt solutions on biological control of postharvest diseases of apple with *Candida* sp. *Phytopathology*, **80**: 456–461.

Narayanasamy, P. 2002. *Microbial Plant Pathogens and Crop Disease Management.* Science Publishers, Inc., Enfield, New Hampshire, USA.

Nunes, C. Usall, J., Teixidó, N., Abadias, M. and Viñas, I. 2002. Improved control of postharvest decay of pears by the combination of *Candida sake* (CPA-1) and ammonium molybdate. *Phytopathology*, **92**: 281–287.

Nunes, C., Usall, J., Teixidó, N. de Eribe, X.C. and Viñas, I. 2001. Control of postharvest decay of apples by preharvest and postharvest application of ammonium molybdate. *Pest Management Science*, **57**: 1093–1099.

Om Prakash and Pandey, B.K. 2000. Control of mango anthracnose by hot water and fungicide treatment. *Indian Phytopathology*, **53**: 92–94.

Park, D.L. and Stoloff, L. 1989. Aflatoxin control – how a regulatory agency managed risk from an unavoidable natural toxicant in food and feed. *Regulation of Toxicology and Pharmacology*, **9**: 109.

Pirgozliev, S.R., Edwards, S.G., Hare, M.C. and Jenkinson, P. 2003. Strategies for the control of *Fusarium* head blight in cereals. *European Journal of Plant Pathology*, **109**: 731–742.

Plaza, P., Usall, J., Smilanick, J.L., Lamarca, N. and Viñas, I. 2004. Combining *Pantoea agglomerans* (CPA-2) and curing treatments to control established infections of *Penicillium digitatum* on lemons. *Journal of Food Protection*, **67**: 781–786.

Porat, R., Daus, A., Weiss, B., Cohen, L. and Droby, S. 2002. Effect of combining hot water, sodium bicarbonate and biocontrol on postharvest decay of citrus fruit. *Journal of Horticultural Science and Biotechnology*, **77**: 441–445.

Prusky, D., Fuchs, Y., Kobiler, I., Roth, I., Weksler, A., Shalom, Y., Fallik, E., Zauberman, B., Pesis, E., Akerman, M., Ykutiely, O., Weisblum, A., Regev, R. and Artes, L. 1999. Effect of hot water brushing, prochloraz treatment and waxing on the incidence of black spot decay caused by *Alternaria alternata* in mango fruits. *Postharvest Biology and Technology*, **15**: 165–174.

Prusky, D., Shalom, Y., Kobiler, I., Akerman, M. and Fuchs, Y. 2002. *Alternaria alternata* in mango fruits at harvest determines the postharvest treatment applied for the control of rots during storage. *Postharvest Biology and Technology*, **25**: 339–347.

Qin, G., Tian, S. and Xu, Y. 2004. Biocontrol of postharvest diseases on sweet cherries by four antagonistic yeasts in different storage conditions. *Postharvest Biology and Technology*, **31**: 51–58.

Qin, G.Z. and Tian, S.P. 2005. Enhancement of biocontrol activity of *Cryptococcus laurentii* by silicon and the possible mechanism involved. *Phytopathology*, **95**: 69–75.

Rodov, V., Agar, T., Peretz, J., Nafussi, B. Kim, J.J. and Ben-Yehoshua, S. 2000. Effect of combined application of heat treatments and plastic packaging on keeping quality of "Oroblanco" fruit (*Citrus grandis* L. X *C. paradisi* Macf.). *Postharvest Biology and Technology*, **20**: 287–294.

Smilanick, J.L., Margosan, D.A. and Hanson, D.J. 1995. Evaluation of heated solutions of sulfur dioxide, ethanol and hydrogen peroxide to control postharvest green mold of lemons. *Plant Disease*, **79**: 742–747.

Smilanick, J.L., Mackey, B.E., Reese, R. Usall, J. and Margosan, D.A. 1997. Influence of concentration of soda ash, temperature and immersion period on the control of postharvest green mold of oranges. *Plant Disease*, **81**: 379–382.

Smilanick, J.L., Margosan, D.A., Milkota, F., Usall, J. and Michael, I. 1999. Control of citrus green mold by carbonate and bicarbonate salts and influence of commercial postharvest practices on their efficacy. *Plant Disease*, **83**: 139–145.

Spadaro, D., Garibaldi, A. and Gullino, M.L. 2004. Control of *Penicillium expansum* and *Botrytis cinerea* on apple combining a biocontrol agent with a hot water dipping and acibenzolar-S-methyl baking, soda, or ethanol application. *Postharvest Biology and Technology*, **33**: 141–151.

Spotts, R.A., Cervantes, L.A. and Facteau, T.J. 2002. Integrated control of brown rot of sweet cherry fruit with a preharvest fungicide a postharvest yeast, modified atmosphere packaging and cold storage temperature. *Postharvest Biology and Technology*, **24**: 251–257.

Spotts, R.A., Cervantes, L.A., Facteau, T.J. and Chandgoyal, T. 1998. Control of brown rot and blue mold of sweet cherry with preharvest iprodione, postharvest *Cryptococcus infirmo-miniatus* and modified atmosphere packaging. *Plant Disease*, **82**: 1158–1160.

Teixidó, N., Usall, J., Palou, L., Asensio, A., Nunes, C. and Viñas, I. 2001. Improving control of green and blue molds of oranges by combining *Pantoea agglomerans* (CPA-2) and sodium bicarbonate. *European Journal of Plant Pathology*, **107**: 685–694.

Ting, Z., Northover, J., Schnieider, K.E. and Lu, X.W. 2002. Interactions between *Pseudomonas syringae* MA-4 and cyprodinil in the control of blue and gray mold of apples. *Canadian Journal of Plant Pathology*, **24**: 154–161.

Tonini, G., Zilibotti, M., Barberini, K., Proni, R. and Bassi, F. 2000. Control of *Botrytis cinerea* rot in kiwifruit: integrated physical and chemical systems. *Atti, Giornate Fitopatologiche*, Vol. 2, 75–80, *Universita di Bologna*, Italy.

Ukuku, D.O., Pilizota, V. and Sapers, G.M. 2004. Effect of hot water and hydrogen peroxide treatments on survival of *Salmonella* and microbial quality of whole and fresh-cut cantaloupe. *Journal of Food Protection*, **67**: 432–437.

Usall, J., Teixidó, N., Torres, R., Ochoa de Eribe, X. and Viñas, I. 2001. Pilot tests of *Candida sake* (CPA-1) applications to control postharvest blue mold on apple fruit. *Postharvest Biology and Technology*, **21**: 147–156.

Wareing, P. 1999. *The application of the Hazard Analysis Critical Control Point (HACCP) approach to the control of mycotoxins in foods and feeds*. Postharvest convention, 1999, Cranfield University, Bedford, UK.

Wilson, C.L., El Ghaouth, A., Chalutz, E., Droby, S., Stevens, C. et al. 1994. Potential of induced resistance to control postharvest diseases of a fruit and vegetables. *Plant Disease*, **78**: 837–843.

Wisniewski, M., Droby, S., Chalutz, E. and Eilam, Y. 1995. Effects of Ca^{2+} and Mg^{2+} on *Botrytis cinerea* and *Penicillium expansum in vitro* and on the biocontrol activity of *Candida oleophila*. *Plant Pathology*, **44**: 1016–1024.

ADDENDUM: BASIC METHODS

1 CHARACTERIZATION OF CAUSAL AGENTS OF POSTHARVEST PATHOGENS

1.1 Fungal Pathogens of Fruits (Xiao and Rogers, 2004)

A. Isolation of Fungal Pathogens

i. Spray decayed fruits with 70% ethanol and air dry in a laminar flow hood.
ii. Cut out small pieces of fruit tissues (approximately $5 \times 5 \times 3\,mm^3$) using sterilized scalpel from the margins of decayed and healthy tissues.
iii. Transfer the fruit tissues to acidified potato dextrose agar (APDA) in Petri plates, prepared by addition of 4.0 ml of a 2.5% solution of lactic acid per liter of the potato dextrose agar (PDA) medium, and incubate at room temperature (20 to 22°C) for 3 to 14 days.
iv. Record the growth characteristics of fungi at 24-h intervals.
v. Transfer the individual fungus to fresh APDA.
vi. Purify cultures by adopting standard single spore or hyphal tip isolation procedures.
vii. Pure cultures in PDA are stored in sterile water at 4°C or in 15% glycerol at −18°C.

Postharvest Pathogens and Disease Management, by P. Narayanasamy

B. Pathogenicity Test

i. Surface-sterilize the fruits with 0.5% sodium hypochlorite (NaOCl); rinse with sterile distilled water for 5 min, three times, and air dry.

ii. Using sterile nail head (4 mm diameter) make wounds on the surface of each fruit to a depth of 4 mm.

iii. Place the mycelial plug (4 mm diameter) prepared from a 4-day old PDA culture in the wound of each fruit; cover the inoculation site with a piece of sterile moist cheese cloth ($2 \times 2\,cm^2$) and wrap the inoculated fruit gently with sterile aluminum foil.

iv. Incubate the inoculated fruit placed in fiberboard trays wrapped in perforated polyethylene bags at appropriate temperature required for the development of the fungus.

v. Maintain adequate number of replicates (at least 10).

vi. Evaluate the extent of decay at 1or 2 weeks after inoculation depending on the incubation temperature.

vii. Isolate the fungus from the inoculated fruit; compare the characteristics of the fungus in stock culture and the reisolated fungus to confirm the Koch's postulates and repeat the experiment twice.

1.2 Fungal Pathogens of Vegetables (Lewis Ivey et al., 2004)

A. Isolation of Fungal Pathogen

i. Cut the leading edge of the lesion (anthracnose) into small sections using sterile scalpel and surface sterilize in 70% ethanol.

ii. Transfer the lesion sections aseptically to potato dextrose agar (PDA) medium (Difco Laboratories, Detroit, MI) and incubate at appropriate temperature to allow the development of the fungus from the lesion sections until conidia are formed.

iii. Prepare a suspension of conidia at the required dilution and follow the standard procedure for single spore isolation and maintain the culture on PDA slants at 10°C.

B. Pathogenicity Test

i. Wash the (tomato or pepper) fruits thoroughly and surface sterilize them in sodium hypochlorite (0.3%) for 2 min and rinse in distilled water twice.

ii. Place the fruits on closed containers on wire mesh screens and label with a permanent marker, making a circle to indicate the point of inoculation on each fruit.

iii. Make wound (approximately 2 mm deep) using a 27.5 gauge sterile needle and dispense conidial suspension (20 μl) per wound.

iv. Use sterile water for inoculating control.

v. Add sterile hot water daily to the bottom of each container and cap it tightly.

vi. Determine the lesion diameter at 0, 3, 4, 5, 6, and 10 days after inoculation.
vii. Reisolate the fungus from the lesions and study the characteristics in comparison to that of the original culture.

2 CHARACTERIZATION OF MICROBIAL BIOCONTROL AGENTS

2.1 Fungal Antagonists (Janisiewicz, 1991; Janisiewicz et al., 2001)

A. *Isolation of Microorganisms*

i. Make wounds by removing the skin (3 cm^2 and 2 mm deep) on intact fruits (apples) 1 week before harvest and remove the wounded areas on fruits using a cork borer (1 cm diameter by 1 cm deep).
ii. Grind the samples with phosphate buffer using a mortar and pestle; dilute the macerate to required levels and plate dilutions on nutrient yeast dextrose agar (NYDA, Difco Laboratories, Detroit, MI) medium.
iii. Alternatively, grind the samples in water (4.5 ml); add the slurry to 10% apple juice (200 ml) in 500 ml Erlenmeyer flasks; shake the flasks at 150 rpm at 26°C; transfer 2 ml aliquots to fresh apple juice at 42 and 66 h and dilution-plate by taking samples after 0, 24, 42, 66, and 140 h.
iv. Purify the colonies of fungi that develop on NYDA.

B. *Screening of Microorganism for Biocontrol Potential*

i. Make two wounds (3 × 3 × 3 mm) in each fruit and transfer 20 μl of water suspension of test microorganism to each wound.
ii. Dispense 20 μl of suspension of the pathogen (10^4 conidia/ml) in a similar manner to each wound within 20 min.
iii. Evaluate each fruit for rot development after incubation for 5 to 7 days at 24°C by using the criteria:
   a. reduction in number of wounds infected to less than 50% and
   b. inhibition of rot expansion by more than 75%.
iv. Select the microorganism(s) based on the two criteria listed above.

2.2 Bacterial Antagonists (Nunes et al., 2001)

A. *Isolation of Microorganisms*

i. Wash the fruits and leaves in 200 ml of sterile 0.05 M phosphate buffer [0.2 M KH_2PO4 (70 ml), 0.2 M K_2HPO4 (30 ml) and deionized water (300 ml)], pH 6.5 on a rotary shaker for 10 min at 150 rpm and pour off the washings.
ii. Wash again for 10 min in an ultrasonic bath (Selecta, Abrera, Barcelona, Spain); plate the sonicated samples (0.1 ml) on nutrient yeast dextrose agar medium [NYDA: nutrient broth (8 g/l), yeast extract (5 g/l), dextrose (10 g/l), and agar (15 g/l)] and incubate for 24 to 48 h at 25°C.

iii. Isolate the colonies with different visual characteristics and purify by single colony isolations after triple streaking on NYDA medium.

B. Screening Microorganisms for Biocontrol Potential

i. Prepare the suspensions of test microorganisms by growing cultures in nutrient yeast dextrose broth (NYDB)-NYDA without agar for 24 to 48h at 25 ± 1°C with shaking at 150rpm; centrifuge at 8315 × g for 10min and resuspend the cells in deionized water.

ii. Prepare appropriate dilutions and determine the optical density at 420nm with a spectrophotometer.

iii. Evaluate the efficacy of test microorganisms based on criteria mentioned above (2.1.B).

3 DETERMINATION OF FUNGICIDAL ACTIVITY OF CHEMICALS

3.1 Poisoned Food Technique or Radial Growth (RG) Test (Carpenter, 1942; Golembiewski et al., 1995)

i. Cultivate the test pathogen(s) in potato dextrose agar medium (PDA) or any medium required for the rapid development of the pathogen and incubate for 7 to 10 days as required.

ii. Prepare different concentrations of the test fungicide in water or organic solvent such as acetone; mix the fungicide solution with about 20ml of molten PDA at about 60°C to give a final concentration of 0.1% (v/v); pour the mixture of the medium and the fungicide into each sterile Petri plate and allow the medium to set.

iii. Prepare the agar disks using a sterile cork borer (7 or 8mm) from the actively growing peripheral zones in the culture plates in which the pathogen is cultivated (step (i)); transfer the agar disks of fungal culture to the centre of the plates containing the medium amended with the fungicides and incubate the plates at a constant temperature (about 20 to 22°C) for optimal growth of the pathogen, for 4 to 7 days.

iv. Measure the colony diameter at two position at a right angle to each other and calculate the mean diameter of the colony in each plate.

v. Calculate the relative growth in different treatments using the formula:

$$\frac{\text{Mean colony diameter in fungicide amended medium}}{\text{Mean colony diameter in unamended medium (control)}} \times 100$$

vi. Calculate the effective concentration (EC_{50}) at which the colony diameter is reduced to 50% of the colony diameter in the unamended medium.

3.2 Assay of Antimicrobial Activity of Chemicals by Automated Quantitative Assay (Raposo et al., 1995)

i. Grow the fungal pathogen in potato dextrose agar (PDA) or other suitable medium in sterile Petri plates in a growth chamber set at 24 ± 1°C with a 16-h photoperiod for 7 days.
ii. Flood the dish with sterile water and gently scrape the surface of the culture with sterile nichrome wire loop; filter the conidial suspension through cotton wool and adjust the concentration of conidia to required level using a hemocytometer.
iii. Dispense the conidial suspension to tubes (15 × 6.5 cm) containing 3 ml of double-concentrated sterile PDA and incubate at 20°C in darkness to allow germination of conidia.
iv. Prepare different dilutions of the test chemical (fungicide); amend the PDA medium with different dilutions of the chemical and transfer the amended medium (100 µl) to each well of a 96-well microplate.
v. Dispense equal volumes (100 µl) of conidial suspension into each well containing medium amended with fungicide dilutions and maintain a minimum of four replicates for each experiment.
vi. Determine absorbance values at 494 nm using a microplate reader immediately after the addition of conidial suspension and again after incubation for 46 h at 20°C in darkness.
vii. Calculate the final absorbance value by subtracting the values of initial measurements from the second measurements.

4 ASSESSMENT OF ANTIFUNGAL ACTIVITY OF PLANT EXTRACTS (THANGAVELU ET AL., 2004)

i. Wash the leaves/plant parts thoroughly with water and grind the tissues in 0.1 M sodium phosphate buffer with pestle and mortar on ice.
ii. Centrifuge the extract at 10,000 rpm for 20 min at 4°C and filter-sterilize the supernatant by passing through 0.22 µm Millipore filter.
iii. Grow the test fungal pathogen (*Colletotrichum musae*) in potato dextrose agar (PDA) medium in Petri plates.
iv. Place the mycelial discs from a 7-day old culture of the pathogen (step (iii) above) at the center of Petri plates containing PDA and incubate the plates at room temperature (28 ± 2°C) for 2 days.
v. Place sterile filter paper discs (6 mm diameter) on the agar surface at 1 cm from the inner edge of the plate; dispense plant extracts (80 µl) at the center of each filter paper disc and incubate for 72 h at room temperature.
vi. Measure the diameter of inhibition zone where the fungal growth is arrested.

5 ASSESSMENT OF SENSITIVITY OF FUNGAL PATHOGENS TO FUNGICIDES (SCHNABEL ET AL., 2004; PERES ET AL., 2004)

i. Scrap the conidia of the test fungal pathogen from infected fruit (pear or citrus); transfer to 1 ml of sterile water and vortex the conidial suspension.
ii. Distribute 50 μl of the conidial suspension evenly into water agar amended with streptomycin sulfate (100 μg/ml) in Petri plate; transfer single hyphal tip from germinating conidium aseptically to potato dextrose agar (PDA, Difco Laboratories, Sparks, MD) and use one single-spore culture per fruit for further experimentation.
iii. Prepare fungicide dilutions (0.001, 0.005, 0.01, 0.05, 0.1, 0.5, and 1.0 μg/ml)
iv. Spread aliquots of 100 μl of conidial suspension uniformly on PDA amended with different dilutions of the fungicide and incubate the plates at 24°C for 3 days with a photoperiod of 12 h.
v. Count the number of fungal colonies and measure their diameter.
vi. Calculate the concentration of fungicide required to suppress mycelial growth by 50% (EC_{50}).
vii. Calculate the sensitivity factor of fungal population by dividing the highest EC_{50} values by the lowest EC_{50} values of isolates within a population.

REFERENCES

Carpenter, J.B. 1942. A toximetric study of some eradicant fungicides. *Phytopathology*, **32**: 845–856.

Golembiewski, R.C., Varges, J.M.Jr., Jones, A.L. and Detweiler, A.R. 1995. Detection of demethylation inhibitor (DMI) resistance in *Sclerotinia homeocarpa* populations. *Plant Disease*, **79**: 491–493.

Janisiewicz, W.J. 1991. Control of postharvest diseases of fruits with biocontrol agents In: *The Biocontrol of Plant Diseases*, Food and Fertilizer Technology Center for the Asian and Pacific region, Taipei, Taiwan.

Janisiewicz, W.J., Tworkoski, T.J. and Kurtzman, C.P. 2001. Biocontrol potential of *Metchnikowia pulcherrima* strains against blue mold of apple. *Phytopathology*, **91**: 1098–1108.

Lewis Ivey, M.L., Nava-Diaz, C. and Miller, S.A. 2004. Identification and management of *Colletotrichum acutatum* on immature bell peppers. *Plant Disease*, **88**: 1198–1204.

Nunes, C., Usall, J., Teixidó, N. and Viñas, I. 2001. Biological control of postharvest pear diseases using a bacterium *Pantoea agglomerans* CPA-2. *International Journal of Food Microbiology*, **70**: 53–61.

Peres, N.A.R., Souza, N.L., Peever, T.L. and Timmer, L.W. 2004. Benomyl sensitivity of isolates of *Colletotrichum acutatum* and *C. gloeosporioides* from citrus. *Plant Disease*, **88**: 125–130.

Raposo, R., Colgan, R., Delcan, J. and Melgarejo, P. 1995. Application of an automated quantitative method to determine fungicide resistance in *Botrytis cinerea*. *Plant Disease*, **79**: 294–296.

Schnabel, G., Bryson, P.K., Bridges, W.C. and Brannen, P.M. 2004. Reduced sensitivity in *Monilinia fructicola* to propiconazole in Georgia and implications for disease management. *Plant Disease*, **88**: 1000–1004.

Thangavelu, R., Sundararaju, P. and Sathiamoorthy, S. 2004. Management of anthracnose disease caused by *Colletotrichum musae* using plant extracts. *Journal of Horticultural Science and Biotechnology*, **79**: 664–668.

Xiao, C.L. and Rogers, J.D. 2004. A postharvest fruit rot in d'Anjou pears caused by *Sphaeropsis pyriputescens* sp. nov. *Plant Disease*, **88**: 114–118.

INDEX

Postharvest Pathogens and Disease Management, by P. Narayanasamy